NUCLEAR ENERGY

NUCLEAR ENERGY

PRINCIPLES, PRACTICES, AND PROSPECTS

SECOND EDITION

David Bodansky

 Springer

David Bodansky
Department of Physics
University of Washington
Seattle, WA 98195
USA

Library of Congress Cataloging-in-Publication Data
Bodansky, David.
  Nuclear energy : principles, practices, and prospects / David Bodansky.—2nd ed.
    p. cm.
  Includes bibliographical references and index.
  ISBN 0-387-20778-3 (hc : alk. paper)
    1. Nuclear engineering.   I. Title.
  TK9145.B54 2003
  333.792′4—dc22   2003070772

ISBN-10: 0-387-20778-3   Printed on acid-free paper.
ISBN-13: 978-0387-20778-?

9 8 7 6 5 4 3

springer.com

Preface to the Second Edition

This second edition represents an extensive revision of the first edition, although the motivation for the book and the intended audiences, as described in the previous preface, remain the same. The overall length has been increased substantially, with revised or expanded discussions of a number of topics, including Yucca Mountain repository plans, new reactor designs, health effects of radiation, costs of electricity, and dangers from terrorism and weapons proliferation.

The overall status of nuclear power has changed rather little over the past eight years. Nuclear reactor construction remains at a very low ebb in much of the world, with the exception of Asia, while nuclear power's share of the electricity supply continues to be about 75% in France and 20% in the United States. However, there are signs of a heightened interest in considering possible nuclear growth. In the late 1990s, the U.S. Department of Energy began new programs to stimulate research and planning for future reactors, and many candidate designs are now contending—at least on paper—to be the next generation leaders. Outside the United States, the commercial development of the Pebble Bed Modular Reactor is being pursued in South Africa, a French-German consortium has won an order from Finland for the long-planned EPR (European Pressurized Water Reactor), and new reactors have been built or planned in Asia.

In an unanticipated positive development for nuclear energy, the capacity factor of U.S. reactors has increased dramatically in recent years, and most operating reactors now appear headed for 20-year license renewals. In a negative development, the German and Dutch governments have announced plans to phase out nuclear power and Sweden continues its earlier, but considerably delayed, program to do the same. Further, it remains unlikely that private U.S. companies will find it financially prudent to order new reactors without incentives from the federal government.

Significant uncertainties remain in important areas, including the fate of the Yucca Mountain nuclear waste repository project, the degree to which the U.S. government will act to further the construction new reactors, the outcome of on-going debates on the effects of low doses of ionizing radiation, and the extent to which nuclear weapons proliferation and nuclear terrorism can be restrained. In the broader energy picture, concern about climate change caused by fossil fuel combustion has intensified, with increased interest in the potential of sequestering carbon dioxide after it is produced and in finding alternatives to fossil fuels.

Given the uncertainties facing nuclear energy, including the overriding uncertainty as to the extent that it may expand or contract, a new look at its current status seems warranted. This book seeks to provide background for considering the role that nuclear energy might play in addressing the overall energy dilemmas facing the United States and other countries throughout the world. It also briefly discusses alternatives to nuclear energy, without attempting a comparative evaluation of the competing, or complementary, possibilities.

The preface to the first edition stated the hope that "the book will be useful to readers with a wide variety of backgrounds who have an interest in nuclear energy matters." This was meant to include readers with technical backgrounds and those without such backgrounds. With the latter readership in mind, the somewhat mathematically oriented material has been slightly reduced for this edition. I hope that where uncongenial equations are found (now mostly confined to Chapter 7), readers will be able to skip over them without too much loss of basic content.

Again, I am indebted to many individuals, at the University of Washington and elsewhere, for much appreciated help. The debts that were acknowledged in the first edition remain. For this edition, assistance from a number of additional individuals calls for special mention. Robert Albrecht, at the University of Washington, has read and discussed many parts of the book with me, and has given me the benefit of his deep understanding of nuclear matters. Robert and Susan Vandenbosch, also in Seattle, have reviewed virtually the entire manuscript and have made numerous helpful suggestions. Edwin Kolbe, the Project Manager for Radioactive Materials at the Swiss National Cooperative for the Disposal of Radioactive Waste (NAGRA) and a 2002 visitor at the Institute for Nuclear Theory at the University of Washington, kindly offered to carry out ORIGEN calculations that give the yield of radionuclides in "typical" spent fuel. Abraham Van Luik, with the Yucca Mountain Project, has provided valuable help in elucidating the DOE's planning and analyses for the project.

Many other colleagues have read drafts of one or more chapters and I am grateful to them for their comments on those chapters, and in many cases, on other aspects of the book. I here thank: Chaim Braun, Bernard Cohen, Stanley Curtis, J. Gregory Dash, David Hafemeister, Isaac Halpern, Robert Halvorsen, William Sailor, Luther Smith, and Gene Woodruff. I also am grateful to Edward Gerjuoy, Phillip Malte, Jeffrey Schneble, and Donald Umstadter for comments on the first edition.

Preface to the Second Edition

This second edition represents an extensive revision of the first edition, although the motivation for the book and the intended audiences, as described in the previous preface, remain the same. The overall length has been increased substantially, with revised or expanded discussions of a number of topics, including Yucca Mountain repository plans, new reactor designs, health effects of radiation, costs of electricity, and dangers from terrorism and weapons proliferation.

The overall status of nuclear power has changed rather little over the past eight years. Nuclear reactor construction remains at a very low ebb in much of the world, with the exception of Asia, while nuclear power's share of the electricity supply continues to be about 75% in France and 20% in the United States. However, there are signs of a heightened interest in considering possible nuclear growth. In the late 1990s, the U.S. Department of Energy began new programs to stimulate research and planning for future reactors, and many candidate designs are now contending—at least on paper—to be the next generation leaders. Outside the United States, the commercial development of the Pebble Bed Modular Reactor is being pursued in South Africa, a French-German consortium has won an order from Finland for the long-planned EPR (European Pressurized Water Reactor), and new reactors have been built or planned in Asia.

In an unanticipated positive development for nuclear energy, the capacity factor of U.S. reactors has increased dramatically in recent years, and most operating reactors now appear headed for 20-year license renewals. In a negative development, the German and Dutch governments have announced plans to phase out nuclear power and Sweden continues its earlier, but considerably delayed, program to do the same. Further, it remains unlikely that private U.S. companies will find it financially prudent to order new reactors without incentives from the federal government.

Significant uncertainties remain in important areas, including the fate of the Yucca Mountain nuclear waste repository project, the degree to which the U.S. government will act to further the construction new reactors, the outcome of on-going debates on the effects of low doses of ionizing radiation, and the extent to which nuclear weapons proliferation and nuclear terrorism can be restrained. In the broader energy picture, concern about climate change caused by fossil fuel combustion has intensified, with increased interest in the potential of sequestering carbon dioxide after it is produced and in finding alternatives to fossil fuels.

Given the uncertainties facing nuclear energy, including the overriding uncertainty as to the extent that it may expand or contract, a new look at its current status seems warranted. This book seeks to provide background for considering the role that nuclear energy might play in addressing the overall energy dilemmas facing the United States and other countries throughout the world. It also briefly discusses alternatives to nuclear energy, without attempting a comparative evaluation of the competing, or complementary, possibilities.

The preface to the first edition stated the hope that "the book will be useful to readers with a wide variety of backgrounds who have an interest in nuclear energy matters." This was meant to include readers with technical backgrounds and those without such backgrounds. With the latter readership in mind, the somewhat mathematically oriented material has been slightly reduced for this edition. I hope that where uncongenial equations are found (now mostly confined to Chapter 7), readers will be able to skip over them without too much loss of basic content.

Again, I am indebted to many individuals, at the University of Washington and elsewhere, for much appreciated help. The debts that were acknowledged in the first edition remain. For this edition, assistance from a number of additional individuals calls for special mention. Robert Albrecht, at the University of Washington, has read and discussed many parts of the book with me, and has given me the benefit of his deep understanding of nuclear matters. Robert and Susan Vandenbosch, also in Seattle, have reviewed virtually the entire manuscript and have made numerous helpful suggestions. Edwin Kolbe, the Project Manager for Radioactive Materials at the Swiss National Cooperative for the Disposal of Radioactive Waste (NAGRA) and a 2002 visitor at the Institute for Nuclear Theory at the University of Washington, kindly offered to carry out ORIGEN calculations that give the yield of radionuclides in "typical" spent fuel. Abraham Van Luik, with the Yucca Mountain Project, has provided valuable help in elucidating the DOE's planning and analyses for the project.

Many other colleagues have read drafts of one or more chapters and I am grateful to them for their comments on those chapters, and in many cases, on other aspects of the book. I here thank: Chaim Braun, Bernard Cohen, Stanley Curtis, J. Gregory Dash, David Hafemeister, Isaac Halpern, Robert Halvorsen, William Sailor, Luther Smith, and Gene Woodruff. I also am grateful to Edward Gerjuoy, Phillip Malte, Jeffrey Schneble, and Donald Umstadter for comments on the first edition.

It is not possible to give a full listing of all the other individuals who have assisted me with information, advice, and documents. In this regard, in addition to those acknowledged above and in the first edition, I want at least to thank Joseph Beamon, James Beard, Mario Carelli, Yoon Chang, Raymond Clark, Paul Craig, George Davis, Herbert Ellison, Rodney Ewing, Tom Ferriera, Steve Fetter, Brittain Hill, Mark Jacobson, John Kessler, Kristian Kunert, Edward Miles, Thomas Murley, Richard Poeton, Jerome Puskin, Lowell Ralston, Stanley Ritterbusch, Finis Southworth, John Taylor, Ronald Vijuk, David Wade, Kevan Weaver, Ruth Weiner, Bruce Whitehead, Bertram Wolfe, and Joseph Ziegler.

Again, as in the first edition, my thanks and apologies are extended to the many others, not named above, who have generously given me their help. I appreciate the willingness of the University of Washington and the Department of Physics to provide space, facilities, and a congenial working environment. Finally, again, I wish to thank my wife, Beverly, for her patience and support during the long continuation of an effort that seemed at times to belie the concept of retirement.

Seattle, Washington   *David Bodansky*
May 2004

Preface to the First Edition

This book has evolved from notes prepared for students in a physics course designed to cover the major aspects of energy production and consumption. About one-third of the course dealt with nuclear energy, and the notes for that segment were revised and expanded for the present book.

The course assumed that the students had at least one year of college-level physics, thus permitting the inclusion of some technical discussions. The present book, in its occasional use of equations and technical terminology, somewhat reflects the nature of that original audience. Readers with relatively little background in physics and engineering may find it useful to refer to the Appendix on "Elementary Aspects of Nuclear Physics," and to the Glossary.

I have sometimes been asked: "For whom is the book written?" One difficulty in addressing this question has already been touched on. Some of the technical discussions include equations, which is not customary in a book for a "lay audience." Other parts are more elementary than would be the case were this a textbook on nuclear engineering. Nonetheless, most of the key issues can be constructively discussed using little or no mathematical terminology, and I therefore hope that the book will be useful to readers with a wide variety of backgrounds who have an interest in nuclear energy matters.

A more fundamental difficulty lies in the fact that such interest is now at a low ebb. In fact, it is often believed that the era of nuclear fission energy has passed, or is passing. While most informed people are aware that France is highly dependent on nuclear energy, this is ignored as an aberration, holding little broader significance. It is not widely realized that nuclear energy, despite its stagnancy in the United States and most of Europe, is expanding rapidly in Asia. Further, many people who are otherwise well-informed on issues of public policy are surprised to learn that the United States now obtains more than 20% of its electricity from nuclear power.

This book has been written in the belief that it is premature and probably incorrect to assume that there is to be only one era of nuclear power and that this era has passed. The future pattern of nuclear energy use will depend on developments in a variety of energy technologies and on public attitudes in differing countries. There can be little certainly as to how these developments will unfold. However, the demands of a growing world economy and the pressures of declining availability of oil will inevitably force a realignment and reassessment of energy options. The goal of this book is to provide basic information to those who want to gain, or refresh, an introductory familiarity with nuclear power, even before broad new reassessments of energy policy are made in the United States and elsewhere.

The preparation of the book has been aided by contributions from many individuals. Among these, I would like especially to acknowledge three. Since I first became interested in energy issues some twenty years ago and continuing until his death in 1991, my understanding of these issues and of nuclear energy in particular benefited greatly from discussions and collaborative writing with my colleague Fred Schmidt. Over the years, I have also gained much from the wisdom of Alvin Weinberg, who has made unique contributions to nuclear energy and its literature and, most recently, has very kindly read and commented on much of this manuscript. I am also grateful to Peter Zimmerman who served the publisher as an anonymous reviewer of a preliminary draft of this book and who subsequently, anonymity discarded, has been a very constructive critic of a revised draft.

In addition, I am heavily indebted to many other individuals at the University of Washington, in government agencies, in industry, and elsewhere. Some have been generous in aiding with information and insights, some have commented on various chapters as the book has evolved, and some have done both. Without attempting to distinguish among these varied contributions, I particularly wish to thank Mark Abhold, Thomas Bjerdstedt, Robert Budnitz, Thomas Buscheck, J. Gregory Dash, Kermit Garlid, Ronald Geballe, Marc Gervais, Emil Glueckler, Lawrence Goldmuntz, Isaac Halpern, Charles Hyde-Wright, William Kreuter, Jerrold Leitch, Norman McCormick, Thomas Murley, James Quinn, Maurice Robkin, Margaret Royan, Mark Savage, Jean Savy, Fred Silady, Bernard Spinrad, Ronald Vijuk, and Gene Woodruff.

This list is far from exhaustive and I extend my thanks and apologies to the many others whom I have failed to mention. I am also grateful to the University of Washington and the Department of Physics for making it possible for me to teach the courses and devote the time necessary for the development of this book. Finally, I must express my appreciation to my wife, Beverly, for her support and encouragement as the book progressed.

Contents

1

The Motivation for Nuclear Energy

1.1 The Need for Energy Sources

1.1.1 The Importance of Energy

The discovery and exploitation of new sources of energy has been central to human progress from the early struggle for biological survival to today's technological world. The first step was learning to control fire, with wood or other biomass as the fuel. This was followed by the harnessing of wind for ships and windmills, the use of water power from rivers, and—mostly much later—the exploitation of chemical energy from the burning of coal, oil, and natural gas. Nuclear energy, which first emerged in the middle of the 20th century, is the latest energy source to be used on a large scale.

It is often pointed out that this has not all been "progress." Some human activities are harmful to other people, to other species, and to the environment, and technological advances enable us to inflict damage more rapidly and on a larger scale than would otherwise be possible. It is also sometimes argued that our lives would be more satisfying if our material surroundings were less complex and changed less rapidly.

Nonetheless, most people in the developed countries gladly accept the fruits of technological advances, and people in less prosperous countries aspire to catch up. While the burden of inefficient or unnecessary energy consumption may be reduced, it is unlikely that there will be a consensus favoring a substantial reduction in energy use in most of the developed countries or a voluntary stemming of the rise of energy use in the developing ones, with their growing population and—it is to be hoped—improved living standards. Thus, the world will demand increasing supplies of energy during the 21st century. Nuclear power provides one option for supplying this energy, albeit a controversial one.

Table 1.1. Commercial energy sources: World consumption in 2001 and U.S. consumption in 2002.

Source	World (2001)		United States (2002)	
	Quads	Percent	Quads	Percent
Fossil fuels				
Petroleum	156	39	38.4	39
Coal	96	24	22.2	23
Natural gas	93	23	23.2	24
All fossil fuels	346	86	83.8	86
Renewable sources[a]				
Hydroelectric	27	6.6	2.6	2.6
Other renewable	3[b]	0.8	3.2	3.3
All renewable	30	7.4	5.8	6
Nuclear	26	6.5	8.1	8.3
TOTAL	397	100	98	100

[a]The reported renewable energy is used mainly for electricity generation, calculated (except for geothermal power) at the average primary energy rate for fossil-fuel steam plants (10,201 BTU/kWh) [1].
[b]The world total for "other renewable" energy is an underestimate, because it includes only energy used for electricity generation, and thus omits other uses, particularly the burning of biomass for heat (wood and wastes).
Source: Refs. [1] and [2].

1.1.2 Energy Use Patterns

Sources of Energy

For well over 100 years, the dominant energy sources in the industrialized world have been fossil fuels—coal, oil, and natural gas—and these now dominate in most of the developing world as well. Other major contributors, of varying importance in different countries, include hydroelectric power, nuclear power, and biomass.[1]

Table 1.1 indicates the main sources of energy for the United States and the world.[2] The dominance of fossil fuels is brought out in these data. They provide 86% of the primary energy for both the United States and the world. The remainder is divided between nuclear and renewable sources. The most

---

[1] The magnitude of biomass consumption is difficult to establish accurately, because much of it involves the collection of wood and wastes on an individual or small-scale basis, outside of commercial channels. Its use is therefore less well documented than is the use of fuels that are purchased commercially.

[2] Here, we follow the practice of U.S. Department of Energy publications, which report energy in BTU or quads, where 1 BTU = 1055 joules (J) and 1 Quad = 10^{15} BTU = 1.055×10^{18} J = 1.055 exajoule (EJ).

important renewable source in commercial energy channels is hydroelectric power. Biomass is included in the U.S. renewable data, but for the world data the only biomass included is the small amount used in electricity generation.

Disparities Among Countries

The disparities among countries are great. In 2001, the per capita consumption of energy for industrialized countries such as France and Japan was about 14 times that of India and almost 50 times that of Bangladesh, whereas it was only about one-half that of the United States [3]. It might be desirable and practical for the United States to reduce its per capita energy use, but in many countries there is a need for more energy. In fact, although the gap between the extremes is still very large, some progress in reducing it has been made in recent years. Thus, U.S. energy consumption per capita hardly changed from 1980 to 2001, whereas per capita consumption more than doubled for India and rose over 150% for Bangladesh—a considerable accomplishment, especially considering the substantial population growth in those countries.

An "overnight" doubling of world energy consumption (i.e., a doubling with no increase in population) would still leave the world's average per capita rate less than 40% of the U.S. rate, with many countries well below the new average. To accommodate an increasing population and an increased per capita demand in much of the world, world energy production may have to more than double over the next 50 years [4, 5]. If present trends continue, most of this additional energy will come from fossil fuels.

Growth of Energy Use in the United States

The history of energy use in the United States since World War II can be divided into two epochs: a period of rapid and unconcerned rise until the oil embargo of 1973 and a subsequent period of much slower growth. Overall, the entire period has been marked by a substantial increase in energy use and an even more rapid increase in electricity use. Figure 1.1 shows the growth from 1949 to 2002 in U.S. population, total energy use, gross domestic product (in constant dollars), and electricity use.

From 1949 to about 1975, total energy consumption closely tracked the gross domestic product (GDP) but in subsequent years it lagged GDP substantially, apparently due to more efficient use of energy and the lessening relative importance of heavy industry in the U.S. economy.

During the same early period, the growth in electricity demand outstripped that of energy and GDP, with average annual growth rates of about 10% from 1949 to 1959 and 7% from 1959 to 1973. Since 1973, electricity growth has continued, but at the relatively modest average annual rate of 2.7% for the 1973–2002 period. This rate is close to, but slightly below, the rate for GDP growth in this period (2.9%) and substantially exceeds the rate for energy consumption (0.9%).

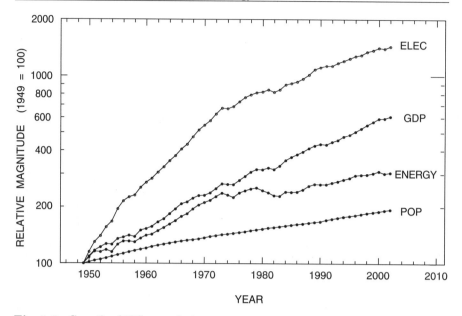

Fig. 1.1. Growth of U.S. population, energy consumption, gross domestic product (in constant dollars), and electricity use, 1949–2002 (normalized to 1949 = 100). (*Sources:* Refs. [2] and [6].)

Before 1975, it was widely accepted that GDP and energy use were tightly coupled, as evidenced by the parallel growth rates seen in Figure 1.1. That link was clearly broken after 1975, to be replaced by an apparent link between electricity and GDP. It is difficult to know how this pair will track in the next decade or two. In a largely unchanging world, incremental improvements in the efficiency of energy use would tend to cause electricity demand to lag GDP. However, expanded applications of electricity can lead to increased total consumption, even if the use rate in certain applications continues to decrease—for example, for lighting and refrigerators. As a somewhat extreme example, the switch from vacuum tube to solid state technology greatly reduced the electricity consumption of an individual radio or television set. On the other hand, the average home today has many more electronic devices than it had before the solid state revolution began.

1.1.3 The Role of Electricity

The Growth of Electrification

The growth in electricity use in the United States has been duplicated in the rest of the industrialized world. Overall, in the 20th century, electrification has been almost synonymous with modernization. It has changed the mechanics of the home, with convenient lighting, refrigeration, and motor-driven appliances, plus expanded entertainment and cultural resources. In industry, electrical motors allow machines to be used where and when they are needed, and electric equipment can deliver heat in highly controllable forms—for example, with electric arcs, laser beams, and microwaves. Medical diagnosis and treatment has been transformed by use of equipment ranging from X-rays and high-speed dental drills to lasers and magnetic resonance imaging.[3] Electricity has made possible entirely new modes of communication, as well as the development of computers and the associated means for exchanging and storing information.

Electricity plays a central role in virtually all spheres of technological life, with the exception of transportation, and even in transportation the rapid trains of Japan and Europe may be pointing to future advances. In the industrialized countries, as represented by the Organization of Economic Cooperation and Development (OECD),[4] electricity consumption rose by 117% in the 1973–2001 period [7].[5] This far outstripped the growth, during the same period, in population (26%) and in total primary energy supply (42%). It was close to but slightly greater than the increase in GDP (111%, in constant dollars). The developing countries—almost by definition—lagged for many years in the use of electric power, but there has been rapid recent growth in some countries. In China, electricity consumption rose from 69 gigawatt-years (GWyr) in 1991 to 138 GWyr in 2000, an average annual increase of 8.1% [8, p. 64]. The growth rate in South Korea in this period was still higher, averaging 10.6% per year.

It appears inevitable that the demand for electricity will increase on a worldwide basis, even if conservation restrains the growth in some countries. This increase will be driven by (a) increasing world population, (b) increased per capita use of energy, at least in successful developing countries, and (c) an increase in electricity's share of the energy budget due to the convenience of electricity in some applications, its uniqueness in others, and its cleanliness

---

[3] The technique was originally called "nuclear magnetic resonance," because magnetic properties of nuclei were being exploited. The word "nuclear" was dropped because of what were felt to be unfavorable connotations.

[4] The OECD was formed in 1960 with 20 members: the 16 countries of western Europe (including Iceland) plus Canada, Greece, Turkey, and the United States. From 1960 to 2002, it was enlarged to 30 member countries with the addition of Australia, the Czech Republic, Finland, Hungary, Japan, Korea, Mexico, New Zealand, Poland, and the Slovak Republic.

[5] Consumption increased from 471 GWyr in 1973 to 1024 GWyr in 2001, corresponding to an average annual growth rate of 2.7%. (Note: Electricity consumption is somewhat less than generation, as in Table 1.2, due to losses in transmission.)

in end use. Additional demands for electricity may arise for the production of
hydrogen and for the desalination of seawater (see Chapter 20).

Present Sources of Electricity

Most electricity is now being generated by the combustion of fossil fuels. World
generation in 2001 was roughly two-thirds from fossil fuels and one-sixth each
from hydroelectric and nuclear power. More detailed numbers are given in
Table 1.2 for the world, the OECD countries, and the United States. The electricity generated by renewable sources other than hydroelectric power is not
always reliably reported, because it is in large measure produced by entities
other than utilities and the accounting is not as reliable as for utility generation. These sources include biomass energy (wood and wastes), geothermal
energy, wind energy, and direct forms of solar generation. Of these, at present
only biomass makes an appreciable contribution, and that varies greatly from
country to country. For the United States in 2002, wind, geothermal energy,
and direct solar energy (e.g., photovoltaic) together accounted for only about
0.6% of the electricity output.

Table 1.2. Total electricity generation and percent shares by source, for world,
OECD, and United States.

	World 2001	OECD 2001	United States 2002
Total (TWh)	14,851	9,490	3,839
Total (GWyr)	1,695	1,083	438
Fossil fuel (%)			
Coal		37.9	50.2
Gas		16.8	17.9
Petroleum		5.9	2.3
Total	64.0	60.6	70.4
Nuclear (%)	17.0	24.1	20.3
Hydroelectric (%)	17.3	13.0	6.6
Other renewable (%)			
Wood and waste		1.33	1.55
Geothermal		0.34	0.35
Wind		0.36	0.27
Solar		0.01	0.01
Tide/wave/ocean		0.01	
Total	1.7	2.0	2.2
Other (%)		0.2	0.5
TOTAL (%)	100	100	100

Source: Refs. [1], [7], and [9].

1.2 Problems with Fossil Fuels

1.2.1 The Need to Replace Fossil Fuels

The long-standing impetus for the development of nuclear power has been the eventual need to replace fossil fuels—oil (or petroleum), coal, and natural gas.[6] Their supply is finite and eventually, at different rates for the different fuels, the readily available resources will be consumed, although experts disagree as to the rate at which this will happen. Some warnings of the imminent exhaustion of supplies have been premature, and concern over oil was less visible in the 1990s than it had been in the late 1970s. However, if oil shortages have been deferred, they cannot in the long run be avoided. Known and projected resources of oil are heavily concentrated in the Persian Gulf region, and unless substitutes for oil are found, the world will face a continuing series of economic and political crises as countries compete for the dwindling supplies.

Natural gas resources may exceed those of oil, measured in terms of total energy content, and the present world consumption rate is less for gas than for oil. Therefore, global shortages are somewhat less imminent. Nonetheless, gas is also a limited resource, with reliance on unconventional resources speculative.

Coal resources are much more plentiful than those of either oil or natural gas, but coal is the least environmentally desirable. Its use was banned in England in the 13th century by King Edward I due to the "intolerable smell" and the injury to the health of "magnates, citizens, and others" [10, p. 5]. With no alternatives other than dwindling supplies of wood, coal became important again in England by the 17th century, and in many countries it is now the leading fuel. It has not had a clean history, with chronic pollution punctuated by severe incidents such as 4000 deaths in the London smog of 1952 [11, p. 297]. However, output of chemical pollutants from coal, particularly sulfur dioxide, can be greatly reduced by "cleaner" burning of the coal, at a moderate additional cost.

The production of carbon dioxide in the combustion of fossil fuels presents a more difficult problem. Unless much of this carbon dioxide can be captured and sequestered, the resulting increase in the concentration of carbon dioxide in the atmosphere carries with it the possibility of significant global climate change.

Among the fossil fuels, natural gas has significant advantages. It is the least environmentally damaging, in terms of both chemical pollutants and carbon dioxide production. If the hypothesized supplies of "unconventional" natural gas live up to some of the projections, then supply difficulties may

---

[6] The terms *oil* and *petroleum* are sometimes used as synonyms, but there is no consistent practice in the literature. In United States DOE compilations, the term "petroleum" is used to embrace "crude oil" and "petroleum products" (see, e.g., Ref. [6], Table 5.3). We will here follow common practice and use the term "oil" in discussing resources.

be postponed for many decades. Further, natural gas can be used in highly efficient combustion turbines operating in a combined cycle mode, in which much of the (otherwise) waste heat from the combustion turbine is used to drive a steam turbine.

Nonetheless, reliance on natural gas as more than a short-term stop gap involves two significant uncertainties or problems. First, gas supplies may be limited to the standard conventional resources, advancing the time at which the availability of gas will become a problem and prices will rise substantially. Second, although preferable to coal in this regard, natural gas is still a source of greenhouse gases, primarily carbon dioxide from combustion and secondarily methane from leaks.[7]

1.2.2 Limitations on Fossil Fuel Supplies

Hubbert's Model

The fossil fuels are generally believed to have been formed in the distant past from the decay of organic matter. Although the supplies are large, especially for coal, they are limited.[8] The emphasis on the finite nature of fossil fuel resources, particular for oil and gas, stems from the work of the geologist M. King Hubbert, who in the 1950s predicted that U.S. oil production would peak in about 1969 if one took the higher of two estimates of total resources (see, e.g., Ref. [13], Figure 22). This prediction departed from the prevailing views of the time, which were more optimistic, but was vindicated when U.S. production actually reached a peak in 1970. By 2001, production in the lower 48 states was only about one-half of the 1970 rate [6, p. 139].

Hubbert's model is very simple. It assumes a finite resource. Exploitation of the resource rises as more uses are found for it. Eventually, the most easily extracted supplies are depleted, extraction costs rise, and higher prices lead to a reduced demand. Therefore, a graph of production as a function of time will show a rise and fall, assumed to follow a bell-shaped curve. The area under the curve corresponds to the magnitude of the total resource. If this magnitude is known and the initial use rate is observed and if one also assumes (as Hubbert did) that the curve is symmetric, then it is possible to determine the height and timing of the peak.

The early success of his prediction and the simplicity of the model behind it brought Hubbert's thinking to the fore of many analyses. The curve is

---

[7] Natural gas is primarily composed of methane (CH_4). Per unit volume (or equivalently, on a per molecule basis), methane is considerably more effective than carbon dioxide as a greenhouse gas.

[8] A maverick opinion, advanced by the astrophysicist Thomas Gold, holds that the gas and oil are largely primordial, i.e., that they were created by processes in the interior of the Earth as it was being formed [12]. If true, this would imply much larger supplies, at greater depths, than have as yet been found. However, this hypothesis is discounted by most analysts.

known as "Hubbert's Peak" [14]. One implication of the model is that the time at which peak production is reached is not very sensitive to the magnitude of the resource, because the peak rate of consumption is higher if the peak comes later. Some caution is needed in applying this model, however. In its simplest form, the model appears to ignore the possibility that if the price rises sufficiently, additional large resource categories may become available. This does not argue against the underlying point of the model, but suggests that the curve of production versus time may not be symmetric or fully predictable.

Resource Estimates

In estimating oil and gas resources, a distinction is made between "conventional" and "unconventional" resources. Conventional resources are those that are found in typical geologic formations and that can extracted by what have become standard methods. Unconventional resources are those that are located in different sorts of formations, requiring special extraction techniques, and may be available only at higher prices. Table 1.3 gives a representative set of resource estimates from a summary by Hans-Holger Rogner in *World Energy Assessment* [15].

Unconventional oil resources include the heavy crude oil found in Venezuela, the tar sands found in Canada, and the oil shale found in the western United States. Unconventional natural gas resources include deposits in coal beds and low-permeability rocks ("tight gas" formations), as well as potentially enormous but highly speculative resources of methane hydrates in permafrost and ocean sediments and of gases at high pressure in deep aquifers [15, p. 147].

The far right column in Table 1.3 gives the ratio of conventional resources to annual (1998) consumption. If consumption and production capabilities both remained constant—which they will not—this ratio would represent the time before the resource is exhausted. The difference between the ratio for oil

Table 1.3. Estimates of world fossil fuel resources, as presented in *World Energy Assessment.*

Fossil Fuel	Resource Base[a] (EJ)		Consumption (EJ/yr), 1998	Ratio[b]
	Conventional	Unconventional		
Oil	12.1×10^3	20.3×10^3	142	85
Natural gas	16.6×10^3	33.2×10^3	84	≈ 200
Coal	200×10^3		92	≈ 2200

[a]The resource base is the remaining resource, omitting less well-established "additional occurences" of unconventional resources. [Note: 1 exajoule (EJ) = 10^{18} J = 0.978 quad.]

[b]This is the ratio of the resource base for conventional resources to the annual consumption.

Source: Ref. [15], p. 149.

and for natural gas may be misleading, because there are pressures at work that may increase gas consumption more rapidly than oil consumption—for example, the efforts in the United States to replace oil in transportation and to emphasize natural gas in future electricity generation. It is clear however, whatever the situation with oil and gas, that there will be no shortage of coal for many hundreds of years. However, the combustion of all this coal would release about 5000 gigatonnes of carbon. If one-half remained in the atmosphere, the pre-industrial atmospheric concentration of CO_2 would be more than quadrupled.

Oil has been the most intensively studied resource. Its resources are usually couched in terms of billions of barrels of oil (bbo), not exajoules, where 1 bbo is equivalent to about 6.0 EJ. The remaining conventional oil resource, as can be seen from Table 1.3, is 2000 bbo and past world consumption of conventional oil (through 1998) was 4.9×10^3 EJ or 800 bbo, giving an estimated "ultimate recovery" of 2800 bbo in the Rogner analysis. This lies between estimates made in the 1990s, cited by James MacKenzie, that cluster around 2000 bbo [16] and a later estimate of about 3000 bbo made by the U.S. Geological Survey [17]. If one adopts the Hubbert curve viewpoint, even with an ultimate resource of 3000 bbo, the peak in world oil production will be reached in the year 2019 [18]. This suggests that, despite the seemingly large conventional oil resource, the squeeze on its supplies will be globally felt within a decade or two. The supplies can be augmented by the unconventional resources, perhaps at higher prices.

The situation is more pressing when regional differences in oil resources are considered. Almost 30% of the world's oil has come from the Persian Gulf region in recent years, and the countries of this region have a disproportionately large share of the remaining resources [9]. As other countries gradually use up their resources, the abundant reservoirs of the Persian Gulf will become proportionally more important, further increasing the political sensitivity of the region.

The United States, with its past high rate of production, has already used up a substantial fraction of its domestic low-cost oil resources, and in 2002, it relied on (net) imports for 53% of its oil supply at a net import cost of $94 billion [9]. This share has been rising over the past two decades and can be expected to continue to rise unless oil consumption is curtailed. The United States and world dependence on oil from the Persian Gulf region has led to considerable political and military unrest. In response, there have been intermittent attempts by the U.S. government to lessen the country's dependence on oil imports. The potential for increased domestic supplies is limited, and the most promising avenue is reduced consumption.

Nuclear power could make only limited additional contributions here, at least in the short run. The use of oil for electricity generation has been greatly reduced since the 1970s and its use in the residential and commercial sectors is also reduced. By 2001, these three sectors accounted for only 9% of U.S. petroleum product consumption [6, Table 5.12]. Virtually all of this use could be replaced by natural gas or electric power. A further gain would come if gas and electricity partially replaced petroleum in industry.

However, two-thirds of all petroleum use in the United States is for transportation, and, here, change cannot be accomplished quickly because our living patterns depend heavily on automobiles and trucks. Moving away from this use of petroleum fuels would require a new (and still unproven) fleet of vehicles and the supporting infrastructure. A more effective approach for the near term would be to increase the average fuel efficiency (in miles per gallon) of conventional vehicles. Further gains could be made by increased use of mass transportation, especially electrified mass transportation. The replacement of petroleum-based fuels by alternatives, particularly hydrogen, is a possibility for the further future. Here, nuclear power could play a role as an energy source for hydrogen production (see Chapter 20).

1.2.3 Global Climate Change

Production of Carbon Dioxide

If the Earth had no atmosphere, its average surface temperature would be about $-18°C$. The Earth is kept at its relatively warm temperature by molecules in the atmosphere, including water molecules and carbon dioxide molecules, that absorb some of the infrared radiation emitted by the Earth and prevent its escape from the Earth's environment.[9] This is the natural "greenhouse effect." Since the beginning of the industrial era, additional gases have been emitted into the atmosphere—particularly carbon dioxide (CO_2)—which add to this absorption and are believed to further increase the Earth's temperature.[10] This increment is referred to as the anthropogenic greenhouse effect. Warnings about the effects of CO_2 emissions date to the 19th century, but they have become a matter of widespread concern only since the 1970s. The anticipated consequences are described as "global warming" or, more broadly, as "global climate change."

The production of CO_2 is the inevitable accompaniment of any combustion of fossil fuels. The amount released per unit energy output varies for the different fuels, due largely to differences in their hydrogen content. Natural gas is primarily methane (CH_4) and a considerable fraction of its combustion energy comes from the chemical combination of hydrogen and oxygen. Its ratio of carbon dioxide production to energy production is the lowest among the fossil fuels.

The releases are usually specified in terms of the mass of carbon (C), not CO_2. Even for a given fuel type, the amount produced per unit energy is

---

[9] Every object radiates energy at an average wavelength determined by its temperature. For the Earth, which is considerably cooler than typical hot glowing objects, the radiation is at wavelengths longer than those of the visible spectrum—namely, in the infrared region.

[10] Other greenhouse gases include methane, chlorofluorocarbons, and nitrous oxide.

not a constant because the chemical composition of the fuels is nonuniform. However, for many purposes, approximate average values are adequate. Approximate coefficients, in megatonnes (Mt) of carbon per exajoule (EJ) of energy, are 24.6 Mt/EJ for coal, 18.5 Mt/EJ for petroleum, and 13.7 Mt/EJ for natural gas.[11] These numbers illustrate the benefit of switching from coal to natural gas, when possible.

The Effects of Greenhouse Gases

The increases during the past century in the atmospheric concentrations of greenhouse gases, especially carbon dioxide, has been unambiguously established. The potential consequences of these increases are controversial and the appropriate policy responses are even more controversial. The conclusions of the Intergovernmental Panel on Climate Change, as put forth in 2001 in its Third Assessment, represent the "conventional wisdom" of the world community of atmospheric scientists—although not a unanimous opinion.[12] The conclusions depend on both the scenario assumed for energy production during the coming century and the results of complex computer models of the response of the environment to the input of greenhouse gases. There is a wide range in the quoted effects, reflecting uncertainties in the atmospheric models and in future rates of greenhouse gas production. The projected effects for the period until 2100 include the following:

- An increase in global average temperature on the Earth's surface of 1.4°C to 5.8°C (2.5–10.4°F). About one-half of this rise is anticipated to take place by 2050.
- Increased average global precipitation.
- A rise in the average sea level due to the melting of glaciers and the thermal expansion of the oceans, by an amount projected to lie in the broad interval of 9–88 cm.
- Increased frequency and intensity of "extreme events," including "more hot days, heat waves, heavy precipitation events, and fewer cold days," with possibly "increased risks of floods and droughts in many regions" [20, p. 14].

The IPCC warns that "large-scale, high-impact, nonlinear, and potentially abrupt changes" could be caused by the greenhouse gases [20, p. 14]. These changes might be irreversible, locked in by positive feedbacks associated, for example, with greater emissions of greenhouse gases from the soil when the temperature rises.

Most of these possibilities are stated in broad terms, reflecting the uncertainties. Additional warnings are contained in a 1997 book by Sir John

---

[11] These values are based on Refs. [19], Appendix B, and [6], Appendix C, which provide more detailed information.

[12] A condensed statement of these conclusions is given in a *Summary for Policy Makers* of the IPCC Synthesis Report [20].

Houghton, then co-chairman of the Scientific Assessment Working Group of the IPCC and chairman of the United Kingdom's Royal Commission on Environmental Pollution. He pointed out that storms and floods claimed over 700,000 lives in the period from 1947 to 1980 [21, p. 3] and suggested that the projected climate changes are likely to lead to more frequent and severe floods (and droughts). Some models suggest an increased intensity of storms as well, but this effect is not well established [21, p. 101]. Even a slight increase in the frequency or severity of floods and storms could mean many additional casualties.

A potentially devastating, but also quite uncertain, possibility is the collapse of the West Antarctic Ice Sheet. If it occurs, it would cause a 5-m rise in sea level, affecting millions of people living in low-lying coastal regions. In the words of Houghton, "there is no reason to suppose there is a danger in the short-term (for instance, during the next century) of the collapse of any of the major ice sheets" [21, p. 110]. The IPCC report suggests a somewhat longer time scale, indicating that "after sustained warming the ice sheet could lose significant mass and contribute several meters to the projected sea-level rise over the next 1000 years" [20, p. 15].

In ordinary thinking, a danger postponed 1000 years is not a matter of much concern. However, the discussions of nuclear waste disposal—where EPA regulations establish a 10,000-year period of responsibility and suggest considering a longer one—point up the question of our responsibility to future generations. How concerned should we be if our actions today may impact people hundreds or thousands of years from now? We return to this question in Chapter 13, in the context of nuclear waste disposal, but the broad issues discussed there are pertinent to any actions or inactions we take today, including those related to global climate change.

Sources of Carbon Dioxide Emissions in the United States

Fossil fuel combustion in the United States in 2001 caused the emission into the atmosphere of 1.56 gigatonnes of carbon (GtC) [19, p. 32].[13] In recent years, the United States has been responsible for about one-quarter of the world CO_2 emissions [6, p. 315]. In 2000, the U.S. share was 24%. CO_2 emissions continue to rise in much of the world, despite calls for their curtailment. From 1991 to 2000, the increase was 17% for the United States and 10% for the world as a whole. There was a small decrease in U.S. CO_2 emissions from 2000 to 2002 due to lower coal and natural gas consumption, but it would be premature to assume that this a trend [9].

The amounts of CO_2 produced in the United States, classified by economic sector and fuel type, are shown in Figure 1.2. There are two major contributors, each responsible for almost one-third of the total emissions: coal in electricity generation and petroleum in transportation.

---

[13] Another 0.032 GtC were produced in other activities, including 0.011 GtC in cement production.

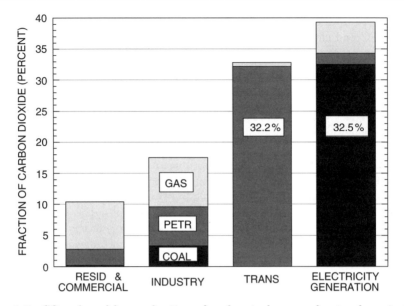

Fig. 1.2. CO_2 released by combustion of coal, petroleum, and natural gas in the various sectors of the U.S. energy economy in 2001, expressed as percents of total. (*Source*: Ref. [19].)

The elimination or reduction of carbon dioxide emissions from coal-fired electricity generation is straightforward, if one is willing to pay the costs. It could be done by substitution of nuclear power or renewable sources for coal or, if the technique proves practical on a large scale, by sequestration of the carbon dioxide emitted from coal-fired power plants. The reduction of carbon dioxide emissions in transportation is more difficult. There is now considerable speculation about hydrogen as an energy carrier for use in transportation, but its practicality has not been established. These possibilities are discussed at greater length in Sections 20.2 and 20.3.

1.3 Nuclear Power as a Substitute for Fossil Fuels

1.3.1 Alternatives to Fossil Fuels

The Range of Alternatives

The challenge in energy policy is to reduce CO_2 emissions and the world's dependence on oil while satisfying a substantially increased demand for energy. Putting aside the still-speculative possibility of sequestering carbon dioxide (see Section 20.2.2), this challenge reduces to that of using energy more efficiently and finding substitutes for fossil fuels. Alternatives to fossil fuels fall into two broad categories:

♦ *Renewable sources.* Most of these sources—including hydroelectric power, wind power, direct solar heating, photovoltaic power, and biomass—derive their energy ultimately from the Sun and will not be exhausted during the next billion years. Geothermal energy and tidal energy are also renewable, in this sense, although they do not rely on the sun.[14] However, there is almost an inverse correlation between the extent to which the source is now being used (see Table 1.2) and the size of the potentially tappable resource. Thus, expansion of hydroelectric power (which is substantially used) is constricted by limited sites and environmental objections, whereas wind (for which the resource is large) is as yet less used and thus is not fully proven as a large-scale contributor.

♦ *Nuclear sources.* The two nuclear possibilities are fission and fusion. The latter would be inexhaustible for all practical purposes, but developing an effective fusion system remains an uncertain hope. Fission energy would also have an extremely long time span if breeder reactors are employed, but with present-day reactors limits on uranium (or thorium) resources could be an eventual problem. At present, fission power faces problems of public acceptance and economic competitiveness.

The broad alternatives of renewable energy and nuclear energy can be considered as being in competition, with one or the other to be the dominant choice, or complementary, with both being extensively employed.

Early Consideration of the Alternatives

When the possibility of nuclear energy first was recognized in the 1930s and early 1940s, it had the attraction of offering very large amounts of energy from very small amounts of material. This excited the imagination of scientists and writers, and fission was looked upon as a very promising potential energy source. Following the technological success of the World War II atomic bomb program, it appeared likely that commercial nuclear energy would prove to be practical.

The development of nuclear power in the United States was the responsibility of the Atomic Energy Commission (AEC), an agency established in 1946 to oversee both military and civilian applications of nuclear energy. Although the desirability of moving ahead with nuclear energy was widely accepted, the AEC undertook in 1949 to sponsor a more analytic study, described as

> ... a study of the maximum plausible world demands for energy over the next 50 to 100 years. The study was envisaged as background

---

[14] Geothermal sources may be temporarily depleted on a local basis, but are replenished from elsewhere in the Earth.

for the Commission's consideration of the economic and public policy problems related to the development and use of machines for deriving electrical power from nuclear fuels.[15]

The AEC asked Palmer C. Putnam, a consulting engineer with broad interests, to carry out the study.[16] The results of Putnam's study appeared in 1953 in the book *Energy in the Future* [22]. In retrospect, the book is a prophetic masterpiece. It started with the consideration of future increases in population, in demand for energy, and in the efficiency of delivering energy. Putnam then addressed the issues of fossil fuel reserves, concluding that we could not live "much longer" off fossil fuels, which he termed "capital energy." He also pointed out the possible dangers of climate change from carbon dioxide produced in the combustion of fossil fuels [22, p. 170].

Putnam next turned to the potential of what we now call renewable energy, which he termed "income energy." This is primarily solar energy, in all its forms. He concluded that the world could not expect to obtain "more than 7 to 15 percent of the maximum plausible demands for energy from 'income' sources at costs no greater than 2 times present costs" [22, p. 204].

This led Putnam to the conclusion that a new "capital" source of energy would be required, i.e., nuclear energy. With breeder reactors, he indicated that world uranium supplies would suffice for "many centuries" [22, p. 250]. However, he pointed out that nuclear energy could only make a decisive contribution if transportation and home heating were electrified to a much greater extent than was the case in the early 1950s. In summary, Putnam urged the prompt development of nuclear power, the exploration of nuclear fusion, and "as our ultimate anchor to windward," exploration of ways to obtain solar energy "in more useful forms and at lower costs than now appear possible" [22, p. 255].

The alternatives to fossil fuels that Putnam contemplated 50 years ago remain the alternatives today: nuclear fission, nuclear fusion, and solar energy. There is considerable disagreement today on both the immediate and ultimate potential of solar energy. One view is that it is not possible to obtain adequate amounts of energy from renewable sources, either now or in the predictable future. An opposing view holds that a combination of renewables and conservation could, in a matter of decades, make fossil fuels and nuclear energy unnecessary.

Lacking confidence that renewable energy alone will suffice to replace fossil fuels, the U.S. government has adopted a policy in recent years of keeping the nuclear option alive, without a major investment in fostering its growth. In

---

[15] From Foreword to Ref. [22].

[16] In selecting Putnam, the AEC does not appear to have attempted to stack the deck in favor of nuclear power. He had previously written books with titles such as *Chemical Relations in the Mineral Kingdom*, *Power From the Wind*, and *Solar Energy*, and he was one of the designers of the giant Smith–Putnam windmill built at Grandpa's Knob in Vermont in the 1940s.

the remainder of the book, this option will be examined in more detail. At the moment, it is not clear whether nuclear power is headed toward a continued hiatus, gradual abandonment, or a renewed rapid expansion. (We will return to these general questions at greater length in Chapter 20.)

1.3.2 The Potential Role of Nuclear Energy

Here, we consider the contribution nuclear power could make to solving the world's energy problems, given a decision to expand its use. In principle, renewable energy could do much the same if it could be made available on the same scale, but we will defer further consideration of its potential until brief mention in Section 20.2.3.

For the developed countries, where the increase in energy demand over the next 50 years could be fairly small if conservation measures are vigorously implemented, the most important contribution would be in direct displacement of fossil fuel sources. Potential measures include the following:

♦ The gradual replacement of present coal-fired power plants by nuclear plants. Both coal and nuclear plants are used primarily for baseload generation; their roles are interchangeable.

♦ The use of nuclear power rather than natural gas when new capacity is needed. This would free natural gas to replace oil or coal in heating and other applications.

♦ The replacement of petroleum in transportation. As already discussed in the context of resources, this change is more difficult to implement. Looking ahead several decades, nuclear energy could contribute by providing power for electric vehicles, hydrogen production (see Section 20.3.2), and electrified mass transportation. More immediately, the most effective remedy is to increase the efficiency of motor vehicles (i.e., improve the "gas mileage"), in which nuclear power would have no role.

♦ The replacement of fossil fuels by electricity for heating. In industry, electric heating can be applied at the desired location and time, with unique precision. In homes and commercial buildings, efficient use of electricity can be achieved with heat pumps or controlled zone heating, although not with electric central furnaces.

For the developing countries, which hope to increase their energy consumption substantially, the expanded use of nuclear power faces the problem of limited capital resources and, in some cases, an inadequate technical base. However, the two largest countries in this category—China and India—have considerable nuclear sophistication and to the extent capital is available, they could turn to nuclear power instead of coal or natural gas to fuel their electricity expansion.

1.3.3 The Example of France

The changes suggested above could not be implemented immediately, but a significant part could be accomplished on the time scale of decades, as illustrated by the history of the French energy economy since the early 1970s. Nuclear advocates cite this history as an example of the contribution nuclear energy can make in reducing carbon dioxide emissions and, in some situations, reducing the demand for oil. Table 1.4 summarizes the changes in the French energy economy from 1970 to 1995—a period during which nuclear's share of electricity generation rose from 6% to 77%. The parallel record for the other European OECD countries (i.e., excluding France) is shown for comparison.

During this period, electricity generation in France more than tripled and total energy supply rose 56%, while carbon emissions and petroleum use each dropped 16%. At the same time both population and GDP rose. Some of these accomplishments can be attributed to the increased use of natural gas, but nuclear power deserves the lion's share of the credit. The replacement of fossil fuels by nuclear energy in the generation of electricity was particularly noteworthy. The fossil fuel share of electricity generation dropped from 62% to 8% in this period while the nuclear share increased from 6% to 77%. The drop in petroleum use also meant lower oil imports.

The rest of OECD-Europe did not experience changes of this magnitude (and sometimes not even in the same direction). Although per capita production of CO_2 decreased slightly, total carbon dioxide production rose. Even here, the rise was somewhat moderated by an increased use of nuclear power. If the other European OECD countries had France's profile of energy sources in 1995 (with no change in their total energy supply), their carbon emissions would have been 550 MtC instead of the actual 866 MtC total, a reduction of 36%. Had their per capital energy use risen at the same time to equal that of France, their carbon emissions still would have been 18% below the actual 1995 levels.

1.3.4 The Status of Nuclear Energy

Initial Optimism and Later Reality

General perceptions of nuclear energy, among both the public and policy makers, have undergone dramatic shifts in the past 50 years. As nuclear energy emerged in 1945 from scientific obscurity and military secrecy, it began to be talked of in speculative terms as an eventual power source. Within a decade, an enthusiastic vision developed of a future in which nuclear power would provide a virtually unlimited solution for the world's energy needs.

It was not difficult to picture nuclear power as the ideal energy source. With the use of breeder reactors, it would be ample in supply. As experience was gained in reactor construction, it would become economical; and because a nuclear reactor would emit virtually no pollutants, it would be clean, especially

Table 1.4. Comparison between changes (Δ) in energy indicators for France and for other European OECD countries (taken together), 1971 to 1995.

	France			Other OECD Europe		
	1971	1995	Δ (%)	1971	1995	Δ (%)
Population (millions)	51.3	58.1	13	359	408	14
GDP (billion $US)[a]	622	1064	71	3307	5901	78
Total energy supply (EJ)[b]	6.5	10.1	56	41.7	55.0	32
Electricity generation (GWyr)	17.8	55.9	214	133	250	88
Per capita indicators						
GDP (1000 $)	12.1	18.3	51	9.2	14.5	57
Total energy supply (GJ)	126	174	38	116	135	16
Electricity generation (MWh)	3.0	8.4	177	3.3	5.4	65
Carbon emissions (tonnes C)	2.34	1.73	−26	2.23	2.12	−5
Energy supply[c] (EJ)						
Petroleum	4.31	3.63	−16	23.0	23.6	3
Coal	1.40	0.67	−52	13.4	11.2	−17
Natural gas	0.41	1.24	205	3.2	11.4	257
Fossil fuel total	6.12	5.54	−10	39.6	46.1	17
Hydroelectric[d]	0.18	0.26	46	1.0	1.5	53
Nuclear	0.10	4.12		0.5	5.3	
Carbon emissions[e] (MtC)						
Petroleum	80	67	−16	425	437	3
Coal	35	17	−52	330	274	−17
Natural gas	6	17	205	44	156	257
Fossil fuel total	120	101	−16	799	867	9
Electricity generation[c] (%)						
Petroleum	28	2		23	10	
Coal	29	5		43	35	
Natural gas	5	1		6	11	
Fossil fuel total	62	8		73	57	
Nuclear	6	77		4	22	
Hydroelectric	31	15		23	19	

[a] GDP values are in 1990 U.S. dollars, based on purchasing power parities, as in Ref. [23].
[b] Electricity exports are subtracted from total energy supply.
[c] Contributions from non-hydroelectric renewable sources are not listed here but are included in the totals.
[d] The energy supply corresponding to hydroelectricity is calculated by the OECD assuming a 100% conversion efficiency; this leads to a quoted hydroelectric energy supply that is about one-third that given in U.S.DOE documents, where a conversion efficiency is used that is equal to the efficiency of steam-electric plants (about 33%).
[e] Carbon emissions, in megatonnes of C (MtC), are calculated using approximate average conversion factors: petroleum, 18.5 MtC/EJ; coal, 24.6 MtC/EJ; natural gas, 13.7 MtC/EJ (based on Refs. [19], Appendix B and [6], Appendix C).
Source: Ref. [23], using conversion factor: 1Mtoe = 0.04187 EJ.

in contrast to coal. More generally, nuclear energy was favored by an almost romantic image of it as a source of abundant, clean energy, by the possible technological imperative to move ahead because it was possible to do so, and by the correct, even if not well quantified, recognition of an eventual limit to fossil fuels.

There was also a negative side, as some doubters pointed out from the first. Very large amounts of radioactivity would be produced. In principle, practically none need escape, but the possibility of mishaps could not be totally excluded. Further, benign nuclear power had a malign older sibling in nuclear weapons. Although many people understood that a reactor itself could not explode like a bomb, some members of the public feared that in some way controlled nuclear energy might go out of control.

The optimists prevailed for two decades, into the early 1970s, and many nuclear reactors were designed, built, and put into operation in the United States and Europe. Part of the motivation for this development was the desire of countries to reduce their heavy dependence on oil, which they expected would eventually be in short supply. The first oil crisis came in 1973, even sooner than had been anticipated. Just as the nuclear buildup was gaining momentum, an oil embargo was imposed by the Organization of Petroleum Exporting Countries (OPEC), as a sequel to the October 1973 war between Egypt and Israel.

An immediate impulse was to rely even more on nuclear power as a substitute for oil. This was especially true in the United States, where nuclear energy had already appeared to many, including the federal government, as an important key to "energy self-sufficiency."[17] However, the embargo had unanticipated effects. It focused new attention on the possibility of reducing *all* energy consumption, and the rising oil prices slowed the pace of economic growth. These factors sharply reduced the demand for electricity and, therefore, the pressure to add new nuclear power plants. At the same time, the economic costs of nuclear power and fears about nuclear power both began to grow. The Three Mile Island accident in 1979 and the Chernobyl accident in 1986 hit a world becoming more attuned to believing the worst about nuclear power.[18]

Nuclear energy development was stopped or brought to a crawl in all but a few countries during the 1980s and 1990s. Contributing factors to this decline included a gradual reduction in oil and gas prices, rising nuclear costs, the sluggishness of the growth in energy demand, general fears of nuclear power,

---

[17] Just prior to the embargo, an Atomic Energy Commission policy statement was formulated in the report, *The Nation's Energy Future*, based on studies carried out during the summer of 1973 and submitted to President Nixon on December 1, 1973 [24]. To eliminate the need for oil imports, it called for major contributions from coal, nuclear power, conservation, and domestic oil and gas.

[18] The antinuclear movie, *The China Syndrome*, had reached the theaters slightly before the Three Mile Island accident, indicative of the growing negative image of nuclear power in popular culture.

and, in some countries, determined campaigns against it. By 2000, it was easy to think that the age of nuclear power was coming to an end. The apparent rise and fall of nuclear power came quickly: Virtually unheard of in 1940, it was a panacea in the early 1970s, and a pariah by the 1990s.

However, this is a very incomplete picture. There was always less unanimity than this description suggests. Not only were there scientists who were aware of nuclear power's potential in 1940, but, more significantly, there were skeptics from the first, and others who have remained enthusiasts throughout. The final verdict has not been given, and the current trends are different in different countries. Nuclear power may seem dormant or dying in Western Europe, but it remains a vital activity in parts of Asia. The actual picture varies from country to country, with considerable long-term uncertainty in most industrialized countries.

Problems Facing Nuclear Energy

The decline in the growth of nuclear power in recent years can be attributed to both a less favorable economic environment and concerns about its overall safety. The nuclear decline is not universal, with a very heavy reliance on nuclear power in some counties—most notably, France—and continuing construction programs in others, especially in Asia. Nevertheless, for the world as a whole, there is a dramatic difference between the expectations of 1970 and the reality of today.

Some of the reasons have been economic. A slowing in the overall growth in energy consumption has lessened the strain on fossil fuel supplies. That strain was further relieved by additional oil and gas production in many parts of the world and partial replacement of fossil fuels by nuclear power. In this changed balance between supply and demand, the cost to utilities of fossil fuels dropped substantially after the early 1980s. In 2001, despite a very high price for natural gas during the first part of the year, the average fossil fuel cost to utilities—dominated by the price of coal—was less than one-half the cost in 1981, expressed in constant (i.e., inflation-adjusted) dollars.[19] During part of this period, nuclear power costs rose. Thus, in many countries, including the United States, nuclear power lost its cost advantage over coal and even natural gas.

However, economic factors are only part of the story. There has also been widespread concern over the possibility of reactor accidents, the disposal of radioactive wastes, and the possibility that nuclear power could help some countries produce nuclear bombs. Since the terrorist attacks of September 11, 2001, there has also been heightened fear that nuclear facilities might be vulnerable targets for future attacks.

---

[19] The average cost in 2001 was 77% of the cost in 1981 in current dollars [9, Table 9.10] and the value of the dollar was 57% of its 1981 value (as measured by the implicit price deflator [6, Appendix E].

In counterpoint, there is also the widely expressed belief that these concerns are exaggerated and that it is less dangerous to use nuclear power than to try to get by without it. The goals of reducing CO_2 production and world dependence on oil will be hard to achieve under any circumstances. They will be all the harder to achieve without taking advantage of all effective energy sources—including, in this view, nuclear power.

It is impossible to resolve these controversies in a universally convincing fashion. Instead of attempting such a task, the remainder of the book discusses basic aspects of nuclear power—primarily technical aspects. The emphasis will be on describing nuclear reactors and the nuclear fuel cycle and examining the associated problems. Issues of history, policy, and public attitudes will be addressed to some extent, but the main focus will be on the physical functioning of nuclear power, rather than its social functioning. The question of nuclear power's future will be discussed further in Chapter 20.

References

1. U.S. Department of Energy, *International Energy Annual 2001*, Energy Information Administration Report DOE/EIA-0219(2001) (Washington, DC: U.S. DOE, March 2003).
2. U.S. Department of Energy, *Monthly Energy Review, August 2003*, Energy Information Administration Report DOE/EIA-0035(2003/08) (Washington, DC: U.S. DOE, August 2003).
3. U.S. Department of Energy, "Table E1C. World Per Capita Primary Energy Consumption, 1980–2001," Energy Information Administration, International Energy Database (February 2003). [From: http://www.eia.doe.gov/pub/ international/iealf/tablee1c.xls]
4. Nebojusa Nakićenović, Arnulf Grübler, and Alan McDonald, eds., *Global Energy Perspectives* (Cambridge: Cambridge University Press, 1998).
5. William C. Sailor, David Bodansky, Chaim Braun, Steve Fetter, and Bob van der Zwaan, "A Nuclear Solution to Climate Change?" *Science* 288, 2000: 1177–1178.
6. U.S. Department of Energy, *Annual Energy Review 2001*, Energy Information Administration Report DOE/EIA-0384(2001) (Washington, DC: U.S. DOE, 2002).
7. International Energy Agency, *Energy Balances of OECD Countries 2000–2001* (Paris: OECD/IEA, 2002).
8. U.S. Department of Energy, *International Energy Annual 2000*, Energy Information Administration Report DOE/EIA-0219(2000) (Washington, DC: U.S. DOE, May 2002).
9. U.S. Department of Energy, *Monthly Energy Review, June 2003*, Energy Information Administration Report DOE/EIA-0035(2003/06) (Washington DC: U.S. DOE, June 2003).
10. Harold H. Schobert, *Coal: The Energy Source of the Past and Future* (Washington, D.C.: American Chemical Society, 1987).
11. Michael Allaby, *Dictionary of the Environment*, 2nd edition (New York: New York University Press, 1983).

12. T. Gold, "The Origin of Natural Gas and Petroleum and the Prognosis for Future Supplies," *Annual Review of Energy* 10 (1985): 53–77.

13. M. King Hubbert, *Nuclear Energy and the Fossil Fuels*, Publication No. 95, Shell Development Company (Houston, TX: Shell, June 1956).

14. Kenneth S. Deffeyes, *Hubbert's Peak: The Impending World Oil Shortage*, (Princeton, NJ: Princeton University Press, 2001).

15. Hans-Holger Rogner, "Energy Resources," in *World Energy Assessment: Energy and the Challenge of Sustainability* (New York: United Nations Development Program, 2000): 135–171.

16. James J. MacKenzie, "Heading Off the Permanent Oil Crisis," *Issues in Science and Technology* XII, no. 4 (Summer 1996): 48–54.

17. USGS World Energy Assessment Team, *U.S. Geological Survey World Petroleum Assessment 2000—Description and Results*, Executive Summary, USGS Digital Data Series—DDS-60 (Denver, Colorado: USGS, 2000). [From: http://greenwood.cr.usgs.gov/energy/WorldEnergy/DDS-60/ESpt4.html#table]

18. Albert A. Bartlett, "An Analysis of U.S. and World Oil Production Patterns Using Hubbert-Style Curves," *Mathematical Geology* 32, no. 1 (2000): 1–17.

19. U.S. Department of Energy, *Emissions of Greenhouse Gases in the United States 2001*, Energy Information Administration Report DOE/EIA-0573(2001) (Washington, DC: U.S. DOE, December 2002).

20. Intergovernmental Panel on Climate Change, *Climate Change 2001: Synthesis Report, Summary for Policy Makers* (Cambridge, England: Cambridge University Press, 2001). [From: http://www.ipcc.ch/pub/syreng.htm]

21. John Houghton, *Global Warming, The Complete Briefing*, 2nd edition (Cambridge: Cambridge University Press, 1997).

22. Palmer C. Putnam, *Energy in the Future* (New York: D. Van Nostrand, 1953).

23. International Energy Agency, *Energy Balances of OECD Countries 1994–1995* (Paris: OECD, 1997).

24. U.S. Atomic Energy Commission, *The Nation's Energy Future, A Report to Richard M. Nixon, President of the United States*, WASH-1281, submitted by Dixy Lee Ray, Chairman, U.S. AEC, December 1, 1973.

2

Nuclear Power Development

2.1 Present Status of Nuclear Power

Nuclear reactors are now being used in 31 countries for the generation of electricity. Overall, they supply about one-sixth of the world's electricity [1]. As of November 2003, there were 440 operating nuclear power reactors in the world, with a combined net generating capacity of 360 gigawatts-electric (GWe) [2].[1] Summary data on these reactors are presented in Table 2.1, which lists for each country the total nuclear generation and nuclear power's fraction of the total electricity generation.

The United States is the leader in total generation and France, among the larger countries, is the leader in the fraction of electricity obtained from nuclear power. By and large, use of nuclear power is concentrated in the industrialized countries of the Organization for Economic Co-operation and Development (OECD) and of Eastern Europe. However, some countries in the OECD have no nuclear power (for example, Italy) and some Asian countries that are not OECD members are increasing their nuclear programs (e.g., China and India).

The listing in Table 2.1 includes only reactors used for electricity generation. In addition, there are a large number of reactors throughout the world used for research and the production of radioisotopes for medical and industrial applications. Beyond these civilian reactors, an undisclosed number of

---

[1] We here use data compiled by the International Atomic Energy Agency (IAEA) and posted, with periodic updating, on its website [2, 3].

Table 2.1. World nuclear status: Operating power reactors (November 2003), net generation (2002), nuclear share of electricity generation (2002).

Country	Status 11/17/03		Generation, 2002		Percent Nuclear 2002
	Number of Units	Capacity (Net GWe)	Net TWh	Net GWyr	
United States	104	98.2	780	89.1	20
France	59	63.1	416	47.4	78
Japan	54	44.3	314	35.8	34
Germany	19	21.3	162	18.5	30
Russia	30	20.8	130	14.8	16
South Korea	18	14.9	113	12.9	39
United Kingdom	27	12.1	81	9.3	22
Ukraine	13	11.2	73	8.4	46
Canada	16	11.3	71	8.1	12
Sweden	11	9.4	66	7.5	46
Spain	9	7.6	60	6.9	26
Belgium	7	5.8	45	5.1	57
Taiwan	6	4.9	34	3.9	21
Switzerland	5	3.2	26	2.9	40
China	8	6.0	23	2.7	1.4
Finland	4	2.7	21	2.4	30
Bulgaria	4	2.7	20	2.3	47
Czech Republic	6	3.5	19	2.1	25
Slovakia	6	2.4	18	2.0	55
India	14	2.5	18	2.0	4
Brazil	2	1.9	14	1.6	4
Lithuania	2	2.4	13	1.5	80
Hungary	4	1.8	13	1.5	36
South Africa	2	1.8	12	1.4	6
Mexico	2	1.4	9.4	1.1	4
Argentina	2	0.9	5.4	0.6	7
Slovenia	1	0.7	5.3	0.6	41
Romania	1	0.7	5.1	0.6	10
Netherlands	1	0.5	3.7	0.4	4
Armenia	1	0.4	2.1	0.2	40
Pakistan	2	0.4	1.8	0.2	2.5
WORLD TOTAL	440	360	2574	294	

Note: IAEA data are used for this table. Other compilations differ somewhat, due to differences in defining the status of some reactors and in dates (e.g., *Nuclear News* lists 444 reactors worldwide at the end of 2002, with a capacity of 364 GWe [4]). *Source*: Data from Refs. [2] and [3].

reactors have been used for the production of plutonium and the propulsion of ships.

2.2 Early History of Nuclear Energy

2.2.1 Speculations Before the Discovery of Fission

Atomic energy, now usually called nuclear energy,[2] became a gleam in the eye of scientists in the early part of the 20th century. The possibility that the atom held a vast reservoir of energy was suggested by the large kinetic energy of the particles emitted in radioactive decay and the resultant large production of heat. In 1911, 15 years after the discovery of radioactivity, the British nuclear pioneer Ernest Rutherford called attention to the heat produced in the decay of radium, writing:

> This evolution of heat is enormous, compared with that emitted in any known chemical reaction.... The atoms of matter must consequently be regarded as containing enormous stores of energy which are only released by the disintegration of the atom. [5]

At this time, however, Rutherford had no concrete idea as to the source of this energy. The magnitudes of the energies involved were gradually put on a more quantitative basis as information accumulated on the masses of atoms, but until the discovery of fission in 1938, there could be no real understanding of how this energy might be extracted. In the interim, however, important progress was made in understanding the basic structure of nuclei. Major steps included the discovery by Rutherford in 1911 that an atom has a nucleus, the discovery in 1932 of the neutron as a constituent of the nucleus, and a series of experiments undertaken in the 1930s by Enrico Fermi and his group in Rome on the interactions between neutrons and nuclei, which eventually led to the discovery of fission.[3]

As these developments unfolded, there was general speculation about atomic energy. One of the earliest scientists to have thought seriously about the possibilities was Leo Szilard, who was later active in efforts to initiate the U.S. atomic bomb program and then to limit it. He attributed his first interest in the extraction of atomic energy to reading in 1932 a book by the British novelist H. G. Wells. Writing in 1913, Wells had predicted that artificially

---

[2] Although the two terms mean the same thing, there has been some shift in usage with time. Originally, "atomic energy" was the more common designation, but it has been largely replaced by "nuclear energy." For example, the U.S. *Atomic* Energy Commission was established in 1946 and the U.S. *Nuclear* Regulatory Commission in 1975.

[3] Historically oriented accounts of these developments and the discovery of fission are given in Refs. [6]–[8]. A discussion of basic nuclear physics concepts and terminology is presented in Appendix A.

induced radioactivity would be discovered in 1933 (he guessed the actual year of discovery correctly!) and also predicted the production of atomic energy for both industrial and military purposes. In Szilard's account, he at first "didn't regard it as anything but fiction." A year later, however, two things caused him to turn to this possibility more seriously: (1) He learned that Rutherford had warned that hopes of power from atomic transmutations were "moonshine"[4] and (2) the French physicist Frederic Joliot discovered artificial radioactivity as predicted by H. G. Wells[5] [9, pp. 16–17].

Szilard then hit upon a "practical" scheme of obtaining nuclear energy. At the time, it was thought that the beryllium-9 (^{9}Be) nucleus was unstable and could decay into two alpha particles and a neutron. This was a misconception, based on an incorrect value of the mass of the alpha particle. Actually, ^{9}Be is stable if only by a relatively small margin. In any event, Szilard thought that it might be possible to "tickle" the breakup of ^{9}Be with a neutron and then use the extra neutron released in the breakup to initiate another ^{9}Be reaction. In each stage, there is one neutron in and two neutrons out. This is the basic idea of a chain reaction. This particular chain reaction cannot work, as was soon realized, because too high a neutron energy is required to cause the breakup of ^{9}Be, and even then, there is a net loss of energy in the process, not a gain.

Szilard tried to find other ways to obtain a chain reaction, but his efforts failed. Nonetheless, in the interim, he went so far as to have a patent on neutron-induced chain reactions entrusted to the British Admiralty for secret safekeeping, the military potential of nuclear energy being important in his thinking [8, p. 225]. However, the key to a realizable chain reaction—fission of heavy elements—eluded Szilard as well as all others.

In these early speculations, there was an awareness of the potential of atomic energy for both military and peaceful applications, and the latter loomed large in the thinking of some scientists. For example, in a document dated July 1934, Szilard explained planned experiments that, if successful, would lead to

> power production...on such a large scale and probably with so little cost that a sort of industrial revolution could be expected; it appears doubtful for instance whether coal mining or oil production could survive after a couple of years. [9, p. 39]

---

[4] It is not clear from the Szilard reference whether the word "moonshine" was Rutherford's own or whether it was a paraphrase appearing in *Nature* in a summary of Rutherford's talk.

[5] Joliot, later Joliot-Curie, shared with his wife, Irene Joliot-Curie, the 1935 Nobel Prize in Chemistry for their discovery in 1933 of artificial radioactivity. They were the son-in-law and daughter of Marie and Pierre Curie, recipients (with Becquerel) in 1903 of the Nobel Prize in Physics, awarded for the discovery of radioactivity.

Along the same lines, Joliot prophesied in his 1935 Nobel Prize acceptance speech:

> scientists, disintegrating or constructing atoms at will, will succeed in obtaining explosive nuclear chain reactions. If such transmutations could propagate in matter, one can conceive of the enormous useful energy that will be liberated. [10, p. 46]

2.2.2 Fission and the First Reactors

Fission of uranium was discovered—or, more precisely, recognized for what it was—in 1938.[6] Scientists quickly recognized that large amounts of energy are released in fission and that there was now, in principle, a path to a chain reaction. By early 1939, it was verified that neutrons are emitted in fission, and it soon became apparent that enough neutrons were emitted to sustain a chain reaction in a properly arranged "pile" of uranium and graphite.[7] It took several more years to demonstrate the practicality of achieving a chain reaction. This work was led by Fermi, who had left Italy for the United States, and it culminated in the development and demonstration of the first operating nuclear reactor on December 2, 1942 at an improvised facility in Chicago.

The discovery and preliminary understanding of fission came at a time when the prospect of war was much on people's minds. The start of World War II in Europe in August 1939 ensured that military, rather than civilian, applications of atomic energy would take primacy, and the early work was heavily focused on the military side, in both thinking and accomplishments. A major goal of the nuclear program was the production of plutonium-239 (^{239}Pu), which was recognized to be an effective material for a fission bomb. The ^{239}Pu was to be produced in a reactor, by neutron capture on uranium-238 (^{238}U) and subsequent radioactive decays.

The first reactor in Chicago was very small, running with a total power output of 200 W. However, even before the successful demonstration of a chain reaction in this reactor, plans had started for the construction of the much larger reactors required to produce the desired amounts of plutonium. A pilot plant, designed to produce 1 MW, was completed and put into operation at Oak Ridge, Tennessee in November 1943 [11, p. 392]. A full-size 200-MW reactor began operating at the Hanford Reservation in Washington state in September 1944—a millionfold increase in power output in less than 2 years.[8]

---

[6] See Sections 6.1 and 6.2 for a brief description of fission and its discovery.

[7] The uranium was the fuel. The graphite served as a "moderator," to slow the neutrons down to energies where they were most effective in producing fission in uranium. See Section 7.2 for a discussion of moderators.

[8] Power is commonly measured in watts (W), kilowatts (kW), megawatts (MW), and gigawatts (GW), where 1 kW = 10^3 W, 1 MW = 10^6 W, and 1 GW = 10^9 W. The basic energy output of a reactor is in the form of heat and often this is explicitly recognized by specifying the power in *megawatts-thermal* (MWt).

The laboratories at Oak Ridge and Hanford were new wartime installations. The Hanford site was not selected until early 1943, and construction of the first reactor began in June 1943. Workers completed the reactor within about 15 months, despite the absence of any directly relevant prior experience.[9] The speed with which the program was pursued is breathtaking by present standards, but it can be understood in the context of the exigencies of World War II.

The pressures of wartime bomb development pushed work on peaceful applications largely into the background, but there was still considerable thinking about future civilian uses. An official report on the development of the atomic bomb was prepared by Henry Smyth, a Princeton physicist. It was published in 1945, shortly after the end of the war, to inform the public about the bomb project. In a closing section, entitled "Prognostication," Smyth pointed out:

> The possible uses of nuclear energy are not all destructive, and the second direction in which technical development can be expected is along the paths of peace. In the fall of 1944 General [Leslie] Groves appointed a committee to look into these possibilities as well as those of military significance. This committee...received a multitude of suggestions from men on the various projects, principally along the lines of the use of nuclear energy for power and the use of radioactive byproducts for scientific, medical, and industrial purposes. [12, pp. 224–225]

With or without such a committee, it was inevitable that imaginative scientists would consider ways of using nuclear energy for electricity generation. The possibility, for example, of power production from a reactor that used water at high pressure for both cooling the reactor and moderating the neutron energies was suggested as early as September 1944, in a memorandum written by Alvin Weinberg who was then working closely with Eugene Wigner on reactor designs [13, p. 43]. This is the basic principle of the dominant reactor in the world today, the *pressurized light water reactor* (PWR).[10]

For reactors designed to produce electricity, the more interesting quantity is the output in *megawatts-electric* (MWe). (For a reactor operating at an efficiency of 33%, the output in MWt is three times the output in MWe.) The maximum electrical power of a nuclear reactor is its *capacity*. When the specification is not explicit, the term "megawatt" usually means MWe for reactors used to produce electricity and MWt for reactors used for other purposes, including the early WWII reactors.

[9] An early summary of this chronology is given in Ref. [11], Chapter XIV.

[10] *Light water reactor* (LWR) refers to a reactor cooled and moderated with ordinary water, the word "light" being used to differentiate these reactors from those cooled or moderated with heavy water [i.e., water in which the hydrogen is primarily in the form of deuterium (^2H)]. The PWR is one version of the LWR. Characteristics of LWRs are discussed at length in later chapters.

2.3 Development of Nuclear Power in the United States

2.3.1 Immediate Postwar Developments

Interest in Commercial Nuclear Power

In the years immediately following World War II, the main activities of the American nuclear authorities continued to be directed toward further military developments, but increased attention was turned to electricity generation. A somewhat guarded assessment of the future was presented in the Smyth report:

> While there was general agreement that a great industry might eventually arise...there was disagreement as to how rapidly such an industry would grow; the consensus was that the growth would be slow over a period of many years. At least there is no immediate prospect of running cars with nuclear power or lighting houses with radioactive lamps although there is a good probability that nuclear power for special purposes could be developed within ten years. [12, p. 225]

This turned out to be rather close to the mark, although the words "slow" and "immediate" may have had different connotations in 1945—in the wake of the rapid pace of wartime development—than they do today.

There were, however, impediments to quick progress. For one, fossil fuels were plentiful in the 1950s, so no immediate urgency was felt. Nuclear facilities and technical knowledge were under the tight control of the AEC, with many aspects kept secret because of the military connections. Further, there was indecision as to the relative roles to be played by the government and private utilities in the development of nuclear power. Finally, it was not clear which type or types of reactor should be built.

The U.S. Navy made a decision first and, under the leadership of Hyman Rickover, built and began tests of pressurized light water reactors by the first part of 1953 [14, p. 188]. These reactors became the foundation of the U.S. nuclear submarine fleet. After some hesitation and consideration of alternative designs for a reactor for the generation of commercial electric power, the AEC announced in autumn 1953 that it would build a 60-MWe power plant [14, p. 194]. Participation by utilities was sought. The general reactor configuration was to be the same as that used by the navy, namely a pressurized light water reactor. A Pennsylvania utility won the competition to participate in this project, contributing the land and buildings and undertaking to run the facility when completed. The reactor was built at Shippingport, Pennsylvania and was put into operation at the end of 1957 [14, pp. 419–423].

Early Enthusiasm: "Too Cheap to Meter"?

Despite the considerable caution in initiating the U.S. nuclear power program, many euphoric statements about the future of nuclear power were made in the 1940s and 1950s. Such statements even go back as far as H. G. Wells in 1913. Nuclear power was to be abundant, clean, and inexpensive. Of course, thoughtful observers modulated and qualified even their optimistic prophecies, but some of the optimism was quite unbridled. For example, an article in *Business Week* in 1947 stated, "Commercial production of electric power from atomic engines is only about five years away.... There are highly respected scientists who predict privately that within 20 years substantially all central power will be drawn from atomic sources" [15].

However, among all the enthusiastic quotations from that era, one in particular has come to haunt nuclear advocates. In the 1980s, as nuclear power became more expensive than electricity from coal, an earlier phrase, "too cheap to meter," was thrown back in the faces of proponents of nuclear power as illustrating a history of false promises and overweening and foolish optimism.

The phrase originated with Lewis L. Strauss, the chairman of the Atomic Energy Commission (AEC) in the 1950s. Speaking at a science writers meeting, he stated: "It is not too much to expect that our children will enjoy electrical energy in their homes too cheap to meter." The phrase was used in a *New York Times* headline on September 17, 1954, the day after the speech.[11] A fuller version of Strauss' remarks, as they appeared in his prepared text, indicated a broadly euphoric technological optimism about nuclear energy and its applications [16, p. 2].

It is doubtful that many professionals shared this euphoria at the time. A more official version of the AEC position, expressed in Congressional testimony in June 1954, held out the hope that

> ...[nuclear power] costs can be brought down—in an established nuclear power industry—until the cost of electricity from nuclear fuel is about the same as the cost of electricity from conventional fuels, and this within a decade or two. [16, p. 4].

Overall, the history appears to have been one of hesitation and examination, followed by a conviction developing in the 1960s that nuclear power would indeed be less expensive than the fossil fuel alternatives. For a period, in the 1970s, this expectation was fulfilled in the United States, and it is still partially fulfilled in some countries. The failure of this expectation in later

---

[11] An account of the history of the remark is given in a brief report prepared by the Atomic Industrial Forum (AIF), a nuclear advocacy organization [16]. There is a good chance that Strauss was thinking of fusion power, not fission power, although he could not be explicit because the practicalities of fusion were secret in 1954, with the development of the hydrogen bomb only recently started. The AIF report quotes Lewis H. Strauss, the son of Lewis L. Strauss and himself a physicist: "I would say my father was referring to fusion energy. I know this because I became my father's eyes and ears as I travelled around the country for him."

years in the United States was a surprise and disappointment to nuclear proponents as well as to many neutral analysts. There was a serious misjudgment, but not as egregious a folly as connoted by the "too cheap to meter" phrase.

2.3.2 History of U.S. Reactor Orders and Construction

The First Commercial Reactors

The Shippingport reactor was a unique case. Although used to supply commercial electricity, it was largely financed by the federal government and built under navy leadership. Following the order of the Shippingport reactor in 1953, there was a fitful pattern of occasional orders by utilities during the next 10 years. This early period of reactor development was characterized by extensive exploration, with a wide variety of reactors being developed for military and research applications and for electricity generation. For the latter, a total of 14 reactors were ordered in the period from 1953 through 1960 [17]. They included nine light water reactors (LWRs), not identical by any means, plus five other reactors with a wide variety of coolants and moderators.[12] With three exceptions, these reactors all had capacities under 100 MWe.

The three reactors ordered in this period that had capacities above 100 MWe were the 265-MWe Indian Point 1 (New York), the 207-MWe Dresden 1 (Illinois), and the 175-MWe Yankee Rowe (Massachusetts) reactors, which were ordered in 1955 and 1956 [17].[13] These were all LWRs. The first to go into operation was Dresden 1, in 1960.

Growth Until the Mid-1970s

The exploratory period ended quickly. There was a brief lull in reactor orders after 1960, with only five more orders until 1965, and then a period of rapid increase from 1965 through 1974. The dominance of LWRs in U.S. reactor orders was complete after 1960, the only exception being the gas-cooled Fort St. Vrain reactor, ordered in 1965.

None of the reactors ordered before 1962 had a capacity as large as 300 MWe. After that, there was a substantial escalation in reactor size, in an effort to gain from expected economies of scale. Some critics believe that the growth in size was too fast to permit adequate learning from experience. The mean size of reactors ordered in 1965 was about 660 MWe, but by 1970, the mean size exceeded 1000 MWe, with some above 1200 MWe. The largest reactors completed and licensed to date in the United States have a (net) capacity of 1250 MWe.

---

[12] These were a fast breeder reactor, a sodium graphite reactor, a high-temperature gas-cooled reactor, an organic moderator reactor, and a heavy water reactor. The variety of reactor types is discussed in Chapter 8.

[13] These reactors have all been shut down, most recently Yankee Rowe in 1991.

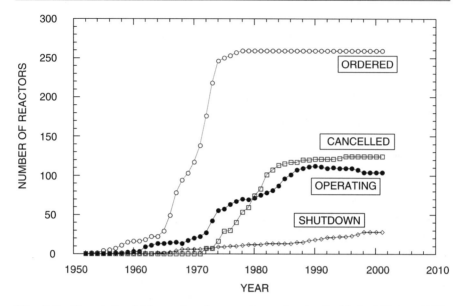

Fig. 2.1. Cumulative history of nuclear reactor orders in the United States, 1953–2001, including cancellations and shutdowns. (Data from Ref. [18], p. 253.)

The history of nuclear reactor deployment from 1953 through 2001 is summarized in Figure 2.1, which shows the cumulative pace of reactor orders as well as the number of reactors that were in operation, that have been canceled, or that have been shut down. There was a period of rapid growth in the number of orders from 1965 to 1975. At first, reactors were completed within about 6 years, and by the early 1970s nuclear power had begun to assume significant proportions. At the end of 1974, there were about 55 reactors in operation, with a total capacity of almost 32 GWe, providing 6% of U.S. electricity [18, p. 255].

Reactor Deployment Since the Mid-1970s

The picture changed abruptly in the mid-1970s. In contrast with the 4-year period from 1971 to 1974, when 129 reactors had been ordered, only 13 reactors were ordered during 1975–1978. After 1978, new orders ceased entirely and all reactors ordered after 1973 have been canceled. The de facto moratorium on commercial reactor orders has continued in subsequent years, and at the time of this writing (early 2004), no new orders are in clear prospect.

A major factor in this change was a sharp drop in the growth in electricity demand, as reflected in Figure 1.1. The growth in electricity consumption averaged over 7% per year from 1953 to 1973—a doubling time of under 10 years. This growth ended in 1974, with an actual drop of 0.4% that year, precipitated by the economic shock of OPEC's oil embargo in late 1973. After 1974, with a reduced rate of economic growth and a new emphasis on conservation, electricity sales grew at a much slower rate than in preceding decades, averaging about 2.7% per year during 1975–2000. Utilities that had previously placed orders for generating facilities found themselves facing a surplus of planned capacity. A first response was to stop plans for expansion.

Nuclear power was particularly impacted by the lessening demand for electricity, because it was additionally confronted by growing opposition and by increasing costs of reactor construction. These factors all worked to slow nuclear deployment after 1974, although reactors continued to go on line until the Three Mile Island accident in March 1979. The accident led to a 1-year hiatus, followed by a gradual resumption of deployment of reactors that were already in the pipeline. Overall, 51 new reactors have been put into commercial operation since 1979. The last of these was in 1996—the Watts Bar I reactor operated by the Tennessee Valley Authority (TVA).

At the end of 2003, there were 104 operating reactors in the United States with a net capacity of 98 GWe.[14] All are light water reactors. Total net generation in 2002 was 772 billion kilowatt-hours, or, in alternative units, 88 gigawatt-years (GWyr) where 1 gigawatt-year $= 8.76 \times 10^9$ kWh [20]. The fraction of electricity provided by nuclear power has risen to about 20% in recent years. Until 1990, the driving force in the increased nuclear output was the addition of new reactors. Since 1990, there has been little change in nuclear capacity, with 7 new reactors having gone into operation and 11 reactors (in general smaller) shut down. Nonetheless, there was an increase of 34% in total nuclear generation due to improved operation of existing reactors (see Section 2.4.2).

Three additional reactors, with capacities of roughly 1200 MWe each, are in something of a state of limbo [4]. They are each listed as more than 50% completed, but construction has been halted for many years, and as of mid-2003, there were no announced plans to resume construction.[15] They all belong to the TVA, which has six licensed reactors, including Browns Ferry 1.

No further nuclear reactors are on the immediate horizon in the United States, and for the time being, nuclear power in the United States is at a plateau. However, the federal government in recent years has looked with more favor on nuclear power. The U.S. Department of Energy in 1998, acting upon the recommendation of the President's Committee of Advisors on Science and Technology (PCAST), launched a Nuclear Energy Research Initiative (NERI) designed to stimulate innovative thinking on topics such as:

---

[14] One of these reactors, the 1065-MWe Browns Ferry 1 reactor, operated by the TVA, was shut down in 1985 along with four other TVA reactors that have since gone back into service. The TVA Board voted in May 2002 to restart Browns Ferry 1 [19]. Major changes are to be made in the reactor building and contents, and the restart will not occur for several years. (During the shutdown period, Browns Ferry 1 has been included in most U.S. government compilations as an "operating reactor" because it has an operating license.)

[15] The three reactors are Bellefonte 1 & 2 and Watts Bar 2.

proliferation-resistant reactors or fuel cycles; new reactor designs with higher efficiency, lower cost, and improved safety to compete in the global market; low-power units for use in developing countries; and new techniques for on-site and surface storage and for permanent disposal of nuclear waste. [21, p. 5–13]

The DOE in 1998 also established the Nuclear Energy Research Advisory Committee (NERAC) to advise on nuclear technology programs. The work of NERI has led to a number of imaginative proposals for new reactors, and NERAC has developed plans to encourage the deployment of new reactors by 2010 as well as the development of advanced designs for later deployment. These initiatives are discussed further in Chapter 16.

2.3.3 Reactor Cancellations

Quantitative detail on the history of orders and cancellations is given in Figure 2.2, which indicates the numbers of reactors ordered in each year and the number of these orders that were not canceled. Overall, almost all reactors now in operation were ordered in the period from 1965 through 1973. This is a very compressed interval, and one that did not allow much opportunity for the manufacturers and utilities to learn from experience.

A striking feature of the data shown in Figure 2.2 is the large number of cancellations of reactors after they had been ordered. In fact, the number of canceled reactors exceeds the number of reactors that are still in operation. The wave of orders in the 1965–1974 period was followed by a wave of cancellations in 1974–1984. The overall record is summarized in Table 2.2, which gives the total nuclear reactor orders from the 1950s to the end of 2002 and a summary of the disposition of the orders.

As seen in Table 2.2, almost one-half of the reactors that were ordered since 1953 have had their orders canceled. In many cases, although not most, cancellation came after construction had started. The other half of the ordered reactors were completed and put into operation, but some of these have since been permanently shut down. The large number of cancellations reflects the same forces that led to the cessation of new orders: the decline in the electricity growth rate, public or political opposition to nuclear power, and increased costs.

2.4 Trends in U.S. Reactor Utilization

2.4.1 Permanent Reactor Closures

One of the threats to the nuclear industry has been the prospect of the permanent shutdown of reactors before the expiration of their operating licenses (commonly issued for 40 years). This has happened to 28 reactors out of 132

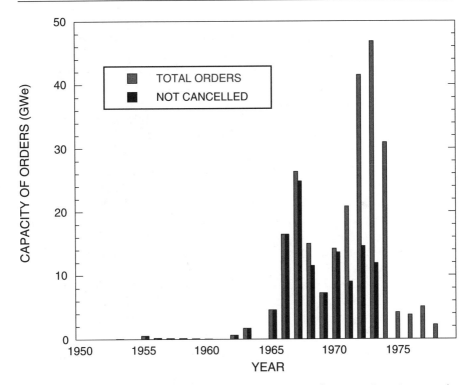

Fig. 2.2. Reactor orders in the United States, 1953–1978 (in GWe of total capacity). Annual figures are given for all orders and for those that were not subsequently canceled. The "not canceled" category includes operating reactors, reactors that have been shut down, and three partially completed reactors. (Data from Ref. [17]; the plot does not reflect several additional cancellations since 1994.)

Table 2.2. Cumulative record of nuclear power in the United States, 1953–2003.

Status	Number of Reactors
Orders for power reactors	259
Construction permit issued	177
Order canceled before reactor operated	124
Construction halted, but not canceled	3
Put into operation	132
Operable as of 12/31/03	104
Permanently shut down, before 12/31/03	28

Source: Ref. [18], p. 253, for period through 12/31/01; there were no changes during 2002 or 2003.

that have gone into operation.[16] However, the large majority of the closed reactors began operating in the 1960s, including 12 with capacities under 80 MWe that were shut down before 1970. Some of the larger of the 1960s reactors remained in operation considerably longer, with several lasting until the 1990s. The largest of the 1960s reactors was the Hanford N reactor (860 MWe), which was built in part for plutonium production and was operated by the U.S. Department of Energy, not by an electric utility. It was shut down in 1988.

In addition to the closing of these older reactors, nine reactors that went into operation after 1970 have been shut down. These are listed in Table 2.3. Their total capacity was 7.5 GWe. They fall into several categories:

- Three Mile Island 2, which went into operation in 1978. It was severely damaged in 1979 in one of the world's two major reactor accidents (the other being the much worse accident at Chernobyl). The damage to the reactor and resulting contamination to the building made it impossible to put the reactor back into operation.
- Shoreham in Long Island, New York, which operated briefly at very low power after issuance of a low-power license in 1985. Local and state opposition prevented the issuance of a license to operate at full power and,

Table 2.3. U.S. reactors that went into operation after 1970 and that have been shut down.[a]

Reactor	Capacity (MWe)	Year of Shutdown	Year of Initial Operation
Millstone 1	660	1998	1971
Zion 1	1040	1998	1973
Zion 2	1040	1998	1974
Maine Yankee	860	1997	1972
Trojan	1095	1992	1976
Shoreham[b]	809	1989	—
Fort St. Vrain	330	1989	1979
Rancho Seco	913	1989	1975
Three Mile Island 2	792	1979	1978

[a]Status as of December 31, 2003.
[b]The Shoreham reactor received a low-power license but was never put into full-scale operation.
Source: Ref. [4].

---

[16] The precise numbers of reactors in these categories is a matter of definition. The DOE tabulation of Ref. [18] indicates that 28 reactors have been shut down. A tabulation in *Nuclear News* names 24 closed reactors [4], and an IAEA listing names 22 closed reactors [3]. The latter list includes three reactors, none larger than 22 MWe, that were shut down in the 1960s that are not in the *Nuclear News* list. The discrepancies in the listings all involve special cases, where the question of whether the reactor was ever in "commercial operation" is blurred.

after lengthy negotiations, the reactor was closed, with neither its st.
as having been in operation nor its date of "shutdown" crisply defined.

♦ Three reactors that were shut down by 1992 that had a history of poor
performance and low capacity factors, which weakened their economic and
political viability (Rancho Seco, Fort St. Vrain, and Trojan). Considerable
public controversy surrounded the decisions as to their fates.

♦ Four reactors that were shut down in 1997 and 1998, in financial decisions
by the utilities that operated them. From the utility standpoint, it was
less costly to shut down the reactor than to continue to operate it, and
these actions elicited little controversy.

Each of these categories carries a cautionary note. In the case of Three
Mile Island it is obvious: A serious accident must be avoided. The case of the
Shoreham reactor has probably been the most worrisome to utilities, because
in this instance a reactor was completed, but public hostility prevented its
use. The weak performers in the third group are a reminder that a reactor is
vulnerable, for both economic and political reasons, if its performance is poor.
The fourth group provides additional warning on the need to be economically
competitive.

The reactor closures of the 1990s were interpreted by some as the fore-
runner of a cascade of further closures, but, in fact, they stopped after 1998.
Instead, as discussed in Section 2.4.5, there has been a switch in the opposite
direction, with a wave of applications for operating license renewals.[17]

2.4.2 Capacity Factors

The cessation of reactor orders, the absence of any new orders on the horizon,
and a spate of closures have represented discouraging trends for U.S. nuclear
power. A bright spot, however, has been the recent improvement in operating
performance. We discuss a number of specific indicators of improved performance in Section 14.4. An overall, more immediate measure of how well a
reactor is performing is given by its *capacity factor*. The capacity factor for
a reactor in any period is the ratio of its total electrical output during that
period to the output if it had run continuously at full power. Capacity factors
are usually expressed in percent, the ideal being 100%.

Although a capacity factor above 90% for an individual reactor in a given
year was not uncommon before the late 1990s,[18] the capacity factors were
usually well below 90% over a more extended period due to scheduled and

---

[17] This does not mean that no reactors will be shut down before their 40 years of
operation are completed. For example, there is pressure from some local environmental groups and political bodies to shut down the Indian Point nuclear plant
located north of New York City. They argue that terrorist attacks could lead to
large releases of radioactive material from the plant and that emergency plans
to evacuate people within a 10-mile radius are inadequate. The plant has two
PWRs, each rated at about 1000 MWe.

[18] For example, the 90% level was exceeded by 29 U.S. reactors in 1994 [23].

unscheduled maintenance and variations in demand. For many years, a capacity factor greater than 80% was considered to be very good. However, in a development that probably even took most nuclear enthusiasts by surprise, the average capacity factor for U.S. reactors has been above 85% since 1999.

The average capacity factor for U.S. reactors since 1973 is plotted in Figure 2.3. After hovering around 50%, it gradually rose to over 60% in 1977 and 1978, but dropped following the Three Mile Island accident in 1979, which precipitated a period of precautionary repairs and modifications. Some of these entailed long periods during which a reactor was out of operation. With the completion of most of these modifications, there has been a significant increase in the annual average capacity factor beginning in 1988. The capacity factor reached 66% in 1990 and then rose to 77% in 1995 and to 90% in 2002 [20, Table 8.1].

The improvement in capacity factors has been attributed by the nuclear industry to concerted efforts at improvement, including modifications in equipment and operator training, as well as better communications within the nuclear industry to facilitate learning from experience. These efforts resulted in fewer unintended shutdowns and a shortened period for planned shutdowns, such as those for refueling the reactor. Some of the capacity factor increase may also be due to the abandonment of reactors that had been below average in performance.

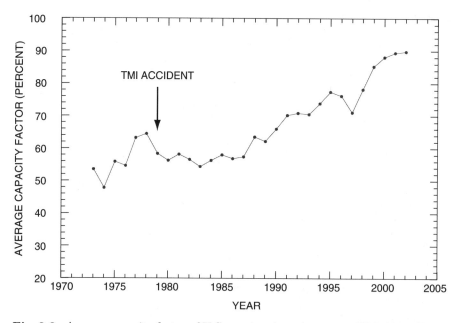

Fig. 2.3. Average capacity factor of U.S. reactors in a given year, 1973–2002. (Data from Ref. [22].)

The improvement in capacity factors has important implications for the economic position of nuclear power. An increase from, say, 60% to 90% corresponds to an increase in output of 50%—equivalent to the addition of roughly 50 reactors—and leads to a large reduction in the cost of electricity. The expenses of a nuclear utility do not change greatly as its electricity output rises or falls, because the capital costs are fixed and the operating costs are insensitive to changes in output. The cost of electricity from a given reactor is therefore reduced by almost one-third if its capacity factor increases from 60% to 90%. A higher capacity factor also reflects a lesser need for repairs or modifications in the reactor, making further savings likely.

2.4.3 Consolidation in the U.S. Nuclear Industry

One of the trends in the U.S. nuclear industry in recent years has been a consolidation of companies that operated nuclear reactors. In early 1998, there were 105 reactors in the United States being operated by 44 separate entities [24]. By the beginning of 2003, the number of operating entities had been reduced to 31, with 104 reactors [4]. The largest of the operators was Exelon Generating, with eight reactors formerly run by Commonwealth Edison and four reactors formerly run by Philadelphia Electric Company. At least in principle, this sort of consolidation allows for greater efficiencies and more concentrated expertise.

Even with the consolidation to date, the U.S. nuclear industry is very fragmented compared to other countries. France is at the other extreme. All but one of its 59 reactors is operated by Electricité de France. The one exception is the Phenix fast breeder reactor, operated by the French atomic energy commissariat.

2.4.4 Renewal of Reactor Operating Licenses

Reactors in the United States operate under Nuclear Regulatory Commission licenses that are issued for a period of 40 years. This period was specified by Congress, according to the Nuclear Energy Institute, because it "was a typical amortization period for an electric power plant," not because of "safety, technical or environmental factors" [25]. Experience with reactors to date indicates no compelling reason to shut them down at the end of 40 years, although, in some cases, retrofits and replacement of components may be necessary.

If a reactor continues to run safely, without the need for a major overhaul, the incentive to extend its operating life is obvious. The capital costs of the reactor have been paid off and the remaining costs of operating the plant are relatively low. Extending the license means relatively inexpensive power for the continued life of the plant. The requirements for obtaining a license renewal are set forth by the Nuclear Regulatory Commission (NRC) and include a study of the effects of aging on reactor safety [26, Part 54]. If granted, a license renewal is specifically limited to 20 years.

The first applications for license renewals were received by the NRC in 1998. By October 2003, the NRC had approved 18 applications and was reviewing 12 more [27]. Looking ahead, the NRC indicated that it expected to receive applications for more than 25 additional reactors by 2006. In view of the financial benefits of continued operation, applications for renewal may eventually be made for most presently operating reactors.

An application for license renewal indicates an intent but not an irrevocable commitment. Just as some reactors have been shut down before the expiration of their original 40-year licenses, it is possible that some renewal applications will be withdrawn, denied by the NRC, or not implemented by the utility. Nonetheless, the rather sudden increase in activity in this area indicates a revived optimism in the U.S. nuclear industry.

2.5 Worldwide Development of Nuclear Power

2.5.1 Early History of Nuclear Programs

The above discussion has emphasized the U.S. nuclear program. However, the United States was not alone in having an early interest in nuclear energy. Other countries had similar interests, although their development lagged because they lacked the head start provided by the U.S. World War II atomic bomb program and they had smaller technological and industrial bases.

For the countries that wanted nuclear weapons, the priority was the same as that of the United States. The bomb came first and peaceful nuclear energy later. Thus, the construction and testing of nuclear weapons was achieved by the USSR in 1949, Britain in 1952, France in 1960, and China in 1964. Commercial nuclear electricity followed: The USSR started with several 100-MWe reactors in 1958[19], Britain with a 50-MWe reactor at Calder Hall in 1956 (preceding the U.S. reactor at Shippingport), France with a 70-MWe prototype reactor at Chinon in 1964, and China with three reactors that went into commercial operation in 1994 [4].

Additional countries had no intention of building nuclear weapons and went directly to nuclear reactors for electric power. These included other countries of Western Europe, beyond France and Britain, as well as Japan. As in the United States, the reactors put into operation in the late 1950s for electricity generation were relatively small in size and few in number, and exploitation of nuclear power remained on a relatively modest scale until the 1970s.

---

[19] A 5-MWe electric plant was put into operation in Obinsk in June 1954, which has been cited by some Soviet authors as being the "first nuclear power plant in the world" (see Ref. [28]).

2.5.2 Nuclear Power Since 1973

Trends in Nuclear Growth

Worldwide nuclear power generation grew rapidly from the 1970s through the 1980s and then slowed. Details on the growth from 1973 to 2000 are shown in Table 2.4 for the world, Western Europe, and the Far East, as well as for the three countries with the greatest individual outputs of nuclear energy.

Looked at broadly, one sees a rapid expansion in the 1970s and 1980s followed by a marked slowdown in the pace of growth in the 1990s. World generation [excluding the former Soviet Union (FSU) and Eastern Europe] increased by more than a factor of 9 between 1973 and 1990—an average increase of 14% per year. Even in the latter part of this period, from 1980 to 1990, the rate was 11% per year. However, it dropped to 2.5% for these countries for 1990 to 2000. For the world as a whole (including all countries), the average annual growth was 2.4% in the period from 1992 to 2001.[20]

The growth rate has become even slower in the most recent years. Six new reactors, with a total capacity of 5.0 GWe, are listed by the IAEA as having been connected to the local electrical grids in 2002 [3]. This represents a 1.4% increase in the world's total nuclear capacity. Five of these additions were in Asia, continuing the recent trend of Asian leadership seen in Table 2.4.[21]

Table 2.4. World growth of nuclear power, 1973 to 2000.

	World (inc.)[a]	West Europe	Asia	United States	France	Japan
Gross generation (GWyr)						
1973	22	8.4	1.4	10	1.7	1.1
1980	71	24	11	30	7.0	9.5
1990	202	84	32	69	36	22
2000	260	102	57	89	47	37
Average increase (% per year)						
1973–1980	18	16	34	17	23	36
1980–1990	11	13	11	8.6	18	8.8
1990–2000	2.5	1.9	5.7	2.6	2.8	5.2

[a]The world figures are incomplete; they exclude Eastern Europe and the former Soviet Union. Nuclear generation for these countries was 32 GWyr in 2000, raising the world total to 292 GWyr.

Source: Ref. [20]; data not included for Eastern Europe and the former Soviet Union (FSU) prior to 1992.

---

[20] These data are from the U.S. DOE publication *Monthly Energy Review*, which does not include data for the FSU and Eastern Europe for the years prior to 1992 [20].

[21] The five reactors in Asia included four in China (Qinshan 2-1 and 3-1 and Lingao 1 and 2), and one in South Korea (Yonggwang 6). The other reactor was in the

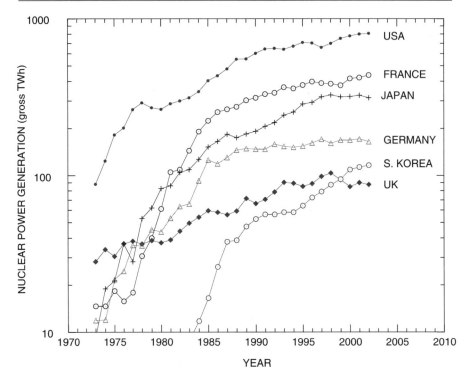

Fig. 2.4. Growth of annual nuclear power generation in selected countries, 1973–2002. (Data from Ref. [20].)

The history of nuclear generation from 1973 to 2002 is shown in Figure 2.4 for several countries with relatively large nuclear programs. France, Japan, and Germany began large-scale use of nuclear power after the United States, but then had larger fractional growth rates, especially in the 1975–1985 period. South Korea, the latest major entry, has had rapid growth in the past 15 years, but its overall program is still relatively small. The United Kingdom, one of the original leaders in nuclear power, later lagged substantially.

Recent Developments in Individual Countries

Despite the overall negative recent history for nuclear power, there were favorable developments in some countries:

♦   France added 10 new reactors from 1990 through 2002. These were large
    reactors, each 1300 MWe or more. In most of this period, nuclear power

Czech Republic (Temelin-2). Construction had started in 1996 or later on the Asian reactors and in 1987 on the Czech reactor.

accounted for more than 75% of France's electricity generation, and France was an exporter of electricity to its neighbors. However, no further reactors are presently in the pipeline.

♦ Japan added 16 new reactors in the same period, most with capacities above 1000 MWe.

♦ South Korea added nine new reactors in this period, more than doubling its nuclear generation.

♦ The average capacity factor of U.S. reactors grew from 66% in 1990 to 90% in 2002, giving a large increase in total output, despite a slight decrease in total capacity.

♦ In Switzerland, voters rejected proposals to phase out nuclear power, in referenda held in May 2003.

On the other side of the ledger, there has been a turning away from nuclear power among some countries that used it. The first to take this step was Italy, which by 1990 had completely shut down its small reactor program, consisting of four reactors with a combined capacity of 1.4 GWe [3]. Since then, decisions to phase out nuclear power has been taken by the governments of Sweden, Germany, and Belgium, each of which obtain a large fraction of its electricity from nuclear power (see Table 2.1). The phasing out is planned to be gradual. By the end of 2003 these decisions had led to two closures, that of the 615-MWe Barsebaeck-1 reactor in Sweden at the end of 1999 and the 640-MWe Stade reactor in Germany in late 2003. The policy was adopted for generalized political or environmental reasons, rather than specific problems with the existing reactors, and could be modified or reversed.

Beyond these broad decisions, there have been several major closures of nuclear power plants in Europe involving special circumstances:

♦ After the reunification of Germany, the government in 1990 shut down five Soviet-manufactured reactors (1.7 GWe total).

♦ The four reactors at Chernobyl have all been shut down, the first by the accident in 1986 and the last by the decision of the Ukrainian government in December 2000 (3.8 GWe total).

♦ France closed its 1200-MWe breeder reactor, Superphenix, in 1998 following a troubled operational history and a lack of an immediate need for a breeder program.

There have been other reactor closures throughout the world, but in most cases, these have involved relatively old and small reactors.

2.5.3 Planned Construction of New Reactors

The pace of reactor construction is now quite slow. According to an IAEA compilation, 32 reactors were under construction as of November, 2003 with a total capacity of 26 GWe. Completion of all these reactors would add 7% to the then existing capacity of 360 MWe, and if accomplished within 5 years, the

Table 2.5. Nuclear reactors under construction as of November, 2003.

Country	No. of Units	Capacity (GWe)	Mean Cap. (MWe)
Asia			
Japan	3	3.70	1232
India	8	3.62	453
China	3	2.61	870
Taiwan	2	2.70	1350
Iran	2	2.11	1056
South Korea	2	1.92	960
North Korea	1	1.04	1040
TOTAL ASIA	21	17.70	843
Eastern Europe			
Ukraine	4	3.80	950
Russian Federation	3	2.83	942
Slovak Republic	2	0.78	388
Romania	1	0.66	655
TOTAL E. EUR.	10	8.06	806
Other: Argentina	1	0.69	692
WORLD TOTAL	32	26.45	827

Source: Ref. [2].

average annual rate of increase would be about 1.4%. Details of the construction plans, subdivided by country and region, are summarized in Table 2.5.

Consistent with the small number of reactors under construction, the IAEA estimated in 2002 that worldwide nuclear capacity will grow at an average rate of only 0.8–1.6% per year for the period from 2001 to 2010 [29, p. 17]. A 2002 U.S. DOE summary projected even slower growth, with world capacity changing from 350 GWe in 2000 to 385 GWe in 2010 in the "high growth case" and to 340 GWe in 2010 in the "low growth case" [30, p. 93]. The "high" figure represents an average annual growth rate of 1.0%.

Table 2.5 clearly shows Asia's dominance in the planned construction. About two-thirds of the reactors listed as being constructed are in Asia. The remainder are almost entirely in Eastern Europe. Even these reactors are not necessarily a sign of vitality in their nuclear programs. Construction on the Eastern Europe reactors was begun in 1987 or earlier, and the current plans therefore reflect projects that have been long delayed and, in most cases, have no announced completion dates [3]. In contrast, with the exception of the Iranian reactors, all the reactors in Asia were started after 1996. Typically, their expected time from the start of construction to commercial operation is 7 years or less.

The only definite move in recent years to start construction of a new reactor outside Asia was in Finland, where the Parliament in May 2002 voted to build a fifth reactor. This plant is not reflected in the IAEA listing in Table 2.5 because actual construction had not yet been started. It is noteworthy that no new reactors are presently slated for the United States or Western Europe. Here, there is either a pause or a planned retraction. Among the leading countries in this group, the United States, France, and the United Kingdom are standing pat, whereas Germany, Sweden and Belgium are scheduled to phase out nuclear power, absent a change in national policy.

However, this may be an incomplete picture. In a number of countries, there are reactors that have been ordered but are not now actively under construction. Some of them could be completed before 2010, given a decision to move ahead. In addition, with a 5-year construction time, additional reactors could be ordered and completed by 2010. For example, Russia has announced plans that call for putting nine additional plants into operation by 2010, primarily by completing delayed projects (see Section 2.6.3), and the U.S. DOE has adopted the goal of having a new commercial nuclear reactor completed by 2010. Conversely, additional plants may be shut down. In particular, there is pressure to shut down two Chernobyl-type reactors that are still operating in Lithuania.

2.6 National Programs of Nuclear Development

2.6.1 France

France is widely cited as the leading success story for nuclear power. Each year since 1986, nuclear power has accounted for over 70% of electricity generation in France. Another large fraction, averaging about 15%, but with considerable year-to-year variations, has come from hydroelectric power. Therefore, France is virtually saturated in terms of the replacement of fossil fuels. Taking advantage of its ample nuclear capacity, France now exports substantial amounts of electricity to its neighbors.

The chief avenues for further major nuclear growth are through increased exports or a greater electrification of the French energy economy, but there are no visible indications that this is a high priority for France. The last new reactors to go into operation in France were four large PWRs each with a capacity of about 1450 MWe, and no new reactors are under construction. The French and German nuclear industries are continuing to consider a large next-generation reactor, the European Pressurized Water Reactor (EPR), a version of which has been ordered as Finland's fifth reactor.

Over the past two decades, nuclear power has greatly changed the structure of French energy supply. It was an almost negligible contributor in 1970, but in 2001, nuclear power provided 39% of all primary energy in France [1]. The period from 1971 to 1995, in which most of the nuclear growth was achieved, has been described in some detail in Section 1.3.3. It was marked by decreases in fossil fuel use and CO_2 emissions while gross domestic product and energy consumption rose. Total electricity generation more than tripled, accompanied by a change from almost total reliance on fossil fuels and hydroelectric power to a dominant reliance on nuclear power (see Table 1.3). The French experience has been pointed to as demonstrating the impact that nuclear power can have in curbing dependence on fossil fuels and on the emission of greenhouse gases.

There have been a number of attempts to explain why French nuclear history has developed so differently from that of other major countries. Several factors have been advanced, although many of them need not have been unique to France:

♦ France is poor in fossil fuels, and nuclear power was seen as the most expeditious way to reduce French dependence on oil and coal imports.
♦ There has been a concentration on a single type of reactor (a PWR originally modeled on Westinghouse designs), operated under the aegis of a single utility, Electricité de France, and built by a single reactor manufacturer, Framatome. This has allowed for standardization to a few PWR types, with resulting economies in design, construction, and operation.
♦ Although France has open political debate, there were few mechanisms whereby opposition to nuclear power could impede its development, short of changing the policy of the central government. In the United States, on the other hand, individual state governments have considerable independent authority and often have taken positions against nuclear development. Further, fewer opportunities for intervention through the courts are available in France than in the United States, especially as the state structure in the United States provides an extra layer of courts.
♦ The Communist party, which in the early years of nuclear power was still an important political force in France, supported nuclear power. In most other European countries, the political left opposed nuclear power.
♦ With the initial success of the French nuclear program, the maintenance of French leadership became a matter of national pride.

Despite these factors, there exists some opposition to nuclear power in France. There is no suggestion that this opposition will reverse the French use of nuclear power, but it may inhibit France from increasing its electricity exports to other European countries.

2.6.2 Japan

Japan is another country with a continuing strong nuclear program, although a smaller one than that of France, especially when adjusted for its larger population and economy. Its nuclear program is particularly driven by Japan's dependence on energy imports, which account for virtually all of Japan's fossil fuels. The chief domestic resources are nuclear energy and hydroelectric power.[22] An increase in the nuclear share of the total energy budget is the most direct way of moving toward greater energy independence. As part of this general strategy, Japan is taking steps toward a self-contained fuel cycle, which could eventually rely on breeder reactors.

To date, Japan has depended for reprocessing on a small domestic plant and reprocessing contracts with facilities in France and Great Britain. Japan's own reprocessing facilities will be greatly expanded with the construction of the Rokkasho-mura plant, started in 1993 and now scheduled for completion in 2005 (see Section 9.4.2). The plant will have a capacity of 800 tonnes per year, sufficient to handle the output of about 25 1000-MWe reactors, each discharging about 30 tonnes of fuel per year. A uranium enrichment plant at Rokkasho-Mura began operation in 1992 and now has an annual capacity sufficient to produce about one-fifth of Japan's requirements of enriched uranium [31, pp. 30–31].

The pace of deployment of new reactors has slowed since the 1990s. However, reactor construction times are relatively short in Japan, averaging under 5 years, so that rapid changes in the program are possible [3, p. 50]. Conversely, political opposition to nuclear power in Japan could interfere with present plans. This opposition has been growing in recent years, exacerbated by incidents that have reflected adversely on the management of Japan's nuclear enterprise.[23]

2.6.3 Other Countries

The Former Soviet Union

The breakup of the Soviet Union left nuclear reactors in a number of the new states, mostly in Russia and Ukraine, but also two in Lithuania and one each in Armenia and Kazakhstan. The 1986 Chernobyl accident led to a cutting back of reactor construction in Russia. Nonetheless, six reactors that had been started before the accident were connected to the grid in Russia between the time of the accident and the end of 2001—five of which were 950-MWe PWRs and one was a 925-MWe RBMK reactor (Smolensk-3 in 1990) [3].[24] From 1988 to 1990, nine older reactors with capacities of 100 MWe to 336

---

[22] In this discussion, use of imported oil to generate electricity is not counted as indigenous production, but electricity generated from imported uranium is counted as indigenous. This is not symmetric, but it is not unreasonable given that the cost of the fuel in the case of nuclear energy is a small fraction of the total cost.

[23] The most notable of these was an accident at the Tokaimura facility in 1999, when two workers died of radiation exposure due to carelessness in their handling of enriched ^{235}U that was to be used in fuel for a research reactor.

[24] The Chernobyl reactor was an RBMK reactor. These reactors have design features that make them more vulnerable to accidents than are the light water reactors more commonly used (see Sections 8.1.4 and 15.3.1).

MWe were taken out of operation. There was also a considerable scaling down of plans. The number of operating plants in Russia rose from 24 in 1992 to 27 in 2002, but over the same period, the number of plants listed as being under construction or on order dropped from 16 to 6 [4, 32].[25] Of the six, two were 750-MWe breeder reactors and it is not clear when, if ever, these will be built.

It is possible that there will be a substantial development of nuclear power in Russia during the next two decades. Several reactors were under construction in Russia in 2003, and Russia is actively engaged in the export market, including reactors under construction in China, India, and Iran. Following a plan formulated by the Russian Ministry of Atomic Energy (MiniAtom) in 2000, the Russian government has adopted a strategy to raise nuclear capacity (which is now 21 GWe) to a level of 52 GWe by 2020 in their "optimistic" economic scenario and to about 35 GWe in their "pessimistic" scenario [33, pp. 171–190]. In the first instance, this will be accomplished by extending the lives of existing plants beyond the originally scheduled 30 years and completing some projects begun earlier. The next phase calls for the construction of 31 GWe of new capacity [33, p. 187]. However Russia is rich in fossil fuels and, for reasons of geography, may be less concerned than most countries about the effects of global warming.

Ukraine was the site of the 1986 Chernobyl accident. The response has been to shut down the three undamaged Chernobyl reactors—the last not until December 2000—but continue with nuclear power. Six PWRs that had been under construction at the time of the accident were completed by 1989, before the Ukraine separated from the USSR. One additional PWR went into commercial operation in 1996, giving a total of 13 operating reactors by the beginning of 2003. Four more reactors are officially under construction (see Table 2.5).

Lithuania led the world in dependence on nuclear electricity in 2002, obtaining 80% of its electricity from two 1380-MWe RBMK reactors that went into operation in 1985 and 1987. However, it has been under pressure to shut these reactors down, given the fears created by their design similarity to the Chernobyl reactors. Lithuania's admission to the European Union, scheduled for 2004, has been made contingent on the subsequent closure of these reactors.

The two other FSU countries with nuclear reactors at the time of Chernobyl were Armenia and Kazakhstan. Armenia had two 376-MWe PWRs, one of which was shut down in 1989 (before the breakup of the FSU) while the other continues to operate, providing 40% of Armenia's electricity in 2002. The one Kazakhstan reactor was a 50-MWe fast breeder that was shut down in 1999, leaving no operating reactors in that country.

---

[25] Note: These numbers, from *Nuclear News*, differ from the IAEA data of Table 2.1. A major reason for the difference is that the IAEA, but not *Nuclear News*, includes four Russian 11-MWe reactors.

Former Soviet Bloc Countries

With only a few exceptions, nuclear power has been at a standstill in Europe since the late 1980s. The main exception, other than France, has been in countries in which there has been construction directed toward completing reactors started in the 1980s under Soviet influence. The operating reactors in the former Soviet bloc as well as those under construction are primarily PWRs of the Russian VVER series. The quality of construction and performance of these reactors has varied among countries, and some of these reactors have been viewed as constituting safety hazards. However, problems of air pollution from coal burning are severe in Eastern Europe, and this may favor nuclear power. Further, a number of these countries are heavily dependent on nuclear energy for their electricity, and it would be difficult to abandon it. For example, the share of electricity in 2002 from nuclear power was 55% in the Slovak Republic, 47% in Bulgaria, and 36% in Hungary (see Table 2.1).

Each of these countries has followed its own path, and we describe these developments briefly (with the number of operating reactors, as of the end of 2003, shown in parentheses):

- *East Germany* (0). The most draconian remedial measures were taken following the merger of East and West Germany in 1990. The East German reactors—primarily VVERs—were all removed from service, as not meeting Western safety standards.
- *Czech Republic* (6). The division of Czechoslovakia into the Czech Republic and Slovakia left four operating reactors in each, and two reactors under construction in the Czech Republic (each about 900 MWe). One of these reactors, Temelin 1, was connected to the grid in 2000. It was located near the Austrian border and its operation was a source of contention with Austria. The second, Temelin 2, was connected to the grid in December 2002.
- *Slovakia* (6). Slovakia completed construction of two 388-MWe VVERs in the late 1990s and has put them both into commercial operation. Two more remain nominally under construction.
- *Bulgaria* (4). Four 440-MWe PWRs of Soviet design were in operation in Bulgaria at the time of Chernobyl, and two new 953-MWe Soviet-type PWRs were brought on line in 1988 and 1992. The two oldest were permanently shut down in December 31, 2002, reducing the total number of reactors to four.
- *Romania* (1). Romania is an anomaly in Eastern Europe, and for that matter a rarity in the world, in opting exclusively for Canadian-type heavy water reactors. Five 625-MWe pressurized heavy water reactors of the CANDU design were under construction in the early 1990s. One went into operation in 1996, but only one of the other four reactors was listed as being under construction in 2003 [3].

Other European Countries

Beyond this activity in Eastern Europe—essentially a cleanup of long-standing projects—nuclear power development has almost stopped, or is in regression, in Europe. Outside of France, the one new reactor built in Western Europe in the 1990s was a 1188-MWe PWR (Sizewell B) that started operation in the United Kingdom in 1995. The only other further step toward European development of nuclear power was the decision of Finland in May 2002 to build its fifth reactor. There have been no tangible signs of other countries following suit, as of early 2004.

Western Hemisphere (Other Than the United States)

Canada embarked on a large program of nuclear construction, mostly in the late 1970s, and in 1995 had 22 operating reactors, all based on the CANDU reactors, which have made Canada a leading country in nuclear capacity and generation. However, in the 1990s, 8 of the 22 reactors in Canada were put into a temporary "laid-up" status, one in 1995 and 7 in late 1997 and early 1998. These eight were the oldest ones in operation, all having started up in the 1970s. This step was taken at the behest of Canada's nuclear regulatory authority to avoid a degradation of the "long-term safety and performance" of Canada's nuclear fleet [34, p. 248]. However a 2003 IAEA summary suggested that the eight closed plants "might re-start in the future" [3, p. 46], and three of these reactors were back in service by early 2004 (Bruce 3 and 4 and Pickering 4). There are also nuclear reactors in use in Argentina, Brazil, and Mexico, but these are small programs, with no imminent expansion plans except for one reactor nominally under construction in Argentina.

Other Asian Countries

Beyond Japan, there are active ongoing nuclear programs in other Asian countries (with number of operating reactors in parentheses):

◆ *South Korea* (18). South Korea is a relative late-comer to nuclear power, but in recent years, it has had rapid growth. Since 1990, it has been second only to Japan in the addition of new nuclear capacity and it plans further expansion. It uses PWRs and heavy water reactors.

◆ *India* (14). India has followed a path of its own, and now has 12 small heavy water reactors in operation—the largest with a capacity of 202-MWe— and two small BWRs. Its eight plants now under construction include four reactors similar to the existing ones and four that are larger, including—in a new departure—two Russian designed 905-MWe PWRs.

◆ *China* (8). China's civilian nuclear power program started comparatively late. Its first three reactors went into operation in 1994 and its next five in 2002 and 2003. China has drawn on many different foreign companies in these first stages, but is developing its own national capabilities.

- *Taiwan* (6). Taiwan's six operating reactors are a mix of PWRs and BWRs with capacities ranging from 604 to 948 MWe [4]. They were all in operation by 1985, giving Taiwan a relatively early nuclear program. There was a subsequent lull, but construction is now underway on two 1350-MWe advanced BWRs.
- *Iran and North Korea* (0). The plans to build civilian nuclear power plants in these countries are intertwined with nuclear weapons proliferation issues, and we discuss them further in Section 18.2.3.

Although recent construction activity in Asia has far exceeded that in Europe and North America, it has not been very rapid compared either to the size of the populations involved or to earlier history elsewhere. The United States and Western Europe each still has a larger total nuclear capacity than does Asia. The Asian countries that are pursuing nuclear power appear to be doing this in a deliberate manner, not in the spirit of "crash programs." However, Asian nuclear leaders have an optimistic view of the future Asian role (see Section 20.4.3).

2.7 Failures of Prediction

In the preceding discussions, the trends of events for nuclear power has been reviewed, and such discussions may appear to suggest the nature of future trends. However, any suppositions as to how nuclear power will develop over

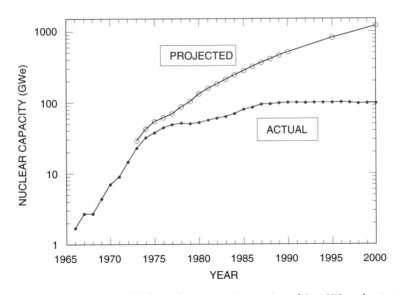

Fig. 2.5. Comparison of U.S. nuclear capacity, projected in 1972 and actual.

the next several decades should be viewed with caution. Past predictions have been rife with errors, and there is no reason to assume we will do better today.

The changing fortunes of nuclear power in the United States offers a cautionary lesson. In Figure 2.5, we compare an early projection for the growth of U.S. nuclear power with the actual subsequent developments. In 1972, the Forecasting Branch of the AEC's Office of Planning and Analysis made a projection for future growth in nuclear capacity, based on past trends in energy use and electricity capacity [35]. Its forecast for the "most likely" case are shown in Figure 2.5, together with the actual history. The projected capacities for 1990 and 2000 were 508 GWe and 1200 GWe, respectively. The actual capacities for 1990 and 2000 were 100 GWe and 98 GWe, respectively [20, p. 113].

This was a spectacular failure of prediction, but one that was in tune with the conventional wisdom of the time. It is natural to speculate on the implications for today. Alternatively, we can believe we now have the wisdom to avoid comparable errors, or we can wonder what new predictive errors are being made. We will return to considering the future of nuclear power in Chapter 20.

References

1. U.S. Department of Energy, *International Energy Annual 2001*, Energy Information Administration report DOE/EIA-0219(2001) (Washington, DC: U.S. DOE, March 2003).
2. International Atomic Energy Agency, "Nuclear Power Plants Information: Operational & Under Construction Reactors by Country" (updated 11/17/03). [From: http://www.iaea.org/cgi-bin/db.page.pl/pris.reaopucct.htm]
3. International Atomic Energy Agency, *Nuclear Power Reactors in the World*, Reference Data Series No. 2, April 2003 edition (Vienna: IAEA, 2003).
4. "World List of Nuclear Power Plants," *Nuclear News* 46, no. 3, March 2003: 41–67.
5. Ernest Rutherford, "Radioactivity," in *The Encyclopaedia Britannica*, 11th edition, Vol. 22, (New York: The Encyclopedia Britannica Company, 1910): 802.
6. Charles Weiner, "1932—Moving into the New Physics," in *History of Physics*, Spencer R. Weart and M. Phillips, eds. (New York: American Institute of Physics, 1985).
7. Emilio Segrè, *From X-rays to Quarks: Modern Physicists and Their Discoveries* (San Francisco: W.H. Freeman, 1980).
8. Richard Rhodes, *The Making of the Atomic Bomb* (New York: Simon and Schuster, 1986).
9. *Leo Szilard: His Version of the Facts*, Spencer R. Weart and Gertrud Weiss Szilard, eds. (Cambridge: MIT Press, 1978).
10. Bertrand Goldschmidt, *Atomic Rivals*, translated by George M. Temmer (New Brunswick: Rutgers University Press, 1990).
11. Samuel Glasstone, *Sourcebook on Atomic Energy* (New York: Van Nostrand, 1950).

12. Henry D. Smyth, *Atomic Energy for Military Purposes* (Princeton, NJ: Princeton University Press, 1945).

13. A.M. Weinberg to R.L. Doan, memorandum, September 18, 1944, in Alvin M. Weinberg, *The First Nuclear Era: The Life and Times of a Technological Fixer* (New York: American Institute of Physics Press, 1994).

14. Richard G. Hewlett and Jack M. Holl, *Atoms for Peace and War: 1953–1961* (Berkeley: University of California Press, 1989).

15. "What Is the Atom's Industrial Future," *Business Week*, March 8, 1947, pp. 21–22, as cited in *The American Atom*, Robert C. Williams and Philip L. Cantelon, eds. (Philadelphia: University of Pennsylvania Press, 1984): 97–104.

16. Atomic Industrial Forum, *"Too Cheap to Meter?" Anatomy of a Cliché*, Special Report (1980).

17. Nuclear Energy Institute, *Historical Profile of U.S. Nuclear Power Development, 1994 edition*, (Washington, DC: NEI, 1994).

18. U.S. Department of Energy, *Annual Energy Review 2001*, Energy Information Administration report DOE/EIA-0384(2001) (Washington, DC: U.S. DOE, November 2002).

19. "Browns Ferry: Unit 1 Restart on Schedule, Budget," *Nuclear News* 46, no. 4, April 2003: 20.

20. U.S. Department of Energy, *Monthly Energy Review, March 2003*, Energy Information Administration report DOE/EIA-0035(2003/03) (Washington, DC: U.S. DOE, April 2003).

21. The President's Committee of Advisors on Science and Technology, *Federal Energy Research and Development for the Challenges of the Twenty-First Century* (Washington, DC: Executive Office of the President, November 1997).

22. U.S. Department of Energy, *Monthly Energy Review, August 2003*, Energy Information Administration Report DOE/EIA-0035(2003/08)(Washington, DC: U.S. DOE, 2003).

23. Nuclear Energy Institute, "U.S. Nuclear Power Plants Top 75-Percent Capacity," *Nuclear Energy Insight*, March 1995: 8.

24. "World List of Nuclear Power Plants," *Nuclear News* 41, no. 3, March 1998: 39–54.

25. Nuclear Energy Institute, *Fact Sheet: Nuclear Plant License Renewal*, February 2002. [From: http://www.nei.org/doc.asp?catnum=3&catid=615]

26. *Energy, U.S. Code of Federal Regulations*, title 10.

27. "Status of license renewal applications in the United States," *Nuclear News* 46, no. 12, November 2003: 24.

28. A.M. Petrosyants, *From Scientific Search to Atomic Industry*, translated from the 1972 Russian edition (Danville, IL: The Interstate Printers & Publishers, 1975).

29. International Atomic Energy Agency, *Energy, Electricity and Nuclear Power Estimates for the Period up to 2020*, Reference Data Series No. 1 (Vienna: IAEA, 2002).

30. U.S. Department of Energy, *International Energy Outlook 2002*, Energy Information Administration Report DOE/EIA-0484(2002) (Washington, DC: U.S. DOE, 2002).

31. Organization for Economic Co-operation and Development, Nuclear Energy Agency, *Nuclear Energy Data 2001* (Paris: OECD, 2001).

32. "World List of Nuclear Power Plants," *Nuclear News* 36, no. 3, March 1993: 41–60.

33. International Energy Agency, *Russia Energy Survey 2002* (Paris: OECD/IEA, 2002).

34. International Energy Agency, *Nuclear Power in the OECD* (Paris: OECD/IEA, 2001).

35. U.S. Atomic Energy Commission, *Nuclear Power 1973–2000*, Report Wash-1139(72) (Washington, DC: AEC, 1972).

3

Radioactivity and Radiation Exposures

3.1 Brief History

Radioactivity and the associated radiation exposures are sometimes thought of as environmental problems that have been created by modern science and technology.[1] However, substantial amounts of radioactivity exist in nature and have existed on Earth since its original formation. All biological species evolved on Earth, for better or worse, in this radioactive environment. Radioactivity could be plausibly termed the oldest "pollutant" if one chooses to describe an integral part of the natural world as a pollutant.

With the advent of controlled nuclear fission, we have obtained the ability to create concentrations of radioactivity that far exceed those encountered in the natural environment. An effort is made to minimize human exposure by isolating this material or surrounding it with shielding, but some exposure nonetheless occurs. To gain perspective on the seriousness of the potential resulting problems, we consider in this chapter the sources and amounts of artificial and natural radioactivity. Natural radioactivity here provides a reference or benchmark by which to gauge the significance of the man-made radioactivity.

Human awareness of the existence of ionizing radiation dates only to the period around 1900. Wilhelm Roentgen discovered X-rays in 1895, and within the next 5 years, Henri Becquerel and Marie and Pierre Curie discovered the

---

[1] In this and succeeding chapters, we are only concerned with *ionizing radiation* (i.e., those radiations for which the individual particles are energetic enough to ionize atoms of the material through which they pass). The environmental effects, if any, of nonionizing radiations, such as those created by microwave generators or 60-cycle electrical transmission lines, are not considered here.

previously unsuspected ionizing radiations from uranium ore. These were the newly named alpha particles, beta particles, and gamma rays from the radioactive decay of uranium and its associated products. The nature of these radiations and of radioactivity was quickly elucidated by investigators in France, Great Britain, and elsewhere.

The particles from radioactivity and cosmic rays constitute natural ionizing radiation. "Man-made" ionizing radiation comes from X-ray machines, nuclear accelerators, and nuclear reactors, as well as from artificial radionuclides produced by accelerators and reactors. The significant safety and environmental issues that arise in the consideration of nuclear power are related to the radionuclides produced in nuclear reactors.

The benefits of X-rays for medical diagnostic purposes were recognized almost immediately after their discovery and the purported benefits of radium, extracted from uranium ore, were proclaimed soon after. The belief that radium had curative properties continued into the 1920s, with horrors such as the sale of "medicines" laced with radium [1]. Surprisingly, underground sites that feature enhanced exposure to radon persist in the United States and elsewhere to this day.[2] At the same time, radium—and later other radionuclides[3]—have been used for better justified purposes, including many applications in medicine and industry.

Along with a growing use of X-rays and natural radioactive materials, there arose a recognition of the health risks of excessive exposure to ionizing radiation. These were recognized following exposures of radiologists, X-ray technicians, and, in a particular tragedy, radium watch dial painters. The dangers were first widely recognized in the 1920s and serious attempts to establish safety standards date from the late 1920s, starting with occupational hazards. Since the 1950s, the setting of standards has been expanded to apply to the general population, usually with criteria that are substantially stricter than those for occupational exposures. The overall issue of radiation risks and standards is discussed in Chapter 4.

3.2 Radiation Doses

3.2.1 Radiation Exposure and Radiation Dose

Radiation hazards are related to the magnitude of the *radiation exposure* or *radiation dose* incurred. The two terms are often treated in casual usage as interchangeable, but, in fact, their definitions are quite different. The dose is a quantitative measure of the impact of radiation, closely related to the energy deposited by incident radiation. Exposure is now used in two senses: (1) in a specialized sense in connection with the roentgen unit (see Section 3.2.2)

---

[2] Such "health" spas can be readily located on the Internet.

[3] The term *radionuclide* denotes a radioactive nuclear species. It is used almost interchangeably with the terms *radioactive isotope* and *radioisotope*.

and (2) as a general qualitative term to indicate "the incidence of *radiation on living or inanimate matter*" [2, p. 46]. Aside from the special case of the roentgen, *dose* is the appropriate term for quantitative descriptions, whereas *exposure* describes a general qualitative situation. Thus, for example, reports of the National Council on Radiation Protection and Measurements (NCRP) carry descriptive titles such as *Ionizing Radiation Exposure of the Population of the United States* [3], whereas the quantitative results presented are given in terms of the doses resulting from the exposures.

3.2.2 Basic Units of Exposure and Dose

Ionization Measure of Exposure: The Roentgen

Historically, radiation exposures were first expressed in terms of the *roentgen* (R). The definition was based on the ionization produced by X-rays or gamma rays in air.[4] This definition made one R equivalent to the deposition of 87 ergs per gram of air. Historically, the roentgen was a convenient unit to introduce because early work with radiation was concerned primarily with X-rays, and radiation was commonly detected with instruments that measured the amount of ionization produced in air.

Absorbed Dose: Rad and Gray

Radiation dosimetry is primarily concerned with radiation effects in human or animal tissue, and deposition of energy in tissue is more relevant than ionization in air. Therefore, radiation exposures are usually described using units other than the roentgen. The basic radiation dose in present usage is the *absorbed dose*, sometimes called the *physical dose*. The units of absorbed dose are the *rad* and, in the SI system, the *gray* (Gy).[5] The definition of these units defines the meaning of the absorbed dose:

---

[4] More specifically, the roentgen was defined as the amount of X-ray or gamma-ray radiation that produces one electrostatic unit of charge (esu) in 1 cm^3 of air (at standard temperature and pressure). The corresponding energy deposition is $E = I/\rho e = 87$ ergs/g (in air), where the average energy expenditure per ion pair is $I = 33.8$ eV $= 5.41 \times 10^{-11}$ ergs [4, p. 18], the electron charge is $e = 4.80 \times 10^{-10}$ esu, and the density of air is $\rho = 0.001293$ g/cm^3. The roentgen has more recently been redefined in terms of SI units to be the amount of X-ray or gamma-ray radiation that produces 2.58×10^{-4} coulomb (C) of charge in 1 kg of air. This formal redefinition is numerically equivalent to the earlier definition.

[5] The situation for dose units is similar to that for other SI units competing with older units. The SI dose units, the gray and the sievert (see next subsection), are used in most foreign countries and in much of the U.S. scientific literature. The older units, the rad and the rem, nonetheless remain in common usage in the United States.

One gray is the absorbed dose corresponding to the deposition of 1 joule/kilogram in the absorber.

One rad is the absorbed dose corresponding to the deposition of 100 ergs/gram in the absorber.

From these definitions, it follows that

$$1 \text{ gray} = 100 \text{ rad.}$$

The rad and the gray apply to any absorber, with the absorbers of greatest interest being human tissues. These units have the advantage over the roentgen in not being restricted to photons, of being defined in terms of fundamental units of energy and mass, which will not require revision if constants are better determined (e.g., the average ionization energy of electrons in air), and of being directly applicable to any tissue. A gamma-ray flux that delivers an exposure of 1 R in air would deliver a dose of approximately 0.96 rads to human tissue.[6] Thus, the roentgen and rad are numerically roughly equivalent.

Dose Equivalent: Rem and Sievert

Biological harm is not determined solely by the total energy deposition per unit mass. It is found also to depend on the *linear energy transfer* (LET) rate, where the LET rate is the ratio of energy deposition by an individual particle to the distance it traverses.[7] It is alternatively referred to as the density of ionization. Greater density corresponds to more localized damage to the tissue and leads to more harm per unit energy (except at very high density where there is a saturation effect).

The energy deposition per unit distance traversed is much greater for alpha particles than for electrons.[8] As a result, the deposition of 100 ergs by alpha particles in a given tissue mass is more injurious than the deposition by electrons of the same amount of energy. More generally, radiations are classified as low LET and high LET. The low-LET radiations are X-rays, gamma rays, and beta particles. X-rays and gamma rays, being neutral, are not themselves ionizing particles, but they transfer their energy to electrons. In all of these cases, the ionizing particle is an electron for which the rate of energy loss per unit path length is small. The high-LET radiations include alpha particles and neutrons. Neutrons themselves are not ionizing particles,

---

[6] This follows from the fact that the mass absorption coefficient of gamma rays is about 10% greater in tissue than in air.

[7] This rate changes as the particle loses energy, and the rate is properly defined as the ratio $\Delta E / \Delta x$, where ΔE is the magnitude of the energy lost by a particle (and transferred to the material) as the particle traverses a small segment of path length Δx.

[8] As a corollary, the distance traversed in matter by alpha particles is much less than the distance traversed by electrons of roughly comparable energy.

but their interactions with nuclei of the material produce high-LET ionizing particles, for example, in elastic scattering with hydrogen nuclei (protons).

The *dose equivalent* is introduced to take into account the *relative biological effectiveness* of different types of radiation. It is defined so that equal dose equivalents have an equal chance of producing the somewhat random biological effects that occur at low and moderate dose levels, independent of the type of radiation [5, p. 15].[9] The dose equivalent H is related to the absorbed dose D by a dimensionless parameter commonly known as the *quality factor* Q, which takes on different values for different radiations:

$$H = QD. \tag{3.1}$$

The unit of dose equivalent corresponding to the rad is the *rem*, an abbreviation for roentgen-equivalent-man. The analogous unit corresponding to the gray is the *sievert* (Sv). Thus, 1 sievert = 100 rem.

Strictly speaking, the relative biological effectiveness of different radiations depends on the tissue exposed, the type of effect in question, the energy of the radiation, and the intensity of the exposure. However, there is usually insufficient information about biological effects to make such precise distinctions. Instead, Q is commonly specified as an approximate average conversion factor—loosely speaking, an average relative biological effectiveness.[10]

Average values of the quality factor for radiations of interest are given in Table 3.1. These correspond to the standard values recommended in International Commission on Radiological Protection (ICRP) Publication 60 [5, p. 6] and are widely accepted by other bodies (see, e.g., Refs. [6], §20.1004, and [7], p. 56). As seen from these data, an absorbed dose of 1 rad from exposure to

Table 3.1. Average quality factors and the absorbed dose corresponding to a dose equivalent of unity for selected radiations.

Type of Radiation	Quality Factor[a] (Q)	Absorbed Dose[b] [D (for $H = 1$)]
X-rays and gamma rays	1	1
Electrons (including beta particles)	1	1
Neutrons (depending on energy)	5–20	0.2–0.05
Alpha particles and fission fragments	20	0.05

[a]From Ref. [5], p. 6, where this quantity is termed the "radiation weighting factor."
[b]For H in rem or Sv, D is in rad or gray, respectively.

---

[9] These are so-called *stochastic* effects. The distinction between stochastic and deterministic effects is discussed in Section 4.2.

[10] In recognition of the imprecision in usage, the International Commission on Radiological Protection has recently recommended that the parameter of Eq. (3.1) be termed the *radiation weighting factor* rather than the quality factor [5, p. 5], but we will use quality factor here, as it is still commonly employed in the literature.

alpha particles is believed to have about the same overall biological impact, in terms of cancer induction, as an absorbed dose of 20 rad from gamma rays.

A word of caution is in order here. In the example cited, for a dose equivalent of 200 mSv the potential impact is a stochastic one, and the use of the dose equivalent is appropriate (see Section 4.2). However, at considerably higher dose levels, when deterministic effects dominate, use of the quality factors of Table 3.1 for neutrons and alpha particles would overestimate the effect [5, p. 15]. It is therefore common practice to describe high doses in terms of the absorbed dose, in rads or grays.

3.2.3 Effective Dose Equivalent or Effective Dose

It is, of course, more serious if the entire body receives a given exposure than if only a single organ receives that exposure. Thus, when a dose equivalent is specified, it is important to know whether the reference is to a limited region of the body or the whole body. If only a small part of the body is irradiated, it is possible to translate an organ dose into the whole-body dose that would produce the same overall risk of a fatal cancer, with the use of appropriate weighting factors. This translated dose is called the *effective dose equivalent* H_E or often just the *effective dose* E.[11]

The effective dose equivalent H_E is related to the individual tissue doses H_T and D_T by tissue weighting factors w_T and a summation over tissues:

$$H_E = \Sigma w_T H_T. \tag{3.2}$$

Standard values for the weighting factors w_T are 0.20 for the gonads, 0.12 for the bone marrow, colon, lungs, and stomach (each), 0.05 for the bladder, breast, liver, esophagus, and thyroid (each), 0.01 for the skin and bone surface (each), and 0.05 for the remainder of the body [5, p. 8]. The sum of these weighting factors is unity. Therefore, when the dose is uniform over the entire body, the effective dose and the "whole-body dose equivalent" are the same.

In judging the probability of cancer induction or genetic damage from a given exposure, the effective dose (or the effective dose equivalent, in U.S. regulatory usage) is the relevant quantity. It is an appropriate dose for both describing exposures and setting radiation standards. It is to be noted that very often this term is informally shortened to "dose," with the context indicating the intended meaning.

---

[11] The "effective dose equivalent" is the earlier usage. In 1990, in its Publication 60, the ICRP adopted the term "effective dose" as a simpler name [5, p. 7], and this terminology is now employed in the UNSCEAR reports [7, p. 21]. The earlier terminology is retained in many U.S. government publications, including those that set radiation standards (see, e.g., Ref. [8], §197.2). NCRP Report No. 116 (unlike ICRP Publication 60) indicates a subtle difference between the definitions of E and H_E [9, p. 21], but any such distinction is commonly ignored (see, e.g., Ref. [10], p. xxiv.

Collective Dose Equivalent or Collective Dose

The *collective dose equivalent* (in person-Sv), or more simply the *collective dose*, to a population is the sum of the effective doses to the individuals in that population. It is intended to be a measure of the overall impact. Use of the collective dose is appropriate if the health effects are linearly proportional to the dose (i.e., if the *linearity hypothesis* is valid) (see Section 4.3.3). Then, for example, the same number of cancers are expected if 1000 people each receive a dose of 200 mSv or if 100,000 people each receive a dose of 2 mSv. In each case, the collective dose is 200 person-Sv. The use of collective dose is controversial, especially when a large collective dose results from small doses to many people, because the underlying premise of linearity is controversial. We will return to this issue in the specific context of the collective dose from ^{14}C (in Section 13.2.2).

Dose Commitment

The concept of the *dose commitment* is particularly relevant to the intake of radionuclides into the body. If the radionuclide has a long residence time in the body, there will be an exposure lasting for many years. Thus, specifying the dose in a given year gives an incomplete picture. The dose commitment is the cumulative dose over the relevant future time (perhaps 50 years on average), allowing for a decrease in annual dose due to radioactive decay and the gradual elimination of the radionuclide from the body. This decrease can be described quantitatively in terms of the radionuclide's radioactive and biological half-lives. In the same spirit as above, it is possible to consider both an individual dose commitment and a collective dose commitment.[12]

3.3 Radioactive Decay

3.3.1 Half-life and Mean Life

The "activity" of a radioactive sample is equal to the number of decay events in the sample per unit time. Each radionuclide is characterized by a *half-life T*—the time interval during which one half of the sample decays and the rate of radioactive decay is halved. For example, in 10 half-lives, the activity drops by a factor of $2^{10} = 1024$. Half-lives of radionuclides vary greatly, from a small fraction of a second to billions of years.

As discussed in more detail in the Appendix, the activity of a sample decreases exponentially with time, proportional to $e^{-\lambda t}$, where λ is the *decay constant* and t is the time. The relationship among λ, the *mean life* (τ), and

---

[12] Here, "dose commitment" almost always refers to the effective dose, but the EPA, for example, avoids any possible ambiguity by using the full term "committed effective dose equivalent" [8, §197.2].

the *half-life* (T) is

$$T = \frac{\ln 2}{\lambda} = \tau \ln 2. \tag{3.3}$$

3.3.2 Units of Radioactivity

The original unit for measuring the amount of radioactivity was the *curie* (Ci), with derivative units in common use ranging from the picocurie (pCi) to the megacurie (MCi).[13] The curie was defined in 1910 to be the amount of radon (^{222}Rn) in equilibrium with 1 g of radium (^{226}Ra) [11, p. 448], but the usage evolved to become the amount of any radionuclide that decays at the same rate as 1 g of ^{226}Ra. Using current numbers for the half-life ($T = 1600$ years) and atomic mass ($M = 226.03$ u) of ^{226}Ra, this rate is 3.66×10^{10} disintegrations/second.

With this definition, the numerical value of the curie would have to be modified whenever a more precise value for T (or M) is found for ^{226}Ra. The curie was subsequently redefined in 1950 as an exact unit, namely "the quantity of any radioactive nuclide in which the number of disintegrations per second is 3.700×10^{10}" [12, p. 472]:

1 curie $= 3.7 \times 10^{10}$ disintegrations per second.[14]

In international usage, and gradually in much of U.S. usage, the curie is being supplanted by the becquerel (Bq), introduced along with other SI units in 1975. The becquerel is defined as the activity corresponding to one disintegration per second:

1 becquerel $= 1$ disintegration per second $= 2.703 \times 10^{-11}$ Ci.

The magnitude of the activity (in Bq) is λN, where λ is the decay constant (in s^{-1}) of the radionuclide and N is the number of nuclei in the sample.

It is difficult to obtain a sense of scale (e.g., to know when an amount of radioactivity is "large"). Indoor air typically has a radon concentration slightly more than 1 pCi (0.037 Bq) of ^{222}Rn per liter. A house with a floor area of 2000 ft^2 therefore contains roughly 0.5 μCi (2×10^4 Bq) of radon. A 70-kg person has a potassium-40 (^{40}K) body content of about 0.1 μCi (3.7×10^3 Bq). Intense sources used for cancer therapy have ranged in size up to about 10,000 Ci (3.7×10^{14} Bq). The core of a typical commercial nuclear reactor has an activity of about 20 billion Ci (7×10^{20} Bq), just after being turned off following a

---

[13] The derivative units use the standard prefixes: *femto* (f) $= 10^{-15}$, *pico* (p) $= 10^{-12}$, *nano* (n) $= 10^{-9}$, *micro* (μ) $= 10^{-6}$, *milli* (m) $= 10^{-3}$, *kilo* (k) $= 10^{3}$, *mega* (M) $= 10^{6}$, *giga* (G) $= 10^{9}$, and *tera* (T) $= 10^{12}$.

[14] There may be a minor semantic problem here, in that the *curie* is used to denote both an amount of material and a number of disintegrations (see, e.g., Ref. [2]). However, this rarely, if ever, leads to an ambiguity.

period of prolonged operation. These amounts, per se, have little meaning. The important quantity is the resulting radiation dose, which also depends on the type of activity and the extent of human contact with it.

3.3.3 Specific Activity

A radioactive substance can be characterized in terms of its *specific activity*, S, defined as the activity per unit mass. The specific activity of a radioactive species (in Bq/g) is

$$S = \lambda N = \frac{\ln 2}{T} \frac{N_A}{M}, \tag{3.4}$$

where N is the number of nuclei per gram, M is the atomic mass of the radionuclide (in atomic mass units, u), T is its half-life (in seconds), and N_A is Avogadro's number. Specific activities for selected radionuclides are given in Table 3.2.

As follows from its definition, and as exhibited in Table 3.2, high specific activity is associated with short half-life. Thus, ^{238}Pu has a specific activity that is 275 times that of ^{239}Pu, corresponding to a half-life that is smaller by about a factor of 275. The radionuclides strontium-90 (^{90}Sr) and cesium-137 (^{137}Cs) have the highest specific activities listed in Table 3.2 because they have the shortest half-lives. Radionuclides with much shorter half-lives have still higher specific activities but often are less of a hazard—for example, as constituents of nuclear waste—because they disappear quickly.

If one is considering an element with more than one significant isotope, as is often the case in discussing natural radioactivity, then Eq. (3.4) must be modified to take into account the contributions of all the isotopes. For the element, the specific activity is the sum of the individual contributions S_i,

Table 3.2. Specific activities S of selected radionuclides.

Nuclide[a]	Isotopic Abund. (%)	Atomic Mass (u)	Half-life, T (years)	S (Bq/g)	S (μCi/g)
^{239}Pu	—	239.05	2.411×10^4	2.30×10^9	6.20×10^4
^{238}Pu	—	238.05	87.7	6.34×10^{11}	1.71×10^7
^{237}Np	—	237.05	2.14×10^6	2.60×10^7	7.03×10^2
^{238}U	99.27	238.05	4.468×10^9	1.24×10^4	0.336
^{235}U	0.72	235.04	0.704×10^9	8.00×10^4	2.16
^{232}Th	100.00	232.04	14.05×10^9	4.06×10^3	0.110
^{226}Ra	—	226.03	1600	3.66×10^{10}	9.89×10^5
^{137}Cs	—	136.91	30.07	3.21×10^{12}	8.68×10^7
^{90}Sr	—	89.91	28.79	5.11×10^{12}	1.38×10^8
^{87}Rb	27.83	86.91	47.5×10^9	3.20×10^3	0.087
^{40}K	0.0117	39.96	1.277×10^9	2.59×10^5	7.01

Source: Data from Ref. [13].

where

$$S_i = \frac{\ln 2}{T_i} \frac{f_i N_A}{M_E}. \tag{3.5}$$

Here, M_E is the atomic mass of the element and, for isotope i, f_i is the fractional isotopic abundance (by number of atoms) and T_i is the half-life. For example, the specific activity of natural uranium is 2.53×10^4 Bq/g, more than twice that of pure ^{238}U. The other contributors are ^{235}U, which adds little due to its low isotopic abundance and relatively long half-life, and ^{234}U ($T = 2.455 \times 10^5$ years), which is in secular equilibrium (see Section 3.4.2) with ^{238}U and has the same ratio f_i/T_1.

3.4 Natural Radioactivity

3.4.1 Origin of Natural Radioactivity

Nucleosynthesis

The existence of natural radioactivity (i.e., the formation of the radionuclides found in nature) cannot be understood in isolation, apart from the understanding of the formation of the stable nuclides. They were produced in the same processes of nucleosynthesis, primarily in the evolution and explosion of stars. Only matters of relatively fine nuclear detail make some species stable and some unstable. A sketch showing the relationship between the atomic number and atomic mass number for the nuclides found in nature is shown in Figure 3.1. These are the "survivors" among the many radionuclides formed in nucleosynthesis, most of which have since decayed.

The first phases of nucleosynthesis, which build the elements up to the general region of atomic mass number $A = 60$, proceed through a series of reactions initiated primarily by protons and alpha particles. Some of the nuclides produced are stable and some are radioactive. The latter are almost all beta-particle emitters, with half-lives that are short compared to the age of the Earth. In consequence, there are very few natural radioactive isotopes in this mass region. The one important exception is ^{40}K.

Beyond the $A \approx 60$ region, nucleosynthesis proceeds primarily by neutron capture. The neutron-rich products of neutron capture decay by β emission (more specifically, by β^- emission) to more stable isobars, and the buildup to heavy elements continues by a sequence of neutron capture and beta decay events. Again, most of the radioactive nuclei that were formed in this fashion have had ample time to decay since they were formed. The most prominent exception is rubidium-87 (^{87}Rb).

As the buildup by neutron capture proceeds up to and past uranium (atomic number $Z = 92$), the energy corresponding to the repulsive electrical force between protons in the nucleus force grows more rapidly than the energy corresponding to the attractive nuclear force, and alpha-particle decay and fission become possible. In several important cases, nuclei are formed that

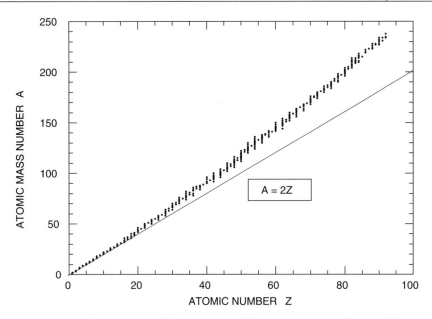

Fig. 3.1. Nuclides in nature: the trajectory of atomic mass numbers and atomic numbers for nuclides that are stable (or very long lived) or members of the natural radioactive decay series. The path $A = 2Z$ is shown for reference (solid line).

are stable against beta-particle decay, but that can decay by alpha-particle emission with half-lives comparable to the age of the Earth. These nuclei are responsible for the radioactive decay series discussed in the next section. For nuclei of very high mass, fission terminates the neutron capture chain and adds to the abundances of nuclei of intermediate mass (in the $A = 80$ to $A = 160$ region). Buildup can begin again in a new neutron-capture chain, with these fission fragments as starting points.

Radioactive Products of Nucleosynthesis

The material of the solar system was formed by several generations of such synthesis processes, possibly including the explosion of a supernova near the protosolar nebula shortly before the formation of the Sun and the condensation of the planets. In the 4.6 billion years since the formation of the Earth, any nucleosynthesis products with short half-lives have disappeared. The only naturally radioactive relics of nucleosynthesis are nuclei that have very long half-lives or are decay products of such nuclei. The heaviest such nuclide is uranium-238 (^{238}U).

One group of natural radioactive nuclides are the intermediate-mass beta-particle emitters. As mentioned earlier, a very large number of beta-unstable nuclei were formed in nucleosynthesis, but few remain today because typical lifetimes for beta-particle emission range from seconds to days. The existence of some long-lived exceptions is due to special circumstances in the nuclear properties of these nuclides and their decay products. With many beta-unstable nuclides having been formed, the chance is enhanced that a few will have the requisite unusual properties.

The two beta-particle emitting nuclides cited earlier are of particular environmental relevance: ^{40}K ($T = 1.28 \times 10^9$ years) and ^{87}Rb ($T = 47.5 \times 10^9$ years). They produce non-trivial radiation exposures to the human body because of the potassium and rubidium naturally present in the tissues. The calculated exposure due to the ^{40}K is considerably greater than that due to the ^{87}Rb, so sometimes only ^{40}K is considered.

At higher atomic masses, the important radionuclides are the members of three radioactive decay series, headed by long-lived parent nuclei. These are discussed in greater detail in the following section.

In addition to the nucleosynthesis products, some radioactive nuclei are being continually produced by cosmic rays, but the amounts are so small that they have little practical significance, except as historical markers for determining ages of samples in which they are found. The most important of these is carbon-14 (^{14}C), produced in the interaction of cosmic-ray neutrons with nitrogen-14 (^{14}N) in the atmosphere. It is the radionuclide that makes possible *carbon dating*, as used in archeology. It has also had a curious role in the consideration of safety criteria in nuclear waste disposal (see Section 13.2.2).

3.4.2 Radioactive Series in Nature

The Occurrence of Three Series

The radioactive decay series are groups of radioactive nuclei that arise from the production in nucleosynthesis of long-lived alpha-particle emitters. The "head" of each series is either a direct product of nucleosynthesis or a remnant of the synthesis of short-lived parents. As it decays, it feeds a set of radioactive progeny which decay by either alpha-particle or beta-particle emission. In alpha-particle decay, the atomic mass number A is reduced by 4. In beta-particle decay, A is unchanged.

This means that there are, in principle, four possible series of radioactive nuclei. They can be classified by the magnitude of the parameter δ, if the atomic mass number is expressed as $A = 4n + \delta$. Here, n is an integer that decreases by unity in alpha-particle decay and δ is a fixed constant for each series, with possible magnitudes of 0, 1, 2, or 3. The observed series in nature are given in Table 3.3.

It is to be noted that there is no series in Table 3.3 for $\delta = 1$. The three listed series are headed by nuclei with half-lives of the order of 1–10 billion years. For a half-life that is very much longer, the nuclide might not be considered "radioactive," but none of the four possible series is headed by a nuclide

Table 3.3. Natural radioactive decay series.

Series Type	Series Name	Parent Nucleus	Half-life $(10^9$ years$)$	Stable Nucleus	Number of decays $(N)^a$	
					α	β^-
$A = 4n$	Thorium	^{232}Th	14.05	^{208}Pb	6	4
$A = 4n + 2$	Uranium	^{238}U	4.47	^{206}Pb	8	6
$A = 4n + 3$	Actinium	^{235}U	0.704	^{207}Pb	7	4

$^a N_\alpha = (A_i - A_f)/4$; $N_{\beta^-} = 2N_\alpha - (Z_i - Z_f) = (A_i/2 - Z_i) - (A_f/2 - Z_f)$, where i and f denote *initial* (parent) and *final* (stable), respectively.

with so long a half-life. If the half-life were much shorter, the series would have disappeared. This is the case for the $A = 4n + 1$ series, which is not found in nature because no member of the series has a sufficiently large half-life. For example, the longest-lived $A = 237$ isobar is ^{237}Np, with a half-life of 2.1×10^6 years, and there exists no alternative long-lived parent at higher or lower A for a $4n + 1$ series.[15]

The Uranium Series

The uranium series, headed by ^{238}U, includes two radionuclides of some special importance, in addition to ^{238}U itself. One of the progeny, ^{226}Ra, was the source of the intense activity found in uranium ore in the early studies carried out by the Curies. It appears as a strong source (i.e., one with a high specific activity), because of its intermediate half-life of 1600 years. A longer half-life would reduce the specific activity. A much shorter half-life would have made it harder to notice in slow chemical separations (as practiced by the Curies), as well as of little practical use.

No other member of any of the natural radioactive series has a half-life lying within a factor of 20 of the ^{226}Ra half-life. Its intermediate half-life, coupled with the relatively high abundance of the parent, ^{238}U, made ^{226}Ra unique for practical applications of radioactivity, until an array of "artificial" radioisotopes became available as products of nuclear reactions using acelerators or reactors.

Radium-226 decays by alpha-particle emission to ^{222}Rn ($T = 3.8$ days). This is the nuclide responsible for the radiation exposures arising from indoor radon. Radon is a noble gas and reacts little with other substances. Therefore, it itself is not trapped in the lungs when inhaled. However, ^{222}Rn has a number of alpha-particle emitting progeny with short half-lives. These are present in the air as decay products of ^{222}Rn and are to a considerable extent retained in the lungs when inhaled. They therefore lead to significant radiation doses to the lungs (see Table 3.5).

---

[15] The heaviest stable nucleus in this series is bismuth-209.

Steady-State Relation in Radioactive Decay Series

In a radioactive decay series, each decay product is itself radioactive until the end of the series is reached when a stable nuclide is formed. In each of the three natural series, the longest half-life is that of the original parent, or head of the series, and the series reaches a steady state.

If the radioactive series is in an isolated volume of material, then in the steady state, each member of the series decays at a virtually constant rate. We say "virtually" constant because there is a very slow change governed by the very slow decay of the head of the series. In the steady state, the number of decays per unit time is the same for each member of the series:

$$\lambda_1 N_1 = \lambda_2 N_2 = \lambda_3 N_3 = \cdots, \tag{3.6}$$

where N_i and λ_i are the number of nuclei and the decay constant, respectively, for the ith member of the series.

The steady-state condition for a radioactive decay series is also known as *secular equilibrium*. The secular equilibrium expression of Eq. (3.6) can be derived rigorously from the equations that describe a series of decaying nuclei, for the limit that the head of the series has a decay constant λ small compared to that of the other members of the series or, equivalently, a half-life T large compared to that of the other members.

The result can also be seen to be qualitatively reasonable. Consider the relationship between the first (long-lived) and second (shorter-lived) member of the series. If the condition of Eq. (3.6) is violated so that, for example, $\lambda_2 N_2 > \lambda_1 N_1$, then the loss from the "pool" of species 2 exceeds the flow into it, and N_2 will decrease until the inequality is removed. A similar restoration to the condition of Eq. (3.6) occurs if $\lambda_2 N_2 < \lambda_1 N_1$. By an extension of the same argument, secular equilibrium is established throughout the radioactive series.

The condition described above applies to isolated material. It does not apply exactly to the natural radioactive series in all environments, due to chemical and physical processes that transfer different elements from one locale to another at different rates. Fractionation can be important in the dissolution and removal of radionuclides by water flowing through rocks and in the transfer of radionuclides from the ocean to ocean sediments.

3.4.3 Concentrations of Radionuclides in the Environment

Radionuclides in the Earth's Crust

There are substantial variations in the concentrations of chemical elements and, hence, of radionuclides in different parts of the Earth's crust. For example, for the crust as a whole, the uranium abundance by mass is about 3 parts per million (ppm) (i.e., 0.0003%), with common rocks having uranium concentrations ranging from 0.5 to 5 ppm [14, p. 61]. Concentrations in uranium ores commonly exceed several hundred ppm, and there are deposits that are reported to have concentrations as high as 65% of U_3O_8, equivalent to 550,000 ppm of U [15, p. 409].

For a detailed picture of radionuclides in nature, it is useful to separate rocks by rock type. It is found, for example, that the abundance of radionuclides is greater in granite than in carbonate rocks such as limestone [14, p. 61]. In some cases, these differences are of crucial importance, as for uranium ores. However, for present purposes of providing perspective, it is of interest to consider the broad average for the continental crust. Average values for the upper continental crust have been presented by a number of authors; we follow Ref. [14] in adopting those presented by Taylor and McLennan [16, p. 46] for use in Table 3.4.

It is seen from Table 3.4 that the magnitude of the activity for the beta-particle emitters, ^{40}K and ^{87}Rb, is significantly greater than that for the listed alpha-particle emitters. However, the disparity in overall effect is much less than might appear because uranium-238 (^{238}U) and thorium-232 (^{232}Th) each head long series of radionuclides and the activities of the progeny should

Table 3.4. Average abundances of natural radionuclides in the upper continental crust, the oceans, and ocean sediments.

	^{40}K	^{87}Rb	^{232}Th	^{238}U
Upper continental crust				
Elemental abundance (ppm)	28,000	112	10.7	2.8
Activity[a] (Bq/kg)	870	102	43	35
Activity (nCi/kg)	23	2.7	1.2	0.9
Activity[b] (kCi/km^3)	66	8	3.3	2.6
Oceans				
Elemental concentration (mg/L)	399	0.12	1×10^{-7}	0.0032
Activity[a] (Bq/liter)	12	0.11	4×10^{-7}	0.040
Activity (nCi/liter)	0.33	0.003	1×10^{-8}	0.0011
Ocean sediments				
Isotopic abundances (mg/kg)	≈ 2		5.0	1.0
Elemental abundance[c] (ppm)	17,000		5.0	1.0
Activity[a] (Bq/kg)	500		20	12
Activity (nCi/kg)	14		0.5	0.3

[a]Calculated from elemental abundances, using Eq. (3.5) with parameters of Table 3.2.
[b]Calculated for a density of 2.8 g/cm^3.
[c]Calculated from isotopic abundances.
Source: Elemental abundance data for the continental crust are from Ref. [16], p. 46. Concentrations of K and Rb in the oceans are from Ref. [16], p. 15, those for Th and U are from Ref. [17]. Isotopic abundances in ocean sediments are from Ref. [18], pp. 23–24.

be added.[16] For example, each ^{238}U decay is followed by seven additional alpha-particle decays and six beta-particle decays (see Table 3.3).[17] Further, alpha-particle emitters, in general, release substantially more energy per disintegration than do beta-particle emitters.

We can get a sense of the magnitudes involved by estimating the total activity in the Earth's crust down to some appropriate depth. The average density of the Earth's continental crust is 2.8 g/cm^3 and the average depth is approximately 17 km [19, p. 11]. Arbitrarily taking 1 km as a depth of interest, it being within relatively easy accessibility, the mass per unit area is 2.8×10^{12} kg/km^2. For this depth, the ^{238}U and ^{232}Th activities are therefore each roughly 3×10^3 Ci/km^2 of surface (see Table 3.4).

The activity represented by the entire series is about a factor of 10 greater, corresponding to roughly 3.5×10^4 Ci/km^2 for each series. Adding these activities and the decay of ^{40}K and ^{87}Rb gives a total of roughly 1×10^5 Ci/km^2, down to a 1-km depth. The land area of the continental United States (48 contiguous states) is about 8×10^6 km^2. Therefore, the total activity over the area of the United States, again down to a depth of 1 km, is roughly 1×10^{12} Ci. If employed with suitable caution, such numbers may help provide perspective when one considers the disposal of radioactive nuclear wastes in the ground (see Section 10.3.3).

Radionuclides in the Oceans

Over the billions of years since the Earth was formed, there has been a substantial interchange of material between the land and oceans. For example, flowing water erodes rocks and carries material from the rocks into the oceans. Therefore, it is to be expected that the same elements that are present in the Earth's crust will also be present, to one degree or another, in the oceans.

However, a high concentration of an element in the Earth's crust does not necessarily mean a high concentration in the ocean. Table 3.4 gives data on the concentrations of some natural radionuclides in the oceans and ocean sediments. It is seen, for example, that the ratio of the concentration (in parts per million by mass) in the oceans to the concentration in the Earth's crust is of the order of 10^{-2} for potassium and about 10^{-8} for thorium. These differences are in part due to differences in the solubility of the compounds in which the elements are normally found in the crust and in their rates of removal by erosion. They are due also to large differences in the average

---

[16] Uranium-235 and the actinium series are omitted in Table 3.4 because the abundance of ^{235}U is small—only 0.7% that of ^{238}U.

[17] At any particular location, the magnitudes of the activities of different species in a decay series may not be the same, due to differences in chemical and physical properties, which cause different movements of these elements. Globally, however, the series are in secular equilibrium, and the total activities are the same for each member.

residence time of elements in the ocean, before transfer to ocean sediments in the ocean floor. For example, the residence time of potassium is about 10 million years, whereas the residence time of thorium is less than 10 years [16, pp. 15–16].

The volume of the oceans is 1.4×10^{21} L [18, p. 22]. This means that the total activity of ^{40}K in the oceans is about 5×10^{11} Ci and the total activity of ^{238}U in the oceans is about 1×10^9 Ci. Thus, in the oceans, ^{40}K is the dominant species in terms of total activity. Its ratio to ^{238}U (and, much more dramatically, ^{232}Th) is greater in the ocean than in the Earth's crust. Fractionation processes, such as rapid removal of ^{230}Th, reduce the abundances of many of the ^{238}U series progeny in the oceans to levels well below secular equilibrium with ^{238}U (see, e.g., Ref. [18]).

3.5 Survey of Radiation Exposures

3.5.1 Natural Sources of Radiation

Overview

Average radiation exposures resulting from natural sources are far greater than those that result from human technology (especially if one omits medical exposures), although the latter usually attract greater attention. The important natural sources include inhaled radon and its progeny, radionuclides in the Earth's crust, radionuclides in the body, and cosmic rays. The resulting doses are summarized in Table 3.5, in terms of averages for the United States and the world.[18] The two sets of estimates differ partly because they refer to different populations and partly because of differences in the methods of calculation.

In aggregate, the dose from natural sources, excluding radon, averages about 1 mSv/yr. Until the early 1980s, many analyses ignored radon and treated this as the *total* natural dose. The actual average, with radon, is substantially greater. U.S. analyses indicate it to be in the neighborhood of 3 mSv/yr, but, as discussed below, there are substantial differences between this value and that estimated in the UNSCEAR report.

More important than the differences in the results of alternative calculations are the very wide variations in the actual doses received by individuals. For example, in parts of Colorado, the terrestrial radiation dose and the cosmic-ray dose each are about triple the U.S. mean. Even greater excursions occur for radon exposures, with indoor radon levels in some houses far above the U.S. average. Thus, the data of Table 3.5 are indicators of overall impact and general scale, rather than a guide to the dose for any particular individual.

---

[18] The values are taken from reports of the National Council on Radiation Protection [3] and the United Nations Scientific Committee on the Effects of Atomic Radiation [7].

Table 3.5. Average radiation exposures for the United States and world: effective dose equivalents (in mSv/yr).

Radiation Source	United States NCRP	World Average UNSCEAR
Natural sources		
Inhalation (mostly radon)	2.0	1.26
Terrestrial radiation	0.28	0.48
Radionuclides in body	0.39	0.29
Cosmic rays	0.27	0.38
Cosmogenic	0.01	0.01
Total of natural sources (rounded)	3.0	2.4
Medical diagnosis[a]		
X-rays	0.39	0.4
Nuclear medicine	0.14	0.03
Other		
Consumer products[b]	0.1	—
Nuclear weapons testing[c]	< 0.01	0.005
Nuclear fuel cycle[d]	0.0005	0.0002
Chernobyl accident	—	0.002
TOTAL (rounded)	3.6	2.9

[a]Medical exposures vary widely among individuals. The quoted numbers are averages over the entire population; the U.S. average is for 1980 and the world average is for 1991–1996.
[b]Major sources include building materials and the domestic water supply.
[c]The U.S. average is for 1987 and the world average is for 2000.
[d]Based on calculated exposure of people within 50 miles of nuclear facilities and averaged over entire population (for 100 reactors).
Sources: Data for United States from Ref. [3]: p. 15 for natural sources, p. 47 for diagnostic X-rays (1980) and nuclear medicine (1982), and pp. 40 and 53 for other. Data for the world from Ref. [7]: p. 140 for natural sources, p. 464 for medical diagnosis (1991–1996), and p. 8 for other (2000).

The next largest sources of radiation exposures are medical treatments, primarily diagnosis. Other anthropogenic sources, including nuclear power, have made comparatively little contribution to the overall dose received by the average individual.

Radon

Radon is the largest single contributor to the average radiation dose received by members of the public (see Table 3.5). Radon concentrations are higher indoors, but there is an appreciable concentration outdoors as well. In addition to impacting the public, radon was responsible for many cancer deaths among uranium miners, due to high exposures before the hazards were recognized. Overall, radon and the Chernobyl accident represent the only firmly established causes of large-scale radiation casualties for workers in the nuclear industry.[19]

The key radon isotope is ^{222}Rn. It is formed in the alpha-particle decay of ^{226}Ra, a member of the ^{238}U decay chain. Uranium, and with it radium and radon, is present in all soils. Another isotope of radon, ^{220}Rn—called *thoron* because it is a member of the series headed by ^{232}Th—also contributes, but it typically adds less than 10% to the total dose [7, p. 140]. Therefore, it is often ignored in discussions of "radon." Radon is a noble gas and readily escapes from the ground, entering houses through openings in their structure. The indoor concentrations vary greatly, depending on the radium content of the underlying soil, the soil porosity, and the house construction.

Radon-induced lung cancer is due to the inhalation of radon and its progeny, and the irradiation of lung tissues by alpha particles that are emitted by these radionuclides—particularly the short-lived progeny of ^{222}Rn. The absorbed dose to lung tissues can be calculated from the radionuclide concentration in the air, the dust content of the air, and the movement of the radionuclides into and within the lung. The result can be expressed as a dose in grays per unit radon concentration in air and can be converted into an effective dose equivalent, using estimated values of the tissue weighting factors for the lung and the alpha-particle quality factor. Calculations of this sort led to the NCRP estimate of 2 mSv/yr for average radon concentrations [3, p. 15]. The UNSCEAR report gives numbers that imply a similar value, if they adopt this dosimetric approach to calculating the dose [7, p. 107].

However, the UNSCEAR report also takes cognizance of the fact that this calculated dose, used in conjunction with standard estimates of risk per unit dose, implies a substantially larger risk to the general population than is found by extrapolating from the experience of uranium miners (see Chapter 4).[20] The discrepancy is about a factor of 2 or 3. The adopted UNSCEAR number (as in Table 3.5) represents a compromise between an estimate based on straight dose calculations and one based on the extrapolation from uranium miners.[21] The difference between the UNSCEAR and NCRP results, therefore, is primarily due to the difference in the approaches taken toward the calculation of the dose. However, this disparity in the quoted average dose is small compared to the differences in doses for different houses. For example, an EPA study finds that well over one million people in the United States live in homes with radon concentrations more than eight times the U.S. mean [20].

---

[19] In addition, there have been isolated accidents, including the death of three army technicians in an accident in 1961 at the National Reactor Testing Laboratory in Idaho and the death of two workers in 1999 at a fuel processing facility in Tokaimura, Japan (see Section 15.1).

[20] The extrapolation assumes that the ratio of the number of cancers to the magnitude of the dose from radon exposures is the same for the general public as for uranium miners (at substantially higher doses).

[21] See, e.g., Ref. [7], Table 26.

Terrestrial Radiation

Many radionuclides in the ground—in particular, ^{40}K and members of the uranium and thorium series—emit gamma rays and produce an appreciable radiation exposure. This "terrestrial" dose includes doses from both the soil and building materials, with substantial variations in the magnitude of the dose from location to location. For example, limestone rocks and wood buildings have much lower radionuclide content than granite rocks and buildings of brick or granite. The outdoor dose is due to the ground alone, whereas the indoor dose is produced by radiation from both the ground (reduced by the structure's shielding effect) and building materials.

The NCRP summaries assume, for simplicity, the indoor dose to be equal to the outdoor dose [14, pp. 84 and 89], but account to some extent for building materials in the "consumer products" category (see Table 3.5). The UN-SCEAR data are based on measurements of both outdoor and indoor dose rates, so the latter include the effects of both the ground and the building. For most countries, the indoor doses are found to somewhat exceed the outdoor doses [7, p. 117].

The reported terrestrial doses for the United States and the world, 0.28 mSv/yr and 0.48 mSv/yr, respectively, differ substantially. Unlike the case for radon exposures, this appears to be a real difference, because the same UNSCEAR report indicates that the average U.S. indoor dose is less than one-half the world average.[22] The average United States dose rate varies from 0.16 mSv/yr in the coastal plane (from SE Texas to New England) to 0.63 mSv/yr in the area near Denver, Colorado [14, p. 89].

Radionuclides in the Body

The largest contributor to the internal dose from radionuclides in the body is ^{40}K, which is unavoidably present due to the presence of ^{40}K in all natural potassium and the large concentration of potassium in human tissue.[23] There are 140 g of potassium in the 70-kg "reference man," considered a prototype by the International Commission on Radiological Protection [14, p. 109]. This corresponds to an activity of 4340 Bq or 0.12 μCi. Other radionuclides in the body, present in lesser amounts, include members of the ^{238}U and ^{232}Th series, ^{14}C, and ^{87}Rb. There is a steady intake of radionuclides into the body, due to

---

[22] No explanation is given for the low U.S. dose rate. It could be understood if U.S. buildings make greater use of wood or if their construction provides more isolation from gamma rays from the soil. NCRP Report 94 attributes a previously noted difference of this sort "to the higher indoor exposure estimated for European masonry structures" [14, p. 89].

[23] Potassium-40 accounts for about 50% of the internal dose in the NCRP estimates and about 60% in the UNSCEAR estimate. The discrepancy in the two results in Table 3.5 is due largely to differences in the estimate for the other radionuclides.

their presence in foods. For example, in a typical U.S. diet, ^{238}U, ^{226}Ra, and ^{210}Po are each ingested at a rate in the neighborhood 1 pCi/day [14, p. 110].

Cosmic Rays

Cosmic rays impinge on the top of the Earth's atmosphere, both from the Sun and from outside the solar system. The "cosmic rays" reaching sea level are primarily secondary particles produced when energetic galactic cosmic rays collide with molecules in the atmosphere. They include muons, neutrons, gamma rays, and electrons. The total flux (and dose) varies greatly with altitude and, to some extent, with latitude. Averaging over latitude, the population-weighted mean outdoor dose at sea level is 0.32 mSv/yr [7, p. 113]. Adjusting for altitude, the population-weighted average is 0.46 mSv/yr outdoors. Overall, due to the reduction caused by shielding within buildings, the worldwide average dose from cosmic rays is 0.38 mSv/y (see Table 3.5.)

The U.S. average value of 0.27 mSv/yr is substantially lower than the world average, largely because, on a population-weighted basis, the average altitude is less. The average dose is higher in mountainous regions [e.g., 0.50 mSv/yr in Denver (altitude of 1600 m) and 1.25 mSv/yr in Leadville, Colorado (3200 m)] [14, p. 23]. At the altitude of jet airplanes, the dose rate is more than 100 times the sea-level rate and, for example, a transatlantic flight between Europe and the U.S. east coast leads to an effective dose of about 0.04 mSv (each way) [7, p. 88].

Cosmogenic Radionuclides

Cosmogenic radionuclides are continuously produced in the atmosphere by incident cosmic rays, most importantly carbon-14 (^{14}C) from the interaction of cosmic-ray neutrons with nitrogen-14 atoms. Actually, tritium (^{3}H) is produced at a still higher rate, but tritium ($T = 12.3$ years) has a much shorter half-life than ^{14}C ($T = 5730$ years) and, therefore, its inventory in the atmosphere is much less. As a result, almost all of the effective dose of 0.01 mSv/yr from cosmogenic radionuclides is due to ^{14}C.

3.5.2 Radiation Doses from Medical Procedures

X-rays have been used for medical diagnosis for almost 100 years. The trend with time has been to more examinations per capita and, with better defined beams and more sensitive detection, smaller doses per examination. The world average dose per capita for 1991–1996 was 0.4 mSv/yr. Variations among countries were great, with the more prosperous countries in general having higher level of health care, including a greater per capita frequency of X-ray examinations and a higher average radiation dose [7, pp. 396 and 401]. According to the UNSCEAR data, the United States frequency of X-rays was about the same as the average for medically advanced countries, but the per capita dose was below their average because of lower doses per examination [7, pp. 393–401].[24] An additional contribution to the average radiation dose comes from the use of radionuclides, again primarily for diagnostic procedures. The average effective dose for the advanced countries in 1991–1996 was 0.08 mSv/yr and the world average was 0.03 mSv/yr [7, p. 464].

X-rays and radionuclides are also used for cancer therapy but the number of such treatments is small compared to the number of diagnostic procedures.[25] Although some individuals receive very high doses from therapy treatments, the contribution to the population average is small. For that reason and in view of the extreme disparities among individuals, these doses are not included in Table 3.5.

It should be noted that the averages given in Table 3.5 mask variations among individuals, some of whom have more frequent and intense exposures than others. For the most part, these radiation exposures are accepted by the individuals involved, because the benefits from the radiation are expected to outweigh the potential harm.

3.5.3 Other Sources of Radiation

Consumer Products

A variety of other human activities involve radiation exposures. One broad group is listed in Table 3.5 as "consumer products." These are considered by the NCRP but not included in the UNSCEAR compilation. Sources of exposure in this category most notably include natural radionuclides in the water supply and in building materials [3, p. 31].

Another large but uncertain source of exposure is tobacco. Tobacco leaves collect lead-210 (^{210}Pb) from the air and this radionuclide and its immediate decay product, polonium-210 (^{210}Pb), are inhaled with the tobacco smoke.[26] One estimate placed the average resulting effective dose for 50,000,000 U.S. smokers at 13 mSv/yr, corresponding to an average U.S. population dose of roughly 2.5 mSv/yr [21, p. 310]. Although this dose is larger than any of the individual doses listed in Table 3.5, it is omitted from most dose summaries— presumably because the dose is unevenly distributed, the calculation is un-

---

[24] The difference between the 0.5 mSv/yr from Ref. [7] and the U.S. average of 0.39 mSv/yr reported by the NCRP for 1980 is consistent with the increased frequency of X-ray examinations.

[25] For the United States in 1991–1996, the number there were 960 diagnostic X-rays per year per 1000 people and less than 2 X-ray therapy treatments [7, pp. 396 and 428].

[26] The half-life of ^{210}Pb is 22.3 years, making it the longest-lived of any of the radioactive progeny of ^{222}Rn. ^{210}Pb decays to ^{210}Po ($T = 138$ days), an alpha-particle emitter. Both radionuclides are present in the air due to the escape of radon from the ground and its subsequent decay.

certain, and other effects of tobacco smoke are thought to be more important than the radiation effects.

It is to be noted that the radionuclides in the water supply, building materials, and tobacco are natural radionuclides, although in the case of building materials and tobacco, they are "technologically enhanced" (i.e, the doses are greater than they would be without houses and cigarettes).

Nuclear Weapon Tests

Extensive aboveground nuclear weapons tests were carried out by the United States, the Soviet Union, and other countries, particularly in the period from 1952 to 1962.[27] The resulting worldwide average effective dose peaked at 0.11 mSv/yr in 1963 and dropped to 0.0055 mSv/yr by 1999 [7, pp. 228–230]. The average cumulative dose from 1945 to 1999 (i.e., the sum of the annual average doses over the 55-year period) was 1.1 mSv in the northern hemisphere, 0.3 mSv in the southern hemisphere, and 1.0 mSv for the world population as a whole.

Doses above the general averages were received by people in the vicinity of nuclear tests. Particularly large exposures were caused by the unexpectedly high fallout from a 1954 U.S. nuclear test at Bikini Atoll in the Pacific. In the United States, there were significantly elevated dose levels in the vicinity of the Nevada test site.[28]

Nuclear Fuel Cycle

The civilian nuclear power fuel cycle, involving mining, fuel fabrication, and reactor operation, contributes a negligible dose to the general public. If calculated on the basis of 1980s practices, as in the NCRP estimate in Table 3.5, it averages about 0.0005 mSv/yr for the United States.[29] The world average for the 1990s reported in Table 3.5 is even lower. The largest contributions to the dose are from uranium mining and processing operations, including the

---

[27] The United States, USSR, and United Kingdom stopped atmospheric testing after the adoption of a limited test ban treaty in 1963. China and France continued atmospheric testing after this date, but the scale was relatively small.

[28] There were also intentional releases of ^{131}I from plants at the Hanford reservation where plutonium was produced for nuclear weapons. These releases ended after the early 1950s. The effects of exposures from these releases, as well as from weapons test fallout, are of historical interest, and extensive efforts were undertaken to assess their consequences, but they are not relevant to current or anticipated future exposures.

[29] The collective effective dose equivalent for all the operations associated with a large reactor (with a capacity of 1 GWe and operating with a capacity factor of 80%) was estimated by the NCRP to be 1.36 person-Sv/yr [22, p. 160]. For a U.S. capacity of 100 GWe, this would imply a national collective dose of 136 person-Sv/yr, or an average individual dose of about 5×10^{-7} Sv/yr.

release of radon from the mill tailings, which are the unused residue of the processes by which uranium is extracted for use as a reactor fuel.

In either estimate, the exposures of the public from the nuclear fuel cycle have been exceedingly small. However, the picture is incomplete. One omission, in principle, is the potential exposures from the nuclear waste disposal program. However, there appears to be little possibility for appreciable population doses from this source (see Chapter 12). A second omission is nuclear accidents, considered separately below.

Nuclear Accidents

Two important accidents have occurred in civilian nuclear power plants: the 1979 Three Mile Island accident and the 1986 Chernobyl accident. Neither is included in the NCRP tabulation because the total release at Three Mile Island was very small and the releases from the accident at Chernobyl had, in the words of the NCRP, "virtually no impact on the population of the United States" [3, p. 28]. However, the Chernobyl accident had an impact in parts of Europe. The total exposures, primarily in regions of Europe close to Chernobyl, translate into a global average of 0.002 mSv/yr, as displayed in Table 3.5. This quoted number is very low because it is a world average and refers to 2000, when dose levels were appreciably below their earlier values. The doses near Chernobyl were much larger, especially in the year following the accident, as discussed in Section 15.3.

Occupational Exposures

The above discussion has focussed on radiation exposures for the general population. Occupational exposures can add appreciably to the dose for some individuals. Here, one can distinguish between past exposures that are of mainly historical interest, present-day exposures that are viewed as "ordinary" risks of the workplace, and accidents that serve as warnings for the future.

In the early history of work with radioactive and fissionable materials, high radiation doses were, on occasion, incurred under circumstances that reflect an ignorance or lack of precautions that seem inconceivable today. This was true for very early investigators of X-rays and radioactivity—for example, Marie and Pierre Curie who discovered radium in 1898. In the 1920s, the use of radium on watch and instrument dials led to very heavy exposures of workers who painted the dials (see Section 4.5.1). In 1945 and 1946, shortly after the end of World War II, two workers at the Los Alamos Laboratory died from acute radiation exposures received in separate accidents involving runaway chain reactions initiated in experimental work with plutonium [23, p. 341]. Somewhat later, in 1961, three technicians died from radiation exposure in an accident at army test reactor (see Section 15.1). When uranium mining began on a large scale in the 1940s, inadequate attention was paid to mine ventilation, and the resulting high radon levels led to several hundred lung cancer deaths in the United States and many elsewhere as well.[30]

The present understanding of radiation hazards and the range of permitted practices make it very unlikely that similar mistakes will occur again. However, accidents can still happen, as illustrated by the events at Chernobyl and Tokaimura (see Chapter 15). In both cases, one can point to egregious errors that led to the accidents, but the possibility of accidents remains an occupational hazard.

Much less dramatically, occupational radiation exposures occur in the routine activities of many workers. For the United States in the early 1980s, the average annual dose was 5.6 mSv for nuclear power plant workers, 1.9 mSv for exposed physicians, and 1.7 mSv for commercial airplane flight crews [24, pp. 65–70]. Within these average numbers, there are wide variations in work environments and in the resulting individual doses. For the 1.5 million workers in all the activities considered, the annual collective dose was was 2300 person-Sv, corresponding to an average of about 1.5 mSv/yr.

Nuclear power plants accounted for roughly one-quarter of the collective occupational dose in the United States in the early 1980s. This dose amounted to about 10 person-Sv per gigawatt-year of electrical output in 1985 [24, p. 38]. Between 1985 and 2001, the average collective dose per U.S. power plant has dropped by more than a factor of five [25, p. 29]. Reported world averages are somewhat similar and also show an improvement with time. For the entire nuclear fuel cycle, the average occupational doses were 18 person-Sv/GWyr for 1980–1984 and 9.8 person-Sv/GWyr for 1990–1994 [7, p. 584].

3.5.4 Summary

Radioactivity is ubiquitously present on Earth, primarily from natural radionuclides. As a result, the average person receives an effective dose in the neighborhood of 3 mSv/yr, with wide geographic variations. Radon and its progeny, particularly indoor radon, are the largest contributors to this dose.

The only human activities that produces large doses are medical diagnosis and treatment (primarily from X-rays, not radionuclides), which give an average dose of about 0.4 mSv/yr—with very large variations from individual to individual. Other anthropogenic sources contribute little. In the absence of accidents, the average radiation dose from nuclear power is in the neighborhood of 0.01% or 0.02% of the natural dose. Although the Chernobyl accident had severe local effects, it added little to the average global radiation exposure (see Section 15.3.6).

---

[30] In the 1960s, mine ventilation was improved and the radiation exposure of miners was substantially reduced.

References

1. Roger M. Macklis, "The Great Radium Scandal," *Scientific American* 269, no. 2, August 1993: 94–99.
2. American Nuclear Society, *Glossary of Terms in Nuclear Science and Technology* (LaGrange Park, IL: ANS, 1986).
3. National Council on Radiation Protection and Measurements, *Ionizing Radiation Exposure of the Population of the United States*, NCRP Report No. 93 (Washington, DC: NCRP, 1987).
4. Gunrun A. Carlsson, "Theoretical Basis for Dosimetry," in *The Dosimetry of Ionizing Radiation*, K.R. Kase, B.E. Bjärngard and F.H. Attix, eds. (Orlando, FL: Academic Press, 1985).
5. International Commission on Radiological Protection, "1990 Recommendations of the International Commission on Radiological Protection," ICRP Publication 60, *Annals of the ICRP* 21, nos. 1–3 (Oxford: Pergamon Press, 1991).
6. *Energy, U.S. Code of Federal Regulations*, Title 10, Part 20 (1993).
7. United Nations Scientific Committee on the Effects of Atomic Radiation, *Sources and Effects of Ionizing Radiation, Volume 1: Sources, UNSCEAR 2000 Report*, (New York: United Nations, 2000).
8. *Protection of Environment, U.S. Code of Federal Regulations*, title 40, part 197 (2001).
9. National Council on Radiation Protection and Measurements, *Limitation of Exposure to Ionizing Radiation*, NCRP Report No. 116 (Washington, DC: NCRP, 1993).
10. Dade W. Moeller, "Radiation Units," in *Health Effects of Exposure to Low-Level Ionizing Radiation*, W.R. Hendee and F.M. Edwards, eds. (Bristol, UK: Institute of Physics Publishing, 1996): xix–xxvi.
11. Samuel Glasstone, *Sourcebook on Atomic Energy* (New York: Van Nostrand, 1950).
12. Robley D. Evans, *The Atomic Nucleus* (New York: McGraw-Hill, 1955).
13. Jagdish K. Tuli, *Nuclear Wallet Cards* (Upton, NY: Brookhaven National Laboratory, 2000).
14. National Council on Radiation Protection and Measurements, *Exposure of the Population in the United States and Canada from Natural Background Radiation*, NCRP Report No. 94 (Washington, DC: NCRP, 1987).
15. DeVerle P. Harris, "World Uranium Resources," *Annual Review of Energy* 4, (1979): 403–432.
16. Stuart Ross Taylor and Scott M. McClennan, *The Continental Crust: Its Composition and Evolution* (Oxford: Blackwell Scientific Publications, 1985).
17. J.H. Chen, R. Lawrence Edwards, and G.J. Wasserburg, " ^{238}U, ^{234}U, and ^{232}Th in Seawater," *Earth and Planetary Science Letters* 80, (1986): 242–251.
18. R. Kilho Park, Dana K. Lester, Iver W. Duedall, and Bostwick H. Ketchum, "Radioactive Wastes and The Ocean: An Overview," in *Wastes in the Ocean, Volume 3* (New York: Wiley, 1983), 3–46.
19. P.A. Cox, *The Elements: Their Origin, Abundance and Distribution* (New York: Oxford University Press, 1989).
20. Frank Marcinowski, Robert M. Lucas, and William M. Yeager, "National and Regional Distributions of Airborne Radon Concentrations in U.S. Homes," *Health Physics* 66 (1994): 699–706.

21. Dade W. Moeller, "Radiation Sources: Consumer Products," in *Health Effects of Exposure to Low-Level Ionizing Radiation*, W.R. Hendee and F.M. Edwards, eds. (Bristol, UK: Institute of Physics Publishing, 1996): 287–313.

22. National Council on Radiation Protection and Measurements, *Public Radiation Exposure from Nuclear Power Generation in the United States*, NCRP Report No. 92 (Washington, DC: NCRP, 1987).

23. Lillian Hoddeson, Paul W. Henriksen, Roger A. Meade, and Catherine Westfall, *Critical Assembly* (Cambridge: Cambridge University Press, 1993).

24. National Council on Radiation Protection and Measurements, *Exposure of the U.S. Population from Occupational Radiation*, NCRP Report No. 101 (Washington, DC: NCRP, 1989).

25. "Performance Indicators: Another Successful Year in Performance, Safety," *Nuclear News* 44, no. 6, May 2002: 28–30.

4

Effects of Radiation Exposures

4.1 The Study of Radiation Effects

4.1.1 Agencies and Groups Carrying out Radiation Studies

In recognition of the importance of determining the dangers that ionizing radiation may pose for humans, the health consequences of radiation exposures have been studied almost since the first discovery of X-rays and radioactivity. The studies greatly intensified during and after World War II, with contributions from many individuals and groups throughout the world.

Official advice as to radiation protection is provided internationally by the International Commission on Radiological Protection (ICRP) and in the United States by the National Council on Radiation Protection and Measurements (NCRP).[1] Each group issues a series of reports on both specific and general radiation topics, sometimes including explicit recommendations for radiation protection. Other influential general reports are the so-called BEIR Reports, which are prepared by the Committee on the Biological Effects of Ionizing Radiations of the U.S. National Research Council,[2] and the reviews published every several years by the United Nations Scientific Committee on

---

[1] In the discussions to follow, the reports of these organizations and the UNSCEAR and BEIR reports are stressed in an effort to rely on "neutral" summaries of sometimes highly controversial issues. This does not eliminate controversy, however, because some of these reports have been criticized as being themselves biased in the selection of data and weighing of issues (e.g., the critique in Ref. [1]).

[2] The National Research Council is an agency of the U.S. National Academy of Sciences and associated academies.

the Effects of Atomic Radiation (the UNSCEAR reports).[3] All of these reports rely heavily on the work of the Radiation Effects Research Foundation (RERF), an organization jointly supported by Japan and the United States. Since the 1950s, its Life Span Study has been carefully following the medical history of the Hiroshima–Nagasaki atomic bomb survivors, and these analyses provide the largest available base of data on the long-term health effects of ionizing radiation.

United States radiation protection regulations, with force of law, are formulated by the Nuclear Regulatory Commission (NRC) and the Environmental Protection Agency (EPA). The regulations are contained in Titles 10 and 40, respectively, of the *Code of Federal Regulations* (CFR).

The evaluations, recommendations, and regulatory limits are under continuous, but slowly evolving, review. Their scientific basis rests on research activities carried out throughout the world. In the United States, these were supported primarily by the Atomic Energy Commission immediately after World War II and more recently by its successor organization, the Department of Energy, along with other federal agencies.

4.1.2 Types of Studies

Much has been learned from the long array of studies and analyses, but as discussed here, some crucial questions remain inadequately resolved. The most direct means of study is the observation of the effects of radiation on humans. There have also been numerous experiments on animals. Although the human and animal studies have provided considerable information on radiation damage at reasonably high doses, they have not led to clear-cut conclusions at low doses. The lower the dose, the smaller the expected effects, and at some point the possible radiation effects become obscured by the "spontaneous" incidence of the same effects—for example, cancers due to causes other than radiation.

The distinction between "high" and "low" doses has no clear definition. For purposes of qualitative orientation it is reasonable to accept the guide provided by NCRP Report No. 136: "low doses are in the range 0 to 10 mSv, moderate doses are from >10 to 100 mSv, and high doses >100 mSv" [4, p. 17]. No explicit reference is made to the time duration of the exposures, but presumably the natural background of 3 mSv/yr would meet the low-dose criterion, even though the cumulative dose over a lifetime might exceed 200 mSv.

With the growing understanding of the role of DNA, there is some hope that experimental and theoretical studies of radiation damage and repair at the cellular level will eventually establish the nature of the biological con-

---

[3] We draw particularly on the UNSCEAR reports published in 2000 [2, 3]. The first UNSCEAR documents were issued in 1958; major subsequent reports appeared in 1972, 1977, 1982, 1986, 1988, 1993, 1994, 1996, and 2000 [2, p. 15].

sequences of low-dose exposures. However, in the absence of definitive information from studies of animals or cells, heavy reliance is placed on human experience, particularly studies of Hiroshima and Nagasaki atomic bomb survivors, of individuals receiving high exposures in the course of medical treatments, and of uranium and other miners, as well from experiences in isolated accidents.

Aside from the Hiroshima and Nagasaki studies, the database on cancer mortality in humans is not extensive, and these data are often pertinent only to a single organ—for example, lung cancer in uranium miners. Additional information is beginning to come from the study of the effects of the Chernobyl nuclear reactor accident, but, as of the time of the 2000 UNSCEAR report, the only observed effect in the general population has been an increase in the rate of thyroid cancer among children in the areas near Chernobyl [3, p. 517].

4.1.3 Types of Effects: Deterministic and Stochastic

In radiation studies, a distinction is made between *deterministic* and *stochastic* effects.[4] Deterministic effects depend on the killing of many cells over a relatively short period of time. They are induced by intense exposures, and the outcome of this exposure is reasonably well defined. The magnitude of the dose determines the *intensity* of the effect. The most obvious deterministic outcome is the death of the victim within a short time (a few months, or less) after the exposure.

Stochastic effects are by definition of a random nature, with the likelihood of one or another outcome a matter of statistical probability. The modification of a limited number of cells by radiation may lead to cancer in the exposed person or to genetic damage which affects later generations. The magnitude of the dose determines the *probability* of the effect.

The impacts of radiation exposure from nuclear power are primarily stochastic. Even in the accident at Chernobyl, relatively few workers received high enough acute radiation doses to cause either death within a few months or other deterministic effects (see Section 15.3.4). Much larger populations were exposed to lower doses. The predicted appearance of cancer in these populations is delayed, typically by more than 10 years after the time of exposure, and it is not possible to identify specific individual victims.

4.2 Effects of High Radiation Doses

4.2.1 Deterministic Effects

It is well established that high radiation doses are fatal. For a dose in the neighborhood of 3–5 Gy received over a short period of time, there is approx-

---

[4] See, e.g., Refs. [5], p. 4, and [6], p. 8.

imately a 50% chance of death within 60 days, although the probability of death is influenced by the prior health of the individual and the treatment administered.[5] For Hiroshima victims, 50% lethality was produced at about 3 Gy. For Chernobyl victims receiving doses between about 4 and 6 Gy, 7 of 23 died. Doses above 6 Gy were lethal in 21 of 22 cases (see Section 15.3.4).

For doses between 1 and 4 Gy, there are clinical symptoms, including nausea and a depressed white blood cell count, described under the general term of "radiation sickness." There is some chance of fatalities at the top end of the range. For doses below 1 Gy, there are usually no clinical symptoms.

4.2.2 Stochastic Effects: Observational Evidence for Cancer at High Doses

Risk Coefficient

The most serious long-term effect of exposures at doses below the 1 Gy (or 1 Sv) region is an increased risk of cancer. This is a stochastic effect because the chance of cancer depends on the magnitude of the dose. There is strong evidence of increased risk down to 0.1 or 0.2 Sv, but there is considerable uncertainty about the effects of smaller doses—as discussed at length below. To fill in the picture, extensive efforts have been made to determine the cancer rate as a function of radiation dose for various dose levels, cancer sites, and age and gender population subgroups.

The most extensive data for this purpose comes from the Life Span Study of the atomic bomb survivors. A 1996 publication summarizes cancer mortality among the survivors for the period from 1950 through 1990 [8].[6] The "cohort" being studied consists of about 87,000 individuals who were within 2.5 km of the bombings and for whom the resulting radiation dose has been estimated. The background cancer mortality rates were determined by studying a comparison group of people who were within 2.5–10 km of the bombings and are presumed to have received little or no exposure.[7]

---

[5] See Refs. [7], pp. 565–574, and 595 and [5], p. 105.

[6] A reassessment of the radiation exposures at Hiroshima and Nagasaki was largely completed in 2002 and a new dosimetry (called DS02) was scheduled for publication in 2003. It will lead to revised estimates of the dose–response relationship, but these are expected to be published too late for inclusion in this book. It is not obvious from a preliminary look at the analysis whether the calculated radiation risk will go up or down, suggesting that the magnitude of the change probably will not be large [9].

[7] The exposure from radioactive fallout was small at Hiroshima and Nagasaki because the bombs exploded about 500 m above the ground [10, p. 37]. In contrast, a nuclear explosion at or very near the ground will lead to high radiation exposures due to the prompt fallout of radionuclides carried on material ejected from the ground and then deposited over a large area.

For the exposed cohort, there were 7827 cancer fatalities in the period from 1950 to 1990, of which 421 were "excess"—that is, attributed to exposures from the bombs rather than to natural causes, as gauged from the cancer rates in the comparison group. Detailed examination of the data provides results subdivided by gender, age at exposure, and type of cancer.

The array of data from this and other studies is often reduced to a single number: a risk coefficient that specifies the average lifetime risk of fatal cancer per sievert of effective dose. This risk coefficient can be thought of as applying to an "average" individual.[8] It is applicable only over a dose interval within which the cancer risk is linearly proportional to the dose. The risk coefficient estimates are based on analyses of data at high doses, and thus the risk coefficient is in the first instance derived as a high-dose coefficient.

This risk coefficient has been estimated by a number of scientific groups, including the BEIR Committee, the NCRP, the ICRP, and UNSCEAR. These several groups all rely on the same base of data, primarily the results of the RERF studies of Hiroshima–Nagasaki. The UNSCEAR report issued in 2000 presents results from BEIR V [11], ICRP Publication 60 [5], a 1994 UNSCEAR report [12], and its own latest analyses [3, p. 431]. These reports all estimate that the risk coefficient lies in the range from 0.08 per sievert to 0.12 per sievert. Although NCRP Report No. 126 was not cited by UNSCEAR, it is in agreement with these estimates, adopting a risk coefficient of 0.10 per sievert for the general population and 0.08 per sievert for workers (a group limited in age range).

The consistency of the results arises primarily from the use of a common database. The actual uncertainties are larger than implied by the agreement among these estimates. For example, an additional result cited in UNSCEAR 2000 is a study of "expert opinion," which also estimated the risk coefficient to be 0.10 per sievert, but with an uncertainty of about a factor of 3 in each direction [13].

The estimates pertain directly to large exposures received during a short time period. For low dose and dose rates, most (but not all) of the above-cited groups recommend using the high-dose coefficient, reduced by a *dose and dose rate effectiveness factor* (DDREF) of about 2. (These issues are discussed at length in Section 4.3.)

It can be expected that these risk assessments will change with time, as new data and insights become available. For example, the conclusions of the 1990 BEIR V report differed substantially from those of the 1980 BEIR III report [14]. During the period between the two studies, additional excess cancers were observed among the atomic bomb survivors—some occurring 40 years after the bombing—raising the cumulative number of cancer fatalities. In addition, a reexamination of evidence on the magnitude of the exposures

---

[8] More specifically, if the members of a representative population all receive the same dose the total number of expected cancer fatalities is the product of the risk coefficient and the collective dose (in person-Sv).

suggested that the average neutron exposures had been considerably overestimated in earlier work. Both factors had the effect of raising the number of cancer fatalities per unit exposure. It is to be expected that the next version of this report series, BEIR VII, will have further changes.

Risk Models: Absolute and Relative Risk

As suggested earlier, the cancer risk from a given exposure depends on a variety of factors other than the dose itself. For instance, the BEIR V report cites "sex, attained age, age-at-exposure, and time-since-exposure" as relevant parameters in determining risk [11, p. 166]. Further, the risk varies from organ to organ. The risk can be related to dose using either an *absolute* or *relative* risk model. In the absolute risk model, the risk depends on the dose. In the relative risk model, the risk depends on both the dose and the underlying "natural" risk for the group being considered. For example, lung cancer is more common among old people than among young people, and radon creates a higher risk of fatal lung cancer for exposure at age 55 than at age 25.

The BEIR V report quotes an overall risk coefficient for cancer mortality of 0.08 per sievert. A more detailed picture of risks to population subgroups is presented in a BEIR V tabulation of the separate risks for 9 age groups, 4 cancer categories, and both genders [11, p. 175]. Adding up the risks for all cancer categories and taking a population-weighted average over the age groups yields risks of 0.077/Sv for males and 0.081/Sv for females—giving an overall average of 0.08/Sv. Although for detailed studies and for considering the risks for specific individuals, the breakdown of risk into the various categories is useful, a single risk coefficient suffices for most studies of the effects of nuclear energy. Should there be radiation exposures, all subgroups of the population will receive roughly similar doses and a single coefficient suffices for many purposes. Therefore, the distinction between relative and absolute risk models is ignored in this book.[9] Nonetheless, it is important to keep in mind that some subgroups face higher than average risks and require particular protection—most notably pregnant women.

Threshold for Unambiguously Observed Effects

It is convenient to apply a single radiation risk coefficient to broad populations. However, as already noted, this coefficient is based on observations of effects for high doses delivered over a short period of time. To estimate the effects at lower doses, it is necessary to make an extrapolation from the high-dose information, or, in other terminology, to establish the shape of *dose–response* curve. This topic is discussed again in Section 4.3.3, but here we consider a basic preliminary issue: over what range of doses is there clear empirical evidence that radiation causes cancer fatalities?

---

[9] The BEIR V report points out that the distinction has become blurred by complexities that can be introduced into these models [11, p. 201].

The most relevant data for addressing this question is contained in the studies of the Hiroshima–Nagasaki survivors. According to ICRP Publication 60, published in 1991, these data gave clear-cut (i.e., at 95% confidence level) indications of cancer induction only at doses above about 200 mSv, although there is weaker positive evidence down to about 50 mSv [5, p. 16]. In a more recent comprehensive analysis of the accumulating data, Donald Pierce and his RERF colleagues explicitly considered the question of "the lowest dose at which there is a statistically significant excess risk." While warning against "the tendency for the failure to find a significant effect to be equated to 'no effect'," they concluded that a significant effect is seen down to 50 mSv [8, p. 10]. In a still later article, reviewing cancer rates in atomic bomb survivors who died in the 1958 to 1994 period, Pierce and Preston concluded that there is a statistically significant cancer excess in the 0–100-mSv range, and at a 95% confidence level, they put the threshold somewhere between 60 and 100 mSv, depending on the method of analysis [15].

The controversial nature of this issue is reflected in a 1996 Position Statement adopted by the Health Physics Society (HPS), a U.S. professional organization with a membership primarily from the area of radiation protection, and the immediate response of the U.S. Environmental Protection Agency. The HPS statement said, in part:

> There is substantial and convincing scientific evidence for health risks at high dose. Below 10 rem [100 mSv] (which includes occupational and environmental exposures), risks of health effects are too small to be observed or are non-existent. [16, p. 3]

The EPA, in response, cited evidence for effects below 100 mSv and termed the overall HPS statement "seriously deficient in scientific justification, logical coherence, and clarity" [17].

It is unlikely that this and similar disagreements will soon be resolved. There are two thresholds to be considered: the threshold for *clearly observed* cancer fatalities and the threshold for *actual* cancer fatalities. The first threshold appears to be in the neighborhood of 100 mSv, although some observers would place it somewhat higher or lower. The question of the threshold, if any, for *actual* cancer fatalities (whether observed or not) is discussed below as part of the broader topic of low dose effects.

4.3 Effects of Low Radiation Doses

4.3.1 Importance of Low Doses

From a policy standpoint, it is unfortunate that the available information on the effects of low doses is as inadequate as it is. Virtually all of our direct evidence on the harmful stochastic effects of radiation, for studies in both animals and humans, comes from observations at high doses or high dose rates, most notably the Hiroshima–Nagasaki studies. However, most of the exposures of concern, such as those from nuclear weapon tests, nuclear energy, and indoor radon, involve much lower doses and dose rates. For example, even for the Chernobyl nuclear reactor accident most of the calculated collective dose is is made up of lifetime individual doses of several millisievert or less (see Section 15.3.6). Similarly, the doses from indoor radon—about 2 mSv/yr for the average person—are the largest source of collective population exposure in the United States.

Should radiation exposures at such levels be a matter of concern? Without firm knowledge about the effects of low-level radiation, there can be no definitive answer to this crucial question. Nonetheless, it is necessary to have at least a tentative answer, in part to provide a rationale for the establishment of standards for the protection of workers and the public from the effects of radiation.

For the most part, these issues have little direct relevance to nuclear power, because the doses from the normal operation of the nuclear fuel cycle are very small compared to natural background (see Section 3.5). In short, the claim that nuclear power causes little harm to health does not rest on details of the effects of radiation. Rather, it rests on keeping routine doses very small and preventing accidents.

However, there are circumstances in which a correct understanding of low-dose effects becomes very important. The most conspicuous case to date involves people who lived near the site of the Chernobyl nuclear reactor accident (see Section 15.3). Decisions as to the evacuation and eventual return of these people from homes in contaminated regions hinges on the evaluation of radiation effects. A similar problem could arise if terrorists succeed in setting off a "dirty bomb" that distributes radioactive material into the environment. Under such circumstances, it is important to have realistic information on radiation effects, to avoid overreaction or underreaction to the radiation hazards. "Erring on the side of caution" could do more harm than good, if it means unnecessarily barring people from their homes. In another context, excessive caution could lead to the possibly unnecessary expenditure of billions of dollars in the cleanup of contaminated sites, such as those at the Hanford reservation in Washington. Conversely, underestimating or ignoring radiation risks could cause avoidable cancer deaths. To aid in reaching informed and sensible decisions, it is important that vigorous efforts be continued toward gaining a better understanding of the consequences of exposures to low doses of radiation.

4.3.2 Observational Evidence for Cancer at Low Dose Rates

Range of Studies

The atomic bomb survivor data, while the best available for the effects of radiation at high doses and high dose rates, become ambiguous for doses below

50 or 100 mSv, and may not be applicable to doses delivered over a protracted period of time. For that reason, there have been extensive studies of the cancer rates in populations other than the atomic bomb survivors that have been exposed to radiation at above-average rates. These include populations living in regions of elevated levels of natural background radiation, workers in the nuclear industry, and groups unintentionally exposed in nuclear weapons tests. In no case is there unambiguous evidence of a positive effect, although often the proper interpretation of the available data is in dispute. Examples of such studies are discussed briefly in the succeeding paragraphs.

Populations Living in Regions of High Natural Radioactivity

There are substantial variations in natural radiation background levels, but comparisons between regions of low and high radiation may be misleading in the face of other possible differences between the regions. For example, the cancer death rate in Colorado is substantially below the United States average, whereas the natural radiation level is substantially above. However, unless one can exclude the effects of differences in other aspects of the environment and in individual lifestyles, it is inappropriate to infer a negative correlation from these data.[10]

Studies of aggregate data for various population sectors, which investigate correlations between cancer rates and radiation dose rates, are termed *ecological* studies. Ecological studies are often criticized by epidemiologists on the ground that they are susceptible to many sorts of difficulty, including spurious correlations introduced by "confounding" factors—as would be the case if the low cancer rate in Colorado was due to low levels of pollutants other than radiation.[11] At a minimum, however, it is of interest to note that no cancer excess has been found in studies involving populations living in regions of high natural background.

A somewhat extreme case is the city of Ramsar in Iran, northwest of Teheran, where very high concentrations of radium have been brought to the surface by hot springs. This leads to gamma-ray radiation doses that are estimated to extend up to 260 mSv/year. The most heavily impacted population is too small to develop reliable epidemiological evidence on the cancer mortality rates. However, studies of blood samples showed no statistically significant signs of radiation damage. When blood cells of people in the region were subjected to a subsequent "challenge dose" of 1.5 Gy, fewer chromosomal abnormalities were observed for people living in the high-radiation areas than for those living in areas with less radiation. This was interpreted by the investiga-

---

[10] The estimated cancer death rates in 1987 were 125 per 100,000 population in Colorado in 1987 compared to a U.S. national average of 199 [18].

[11] A summary of possible pitfalls is given, for example, in NCRP Report No. 136 [4, pp. 134–136].

tors as evidence for an *adaptive* effect, in which a prior radiation dose decreases an organs's sensitivity to later radiation insults (see Section 4.3.3) [19].

Indoor Radon

Studies of the effects of indoor radon are important in their own right, as a guide to establishing sensible protective measures against the possible hazards of the source of most of the collective human dose from natural radiation sources. Because the doses vary substantially from place to place, it might be expected that studies of the relation between indoor radon and lung cancer levels would shed light on the general question of the effects of low-dose exposures. However, as discussed further in Section 4.5.2, the leading studies have produced a high-profile dispute, not consensus.

Nuclear Industry Workers

A comprehensive study of cancer mortality workers in the nuclear industry analyzed aggregated data from seven laboratories: three in the United States, three in the United Kingdom, and one in Canada [20]. Among the 95,673 workers included, there were 15,825 deaths of which 3976 were from cancer. The population was subdivided by dose and an "expected" number of deaths was calculated for each group, based on its size, with adjustments for factors such as gender and age. Most of the workers (60%) had doses below 10 mSv and about 20% had doses between 10 mSv and 50 mSv. Fewer than 2% had doses over 400 mSv. Within each group, a comparison was made between the observed and expected deaths.

The data showed a very small, but statistically insignificant, decrease in total cancer death rate with increasing dose.[12] Overall, although this was the most comprehensive study of nuclear workers undertaken, it provided relatively little new information due to the relatively low doses received and the accompanying statistical uncertainties. The authors concluded that their results are consistent with "a range of risks, ranging from close to zero to a risk approximately twice the linear estimates from analyses of atomic bomb survivors" [20, p. 129].

It may be noted that the study is based on internal comparisons among nuclear industry workers. This follows the precedent set in the studies of the Hanford facility in the United States, where it was noted that the cancer rate for exposed workers was lower than that for an age-matched general population. This was attributed to the "healthy worker effect," and to avoid this effect, comparisons were restricted to those between different groups of Hanford workers [21].

---

[12] However, within total cancers, excess deaths were observed for a leukemia category (all leukemias other than chronic lymphocytic leukemia) at doses above 400 mSv (6 instead of an expected 2.3).

Difficulties in Obtaining Observational Evidence
of Low-Dose Effects

In general, there are major difficulties in trying to identify a small number of radiation-induced cancers in the presence of the background of many "natural" cancers. One difficulty is statistical, especially if the data are subdivided to study cancers for particular age groups or in a particular organ. For example, if a population is divided into 6 age groups and one separately considers 5 cancer types and both genders, there are a total of 60 categories. On average, there should be three cases of a departure of two standard deviations or more from the norm, even if there is no true effect. When a two-standard-deviation anomaly is seen under such circumstances, is the effect "significant"?

A second difficulty is that of determining whether an observed high (or low) cancer rate in a given population could be due to some confounding factor—such as smoking or chemical pollution—rather than to the radiation. The expected rate is inferred from a comparison group, but it is often difficult to identify a large and sufficiently similar comparison group.

The difficulty of obtaining information on effects at low doses is reflected in a series of comments in major summary documents. Thus, in ICRP Publication 60, issued in 1990, it was judged:

> Overall, studies at low dose, while potentially highly relevant to the radiation protection problem, have contributed little to quantitative estimates of risk. [5, p. 17]

The 1990 BEIR V report was similarly unable to draw any crisp conclusions from epidemiological studies. The studies indicated that the extrapolations from results at higher doses did not *underestimate* the risk at low doses, but an *overestimate* could not be excluded.

> Studies of the imputed effects of irradiation at low doses and low dose rates fulfill an important function even though they do not provide sufficient information for calculating numerical estimates of radiation risks. They are the only means available now for determining that risk estimates based on data accumulated at higher doses and higher dose rates do not underestimate the effects of low-level radiation on human health. [11, p. 371]

Matters were not much clearer a decade later, as summarized in NCRP Report 136, issued in 2001, which states:

> In general, however, because of limitations in statistical power and the potential for confounding, low-dose epidemiological studies are of limited value in assessing dose–response relationships and have produced results with sufficiently wide confidence limits to be consistent with an increased effect, a decreased effect, or no effect. [4, p. 6]

It remains of interest to attempt to extract what epidemiological evidence is available on the effects of low radiation doses, but as suggested elsewhere in

NCRP Report 136, it "may never be possible" to determine the shape of the dose–response curve at the low-dose end [4, p. 7]. Of course, if the dose is low enough, the effects will be unobservable, no matter what analyses are made.

Partly with this in mind, increased attention has been given in recent years to the study of the underlying biological processes, and these studies in the end may provide the strongest basis for estimating radiation effects at low doses. However, as yet they have not advanced to the point where they suffice to determine the risks of low-level radiation.

4.3.3 The Shape of the Dose–Response Curve

Alternative Models

In the absence of direct evidence on the rate of cancer induction at low doses, estimates are made by extrapolation from observations at high doses. The extrapolation is couched in terms of a *dose–response curve*, namely an expression that relates excess cancer mortality to radiation dose. The curve is anchored at the bottom by definition: There is zero excess if there is zero dose. It is anchored reasonably well at the top (e.g., an excess lifetime cancer mortality risk of 0.10 for a dose of 1 Sv).[13]

Figure 4.1 presents alternative curves for the extrapolation from high to zero doses. The dominant hypothesis is the *linearity hypothesis*, embodied in curve B. Many alternatives to the linearity hypothesis have been proposed. Some of these are also illustrated in Figure 4.1. Overall, the most prominent possibilities are the following:

1. A dose–response curve that corresponds to greater effects at low doses than implied by linearity; this is termed "supralinearity" (curve A).
2. The linearity hypothesis, or linear-nonthreshold hypothesis, according to which the risk of radiation-induced cancer is proportional to the magnitude of the dose (curve B).
3. A linear-quadratic dose response, in which the risk is dependent on the sum of linear and quadratic terms. Depending on the relative importance of the two terms, the extrapolated risk may be substantially depressed at low doses (curve C).
4. A dose–response curve that takes on negative values at low doses, corresponding to a beneficial effect of small radiation doses; this is termed "hormesis" (curve D).
5. The existence of a threshold, below which there is no appreciable rate of cancer induction (not shown).
6. A reduced risk per unit dose at low doses or low dose rates (or both), which can be represented by applying a dose and dose rate effectiveness factor (DDREF) to the predictions of the linearity model (not shown).

---

[13] The caution remains that this is based largely on doses received over a very short period of time.

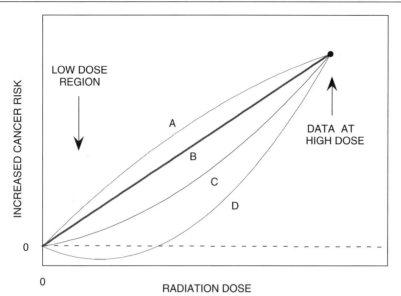

Fig. 4.1. Schematic representation of several alternative assumptions for the extrapolation of the cancer risk versus radiation dose to low dose levels, given a known risk at a high dose: supralinear (A), linear (B), linear-quadratic (C), and hormesis (D).

Alternatives 3 to 6 correspond to reduced effects at low doses. Of these, the only one that has been adopted in major recent studies is possibility 6, the inclusion of a dose and dose rate effectiveness factor (DDREF). At an earlier time, as reflected in the 1980 BEIR III study, the linear-quadratic model 3 had been favored [14]. Alternatives 1 and 4 are outside the mainstream of standard assessments, but this in itself does not prove they are wrong. In fact, as an increasing volume of research is accumulated, individual studies can be found that lend support to each of these nonstandard assumptions. However, they have failed to convince the leading national and international advisory bodies that any of the alternatives is to be preferred to the linearity hypothesis (except for the inclusion of a DDREF).

The relative merits of the linearity hypothesis and of the several alternatives could, in principle, be determined from either conclusive observational data at low radiation doses or a fuller understanding of the damage mechanisms. However, as discussed earlier and implied by the persistence of alternatives, neither is available.

The Linearity Hypothesis

According to the linearity hypothesis, the cancer risk (above the normal rate) is proportional to the magnitude of the excess radiation dose, over the full range from zero dose to high dose, independent of the dose rate.[14] Thus, if a dose of 1 Sv in a brief period corresponds to a 10% risk of a radiation-induced cancer, then a dose of 1 mSv (gradual or sudden) corresponds to a 0.01% chance. More generally, each increase of 1 mSv in dose would add 0.01% to the cancer risk. Sometimes, to be explicit, this model is referred to as the *linear-nonthreshold* (LNT) hypothesis, because the linear behavior persists down to very small doses.

It is to be noted that the key aspect of the LNT hypothesis is that the slope of the entire dose–response curve (see Figure 4.1) is defined by the risk at high dose. All of the curves in Figure 4.1 are linear over any small region (in the same sense that the Earth is flat over a small region), but only for curve B is the slope of each near-linear segment defined by the location of the high-dose "anchor point."

The linearity hypothesis is dominant among the models in the sense that virtually all official bodies use it as a working assumption in the formulation of radiation protection policy. Nonetheless, its validity is strongly contested. Proponents can cite data on cancer incidence and studies of cellular processes in support of the linearity hypothesis, whereas opponents can cite evidence against in both areas. It requires a massive study of all available evidence to form a balanced view, and bodies such as the NCRP that have undertaken these studies have not ended up with crisp statements on the scientific validity of the LNT hypothesis. Perhaps because individual members of these consensus-seeking groups assess the evidence somewhat differently and perhaps because many scientists see it as an unsettled issue, the final documents end up being guarded in their conclusions about linearity, although endorsing it for radiation protection purposes.

Hormesis and Adaptive Response

The simplest criticism of the LNT hypothesis is that it has not been proven, namely, that there is little evidence from studies of exposed populations that directly demonstrates that low doses of radiation cause cancer. In fact, the converse has also been argued—that there are reasons to believe that low doses of radiation do *not* cause cancer.

One argument that is advanced in support of this thesis is based on the existence of biological mechanisms to repair the cellular damage that occurs at low doses. At large enough doses, these repair mechanisms are overwhelmed, but at low doses, they can undo the damage caused by the passage of ionizing radiations. In this description, low doses have little effect, good or bad.

Beyond this "no significant effects" picture, there is evidence that low doses of radiation can stimulate an "adaptive" response, in which the ability of the system to repair cellular damage increases. This effect has been seen

---

[14] Here, a noncontroversial caveat is in order. At extremely high doses, there is saturation. The probability of death cannot exceed unity.

in experiments in which a prior exposure to a relatively small radiation dose lessens the harm caused by a subsequent larger dose. At one time, suggestions that "a little radiation was good for you" elicited reactions that ranged from the amused to the outraged. Now, however, adaptive responses have gained scientific respectability, as illustrated in NCRP Report No. 136:

> There is growing evidence that small doses of radiation can sometimes elicit transient homeostatic responses which may enhance the ability of cells and organisms to withstand subsequent irradiation....
>
> Among the adaptive responses to radiation most studied thus far is a heightened capacity for repair of chromosome damage... [4, p. 202]

Nonetheless, NCRP Report No. 136 raises a series of questions concerning the range of applicability of the adaptive phenomena and concludes that the data are "generally interpreted" to provide "insufficient grounds for rejecting the linear-nonthreshold dose–response model as a basis for assessing the risks of low-level ionizing radiation in radiation protection" [4, p. 205].

The adaptive response in the above-described form provides a limited benefit—it helps cells repair damage and withstand later radiation insults. However, adaptive responses are believed by some scientists to carry implications well beyond the cellular level. They provide a basis for the challenge to the LNT theory presented by the advocates of hormesis—the theory that small radiation doses are beneficial to human health and *decrease* the overall cancer risk. This viewpoint has been especially put forward by T.D. Luckey in books and numerous publications. Believers in hormesis cite examples in which a reduced cancer rate is associated with an increased radiation level, although the cited evidence has not convinced the major organizations such as the NCRP or UNSCEAR to accept hormesis as a strong alternative to linearity.[15]

A particular interesting case is provided by the studies by Bernard Cohen of lung cancer and indoor radon. Although Cohen's basic claim is that his study disproves the LNT hypothesis, not that it establishes hormesis, his work is widely cited by advocates of hormesis [22, 23]. It is also vigorously disputed, especially by some epidemiologists. We will discuss the radon results in more detail in Section 4.5.2.

A rather general challenge to the LNT theory and its application to radiation protection has been presented by Zbigniew Jaworowski, formerly an UNSCEAR chairman [25]. A belief in hormesis enters into his argument, but he criticizes the linearity hypothesis on other grounds. One phase of his analysis draws on the fact that life has evolved in an environment with widely different levels of radiation exposure. He contrasts changes in radiation level (which we cannot detect with our senses) to changes in temperature (which

---

[15] Examples of studies that are interpreted as demonstrating hormesis are presented, for example, in articles by Luckey (e.g., [22]), Myron Pollycove [23], and John Cameron [24].

we can readily detect) and argues that we cannot detect the radiation changes because, although they are large, they are harmless.

> Perhaps we humans lack a specific organ for sensing ionizing radiation simply because we do not need one. Our bodies' defense mechanism provides ample protection over the whole range of natural radiation levels—that is, from below 1 mSv to above 280 mSv/yr. . . .

> In short, conditions in which levels of ionizing radiation could be noxious do not normally occur in the biosphere, so no radiation-sensing organ has been needed in humans and none has evolved. [25, p. 25]

In contrast, having "an organ that can sense heat and cold has been essential for survival." Similarly, our senses of smell and taste help protect us from being poisoned by toxic food.

Jaworowski goes on to explain the original adoption of the LNT theory by the ICRP in 1959 as a convenient choice for purposes of radiation protection and one that was "politically useful" because it "played an important part in effecting first a moratorium and then a ban on atmospheric nuclear tests." He suggests that there is little justification for the transformation of this "working assumption" into what is now often presumed to be a "scientifically documented fact," and he cites as an "absurdity" the use of the LNT theory in estimating the total eventual number fatalities at Chernobyl.[16]

4.3.4 Conclusions of Advisory Bodies on Low-Dose Effects

BEIR V: Linearity with No DDREF

Despite these criticisms, the LNT hypothesis is accepted by most major scientific bodies as a working assumption, sometimes with the minor modification of a DDREF. In its 1990 report (BEIR V), the BEIR Committee adopted the linearity assumption for cancers other than leukemia, with no dose or dose rate reduction factor [11].[17] The risk of excess cancer mortality is taken to be 0.08 per sievert over a wide range of dose levels. The use of this assumption can be illustrated by applying it to the case of a constant low-level lifetime exposure. Consider a population of 100,000, with a representative distribution by age and sex. The "normal" number of eventual cancer deaths in such a population is about 19,000 [11, p. 172]. If this population sustains a 1-mSv/yr dose for a period of 70 years, the cumulative exposure is 70 mSv per person, or 7000 person-Sv, and the calculated excess cancer mortality is 560.[18] This would represent an increase of about 3% in the total cancer rate.

---

[16] This "absurdity" is repeated, with reservations, in Section 15.3.

[17] Leukemia is responsible for roughly 4% of normal cancer deaths and for 12% of excess fatalities for a single exposure of 0.1 Sv or a lifetime exposure of 1 mSv/yr [11, pp. 172].

[18] The actual quoted numbers in BEIR V are 520 per 100,000 for men and 600 per 100,000 for women [11, p. 172].

However, the BEIR V report expresses concern about this sort of extrapolation to doses and dose rates that are far lower than those on which the fundamental risk estimate is based, especially the Hiroshima–Nagasaki data. Thus, while adopting the linear extrapolation as a basis for predicting the risk at low doses, the BEIR V report also indicated the possibility that there is no risk from exposures in the neighborhood of several millisieverts per year:

> Departure from linearity cannot be excluded at low doses below the range of observation. Such departures could be in the direction of either an increased or decreased risk. Moreover, epidemiological data cannot rigorously exclude the existence of a threshold in the millisievert dose range. Thus the possibility that there may be no risks from exposures comparable to external natural background radiation cannot be ruled out. At such low doses and dose rates, it must be acknowledged that the lower limit of the range of uncertainty in the risk estimates extends to zero. [11, p. 181]

ICRP, UNSCEAR, and NCRP Reports: Linearity Modified by a DDREF

Most other major bodies that have evaluated low-dose risks have also accepted the linearity hypothesis, but with the inclusion of a DDREF.[19] Thus, a DDREF of about 2 has been adopted in ICRP, UNSCEAR, and NCRP reports for low doses and dose rates (see Table 4.1). For example, the 2000 UNSCEAR report used a DDREF of 2, with the caution that this factor may be in error by a factor of 2, which means that the true factor could lie between 1 and 4 [2, p. 13]. Many of these recommendations are explicitly limited to low-LET radiation and thus apply to gamma-ray and beta-particle exposures, but not to neutron or alpha-particle exposures [5, p. 18]. It should be noted that the "low" dose designations of the ICRP and UNSCEAR recommendations go into effect at levels much above those encountered under all natural conditions as well as by most people believed to have been impacted by Chernobyl.[20]

---

[19] Even the BEIR V study, which in the end did not adopt a DDREF, suggested that there might a reduction of risk "possibly by a factor of 2 or more" at low dose rates [11, p. 6].

[20] The ICRP requirement is that the dose be below 0.2 Gy, with higher doses acceptable if the rate is below 0.1 Gy/h [5, p. 19]. The 1993 UNSCEAR report sets a dose rate limit of 0.006 Gy/h [27, p. 682]. A dose of 0.2 Gy, which translates to 200 mSv for low-LET radiation, is more than 50 times the U.S. average annual dose of 3.6 mSv (see Table 3.5). Similarly, a dose rate of 0.006 Gy/h (equivalent to 6 mSv/h) means a larger dose in 1 h than normally received in 1 year.

Table 4.1. Estimates of cancer fatality risk for exposures to high and low doses of ionizing radiation.

Source of Estimate	Year	High Dose[a] (per sievert)	Adopted DDREF	Low Dose (per Sv)	
				Indicated[b]	Inferred[c]
General public					
BEIR V [11, p. 175]	1990	0.08	1[d]	0.08	0.08
ICRP 60 [5, p. 20][e]	1991	0.10	2	0.05	0.05
UNSCEAR 1993 [27, pp. 16, 17][e]	1993	0.11	2	0.05	0.05
NCRP Report 115 [28, pp. 111–112]	1993	0.10	2	0.05	0.05
UNSCEAR 1994 [12, p. 3–4]	1994	0.08–0.12	2	0.05	0.05
NCRP Report 126 [29, pp. 69, 71]	1997	0.10	2	0.040[f]	0.05
EPA (FGR 13) [26, p. 179]	1999			0.0575	
UNSCEAR 2000 [3, pp. 358, 431]	2000	0.09–0.12	2		0.05
Occupational					
ICRP 60 [5, p. 20]	1991	0.08	2	0.04	0.04
NCRP Report 115 [28, pp. 111–112]	1993	0.08	2	0.04	0.04
NCRP Report 126 [29, p. 73]	1997	0.08	2	0.037[f]	0.04

[a]The "high dose" here is a single dose of 0.1 Sv or 1 Sv.

[b]The indicated value is one that is stated or implied in reference.

[c]The inferred value is calculated from the given high-dose risk and DDREF.

[d]The (relatively small) leukemia contribution is based on the linear-quadratic model and therefore contains an implicit DDREF.

[e]The adoption of a DDREF (other than unity) is explicitly limited to low-LET radiation.

[f]Mean value, as determined by Monte Carlo calculation.

Summary of Estimates Relating to Cancer Induction at Low Doses

Table 4.1 summarizes recent estimates for radiation risk coefficients. The overall consensus of these estimates suggests that the risk of fatal cancer at a dose of 1 Sv is 0.10 for the general population and that it is appropriate to use a DDREF of 2 at low doses and low dose rates, at least for low-LET radiation. This leads to an overall estimate for the risk to the general population when exposed to low, protracted doses:

Risk of eventual fatal cancer: 0.05 per sievert (0.0005 per rem).

The EPA has derived a virtually indistinguishable factor for low-LET radiation at low doses: 0.0575 per sievert [26, Table 7.3].[21] For a worker population, which omits the young and very old, a slightly lower risk factor is favored in the studies cited above, namely 0.04 per sievert.

In making estimates of cancer risk to the general population (e.g., in considering the Chernobyl accident), we will use the consensus risk factor of 0.05 per sievert. However, consensus should not be mistaken for precision or even basic validity. NCRP Report 126 specifically analyzes the uncertainties involved in making these estimates, within the context of the linearity hypothesis, and calculates a distribution of possible risk coefficients. It finds a mean of 0.0399 per sievert and a 90% "subjective" confidence interval extending from 0.012 to 0.088 per sievert [29, p. 71]. This implies that the 0.05 per sievert coefficient might be wrong by more than a factor of 2—even without considering the more extreme hypotheses of a threshold or of hormesis.

Regulatory Guidelines and Scientific Assessments

As emphasized earlier, risks calculated on the basis of the linearity hypothesis do not rest on a firmly established scientific foundation. Instead, they represent estimates that, as thought appropriate for guidelines to be used in radiation protection, may "err on the side of caution." If the effects of low-level radiation are not well known, it is appropriate for an advisory body that will influence regulations to take a conservative approach.[22]

This position was enunciated, for example, in NCRP Report No. 116, issued in 1993, which explicitly assumed linearity for "radiation protection purposes" while indicating that one could not exclude the "possibility that there is no risk" at very low doses [6, pp. 10 and 13]. The same viewpoint is reflected more recently in NCRP Report No. 136, issued in 2001. As discussed earlier, although it recognized that validity of the LNT assumption at very low doses might never be proved, it found no other model to be "more plausible" [4, p. 7]. Overall, the LNT hypothesis is adopted in setting standards because there is some evidence for it, it is not seen as having been proven untrue, it is simple, and it is generally believed to be conservative.

It is also widely used in estimating the number of cancer fatalities that may result from radiation exposures, as, for example, in assessing the long-term risks for people living near Chernobyl. We will follow this practice in considering potential harm from radiation in the context of nuclear energy. This makes it unlikely that the harm is being underestimated. At the same

---

[21] This risk coefficient, which includes a DDREF of 2, is specified as applying to low-LET radiation. It can also be used for alpha-particle exposures because the relative biological effectiveness factor for alpha particles is taken to be 20 in comparisons with low-dose, low-LET radiation (equivalent to setting the quality factor Q equal to 20) [26, p. 174].

[22] However, as discussed in Section 4.3.1, this is not necessarily a "cost-free" approach.

time, it is to be remembered that the estimates are based on a model that has not been proven to be valid.

4.3.5 Genetic Effects

There is no evidence that directly demonstrates genetic damage to the offspring of people exposed to radiation, although in the 1950s and earlier, there was considerable emphasis on the potential genetic effects of radiation, based on animal studies. Since then, in the words of NCRP Report No. 116 "the genetic risks were found to be smaller and cancer risks larger than were thought at the time" [6, p. 12]. A 2001 UNSCEAR Report summarizes the current evidence as follows:

> No radiation-induced genetic (= hereditary) diseases have so far been demonstrated in human populations exposed to ionizing radiation. However, ionizing radiation is a universal mutagen and experimental studies in plants and animals have clearly demonstrated that radiation can induce genetic effects; consequently, humans are unlikely to be an exception in this regard. [30, p. 84]

One might expect that there would be evidence from the extensive study of the atomic bomb survivors and their children. However, here the evidence is negative: On comparing offspring of heavily irradiated parents with offspring of parents who received little or no radiation, "no statistically significant effects of radiation have been demonstrated to date" [11, p. 95]. In these studies, stillbirths, abnormalities at birth, and early deaths among children were all compared, with no overall statistically significant positive effects.[23] An increased rate of mutations was seen at the cellular level among the atomic bomb survivors themselves, but no increase was seen in similar studies of their children [4, p. 38].

Inferences as to inherited effects therefore come from animal experiments, primarily with fruit flies and mice. The atomic bomb results serve to verify that the deduced effects in humans are not thereby substantially underestimated. In fact, they suggest (but do not establish at a 95% confidence level) that extrapolating from animal data overestimates the genetic effects of radiation in humans.

It has been common in discussing genetic effects to introduce the concept of a *doubling dose*, namely the dose at which the normal incidence of genetic defects is doubled. An evaluation of the overall evidence on radiation induced genetic effects is contained in NCRP Report No. 115:

---

[23] It interesting to speculate why it is the common impression that there was great genetic damage from the atomic bombs. Perhaps the misconception arose in part from the fact that there *was* substantial damage to children of pregnant mothers at Hiroshima and Nagasaki, and in part because responsible people did not wish to risk underestimating the horrors of the bombs.

The doubling dose (DD) for genetic diseases that cause morbidity or mortality in humans is now estimated to be about 1.7 to 2.2 Sv or about 3.4 to 4.4 Sv for exposures at low-dose rates.... Because of the large uncertainties in both estimates and the desire to provide adequate protection, a risk of severe hereditary effects for the general population (for all generations) of 1×10^{-2} Sv^{-1} should provide a reasonable basis for dose limitation [28, p. 3]

Consistent with this last number, the UNSCEAR report estimates the doubling dose for low-dose, low-LET exposures to be 1 Gy [30, p. 87]. The normal frequency of genetic diseases is about 24 per 1000 live births (i.e., 2.4%). [30, p. 19]

Because genetic effects are less frequent and less severe than cancer fatalities, the effects of radiation exposures are often couched in terms of cancer fatalities alone. Thus, for example, in the discussion of the Chernobyl accident in Chapter 15, the focus is on cancer rather than genetic or other possible effects such as mental retardation. These are not unimportant, but whatever the importance, it appears to be less than that of cancer.

4.4 Radiation Standards and Health Criteria

4.4.1 Standards for the General Public

From about 1960 to 1990, the standard established in the United States and internationally was that the average additional exposure for members of the general population should not exceed 1.7 mSv/yr, and for any individual, it should not exceed 5 mSv/yr (excluding radiation workers). This was the maximum "permitted" dose, over and above the dose received from natural and medical sources. Not coincidentally, this limit for additional dose was about equal to the background from natural and medical sources (omitting radon, which had more or less been ignored!)

Coupled with such limits is the general view that it is prudent to incur no excess radiation unless there are compensating benefits. This policy was formally recommended by the ICRP in 1977 in what has come to be known as the ALARA principle: "all exposures shall be kept as low as reasonably achievable, economic and social factors being taken into account" [31, p. 3]. This policy is widely accepted by regulatory agencies. Thus, the U.S. Nuclear Regulatory Commission explicitly instructs its licensees to achieve doses "that are as low as reasonably achievable (ALARA)" [32, §20.1101].

In the general spirit of reducing exposure levels, a lowering of the individual dose limit for the general population from 5 to 1 mSv/yr was recommended by the ICRP in 1990 [5, p. 45] and was adopted by the U.S. Nuclear Regulatory Commission effective January 1, 1994 [32, §20.1301]. This limit refers to exposures from a single facility licensed by the NRC, excluding medical facilities. The reduction does not have much of a practical impact. The nuclear fuel cycle, including reactor operation, falls under an EPA whole-body dose limit of 0.25 mSv/yr for total exposures (other than radon exposures) from all facilities [33, §190.10] and somewhat more restrictive NRC conditions for effluents from light water nuclear reactors [32, Part 50, Appendix I]. In addition, nuclear waste disposal facilities covered by the EPA are required to satisfy a 0.15-mSv/yr limit.[24]

Actual exposures to the public from nuclear power operations and facilities are much lower than the regulatory limit of 0.25 mSv/yr, and the limit is therefore not presently constraining (see Table 3.5). It would be neither violated nor even approached except in the case of an accident, in which case the existence of regulations would be moot. However, the regulatory limits provide a safeguard against negligent operation of existing facilities. Without a history of external pressures from regulatory agencies, radiation exposures from nuclear facilities might not be as low as they now are.

Under the terms of the Safe Water Drinking Act, the EPA has established stringent limits on the contamination of drinking water from a wide variety of pollutants, including radionuclides. For beta-particle and gamma-ray emitters, the radionuclide concentrations in community water systems must not produce a dose above 4 mrem (0.04 mSv) per year for a person drinking 2 liters/day from this source [33, §141.66]. The limit for alpha-particle emitters is stated in terms of maximum concentrations (in picocuries per liter), rather than in terms of a radiation dose, but the intent is to achieve the roughly the same dose limit.

Uranium and radon are explicitly omitted from control in this limit on alpha-particle emitters. The chief ingestion danger for uranium is from its chemical properties, not its radioactive emissions, because the half-life of ^{238}U is large and its rate of decay slow. A concentration limit for uranium is set in terms of mass per liter rather than activity per liter. For radon, there is a special problem. The hazard from radon in water supplies is caused mainly by the escape of the radon from the water into the air, raising the household radon concentration. This leads to an additional dose that often exceeds 0.04 mSv/yr, but that is still small compared to the dose from outdoor air, to say nothing of indoor air (see Table 3.5). The EPA has put forth a proposal to address the question of radon in drinking water, but the proposal remains in abeyance [34].[25]

---

[24] The regulation of nuclear waste disposal facilities is discussed further in Section 13.2.5. The 0.15-mSv/yr limit was established in 1993 for waste disposal facilities other than Yucca Mountain [33, §191.15] and in 2001 for the proposed Yucca Mountain repository [33, §197.20].

[25] A National Research Council study hypothesized (but did not recommend) an eventual standard of 25,000 Bq/m^3 (about 700 pCi/L) [35, p. 182]. Using the ratio of air to water concentrations that is found to apply to a typical house, a 700 pCi/L concentration in water translates to an addition of 0.07 pCi/L to the ambient radon concentration [35, p. 13]. Roughly speaking, this would represent a 7% increase in the typical indoor air concentration, were this standard to be adopted by the EPA.

The EPA has taken an advisory, but not regulatory, position on the larger problem of radon in indoor air. Regulation would be much more difficult here because radon is a natural product and limits would have to be enforced house by house. In any event, although the EPA has the authority to set standards for public drinking water supplies, it does not have the authority to set any mandatory rules for household air pollution. To provide guidance, the EPA suggests that remedial action be taken if the indoor radon concentration exceeds 4 pCi of ^{222}Rn per liter of air (148 Bq/m^3). This concentration corresponds to an annual dose of approximately 8 mSv/yr.[26]

A summary of these standards and recommendations, along with those for occupational exposure, is given in Table 4.2. It is seen that the limits established by the EPA on nuclear facilities are much stricter than the level at which action is recommended to reduce indoor radon levels.

4.4.2 Standards for Occupational Exposures

For many years, the limit for occupational exposures was 50 mSv/yr. The ICRP in 1990 reduced its recommended limit to 100 mSv (10 rem) for occupational exposures over a 5-year period, equivalent to an average of 20 mSv/yr [5, pp. 72–73]. The limit of 50 mSv in any one year is retained. More stringent limitations are recommended for pregnant women. For the United States, as codified by U.S. Nuclear Regulatory Commission, the basic occupational limit

Table 4.2. EPA and NRC standards for U.S. radiation protection, expressed in terms of annual effective dose equivalent.

Source of Exposure	Dose Limit (mSv)	Agency	Authority[a]
Occupational	50	NRC	10 CFR 20.1201
General public			
Any licensed facility	1	NRC	10 CFR 20.1301
Nuclear power facility	0.25	EPA	40 CFR 190.10
Nuclear waste repository	0.15	EPA	40 CFR 191.15
Yucca Mountain repository	0.15	EPA	40 CFR 197.20
Public water supplies	0.04	EPA	40 CFR 141.66
Indoor radon (general public)[b]	8	EPA	

[a] The authority in each case, other than indoor radon, is the *Code of Federal Regulations* [32, 33].
[b] This is not a limit but rather a level at which the EPA recommends remedial action.

---

[26] There has been some consideration within the EPA of reducing the suggested action level from 4 to 2 pCi/L but no action had been taken on this as of the end of 2003. The effective dose equivalent in millisievert is here established using the following approximate equivalences: 1 pCi/L ≈ 0.2 working level months per year ≈ 2 mSv effective dose equivalent per year.

remains 50 mSv/yr [32, §20.1201]. In addition, occupational exposures are explicitly included in the application of the ALARA principle [32, §20.1101]. Were a worker exposed to the maximum dose of 50 mSv/yr over a working lifetime of 40 years, the cumulative total would amount to 2 Sv. Adopting a risk factor of 0.04 per sievert, this would imply an 8% chance of a fatal cancer due to occupational exposure. Average occupational exposures have been far below the limit, except for uranium miners, but nonetheless there have been calls for a formal reduction in the NRC occupational limit.

4.4.3 Alternative Risk Criteria

The cancer risk from the ingestion or inhalation of a radionuclide depends on the amount of the radionuclide taken into the body, its chemical form, the tissues in which it concentrates, the time it remains in the body, the rate of radioactive decay, and the type and energy of the emitted radiations. Together, these factors determine the effective dose and the risk of cancer mortality. Knowledge of these factors is evolving as research progresses, but remains incomplete. Calculations based on the current understanding are updated periodically and provide the basis for compilations of radiation dose and risk factors, most conspicuously in compilations prepared by the ICRP and the U.S. EPA. The findings are presented in a number of different ways, with varying choices for the parameter on which to focus, including the following:

- *Annual limit on intake (ALI).* The ALI for a radionuclide corresponds to the intake that would lead to an effective committed dose equal to the occupational dose limit. Tables of ALIs for inhalation and ingestion have been presented in a series of ICRP reports (including ICRP Publication 61 in 1990 [36]), as well as in Federal Guidance Report No. 11 (FGR-11) issued by the U.S. EPA in 1988 [37]. Tables of ALIs are also given by the U.S. Nuclear Regulatory Commission in Title 10 of the *Code of Federal Regulations* [32, Appendix B to Part 20].[27]
- *Committed dose per unit intake.* A dose calculation logically precedes the calculation of the ALI. Tabulations of the dose per unit of activity of the radionuclide ingested or inhaled (in Sv/Bq) are presented in FGR-11 and later (1995) in ICRP Publication 72 [38].
- *Risk per unit intake.* If the dose–response relation is known, the cancer risk can be calculated from the dose. Tabulations of the risk per unit intake (in becquerels) of cancer incidence and cancer mortality are presented in Federal Guidance Report No. 13 (FGR-13), published by the EPA in 1999 [26]. This report, based on ICRP Publication 72 and the EPA's own calculations, represents a major update of FGR-11 and in some cases differs significantly from it.

---

[27] In each case, the ALI corresponds to the governing occupational dose limit; 20 mSv/yr for the ICRP and 50 mSv/yr for the U.S. agencies.

Table 4.3. Cancer mortality risk for ingestion (in tap water) and inhalation of selected radionuclides, expressed in risk per becquerel and risk per microgram.

Radio-nuclide	Half-life (years)	Risk for Ingestion		Risk for Inhalation	
		per Bq	per μg	per Bq	per μg
^{3}H	1.23×10^1	9.44×10^{-13}	3.36×10^{-4}	4.58×10^{-12}	1.63×10^{-3}
^{14}C	5.73×10^3	2.89×10^{-11}	4.77×10^{-6}	3.68×10^{-13}	6.07×10^{-8}
^{40}K	1.28×10^9	4.30×10^{-10}	1.12×10^{-10}	1.20×10^{-9}	3.11×10^{-10}
^{60}Co	5.27×10^0	2.75×10^{-10}	1.15×10^{-2}	8.02×10^{-10}	3.36×10^{-2}
^{90}Sr	2.88×10^1	1.34×10^{-9}	6.85×10^{-3}	2.65×10^{-9}	1.35×10^{-2}
^{99}Tc	2.11×10^5	4.28×10^{-11}	2.71×10^{-8}	3.49×10^{-10}	2.21×10^{-7}
^{129}I	1.57×10^7	4.07×10^{-10}	2.66×10^{-9}	1.68×10^{-10}	1.10×10^{-9}
^{131}I	2.20×10^{-2}	1.31×10^{-10}	6.03×10^{-1}	5.55×10^{-11}	2.56×10^{-1}
^{137}Cs	3.01×10^1	5.66×10^{-10}	1.82×10^{-3}	2.19×10^{-10}	7.04×10^{-4}
^{210}Pb	2.23×10^1	1.75×10^{-8}	4.95×10^{-2}	6.84×10^{-8}	1.93×10^{-1}
^{226}Ra	1.60×10^3	7.17×10^{-9}	2.62×10^{-4}	2.93×10^{-7}	1.07×10^{-2}
^{232}Th	1.41×10^{10}	1.87×10^{-9}	7.59×10^{-12}	1.10×10^{-6}	4.47×10^{-9}
^{235}U	7.04×10^8	1.21×10^{-9}	9.68×10^{-11}	2.57×10^{-7}	2.06×10^{-8}
^{238}U	4.47×10^9	1.13×10^{-9}	1.41×10^{-11}	2.38×10^{-7}	2.96×10^{-9}
^{237}Np	2.14×10^6	1.10×10^{-9}	2.86×10^{-8}	4.18×10^{-7}	1.09×10^{-5}
^{238}Pu	8.77×10^1	2.75×10^{-9}	1.74×10^{-3}	8.04×10^{-7}	5.10×10^{-1}
^{239}Pu	2.41×10^4	2.85×10^{-9}	6.54×10^{-6}	7.94×10^{-7}	1.82×10^{-3}
^{240}Pu	6.56×10^3	2.85×10^{-9}	2.39×10^{-5}	7.95×10^{-7}	6.68×10^{-3}
^{241}Am	4.32×10^2	2.01×10^{-9}	2.55×10^{-4}	6.59×10^{-7}	8.37×10^{-2}

Source: Data are from Ref. [26], Tables 2.1 and 2.2. The inhalation factors are for the recommended default absorption types, except for ^{14}C, where carbon dioxide is assumed, and for ^{40}K, for which no default recommendation is indicated.

Table 4.3 presents results from FGR-13 for the risk for ingestion and inhalation of selected radionuclides.

The alternative means of presentation are closely related. However, there are discrepancies among sets of tables prepared at different times, most importantly due to changes in the understanding of the behavior of particular radionuclides in the body. The connections between these measures of risk can be illustrated by considering a specific case. According to FGR-11, the effective committed dose equivalent for the ingestion of ^{137}Cs is 1.35×10^{-8} Sv/Bq [37, p. 70]. Therefore, the intake corresponding to the dose limit of 50 mSv/yr is 3.7×10^6 Bq/yr. This result is in agreement with the tabulated ALI of 4×10^6 Bq, also given in FGR-11 [37, p. 70].

Agreement between the results of FGR-11 and FGR-13, which were published more than a decade apart, is not as good. In the calculations of FGR-13, the average risk factor is taken to be 0.0575 per sievert [26, p. 179]. Combining this factor with the dose per becquerel, gives a calculated risk for ^{137}Cs of 7.8×10^{-10} per becquerel. This result is in moderate agreement with the tabulated risk of 5.7×10^{-10} given in FGR-13 for ingestion via tap water (see Table 4.3). However, even this level of agreement between the results presented in FGR-11 and FGR-13 is not uniformly attained, and often there are differences of more than a factor of 10 between FGR-11 and FGR-13. The very large difference in the case of ^{237}Np is discussed in Section 4.5.3.

In terms of its delivery to the body, the "amount" of a contaminant that a person consumes by ingestion or inhalation is represented by the mass or volume of the material, not by its activity. Thus, in considering the effects of consumption of a radionuclide, the risk per unit mass is more relevant than the risk per unit activity. Table 4.3 therefore lists the risk per microgram along with the risk per becquerel. It is seen, for example, that the risk per becquerel for ^{238}U is roughly the same as for many other alpha-particle emitters. However, the risk per microgram is much less for ^{238}U, because its half-life is so long. Similarly, although ^{226}Ra was unconcernedly (and apparently harmlessly) carried by millions of people in their wrist watches and ^{239}Pu has gained the (undeserved) reputation of being uniquely poisonous, ^{226}Ra is considerably more hazardous than ^{239}Pu on a risk-per-microgram basis.

4.4.4 Collective Doses and de Minimis Levels

Problems with Collective Doses

In assessing the impact upon a population of radiation exposures, it is tempting to add up all the individual doses and calculate the collective dose. If the LNT theory is literally true, the collective dose is the measure of the total impact (i.e., the total number of cancer fatalities). Although this may seem a reasonable approach, it can readily lead to unreasonable results when the total effect is the result of summing small doses for many people.

In a particularly egregious case, a collective dose of 300,000 Sv was calculated as the potential consequence of the release of ^{14}C from a nuclear waste repository. This would lead to 15,000 calculated fatalities if one applies the LNT theory. However, this collective dose results from considering a world population of 10 billion people over 10,000 years, with an average individual dose about 3×10^{-6} mSv/yr—about one-millionth of natural background. The history of coping with the conflicting implications of the large collective dose and the miniscule individual doses that then arose is discussed in Section 13.2.2.

The problem was resolved in this case by ignoring the collective dose. However, the issue persists in other less extreme contexts. Another example, already alluded to, is that of Chernobyl. There is no consensus on how to view this matter. One authority, as mentioned earlier, considers calculating the collective dose from Chernobyl an absurdity. Others undoubtedly believe that *not* calculating the collective dose would mean irresponsibly ignoring the global effects of the accident—tantamount to a whitewash.

Adoption of a de Minimis Level

The collective dose issue can be bypassed by applying a cutoff, sometimes termed a *de minimis* level or level of negligible concern. For example, it has often been suggested that exposures less than 0.01 mSv (1 mrem) be ignored. Whatever is believed about the scientific foundations of the LNT theory, it is clear that as a society we are indifferent to variations in natural background of 0.01 mSv/yr—and even of 0.1 or 1 mSv/yr. If an increment of 0.01 mSv is of no concern when it comes from natural radiation, it may be justifiable to ignore it in all circumstances.

A policy embodying this viewpoint is being considered by the ICRP as a part of a major recasting of radiation protection criteria. As discussed in a 2001 document describing the progress of this effort, one of the proposed recommendations is to define bands of radiation dose magnitude, with the action to be taken for each band depending on the dose levels included in it [39, pp. 119–120]. The reference for these bands is the "normal" or "typical natural background" level, taken to be 1–10 mSv/yr. The lowest band in this system is for doses less than 1% of normal (i.e., below roughly 0.03 mSv/yr).[28] Such doses are characterized as being "negligible" and the proposed action at this level is to ignore it ("exclude from the ICRP system of protection"). At a level 10 times higher, where the dose is termed "trivial," it is not to be ignored but "no protective action" is to be taken.

These proposals represent a shift from what has appeared to be a trend of increasing attention to small radiation doses. If these or similar proposals are officially adopted by the ICRP—after what the ICRP looks forward to as a "debate with an iteration of ideas over the next few years" [39, p. 113]—it will imply a considerable change in attitudes toward exposures at very low dose levels.

4.5 Radionuclides of Special Interest

4.5.1 Radium-226

In the early studies of the radioactive decay series in nature, it became obvious that radium was a uniquely valuable radionuclide. The chemical separation of radium from uranium ore yields almost pure ^{226}Ra, with no significant admixture of other isotopes of radium. The half-life of ^{226}Ra is 1600 years, which means that it has a much higher specific activity than, say, ^{238}U or ^{232}Th and yet the activity of a sample of radium will remain almost constant for many decades. Thus, radium became the dominant radionuclide in medical

---

[28] With typical natural background taken as 1–10 mSv/yr, the 1% level is not well defined. For specificity in the discussion, we here use the geometric mean of 3 mSv/yr; this matches the average U.S. dose listed in Table 3.3 and is slightly higher than the listed world average dose.

and industrial applications, with no effective substitute until the advent of artificial radionuclides—first, as a trickle in the late 1930s and then profusely as nuclear reactors and accelerators were built in the decades following World War II.

Radium was used in cancer treatments as early as 1903 [40, p. 453]. This use, along with the use of X-rays, were the forerunners of modern radiation therapy which now employs, in addition to X-rays, a variety of radionuclides and beams of particles. Some use of radium for luminescent dials was also attempted in 1903, but this application did not become widespread until World War I, when improved paints were developed for dials in German submarines [41, p. 17].

By the 1930s, radium dial watches and instruments became the standard for quality equipment. At one time, as many as 3 million radium-dial watches and clocks were sold in the United States each year, and although these watches were no longer being sold after about 1968, an estimated 10 million radium-dial watches were still in use for another decade [42, p. 318]. The amounts of radium used were typically on the scale of 0.1 μCi for watches and 0.5 μCi for clocks [42]. The use of radium for watch dials led to very harmful exposures for U.S. radium dial painters in the early 1920s, when it was the practice to paint radium onto the dials with brushes whose tips were put into the workers' mouths to form sharper points.

The novelty of radium made it a natural for employment in "quack medicine." For example, in one scheme, water was laced with radium and the product sold under the name of *Radithor*. That product is believed to have been responsible for the death in 1932 of at least one user—a person who drank it regularly to take advantage of its advertised broad curative properties [41, p. 13]. The belief in the curative powers of radium finds an echo to this day in spas that advertise treatments by immersion in an atmosphere with high concentrations of radon, a radioactive gas produced by the decay of ^{226}Ra.

4.5.2 Radon-222

Radiation Doses from Radon

As discussed in Section 3.5.1, indoor radon is the largest contributor to the radiation dose that the average person receives from natural or other sources. The UNSCEAR estimate of 1.3 mSv/yr (the lower of the two estimates quoted in Table 3.5) corresponds to an annual collective dose of about 8 million person-Sv for the world population of 6 billion people. This is more than 10 times the estimated global dose commitment from the Chernobyl accident of roughly 600,000 person-Sv over a 70-year period (see Section 15.3.6). Roughly speaking, therefore, the 70-year collective dose from indoor radon is 1000 times greater than that from Chernobyl.

That does not mean, however, that indoor radon is 1000 times more harmful. A collective dose calculation is meaningful only if the LNT theory is valid. Considering only directly established harm to human health, Chernobyl caused 31 deaths among workers and has led to an increase in thyroid cancer in children (see Section 15.3.4), whereas there is no specifically identifiable harm from indoor radon.

Studies of lung cancer incidence among uranium miners have unequivocally demonstrated that inhalation of radon and radon progeny at high concentrations causes lung cancer fatalities. These studies also established a risk factor relating lung cancer to radon concentration. It is possible to calculate the impact of indoor radon on the general population in two ways:

1. Apply the risk factor from the miners' experience to the general population, correcting for differences between mine conditions and normal household conditions (e.g., in breathing rates and dust concentrations).
2. Calculate the dose to the lung for a given radon concentration, using a model for the movement of radon and its progeny into and within the lung.

The calculated fatalities for the second approach, which was the approach used in NCRP Report No. 93 (see Section 3.5.1), are a factor of 2 or 3 times those for the first. The UNSCEAR 2000 report found a discrepancy of a factor of 2.5 between the two approaches, but viewed it as "not a big discrepancy, considering the complex physical and biological issues involved" [2, p. 107]. It addressed the matter by reducing the imputed radiation dose from radon to a level somewhat lower than that given by lung model calculations alone. The reason for the contradiction is not known. Possibilities include (1) the quality factor of 20, assumed for alpha particles in the dose calculation, may be too high, making the calculated effective dose equivalent too high; and (2) if the effects depend on the rate at which the dose is delivered, and not just on the total dose, then a risk coefficient based largely on the atomic bomb exposures is not properly applicable to the protracted exposures received by the miners.

Effects of Radon Exposure in the General Population

The discrepancy discussed in the previous subsection is of lesser immediate importance than the disagreement between the results of two approaches to relating lung cancer rates in the general population to residential radon concentrations. One method for studying the relationship is through *case-control* epidemiological studies. A *case* group, composed of people who have been diagnosed as having lung cancer, is compared to a *control* group of randomly selected people from the same general area. The radon concentrations in the homes of the two groups are compared, with adjustments for demographic factors, including age, gender, and smoking history. Individual studies of this type generally have too few subjects to be statistically conclusive, but an analysis that combines the results of a number of such studies may yield more meaningful results.

A meta-analysis of eight studies reported a "significant exposure–response relationship" with quantitative results that were in general agreement with an extrapolation made from studies of lung cancer in miners [43]. However, the statistical uncertainties in the individual studies were substantial and the authors warned that until further studies were completed, the "results should be interpreted cautiously." A larger pooled study of 18 individual case-control studies is underway. According to a preliminary analysis, some of the studies found a positive association between radon concentration and lung cancer rate, some found no association, and none "has reported a statistically significant negative association" [44].

Taking a different approach, Bernard Cohen has made an extensive study of correlations between county-by-county average indoor radon concentrations and county-by-county lung cancer mortality rates. Such studies are termed "ecological" studies, as distinct from "epidemiological" studies. This study now includes about 1600 counties that contain 90% of the U.S. population. A strong negative correlation is found in the data; that is, the high-radon counties tend to have below average lung cancer rates, extending to radon concentrations that are about five times the U.S. average [45, 46]. Thus, the slope of the curve of cancer fatalities versus household radon concentration is downward, rather than upward as predicted by a linear extrapolation from the miners data. (By chance the magnitudes of the observed downward and predicted upward slopes are about the same.)

The negative slope remained unchanged when the data were analyzed by subdividing them (stratifying) on the basis of potential socio-economic, geographic, and climatic confounding factors. Particular attention was paid to smoking as a possible confounding factor, but this too did not serve to explain the negative slope. Cohen concluded that his results prove that the LNT theory fails, because the predicted positive slope was not found. Many others have cited Cohen's work as support for hormesis, and on occasion Cohen also has suggested a protective effect.

This result, explicated and defended in a long series of detailed articles by Cohen going back to the 1980s, is highly controversial. It been been criticized by many epidemiologists who argue that it, along with all ecological studies, has methodological weaknesses that invalidate it. An obvious potential pitfall is the confounding effects of smoking, and Cohen's analyses—although Cohen has specifically addressed this issue—have not satisfied his critics.

In one criticism, Jerome Puskin has shown that negative correlations similar to that observed by Cohen also exist for other cancers linked to smoking (e.g., oral cancer and cancers of the larynx and esophagus) but are not seen for cancers that are not linked to smoking (e.g., colon, breast, and prostate cancer) [47]. Puskin interpreted these results as demonstrating a negative correlation between radon concentrations and smoking prevalence, and he concluded that the lowered lung cancer rates in high radon regions are due to less smoking. Thus, in this view, smoking is a decisive confounder and Cohen's thesis is invalidated. In response, Cohen argued that even when he assumes an unrealistically extreme correlation between smoking and radon concentrations, the correlation between lung cancer and radon concentrations does not become positive [48]. To explain Puskin's findings, he suggested instead that radon exposures may possibly reduce other cancer rates, beyond just lung cancer.

At present, Cohen's results are not accepted by major consensus bodies, such as the NCRP and UNSCEAR (e.g., Refs. [4, pp. 176–177] and [3, pp. 323–324]). They are not likely to be accepted by the broad epidemiological community unless case-control studies give evidence in support of a negative radon–lung cancer correlation—and, to date, the weight of evidence points in the other direction. On the other hand, scientists who are impressed by the apparent robustness of Cohen's analyses are unlikely to reject his findings, unless a specific error is convincingly identified in the analyses or the case-control evidence becomes incontrovertible.

Radon Spas

For nearly a century, some people have believed that radiation has beneficial health effects and radon spas have flourished. In an interchange in the December 2001 issue of *Health Physics*, one author termed the notion of a "radon health spa" an oxymoron, while another brief paper suggested that radon may be helpful in treating rheumatoid arthritis and that the dangers of exposures to radon at low doses are greatly exaggerated in calculations that assume the validity of the linearity hypothesis.

4.5.3 Neptunium-237

The difficulties in accurately establishing the dose caused by the ingestion of radionuclides is illustrated by the case of ^{237}Np.[29] It is an extreme case because the changes over time in the assessment of ^{237}Np hazards have been exceptionally large. It is of particular interest because in current calculations for the Yucca Mountain nuclear waste repository, ^{237}Np is the largest contributor to the potential radiation dose after 100,000 years.

The dose from the ingestion of a given amount of a radionuclide can be calculated using the ingestion dose coefficient, expressed, for example, as the effective dose equivalent per unit activity ingested (in Sv/MBq). Table 4.4 displays estimates for the ^{237}Np dose coefficient extracted from reports published from 1959 to 1999. Each value is either taken directly from the individual report or calculated from alternative data presented in the report. A striking feature of the data in Table 4.4 is the large changes during this time period.

---

[29] I am indebted to Dr. Lowell Ralston (EPA) for informative communications on neptunium dosimetry.

Table 4.4. History of estimates of the dose coefficient d for ingested ^{237}Np, in units of Sv/MBq.[a]

Year	Reference	Risk Measure ALI[b]	Sv/MBq	d (Sv/MBq)	Ratio to FGR-13[c]
1959	ICRP 2[d]	25 μCi		0.054	3
1980	ICRP 30, Pt. 2 [52, p. 71]	0.005 MBq		10	500
1981	ICRP 30, Pt. 2S [53, p. 739][e]		≈ 11	≈ 11	≈ 600
1983	NAS [54, pp. 249, 341][f]			4	200
1988	ICRP 30, Pt. 4 [50, p. 4]	0.05 MBq		1.0	50
1988	NCRP 90 [49, p. 35]	0.06 MBq		0.8	40
1988	EPA: FGR-11 [37, p. 177]		1.2	1.2	60
1991	ICRP 61 [36, p. 38]	0.03 MBq		0.7	40
1991	NRC [32, Part 20, App. B]	1μCi		1.35	70
1993	ICRP 67 [55, p. 165]		0.11	0.11	6
1995	ICRP 72 [38, p. 41]		0.11	0.11	6
1996	EPRI TSPA [56, pp. 9–11]		0.11	0.11	6
1999	DOE: GENII-S [57, p. 16]		1.4	1.4	70
1999	EPA: FGR-13 [26, pp. 103, 179]			0.019[g]	1

[a]Where multiple values are given in the source document, the result tabulated here is for ingestion via tap water, applies to adults, and uses the stochastic limit (not the bone surface limit).

[b]The dose coefficient d is calculated from the ALI as $d = $ (dose limit)/ALI, where the (occupational) dose limit is 50 mSv/yr, except for ICRP 61, for which it is 20 mSv/yr.

[c]Ratios are rounded off to one significant figure.

[d]Value reported in NCRP Report No. 90 [49, p. 34].

[e]Based on listed individual organ dose equivalents.

[f]To "allow for ICRP-60 corrections," the NAS report multiplied by 200 a coefficient calculated by B.A. Napier (Pacific Northwest Laboratories) of 0.02 Sv/MBq [54, pp. 249, 341]. The Napier result is in excellent agreement with the result in FGR-13.

[g]See text, footnote 30.

The dose coefficient depends on the fraction of ingested Np that enters the bloodstream rather than being excreted (termed f_1), the fraction of the neptunium that is transferred from the blood to each organ (particularly bone and the liver), and the tissue weighting factor for that organ. The greatest variations have come from changes in the estimated value of f_1. This coefficient was assumed to be 0.0001 for the 1959 ICRP Publication 2, but it was abruptly increased by a factor of 100 (to 0.01) in ICRP Publication 30 on the basis of a limited set of animal experiments. It was soon recognized that these data gave a misleadingly high value of f_1 because they were obtained in animal experiments that involved the ingestion of very large amounts of neptunium [49, 51]. The chemical form of neptunium in the stomach depends on the amount present, and at high concentrations, neptunium is in a form more readily transferred to the blood than is the case for the lower concentrations that would plausibly occur if ^{237}Np was unintentionally ingested.

In response to new experimental results that showed how f_1 depended on the amount of ^{237}Np ingested, a value of 0.001 was adopted in NCRP and ICRP reports published in 1988 [49, 50]. The value was further reduced to $f_1 = 0.0005$ in ICRP Reports issued in the 1990s as well as for the EPA's FGR-13 report in 1999 [26, 38]. Other changes in the estimates of the dose coefficient stem from differences in assumptions as to the pathways for neptunium once it enters the bloodstream, including the relative amounts going to the bone and liver, as well as tissue weighting factors that translate the dose to an organ to a whole-body dose.

In the documents referenced in Table 4.4, the hazard was alternatively specified in terms of the annual limit of intake (ALI) of ^{237}Np, the cancer risk per unit intake of ^{237}Np, or the dose per unit intake of ^{237}Np. For comparison purposes, these results are all converted to a dose coefficient in Sv/MBq.[30]

The calculated values for this coefficient have fluctuated markedly over the past four decades. As indicated earlier, ^{237}Np plays a prominent part in nuclear waste disposal calculations. In particular, analyses of the behavior of the protective barriers for the Yucca Mountain repository suggest that radionuclides will be largely kept out of the accessible environment for over 100,000 years but that, eventually, the still remaining radionuclides will reach water supplies that can be tapped by residents of the area and be taken up by plants. The half-life of ^{237}Np $(T = 2.14 \times 10^6$ yr) is long enough for most of the ^{237}Np to still be present, and ^{237}Np is the largest contributor to potential radiation doses in many calculations.

The apparent error caused by using too large a coefficient was particularly great in the calculations of Ref. [54], which reflected the results in ICRP 30, Part 2. More significantly for present concerns, there may be a large overestimate in the calculations used in the studies supporting the DOE's 2002 recommendation to proceed with the Yucca Mountain project (see Chapter 12). For these calculations, the DOE used the dose coefficients obtained from the so-called GENII-S code [57]. The GENII-S calculations are based on the methods of ICRP Report No. 30, Part 4 (1988) and, for the most part, give results that are similar, but not identical, to those of ICRP-30. GENII-S gives a dose conversion factor that is 70 times higher than that deduced from EPA's latest analysis (FGR-13) and about 10 times higher than the factors adopted by the ICRP in the mid-1990s [38, 55].

If doses at times beyond 100,000 years are of interest—and the Yucca Mountain analyses presume that they are—then it is desirable to use the best available information for ^{237}Np. There is no apparent rationale for retaining coefficients based on older ICRP reports, as modified in GENII-S, when pre-

---

[30] When the ALI is specified, the dose coefficient is the ratio of the occupational dose limit to the occupational ALI. For FGR-13, it is the ratio of the tabulated risk per million becquerel (0.0011 MBq^{-1}) to the assumed risk per sievert (0.0575 Sv^{-1}).

sumably more accurate coefficients are available from recent ICRP and EPA reports. It is hard to believe that the DOE would have been able to retain for a decade numbers that *underestimate* the risk, but there apparently have not been significant pressures to correct numbers that probably *overestimate* the risk. Presumably, more up-to-date coefficients will be used in the final calculations that are being developed in support of the DOE's forthcoming application to the NRC for a permit to construct the Yucca Mountain repository.

References

1. Klaus Becker, Book Review of NCRP Report 136, *Health Physics* 82, no. 2, 2002: 257–258.
2. United Nations Scientific Committee on the Effects of Atomic Radiation, *Sources and Effects of Ionizing Radiation, Volume I: Sources*, UNSCEAR 2000 Report (New York: United Nations, 2000).
3. United Nations Scientific Committee on the Effects of Atomic Radiation, *Sources and Effects of Ionizing Radiation, Volume II: Effects*, UNSCEAR 2000 Report, (New York: United Nations, 2000).
4. National Council on Radiation Protection and Measurements, *Evaluation of the Linear-Nonthreshold Dose–response Model for Ionizing Radiation*, NCRP Report No. 136 (Washington, DC: NCRP, 2001).
5. International Commission on Radiological Protection, "1990 Recommendations of the International Commission on Radiological Protection," ICRP Publication 60, *Annals of the ICRP* 21, nos. 1–3 (Pergamon Press, Oxford, 1991).
6. National Council on Radiation Protection and Measurements, *Limitation of Exposure to Ionizing Radiation*, NCRP Report No. 116 (Washington, DC: NCRP, 1993).
7. United Nations Scientific Committee on the Effects of Atomic Radiation, *Sources, Effects, and Risks of Ionizing Radiation*, UNSCEAR 1988 Report (New York: United Nations, 1988).
8. Donald A. Pierce, Yukiko Shimizu, Dale L. Preston, Michael Vaeth, and Kiyohiko Mabuchi, "Studies of the Mortality of Atomic Bomb Survivors," Report 12, Part 1. Cancer: 1950–1990," *Radiation Research* 146, 1996: 1–27.
9. Gen Roessler, "DS02: A New and Final Dosimetry System for A-Bomb Survivor Studies," Health Physics News XXXI, no. 6, 2003: 1, 4–6.
10. Samuel Glasstone and Philip J. Dolan, *The Effects of Nuclear Weapons* (Washington, DC: U.S. Department of Defense and Energy Research and Development Administration, 1977).
11. National Research Council, *Health Effects of Exposure to Low Levels of Ionizing Radiation, BEIR V*, Report of the Committee on the Biological Effects of Ionizing Radiations (Washington, DC: National Academy Press, 1990).
12. United Nations Scientific Committee on the Effects of Atomic Radiation, *Sources and Effects of Ionizing Radiation*, UNSCEAR 1994 Report (New York: United Nations, 1994).

13. M.P. Little, C. R. Muirhead, L.H.J. Goossens, B.C.P. Kraan, and R.M. Cooke, *Probabilistic Accident Consequence Uncertainty Analysis, Late Effects Uncertainty Assessment*, Report NUREG/CR-6555, EUR 16774 (Luxembourg: Office for Official Publications of the European Communities, 1997).

14. National Research Council, *The Effects on Populations of Exposure to Low Levels of Ionizing Radiation: 1980*, Report of the Committee on the Biological Effects of Ionizing Radiations (Washington, DC: National Academy Press, 1980).

15. Donald A. Pierce and Dale L. Preston, "Radiation-Related Cancer Risks at Low Doses Among Atomic Bomb Survivors," *Radiation Research* 154, 2000: 178–186.

16. Kenneth L. Mossman, Marvin Goldman, Frank Massé, William A. Mills, Keith J. Schiager, and Richard J. Vetter,"Health Physics Society Position Statement: Radiation Risk in Perspective," *HPS Newsletter* XXIV, no. 3, 1996: 3.

17. Jerome Puskin, "An EPA Response to Position Statement of the Health Physics Society: Radiation Risk in Perspective," *HPS Newsletter* XXIV, no. 5, 1996: 6.

18. *Cancer Facts & Figures—1987* (New York: American Cancer Society, 1987).

19. M. Ghiassi-nejad, S.M.J. Mortazavii, J. R. Cameron, A. Niroomand-rad, and P. A. Karam, "Very High Background Radiation Areas of Ramsar, Iran: Preliminary Biological Studies," *Health Physics* 82, no. 1, 2002: 87–93.

20. E. Cardis, E.S. Gilbert, L. Carpenter, et al., "Effects of Low Doses and Low Dose Rates of External Ionizing Radiation: Cancer Mortality among Nuclear Industry Workers in Three Countries," *Radiation Research* 142, 1995: 117–132.

21. Ethel S. Gilbert and Sidney Marks, "An Analysis of the Mortality of Workers in a Nuclear Facility," *Radiation Research* 79, 1979: 122–148.

22. T.D. Luckey, "Radiation Hormesis Overview," *Radiation Protection Management*, 16, no. 4, 1999: 22–34.

23. M. Pollycove, "The Issue of the Decade: Hormesis," *European Journal of Nuclear Medicine* 22, no. 5, 1995: 399–401.

24. John Cameron, "Is Radiation an Essential Trace Energy?" *Physics and Society* 30, no. 4, 2001: 14–16.

25. Zbigniew Jaworowski,"Radiation Risk and Ethics," *Physics Today* 52, no. 9, 1999: 24–29.

26. Keith F. Eckerman, Richard W. Leggett, Christopher B. Nelson, Jerome S. Puskin, and Allan C.B. Richardson, *Federal Guidance Report No. 13, Cancer Risk Coefficients for Environmental Exposure to Radionuclides,* Report EPA 402-R-99-001 (Washington, DC: Environmental Protection Agency, 1999).

27. United Nations Scientific Committee on the Effects of Atomic Radiation, *Sources and Effects of Ionizing Radiation*, UNSCEAR 1993 Report (New York: United Nations, 1993).

28. National Council on Radiation Protection and Measurements, *Risk Estimates for Radiation Protection*, NCRP Report No. 115 (Washington, DC: NCRP, 1993).

29. National Council on Radiation Protection and Measurements, *Uncertainties in Fatal Cancer Risk Estimates Used in Radiation Protection*, NCRP Report No. 126 (Washington, DC: NCRP, 1997).

30. United Nations Scientific Committee on the Effects of Atomic Radiation, *Hereditary Effects of Radiation*, UNSCEAR 2001 Report (New York: United Nations, 2001).

31. International Commission on Radiological Protection, "Recommendations of the International Commission on Radiological Protection," ICRP Publication 26, *Annals of the ICRP* 1, no. 3 (Oxford: Pergamon Press, 1977).

32. *Energy, U.S. Code of Federal Regulations*, title 10.
33. *Protection of Environment, U.S. Code of Federal Regulations*, title 40.
34. U.S. Environmental Protection Agency, "40 CFR Parts 141 and 142, National Primary Drinking Water Regulations; Radon-222; Proposed Rule," *Federal Register* 64, no. 211, 1999: 59246–59378.
35. National Research Council, *Risk Assessment of Radon in Drinking Water*, Report of the Committee on Risk Assessment of Exposure to Radon in Drinking Water (Washington, DC: National Academy Press, 1999).
36. International Commission on Radiological Protection, "Annual Limits of Intake of Radionuclides by Workers Based on the 1990 Recommendations," ICRP Publication 61, *Annals of the ICRP* 21, no. 4 (Oxford: Pergamon Press, 1991).
37. Keith F. Eckerman, Anthony B. Wolbarst, and Allan C.B. Richardson, *Federal Guidance Report No. 11, Limiting Values of Radionuclide Intake and Air Concentration and Dose Conversion Factors For Inhalation, Submersion, and Ingestion*, Report EPA 520/1-88-020 (Washington, DC: Environmental Protection Agency, 1988).
38. International Commission on Radiological Protection, "Age-Dependent Doses to Members of the Public from Intake of Radionuclides: Part 5, Compilation of Ingestion and Inhalation Dose Coefficients," ICRP Publication 72, *Annals of the ICRP* 26, no. 1 (Oxford: Pergamon Press, 1996).
39. International Commission on Radiological Protection, "A Report on Progress Towards New Recommendations: A Communication from the International Commission on Radiological Protection," *Journal of Radiological Protection* 21, 2001: 113–123.
40. E. Russell Ritenour and Richard A. Geise, "Radiation Sources: Medicine," in *Health Effects of Exposure to Low-Level Ionizing Radiation*, W.R. Hendee and F.M. Edwards, eds. (Bristol, UK: Institute of Physics Publishing, 1996): 433–467.
41. Ronald L. Kathren, *Radioactivity in the Environment: Sources, Distribution, and Surveillance* (New York: Harwood Academic Publishers, 1984).
42. Merril Eisenbud and Thomas Gesell, *Environmental Radioactivity from Natural, Industrial, and Military Sources*, 4th edition (San Diego, CA: Academic Press, 1997).
43. Jay H. Lubin and John D. Boice, Jr., "Lung Cancer Risk from Residential Radon: Meta-analysis of Eight Epidemiological Studies," *Journal of the National Cancer Institute* 89, no. 1, 1997: 49–57.
44. Daniel Krewski, et al., "A Combined Analysis of North American Case-Control Studies of Residential Radon and Lung Cancer: An Update," *Radiation Research* 158, 2002: 785–790.
45. Bernard L. Cohen, "Test of the Linear-No Threshold Theory of Radiation Carcinogenesis for Inhaled Radon Decay Products," *Health Physics* 68, 1995: 157–174.
46. Bernard L. Cohen, "Updates and Extensions to Tests of the Linear-No Threshold Theory," *Technology* 7, 2000: 657–672.
47. J. S. Puskin, "Smoking as a Confounder in Ecologic Correlations of Cancer Mortality Rates with Average County Radon Levels," *Health Physics* 84, no. 4, 2003: 526–532.
48. Bernard L. Cohen, "The Puskin Observation on Smoking as a Confounder in Ecologic Correlations of Cancer Mortality Rates with Average County Radon Levels," *Health Physics* 86, no. 2, 2004: 203–204.

49. National Council on Radiation Protection and Measurements, *Neptunium: Radiation Protection Guidelines*, NCRP Report No. 90 (Washington, DC: NCRP, 1988).

50. International Commission on Radiological Protection, "Limits for Intakes of Radionuclides by Workers: An Addendum," ICRP Publication 30, Part 4, *Annals of the ICRP* 19, no. 4 (Oxford: Pergamon Press, 1988).

51. Bernard L. Cohen, "Effects of Recent Neptunium Studies on High-Level Waste Hazard Assessments," *Health Physics* 44, no. 5, 1983: 567–569.

52. International Commission on Radiological Protection, "Limits for Intakes of Radionuclides by Workers," ICRP Publication 30, Part 2, *Annals of the ICRP* 4, no. 3/4 (Oxford: Pergamon Press, 1980).

53. International Commission on Radiological Protection, "Limits for Intakes of Radionuclides by Workers," ICRP Publication 30, Supplement to Part 2, *Annals of the ICRP* 5, no. 1–6 (Oxford: Pergamon Press, 1981).

54. National Research Council, Board on Radioactive Waste Management, *A Study of the Isolation System for Geologic Disposal of Radioactive Wastes* (Washington, DC: National Academy Press, 1983).

55. International Commission on Radiological Protection, "Age-Dependent Doses to Members of the Public from Intake of Radionuclides: Part 2, Ingestion Dose Coefficients." ICRP Publication 67, *Annals of the ICRP* 23, no. 3/4 (Oxford: Pergamon Press, 1993).

56. Electric Power Research Institute, *Yucca Mountain Total System Performance Assessment, Phase 3*, Report TR-107191 (Palo Alto, CA: EPRI, 1996).

57. U.S. Department of Energy, Office of Civilian Radioactive Waste Management, *Dose Conversion Factor Analysis: Evaluation of GENII-S Dose Assessment Methods*," Report ANL-MGR-MD-000002 REV 00 (October 1999).

5

Neutron Reactions

5.1 Overview of Nuclear Reactions

5.1.1 Neutron Reactions of Importance in Reactors

The term *nuclear reaction* is used very broadly to describe any of a wide array of interactions involving nuclei. Innumerable types of nuclear reactions can occur in the laboratory or in stars, but in considering energy from nuclear fission interest is limited almost entirely to reactions initiated by neutrons. Here, the important reactions are those that occur at relatively low energies, several million electron volts (MeV) or less, characteristic of neutrons produced in nuclear fission.[1] These reactions are elastic scattering, inelastic scattering, neutron capture, and fission.[2]

Elastic Scattering

In elastic scattering, a neutron and nucleus collide with no change in the structure of the target nucleus (or of the neutron). An example, the elastic scattering of neutrons on carbon-12 (^{12}C), is

$$n + {}^{12}C \rightarrow n + {}^{12}C.$$

Although the structure of the ^{12}C nucleus is unchanged, the neutron changes direction and speed, and the ^{12}C nucleus recoils. The total kinetic energy of

---

[1] See Section A.2.4 of Appendix A for the definition of the *electron volt* (eV) and the related terms keV (equal to 10^3 eV) and MeV (equal to 10^6 eV).

[2] This is not a fully exhaustive list. Other reactions are sometimes possible (e.g., those in which an alpha particle is produced), but the exceptions are only rarely of interest in nuclear reactors.

the system is unchanged, but some of the neutron's energy is transferred to the ^{12}C target nucleus.

Elastic neutron scattering can occur with any target nucleus, but in reactors it is of greatest importance when the target nucleus is relatively light and the loss of kinetic energy of the neutron is therefore relatively large. In such cases, elastic scattering is an effective means of reducing the energy of the neutrons without depleting their number. This is the process of *moderation* (see Section 7.2).

Inelastic Scattering

Inelastic scattering differs from elastic scattering in that the target nucleus is left in an excited state. It decays, usually very quickly, to the ground state, with the emission of one or more gamma rays. An example, the inelastic scattering of neutrons on ^{238}U, is

$$n + {}^{238}U \rightarrow n + {}^{238}U^* \rightarrow n + {}^{238}U + \gamma\text{'s},$$

where the asterisk indicates an excited state of ^{238}U. The total kinetic energy of the incident neutron equals the sum of the kinetic energies of the neutron and ^{238}U nucleus after the scattering plus the excitation energy of the ^{238}U nucleus.

Inelastic scattering is possible only if the incident neutron energy is greater than the excitation energy of the lowest excited state of the target nucleus. Heavy nuclei, such as uranium isotopes, have many excited states well below 1 MeV; thus, inelastic scattering can contribute to the initial slowing down of neutrons in a reactor.

Neutron Capture

In the first stage of many reactions, the neutron combines with the target nucleus to form an excited *compound nucleus*. The term *neutron capture* is usually restricted to those cases where the excited compound nucleus decays by the emission of gamma rays. For example, again taking ^{238}U as the target nucleus, neutron capture leads to the formation of ^{239}U:

$$\text{Formation:} \qquad n + {}^{238}U \rightarrow {}^{239}U^*,$$
$$\text{De-excitation:} \qquad {}^{239}U^* \rightarrow {}^{239}U + \gamma\text{'s}.$$

Here, the compound nucleus is ^{239}U*, where the asterisk again indicates an excited state. The number of gamma rays emitted in the de-excitation of ^{239}U* may vary from one to many, but the gamma rays themselves are not the product of chief interest. The important effect here is the transformation of ^{238}U into ^{239}U.

Neutron capture can occur for almost any target nucleus, although at widely differing rates. Neutron-capture reactions play two general roles in nuclear reactors: (1) They consume neutrons that might otherwise initiate fission, and (2) they transform nuclei into different nuclei, of higher mass number. The specific reaction indicated above is an example of role 2, and is of particular interest because it is the first step in the production of plutonium-239 (^{239}Pu) in reactors. The ^{239}Pu results from the beta decay of ^{239}U ($T = 23.5$ min) to neptunium-239 (^{239}Np), followed by the beta decay of ^{239}Np ($T = 2.355$ days) to ^{239}Pu. ^{239}Pu has a half-life of 2.41×10^4 years and therefore does not decay appreciably while in the reactor.

Nuclear Fission

Energy production in a nuclear reactor derives from fission. In a typical fission reaction, the excited compound nucleus divides into two main fragments plus several neutrons. The fission fragments have high kinetic energies, and it is this energy—quickly converted into heat— that accounts for most of the energy production in nuclear reactors or bombs (see Section 6.4 for more details on the energy release).

A typical fission reaction, here illustrated for a ^{235}U target, is of the general form

$$n + {}^{235}U \rightarrow {}^{236}U^* \rightarrow {}^{144}Ba + {}^{89}Kr + 3n.$$

The products in this example are barium-144 (atomic number $Z = 56$), krypton-89 ($Z = 36$), and three neutrons. Many other outcomes are also possible, always subject to the condition that the sums of the atomic numbers and atomic mass numbers of the products are the same as those of the initial system, 92 and 236, respectively.

Both the ^{144}Ba and ^{89}Kr nuclei are radioactive. For each, its formation is followed by a series of beta decays that continues with the successive emission of beta particles until a stable isobar is reached. In addition, gamma rays are emitted in the de-excitation of the two fission fragments (^{144}Ba and ^{89}Kr), assuming they are formed in excited states, as well as in the de-excitation of the products of the successive beta decays.

Fission is possible for only a very few target nuclei, the most important cases being isotopes of uranium and plutonium.

5.1.2 Reaction Cross Sections

Definition of Cross Section

The neutrons in a reactor may interact with a variety of target nuclei. Possible reactions are elastic scattering, capture, and, in some cases, fission or inelastic scattering. Which of these reactions occurs for an individual neutron is a matter of chance. The probability of each outcome is commonly couched in terms of the reaction *cross section* σ for the event.

The term *cross section* suggests an area, and the reaction cross section for a given reaction can be thought of as the effective cross-sectional or projected area of a nucleus as a target for that reaction. The so-called *geometric cross section* of a nucleus in fact corresponds to a physical area. It is defined as the area of the disk presented to incident particles, namely πR^2, where R is the effective radius of the nucleus.[3]

Although a nucleus does not have a sharp boundary, its density is rather uniform over its volume and falls off fairly rapidly at the exterior. Thus, it is meaningful to speak in terms of a nuclear radius. Nuclear densities are approximately the same for all nuclei, independent of nuclear mass, and the nuclear radius is therefore proportional to the cube root of the nuclear mass. For heavy nuclei such as ^{238}U, the nuclear radius is roughly 9×10^{-13} cm, and the geometric cross section is therefore in the neighborhood of 2.5×10^{-24} cm^2. In common notation, this is expressed as 2.5 barns (b), where $1 \text{ b} = 10^{-24} \text{ cm}^2 = 10^{-28} \text{ m}^2$.

Actual cross sections for specific reactions may be much larger or much smaller than the geometric cross section. Larger cross sections are difficult to understand if one thinks in terms of a classical geometric picture. They are a consequence of the wave properties of moving particles.[4] These properties are most important for neutrons at low energies, and they explain the large reaction cross sections for very low-energy neutrons.

Cross sections are appreciably smaller than the geometric cross section when the particular reaction is relatively improbable. If the reaction cannot occur at all (e.g., if the energy in the initial system is too low to create the final system), then the cross section is zero.

The cross section σ for a given reaction can be defined by relating it to the probability that the reaction occurs. For neutrons traversing a short distance δx, the probability δP of a reaction is given by

$$\delta P = N\sigma\delta x, \tag{5.1}$$

where N is the number of nuclei per unit volume. For a homogenous material, $N = \rho N_A/M$, where M is the atomic mass of the material, ρ is the density, and N_A is Avogadro's number.

The probability δP is an experimentally determined number for a given path length δx. It is the ratio of the number of neutrons undergoing the given reaction to the number of incident neutrons. The number of nuclei per unit

---

[3] The small but finite radius of the neutron can be taken into account in assigning a numerical magnitude to the effective radius.

[4] It would carry the discussion too far afield to discuss the relationship between wave and particle aspects of matter. An elementary introduction is given, for example, in Ref. [1].

volume N is in principle known or measurable. Therefore, Eq. (5.1) defines the cross section σ for that reaction.

Total Cross Section

The *total cross section* σ_T for neutron reactions is the sum of the individual cross sections. If we limit consideration to the reactions considered earlier, then

$$\sigma_T = \sigma_{\text{el}} + \sigma_a = \sigma_{\text{el}} + \sigma_{\text{in}} + \sigma_\gamma + \sigma_f, \tag{5.2}$$

where σ_T is the sum of the elastic scattering cross section (σ_{el}) and the *absorption cross section* (σ_a), and the absorption cross section is the sum of the cross sections for inelastic scattering (σ_{in}), capture (σ_γ), and fission (σ_f).

Mean Free Path

Neutrons do not travel a well-defined distance through material before undergoing interactions. They are removed exponentially, at separate rates for each of the possible reactions.[5] The average distance traversed by a neutron before undergoing a reaction of the type specified is the *mean free path* λ for the reaction. Numerically, it is equal to the reciprocal of $\delta P/\delta x$ in Eq. (5.1), namely

$$\lambda = \frac{1}{N\sigma}. \tag{5.3}$$

Thus, the mean free path is inversely proportional to the magnitude of the cross section. Although the mean free path can be defined for individual types of reactions, the concept is most suggestive when the cross section in Eq. (5.3) is the total cross section σ_T. If the medium has more than one nuclide, then the overall mean free path is found by replacing $N\sigma$ in Eq. (5.3) by the summation $\Sigma\, N_i\sigma_i$ over different nuclear species.

Competition Between Capture and Fission

For neutrons in a nuclear fuel, such as uranium or plutonium, the most interesting reactions are fission and capture. (Elastic scattering in the fuel changes the neutron energy only slightly, due to the high mass of the target nuclei, and has little importance.) If the goal is fission, as it is with ^{235}U or ^{239}Pu, capture has the effect of wasting neutrons. The ratio of the capture cross section to the fission cross section is therefore an important parameter. It is commonly denoted by the symbol α, defined as

---

[5] This is analogous to radioactive decay, as is seen more explicitly by rewriting Eq. (5.1) in the equivalent form: $-(dn/n) = N\sigma\, dx$, where n is the number of neutrons in a neutron beam and $\delta P = -dn/n$.

$$\alpha(E_n) = \frac{\sigma_\gamma(E_n)}{\sigma_f(E_n)}, \tag{5.4}$$

where the dependence of the cross sections and of α on neutron energy E_n is shown explicitly.

5.1.3 Neutron Reactions in Different Energy Regions

As discussed in more detail in Section 7.1.1, a nuclear chain reaction is sustained by the emission of neutrons from fissioning nuclei. The neutrons emitted in fission have a broad energy spectrum, with the peak lying between several hundred keV and several MeV and with a typical central energy in the neighborhood of 1 MeV. Reactions take place at all neutron energies E_n, from several MeV down to a small fraction of 1 eV. We are interested in the main features of the reaction cross sections over this entire region.

Such information is obtained from experimental measurements, and extensive measurements have been carried out. Here, we will consider results for neutron reactions with ^{235}U, partly because of the importance of ^{235}U reactions in nuclear reactors and partly to illustrate general aspects of neutron-induced reactions. The illustrative graphs of cross sections versus neutron energy in Figures 5.1, 5.3, and 5.4 are taken from a Brookhaven National Laboratory report [2].[6]

Experimentally measured cross sections for nuclear fission are shown in these figures for three somewhat distinct energy regions, in which the variation with energy of the cross sections differ significantly. There is no universal terminology to characterize the regions. We will here term them the *continuum* region, the *resonance* region, and the *low-energy* region. Although there are no clear boundaries, the continuum region corresponds to the higher neutron energies, from about 0.01 MeV to 25 MeV. The resonance region extends from about 1 eV to about 0.01 MeV. The low-energy region extends from zero energy up to about 1 eV.[7]

5.2 Cross Sections in the Resonance Region

5.2.1 Observed Cross Sections

We start with a discussion of the resonance region, because phenomena observed here are pertinent to understanding characteristics of the cross sections

---

[6] These figures are copied from the 1965 second edition of Report BNL-325. Although not the latest available data, they bring out in clear fashion the key aspects of the cross sections.

[7] In other usage, what is here termed the continuum region is called the *fast-neutron* region; what is here termed the low-energy region is divided into the *epithermal* and *thermal* regions.

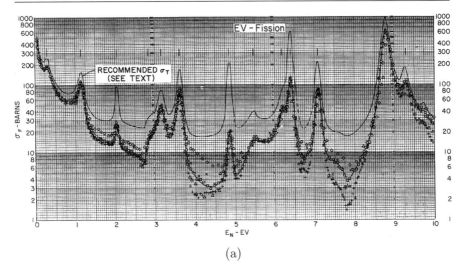

(a)

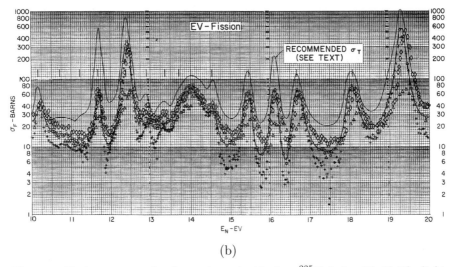

(b)

Fig. 5.1. Fission cross section for neutrons incident on ^{235}U, below 20 eV. The light curve gives the total cross section, σ_T. [From Ref. [2], p. 92-235-40.]

at higher and lower energies. Figure 5.1 shows fission cross sections σ_f in the lower part of this region for neutrons incident on ^{235}U. The cross section for neutron capture in ^{238}U is qualitatively similar. The striking aspect of Figure 5.1 is the rapid variations in cross section as a function of energy. For example, σ_f falls from 500 b at 19 eV to less than 10 b at 17.5 eV. The high peaks are due to resonances in the neutron cross section. Resonances in neutron cross sections are analogous to the resonance absorption of light in a gas at wavelengths characteristic of the gas.[8] If the neutron energy matches the energy difference between the ground state of ^{235}U and an excited state of ^{236}U, then there can be a large cross section for absorption.[9] The product of this absorption is a compound nucleus of ^{236}U. The number of possible resonance lines depends on the number of available states in the compound nucleus. In energy regions where the spacing of these states is not too close, there will be separate, discrete resonance lines, as seen in Figure 5.1.

In Figure 5.1, the total cross section σ_T is plotted along with σ_f. In the energy region covered, the only important reactions are fission, neutron capture, and elastic scattering. Thus, $\sigma_T = \sigma_f + \sigma_\gamma + \sigma_{el}$. Over this region, σ_{el} is fairly constant, at about 10 b, whereas the magnitudes of σ_γ and σ_f and their ratio vary considerably. We give approximate cross sections for one example, for the resonance peak at $E_n = 18.05$ eV: $\sigma_f = 70$ b, $\sigma_\gamma = 50$ b, and $\sigma_{el} = 10$ b, for a total of total cross section $\sigma_T = 130$ b.

5.2.2 Shape of the Resonance Peak

The measured width of resonance peaks, such as those seen in Figure 5.1, depends in part on how monoenergetic the neutron beam is (i.e., on how sharply defined it is in energy). However, if measurements are made of cross section versus energy with more and more closely monoenergetic beams, the peaks do not continue to get appreciably narrower. There is a certain inherent width to the resonance peak that has nothing to do with the measurement technique. It can be parameterized in terms of a *level width* Γ, characteristic of the resonance. Neutrons will be absorbed not only at some central energy, E_o, but over a band of energies, with the absorption strongest in the region from $E_o - \Gamma/2$ to $E_o + \Gamma/2$.

For a single isolated resonance, the shape of this absorption peak is given by the *Breit–Wigner formula*. The cross section for formation of the compound nucleus σ_c, which we take here to be approximately equal to the absorption cross section, σ_a, can be written in a simplified form as[10]

$$\sigma_a = \frac{C/v}{(E_n - E_o)^2 + \Gamma^2/4}, \tag{5.5}$$

where E_n is the kinetic energy of the incident neutron, E_o is the neutron energy at resonance, v is the velocity of the incident neutron, C is a constant, for a given resonance, and Γ is the width for the level, different for each level.

---

[8] This is the phenomenon that produces the so-called Fraunhofer dark lines in the solar spectrum.

[9] For both light and neutrons, strong absorption is not assured if the energies match, but it is made possible. In addition, other criteria must be satisfied, (e.g., those relating to the angular momenta of the states).

[10] See, e.g., Ref. [3] for simple forms of the Breit–Wigner formula as well as a more complete treatment.

When the neutron kinetic energy E_n is very close to the resonance energy E_o, the reaction cross section may be very large, up to hundreds of barns, and the neutron is likely to be absorbed. Neutrons that differ in energy by several tenths of an electron volt, on the other hand, can have much lower cross sections (and larger mean free paths) and are more likely to escape absorption.

5.2.3 Level Widths and Doppler Broadening

The level width Γ, as seen from Eq. (5.5), is the full width at half-maximum of the resonance peak; that is, when $E_n - E_o = \Gamma/2$ the magnitude of the cross section is one-half of its maximum value. There is a minimum width for any level, the *natural* width Γ_n.[11] A typical width for an isolated resonance for neutrons interacting with ^{235}U or ^{238}U nuclei is about 0.1 eV.

At nonzero temperatures, the actual level width is increased over the natural width by the thermal motion of the uranium nuclei in the fuel. This is called *Doppler broadening* because the broadening is due to the motion of the nuclei, analogous to the change in the observed frequency of sound or light due to the motion of the source or receiver. The Doppler broadening increases as the temperature of the medium increases, as illustrated schematically in Figure 5.2.

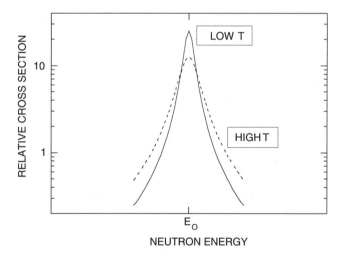

Fig. 5.2. Schematic representation of Doppler broadening of Breit–Wigner resonance at high temperature: Cross section as a function of neutron energy at high and low temperatures.

---

[11] The natural width is related to the mean lifetime τ for the decay of the compound nucleus by the expression $\Gamma_n = h/2\pi\tau$, where h is Planck's constant.

The effect of Doppler broadening is to decrease the neutron-absorption cross section at the center of a resonance peak and to increase the cross section in the wings of the peak. At the center of a resonance, E_o, the cross section is often sufficiently large that almost all the neutrons with energies very close to E_o are absorbed. For example, if the cross section for neutron absorption in ^{238}U is 500 b—a value that is exceeded for a considerable number of resonances—the mean free path of a neutron in uranium oxide (UO_2) fuel is under 0.1 cm. If the cross section is somewhat reduced at E_o, as happens with increased Doppler broadening, most of these neutrons are still absorbed.

However, in the wings of the resonance, at either side of the peak, the cross sections are low enough that many neutrons are not absorbed. At these energies, the increase in the cross section with increasing temperature (see Figure 5.2) means a greater chance of neutron absorption. Summed over all neutron energies, the overall effect of the increased Doppler broadening at higher temperatures is to increase the total resonance absorption in the fuel. This has important implications for reactor safety (see Section 14.2.1).

5.3 Cross Sections in the Continuum Region

For neutrons with energies in excess of tens of kilovolts, the accessible states in the compound nucleus are so numerous that their spacing is small compared to their width. Thus, the levels overlap. A compound nucleus is still formed, but it involves a continuum of overlapping states. The cross section varies relatively smoothly with energy, without the rapid changes characteristic of the resonance region. The cross section for fission in ^{235}U is displayed for this region in Figure 5.3.

For 1-MeV neutrons incident upon ^{235}U, the main decay modes of the ^{236}U compound nucleus are fission, emission of a neutron to an excited state of ^{235}U (inelastic scattering), and γ-ray emission (capture). The cross sections for these reactions at 1 MeV are approximately $\sigma_f = 1.2$ b, $\sigma_{in} = 1.1$ b, and $\sigma_\gamma = 0.11$ b, respectively [2]. The sum of the cross sections for the possible reactions (excluding elastic scattering) is 2.4 b, close to the geometric cross section cited in Section 5.1.2 for the neighboring nucleus, ^{238}U.

At somewhat lower energies, the relative importance of the different decay modes changes. For example, at $E_n = 0.01$ MeV, $\sigma_f = 3.2$ b and $\sigma_\gamma = 1.15$ b. The cross section for inelastic scattering is very small, in part because of the lack of suitable low-lying excited states of ^{235}U. The total absorption cross section $\sigma_f + \sigma_\gamma$ is 4.4 b, well above the geometric cross section.

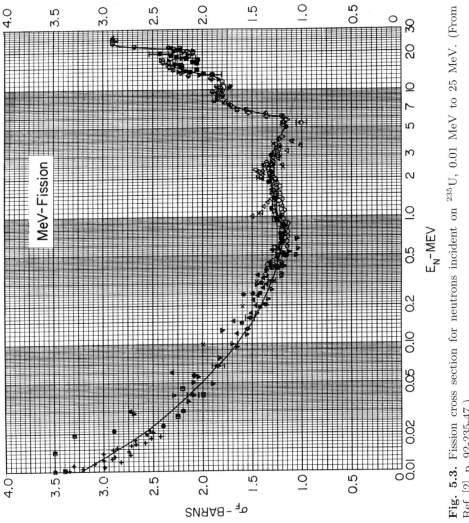

Fig. 5.3. Fission cross section for neutrons incident on ^{235}U, 0.01 MeV to 25 MeV. (From Ref. [2], p. 92-235-47.)

5.4 The Low-Energy Region

5.4.1 Low-Energy Region and the $1/v$ Law

For slow neutrons, below 1 eV, one is interested in variations of cross section with energy over a very small energy region. For example, from 0.1 to 0.025 eV the neutron velocity drops by a factor of 2, but the energy change may be small compared to the width of a resonance, typically of the order of 0.1 eV.[12] The behavior of the cross section over so narrow a region is often dominated by the tail of a single resonance. If the denominator in Eq. (5.5) does not change appreciably over this interval, the cross section is proportional to $1/v$, where v is the neutron velocity.

The $1/v$ dependence is contingent upon the presence of an appropriately located resonance (or resonances), and it characterizes many, although not all, neutron-absorption cross sections at very low energies. It is exhibited in Figure 5.4, which shows the fission cross section for ^{235}U below 1 eV. Were the cross section σ itself plotted, it would exhibit a steady rise as E_n falls, roughly proportional to $1/v$. Instead, the curve in Figure 5.4 is plotted with the ordinate as $\sigma_f \times E^{0.5}$. There is not much more than a factor of 2 difference between the highest and lowest value of the ordinate in Figure 5.4, although the energy varies by a factor of almost 100 and the velocity by a factor of almost 10. Thus, the $1/v$ dependence of σ holds reasonably well. If it held perfectly, the curves in Figure 5.4 would be horizontal lines.

The cross section itself can be found from Figure 5.4 by dividing the observed ordinate by $\sqrt{E_n}$, where E_n is the neutron energy in electron volts. At 1 eV, $\sigma_f = 65$ b, whereas at 0.0253 eV, $\sigma_f = 92/\sqrt{0.0253} \doteq 580$ b.

5.4.2 Thermal Neutrons

In many reactors, a succession of elastic collisions with the nuclei of the so-called moderator (see Section 7.2) reduces the neutron energies to lower and lower values as the neutrons impart some of their kinetic energy to the target nuclei. Ultimately, the thermal motion of the nuclei in the reactor cannot be neglected, and the neutron energies then approach a Maxwell–Boltzmann distribution (qualitatively, a "bell-shaped" curve), characteristic of the temperature of the fuel and moderator. This is the *thermal* region.[13]

---

[12] Although the discussion has been couched in terms of a single resonance, a $1/v$ dependence can also occur with several resonances contributing to the cross section.

[13] Thermalization of the neutrons is not complete because absorption of neutrons removes some of them before equilibrium is reached. Thus, the actual velocity or energy distribution of the neutrons has an excess high-energy tail and a suppression of low-energy neutrons, as compared to an exact Maxwell–Boltzmann distribution [4, p. 332 ff.].

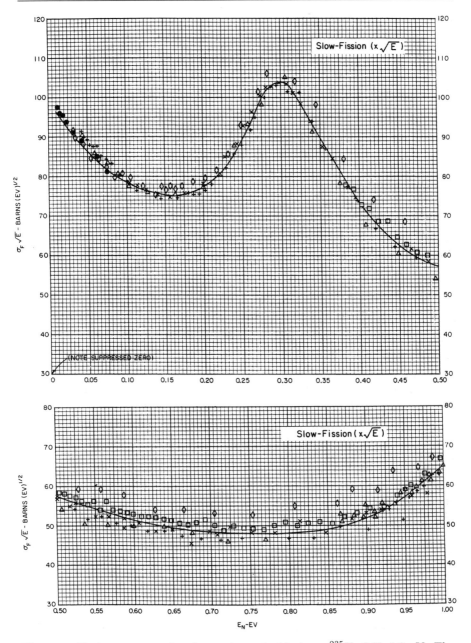

Fig. 5.4. Fission cross section for neutrons incident on ^{235}U, 0.01–1.0 eV. The ordinate is the product $\sigma_f E^{0.5}$, where E is in eV. [From Ref. [2], p. 92-235–32.]

So-called thermal neutron cross sections have been measured in the laboratory using monoenergetic neutrons with a velocity v of 2200 m/s. At this velocity, $E_n = 0.0253$ eV. This is the energy corresponding to the most probable velocity of a Maxwell–Boltzmann velocity distribution at a temperature of $20°$C.[14] Thus, v here characterizes thermal neutrons at $T = 20°$C. The actual neutrons in a reactor are characterized by higher temperatures, roughly $300°$C, and are distributed in energy rather than having a single value. Nonetheless, use of these nominal thermal energy cross sections gives quite accurate results.

This fortuitously simple result follows from the $1/v$ dependence of the cross section. The rate at which reactions occur is proportional to the product σv of the cross section and the velocity.[15] Over the region where σ is proportional to $1/v$, the product σv is a constant, and the number of reactions per second is independent of the neutron energy. The measured result at 2200 m/s can then be used to characterize the full thermal region, if approximate results suffice. Thus, in subsequent chapters, we will follow what is standard practice in the nuclear reactor literature, namely to characterize thermal neutron interactions in the fuel in terms of the cross sections at a single energy.

Departures from the $1/v$ dependence of neutron cross section and from a thermal distribution of neutron energies cannot be neglected if accurate calculations are desired. Correction factors to take into account deviations from the idealized $1/v$ dependence are known as Westcott parameters, and tables of these parameters are found in standard nuclear engineering references (see, e.g., Refs. [5] and [6]). However, these corrections are not large in the cases of greatest interest and are often ignored in qualitative discussions. In nonthermal reactors, where the $1/v$ dependence does not describe the neutron cross sections, none of these simplifications is relevant and it is necessary to examine $\sigma(E)$ in detail.

References

1. David Halliday, Robert Resnick, and Kenneth S. Krane, *Physics, Volume Two, Extended Version*, 4th edition (New York: Wiley, 1992).
2. J.R. Stehn, M.D. Goldberg, R. Wiener-Chasman, S.F. Mughabghab, B.A. Magurno, and V.M. May, *Neutron Cross Sections, Volume III, Z = 88 to 98*, Report No. BNL-325, 2nd edition, supplement No. 2 (Upton, NY: Brookhaven National Laboratory, 1965).

---

[14] Note: At 0.0253 eV, $T = E_n/k = 0.0253(1.602 \times 10^{-19})/1.381 \times 10^{-23} = 293$ K $= 20°$C, where k is Boltzmann's constant.

[15] We are speaking as if the target nuclei are at rest. This is not true at thermal equilibrium, where all nuclei are in motion. However, if the target has high atomic mass M, such as for U or Pu, it is a reasonably good description, because at fixed temperature, the average velocity of nuclei varies as $M^{-0.5}$. For targets of any mass, the present analysis is valid if v is interpreted to be the *relative* velocity of the neutron and target.

3. Emilio Segrè, *Nuclei and Particles*, 2nd edition (Reading, MA: W.A. Benjamin, 1977).

4. Alvin M. Weinberg and Eugene P. Wigner, *The Physical Theory of Neutron Chain Reactors* (Chicago: University of Chicago Press, 1958).

5. Manson Benedict, Thomas H. Pigford, and Hans Wolfgang Levi, *Nuclear Chemical Engineering*, 2nd edition, (New York: McGraw-Hill, 1981).

6. John R. Lamarsh, *Introduction to Nuclear Engineering*, 2nd edition (Reading, MA: Addison-Wesley, 1983).

6

Nuclear Fission

6.1 Discovery of Fission

Prior to the actual discovery of the neutron in 1932, it had long been suspected, in part due to a suggestion by Ernest Rutherford in 1920, that a heavy neutral particle existed as a constituent of the atomic nucleus.[1] The discovery of the neutron, as with many of the early discoveries in nuclear physics, was accidental. In the bombardment of beryllium by alpha particles from a naturally radioactive source, a very penetrating radiation had been observed in experiments begun in 1928. In view of its penetrating power, it was first thought to be gamma radiation, but the radiation was found to eject energetic protons from paraffin, which would not have been possible with gamma rays.

Although in retrospect the explanation is obvious, it was not until early 1932 that James Chadwick demonstrated that the radiation was made up of particles (already recognized to be neutral) with a mass close to that of a proton. The ejected protons were the recoil products of elastic scattering between the neutron and hydrogen nuclei in the paraffin. Thus, the previously suspected neutron had been found. The reaction responsible for the neutron production was

$$^4\mathrm{He} + {}^9\mathrm{Be} \rightarrow {}^{12}\mathrm{C} + \mathrm{n}.$$

With a source of neutrons readily available, it was natural to explore what reactions the neutrons might induce in collisions with nuclei of various elements. There was particular interest in doing this because, although protons and alpha particles of the energies then obtainable had too little kinetic energy to overcome the Coulomb repulsion of the nuclei of heavy elements, there

---

[1] For a brief account of this history, see, e.g., Refs. [1] and [2].

would be no Coulomb force acting upon the neutrons.[2] Enrico Fermi in Italy quickly became the leader in this work, undertaking a systematic program of bombardment by neutrons of nuclei throughout the periodic table.

The bombardment of uranium, undertaken in 1934, proved to be particularly interesting because many new radioactive products were discovered. Fermi at first believed that this was due to the formation of a transuranic element (atomic mass number $A = 239$) through neutron capture in ^{238}U, followed by a chain of radioactive decays. In fact, when the decision was made to award the 1938 Nobel Prize in physics to Fermi for his discoveries of radioactivity produced with neutrons, it was still thought that the most interesting of these were transuranic elements and their radioactive progeny.[3] Efforts were made to isolate the hypothesized new elements chemically, but it proved impossible to demonstrate properties that differed from those of known lighter elements.

Finally, in 1938, Otto Hahn and Fritz Strassmann in Germany demonstrated that a residue of neutron bombardment of uranium that had previously been thought to be radium (atomic number $Z = 88$), a presumed decay product of the radioactive chain, was in fact barium ($Z = 56$). This result was communicated by them to Lise Meitner, a former colleague of Hahn who was then a refugee in Sweden, and she and her nephew Otto Frisch formulated the first overall picture of the fission process. From known nuclear binding energies, it was clear that there should be a large accompanying energy release, and Frisch quickly demonstrated this release experimentally. Thus, fission was hypothesized and demonstrated in late 1938 and early 1939.

The potential practical implications of fission were immediately recognized, and work on fission quickly intensified in Europe and the United States. The first detailed model of the fission process was developed by Niels Bohr and John Wheeler, during a visit by Bohr to Princeton University in early 1939—the so-called liquid drop model. The essential features of this model are still accepted.

Among the early insights from this work was the explanation of the puzzling fact that fission was copiously produced in natural uranium by both very slow neutrons (under about 0.1 eV in energy) and moderately fast neutrons (above about 1 MeV), but not to any appreciable extent by neutrons of intermediate energy (say, 0.5 MeV). Bohr recognized, in what Wheeler suggests was a rather sudden inspiration, that the slow neutrons cause fission in the relatively rare isotope ^{235}U, for which the fission cross section is high.[4] Fast neutrons (above 1 MeV) are required to produce fission in the abundant isotope ^{238}U.

---

[2] "Coulomb force" is the standard name for the electrical force between two charged particles, which is described quantitatively by Coulomb's law.

[3] The suggestion made in 1934 by the German chemist Ida Noddack that lighter elements were being produced was largely ignored, because she lacked the eminence and the unambiguous evidence to gain credence for her radical suggestion.

[4] A brief account of this history is given in Ref. [3].

6.2 Simple Picture of Fission

6.2.1 Coulomb and Nuclear Forces

The energy available for fission (or other processes) within the nucleus is the result of the combined effects of the Coulomb force between the protons, which is repulsive, and the nuclear force between nucleons (neutrons or protons), which is attractive. In a stable nucleus, the nuclear force is sufficiently strong to hold the nucleus together. In alpha-particle emitters, the Coulomb repulsion causes eventual decay, but on a slow time scale. Fission represents a more extreme "victory" for the Coulomb force.

The fact that fission occurs for heavy nuclei follows from the dependence of the forces and the corresponding potential energies on nuclear charge and mass. For the repulsive Coulomb force, the potential energy is positive and increases as Z^2/R, where the nuclear radius R is proportional to $A^{1/3}$. The atomic number Z is roughly proportional to the atomic mass number, A, and the Coulomb potential energy therefore rises roughly as $A^{5/3}$.

For the attractive nuclear force, the potential energy is negative. In light of the saturation of nuclear forces, which limits the number of nucleons with which each nucleon interacts, the magnitude of the nuclear potential energy is proportional to A. Thus, the ratio of the magnitudes of the Coulomb and nuclear potential energies rises with increasing A and, eventually, the Coulomb force dominates. This limits the maximum mass of stable nuclei.

The existence of ^{235}U and ^{238}U in nature is evidence that the balance between their nuclear and Coulomb forces is not such as to cause instability, beyond alpha-particle decay at a slow rate.[5] However, the fact that no elements of higher atomic number are found in nature suggests that these uranium isotopes are "vulnerable" and that fission is possible if they are sufficiently excited. As described in the liquid drop model, the excitation can take the form of a shape oscillation, in which the nucleus deforms into a dumbbell shape (where the two ends of the dumbbell are usually not of equal size). The Coulomb repulsion between these two parts may be sufficient to overcome the remaining attraction from the nuclear force, and the parts move further apart, culminating in fission.

6.2.2 Separation Energies and Fissionability

The striking qualitative fact that thermal neutrons can produce fission in ^{235}U but not in ^{238}U is explained by rather small differences in the excitation energies of the nuclei ^{236}U and ^{239}U, produced when neutrons are absorbed in ^{235}U and ^{238}U, respectively. Consider, for example, the neutron-capture process in which the excited ^{236}U nucleus is formed, followed by gamma-ray

---

[5] In addition, spontaneous fission occurs for both ^{235}U and ^{238}U, but at rates that are very slow; the respective half-lives are are 1.0×10^{17} and 8×10^{13} years.

emission to bring ^{236}U to its ground state:

$$n + {}^{235}U \rightarrow {}^{236}U^* \rightarrow {}^{236}U + \gamma\text{'s.}$$

Here, the energy carried off by the gamma rays is equal to the excitation energy E of the excited compound nucleus ^{236}U*. It is given by the difference in the masses of the initial and final constituents.[6] Assuming the kinetic energies to be zero (or negligible), the excitation energy E equals the energy required to remove a neutron from ^{236}U to form ^{235}U $+ n$, the so-called *separation energy* S.

In terms of the atomic masses M and the mass of the neutron m_n, these energies are

$$E = S = [M({}^{235}U) + m_n - M({}^{236}U)] c^2,$$

where c is the velocity of light. Expressing masses in atomic mass units (u) and noting that 1 u corresponds to 931.49 MeV,

$$E = S = (235.043923 + 1.008665 - 236.045562) \times 931.49 = 6.54 \text{ MeV.} \quad (6.1)$$

The meaning of Eq. (6.1) can be stated as follows: (a) The ^{236}U excitation energy E following the capture of a thermal neutron in ^{235}U is 6.54 MeV, or (b) to separate a neutron from a ^{236}U nucleus (originally in its ground state) and form $n + {}^{235}$U, an energy must be supplied that is at least as great as the separation energy S of 6.54 MeV.

For the capture reaction described above, the excitation energy is carried off by gamma rays. However, the absorption of thermal neutrons can, in some cases, lead to fission. For ^{235}U, in particular, fission is the dominant process. In general, thermal neutron absorption leads to fission if the excitation energy is sufficiently high. Relevant excitation energies are listed in Table 6.1 for thermal neutron absorption in a number of nuclei of interest.

Table 6.1. Excitation energies E* for thermal neutron capture.

Target Nucleus	Compound Nucleus	Excitation Energy (MeV)	Fissile Target?
^{232}Th	^{233}Th	4.79	No
^{233}U	^{234}U	6.84	Yes
^{235}U	^{236}U	6.54	Yes
^{238}U	^{239}U	4.81	No
^{239}Pu	^{240}Pu	6.53	Yes
^{240}Pu	^{241}Pu	5.24	No
^{241}Pu	^{242}Pu	6.31	Yes

---

[6] In the standard nuclear terminology, this is the Q *value* of the reaction.

If an appreciable fission yield is produced by thermal neutrons incident upon a target nucleus, that nuclear species is termed *fissile*. It is seen in Table 6.1 that the fissile nuclei are all odd-A nuclei, and the resulting compound nuclei all have excitation energies exceeding 6.3 MeV. For the nonfissile nuclei, the excitation energies are about 5.0 MeV. However, some of the nonfissile nuclei can play an important role in nuclear reactors. Neutron capture in ^{232}Th and ^{238}U leads, with intervening beta decays, to the production of the fissile nuclei ^{233}U and ^{239}Pu. Therefore, ^{232}Th and ^{238}U are termed *fertile*.

The question arises: How much excitation energy is required for fission to occur? An alternate means of exciting nuclei is by bombardment with gamma rays. If fission then occurs, it is called *photofission*, to distinguish the process from the more common neutron-induced fission. It was found early that the thresholds for photofission in uranium and thorium isotopes lie in the range from 5.1 to 5.4 MeV [4, p. 492]. This means that an excitation energy between about 5.1 and 5.4 MeV is required for photofission. For all fissile nuclei, the excitation energies following neutron capture are higher than the photofission threshold. For the nonfissile nuclei, they are below the threshold.

The association between high separation energies and odd-A target nuclei (i.e., even-A compound nuclei) results from a special property of nuclear forces, the pairing energy. In nuclei where both Z and A are even (e.g., ^{234}U, ^{236}U, and ^{240}Pu), there are even numbers of both protons and neutrons. There is a stronger binding for members of a neutron pair than for unpaired neutrons (as in ^{239}U, for example). Thus, the neutron separation energy is high for these even–even nuclei, and, correspondingly, their excitation energy is high when they are formed by neutron capture. The fact that all fissile target nuclei are odd-A nuclei is not an accident, but rather a manifestation of pairing interactions in nuclei.

6.2.3 Fission Cross Sections with Fast and Thermal Neutrons

For orientation purposes, the cross sections for fission in ^{235}U, ^{238}U, and ^{239}Pu are shown in Figure 6.1. (The data are extracted from detailed plots given in Ref. [5].) In the thermal region, the cross sections in ^{235}U and ^{239}Pu are hundreds of barns. In general, they then fall with increasing energy, with very rapid fluctuations in the resonance region. In the 1–10 MeV region, the fission cross sections are in the vicinity of 1 or 2 b for these nuclei.

The cross section for ^{238}U fission is negligible at low neutron energies. For neutrons with energies of several electron volts, σ_f is typically in the neighborhood of 10^{-5}–10^{-4} b [5], and it is still under 0.02 b at $E_n = 1.0$ MeV. However, it then rises rapidly to a plateau of about 0.6 b, which extends from 2 to 6 MeV. In this region, σ_f in ^{238}U is about one-half the magnitude of σ_f in ^{235}U. Thus, although ^{238}U is not "fissile" in the sense that the term is used—namely having a significant fission cross section for thermal neutrons— the ^{238}U fission cross sections becomes appreciable for neutron energies above about 1 MeV.

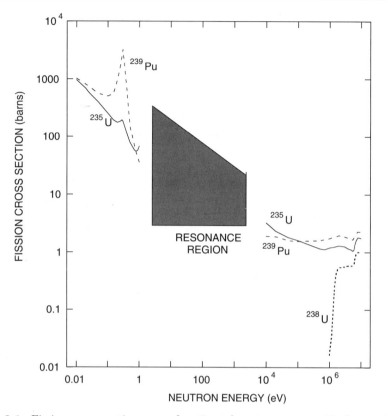

Fig. 6.1. Fission cross sections as a function of neutron energy E_n for neutrons on ^{235}U, ^{238}U, and ^{239}Pu. The shaded region is a schematic of the resonance region where the cross section varies rapidly. [The top and bottom boundaries of the shaded region are *not* limits on the peaks and valleys of the resonance cross sections, nor are there sharp boundaries at low and high energies.]

6.3 Products of Fission

6.3.1 Mass Distribution of Fission Fragments

The fission reaction, again taking the example of ^{235}U, can be written in the somewhat general form[7]

$$\text{n} + {}^{235}\text{U} \rightarrow {}^{236}\text{U}^* \rightarrow (A_1, Z_1) + (A_2, Z_2) + N\text{n} + \gamma\text{'s}.$$

This expression represents the fission of ^{236}U into two fragments (with atomic mass numbers A_1 and A_2) and N neutrons. From conservation of charge and of number of nucleons, $Z_1 + Z_2 = 92$ and $A_1 + A_2 + N = 236$. N is a

---

[7] Here, we ignore relatively rare additional particles, most notably alpha particles.

small number, typically 2 or 3. If $A_1 = A_2$, or nearly so, the fission process is called *symmetric fission*. Otherwise, it is *asymmetric fission*. The relative yields of fission fragment nuclei of different atomic mass number are plotted in Figure 6.2.[8]

As seen in Figure 6.2, thermal fission of ^{235}U leads overwhelmingly to asymmetric fission.[9] The fission yield is dominated by cases where one fragment has a mass number A between about 89 and 101, and the other has a mass number between about 133 and 144. The fission yields, namely the fraction of the events that produce nuclei of a given A, lie between about 5% and 7% in this region, with a single anomalously high value at $A = 134$. For symmetric fission (A of about 116), the fission yield at a given value of A is only about 0.01%. The integral of the area under the curve of Figure 6.2 is 200%, as there are two fragments per fission.

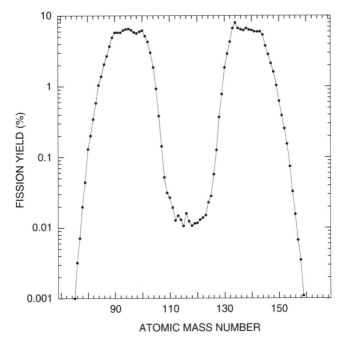

Fig. 6.2. Yield of fission fragments as a function of atomic mass number A for thermal fission of ^{235}U (in percent per fission).

---

[8] Data for Figure 6.2 are the yields after delayed neutron emission, as listed in Ref. [6, Table I].

[9] This fission process for neutrons incident on ^{235}U is commonly referred to as "fission of ^{235}U," although, of course, the fissioning nucleus is ^{236}U. Usually, this does not lead to ambiguity, because the meaning is clear from the context.

Although in most cases the massive products from the breakup of the fissioning nucleus are the two fission fragments and neutrons, there is also a small possibility of emission of other nuclei, especially alpha particles. Alpha-particle emission, with energies up to 20 MeV, occurs in 0.2% of fission events [7, p. 590]. While of some interest in its own light, this contribution is too small to be of any importance in considering nuclear reactors.

The fission fragment yields for other fissile targets, in particular ^{233}U and ^{239}Pu, are similar to those in Figure 6.2, although the minima are not quite as deep and there are small displacements in A due to the differences in nuclear mass number. Fast fission yields also have the same general character, for both fissile and fertile targets [8]. Again, asymmetric fission dominates, with peak yields primarily in the range between 5% and 7%, although there are some exceptions and differences in detail.

6.3.2 Neutron Emission

Number of Neutrons as a Function of Neutron Energy

The average number of neutrons emitted in fission ν is a crucial parameter in establishing the practicality of a chain reaction. The number of neutrons varies from event to event, ranging from 0 to about 6 [7, p. 587]. Typically, the number is 2 or 3. Because of its importance, the magnitude of ν has been determined with high precision. It increases with neutron energy and can be described approximately by the relations [9, p. 61]

$$^{235}\text{U}: \nu = 2.432 + 0.066E, \qquad \text{for } E < 1 \text{ MeV}$$
$$^{235}\text{U}: \nu = 2.349 + 0.15E, \qquad \text{for } E > 1 \text{ MeV}$$
$$^{239}\text{Pu}: \nu = 2.874 + 0.138E, \qquad \text{for all } E,$$

where the neutron energy E is in MeV. It is seen that ν is significantly higher for ^{239}Pu than for ^{235}U (e.g., 3.0 versus 2.5, at 1 MeV).

Energy Spectrum of Neutrons

The neutrons are emitted with a distribution of energies, with a relatively broad peak centered below 1 MeV. Several approximate semiempirical expressions are used to represent the actual neutron spectrum. One, that is in a form which lends itself easily to calculation, is (see, e.g., Ref. [7, p. 586])

$$f(E) = 0.770\sqrt{E}e^{-0.775\,E}, \tag{6.2}$$

where E is the neutron energy in MeV and $f(E)\,dE$ is the fraction of neutrons emitted with energies in the interval E to $E + dE$. An alternative expression for $f(E)$ has the slightly different form (see, e.g., Ref. [10, p. 75])

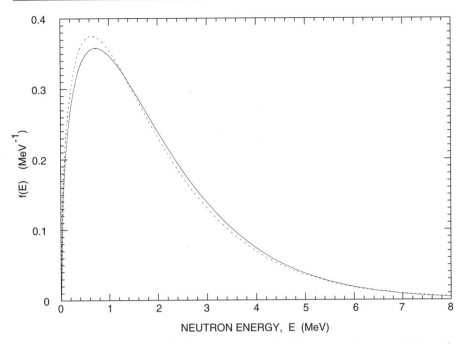

Fig. 6.3. Energy spectrum $f(E)$ of neutrons emitted in thermal neutron fission of ^{235}U. The dotted curve corresponds to Eq. (6.2) and the solid curve to Eq. (6.3).

$$f(E) = 0.453\, e^{-1.036E} \sinh[(2.29E)^{0.5}], \qquad (6.3)$$

where again E is in MeV.

The spectra corresponding to Eqs. (6.2) and (6.3) are plotted in Figure 6.3. It is seen that there is little difference between the curves for the two expressions. In both cases, there is a peak at roughly 0.7 MeV. At $E = 1$ MeV, $f(E)$ is only slightly below its peak value, and at $E = 2$ MeV, it is still above 60% of the peak value. The median neutron energy is about 1.5 MeV. There is a substantial tail extending to higher energies, and the "average" neutron energy is commonly taken to be about 2 MeV.

Delayed Neutrons

Most of the neutrons produced in fission are emitted promptly, essentially simultaneously with the fission itself. However, some neutrons are emitted from nuclei produced in the beta decay of the original fission fragments. These are called *delayed neutrons*. The delay arises from the beta-decay process and has a time scale determined by the beta-decay half-life of the precursor to the neutron emitter. The neutron emission itself follows virtually immediately after the beta decay.

An example of delayed neutron emission occurs at $A = 87$. Bromine 87 (^{87}Br, $Z = 35$) is formed in about 2% of ^{235}U fission events. Its beta decay, with a half-life of 56 s, proceeds to a number of different states of krypton-87 (^{87}Kr, $Z = 36$). In 2.3% of the cases, the resulting ^{87}Kr state decays by neutron emission to ^{86}Kr. In aggregate, for every 100 ^{235}U fission events, there are 0.05 delayed neutrons with a ^{87}Br precursor. Considering all delayed neutron paths, with other precursors, there are 1.58 delayed neutrons per 100 fissions [7, p. 589]. Given an average of 2.42 neutrons per fission event, this means that the fraction of fission neutrons that are delayed, conventionally designated as β, is about 0.0158/2.42. Thus, for ^{235}U about 0.65% of fission neutrons are delayed neutrons. The precursors for these neutrons have half-lives ranging from 0.2 to 56 sec, with the largest group having a half-life of about 2 s.

This might seem to be a small point—either a small detail introduced for completeness or a minor nuisance. It is neither. As discussed in Section 7.3.2, the existence of these delayed neutrons is crucial for the stable operation of nuclear reactors.

6.3.3 Decay of Fission Fragments

Gamma Ray Emission from Excited Nuclei

If the initial fission fragments are formed in excited states, as is usually the case, they will decay by gamma-ray emission to the ground state before beginning a beta-decay chain. Although the half-lives for beta decay are typically seconds or longer, gamma decay usually occurs within about 10^{-14} s of the fission event. The gamma rays have an average energy of about 1 MeV, and, overall, about 7 MeV are emitted per fission event in the form of gamma rays [7, p. 590]. In a reactor, this energy is almost entirely recovered because the gamma rays are absorbed in a distance small compared to the dimensions of the reactor.

Beta Decay

For the stable nuclei that exist in nature with atomic mass number up to $A = 40$, the ratio Z/A is close to 1/2. Beyond this, as one goes to heavier nuclei, Z/A gradually decreases. In fission, aside from the few emitted neutrons, the average value of Z/A of the initial fragments must be the same as that of the fissioning nucleus. Thus, for their mass region, the fission fragments will be proton poor, or equivalently, neutron rich. For example, if we consider the nucleus at $A = 234/2 = 117$, the stable nucleus is tin-117 (^{117}Sn, $Z = 50$), whereas the fragments of symmetric fission of ^{235}U (with the emission of two neutrons) are two palladium-117 (^{117}Pd, $Z = 46$) nuclei. Each of the ^{117}Pd fragments undergoes four beta decays to reach the stable ^{117}Sn nucleus.

As a more typical example, where the fission is asymmetric, we can consider the following fission event:

$$n + {}^{235}U \rightarrow {}^{144}_{56}Ba + {}^{89}_{36}Kr + 3n.$$

This is followed by the beta-decay chains (half-lives in parentheses):[10]

$$A = 144: Ba(11\ s) \rightarrow La(41\ s) \rightarrow Ce(285\ d) \rightarrow Pr(17\ m) \rightarrow Nd$$
$$A = 89: Kr(3\ m) \rightarrow Rb(15\ m) \rightarrow Sr(51\ d) \rightarrow Y$$

The "final" products are the (virtually) stable nuclei ${}^{144}_{60}Nd$ and ${}^{89}_{39}Y$.

Some qualifications should be made on the designations "final" and "stable." Neodymium-144 (${}^{144}Nd$) is not truly stable. It has a half-life of 2.3×10^{15} years. Of course, this is stable enough for any practical purposes. A much more significant point is illustrated in the $A = 144$ chain by the role of cerium-144 (${}^{144}Ce$), with a half-life of 285 days. Only part of the ${}^{144}Ce$ decays before the fuel is discharged from the reactor, which, on average, takes place 1 or 2 years after the fission event. Thus, only part of the ${}^{144}Ce$ decay energy should be included in considering the energy budget of the reactor. Further, the remaining activity will appear in the radioactivity inventory of the spent fuel. More generally, although most of the radioactive fission products decay quickly enough to have disappeared before the spent fuel is removed, there are some longer-lived species that remain. This is the origin of the nuclear waste disposal problem.[11]

6.4 Energy Release in Fission

6.4.1 Energy of Fission Fragments

The energy release in fission can be calculated in two ways: (1) from the mass differences and (2) from the Coulomb potential. Method (1) gives the more accurate result. Method (2) is the more physically instructive in terms of mechanism. We will perform the calculation, using both methods, for the same case that we have been citing:

$$n + {}^{235}U \rightarrow {}^{144}_{56}Ba + {}^{89}_{36}Kr + 3n.$$

In terms of the atomic masses, the energy release is

$$Q = [M({}^{235}U) - M({}^{144}Ba) - M({}^{89}Kr) - 2m_n]c^2 = 173\ \text{MeV},$$

---

[10] Here, we use the common abbreviations s = seconds, m = minutes, and d = days.

[11] Actually, nuclides with half-lives of less than 1 year, such as ${}^{144}Ce$, are not the main problem, because they are essentially gone after a decade or two. It is the longer-lived nuclides, with half-lives of many years, that create the demand for repositories that will be safe for thousands of years.

where the atomic masses M of ^{235}U, ^{144}Ba, ^{89}Kr, and the neutron are 235.0439 u, 143.9229 u, 88.9176 u, and 1.0087 u, respectively.

Thus, on the basis of mass differences, the energy carried by the fission fragments and neutrons is 173 MeV. For the fission fragments alone, it is about 167 MeV. This is for only one particular pair of fission fragments. For other pairs, the energy release is slightly different, although the process is qualitatively similar. The calculation tacitly assumes that the fragments are formed in their ground state. However, they are commonly formed in excited states, and some of the available energy is carried off by de-excitation gamma rays rather than particle kinetic energy. However, either way, the energy is part of the total energy release.

A rough alternate calculation of the energy release can be made from consideration of the Coulomb repulsion, with the aid of some gross simplifications, including the approximation that the entire energy release is given by the kinetic energy of the two fragments. We assume that at some stage in the fission process, the two fragments have become spheres that are just touching, and we neglect the energy changes in getting to this point. We further assume that at this point, the nuclear force between the fragments will have become vanishingly small and that one only need consider the Coulomb repulsion between them. If we take the nuclear radius to be given by $r = r_o A^{1/3}$, with $r_o = 1.4 \times 10^{-13}$ cm, then the radii of the ^{144}Ba and ^{89}Kr nuclei are 7.34×10^{-13} cm and 6.25×10^{-13} cm, respectively, for a total separation between the centers of these fragments of $R = 13.6 \times 10^{-13}$ cm.

The potential energy for two uniformly charged spheres, carrying charges Q_1 and Q_2 and with centers separated by a distance R, is

$$\mathrm{PE} = \frac{1}{4\pi\epsilon_o} \frac{Q_1 Q_2}{R}, \tag{6.4}$$

where $\epsilon_o = 8.85 \times 10^{-12}$ (in SI units). Given that $Q = Ze$, Eq. (6.4) gives the result that $\mathrm{PE} = 3.42 \times 10^{-11}$ J $= 214$ MeV. All of this potential energy is converted into kinetic energy of the fission fragments; thus, the fission fragment energy is about 214 MeV.

This result is about 25% greater than the much more accurate result from mass excesses. It is impressive that so simple a model captures enough of the essential features of the fission process to give an approximately correct result.

6.4.2 Total Energy Budget

The total energy release in the fission of ^{235}U can be divided into two components. We list these below and give energies associated with each contribution, averaged over the various possible outcomes of the fission process (from Ref. [7, p. 591]):

1. *The energy released in the initial fission process (180.5 MeV).* As discussed earlier, this energy can be determined from the mass differences.

Contributions include the following:
- The kinetic energy of the fission fragments (168.2 MeV).
- The kinetic energy of the prompt neutrons (4.8 MeV).
- The energy carried off by prompt gamma rays (7.5 MeV).

2. *The energy released in the decay of the fission fragments to more stable nuclei (14.6 MeV).* This component includes the following:
- The kinetic energy of the emitted beta rays (7.8 MeV), which is almost all recovered in the reactor.
- The energy of the neutrinos emitted in beta decay (12 MeV). This energy is lost to the reactor and is omitted from the indicated total energy release. It is somewhat greater than the beta-particle energy because the neutrino typically carries away more than half of the available energy in these decays.
- The energy of the gamma rays accompanying beta decay (6.8 MeV), which is almost all recovered in the reactor.
- The energy associated with the delayed neutrons, which is negligible because delayed neutrons constitute less than 1% of all fission neutrons.

For ^{235}U, the sum of the above contributions gives an average total energy recovery of 195 MeV per fission event. For ^{239}Pu, the corresponding energy is 202 MeV [7, p. 591].

This accounts for the energy release in a fission event and its aftermath, but is not a full accounting of the energy produced in the reactor. In addition, some neutrons do not cause fission, but are captured in the ^{235}U and ^{238}U of the reactor fuel as well as in other parts of the reactor. This capture is accompanied by the emission of gamma rays, whose energy adds roughly 2% to the energy from the fission itself [11, p. 46]. Thus, the average energy deposition in the reactor is very close to 200 MeV per ^{235}U fission event, when these capture events are included.[12]

References

1. Charles Weiner, "1932—Moving Into The New Physics," in *History of Physics*, Spencer R. Weart and Melba Phillips, eds. (New York: American Institute of Physics, 1985).
2. E. Segrè, *From X-Rays to Quarks: Modern Physicists and Their Discoveries* (San Francisco: W.H. Freeman, 1980).
3. John A. Wheeler, "Fission in 1939: The Puzzle and the Promise," *Annual Review of Nuclear and Particle Science* 39, 1989: xiii–xxviii.
4. Irving Kaplan, *Nuclear Physics* (Cambridge, MA: Addison-Wesley, 1954).

---

[12] This accounting considers only events in uranium. It omits fission in the plutonium produced in a uranium-fueled reactor, which can add importantly to the reactors's total energy output (see Section 9.3.3).

5. Victoria McLane, Charles L. Dunford, and Philip F. Rose, *Neutron Cross Section Curves*, Volume 2 of *Neutron Cross Sections* (New York: Academic Press, 1988).

6. Arthur C. Wahl, "Nuclear-Charge Distribution and Delayed-Neutron Yields for Thermal-Neutron-Induced Fission of ^{235}U, ^{238}U, and ^{239}Pu and for Spontaneous Fission of ^{252}Cf," *Atomic Data and Nuclear Data Tables* 39, 1988: 1–156.

7. E. Segrè, *Nuclei and Particles*, 2nd edition (Reading, MA: W.A. Benjamin, 1977).

8. Manson Benedict, Thomas H. Pigford, and Hans Wolfgang Levi, *Nuclear Chemical Engineering*, 2nd edition (New York: McGraw-Hill, 1981).

9. James J. Duderstadt and Louis J. Hamilton, *Nuclear Reactor Analysis* (New York: Wiley, 1976).

10. John R. Lamarsh, *Introduction to Nuclear Engineering*, 2nd edition (Reading, MA: Addison-Wesley, 1983).

11. Ronald Allen Knief, *Nuclear Engineering: Theory and Technology of Commercial Nuclear Power*, 2nd edition (Washington, DC: Hemisphere Publishing Company, 1992).

7

Chain Reactions and Nuclear Reactors

7.1 Criticality and the Multiplication Factor

7.1.1 General Considerations

Effective Multiplication Factor

Release of significant amounts of energy from nuclear fission requires a *chain reaction*. For the buildup of a chain reaction, each generation must have more fission events than the preceding one. This means that the average number of neutrons produced in a fission event must be significantly greater than unity. This excess is necessary because some neutrons will be lost, instead of inducing fission in the next generation. For example, some neutrons will be lost to capture reactions, and others will exit the region in which the chain reaction is to be established.

The condition for establishing a chain reaction in a nuclear reactor is commonly expressed as the achievement of *criticality*. This can be quantified in terms of the *criticality factor* or *effective multiplication factor* k, defined as the ratio of the number of neutrons produced by fission in one generation to the number in the preceding generation.[1] A system is termed *critical* if $k = 1$. It is *subcritical* if $k < 1$ and *supercritical* if $k > 1$.

---

[1] In some notations, the effective multiplication factor is denoted k_{eff}.

153

Number of Neutrons per Fission Event, ν

A quantity of key importance in establishing the possibility of a fission chain reaction is the magnitude of ν, defined as the average number of neutrons produced per fission event. Before a nuclear chain reaction could be envisaged as being achievable, it was necessary to establish that ν was sufficiently large. Depending on the magnitude of ν, achievement of criticality might be easy, impossible, or possible but difficult.

Measurements to establish ν for uranium were undertaken almost immediately after the discovery of fission. In 1939, an erroneously high value of 3.5 was reported by a French group headed by Frédérick Joliot [1, p. 48], and a more accurate value, about 2.5, was reported by Fermi's group at Columbia University [2]. Even the latter value was high enough to make the achievement of a chain reaction plausible. (See Section 6.3.2 for more recent values of ν.)

Chain Reactions in Uranium and the Use of Moderators

A typical fission neutron has an energy in the neighborhood of 1 or 2 MeV. For ^{238}U, the fission cross section σ_f is small below 2 MeV (see Figure 6.1), and over much of the relevant neutron energy region the capture cross section σ_γ is appreciable [3]. More specifically, below 1.3 MeV, $\sigma_\gamma/\sigma_f > 1$, and below 1 MeV, $\sigma_\gamma/\sigma_f \gg 1$. With these cross sections, a fast-neutron chain reaction cannot be sustained in ^{238}U.

For ^{235}U, the situation is more favorable. At 1 MeV, $\sigma_f = 1.2$ b and σ_γ is only one-tenth as great. However, the isotopic abundance of ^{235}U in natural uranium is 0.72%, while the abundance of ^{238}U is 99.27%. Thus, natural uranium is not much better than pure ^{238}U in the ability to sustain a chain reaction with fast neutrons.

There are several possible solutions, if one is restricted to uranium: (a) increase the fraction of ^{235}U, (b) reduce the energy of the neutrons to a region where the cross sections are more favorable,[2] or (c) both. Solution (a) was the basis of uranium-fueled nuclear bombs. Solution (b) was used by Fermi for the first operating chain reactor. Solution (c) is the approach adopted for most present-day reactors used for electricity generation, although the ^{235}U fraction is much lower for reactors than for bombs.

The "favorable" energy for fission is the thermal energy region. At the nominal thermal energy of 0.0253 eV, the ratio of the cross section for fission in ^{235}U to capture in ^{238}U is greater than 200. Under these circumstances, a chain reaction is possible, even in natural uranium. In so-called thermal reactors, which have been and remain the dominant type of nuclear reactor, the neutron energy is reduced from the MeV region to the thermal region by successive

---

[2] To achieve criticality with natural uranium, the moderator is physically separated from the uranium (in a carefully designed arrangement). This defines a *heterogeneous* reactor.

elastic collisions with light nuclei (i.e., nuclei with low atomic mass), possibly preceded by inelastic scattering in uranium. This is the process of *moderation*, and the material with low atomic mass is the *moderator*. Moderators will be discussed at greater length in Section 7.2.

7.1.2 Formalism for Describing the Multiplication Factor

Number of Neutrons per Absorption Event, η

In this section, we consider chain reactions in thermal reactors. As mentioned earlier (Section 5.4.2), the term "thermal" is commonly used to refer specifically to room temperature, $T = 293$ K and $kT = 0.0253$ eV, because cross sections are normally measured in the laboratory with the sample at this temperature. When neutrons are thermalized, they come into equilibrium with the local medium. The thermalization occurs primarily in the moderator; therefore, the moderator temperature is the relevant one. It is higher than room temperature, but kT is still well under 0.1 eV.[3] In this section, we assume that a means is found to reduce the neutron energy from the 1-MeV region to below 0.1 eV, and in that context, we will discuss the formalism used to describe the multiplication factor.

As a preliminary, it is useful to define a parameter η:

η = average number of fission neutrons produced per thermal neutron *absorbed* in fuel.

This parameter is related to, but distinct from, the parameter ν defined in Section 6.3.2—the average number of neutrons produced per fission event. When a neutron is absorbed in, say, ^{235}U, either fission or capture may result. Only fission leads to neutron emission. Therefore, there are more neutrons per fission event than per absorption event (i.e., $\nu > \eta$).

The relationship between these parameters can be quantitatively described in terms of cross sections. The absorption cross section σ_a is the sum of the fission cross section σ_f and the capture cross section σ_γ.[4] For a fissile nuclide (e.g., ^{235}U), it follows that

$$\frac{\eta_o}{\nu} = \frac{\sigma_f}{\sigma_a} = \frac{1}{1+\alpha}, \tag{7.1}$$

where η_o is the magnitude of η for pure ^{235}U, $\alpha = \sigma_\gamma/\sigma_f$ [as in Eq. (5.4)], and $\sigma_a = \sigma_f + \sigma_\gamma$.

---

[3] For example, the moderator temperature, namely the water temperature, in a pressurized water reactor is about 300°C (573 K), corresponding to $kT = 0.05$ eV [4, p. 81].

[4] Inelastic nuclear scattering, which was included, for example, in Eq. (5.2) does not occur at thermal energies.

For a mixture of isotopes, the fission and capture cross sections must be summed for the various components. For the case of uranium as the fuel, if the isotopic fraction (by number of atoms) of ^{235}U is x,[5]

$$\frac{\eta}{\nu} = \frac{x\,\sigma_f(235)}{x\,\sigma_a(235) + (1-x)\,\sigma_a(238)}, \tag{7.2}$$

where η refers to the mixture of uranium isotopes in reactor fuel. For ^{238}U, $\sigma_f \doteq 0$ and $\sigma_a = \sigma_\gamma$. The ratio of η to η_o follows from Eqs. (7.1) and (7.2):

$$\frac{\eta}{\eta_o} = \frac{x\,\sigma_a(235)}{x\,\sigma_a(235) + (1-x)\,\sigma_a(238)} \tag{7.3}$$

For $x = 1$, as in pure ^{235}U, Eq. (7.3) reduces to $\eta = \eta_o$.

Thus, despite the early emphasis on the quantity ν, the key quantity in determining the progress from generation to generation in a chain reaction is η. This is the average number of neutrons produced per absorption event in uranium. To summarize, η is less than ν for two reasons: (1) Some of the thermal neutron absorptions in ^{235}U result in capture, not fission; (2) some of the thermal neutrons absorbed in uranium are absorbed in ^{238}U, not ^{235}U.

Five-Factor and Four-Factor Formulas

The effective multiplication factor can be expressed as a product of factors:

$$k = \eta f p \epsilon P_L, \tag{7.4}$$

where η is the number of fission neutrons per thermal neutron absorbed in the fuel, p is the fraction of fission neutrons that avoid capture and reach thermal energies, f is the fraction of thermal neutrons captured in the fuel, ϵ is a factor that accounts for the contribution from fast neutrons, and P_L is the probability that the neutron will not leak out of the reactor.

A neutron faces two main hurdles before it can initiate thermal fission. First, as quantified in the factor p, it must be thermalized before capture occurs. If the reactor fuel is primarily ^{238}U—as is the case for today's reactors—then capture will occur mostly when the neutron energy matches resonance energies for neutron capture in ^{238}U. The factor p is commonly termed the *resonance escape probability*.

Second, as quantified in the factor f, once the neutron is thermalized, it must be absorbed in the uranium fuel (in either ^{235}U or ^{238}U), rather than in something else. All of the other materials of the reactor offer competition for absorption of thermal neutrons. These materials include the moderator, the

---

[5] For simplicity, we ignore in Eq. (7.2) the presence of plutonium isotopes. In practice, as uranium fuel is consumed in a reactor, the ^{235}U fraction (x) decreases and ^{239}Pu and other plutonium isotopes are produced, as taken into account in more complete treatments.

reactor coolant (if different from the moderator), and structural materials. The factor f is commonly termed the *thermal utilization factor*.

This description is based on the conceptual picture that fission occurs at thermal energies and therefore only in ^{235}U (in a uranium reactor). This is a good approximation, but not a perfect one, because fast neutrons can also initiate fission, albeit with a smaller cross section. The enhancement factor ϵ compensates for that oversimplification. Typically, ϵ is about 1.02.

The loss factor P_L describes the escape of neutrons from the reactor. If the reactor system is too small, the losses will be large and a chain reaction is impossible. Often, the other extreme is considered, at least for conceptual purposes, and an infinite system is considered. In that case, $P_L = 1$, and Eq. (7.4) reduces to the so-called *four-factor formula*:

$$k_\infty = \eta f p \epsilon \qquad (7.5)$$

This multiplication factor is commonly designated as k_∞ (where $k = k_\infty P_L$).[6]

7.1.3 Numerical Values of Thermal Reactor Parameters

Table 7.1 lists cross sections and other pertinent parameters for important nuclear fuel constituents. The *fissile* nuclei are those with high fission cross sections at thermal energies. Several of the remainder, most importantly ^{232}Th and ^{238}U, are termed *fertile* nuclides. Although they have negligible cross

Table 7.1. Parameters at thermal energies (0.0253 eV) for selected actinides (cross sections are in barns).

Nuclide	σ_f	σ_γ	σ_a	α	ν	η_o^a
Fissile						
^{233}U	529	45.5	575	0.0861	2.493	2.296
^{235}U	583	98.3	681	0.1687	2.425	2.075
^{239}Pu	748	269	1017	0.360	2.877	2.115
^{241}Pu	1011	358	1369	0.354	2.937	2.169
Fertile						
^{232}Th	$< 3 \times 10^{-6}$	7.37	7.37			
^{238}U	4×10^{-6}	2.68	2.68			
^{240}Pu	0.06	290	290			

[a] The designation η_o (rather than η) is used to emphasize that the values are for a single isotope, not a mixture.
Source: Data are from Ref. [6], part B.

---

[6] Strictly speaking, the magnitudes of the parameters on the right-hand side of Eq. (7.5) differ from those in Eq. (7.4) because of differences in the conditions in an infinite reactor and in a finite one [5]. However, these small differences will be ignored.

sections for fission at thermal energies, they capture neutrons leading to the formation of the fissile nuclides, ^{233}U and ^{239}Pu. The fissile nuclei all have odd atomic mass numbers A and the fertile nuclei all have even A, for reasons discussed in Section 6.2.2.

The magnitudes of η given in Table 7.1 are for pure isotopes. For natural uranium, $x = 0.0072$. Then, from Eq. (7.3), $\eta/\eta_o = 0.648$ and $\eta = 1.35$, using the parameters of Table 7.1. For uranium enriched to $x = 0.03$, $\eta = 1.84$.

As the reactor fuel is consumed, the ^{235}U fraction in the fuel will decrease, and it is to be expected that η will be lower than the 1.84 calculated for a 3% ^{235}U enrichment. This is illustrated in typical quoted values of the parameters in Eq. (7.4) for a thermal reactor: $\eta = 1.65$, $p = 0.87$, $f = 0.71$, $\epsilon = 1.02$, and $P_L = 0.96$ [7, p. 83]. These give the product $k = 1.00$.

It is clear from the preceding discussion that greater enrichment in ^{235}U will increase η. It will also increase the thermal utilization factor f, because the large cross section for fission in ^{235}U increases the relative probability of absorption in the fuel, rather than in other materials.

7.2 Thermalization of Neutrons

7.2.1 Role of Moderators

Kinematic Relations

As explained in Section 7.1.1, fission in natural uranium will not result in a chain reaction if the neutrons interact primarily at energies close to those at which they are emitted, in the vicinity of 1 or 2 MeV. The neutrons are reduced in energy from this region to the more favorable thermal energy region by elastic collisions with the nuclei of the *moderator.*

The energy transfer in an elastic nuclear collision depends on the angle at which the incident neutron is scattered. The energy of the scattered neutron E_s can be calculated from the kinematics of "billiard ball" collisions (i.e., by the application of conservation of energy and momentum). If a neutron of initial energy E_i is scattered at 0° by a nucleus of the moderator material— namely it continues in the forward direction—then the energy of the scattered neutron will be $E_s = E_i$. In brief, nothing is changed. If the neutron is scattered at 180°, then the energy of the scattered neutron can be shown to be

$$E_s = \left(\frac{M - m_n}{M + m_n}\right)^2 E_i = \left(\frac{A - 1}{A + 1}\right)^2 E_i, \qquad (7.6)$$

where M is the atomic mass of the moderator, m_n is the mass of the neutron, and the approximation has been made that $M = A$ and $m_n = 1$ (both in atomic mass units, u). Equation (7.6) describes the condition for minimum neutron energy or maximum energy loss. If we define δ to be the ratio of *average* scattered neutron energy to E_i and assume this average to be the mean of the 0° and 180° cases, it follows that

$$\delta = \frac{1}{2}\left[1 + \left(\frac{A-1}{A+1}\right)^2\right] = \frac{A^2+1}{A^2+2A+1}. \qquad (7.7)$$

For example, $\delta = 0.50$ for hydrogen ($A = 1$), $\delta = 0.86$ for carbon ($A = 12$), and $\delta = 0.99$ for uranium ($A = 238$).

Obviously, Eq. (7.7) could not give a correct average if the scattering events were either predominantly forward or predominantly backward. Less obviously, it gives an exactly correct result if the scattering is isotropic, namely an equal number of neutrons scattered in all directions.[7] This isotropy condition is well satisfied for neutrons incident on light elements such as hydrogen and carbon, at the energies relevant to fission; therefore, Eq. (7.7) is a good approximation.

An effective moderator (i.e., one that reduces the neutron energy with relatively few collisions) requires a low value of the parameter δ. As seen from Eq. (7.7) and illustrated by the above examples, this means that the moderator must have a relatively low atomic mass number A. This rules out using uranium as its own moderator. Instead, hydrogen in water (H_2O), deuterium in heavy water (D_2O), or carbon in graphite (C) are commonly used as moderators.

Number of Collisions for Thermalization

The factor δ can be used to calculate the number of collisions n required to reduce the *mean* neutron energy by any given factor. For example, the reduction of the mean neutron energy from 2 MeV to 1 eV (a factor of 2×10^6) corresponds to $\delta^n = 5 \times 10^{-7}$. For hydrogen ($\delta = 1/2$), this gives $n = 21$. However, the distribution of neutron energies becomes very skewed in successive collisions, with many neutrons having very small energies and a few neutrons having high energies. These relatively few high-energy neutrons elevate the mean energy to a level considerably above the more representative *median* energy. Thus, the decrease in mean energy provides only a crude characterization of the extent to which neutrons have lost energy, and δ, although simple to calculate, is not widely used.

It is more common to characterize the evolution of the neutron energy distribution in terms of a quantity that is less perturbed by a few remaining high-energy neutrons, namely the logarithm of the energies. The mean *logarithmic energy decrement* ξ is defined as[8]

---

[7] Here, angles are taken to be in the center-of-mass system (i.e., a system moving at the velocity of the center of mass).

[8] In the nuclear engineering literature, ξ is the mean gain in *lethargy* per generation. We will not undertake here the detailed calculations that make it desirable to employ the concept of lethargy.

$$\xi = \left\langle \ln\left(\frac{E_i}{E_s}\right) \right\rangle. \tag{7.8}$$

Here, ξ is the mean of the logarithm of the ratio of the incident and scattered neutron energies in successive generations. The average value of the logarithm of the neutron energy therefore drops by ξ in successive generations, or—equivalently but less precisely—the "average" neutron energy changes by a factor of $e^{-\xi}$. To reduce this neutron energy by a factor of 2×10^6 requires n generations, where

$$e^{n\xi} = 2 \times 10^6. \tag{7.9}$$

It can be shown that for an isotropic distribution in scattering angle (see, e.g., Ref. [8, p. 624])[9]

$$\xi = 1 - \frac{(A-1)^2}{2A} \ln\left(\frac{A+1}{A-1}\right), \tag{7.10}$$

where A is the atomic mass number, and the small difference between A and the atomic mass M is ignored.

It is desirable that thermalization be accomplished in as few collisions as possible, to minimize the loss of neutrons by absorption in ^{238}U and other materials of the reactor. For hydrogen, $\xi = 1$ by Eq. (7.10) (noting that the second term approaches zero as A approaches 1) and $n = \ln(2 \times 10^6) = 14.5$. The same analysis shows that $n = 92$ for ^{12}C ($\xi = 0.158$) and $n = 1730$ for ^{238}U ($\xi = 0.0084$). This highlights the key reason hydrogen is useful as a moderator and indicates the impossibility of accomplishing thermalization without a moderator of a material that is much lighter than uranium.

7.2.2 Moderating Ratio

In a moderator, there is competition between elastic scattering, which produces the desired reduction in energy, and capture, which means the loss of the neutron. The competition depends on the relative magnitudes of the elastic scattering cross section σ_{el} and the capture cross section σ_c. A figure of merit for a moderator is the *moderating ratio*, defined as

$$\mathrm{MR} = \frac{\xi\sigma_{el}}{\sigma_c}. \tag{7.11}$$

A good moderator is one in which ξ and σ_{el} are high and σ_c is low. Properties of various moderators are presented in Table 7.2.[10]

---

[9] Again, isotropy refers to the center-of-mass system.
[10] The moderating ratio should be calculated as an average over the energies of the slowing neutrons.

Table 7.2. Properties of alternative moderators.

Material[a]	A	ξ	n[b]	MR
H	1	1.000	14	
D	2	0.725	20	
H_2O		0.920	16	71
D_2O		0.509	29	5670
Be	9	0.209	69	143
C	12	0.158	91	192
^{238}U	238	0.008	1730	0.009

[a]Here, H denotes ^1H and D denotes ^2H.
[b]n is the number of collisions to reduce the energy by factor of 2×10^6 [see Eq. (7.9)].
Source: Data are from Ref. [7, p. 324].

An obvious choice for use as a moderator is hydrogen, in the form of ordinary ("light") water (^1H_2O). The oxygen in the water is relatively inert, having a very low capture cross section and being much less effective than hydrogen as a moderator. However, ordinary hydrogen has the major drawback of having a large cross section (0.33 b at 0.0253 eV) for the capture reaction

$$n + {}^1H \rightarrow {}^2H + \gamma.$$

The moderating ratio for light water, as defined in Eq. (7.11), is 71 (see Table 7.2).

The moderating ratio for heavy water, D_2O (^2H_2O), is almost 100 times greater, despite the larger mass of deuterium.[11] This is because the neutron-capture cross section is much less in deuterium than in hydrogen (0.0005 b at 0.0253 eV). It was recognized early in the nuclear weapons programs of World War II that heavy water would be a very valuable substance in reactor development, and considerable effort was devoted to attacking heavy water production facilities in Norway, which were then under German control. The choice between light and heavy water remains unclear, with the former having the advantage of being very easily obtainable and the latter making it possible to use natural uranium as the reactor fuel.

Carbon, in the form of pure graphite, is a more effective moderator than light water, with a moderating ratio of 192, despite the higher A and lower ξ for ^{12}C. The reason for this, again, is a low neutron-capture cross section (0.0035 b at 0.0253 eV). However, in contrast to light water and heavy water, graphite cannot be used as a coolant as well as a moderator.

---

[11] The actual moderating ratio for a heavy water reactor depends on the magnitude of the light water impurity in the heavy water.

7.3 Reactor Kinetics

7.3.1 Reactivity

If the multiplication exceeds unity by more than a small amount, the reactor power will build up at a rapid rate. For constant power, the effective multiplication factor must be kept at unity. Thus, a key quantity is the difference $k - 1$. This quantity is usually expressed in terms of the *reactivity* ρ, where, by definition,

$$\rho = \frac{k - 1}{k} \tag{7.12}$$

and

$$k = \frac{1}{1 - \rho}. \tag{7.13}$$

A runaway chain reaction is one in which k rises appreciably above unity or, equivalently, the reactivity ρ is appreciably greater than zero. A major aspect of reactor safety is the avoidance of such an excursion.

7.3.2 Buildup of Reaction Rate

The rate at which a reactor delivers power is proportional to the number of fission events per unit time, which, in turn, is proportional to the neutron flux, here designated as ϕ. To describe changes in the reaction rate, it is convenient to introduce the *mean neutron lifetime l*, defined as the average time between the emission of a fission neutron and its absorption. Fission follows absorption with no time delay, and l gives the average time between successive fission generations.[12] The change in the neutron flux per generation is $(k - 1)\phi$, and the time rate of change in neutron flux is therefore

$$\frac{d\phi}{dt} = \frac{(k - 1)\,\phi(t)}{l} = \frac{\phi(t)}{T}, \tag{7.14}$$

where T is the *reactor period*, defined as

$$T = \frac{l}{k - 1}. \tag{7.15}$$

The solution of Eq. (7.14) is an exponential buildup:

$$\phi(t) = \phi(0)\,e^{t/T}. \tag{7.16}$$

---

[12] The time between generations and the mean neutron lifetime l are equal in a reactor where $k = 1$ and are approximately equal if $k \approx 1$.

The flux in the reactor grows exponentially if the period T is not infinite. It is important for the safety of the reactor that the growth rate not be too rapid (i.e., the period T not be small). As is obvious qualitatively and follows from Eq. (7.16), this is achieved by keeping the multiplication constant k close to unity.

The reactor period also depends on the mean neutron lifetime l. In normal reactor operation, the achievement of criticality requires the small, but crucial, number of fission events that are initiated by delayed neutrons. For ^{235}U, the delayed neutron fraction is 0.65% [i.e., $\beta = 0.0065$ (see Section 6.3.2)]. The mean lifetime l can be expressed as the sum of terms for the prompt and delayed components:

$$l = (1 - \beta)\, l_p + \beta \tau_e, \qquad (7.17)$$

where l_p and τ_e are mean lifetimes for prompt and delayed neutrons, respectively.

The prompt neutron lifetime is the sum of the time required for moderation to thermal energies and the time spent by the neutron at thermal energies before being absorbed. The magnitude of this time depends on the type of reactor. For a water-moderated, uranium-fueled reactor, the mean neutron lifetime l_p is typically about 0.0001 s (see, e.g., Ref. [7, p. 236]).

The effective mean lifetime for delayed neutrons τ_e is found from the beta-decay lifetimes of the individual contributing nuclides, averaged over all precursors of the delayed neutrons (ignoring the slowing down of the delayed neutrons themselves).[13] For thermal neutron fission in ^{235}U, τ_e is about 10 s (see, e.g., Ref. [9, p. 76]).

Together, the two components give an average mean lifetime for all neutrons of $l \approx 0.1$ s. This mean lifetime is almost 1000 times greater than the mean lifetime l_p for prompt neutrons alone. The importance of this difference can be seen from a simple numerical example. If k were to rise, say, from 1.000 to 1.001, then the period T would be 100 s and the chain reaction would build up relatively slowly. On the other hand, were there no delayed neutron component (i.e., $\beta = 0$), the period would be 0.1 s and, by Eq. (7.16), the flux would double every 0.07 s (in the absence of other changes). It is important to the safe operation of the reactor that changes not be too rapid (i.e., the period T not be too small). Delayed neutrons are crucial to achieving this.

A more complete analysis of the buildup rates with delayed neutrons shows that it is only valid to use Eq. (7.17) in the special case where the reactivity ρ is less than β [7, p. 246]. When ρ is large compared to β, then, as might be expected, the fact that some of the neutrons are delayed makes little difference, and the period is determined by the prompt lifetime l_p alone.

An important distinction can be made between the case where the reactor is critical without delayed neutrons and the case where delayed neutrons are

---

[13] More specifically, $\beta \tau_e = \Sigma \beta_i \tau_i$, where τ_i and β_i are the mean lifetime and fraction of emitted neutrons, respectively, for the ith delayed neutron group.

required for criticality. The cases are distinguished as follows:[14]

$\rho = 0$: *critical* or, alternatively, *delayed critical*

$0 < \rho < \beta$: *supercritical*[15]

$\rho = \beta$: *prompt critical*

$\rho \geq \beta$: *super prompt critical*

As long as the reactor operates below the prompt critical threshold, it is protected against rapid changes by the dependence of the chain reaction on delayed neutrons.

7.4 Conversion Ratio and Production of Plutonium in Thermal Reactors

With uranium fuel, ^{239}Pu is produced by the capture of neutrons in ^{238}U (see Section 5.1.1). This provides a means of obtaining ^{239}Pu for possible use in weapons or in other reactors. It also contributes fissile material which is consumed in the reactor before the fuel is removed, supplementing the original ^{235}U in the fresh fuel. The rate of ^{239}Pu production is described in terms of the *conversion ratio*, which is defined in general terms as the ratio of the number of fissile nuclei produced to the number of fissile nuclei destroyed.

If we specialize to uranium fuel, the initial conversion ratio (i.e., the conversion ratio before there is an appreciable buildup of ^{239}Pu) is given by

$$C = \frac{N(238)}{N(235)}, \tag{7.18}$$

where $N(238)$ is the number of capture events in ^{238}U, producing ^{239}Pu, and $N(235)$ is the number of absorption events in ^{235}U, destroying ^{235}U by fission or neutron capture.

To evaluate Eq. (7.18), it is convenient to normalize to a single absorption event in ^{235}U, setting $N(235)$ at unity. For simplicity, we ignore the destruction of ^{239}Pu by fission or capture. Then, Eq. (7.18) reduces to

$$C = N(238) = N_{\text{fast}} + N_{\text{th}}, \tag{7.19}$$

where N_{fast} and N_{th} are the number of capture events in ^{238}U for fast (nonthermal) neutrons and thermal neutrons, respectively, per absorption in ^{235}U.

---

[14] It can be seen that $\rho = \beta$ is a dividing point by separating the multiplication factor k into a prompt and delayed part, $k = k_p + k_d$, where $k_d = \beta k$. Then, $k_p = k(1 - \beta)$ and, by Eq. (7.13), the condition $\rho = \beta$ is equivalent to the condition $k_p = 1$.

[15] Usage is not uniform, and some authors call this region "delayed critical" rather than "supercritical" [7, p. 246].

To calculate C, we follow the consequences of an absorption event in ^{235}U. Each absorption leads to the production of η_o fission neutrons, where η_o is the value for pure ^{235}U. For simplicity, we will ignore the fast-fission enhancement factor, ϵ, assuming it to be close to unity [see the four-factor formula, Eq. (7.5)]. As a further simplification, we will assume that all absorption in ^{235}U occurs at thermal energies. The multiplication factor of Eq. (7.5) then reduces to

$$k = \eta f p. \tag{7.20}$$

It is a good approximation to assume that all neutrons that do not reach thermal energies are captured in ^{238}U in the resonance region. The number of fast-neutron ^{238}U capture events can then be expressed in terms of the resonance escape probability p:

$$N_{\text{fast}} = (1 - p)\eta_o. \tag{7.21}$$

The remaining neutrons reach thermal energies, where the fraction f is captured in uranium. The number of these captured neutrons is therefore $\eta_o f p$. From Eq. (7.3), the fraction of these that are captured in ^{235}U is η/η_o and, therefore, the fraction captured in ^{238}U is $(1 - \eta/\eta_o)$. Therefore, the number of thermal neutron capture events in ^{238}U is

$$N_{\text{th}} = \eta_o f p \left(1 - \frac{\eta}{\eta_o}\right) = \eta f p \left(\frac{\eta_o}{\eta} - 1\right) = \frac{\eta_o}{\eta} - 1, \tag{7.22}$$

where we have used the requirement for criticality ($k = 1$) that $\eta f p = 1$ [see Eq. (7.20)]. The total conversion ratio C is given by adding the components in Eqs. (7.21) and (7.22):

$$C = (1 - p)\eta_o + \left(\frac{\eta_o}{\eta} - 1\right). \tag{7.23}$$

As a specific example, we adopt the parameters cited earlier (Section 7.1.3) for a conventional light water reactor ($p = 0.87$, $f = 0.71$, $\eta_o = 2.075$, and $\eta = 1.65$). It then follows that

$$C = 0.27 + 0.26 = 0.53.$$

Thus, for this case, roughly one ^{239}Pu nucleus is produced for every two ^{235}U nuclei consumed.

The actual conversion ratio at a given time and place in the reactor will depend on the magnitude of the ^{235}U fraction x. As this quantity decreases from its initial value, absorption in ^{238}U becomes relatively more important for thermal neutrons, η decreases, and C rises. The example cited is for an intermediate value of x. A more complete treatment also includes the destruction of ^{239}Pu. A typical overall value for the conversion ratio in a light water reactor is about 0.6 (see, e.g., Ref. [7, p. 86]). Reactors using moderators with a higher moderating ratio (heavy water or graphite) in general will have a high thermal utilization factor f and a lower resonance escape probability, p. This means a greater production of ^{239}Pu (i.e., a conversion ratio C that typically is about 0.8 [7]).

If $C = 1$, the condition for breeding is reached (i.e., as much fissionable material is produced as has been consumed). However, most breeder reactors are *fast breeder reactors*, with fission occurring far above the thermal region of neutron energy, and the above formalism is not applicable to them.

7.5 Control Materials and Poisons

7.5.1 Reactor Poisons

Some of the fission products in a reactor have very high capture cross sections at thermal energies. For example, the thermal neutron-capture cross section is 2.65×10^6 b for xenon-135 (^{135}Xe) and is 4.1×10^4 b for samarium-149 (^{149}Sm) (see, e.g., Ref. [9, p. 643]).[16] Such nuclei are known as *poisons* because their presence in the reactor fuel creates an additional channel for the loss of neutrons and reduces the reactivity of the core. In terms of Eq. (7.5), they lower the thermal utilization factor f and the multiplication factor k.

The buildup of the poisons is limited by both their radioactive decay (if they are not stable) and their destruction in the capture process. Some poisons asymptotically reach a constant level where production of the poison is just balanced by radioactive decay and neutron capture in the poison, whereas for others, the abundance keeps increasing with time. The resulting increase in the total poison content is one of the factors that limits the length of time that fuel can be used in a reactor.

7.5.2 Controls

Control materials are materials with large thermal neutron-absorption cross sections, used as controllable poisons to adjust the level of reactivity. They are used for a variety of purposes:

- To achieve intentional changes in reactor operating conditions, including turning the reactor on and off.
- To compensate for changes in reactor operating conditions, including changes in the fissile and poison content of the fuel.
- To provide a means for turning the reactor off rapidly, in case of emergency.

---

[16] The capture cross sections for these nuclei are not well fit by the $1/v$ law. The cross sections specified here are the measured values for a neutron velocity of 2200 m/s (see Section 5.4.3).

Two commonly used control materials are cadmium and boron. The isotope ^{113}Cd, which is 12.2% abundant in natural cadmium, has a thermal neutron capture cross section of 20,000 b. The isotope ^{10}B, which is 19.9% abundant in natural boron, has a thermal absorption cross section of 3800 b. (In this case, absorption of the neutron leads not to capture but to the alpha-particle emitting reaction, $n + {}^{10}B \rightarrow \alpha + {}^{7}Li$, but the effect on the reactivity is the same as for capture.) Thus, cadmium and boron are effective materials for reducing the reactivity of the reactor.

Cadmium is commonly used in the form of control rods, which can be inserted or withdrawn as needed. Boron can be used in a control rod, made, for example, of boron steel or it can be introduced in soluble form in the water of water-cooled reactors. It is also possible to use other materials.

A control variant is the so-called *burnable poison*, incorporated into the fuel itself. During the three or more years that the fuel is in the reactor, the ^{235}U or other fissile material in the fuel is partially consumed, and fission product poisons are produced. The burnable poison helps to even the performance of the fuel over this time period. It is a material with a large absorption cross section, which is consumed by neutron capture as the fuel is used. The decrease in burnable poison content with time can be tailored to compensate for the decrease in ^{235}U and the buildup of fission fragment poisons. Boron and gadolinium can be used as burnable poisons.

7.5.3 Xenon Poisoning

Reference has already been made to the buildup of poisons as the fuel is used. A particularly interesting example is xenon-135 (^{135}Xe)—the cause of the "xenon poisoning" that for a brief time threatened operation of the first large U.S. reactors in World War II, which were designed to produce ^{239}Pu.[17]

The first large reactor for this purpose went into operation in September 1944. After a few hours of smooth operation, the power level of the reactor began to decrease, eventually falling to zero. It recovered by itself within about 12 h and then began another cycle of operation and decline. The possibility of fission-product poisoning had previously occurred to John Wheeler, who had participated in the project. From the time history of the poisoning cycle and the known high-absorption cross section of xenon, Wheeler and Fermi "within a couple of days...discovered the culprit."[18]

The phenomenon was due to the production of $A = 135$ precursors, including ^{135}I, which decay into the strongly absorbing ^{135}Xe, which, in turn, decays into a stable (or long-lived) product with a small absorption cross section. In anticipation of possible difficulties, the reactor had been built in a way such that more uranium could be added to the core. This enabled it to sustain criticality even in the face of the xenon poisoning. Thus, the wartime pluto-

---

[17] A vivid description of these events is given in Ref. [10, pp. 557–560].
[18] See Ref. [11, pp. 29–30].

nium production was not greatly delayed. Nonetheless, the event dramatized a phenomenon that remains an important consideration in reactor design.

The xenon poisoning can be described in terms of the chain of $A = 135$ isotopes (with their half-lives in parentheses):[19]

$$^{135}\text{Te}(19.0 \text{ s}) \rightarrow {}^{135}\text{I}(6.57 \text{ h}) \rightarrow {}^{135}\text{Xe}(9.14 \text{ h}) \rightarrow {}^{135}\text{Cs}(2.3 \times 10^6 \text{ yr})$$

The main fission product in this chain is tellurium-135 (^{135}Te), which almost immediately decays to iodine-135 (^{135}I). This is the chief source of ^{135}Xe, although a small amount (about 1/20th the amount from ^{135}I) is produced directly. Thus, initially there is a negligible concentration of ^{135}Xe. On a time scale of about 7 h, the concentration begins to build. If this is sufficient to shut down the reactor, as was the case in the above-cited events, then the concentration eventually declines with a 9-h half-life. Reactor operation can then resume and the cycle can repeat.

The effectiveness of ^{135}Xe as a poison, despite the very small amount present, stems from its extremely high cross section: 2.65×10^6 b at 0.0253 eV. In a normally operating reactor, the amount of ^{135}Xe is held in check by both its own radioactive decay and its destruction by neutron capture. If the reactor is shut down suddenly, the ^{135}Xe level will increase for a period, as the ^{135}I decay continues but the destruction by neutron capture stops. The early restart of the reactor is then more difficult. This effect, and imprudent measures taken to overcome it, were important contributors to the Chernobyl nuclear reactor accident (see Section 15.3.2).

References

1. Bertrand Goldschmidt, *Atomic Rivals*, translated by Georges M. Temmer (New Brunswick, NJ: Rutgers University Press, 1990).
2. Rudolf Peierls, "Reflections on the Discovery of Fission," *Nature* 342, 1989: 853–854.
3. Victoria McLane, Charles L. Dunford, and Philip F. Rose, *Neutron Cross Section Curves*, Vol. 2 of *Neutron Cross Sections* (New York: Academic Press, 1988).
4. Anthony V. Nero, Jr., *A Guidebook to Nuclear Reactors* (Berkeley: University of California Press, 1979).
5. Alvin M. Weinberg and Eugene P. Wigner, *The Physical Theory of Neutron Chain Reactors* (Chicago: University of Chicago Press, 1958).
6. S.F. Mughabghab, M. Divadeenam, and N. E. Holden, *Neutron Resonance Parameters and Thermal Cross Sections*, Part A, $Z = 1$–60 and Part B: $Z = 61$–100, Vol. 1 of *Neutron Cross Sections* (New York: Academic Press, 1981 and 1984).
7. James J. Duderstadt and Louis J. Hamilton, *Nuclear Reactor Analysis* (New York: Wiley, 1976).

---

[19] See, for example, Ref. [12, p. 195], for relative fission yields.

8. E. Segrè, *Nuclei and Particles*, 2nd edition (Reading, MA: W.A. Benjamin, 1977).

9. John R. Lamarsh, *Introduction to Nuclear Engineering*, 2nd edition (Reading, MA: Addison-Wesley, 1983).

10. Richard Rhodes, *The Making of the Atomic Bomb* (New York: Simon and Schuster, 1986).

11. Alvin M. Weinberg, *The First Nuclear Era: The Life and Times of a Technological Fixer* (New York: American Institute of Physics Press, 1994).

12. D.J. Bennet and J.R. Thomson, *The Elements of Nuclear Power*, 3rd edition (Essex, UK: Longman, 1989).

8

Types of Nuclear Reactors

8.1 Survey of Reactor Types

8.1.1 Uses of Reactors

The first nuclear reactors were built to produce ^{239}Pu for bombs. Subsequently, reactors have been used for many other purposes, of which electricity generation is now, by far, the most prominent. Further uses have been to propel ships (mostly naval vessels), to produce radioisotopes, and, to a limited extent, to supply heat. Many additional reactors have been built for teaching or research, much of the latter involving the study of the properties of materials under neutron bombardment and the intrinsic properties of neutrons and other subatomic particles.

In some cases, applications have been combined. For example, the N reactor at Hanford and the Chernobyl-type reactors were used for both ^{239}Pu production and electricity generation. There has also been limited use of waste heat from reactors to produce hot water or steam for industrial applications, for heavy water production, and for desalination. These reactors have been primarily in the USSR and Canada [1]. The total thermal capacity of these facilities is much less than that of electricity-generating nuclear reactors, although the use of heat from reactors could increase in the future, for example if used for hydrogen production (see Section 16.6).

The focus in this book is on the use of reactors for electricity generation, but the issue of ^{239}Pu production still arises. If the reactor fuel contains ^{238}U, then ^{239}Pu will inevitably be produced following neutron capture in ^{238}U. In breeder reactors, the production of ^{239}Pu is a central goal, with the ^{239}Pu intended as fuel for further reactors. In nonbreeders, which means almost all of the world's operating reactors, ^{239}Pu is a by-product, but its fission, nonetheless, often contributes significantly to the reactor's total energy output. (We will return in Chapter 17 to the link between power reactors and the possible use of their ^{239}Pu for bombs.)

8.1.2 Classifications of Reactors

Thermal Reactors and Fast Reactors

In previous chapters, reference has been made to the thermalization of neutrons in reactors (i.e., to the slowing of the neutrons to thermal energies). Reactors designed to operate in this fashion are termed *thermal* reactors. However, it is also possible to operate a reactor with "fast" neutrons. There is no moderator in such a reactor and few neutrons reach thermal energies, but many lose energy by inelastic scattering and fission occurs at energies ranging from the MeV region down to the keV region and below. These reactors are called *fast-neutron* reactors or just *fast* reactors. The only prominent example of an operating fast power reactor is the liquid-metal reactor (see Section 8.3.3), although other types of fast reactors have been built for experimental purposes.

Homogeneous and Heterogeneous Reactors

We have been tacitly assuming that the reactors under consideration are what are sometimes known as *heterogeneous*, in which the fuel, coolant, and moderator (if any) are distinct physical entities. All reactors used today for power generation are of this form. However, in the early days of nuclear power, there was considerable exploration of an alternative configuration, the *homogeneous* reactor, defined as "a reactor whose small-scale composition is uniform and isotropic" [2, p. 378]. Homogeneity can be achieved if the fuel is in liquid form, where the liquid is circulated for heat transfer to a steam generator.

One variant of this reactor type was known as the *aqueous homogenous reactor* because the fuel was mixed with water (H_2O or D_2O). In the so-called homogeneous reactor experiment, two small reactors, called HRE-1 and HRE-2, were built at Oak Ridge National Laboratory (ORNL) in the 1950s.[1] For HRE-2, the fluid was uranyl sulfate (UO_2SO_4) in heavy water (D_2O), with the uranium highly enriched in ^{235}U. This program had the potential of developing a thermal breeder reactor, but although HRE-2 operated uninterruptedly for over 100 days at 5 MWe, some difficulties developed, and the program was dropped in favor of alternative liquid-fuel projects.

One alternative was the *molten-salt reactor*. The fluid was a mixture of fluoride compounds, including the fissile component ^{235}UF$_4$ and the fertile component ThF$_4$. After initial operation, ^{233}UF$_4$ was successfully tried as an

---

[1] For a description of the history of this program and the ORNL program on molten-salt breeders, see Chapter 6 of Ref. [3]. Technical aspects of fluid-fuel reactors are also discussed in Ref. [4, pp. 403–413].

alternative to $^{235}UF_4$. Like HRE-2, this reactor was designed to be a thermal breeder reactor. Again, there was initial success in the reactor operation, but a decision was made in the 1960s to abandon development of thermal breeders in favor of fast breeder reactors.[2] A further homogenous reactor approach, a *liquid metal thermal breeder* using uranium compounds in molten bismuth, was also investigated but was abandoned without construction even of a test reactor.

At present, there are no electricity-generating homogenous reactors, and we will not consider them further in the chapter. Instead, we will restrict consideration to heterogeneous reactors, which are so dominant that it is unusual to include the specification "heterogeneous." Nonetheless, some interest remains in homogenous reactors, particularly in molten-salt reactors (see, e.g., Ref. [5]).[3] In fact, one of the reactors selected for possible long-term development under the Generation IV program is a molten-salt reactor (see Chapter 16). Overall, although we focus in this chapter on reactors of the sort in actual use or in immediate prospect, it is well to remember that on a longer time scale, a wide array of variants are possible. Some of these are being tentatively explored in the thinking underway in the United States and elsewhere on nuclear options for the future, as discussed in Chapter 16.

8.1.3 Components of Conventional Reactors

Overall

Any generating plant consists of an array of structural components and a system of mechanical and electrical controls. In a nuclear plant, there are special demands on structural integrity and reliability. In addition, a nuclear reactor is characterized by the use of specialized materials, some aspects of which were already discussed in Chapter 7. In standard reactors, these are the fuel itself, the coolant, the moderator, and neutron-absorbing materials used to control the power level. A main distinction between different reactor types lies in the differences in the choices of fuel, coolant, and moderator.

Fuels

There are few nuclides that can be used as reactor fuels. The paucity of possible candidates can be seen by examining the properties of the naturally occurring heavy elements:

---

[2] The fast breeder reactor program was subsequently sharply reduced, with the centerpiece of the U.S. fast breeder reactor program, the Clinch River Breeder Reactor, abandoned in 1975.

[3] There has been speculation about a quite different sort of molten-salt reactor, driven by a proton accelerator. If pursued, this would represent a radical departure from the sorts of reactors that have been built to date.

- *Uranium (atomic number $Z = 92$).* This is the main fuel in actual use, especially ^{235}U which is fissile. In addition, ^{238}U is important in reactors, primarily as a fertile fuel for ^{239}Pu production, and ^{233}U could be used as a fissile fuel, formed by neutron capture in ^{232}Th.
- *Protactinium (atomic number $Z = 91$).* The longest-lived isotope of protactinium (^{231}Pa) has a half-life of 3.3×10^4 yr, and therefore there is essentially no Pa in nature.
- *Thorium (atomic number $Z = 90$).* Thorium is found entirely as ^{232}Th, which is not fissile (for thermal neutrons). It can be used as a fertile fuel for the production of fissile ^{233}U.

Between thorium ($Z = 90$) and bismuth ($Z = 83$), the isotope with the longest half-life is ^{226}Ra ($T = 1600$ years); therefore, there are no fuel candidates, quite apart from the issue of fissionability. By the time the atomic number is as low as 83, the threshold for fission is much too high for a chain reaction to be conceivable (see, e.g., Ref. [6, p. 574]). Thus, uranium and thorium are the only natural elements available for use as reactor fuels. In addition, ^{233}U and ^{239}Pu can be produced from capture on ^{232}Th and ^{238}U in reactors. This means that the nuclides listed in Table 6.1 exhaust the practical possibilities for reactor fuels. Of these, only ^{235}U is both fissile and found in nature in useful amounts.

Restricting consideration to uranium fuel, there are a number of options as to the form of the fuel. Reactors can operate over a considerable range of enrichments in ^{235}U. Enrichment to a concentration in the neighborhood of 4% is now somewhat typical in the light water reactors that account for most of today's nuclear generation (see Section 8.1.4), with a trend over time toward higher enrichments and greater burnup of the fuel. In today's heterogeneous reactors, the fuel is solid. For the most part, it is in an oxide form, as UO_2, but metallic fuel is a possibility and has been used in some reactors. The fuel usually is in cylindrical pellets with typical dimensions on the scale of 1 cm, but some designs for future reactors are based on fuel in submillimeter microspheres embedded in graphite, with the goal of enhanced ruggedness at high temperatures.[4]

Moderators

As discussed in Section 7.2, a moderator is required if the reactor is to operate at thermal neutron energies. This means that most operating reactors use moderators, with the fast breeder reactor the exception. The options for moderating materials are limited:

- *Hydrogen ($Z = 1$).* The isotopes ^{1}H and ^{2}H are widely used as moderators, in the form of light (ordinary) water and heavy water, respectively.

---

[4] In particular, this fuel is for high-temperature gas-cooled reactors (see Section 16.4.3).

- *Helium (Z = 2)*. The isotope ^{4}He is not used, because helium is a gas and excessive pressures would be required to obtain adequate helium densities for a practical moderator; ^{3}He would be similarly excluded, but, in addition, it is a strong neutron absorber and would be unsuitable as a moderator.
- *Lithium (Z = 3)*. The isotope ^{6}Li (7.5% abundant) has a large neutron-absorption cross section, making lithium impractical as a moderator.
- *Beryllium (Z = 4)*. ^{9}Be has been used to a limited extent as a moderator, especially in some early reactors. It can be used in the form of beryllium oxide, BeO. However, beryllium is expensive and toxic.
- *Boron (Z = 5)*. The very large neutron-absorption cross section in ^{10}B (20% abundant) makes boron impossible as a moderator.
- *Carbon (Z = 6)*. Carbon in the form of graphite has been widely used as a moderator. It is important that the graphite be pure (i.e., be free of elements that have high absorption cross sections for neutrons).

There are no advantages in considering elements heavier than carbon. The effectiveness for moderation decreases with increasing mass, and there are no counterbalancing advantages in other properties. Again, therefore, there is a limited list of candidates: light water, heavy water, graphite, and beryllium. Any of these can be used with uranium enriched in ^{235}U. With natural uranium, it not possible to achieve a chain reaction with a light water moderator, but it is practical to use heavy water or graphite, both of which have high moderating ratios (see Table 7.2).

Coolants

The main function of the coolant in an electricity generating plant is to transfer energy from the hot fuel to the electrical turbine, either directly or through intermediate steps. During power plant operation, cooling is an intrinsic aspect of energy transfer. However, in a nuclear reactor, cooling has a special additional importance, because radioactive decay causes heat production to continue even after the reactor is shut down and electricity generation has stopped. It is still essential to maintain cooling to avoid melting the reactor core, and in some types of reactor accidents (e.g., the accident at Three Mile Island) cooling is the critical issue.[5]

The coolant can be either a liquid or a gas. For thermal reactors, the most common coolants are light water, heavy water, helium, and carbon dioxide. The type of coolant is commonly used to designate the type of reactor. Hence, the characterization of reactors as light water reactors (LWRs), heavy water reactors (HWRs), and gas-cooled reactors (GCRs).

---

[5] If the fuel is designed to operate at high enough temperatures, cooling can be provided by radiation from the fuel, and maintaining the flow of coolant would not be essential under accident conditions. However, at present no operating reactors are designed for such high temperatures.

The coolant may also serve as a moderator, as is the case for LWRs and HWRs. In gas-cooled reactors, the density of the coolant is too low to permit it to serve as the chief moderator, and graphite is used. Fast reactors, in which fission is to occur without moderation to thermal energies, usually use a coolant that has a relatively high atomic mass number (A).[6] Generically, these reactors are termed liquid-metal reactors, because the coolant is a liquid metal, most commonly sodium $(A = 23)$.

Control Materials

As discussed in Section 7.5.2, control materials are needed to regulate reactor operation and provide a means for rapid shutdown. Boron and cadmium are particularly good control materials because of their high cross sections for the absorption of thermal neutrons. These control materials are usually used in the form of rods. Control rods for pressurized water reactors (PWRs) commonly use boron in the form of boron carbide (B_4C) or cadmium in a silver–indium alloy containing 5% cadmium. Control rods for boiling water reactors (BWRs) commonly use boron carbide [7, p. 715]. In addition, boron may be introduced into the circulating cooling water to regulate reactor operation.

8.1.4 World Inventory of Reactor Types

Reactor Sizes

The earliest reactors had generating capacities well below 100 MWe, but there was a rapid transition to 1000-MWe reactors and larger. The move to a larger size was motivated by the desire to capture economies of scale. Some analysts suggest that this escalation proceeded too rapidly, especially in the United States, and was responsible for some of the difficulties encountered in achieving short construction times and reliable operation.

The mean capacity of all reactors in operation worldwide in 2003 was about 820 MWe (see Table 2.1). At the extremes, a class of older British gas-cooled reactors have capacities of 50 MWe, whereas four PWRs that went into operation in France in 2000 each have a capacity close to 1450 MWe [8].[7] While most reactors built in recent years—including in France, Japan, South Korea, the United States, and the United Kingdom (for the one reactor in 1995)—are large, considerable attention is being given to smaller reactors. Although going to smaller reactors means sacrificing economies of

---

[6] It is also possible to use helium as a coolant because there the helium gas is not dense enough to be an effective moderator.

[7] Four 50-MWe units at the Calder Hall power plant in the United Kingdom were officially shut down on March 31, 2003 after operations that began in the late 1950. (Three of these units had actually suspended operations in 2001, but remained listed as "operating" in standard tables [9]). Four other 50-MWe reactors remain in operation at the Chapelcross plant.

scale, some advantages can be regained if a number of identical units are placed at the same site. (Questions of future reactor size are discussed further in Section 16.1.3.)

Types of Reactors

A variety of different reactors are in use in the world today, although there was greater diversity in the early days of reactor design. Table 8.1 lists the types of nuclear power plant in operation in late 2003 as well as those reported to be under construction or on order [10]. The dominant reactor is the light water reactor (LWR), which uses ordinary water as both the coolant and moderator and enriched uranium in UO_2 pellets as the fuel. There are two types of light water reactor: the pressurized water reactor (PWR) and the boiling water reactor (BWR). Together, they account for 88% of the world's present generating capacity and 85% of the capacity nominally being built or on order. The main types of reactors in past or present use for electricity generation are as follows:

PWR. The pressurized water reactor accounts for almost two-thirds of all capacity and is the only LWR used in some countries, for example France, the former Soviet Union, and South Korea.

BWR. The boiling water reactor is a major alternative to the PWR and both are used, for example, in the United States and Japan.

Table 8.1. World totals for nuclear reactors in commercial operation and under construction, November 2003, classified by reactor type: number of reactors and capacity (in GWe).

	Number		Capacity (GWe)					First
Type	Oper	Cons	Oper	Cons	Usual Fuel[a]	Moderator	Coolant	Developed
PWR	263	18	236.0	16.0	UO_2 enr	H_2O	H_2O	USA
BWR[b]	92	5	80.6	6.4	UO_2 enr	H_2O	H_2O	USA
PHWR[c]	39	8	19.3	3.1	UO_2 nat	D_2O	D_2O	Canada
GCR	26	0	10.9	0.0	U, UO_2^d	C	CO_2	UK
LGR	17	1	12.6	0.9	UO_2 enr	C	H_2O	USA/USSR
LMFBR	3	0	1.0	0.0	$UO_2 + PuO_2$	None	Liq Na	Various
TOTAL	440	32	360	26				

[a]Fuel designations: enr = enriched in ^{235}U, nat = natural.

[b]The listing for BWRs includes two ABWRs in operation and four under construction.

[c]Includes one 148-MWe HWLWR in Japan.

[d]Natural U used for GCR; enriched UO_2 used for AGCR.

Source: Capacity data are from Ref. [10]. Fuel and country data are from Ref. [11, p. 67].

ABWR. The advanced boiling water reactor incorporates improvements over earlier BWRs. It is in use in Japan, with additional units under construction in Japan and Taiwan.

PHWR. The pressurized heavy water reactor uses heavy water for both the coolant and moderator and operates with natural uranium fuel. It was developed in Canada and is commonly referred to as the CANDU.[8] Other countries with CANDU units in operation include India, South Korea, and Argentina.

GCR. The gas-cooled, graphite-moderated reactor uses a CO_2 coolant and a graphite moderator. Its use is limited to the United Kingdom; it is sometimes known as the Magnox reactor. A larger second-generation version is the advanced gas-cooled, graphite-moderated reactor (AGCR).

LGR. The light-water-cooled graphite-moderated reactor uses water as a coolant and graphite (in addition to water) for moderation. The world's major currently operating LGRs are the RBMK reactors in the former Soviet Union (11 in Russia and 2 in Lithuania).[9] There were four such units at the Chernobyl plant at the time of the accident there, but they have all been shut down.

HTGR. The high-temperature gas-cooled reactor uses helium coolant and a graphite moderator. The only HTGR that had been operating in the United States (Fort St. Vrain) has been shut down, and there are no HTGRs being used elsewhere for electricity generation, although active studies of variants of the HTGR are underway.

LMFBR. The liquid-metal fast breeder reactor uses fast neutrons and needs no moderator. A liquid metal is used as coolant, now invariably liquid sodium. There are only two LMFBR reactors in operation (one each in France and Russia).[10]

HWLWR. The heavy-water-moderated, light-water-cooled reactor is an unconventional variant of the heavy water reactor, and only one has been in recent operation, a 148-MWe plant in Japan. A new 700-MWe version of the HWLWR is being designed in Canada, the ACR-700 (see Section 16.2.2).

The dominance of light water reactors, both for plants in operation and those under construction, is seen in Table 8.1. These reactors were first developed in the United States, in both the PWR and BWR configurations, and have become the reactors of choice in almost all other major nuclear

---

[8] This acronym stands for Canadian deuterium uranium and has an obvious double meaning.

[9] In addition, there are four 11-MWe LGRs in Russia.

[10] In addition, the 246-MWe Monju reactor in Japan is listed by the IAEA as connected to the grid, but it has been shut down since 1995 after operating for only a few months [8].

countries. The main exceptions are Canada, the United Kingdom, the former Soviet Union (FSU), and India. Even in the United Kingdom and the FSU, the most recently completed reactors are PWRs.

The number of reactors under construction or on order as of late 2003 (32) was small compared to the number in operation (440). The average capacity of these reactors is about 830 MWe—very close to the average for operating plants. They range in size from four 202-MWe PHWRs being built in India [12] to two ABWRs in Japan with capacities near 1300 MWe.

History of Commercial Reactor Development

After World War II, the leading countries in nuclear reactor development were the United States, the United Kingdom, Canada, and the Soviet Union. Each went in a different direction.

The first U.S. power reactors, beyond plutonium-producing or experimental reactors, were built for submarines, not for civilian electricity generation. The earliest were a PWR for the submarine *Nautilus*, commissioned in 1955, and a sodium-cooled reactor for the submarine *Seawolf*. The *Seawolf* reactor had difficulties, and sodium-cooled reactors were abandoned by the navy in 1956 in favor of light water reactors [13, p. 423]. The navy PWR program provided the foundation for the development of PWRs for electricity generation, starting with the 60-MWe reactor at Shippingport, Pennsylvania, in 1957.

As was noted in Section 2.3.2, during the 1950s a varied array of reactors were ordered in the United States. These even included a small fast breeder reactor (Fermi I) in Michigan, which went into operation for a few years starting in 1966. However, after 1967, the only commercial power reactors put into operation in the United States have been PWRs and BWRs, with the sole exception (in 1979) of the trouble-plagued Fort St. Vrain HTGR in Colorado, which has since been shut down. The commercial BWRs were an outgrowth of a program of experimental BWR development carried out in the mid-1950s at Argonne National Laboratory.

The United Kingdom and Canada followed routes that did not require enriched uranium. The United States had a monopoly on uranium enrichment at the time, and although it presumably would have provided enriched uranium to such close allies, there may have been a reluctance on their part to become dependent. The United Kingdom program began very early, with two 50-MWe reactors at Calder Hall in 1956. These were GCRs, with graphite moderation and CO_2 cooling. They differed from most later reactors in the world in that they used uranium metal for the fuel, not uranium dioxide (UO_2). They gained the name Magnox, because the fuel pin cladding material was a magnesium alloy called Magnox [11, p. 165]. The GCRs that were built later had increasing size, up to 420 MWe. From the mid-1970s, with one exception, the few new plants brought on line in the United Kingdom have been AGCRs in the

600–700 MWe range. Like the Magnox reactors, they use graphite moderation and CO_2 cooling, but their fuel is enriched UO_2.

On the whole, after a fast start, the British reactor program has moved fitfully, with indecision abetted by North Sea oil and natural gas. After 1989, only one new reactor was put into commercial operation in the United Kingdom, the 1188-MWe Sizewell B reactor in 1995 [8]. Interestingly, it is a PWR, selected after prolonged study, adding further to the dominance of LWRs in the world nuclear picture.

The Canadian nuclear program offers the main alternative to the LWR among reactors now in operation or under construction. This program has involved only one type of reactor, the pressurized heavy water reactor (PHWR) known as the CANDU (Canadian deuterium uranium). Use of a deuterium moderator enabled Canada to use natural, rather than enriched, uranium. This was an attractive option for a country that had sophisticated scientific and engineering capabilities, including experience with heavy water reactors gained during World War II, but no enrichment facilities. The larger CANDU reactors (greater than 600 MWe) had the best cumulative capacity factors of any reactor type, as of the end of 2001 [14].

The PHWR has made substantial inroads outside of Canada, in particular in India, South Korea, and Argentina, with smaller programs elsewhere. In Canada, these reactors vary relatively little in size, starting at 525 MWe and most recently built at 881 MWe. India has emulated the Canadian example of reliance on PHWRs, although initially at the smaller size of about 200 MWe. The first of these were constructed under Canadian supervision, but, subsequently, India has assumed independent responsibility. South Korea has fewer PHWRs than India (four compared to fourteen) but they are larger— each about 650 MWe.

The other major dissenter from LWRs had been the Soviet Union, but this is now changing. The Soviet Union began with six 100-MWe light-water-cooled, graphite-moderated reactors (LGR) put into operation from 1958 to 1963 [8]. These were followed by larger LGR reactors with capacities of about 950 MWe, the so-called RBMK reactors among which were the Chernobyl reactors. These LGRs were built to produce both plutonium and electricity, as was the now-closed Hanford-N reactor in Washington. The Soviet Union also developed its own PWRs, the WWER series, and these are now the most numerous reactors in Russia and are widely used in much of Eastern Europe [8].

Overall, what worldwide growth there is in nuclear power is now primarily in the form of LWRs, with the Canadian PHWR as the only other significant player. It is not clear whether this is because of intrinsic technical and economic advantages of water-cooled reactors or because of historical and commercial forces. For the future, there is considerable interest in new HTGRs as a relatively near-term option and in liquid-metal-cooled reactors for the longer term (see Chapter 16), but the dominance of water-cooled reactors, and particularly LWRs, has not yet been seriously challenged.

8.2 Light Water Reactors

8.2.1 PWRs and BWRs

The two types of LWR in use in the world are the pressurized water reactor (PWR) and the boiling water reactor (BWR). The difference between them, as embodied in the names, is in the condition of the water used as coolant and moderator. In the PWR, the water in the reactor vessel is maintained in liquid form by high pressure. Steam to drive the turbine is developed in a separate steam generator. In the BWR, steam is provided directly from the reactor. These differences are brought out in the schematic representation of the two reactor types in Figure 8.1.

Under typical conditions in a PWR, temperatures of the cooling water into and out of the reactor vessel are about 292°C and 325°C, respectively, and the pressure is about 155 bar [7, p. 713].[11] For the BWR, typical inlet and outlet temperatures are 278°C and 288°C, respectively, and the pressure is only about 72 bar. The high pressure in the PWR keeps the water in a condensed phase; the lower pressure in the BWR allows boiling and generation of steam within the reactor vessel.

Neither the PWR nor BWR has an overwhelming technical advantage over the other, as indicated by the continued widespread use of both. Among the major LWR users, the United States, Japan, and Germany use both types, while France, South Korea, and Russia use PWRs almost exclusively in the LWR part of their programs. Overall, the number of PWRs in operation is significantly greater than the number of BWRs, and PWRs also have a lead in reactors listed as under construction. The future is not clear-cut, however. For example, in Japan, all three reactors under construction in 2003 were BWRs, including two ABWRs. In the following discussion, we will emphasize the PWR in giving specific illustrations but will consider both to some extent.

8.2.2 Components of a Light Water Reactor

The containment structure and enclosed components for a typical PWR and a typical BWR are shown schematically in Figures 8.2 and 8.3.[12] The most conspicuous difference between them is the absence of a steam generator in the BWR. At the heart of the reactors, literally and figuratively, is the reactor core, contained within the reactor pressure vessel. The pressure vessel encloses three vital components:

♦ The fuel itself, contained in many small fuel rods comprising the reactor core.

♦ The surrounding water, acting as coolant, moderator, and heat-transfer agent.

---

[11] 1 bar = 10^5 newton/m^2 = 0.987 atm.

[12] These diagrams are copied from a draft version of Ref. [15].

Boiling water reactor (BWR)

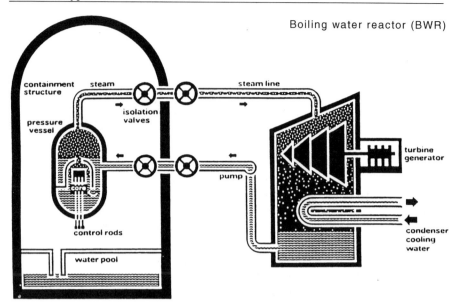

Pressurized water reactor (PWR)

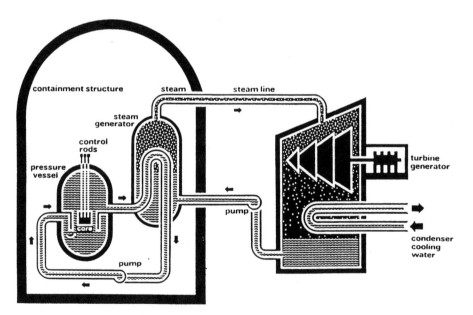

Fig. 8.1. Schematic representation of BWR and PWR systems, emphasizing the difference in the means for providing steam to the steam turbine. [Adapted from figures provided by the U.S. Council on Energy Awareness.]

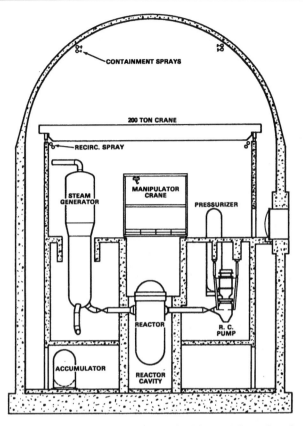

Fig. 8.2. Schematic diagram of containment building with enclosed reactor vessel and steam generator for an illustrative PWR: the 781-MWe Surry plant. The output of the steam generator drives the turbines, external to the containment building. (From [15, p. 4–4].)

♦ Control rods, used to establish the reactivity at the desired level and shut the reactor down in case of an emergency.

The reactor pressure vessel is a massive cylindrical steel tank. Typically for a PWR, it is about 12 m (40 ft) in height and 4.5 m (15 ft) in diameter [16, p. 304]. It has thick walls, about 20 cm (8 in.), and is designed to withstand pressures of up to 170 atm.

A second major component, or set of components, is the system for converting the reactor's heat into useful work. In the BWR, steam is used directly from the pressure vessel to drive a turbine. This is the step at which electricity is produced. In the PWR, primary water from the core is pumped at high pressure through pipes passing through a heat exchanger in the steam generators. Water fed into the secondary side of the steam generator is converted into steam, and this steam is used to drive a turbine. The secondary loop is also

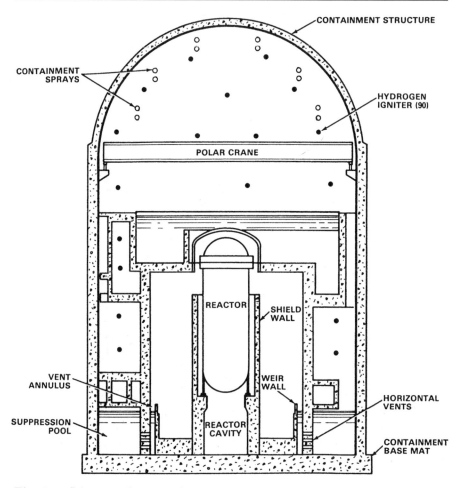

Fig. 8.3. Schematic diagram of containment building with enclosed reactor vessel for an illustrative BWR: the 1142-MWe Grand Gulf plant. Steam from the reactor vessel drives the turbines, external to the containment building. (From [15, p. 4–40].)

closed. The exhaust steam and water from the turbine enter a condenser and are cooled in a second heat exchanger before returning to the steam generator.

The cold side of the condenser heat exchanger represents the tertiary loop for the PWR. In principle, this loop need not be closed, and the condenser cooling water could circulate to and from a river or the ocean. More commonly, this condenser output is circulated through a cooling tower, where it is cooled by evaporation and the ultimate heat sink becomes the air. The part of the water that is lost as steam from the cooling tower is made up by water drawn, say, from a river.

Cooling towers became represented as ominous-looking symbols during the Three Mile Island accident, but they themselves are quite benign. The steam that may be seen rising from a cooling tower is isolated from water passing through the reactor core by the heat exchangers and, thus, is not a source of radioactive emissions, even if the primary cooling water is slightly radioactive. Such cooling towers are not unique to nuclear power plants and are used in other facilities, including coal-fired plants, where it is necessary to dissipate large amounts of heat.

The pressure vessel and the steam generators are contained within a massive structure, the *containment building*, commonly made of strongly reinforced concrete. In some designs, the concrete containment is lined with steel; in others, there is a separate inner steel containment vessel. The containment is intended to retain activity released during accidents and is believed capable of protecting a reactor against external events including some airplane impacts (see Section 17.5.5). In the Three Mile Island accident, the containment very successfully retained the released radioactivity, although it may be noted that the physical structure was not put fully to the test because there was no explosion or buildup of high pressures. At Chernobyl, there was no containment, with disastrous results. In principle, were the possibility of an accident made negligibly small with improved reactor design and if terrorist attacks were not a concern, a containment would be unnecessary. Nonetheless, it is widely considered to be an important safety feature, providing an additional layer of protection.

8.2.3 PWR Reactor Cores

We consider here the specific characteristics of a reactor core based on a Westinghouse Corporation PWR design, but the gross features are similar for all large LWRs.[13] The reactor fuel is in the form of cylindrical uranium oxide (UO_2) pellets, about 0.8 cm in diameter and 1.35 cm in length. The pellets are placed in tubes—called fuel rods or fuel pins—made of zircaloy, a zirconium alloy (98% Zr, 1.5% Sn, and small amounts of other metals [16, p. 234]) selected on the basis of structural strength and low neutron absorption. The zircaloy cladding is thin, 0.06 cm. The fuel rod is typically 3.7 m (12 ft.) in length and 1.0 cm in diameter. There is some free space within the fuel rod to allow for the expansion of the fuel pellets and to accommodate gaseous fission products such as xenon and krypton. The fuel expansion is due to both increased temperature and the replacement in the fuel of one uranium atom by two fission-product atoms. Noble gases produced as fission products may be trapped as bubbles in the fuel or may escape from the fuel into the gap between the fuel and cladding.

A 17 × 17 array of fuel rods forms a "fuel bundle" or *assembly*. Although this would allow 289 fuel rods in an assembly, positions are left open in some

---

[13] Most of the detailed numbers in this paragraph are based on Westinghouse Corporation information, as reported in Ref. [16], especially Table 9.1. They are specific to this particular Westinghouse design; others differ in detail.

assemblies for the insertion of control rods or measuring instrument rods. The total core contains 193 assemblies and 50,952 fuel rods. Some 53 of the assemblies have spaces for clusters of 20 control rods, which can be moved in and out within the assembly. These control rods are made from a silver–indium–cadmium alloy.

Fuel assemblies are replaced periodically, but not all in the same period of reactor refueling. Thus, although a particular fuel assembly may remain in the reactor for 3 years, one-third of the core can be changed every year. (Recently, cores have been designed to have a longer time between fuel changes.) As the ^{235}U is consumed in the reactor, the reactivity of the fuel decreases. This is compensated for in several ways. Burnable poisons in the fuel are consumed, control rods that limit the reactivity are partly withdrawn, or the soluble poison concentration in the cooling water is reduced.

8.3 Burners, Converters, and Breeders

8.3.1 Characterization of Reactors

As discussed in Section 7.1.1, the condition for a chain reaction is that for every neutron initiating fission in one generation, one or more neutrons initiate fission in the next generation. If, in addition, another fissile nucleus is produced for every ^{235}U atom consumed, then there is no decrease in the amount of nuclear fuel available. This is the principle of the breeder reactor.

The *conversion ratio C* (or *breeding ratio B*) is defined as the ratio of the rate of production of fissile nuclei to the rate of consumption of fissile nuclei [see Eq. (7.18)].[14] For uranium fuel, this is the ratio of ^{239}Pu produced to ^{235}U consumed. If the conversion ratio is small, the reactor is sometimes called a *burner*; if the conversion ratio is between about 0.7 and 1.0, it is commonly called a *converter*; and if it exceeds unity, the reactor is called a *breeder* (see, e.g., Ref. [16, p. 458]).

8.3.2 Achievement of High Conversion Ratios in Thermal Reactors

Difficulty of Reaching a Conversion Ratio of Unity with ^{235}U

As discussed in Section 7.4, the limiting condition for a breeder reactor is that the conversion ratio, C, be at least 1. This means that the number η_o of neutrons produced for each neutron absorbed in ^{235}U, must be two or more.[15] For thermal neutrons absorbed in ^{235}U, $\eta_o = 2.075$ (see Table 7.1). Were there no losses, this would suffice for breeding: 1 neutron for continuing the

---

[14] Sometimes, a distinction is made in terminology, with conversion ratio used when $C < 1$ and breeding ratio used when $C > 1$.

[15] The relation among η_o, η, and the commonly cited parameter ν, is discussed in Section 7.1.2.

chain reaction, 1 neutron for production of ^{239}Pu, and 0.08 neutrons free to be "wasted." Such efficient utilization cannot be achieved, however, because there is absorption in the moderator and other nonfuel materials, as well as escape of neutrons from the core. Therefore, a ^{235}U-fueled thermal breeder reactor is not practical. Nonetheless, the production of ^{239}Pu is significant in uranium-fueled thermal reactors because its fission increases the overall energy output from the reactor fuel beyond that gained from the ^{235}U alone.

Potential of ^{233}U for a Thermal Breeder Reactor

The number of neutrons produced is significantly higher for ^{233}U ($\eta_o = 2.296$) than for ^{235}U, and there have been serious suggestions for developing ^{233}U thermal breeders. These date to as early as 1945, in work done by Eugene Wigner's group in Chicago [3, Chapter 6]. A cycle is envisaged in which ^{233}U is produced initially in a reactor with ^{235}U as the fissile fuel and ^{232}Th as the fertile fuel. Subsequently, a ^{233}U–^{232}Th cycle could, in principle, be self-sustaining. Not only is η_o higher for ^{233}U than for ^{235}U, but the capture cross section is significantly higher for ^{232}Th than for ^{238}U at thermal energies, making the conversion ratio higher than for a cycle based on ^{238}U (see Table 7.1).

However, although thermal breeders based on ^{233}U are, in principle, possible and preliminary exploratory work toward their development was done at Oak Ridge National Laboratory in the 1950s, thermal breeders were abandoned in favor of the fast breeder reactor. It is conceivable that interest in thermal breeders could revive, but, to date, the few breeder reactors that have gone into commercial operation have all been fast breeders.

High Conversion Ratios Without Breeding

Before turning to fast breeder reactors, it may be noted that even if breeding is not achieved with thermal reactors, a high conversion ratio can still be desirable. One motivation could be plutonium production. Another motivation is the extension of fuel resources. As the conversion ratio increases, the energy output increases for a given original ^{235}U content. A high conversion ratio means a high ratio of capture in ^{238}U to absorption in ^{235}U. This must be accomplished without losing criticality. Greater losses of neutrons to ^{238}U can be compensated for by smaller losses in the moderator and reactor structure.

The use of carbon instead of light water as a moderator is favorable on two counts if a high conversion ratio is desired (in addition to the advantage that with a carbon moderator it is possible to use natural uranium). Because carbon is a less effective moderator than water, more collisions are required to reach thermal energies; therefore, there is more possibility of neutron capture in ^{238}U at intermediate energies. Further, because of the low neutron-capture cross section in ^{12}C (see Section 7.2.2), the loss of thermal neutrons to absorption will be less for carbon than for light water.[16] Together, this means a higher conversion ratio.

The same general arguments apply to heavy water reactors. The conversion ratio is higher in a heavy water reactor than in a light water reactor due to less effective moderation in heavy water and a lower neutron-capture cross section. For both graphite-moderated and heavy water reactors, there have been suggestions that the ^{232}Th–^{233}U cycle be used, to further increase conversion and extend the life of the uranium fuel, even without breeding.

It may be noted that reactors designed with production of plutonium for weapons in mind, either as the main or as an auxiliary function, have been mostly graphite moderated. Examples include the Windscale plant in England, the plutonium production reactors at Hanford, and the RBMK reactors built in the USSR. The five heavy water reactors at the Savannah River (South Carolina) complex for plutonium production are the major exception.

8.3.3 Fast Breeder Reactors

Plutonium as Fuel for Fast Breeders

A thermal breeder reactor is not possible using ^{239}Pu due to the high ratio α of the capture cross section to the fission cross section for thermal neutrons.[17] However a fast breeder reactor is possible. It relies on a chain reaction in which the neutrons are not thermalized but instead produce fission at relatively high energies. If ^{239}Pu is the fissile fuel, the cycle uses ^{238}U as the fertile fuel and ^{239}Pu is both consumed and produced in the reactor. The cycle is started using ^{239}Pu produced in a uranium-fueled reactor.

Although most of the fission neutrons are emitted with energies above 1 MeV, they can lose energy through inelastic scattering in ^{239}Pu and ^{238}U. Fission is therefore produced at energies extending over a very broad energy region, from above 1 MeV to below 1 keV. In the high-energy part of this region the conditions are very favorable for breeding. For example, for 1.0-MeV neutrons on ^{239}Pu, $\sigma_f = 1.7$ b, and the ratio (α) of the capture cross section to the fission cross section is less than 0.03. The low value of this ratio means that almost all absorption in ^{239}Pu leads to fission. With about three neutrons per fission at 1 MeV, breeding with ^{239}Pu is readily achieved, with 1 neutron for continuing the chain reaction, 1^+ for breeding, and 1^- for losses. At lower

---

[16] In terms of the formalism introduced in Chapter 7, this means that carbon leads to a lower resonance escape probability p and a higher thermal utilization factor f than does light water [see the four-factor formula, Eq. (7.5)]. Criticality can still be maintained ($k = 1$), and C will be greater (see Eq. (7.23)).

[17] For ^{239}Pu, $\alpha = 0.360$ and $\eta_o = 2.115$ at thermal neutron energies. This may be compared to the values for ^{233}U: $\alpha = 0.086$ and $\eta_o = 2.296$. For breeding, the crucial condition is that $\eta_o > 2$. Therefore ^{233}U is significantly better than ^{239}Pu at thermal energies.

neutron energies, the cross section σ_f remains high, above 1.5 b for the most part, but the ratio α rises—reaching about 0.3 at 40 keV—greatly reducing the number of "surplus" neutrons [17, p. 753]. The actual conversion ratio depends on the neutron reactions over the full range of "incident" neutron energies, with the lower-energy neutrons contributing even if not as effectively as the high-energy neutrons.

To avoid thermalization of the neutrons, fast breeder reactors use a coolant with a relatively high mass number A. Liquid metals have the best combination of high A and favorable heat-transfer properties, and the fast breeder reactors in actual use have been *liquid-metal fast breeder reactors* (LMFBR). The standard choice for the coolant is liquid sodium (^{23}Na).

The fuel is made of pellets of mixed plutonium and uranium oxides, PuO_2 (about 20%) and UO_2 (about 80%). Uranium depleted in ^{235}U is commonly used, it being available as a residue from earlier enrichment. The fission cross section for ^{239}Pu is between 1.5 and 2.0 b over virtually the entire fast-neutron region (from 10 keV to 6 MeV), whereas for ^{238}U it is below 0.2 b for $E_n < 1.4$ MeV and falls rapidly at lower E_n (see Figure 6.1 and Ref. [17]). Therefore, fission in ^{239}Pu is much more probable than fission in ^{238}U. The most probable fast-neutron reactions in ^{238}U are inelastic scattering, which produces lower energy neutrons, and capture, which produces ^{239}Pu.

Status of Fast Reactor Programs

The main incentive for the development of fast breeder reactors is the extension of uranium supplies. A fast breeder economy would extract much more energy per tonne of uranium than is obtained from other reactors (e.g., the LWRs). Further, with more energy per unit mass, it becomes economically practical to use more expensive uranium ores, increasing the ultimate uranium resource. A secondary incentive is the easing of the waste disposal problem if plutonium and uranium (and possibly other actinides) are reused in a closed cycle rather than disposed of as waste.

However, during the 1980s and 1990s, growth of nuclear power fell far short of earlier expectations, there was little pressure on uranium supplies, and interest in fast breeder reactors declined. Further, the initial fast breeder reactors proved to be more expensive than alternatives, such as the LWR or HWR. Particularly in the United States, there was also the concern that the large-scale use and availability of ^{239}Pu might increase dangers from terrorism and nuclear weapon proliferation (see Section 9.4.2 and Chapter 17). These factors made breeder programs a vulnerable target, at a time when there was significant opposition to any projects to advance nuclear power. Nonetheless, some development of breeders has continued, in part to maintain the technology as insurance against future needs.

In a later turn of the argument, it has been pointed out that a liquid-metal fast reactor (LMR) can be used to *destroy* unwanted plutonium and other heavy elements which are in weapons stockpiles or nuclear wastes. In this reversal of motivation, the LMR would be used to consume plutonium, rather than to produce plutonium as a fuel. There is flexibility in this, because as LMR technology and facilities are developed, they could be turned to either purpose. However, if the driving fear is concern about misuse of plutonium, it may appear more desirable to dispose of plutonium from weapons stockpiles in ways that do not involve expanding a technology that is closely related to potential plutonium production.

France had led in the development and deployment of breeder reactors, with two completed reactors, the 233-MWe Phenix, put into operation in 1973, and the 1200-MWe SuperPhenix at Creys-Malville, which first generated electricity in 1986 but was finally shut down in 1998 after a troubled history of recurring technical difficulties. Small breeder reactors in Great Britain and Kazakhstan were also shut down in the 1990s, and there now remain only two LMFBRs operating to produce electricity, Phenix in France and a 560-MWe reactor in Russia. However, interest in breeder reactors remains in a number of countries, with ongoing and new activity, and it would be premature to write breeder reactors off as an option for the future.[18]

The United States breeder reactor program has been marked by indecision and opposition, with successive projects started and abandoned. It had started auspiciously, with the operation of the Experimental Breeder Reactor (EBR-I) in Idaho. On December 20, 1951 EBR-I generated the first electricity from a nuclear reactor produced anywhere—enough for four light bulbs. Its output was shortly thereafter increased to 0.1 MWe [19]. Two larger, fast reactors made important research contributions in subsequent years—the Experimental Breeder Reactor II (EBR-II) in Idaho and the Fast Flux Test Reactor (FFTR) in Washington—but both of these projects have been terminated. The most ambitious breeder proposal in the United States was for the Clinch River Breeder Reactor (CRBR)—a project that was under active consideration in the 1970s but was terminated by Congress in 1983.

Following the end of the CRBR project, a major fast reactor development program was undertaken at Argonne National Laboratory as part of the *integral fast reactor* plan (see Section 16.5). Advocates of this program stressed its potential to offer a high degree of safety against reactor accidents and to consume nuclear wastes in an on-line process. The breeding potential was often secondary in these arguments, and the planned LMR need not have operated as a breeder, namely with a conversion ratio greater than unity. Nonetheless, the basic configuration of the system was similar to that of a breeder reactor. Culminating several years of debate, most funding for this project was terminated in the mid-1990s.

---

[18] Current breeder projects include the 246-MWe Monju breeder reactor in Japan, which is now shut down due to operating difficulties but which may be restarted, a 750-MWe LMFBR being planned in Russia (Beloyarsk-4) with completion scheduled for 2009, a prototype 500-MWe breeder being planned in India (Kalpakkam), and a 65-MWt fast neutron reactor being built in China [18].

Despite these difficulties, the long-term argument for breeder reactors remains and breeder development may intensify, especially if the proliferation problems can be satisfactorily addressed. Some of the reactors now being considered under the program of the Generation IV International Forum (see Chapter 16) are designed to operate with a fast-neutron spectrum and thus have the potential of being operated as breeder reactors.

8.4 The Natural Reactor at Oklo

A remarkable discovery was made in 1972 by French scientists analyzing uranium extracted from the Oklo uranium mine in Gabon. The uranium was depleted in ^{235}U, sometimes by large amounts, although, normally, the isotope ratios in uranium are nearly constant over the surface of the Earth. It was soon suspected and then demonstrated that this isotopic anomaly was due to a natural uranium chain reaction occurring more than a billion years ago. Conclusive evidence in support of this explanation was provided by the relatively high abundance of intermediate-mass nuclei, the rare earths, which are characteristic fission products but are not normally found in large abundance in nature (see, e.g. Ref. [20]).

The scenario, as it has been recreated, puts the event about 1.8 billion years ago. At that time, the isotopic abundance of ^{235}U exceeded its present value by the factor $\exp(\Delta\lambda t)$, where $\Delta\lambda$ is the difference in the decay constants of the two isotopes and t is the time since the event. The decay constants of ^{235}U and ^{238}U are 0.985×10^{-9} yr^{-1} and 0.155×10^{-9} yr^{-1}, respectively (see Table 2.1), giving $\Delta\lambda t = 1.49$ for $t = 1.8 \times 10^9$ yr. Therefore. the isotopic abundance of ^{235}U was 4.4 times greater at the time of the Oklo event than it is today, putting the enrichment at slightly above 3%. (This is strikingly close to the enrichment used in modern LWRs.) The intrusion of water, acting as a moderator, apparently initiated a chain reaction, which appears to have simmered for at least several hundred thousand years. In this model of what took place, the reaction did not occur earlier because the concentrated uranium deposits had been only recently formed by the leaching of rocks and the precipitation of the dissolved uranium.

Aside from having posed an intriguing scientific puzzle, with a very interesting explanation, the Oklo event is considered by some to have significance as a test of the motion of fission products through the ground. For the most part, these products have moved very little over a period of more than 1 billion years. This could have implications for the rate of movement of fission products in buried nuclear wastes. The Oklo example cannot be used as an all-embracing guide because differences in the chemical form of the product and in the type of rock formation may vitiate an extrapolation from Oklo to the behavior of an individual modern waste disposal site. However, Oklo illustrates that at least under some circumstances, radionuclides do not migrate appreciably from their initial location.

References

1. H. Barnert, V. Krett, and J. Kupitz, "Nuclear Energy for Heat Applications," *IAEA Bulletin* 33, no. 1, 1991: 21–24.
2. Alvin M. Weinberg and Eugene P. Wigner, *The Physical Theory of Neutron Chain Reactors* (Chicago: University of Chicago Press, 1958).
3. Alvin M. Weinberg, *The First Nuclear Era: The Life and Times of a Technological Fixer* (New York: American Institute of Physics Press, 1994).
4. J. Smith, "Novel Reactor Concepts," in *Nuclear Power Technology*, Vol. 1 of *Reactor Technology*, W. Marshall, ed. (Oxford: Clarendon Press, 1983), pp. 390–415.
5. Uri Gat and H. L. Dodds, "The Source Term and Waste Optimization of Molten Salt Reactors with Processing," in *Future Nuclear Systems: Emerging Fuel Cycles & Waste Disposal Options*, Proceedings of *Global '93* (La Grange Park, IL: American Nuclear Society, 1993), pp. 248–252.
6. Emilio Segrè, *Nuclei and Particles*, 2nd edition (Reading, MA: W.A. Benjamin, 1977).
7. Ronald Allen Knief, *Nuclear Engineering: Theory and Technology of Commercial Nuclear Power*, 2nd edition (Washington, DC: Hemisphere Publishing Company, 1992).
8. "World List of Nuclear Power Plants," *Nuclear News* 46, no. 3, March 2003: 41–67.
9. "United Kingdom: Calder Hall Bows out at the Age of 47," *Nuclear News* 46, no. 6, May 2003: 48.
10. International Atomic Energy Agency, "Nuclear Power Plants Information, Operational & Under Construction Reactors By Type" (updated 11/24/03). [From: http://www.iaea.org/cgi-bin/db.page.pl/pris.reaopucty.htm]
11. D.J. Bennet and J.R. Thomson, *The Elements of Nuclear Power*, 3rd edition (Essex, UK: Longman, 1989).
12. International Atomic Energy Agency, "Latest News Related to PRIS and the Status of Nuclear Power Plants" (updated May 2003). [From: http://www.iaea.org/programmes/a2/]
13. Richard G. Hewlett and Jack M. Holl, *Atoms for Peace and War, 1953–1961* (Berkeley, CA: University of California Press, 1989).
14. International Atomic Energy Agency, *Nuclear Power Reactors in the World*, Reference Data Series No. 2, April 2003 edition (Vienna: IAEA, 2003).
15. U.S. Nuclear Regulatory Commission, *Reactor Risk Reference Document*, NUREG-1150, Vol. 1, Draft for Comment (Washington, DC: NRC, 1987).
16. F.J. Rahn, A.G. Adamantiades, J.E. Kenton, and C. Braun, *A Guide to Nuclear Power Technology* (New York: Wiley, 1984).
17. Victoria McLane, Charles L. Dunford, and Philip F. Rose, *Neutron Cross Section Curves*, Volume II of *Neutron Cross Sections* (New York: Academic Press, 1988).
18. World Nuclear Association, *Information and Issues Briefs; Particular Countries*. [From: http://www.world-nuclear.org/info/info.htm]
19. Argonne National Laboratory West, "ANL-W History—Reactors (EBR 1): EBR-1." [From: http://www.anlw.anl.gov/anlw_history/reactors/ebr_i.html]
20. Michel Maurette, "Fossil Nuclear Reactors," *Annual Review of Nuclear Science* 26, 1976: 319–350.

9

Nuclear Fuel Cycle

9.1 Characteristics of the Nuclear Fuel Cycle

9.1.1 Types of Fuel Cycle

The *nuclear fuel cycle* is the progression of steps in the utilization of fissile materials, from the initial mining of the uranium (or thorium) through the final disposition of the material removed from the reactor. It is called a cycle because in the general case, some of the material taken from a reactor may be used again, or "recycled."

Fuel cycles differ in the nature of the fuel used, the fuel's history in the reactor, and the manner of handling the fuel that is removed from the reactor at the end of the fuel's useful life (known as the *spent fuel*). For uranium-fueled reactors—which means virtually all commercial reactors—a key difference is in the disposition of the plutonium and other actinides that are produced in a chain of neutron captures and beta decays that starts with neutron capture in ^{238}U to produce ^{239}Pu (see Section 7.4).[1] The actinides are important because (1) some, especially ^{239}Pu, are fissile and can be used as nuclear fuel in other reactors or in bombs, and (2) many of the actinides have long half-lives, complicating the problems of nuclear waste disposal. The three broad fuel cycle categories are as follows:

---

[1] The *actinides* are the elements with atomic numbers Z greater than or equal to that of actinium ($Z = 89$). (The terminology is not uniform and, sometimes, actinium is not included among the "actinides.") Neptunium ($Z = 93$), americium ($Z = 95$), and curium ($Z = 96$) are referred to as *minor actinides* in view of their low abundance in spent fuel compared to uranium ($Z = 92$) and plutonium ($Z = 94$). Elements with atomic numbers greater than 92 are termed *transuranic* elements.

♦ *Once-through fuel cycle.* This is sometimes called an *open fuel cycle* or a "throw-away" cycle. It is not really a cycle, in that the spent fuel is treated as waste when it is removed from the reactor and is not used further. The ^{239}Pu and other actinides are part of these wastes.

♦ *Reprocessing fuel cycle.* In the present standard reprocessing fuel cycle, plutonium and uranium are chemically extracted from the spent fuel. The plutonium is used to make additional fuel, often by mixing it with uranium oxides to produce *mixed-oxide fuel* (MOX) for use in thermal reactors. This provides additional energy and changes the nature of the wastes. In potential variants of the reprocessing fuel cycle, the minor actinides would also be extracted, and they and the plutonium would be incorporated in fresh fuel for fast reactors (see Section 9.4.3).

♦ *Breeding cycle.* For this cycle, the reactor is designed so that there is more fissile material (mostly ^{239}Pu) in the spent fuel than there was in the fuel put into the reactor (see Section 8.3). As in the reprocessing fuel cycle, the plutonium can be removed and be used in another reactor. With a sequence of such steps, fission energy is in effect extracted from a substantial fraction of the ^{238}U in uranium, not just from the small ^{235}U component, increasing the energy output from a given amount of uranium by a factor that could, in principle, approach 100.

It may be noted that uranium accounts for most of the mass of the nuclear wastes in the once-through cycle. It is separated out in the reprocessing and breeding cycles for possible reuse in reactor fuel.

At present, all U.S. commercial reactors and the majority of reactors worldwide are operating with a once-through fuel cycle, although some countries, particularly France, have large-scale reprocessing programs with use of plutonium in the form of MOX fuel. It should be noted, of course, that even in the once-through fuel cycle, the potential for eventually using the fuel in a reprocessing cycle remains until the fuel is disposed of irretrievably. No country is employing a breeder cycle at this time, although France appeared on the verge of attempting such a program with its Phenix and Superphenix reactors—but this effort has been abandoned, at least for the time being (see Section 8.3.3).

Although virtually all of the world's commercial reactors have used uranium fuel, there is continuing interest in the use of thorium fuel.[2] In a thorium fuel cycle, the thorium (all ^{232}Th in nature) serves as the fertile fuel. Neutron capture and beta decay result in the production of ^{233}U, which has favorable properties as a fissile fuel. To start the thorium cycle, a fissile material such as ^{235}U or ^{239}Pu is needed, but once begun, it can be sustained if enough ^{233}U is produced to at least replace the initial fissile material. It is often argued that a thorium cycle is preferable to a uranium cycle, because if ^{233}U is ex-

---

[2] The Fort St. Vrain high-temperature, gas-cooled, graphite-moderated reactor in Colorado, which was shut down in 1989, is one of several exceptions to the exclusive use of uranium, having used thorium for part of its fuel [1, p. 41].

tracted from the spent fuel, it can be "denatured" by mixing it with natural uranium to make a fuel that cannot be used in a bomb. Bomb material could be obtained only after the isotopic separation of ^{233}U. In contrast, bomb material can be obtained from a uranium-fueled reactor by chemical separation of the plutonium (see Chapter 17). Isotopic separation is technically more difficult than chemical separation; thus, a thorium fuel cycle could be more proliferation resistant than a uranium fuel cycle unless, in the latter case, the plutonium is well protected from diversion or theft.

9.1.2 Steps in the Nuclear Fuel Cycle

A schematic picture of the fuel cycle is shown in Figure 9.1, which indicates alternative paths, with and without reprocessing [2]. The steps in the fuel cycle that precede the introduction of the fuel into the reactor are referred to as the *front end* of the fuel cycle. Those that follow the removal of the fuel from the reactor comprise the *back end* of the fuel cycle. At present, there is only a truncated back end to the fuel cycle in the United States, as virtually all commercial spent fuel is accumulating in cooling pools or storage casks at the reactor sites.

Implementation of a spent fuel disposal plan, or of a reprocessing and waste disposal plan, would represent the "closing" of the fuel cycle. This closing is viewed by many to be an essential condition for the increased use of nuclear power in the United States—and perhaps even for its continued use beyond the next several decades.

Key aspects of the fuel cycle will be surveyed in the remainder of this chapter. The fuel cycle will be discussed particularly in the context of light water reactors, in view of their dominance among world nuclear reactors. The main aspects are relevant to other types of reactor as well. A more extensive treatment of the crucial step of waste disposal will be given in Chapters 10–13.

9.2 Front End of the Fuel Cycle

9.2.1 Uranium Mining and Milling

Uranium Deposits in the Earth's Crust

The concentration of uranium varies greatly among geological formations. The average concentration in the Earth's crust is about 3 parts per million (ppm) by weight, but extremes extend from under 1 ppm to something in the neighborhood of 500,000 ppm.[3]

Uranium resources are widely distributed, with substantial uranium production in many countries, including Australia, Canada, Kazakhstan, Namibia,

---

[3] For example, one deposit in Canada is identified as having zones of "over 50% U_3O_8," which translates to over 42% uranium [3].

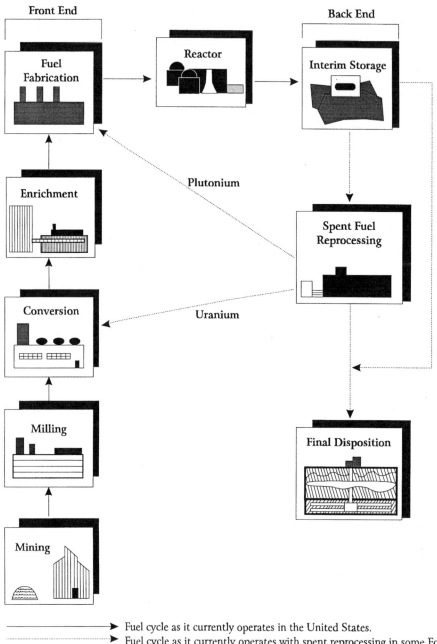

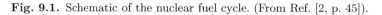

Fuel cycle as it currently operates in the United States.
Fuel cycle as it currently operates with spent reprocessing in some Foreign Countries and/or final waste storage.

Fig. 9.1. Schematic of the nuclear fuel cycle. (From Ref. [2, p. 45]).

Niger, the Russian Federation, South Africa, the United States, and Uzbek-istan [4, p. 36]. Most of the uranium now used in the United States is imported, with Canada being the largest supplier. Through 2000, the United States had been the world's leader in cumulative production of uranium, with Canada a close second. However, the U.S. share of production declined by the 1990s, and now Canada and Australia are the world's main suppliers of uranium. Together, they accounted for about 50% of world production in 2000 [4].

Uranium in rocks is mostly in the form of a uranium oxide, U_3O_8. In "conventional" mining, the rock is extracted from open pit or underground mines, the U_3O_8 is then extracted in the milling process by crushing the rock and leaching with acid, and the U_3O_8 is then recovered from the liquid and dried. The concentrated U_3O_8 is known as *yellowcake*.[4] In an "unconventional" method, appropriate for only certain types of uranium deposits, U_3O_8 is extracted by in situ leaching (i.e., by pumping a leaching agent through the ore without physical removal of the rock).[5]

At low concentrations, the uranium content is expressed in terms of the uranium grade, given in percent by weight of either uranium or U_3O_8. Thus, ore which is 1000 ppm of uranium corresponds to grades of 0.100% U or 0.118% U_3O_8.[6] In an extensive 1983 study of U.S. uranium ores, deposits were listed with U_3O_8 grades ranging from under 0.01% to over 1.8%, with a median of about 0.1% [6, p. 39]. The higher the grade, the less the amount of ore that must be extracted, which, in general, leads to lower costs. Ores below a grade of 0.05% are considered low-grade ores and have not been widely needed. Of course, the ultimate criterion is overall cost, not grade per se, and at one time open-pit mining utilized ores down to 0.04% [7, p. 411].

During 2001, most of the uranium extraction in the United States was done by in situ leaching, using ores ranging in grade from 0.09% to about 0.20% U_3O_8 [8]. Worldwide, conventional mining dominates but is now economically practical only at higher uranium grades ("above a few tenths of a percent") [1, p. 25].

Radon Exposures from Uranium Mining and Mill Tailings

In the early days of uranium mining, little attention was paid to radiation safety. In the Middle Ages, long before uranium had been identified as an element, metal miners in southern Germany and Czechoslovakia contracted lung ailments, called Bergkrankheit ("mountain sickness"). Modern scientists have attributed the ailment to lung cancer caused by a high uranium concentration

---

[4] U_3O_8 is not yellow in its pure form. Yellowcake is about 85% U_3O_8 [5, p. 241], and the yellow color results from another uranium compound in the ore.

[5] The designations "conventional" and "unconventional" correspond to those, for example, of Ref. [1, p. 25].

[6] In international usage, "grade" usually refers to U content, whereas in U.S. DOE documents it refers to U_3O_8 content. Note: U_3O_8 is 84.8% uranium and 15.2% oxygen, by weight.

that, by chance, was in the rock formations being mined. The decay of the radionuclides in the uranium series proceeds from ^{238}U through several steps to ^{226}Ra and then to radon gas (^{222}Rn) and its radioactive progeny. Inhalation of these "radon daughters" can lead to lung cancer (see Section 3.5.1).

As one would expect, the problem of radon exposure is more extreme in uranium mines than in other sorts of mine. It became a particularly serious problem in a number of countries—for example, in the United States, Czechoslovakia, and Canada—when large-scale uranium mining was begun in the 1940s to meet the demands of nuclear weapon and nuclear power programs. By the late 1950s, steps were initiated in the United States to reduce radon exposures, mainly through better ventilation, and by the 1970s, the average exposures of uranium miners had become quite low (lower than that from indoor radon in many homes). However, a good deal of damage had already been done, and there is unambiguous evidence of increased lung cancer fatalities among uranium miners.

The residues of the milling operation, representing the remainder of the ore after extraction of the U_3O_8, are the *mill tailings*. All of the uranium progeny, starting with ^{230}Th, are present in the tailings.[7] The radionuclide ^{230}Th has a half-life of 75,400 years and thus sustains the remainder of the uranium series for a long period of time. This results in the continuous production of radon, some escaping to the atmosphere. Of course, these steps do not increase the rate of radon production above what it would have been without mining, but the radon in the tailings can more readily reach the atmosphere than can radon in underground ore. At one time, this was viewed by some as constituting an important environmental hazard, and it is still deemed necessary to take remedial measures to limit radon emissions from the tailings (using overlying layers of material to impede radon escape). However, interest in the issue has diminished as it has become obvious that exposures from "normal" indoor radon pose a much more serious problem, in terms both of the number of people impacted and the magnitudes of the radon concentrations to which they are exposed.[8]

9.2.2 Enrichment of Uranium

Preparation for Enrichment: Conversion

There are a variety of approaches to the enrichment of uranium, each taking advantage of the small mass difference between ^{235}U and ^{238}U. In the most used of these processes, it is necessary to have the uranium in gaseous form. For that purpose, the U_3O_8 is chemically converted to gaseous uranium hexafluoride, UF_6. This is the compound of choice, because UF_6 is a gas at lower

---

[7] The ^{234}U remains in the yellowcake and the radionuclides between ^{238}U and ^{234}U in the uranium series are short-lived.

[8] For a comparison of the hazards from mill tailings and indoor radon, see, for example, Ref. [9].

temperatures than can be reached by any other uranium compound in gaseous form [10, p. 589].

Degrees of Enrichment

Natural uranium has an isotopic abundance by *number of atoms* of 0.0055% ^{234}U, 0.720% ^{235}U and 99.275% ^{238}U.[9] In the remainder of the discussion of uranium isotopic enrichment, we will follow the standard practice of describing the ^{235}U fraction in terms of *mass* rather than, as is common in many other scientific applications, of number of atoms.[10] For natural uranium, the ^{235}U abundance by mass is 0.711%. The presence of the small amount of ^{234}U is often ignored, because corrections on the order of 10^{-4} or less are irrelevant.

The fissile nuclide in thermal reactors is ^{235}U. For reactors that require uranium with a higher fraction of ^{235}U than is found in natural uranium, *enrichment* is necessary. This is, of course, the case for light water reactors (LWRs). Fuel used in LWRs in past years has been enriched to ^{235}U concentrations ranging from under 2% to over 4%. The anticipated average for the United States, for cumulative production up until about 2010, is 3.0% for BWR fuel and 3.75% for PWR fuel.[11]

The material used in LWRs is known as slightly enriched uranium, in contrast to the highly enriched uranium used for nuclear weapons and submarine reactors. Within the core of a given reactor, enrichments vary with the location of the fuel assemblies. As discussed later in the context of the burnup of fuel, there is a general trend toward using fuel with higher initial enrichments.

The products of the enrichment process are the enriched material itself and the depleted uranium, sometimes called *enrichment tails*. Typically, enrichment tails have in the neighborhood of 0.2% to 0.35% ^{235}U remaining [12, p. 7]. As one goes to lower concentrations of ^{235}U in the tails, the consumption of uranium ore is reduced, but the cost of enrichment is increased. Thus, there is a trade-off.

The depleted uranium is sometimes used in special applications. Its use in armor-piercing shells, where the high density of uranium is advantageous ($\rho \approx 19$ g/cm^3), has led to some public concern about the resulting environmental risks. However, depleted uranium has a lower specific activity than

---

[9] The ^{234}U arises as a member of the ^{238}U series, with an abundance relative to ^{238}U that is inversely proportional to the half-lives of the two isotopes (2.45×10^5 yr and 4.468×10^9 yr, respectively).

[10] These descriptions of isotopic abundance are related by the expression $w = [(1 - \delta)/(1 - x\delta)] x$, where, specialized to the case of uranium, w is the ratio of ^{235}U mass to total uranium mass, x is the ratio of the number of ^{235}U atoms to the total number of uranium atoms, and δ is the ratio of the difference between the ^{238}U and ^{235}U atomic masses to the ^{238}U atomic mass. For low enrichments (with $\delta = 0.0126$ for uranium), $w \doteq 0.987x$, and there is little difference between the two formulations. For natural uranium, $x = 0.00720$ and $w = 0.00711$.

[11] This is the planning basis for the Yucca Mountain nuclear waste repository [11, p. 3–13].

does natural uranium, and there is no evidence of appreciable radiation hazards except for occupants of a closed vehicle that has been struck by a shell that partially vaporizes within it.[12]

Methods for Enrichment

The leading enrichment methods in terms of past or anticipated future use are as follows:[13]

- *Gaseous diffusion.* The average kinetic energy of the molecules in a gas is independent of the molecular weight M of the gas and depends only on the temperature. At the same temperature, the average velocities are therefore inversely proportional to $\sqrt{M}$. For uranium in the form of UF_6, the ratio of the velocities of the two isotopic species is 1.0043.[14] If a gas sample streams past a barrier with small apertures, a few more ^{235}U molecules than ^{238}U molecules pass through the barrier, slightly increasing the ^{235}U fraction in the gas. The ratio of $^{235}U/^{238}U$ before and after passing the barrier is the enrichment ratio α. Its ideal or maximum value is given by the velocity ratio $\alpha = 1.0043$. However, one cannot calculate the number of stages of diffusion needed to achieve a given enrichment merely in terms of powers of α, because the ideal value is not achieved in practice and because it is necessary to continually recycle the less enriched part of the stream. Typically, if one starts with natural uranium (0.71%) and with tails depleted to 0.3%, it is found that about 1200 enrichment stages are required to achieve an enrichment of 4% [15, p. 36].
- *Centrifuge separation.* Any fluid—liquid or gaseous—can be separated in a high-speed centrifuge. The centrifugal action causes the heavier component to become more highly concentrated at large radii. As in gaseous diffusion, only a small gain is made in any one stage, and high enrichments of the UF_6 are reached using multiple centrifuge stages, with the slightly enriched output of one stage serving as the input to the next one. The centrifuges used for uranium enrichment are rotating cylinders. Uranium that is slightly enriched in ^{238}U (and depleted in ^{235}U) can be extracted from the outer region of the cylinder and returned to an earlier stage in the centrifuge cascade. Uranium slightly enriched in ^{235}U can be extracted from regions near the center and used as input to the next higher stage in the array of centrifuge units. High enrichments of the UF_6 are reached using multiple centrifuge stages. The power requirement for a given degree of enrichment is much less for centrifuge separation than for diffusion separation.
- *Aerodynamic processes.* These processes exploit the effects of centrifugal forces, but without a rotating centrifuge. Gas—typically UF_6 mixed with

---

[12] In this case, direct damage from the shell is a still greater concern.

[13] Detailed discussions of these methods are given in, for example, Refs. [13] and [14].

[14] The atomic mass of fluorine (F) is 19.00 u.

hydrogen—expands through an aperture, and the flow of the resulting gas stream is diverted by a barrier, causing it to move in a curved path. The more massive molecules on average have a higher radius of curvature than do the lighter molecules, and a component enriched in ^{235}U is preferentially selected by a physical partition. The process is repeated to obtain successively greater enrichments. The gas nozzle process was developed in Germany as the *Becker* or *jet nozzle* process. A variant with a different geometry for the motion of the gas stream, the so-called *Helikon* process, has been developed and used in South Africa.

♦ *Electromagnetic separation.* When ions in the same charge state are accelerated through the same potential difference, the energy is the same and the radius of curvature in a magnetic field is proportional to $\sqrt{M}$. Thus, it is possible to separate the different species magnetically. This separation can be done with ions of uranium and, so, conversion to UF_6 is not, in principle, necessary. Overall, this approach gives a low yield at a high cost in energy, but it has the advantage of employing a relatively straightforward technology.

♦ *Laser enrichment.* The atomic energy levels of different isotopes differ slightly.[15] This effect can be exploited to separate ^{235}U from ^{238}U, starting with uranium in either atomic or molecular form. For example, in the atomic vapor laser isotope separation (AVLIS) method, the uranium is in the form of a hot vapor. Lasers precisely tuned to the appropriate wavelength are used to excite ^{235}U atoms, but not ^{238}U atoms, to energy levels that lie several electron volts above the ground state. An additional laser is used to ionize the excited ^{235}U atoms.[16] The ionized ^{235}U atoms can be separated from the un-ionized ^{238}U atoms by electric and magnetic fields. An alternative to the AVLIS method is the SILEX process (separation of isotopes by laser excitation). It is based on the selective dissociation of UF_6 (a gas) into UF_5 (a solid) [16]. The costs in energy of laser enrichment are lower than those of other enrichment methods, but a sophisticated laser technology is required, and, to date, there are no commercial facilities for laser enrichment of uranium. Once mastered, the laser technique is expected to be relatively inexpensive. On the negative side, there have been fears that if the technique develops sufficiently, laser separation may make it easy for small countries or well-organized terrorist groups to enrich uranium for nuclear weapons.

---

[15] This "isotope effect" was responsible for the discovery of ^{2}H. It arises for two reasons: (1) the atomic energy levels depend on the reduced mass of the electrons, which differs from the mass of a free electron by an amount proportional to m_e/M, where m_e is the electron mass and M the atomic mass, and (2) the energy levels of heavy atoms depend in a small measure on the overlap between the wave functions of the innermost electrons and the nucleus, with differences between isotopes due to differences in their nuclear radii.

[16] The ionization energy to remove an electron from uranium in its unexcited (ground) state is 6.2 eV.

Adopted Enrichment Practices

During World War II, not knowing which method would be the most effective, the United States embarked on both diffusion and electromagnetic separation, as well as still another method that was later discarded (namely thermal diffusion, which exploits temperature gradients). The electromagnetic separation technique was abandoned in the United States after World War II and was widely considered to be obsolete. However, it was found in 1991, after the Gulf War, that Iraq had been secretly using this approach in an attempt to obtain enriched uranium for nuclear weapons.

In the United States since World War II, the enrichment program has relied on gaseous diffusion, as did early European programs. Since the 1950s, the DOE (and its predecessor agencies, starting with the AEC) operated two large gaseous diffusion enrichment facilities—one in Paducah, Kentucky and one in Portsmouth, Ohio. In 1999, operation of these facilities was privatized under the management of the United States Enrichment Corporation (USEC), and in 2001, the Portsmouth plant—the smaller of the two—was shut down. Outside the United States, there are major enrichment facilities in France and Russia and smaller ones in a number of other countries, including the United Kingdom, Netherlands, Germany, Japan, and China [1, p. 33]. In recent years (1999–2001), most of the enrichment for fuel used in U.S. reactors has been carried out abroad, particularly in Russia [17, p. 28]. The gaseous diffusion method is used in France and the United States, whereas the centrifuge method is used almost everywhere else.

A 2001 OECD report anticipated that most new plants will use the centrifuge method [1, p. 85]. The USEC, which has been the sole U.S. company pursuing enrichment, plans to replace its diffusion plant with a "second-generation" centrifuge plant [18]. To that end, it submitted to the NRC in February 2003 a license application for a preliminary demonstration facility scheduled to be on-line in 2005. A larger, full-scale plant—the so-called American Centrifuge—is planned for later in the decade [19]. In addition, the Louisiana Energy Services Partnership—an organization that includes, among others, Urenco (a major European enrichment company), the Westinghouse Electric Company, and several U.S. utilities is seeking to build a centrifuge enrichment facility in New Mexico [20].

USEC has also worked on developing laser enrichment technology as a "third-generation" option. It originally focused on the AVLIS method and later on the SILEX process, but as of Spring 2003, USEC decided to concentrate on its centrifuge projects to the exclusion of laser options [21].

Separative Work

In a ^{235}U enrichment process, there are three streams of material: the input or *feed*, the output or *product*, and the residue or *tails*. The system operates with a cascade of steps, with the enrichment of the product increasing successively in each step.[17] As the enrichment cascade progresses, the tails from an intermediate stage have a higher ^{235}U concentration than the original feed material, and these tails can profitably be returned to the cascade. There are different strategies for reusing the tails of successive steps to maximize the efficiency of the process, including an "ideal cascade" (see, e.g., Ref. [5]).

The difficulty of carrying out uranium enrichment, as measured, for example, by the relative energy required in the diffusion process, is described by a quantity known as the *separative work*.[18] Separative work has the dimensions of mass and is specified in *separative work units* (SWU), as kg-SWU or tonne-SWU. Figure 9.2 shows the separative work required to produce 1 kg of enriched uranium product and 1 kg of ^{235}U in the form of enriched uranium, for different degrees of final enrichment.

As seen in Figure 9.2, it requires more separative work per kilogram of ^{235}U to enrich uranium from 0.7% to 5% than to carry it the rest of the way to 95% enrichment. Even 3% enriched uranium fuel is more than "halfway" to 95% enrichment. This could make the slightly enriched uranium produced for reactors a somewhat attractive initial material for the production of uranium for weapons (see Chapter 17).

Although the formalism was developed in the context of gaseous diffusion enrichment methods, it is used for other processes as well. Separative work serves as a general measure of what is achieved in the enrichment. For an individual process, it serves also as a measure of relative energy consumption. Different processes vary greatly in their energy consumption. For example, gaseous diffusion uses 2.5 MWh/kg-SWU, while the gas centrifuge uses about 1/50th as much energy and laser isotope separation methods still less [12, p. 530].

It is of interest, in the spirit of what is known as "net energy analysis," to compare the energy required to enrich uranium with the energy obtained from it. The production by the diffusion process of 1 kg of ^{235}U in the form of

---

[17] The logical structure of the system is similar to that of fractional distillation, and some of the formalism was developed in the 19th century by Lord Rayleigh [5, p. 649].

[18] In a separation process, the masses are in the ratios:

$$\frac{M_F}{M_P} = \frac{w_P - w_T}{w_F - w_T} \quad \text{and} \quad \frac{M_T}{M_P} = \frac{w_P - w_F}{w_F - w_T},$$

where M_F, M_P, and M_T are the masses and w_F, w_P, and w_T are the ^{235}U concentrations (by weight) of the feed, product, and tails, respectively. This result follows from the conservation of total mass and ^{235}U mass: $M_F = M_P + M_T$ and $w_F M_F = w_P M_P + w_T M_T$. The separative work in an isotopic enrichment process is

$$\Delta V = M_P V_P + M_T V_T - M_F V_F,$$

where V is the *value function*, defined as $V = (1 - 2w) \ln[(1 - w)/w]$. (For the derivation leading to this result see, e.g., Ref. [5, Chapter 12].)

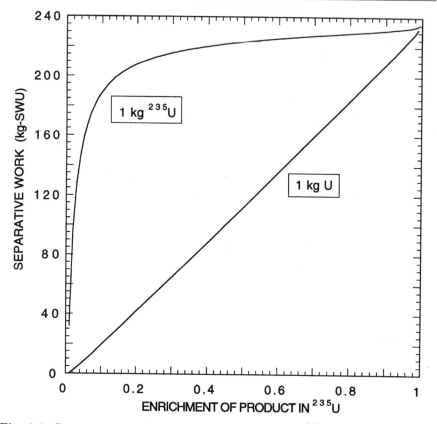

Fig. 9.2. Separative work (in kg-SWU) as a function of ^{235}U enrichment for the production of 1 kg of U and 1 kg of ^{235}U. (The initial feed concentration is $w_F = 0.00711$ and the assumed tails concentration is $w_T = 0.0025$.)

uranium enriched to 3.75% in ^{235}U, with a ^{235}U concentration in the tails of 0.25%, requires 142 kg-SWU (see Figure 9.2) or about 355 MWh. The reactor output from 1 kg of ^{235}U is about 1 MWyr or 8760 MWh (see Section 9.3.2). Therefore, the relatively energy-intensive diffusion process requires roughly 4% of the energy output of a reactor. Although this is the largest single energy input to the production of nuclear power, it does not significantly reduce the net positive energy balance from nuclear power.

9.2.3 Fuel Fabrication

Most nuclear fuel used in light water reactors is in the form of uranium dioxide (UO$_2$), also sometimes called "uranium oxide." This is not a single compound, but a mixture of oxides (UO$_n$), where n typically ranges from 1.9 to 2.1 [22, p. 226]. The UO$_2$ is produced by chemical conversion of the enriched UF$_6$. It is then processed into a fine powder and compacted and sintered to form rugged pellets. During sintering, the oxygen content of the fuel can be adjusted. The pellets are corrected in size, to close tolerances, by grinding.[19] The pellets are loaded into zircaloy fuel pins that are arranged in a matrix to form the fuel assembly. As discussed in Section 8.2.3, the reactor core consists of a large number of such assemblies.

9.2.4 Other Fuel Types

The focus here has been on UO_2, which is the usual fuel for LWRs. Other fuel types are of interest, however, even if not widely used at present (see, e.g., Ref. [1]). Possibilities include the following:

- *Mixed-oxide fuel (MOX)*. MOX fuel, a mixture of uranium and plutonium oxides, uses plutonium in order to exploit its energy content, reduce the stocks of potential weapons materials, or both (see Section 9.4.2).
- *Metal alloy fuels*. Metallic fuel, in the form of alloys of uranium, provide an alternative to oxide fuels that is easier to reprocess.
- *Microsphere fuel particles*. High-temperature gas-cooled reactors utilize uranium or thorium oxide in the form of very small spheres, with multiple layers of outer protection. These are the so-called TRISO fuel particles (see Section 16.4.3).
- *Thorium fuels*. A fuel cycle based on thorium-232 as the fertile fuel and uranium-233 as the fissile fuel could be used to supplement uranium resources. It has the advantage of producing little plutonium and thereby lessening waste disposal and proliferation problems.
- *Molten salt*. Although all operating commercial reactors presently use fuel in solid form, it is also possible to have the fuel as a liquid uranium fluoride, mixed with other liquid fluorides, as would be done in the proposed molten salt reactor (see Section 16.6.1).

9.3 Fuel Utilization

9.3.1 Burnup as a Measure of Fuel Utilization

Thermal Efficiency of U.S. Reactors

The *thermal efficiency* of a reactor is the ratio of the electrical energy produced to the total heat energy produced. Since 1973, the average thermal efficiency of U.S. reactors has ranged between 30.6% and 32.1%, according to DOE compilations [23, Table A6]. There has been a gradual improvement with time, and since 1985 it has been above 31.5%, reaching 32.1% for the years 1996–2002. We will use the approximate figure of 32% as the nominal average efficiency of LWRs.

---

[19] For a discussion of the details of these processes, see Ref. [22, Section 7.5].

Basic Unit for Burnup: GWDT per MTHM

A useful measure of the performance in the nuclear fuel cycle is the energy obtained per unit mass of fuel, known as the fuel's *burnup*. The burnup is commonly specified in megawatt-days or gigawatt-days of thermal output per metric tonne of heavy metal (MWDT/MTHM or GWDT/MTHM). This is a cumbersome notation for repeated use, and we represent GWDT/MTHM in a more compact form as GWd/t (gigawatt-days per tonne). In standard energy units, $1 \text{ GWd} = 8.64 \times 10^{13}$ joules (J).

For U.S. reactors, as well as most reactors elsewhere, the "heavy metal" in the original fuel is uranium.[20] The fuel is in the form of uranium oxide (UO_2). About 12% of the mass of the fuel is oxygen and, therefore, there is a distinction between the mass of heavy metal and the mass of the fuel. The heavy metal in the spent fuel removed from the reactor is still primarily uranium, but it also includes isotopes of plutonium and—to a small extent—other transuranic elements. Typically, the mass of heavy metal is about 3% or 4% less in the spent fuel than in the initial fuel due to fission (including fission of plutonium isotopes).[21]

A 1000-MWe reactor, operating at a typical thermal efficiency of 32%, produces energy at the rate of 3125 MWt (where it is explicitly indicated that this is the thermal output). One gigawatt-year of electric power therefore represents a thermal output of 1141 GWd(t). If, for example, the average burnup in a reactor is 40 GWd/t, the fuel consumption is 28.5 tonnes of enriched uranium per gigawatt-year.

Trends in Burnup of LWR Fuel

Average burnup values for past years are shown in Table 9.1 along with the average projected for the fuel to be deposited at the Yucca Mountain waste repository. Overall, there has been a trend with time toward higher burnup, on average roughly doubling in the 25 years from 1973 to 1998 and projected to continue to rise. Thus, in a 1993 DOE projection, it was expected that the median PWR fuel burnup for standard assemblies would be about 43 GWd/t in the year 2000—a value actually achieved in 1998—and 51 GWd/t

---

[20] The main exception is for reactors that use a mixture of uranium and plutonium in *mixed-oxide* fuels (see Section 9.4.2).

[21] The designation "metric tons of heavy metal" (MTHM) commonly appears in discussions of the utilization and disposal of nuclear fuel. Here, MTHM refers to the heavy metal mass of the initial fuel. Alternatively, this can be made explicit by using the designation "metric tonnes of *initial* heavy metal" (MTIHM). In effect, "MTHM" and "MTIHM" are used interchangeably and mean the same thing. Thus, the U.S. spent fuel inventory at the end of 1995 is given as 31926 MTHM in the 2002 Yucca Mountain EIS [24, Table A-7] and as 31952 MTIHM in a 1996 report [25, Table 1.2]. (The 0.08% difference is insignificant compared to the difference of several percent in the actual heavy metal contents of the initial and spent fuel.)

Table 9.1. Average burnup of U.S. spent fuel, for BWRs and PWRs.

| Year | Burnup (GWd/t) | |
Discharged	BWRs	PWRs
Annual		
1973	12.4	23.7
1978	19.8	26.4
1983	26.7	30.1
1988	24.1	33.4
1993	30.3	38.9
1998	36.4	43.3
Average		
1968–1998	25.1	33.4
Yucca Mountain[a]	33.6	41.2

[a]This includes all commercial spent fuel slated for Yucca Mountain (i.e., all of the fuel discharged through about 2010).
Source: Refs. [26, Table 2], and [11, Table 3–6].

in 2010 [27, p. 144]. It should be noted that Table 9.1 gives only averages. In any year, there is a wide disparity among reactors. Thus, in 1993, when the average PWR burnup was 38.9 GWd/t, 12% of the spent fuel had a burnup in excess of 45 GWd/t [28, p. 25].

An estimated average for the fuel expected to be deposited in the Yucca Mountain repository (i.e., all of the fuel generated in the United States from the start of nuclear power until about 2010) is given in Yucca Mountain planning documents. The indicated average burnups for BWRs and PWRs are 33.6 GWd/t and 41.2 GWd/t, respectively, with average initial ^{235}U enrichments of 3.03% and 3.75% [11, p. 3–13]. The corresponding weighted average for all LWR fuel is 38.6 GWd/t, with an enrichment of 3.5%.

The burnup depends on the power density in the fuel and the length of time the fuel is kept in the reactor. Unless the conversion ratio is very large, a high burnup requires a high initial enrichment in ^{235}U. For example, model calculations carried out at Oak Ridge National Laboratory included a fuel cycle that gives a burnup of 60 GWd/t, with a 4.7% ^{235}U concentration and the fuel kept in the reactor for three 600-day cycles (i.e., 1800 days) [29, p. 2.4–3]. Looking to the future, the term "high burnup" is sometimes used to refer to a burnup in excess of 70 GWd/t [1, p. 88].

High enrichment alone is not sufficient for high burnup. In addition, the fuel and cladding must be able to withstand the added neutron bombardment and the buildup of fission gases. This depends both on the fuel itself and on the composition of the alloy used for the cladding. Further, to compensate for the initial high reactivity with high enrichment, burnable poisons are used with the fuel.

With higher burnup, the mass and volume of spent fuel required for a given energy output are reduced. The consumption of uranium is also reduced, because a larger fraction of the ^{235}U is consumed and more ^{239}Pu is produced and consumed. In addition, the time intervals between refueling operations can be longer, meaning that less reactor time is lost for refueling. High burnup fuel is also less proliferation-prone than fuel with lesser burnup because the concentrations of ^{240}Pu and ^{242}Pu are higher and the fuel is harder to handle due to its high level of radioactivity (see Chapter 17).

9.3.2 Uranium Consumption and Plutonium Production

In uranium-fueled reactors, there is a continual destruction of ^{235}U, through fission and neutron capture, and buildup of plutonium isotopes through neutron capture and beta decay. The plutonium sequence starts with ^{239}Pu, following neutron capture in ^{238}U, and continues to include plutonium isotopes up to ^{242}Pu, as well other heavy radionuclides—for example, ^{241}Am (atomic number $Z = 95$), which is produced primarily from the beta decay of ^{241}Pu ($T = 14.39$ yrs). Similarly, radionuclides of atomic mass numbers 236 and 237 (primarily, ^{236}U and ^{237}Np), are formed through neutron capture in ^{235}U and, for ^{237}Np, subsequent neutron capture and beta decay.

Most of the reactor's energy output comes from the fission of ^{235}U. However, as discussed earlier, production and fission of ^{239}Pu also play a significant part in the energy economy of the reactor. Thus, the total energy output is the sum of the energy from the fissile plutonium isotopes (^{239}Pu and ^{241}Pu) plus the energy from ^{235}U and, to a much lesser extent, ^{238}U.

Table 9.2 gives the masses of the main actinide isotopes in the spent fuel when it is discharged from the reactor, for a representative PWR case, namely UO$_2$ fuel with a 3.75% ^{235}U enrichment, a burnup of 40 GWd/t, and a residence time in the reactor of a little over 3 years. The results in Table 9.2 were calculated with the ORIGEN program, developed at the Oak Ridge National Laboratory, which traces the production and consumption of the nuclides as a function of time, taking into account the simultaneous nuclear processes for all the nuclides, including fission, neutron capture, and radioactive decay.[22] Important qualitative features of the results include the following:

♦ Fission products. The fission product production is about 40 kg, as seen from the 40.4-kg reduction in the heavy metal mass.[23] It may seem sur-

---

[22] I am indebted for the calculation of the activities to Dr. Edwin Kolbe, Project Manager for Radioactive Materials at the Swiss National Cooperative for the Disposal of Radioactive Waste (NAGRA) [31]. The mass of each radionuclide was calculated from the ratio of its activity to its specific activity [see Eq. (3.4)]. These ORIGEN results are also used in Chapter 10.

[23] The precise number is slightly less than 40.4 kg due, in part, to neutron escape from the reactor. The mass equivalent of the energy produced in the reactor is 0.04 kg, representing a very small additional "correction" to the calculated fission product mass.

Table 9.2. Activity and mass of actinides in PWR spent fuel (per MTHM), for burnup of 40 GWd/t and ^{235}U enrichment of 3.75%.

Nuclide	Half-life (years)	Activity (Bq)	Mass (kg)	Isotopic Percent
Input fuel				
^{234}U	2.46×10^5		0.05	0.005
^{235}U	7.04×10^8		37.5	3.75
^{238}U	4.47×10^9		962.4	96.24
Total HM^a			1000	
Spent fuel				
^{234}U	2.46×10^5	4.70×10^{10}	0.2	0.02
^{235}U	7.04×10^8	6.85×10^8	8.6	0.90
^{236}U	2.34×10^7	1.21×10^{10}	5.1	0.53
^{238}U	4.47×10^9	1.16×10^{10}	934.4	98.54
Total U			948.2	100
^{237}Np	2.14×10^6	1.49×10^{10}	0.57	87
^{239}Np	0.0065	7.18×10^{17}	0.08	13
Total Np			0.65	100
^{238}Pu	87.7	1.38×10^{14}	0.22	2.1
^{239}Pu	2.41×10^4	1.28×10^{13}	5.56	53.2
^{240}Pu	6564	2.07×10^{13}	2.46	23.6
^{241}Pu	14.29	5.93×10^{15}	1.54	14.8
^{242}Pu	3.73×10^5	9.66×10^{10}	0.66	6.3
Total Pu			10.4	100
^{241}Am	432.2	6.41×10^{12}	0.05	24
^{243}Am	7370	1.18×10^{12}	0.16	76
Total Am			0.21	100
^{244}Cm	18.1	1.54×10^{14}	0.05	
Total HM			959.6	
Decrease in HM			40.4	

^{a}The initial total heavy metal (HM) mass is the uranium mass.
Source: Half-lives from Ref. [30]; activities are based on ORIGEN calculation (see text) [31]. The mass is calculated from the activity and half-life.

prising that the fission product mass exceeds the mass of the destroyed ^{235}U. However, in addition to fission in ^{235}U, there is fission of ^{239}Pu and ^{241}Pu.

♦ ^{236}U. Neutron absorption in ^{235}U leads to capture in 14% of the events (see Table 7.1). Therefore, the destruction of ^{235}U is accompanied by the production of a significant amount of ^{236}U. The isotope ^{236}U has a modest capture cross section at thermal energies ($\sigma_\gamma = 5.1$ b) and a negligible fission cross section. Its capture product, ^{237}U, decays with a rather short half-life (6.75 days) to ^{237}Np ($T = 2.14 \times 10^6$ years), which has a large capture cross section and a considerably smaller fission cross section. This produces ^{238}Np, followed by beta decay to ^{238}Pu [32]. Overall, ^{236}U is more a weak poison rather than a fissile or fertile fuel.

- ^{238}Pu. With a half-life of 88 years, ^{238}Pu is useful as an energy source for use in "radioisotope thermoelectric generators" (RTGs), in which heat from the radionuclide is used to produce electricity. Its intermediate half-life means that the output from ^{238}Pu changes little in a decade or two, yet its specific activity is relatively high. RTGs based on ^{238}Pu were at one time used in heart pacemakers, but, more recently, their main use has been in the space program (e.g., in the Cassini mission to Saturn launched in 1997).

- ^{239}Pu. There is continuous production of ^{239}Pu following neutron capture in ^{238}U, and also continuous destruction of ^{239}Pu, primarily by fission but also by neutron capture. The final ^{239}Pu abundance reflects the net effect of production and destruction. The relatively low value of its isotopic abundance (53%) is characteristic of "reactor-grade" plutonium (see Section 17.4.1).

- ^{240}Pu. The capture cross section is relatively high in ^{239}Pu, so that neutron absorption results in the production of ^{240}Pu as well as in fission. ^{240}Pu acts as a fertile fuel for the production of ^{241}Pu, but the absorption cross section for ^{240}Pu is less than that for ^{239}Pu, and ^{240}Pu is not consumed as rapidly as is ^{239}Pu.

- Other Pu isotopes. Neutron capture on ^{240}Pu produces ^{241}Pu, which is fissile but which also has an appreciable branch for neutron capture to ^{242}Pu. The buildup of plutonium isotopes stops here because ^{243}Pu, the next in the series, decays with a half-life of 5.0 h to ^{243}Am and further neutron captures and beta decays moves the chain to atomic numbers higher than that of plutonium.

The plutonium isotopes through atomic mass number $A = 242$ have half-lives that are long compared to normal exposure periods in the reactor (the shortest is 14.3 years for ^{241}Pu). Considering those with $A > 238$ as a group, they are fed by neutron capture in ^{238}U (quickly followed by beta decay to ^{239}Np and ^{239}Pu) and are depleted primarily by the fission of ^{239}Pu and ^{241}Pu and the beta decay of ^{243}Pu.

9.3.3 Energy from Consumption of Fuel

Energy per Unit Mass from Fission of ^{235}U

The fission of a ^{235}U nucleus corresponds on average to the release of 200 MeV $(3.20 \times 10^{-11}$ J), including the associated contributions from neutron capture and the decay of fission fragments (see Section 6.4.2). The number of ^{235}U atoms per gram of uranium is wN_A/M, where w is the fraction of ^{235}U in the uranium (by weight), N_A is Avogadro's number, and M is the atomic mass of ^{235}U. Thus, 1 kg of natural uranium ($w = 0.00711$) has 1.822×10^{22} nuclei

Table 9.3. Energy per unit mass from fission of uranium, at different degrees of enrichment.

Category	Enrichment, w (%)	Energy per Unit Mass[a]		Tonne/GWyr(e)[b]
		J/kg	GWd/t	
Natural U	0.711	5.84×10^{11}	6.8	169
Enriched U	3.2	2.63×10^{12}	30.4	38
Enriched U	3.5	2.87×10^{12}	33.3	34
Enriched U	3.75	3.08×10^{12}	35.6	32
Enriched U	5.0	4.10×10^{12}	47.5	24
Pure ^{235}U	100	8.21×10^{13}	950	1.2

[a] Thermal energy, assuming fission of *all* ^{235}U, a release of 200 MeV per ^{235}U fission, and no fission of other nuclides.
[b] Assuming a thermal conversion efficiency of 32%.

of ^{235}U and their complete fission would release 5.8×10^{11} J. The available fission energy per kilogram of uranium, for different degrees of enrichment, is given in Table 9.3 for a few illustrative cases, assuming fission of all ^{235}U and ignoring losses due to capture in ^{235}U and gains from fission in ^{239}Pu and other nuclides.

As indicated earlier, 1 GWyr of electric power corresponds to a thermal output of 1141 GWd(t) or 9.86×10^{16} J. Assuming complete fission of all the ^{235}U and ignoring fission in plutonium, this corresponds to a fuel requirement of 1.20 tonnes of ^{235}U per GWyr(e). For the specific case of a 3.75% enrichment in ^{235}U (as in Table 9.2), the requirement translates to 32 tonnes of uranium. The corresponding thermal burnup is 35.6 GWd/t, as indicated in Table 9.3.

Energy per Unit Mass of Fuel

The discussion in the preceding subsection is incomplete, because it omits many crucial factors that significantly modify the amount of ^{235}U required by a reactor. These include the following:

1. Not all of the ^{235}U is consumed in the reactor. For example, for the case described in Table 9.2, the ^{235}U content per MTHM is 37.5 kg in the fresh fuel and 8.6 kg in the spent fuel (i.e., a consumption of only 77% of the ^{235}U).
2. About 14% of the thermal neutron-absorption reactions in ^{235}U result in capture rather than fission.
3. Fission in ^{239}Pu (and, to a lesser extent, in ^{241}Pu) provides a substantial additional energy source. This reduces the ^{235}U required for a given energy production.
4. There is a small contribution from fast-neutron fission in ^{238}U.

These effects all change the number of fission events, reducing the number of fissions in ^{235}U and adding fission in plutonium isotopes and even ^{238}U. The overall consequence can be crudely estimated by comparing the decrease in the total heavy metal mass—which results almost entirely from fission of uranium and plutonium isotopes—to the original ^{235}U mass. The ratio of these quantities is 1.081, which would suggest that there are about 8.1% more fission events than would be given by complete fission of ^{235}U. Applying this 8.1% "correction" raises the ^{235}U burnup of 35.6 GWd/t to a revised value of 38.5 GWd/t. This estimate still ignores the differences in fission energy yields and atomic masses between ^{235}U and the heavier actinides. Roughly 40% of the fissions are in ^{239}Pu and ^{241}Pu, not ^{235}U. Taking this into account adds roughly 0.6 GWd/t. Gamma-ray emission following neutron capture in the actinides adds further. Together, these approximate corrections bring the total close to the burnup of 40 GWd/t indicated in Table 9.2.

A value of 40 GWd/t is a good representation of recent LWR performance, although average PWR burnups are now higher and future ones are expected to be still higher (see Table 9.1).[24] The burnup in GWd/t can be translated into the fuel requirement per year. For example, for a burnup of 40 GWd/t the total uranium requirement is 28.5 tonnes per GWyr, or 1.07 tonnes of ^{235}U for an enrichment of 3.75%.

The precise uranium requirements for a given energy output depend on details of the fuel cycle and reactor operation. Nonetheless, the following equivalence, as found for the above example, is useful for approximate estimates of the general magnitudes for a once-through LWR fuel cycle:

$$1 \text{ tonne of } {}^{235}\text{U} \rightarrow 1 \text{ GWyr(e)} \quad \text{(approximate)}.$$

9.3.4 Uranium Ore Requirement

The amount of uranium ore required to operate a reactor depends on the burnup achieved, the initial enrichment, and the amount of ^{235}U lost in the enrichment process. The mass M_F of natural uranium used as feed input to the enrichment facility is related to the mass M_P of enriched uranium produced in the fuel by the expression $M_F/M_P = (w_P - w_T)/(w_F - w_T)$, where w_P, w_T, and w_F are the enrichments of the fuel, the tailings, and natural uranium, respectively. For $w_P = 3.75\%$, $w_T = 0.2\%$, $w_F = 0.711\%$, and $M_P = 28.5$ tonnes/GWyr, the natural uranium requirement is 198 tonnes/GWyr. Thus, in round numbers, a once-through LWR fuel cycle requires about 200 tonnes

---

[24] The average burnup is systematically less in BWRs than in PWRs because burnup in the former is less uniform along the length of the fuel rods. The water at the bottom of the BWR tank has a high density and is a better moderator than the steam–water mixture at the top. This means that the maximum burnup is achieved at the bottom of the rod, with a smaller burnup higher up on the rod. The maximum acceptable burnups are about the same for the PWR and BWR, but the PWR has a more uniform profile along the length of the fuel rod and therefore a greater average burnup.

of natural uranium per gigawatt-year. Present world demand for uranium is 60,000 tonnes/yr, corresponding to the requirements for the present annual generation by nuclear power plants of about 300 GWyr [1, p. 17].

9.4 Back End of Fuel Cycle

9.4.1 Handling of Spent Fuel

Initial Handling of Reactor Fuel

Periodically, a portion of the fuel in the reactor is removed and replaced by fresh fuel. In typical past practice, an average sample of fuel remained in the reactor for 3 years, and approximately one-third of the fuel was removed each year, with a shutdown time for refueling and maintenance of up to about 2 months. The trend is now to extend the interval between refueling operations and to reduce the time for refueling.[25] Currently, time intervals of 18 months and shutdowns of 1 month are typical.

When the spent fuel is first removed from the reactor, the level of radioactivity is very high, due to the accumulation of radioactive fission products and radioactive nuclei formed by neutron capture. Each radioactive decay involves the release of energy, which immediately appears as heat, so the fuel is thermally hot as well as radioactively "hot." Independent of the reprocessing question, the first stage is the same, namely allowing the fuel to cool both thermally and radioactively. The cooling of the fuel normally takes place in water-filled cooling pools at the reactor site.

Originally, it was planned to keep the spent fuel at the reactor for roughly 150 days and then to transfer it to handling facilities at other locations. The nature of the next step, in principle, depends on whether the fuel is to be disposed of as waste or reprocessed. However, as yet, this "next step" has been much delayed in the United States because no off-site facilities have been developed. Instead, almost all of the fuel has remained at the reactor sites—in many cases for more than 20 years.

In the absence of alternatives, some U.S. utilities are transferring older fuel rods from cooling pools to air-cooled (dry storage) casks at the reactor site. This may provide a workable temporary solution to the long delay in implementing a national waste disposal program. However, it is only a stopgap because the reactor operator cannot be counted on to be willing and able to supervise the spent fuel for prolonged periods of time (see Section 11.1.3).

Disposal or Storage of Spent Fuel

For many years, it had been assumed that all U.S. civilian nuclear waste would be reprocessed, but U.S. reprocessing plans have been abandoned. Instead,

---

[25] An annual refueling shutdown of 2 months would mean a *maximum* capacity factor of 83%, which is well below the present U.S. average.

official plans now call for disposing of the spent fuel directly, while retaining for many decades the option of retrieving it. The fuel is to remain in solid form and the fuel assemblies eventually placed in protective containers and ultimately moved in secure casks to either a permanent or an interim repository site. In the latter case, the waste would be moved to a permanent repository at a later time.

A distinction is sometimes made been "disposal" and "storage." The former suggest permanence, whereas the latter suggests the possibility that the spent fuel might be later retrieved. This possibility is made explicit in *retrievable* storage systems, where the permanent sealing of the repository is deferred, allowing the spent fuel to be recovered should this be desired at a later time.[26] In this case, the reprocessing option is not foreclosed, and the spent fuel may ultimately not be a "waste."

There are several motivations for maintaining retrievability: (a) It allows for remedial action in case surprises are encountered in the first decades of waste storage that require modifying the fuel package or the repository; (b) it keeps open the option of recovering plutonium from the fuel; and (c) it allows the recovery of other materials deemed useful—for example, fission products for use in medical diagnosis and therapy or in the irradiation of food or sewage sludge. When the placement becomes irreversible, with no prospect of retrieving the fuel, this becomes final disposal.

9.4.2 Reprocessing

Extraction of Plutonium and Uranium

The alternative to disposing of the spent fuel is to reprocess it and extract at least the uranium and plutonium. In reprocessing, the spent fuel is dissolved in acid and the plutonium and uranium are chemically extracted into separate streams, for use in new fuel. The most widely used method for this is the suggestively named PUREX process.

Most early U.S. plans for reprocessing assumed that 99.5% of the U and Pu would be removed. The remainder constitutes the high-level waste. In the traditional plans, the wastes include almost all of the nonvolatile fission products, 0.5% of the uranium and plutonium, and almost all of the minor actinides [i.e., neptunium ($Z = 93$), americium ($Z = 95$), and curium ($Z = 96$)]. The uranium represents most of the mass of the spent fuel, but the fission products contain most of the radioactivity.

Extraction can be more complete than contemplated in the original U.S. thinking. The French program has exceeded the 99.5% goal, separating out more than 99.9% of the uranium and 99.8% of the plutonium [33, p. 28]. There is no essential reason to limit extraction to plutonium and uranium,

---

[26] Plans for the Yucca Mountain repository call for it to remain open for perhaps as much as several hundred years, but the preservation of the reprocessing option does not now appear to be the major motivating factor (see Section 12.2.1).

although the former represents the valuable fuel and the latter represents the bulk of the mass. It is possible to extract other radioisotopes as well, either because they are deemed pernicious as components of the waste or because they are useful in other applications. The minor actinides have been of particular interest. They include long-lived products whose removal would decrease the long-term activity in the waste. One option is to separate them and return them to a reactor where they would be transmuted in neutron reactions.

Of course, if the chief goal is safety, it is necessary to balance the benefits from decreased activity in the wastes against the increased hazards of handling and processing them when they are still very hot. At present, this further separation option has not been adopted in the major reprocessing programs in France and the United Kingdom, and the minor actinides remain with the fission products [34, p. 149].

The residue of reprocessing constitutes the wastes. They are to be put in solid form for eventual disposal. The standard method is to mix the high-level waste with molten borosilicate glass and contain the solidified glass in metal canisters. Although other solid waste forms have been suggested, borosilicate glass has been used in the French nuclear program and had figured prominently in the original U.S. plans for reprocessing commercial wastes. It is being used for the sequestering of already reprocessed U.S military wastes at the Savannah River site in South Carolina and is planned for the wastes at the Hanford reservation in Washington state.

Status of Reprocessing Programs

Until the late 1970s, reprocessing had been planned as part of the U.S. nuclear power program. A reprocessing facility at West Valley, New York was in operation from 1966 to 1972, with a capacity of 300 MTHM/yr. This is enough, roughly speaking, for the output of 10 large reactors. There were plans for further facilities at Morris (Illinois) and Barnwell (South Carolina) which would have substantially increased the reprocessing capacity. However, all these plans have been abandoned.[27]

In part, the abandonment was impelled by technical difficulties. There had been high radiation exposures of workers at West Valley and the plant was shut down in 1972; plans to remodel and expand it were later aborted. When the Morris plant was first tested with nonradioactive materials, it did not perform reliably, and the General Electric Co., which was building the plant, decided there were serious difficulties. The Barnwell plant moved ahead until the early 1980s, but it faced problems of meeting increasingly strict standards on permissible radioactive releases.

These difficulties might have been surmounted had there been a belief that reprocessing was needed. However, the fundamental motivation for re-

---

[27] For a discussion of this history, see Ref. [35].

processing began to slip away. There was no prospective near-term shortage in uranium supply, and uranium prices were low enough to remove the economic incentive for reprocessing. Further, an important body of opinion had developed in the United States against reprocessing, on the grounds that it might make plutonium too readily available for diversion into destructive devices.

This view was expressed in the 1977 report *Nuclear Power Issues and Choices*, sponsored by the Ford Foundation and authored by an influential group of national science policy leaders. In its conclusions on reprocessing, the report stated:

> [T]he most severe risks from reprocessing and recycle are the increased opportunities for the proliferation of national weapons capabilities and the terrorist danger associated with plutonium in the fuel cycle.
>
> In these circumstances, we believe that reprocessing should be deferred indefinitely by the United States and no effort should be made to subsidize the completion or operation of existing facilities. The United States should work to reduce the cost and improve the availability of alternatives to reprocessing worldwide and seek to restrain separation and use of plutonium. [36, p. 333]

Consistent with this thinking, the Carter administration decided in 1977 to "defer indefinitely the commercial reprocessing and recycling of the plutonium produced in U.S. nuclear power programs" [37, p. 54]. Work on U.S. reprocessing plants for commercial fuel was phased out, culminating in the closing of Barnwell at the end of 1983 [35, p. 124].

Nonetheless, reprocessing has been pursued in other countries. The largest reprocessing programs are in France and the United Kingdom, both of which completed major expansions of reprocessing capacity in 1994 to handle both domestic and foreign fuel. In addition, a large facility is being built in Japan. France has the most fully developed fuel cycle. Although much of its present reprocessing capacity is devoted to foreign orders, it also has a program of reprocessing and plutonium recycle of domestic fuel.

Table 9.4 lists the reprocessing plants in operation or under construction. It omits plants that were closed down before 2002, including plants in Belgium, France, Germany, the United Kingdom, and the United States.

Use of Mixed-Oxide Fuel

The fuel manufactured from the output of the reprocessing phase is generally a mixture of plutonium oxides and uranium oxides, with 3% to 7% PuO_2 and the remainder UO_2. It is called a *mixed-oxide fuel* or MOX. At the higher ^{239}Pu enrichments, a burnable poison would be added to the fuel to reduce its initial reactivity. Due to differences in the nuclear properties of ^{239}Pu and ^{235}U, most LWRs are limited to using only about a one-third fraction of MOX

Table 9.4. Reprocessing plants for commercial nuclear fuel in operation or under construction, 2002.

Country	Location	Year of Start-up	Capacity (MTHM/yr)[a]
In operation			
France	La Hague (UP2)[a]	1976	800
France	La Hague (UP3)	1989	800
India	Tarapur	1974	100
India	Kalpakkam	1998	100
Japan	Tokai-mura	1977	90
Russia	Chelyabinsk	1984	400
United Kingdom	Sellafield (B205)	1964	1500
United Kingdom	Sellafield (Thorp)	1994	1200
Under construction			
China	Diwopu	2002 (?)	25–50
Japan	Rokkasho-Mura	2005	800

[a] UP2 was upgraded and redesignated as UP2-800, with full capacity reached in 1994 [40].
Sources: Refs. [1, p. 46]; [38, Table 4]; and [39, p. 119].

in the reactor core, with the remainder ordinary uranium-oxide fuel [41, 42].[28] Some LWRs, however, have been designed to accommodate a full load of MOX fuel.[29]

By 2001, about 20 PWRs in France (out of 58) were using MOX for one-third of their fuel [34, p. 138]. In the United States, the interest in MOX fuel has been motivated by the need to dispose of plutonium from dismantled nuclear weapons (see Section 18.3.3). Toward this end, the DOE is planning to build facilities for conversion of plutonium into MOX fuel at its Savannah River site. At least one nuclear plant operator (Duke Energy) has made a

---

[28] It is more difficult to control a thermal reactor using plutonium than one using uranium. Contributing reasons include (a) the delayed neutron fraction, β, is smaller for ^{239}Pu than for ^{235}U and (b) the fission cross section resonance in ^{239}Pu near 0.3 eV (see Figure 6.1) leads to a positive feedback if the reactor temperature rises. In addition, with ^{239}Pu, the neutron and gamma-ray spectra are more energetic than with ^{235}U, causing more radiation damage [42, p. 119]. As a result, it is necessary to have design changes, including more control rods, if a full load of MOX fuel is used in place of uranium fuel. This cannot be readily accomplished in most LWRs. However, it is possible in the so-called System-80 PWRs. Three such reactors are in operation at the Palo Verde nuclear plant in Arizona, but, at present, no U.S. LWR is licensed by the NRC to operate with MOX fuel.

[29] See Section 18.3 for a further discussion of MOX fuel, in the context of the burning of plutonium from dismantled nuclear weapons.

commitment to use MOX fuel in some of its reactors, beginning in 2007 if plans proceed according to the initial schedule [43].

9.4.3 Alternative Reprocessing and Fuel Cycle Candidates

Advanced Aqueous Process

In the widely used PUREX process, the plutonium and uranium are extracted and the fission products and minor actinides constitute the wastes. The advanced aqueous process is a modification of the PUREX process in which the minor actinides are recovered as well. Uranium is crystallized out at an early stage to reduce the bulk of the material that must be dealt with in the further chemical processing [44, p. 60]. The two product streams, one of uranium and the other of plutonium and the minor actinides, are used to fabricate fuel for use in either thermal or fast reactors.

The UREX Process

An alternative to the advanced aqueous process is the uranium extraction process (UREX and UREX+). It differs in the means of separating out the uranium. Several output streams are specifically identified in this process [45, p. II-3]:

1. Uranium. The uranium is extracted in very pure form ("at purity levels of 99.999 percent"). The leaves it free of highly radioactive contaminants and makes it easy to handle for disposal or reuse in a reactor.
2. Plutonium and minor actinides. Neptunium, americium, and curium are retained with the plutonium. These elements can be incorporated into the reactor fuel.
3. Long-lived fission products. Long-lived fission products (in particular, iodine-129 and technicium-99) are separately extracted, for destruction in a reactor (see Section 11.3.3).
4. Other fission products. These become the wastes. The waste disposal problem is simplified because the long-lived radionuclides have, for the most part, been removed.

Pyroprocessing

The above-discussed chemical reprocessing processes are known as *aqueous* processes. An alternative approach, under active exploration for use in conjunction with future reactors, is the *pyroprocess* or electrorefining process. In this method, the spent fuel is dissolved at very high temperatures in molten cadmium, creating an "electrolytic bath." Groups of chemical elements are separately extracted on the basis of differences in the potentials at which they dissolve and ionize. In particular, ions of the actinides, including uranium, plutonium, and the minor actinides, are attracted to cathodes and are extracted. The actinides are then incorporated in the fabrication of new fuel elements.

Full Actinide Recycle

A fuel cycle based on the nearly complete extraction of plutonium and minor actinides (collectively, the transuranic elements) and their consumption by fission in fast reactors has been sketched in the MIT report *The Future of Nuclear Power* [46]. This cycle envisages a global nuclear economy in 2050 with a capacity of 1500 GWe based on a balanced combination of thermal and fast reactors. The thermal reactors are assumed to be LWRs, fueled by enriched uranium oxide. The fast reactors, undefined as to type, are fueled by transuranics obtained from the LWR spent fuel. Pyroprocessing is used to extract the transuranics from both the thermal and fast reactor spent fuel. For each load of fresh fuel in the fast reactors, 20% of the transuranics are consumed in fission and the remaining 80% are available for recycle.

A balanced system is one in which the spent fuel from the thermal reactors provides the transuranics needed to make up for those consumed in the fast reactors. With the assumptions made in the MIT analysis, this is achieved by having slightly more capacity in the thermal reactors than in the fast reactors (815 GWe and 685 GWe, respectively). A variety of choices exist for the fast reactors, including some of the Generation IV reactors discussed in Section 16.6.

The uranium requirement for the entire fuel cycle is the amount needed to provide for the 815 GWe of LWRs, which is 54% of the amount needed if LWRs accounted for the full 1500-GWe capacity. A further, and perhaps even more important, benefit is the almost complete elimination of plutonium and minor actinides from the stream of wastes that require permanent disposal.

General Features of Reprocessing Options

Any fuel cycle that recycles the fissile components of the spent fuel (mainly the remaining ^{235}U and the plutonium isotopes ^{239}Pu and ^{241}Pu), increases the energy obtained from the existing uranium resources. If the minor actinides are included with the uranium and plutonium in the new fuel, the wastes will have much less long-term radioactivity than wastes in the once-through fuel cycle. The mass of the spent fuel is greatly reduced if the uranium is either returned to the reactor or is separated from other radionuclides to become low-activity depleted uranium. The fission products then constitute the waste product that requires long-term disposal. This greatly reduces the mass of the waste product and the period during which it must be kept isolated from the environment.

All of the reprocessing fuel cycles that have been described in this section achieve these resource extension and waste reduction benefits. The PUREX process accomplishes much of this, but in its standard form the minor actinides are not removed.

A long-standing objection to reprocessing is based on the increased proliferation risks if ^{239}Pu is in wide circulation. The above-described methods lessen the risks in two ways: (1) The presence of the minor actinides increases the activity of the fuel and makes it more difficult to handle, and (2) the reprocessing and fuel fabrication facilities can be located adjacent to the reactor, making theft or diversion of the fuel very difficult without the collaboration of the plant operators. The collocation aspect was particularly stressed in planning documents for the Integral Fast Reactor (see Section 16.5.1) and combined facilities were tested on a small scale using fuel from the experimental breeder reactor in Idaho (EBR-II).

A further objection is based on economics. Given present demand and prices, it is more expensive to reprocess spent fuel than to obtain fuel from newly mined uranium. This is a cogent objection at the present scale of nuclear power use. The advantages of these reprocessing approaches become more relevant in the context of a possible major expansion of nuclear power.

At present, the reprocessing approaches discussed here are in the development and study stage, except for the long-used PUREX process. In general, the pyroprocessing technique is more suitable for use with fuel in metallic form, while the aqueous processes are more suitable for oxide fuels. Thus pyroprocessing was originally studied for use with metallic fuel from a sodium-cooled fast reactor. However, either class of process could be used with a wide variety of fuel forms, given appropriate pretreatment stages.

9.4.4 Waste Disposal

All countries with announced plans for disposing of high-level radioactive wastes are planning on eventual disposal in deep geologic repositories, typically made by excavating caverns or holes in favorable environments. Many of the plans for these permanent disposal facilities include a period during which the waste could still be retrieved.

Deep geologic disposal has been the favored course in U.S. thinking since the first attempts to formulate plans. There have been continuing efforts to locate and design a suitable facility. A site at Yucca Mountain in Nevada was selected in 1987 as the candidate for a U.S. repository and it has been under intense study since. The DOE in 2002, with the subsequent concurrence of the president and Congress, recommended going ahead with the Yucca Mountain project. The announced goal is to have a facility ready to receive wastes by 2010, subject to approval by the Nuclear Regulatory Commission. (Much more extensive discussions of nuclear wastes are presented in Chapters 10–13.)

9.5 Uranium Resources

9.5.1 Price of Uranium

Conventional Units for Amounts of Uranium

The magnitude of uranium resources can be specified in terms of the amount of uranium oxide (U_3O_8) or the amount of natural uranium (U). Commonly, U.S. organizations expressed the resources in short tons of U_3O_8, whereas international organizations, such as the OECD, use tonnes of uranium. The units are related by the equivalence:[30]

$$1 \text{ ton of } U_3O_8 = 0.769 \text{ tonnes of U.}$$

In terms of the units in which uranium prices are usually couched, 1 kg of U is equivalent to 2.60 lbs of U_3O_8; therefore, a price of \$100/lb of U_3O_8 is equivalent to \$260/kg of U.

Uranium Prices and Electricity Costs

Uranium prices are now very low compared to those projected several decades ago. They dropped markedly in recent years, in large measure due to the lag in the expansion of nuclear power. U.S. prices peaked in 1978 at an average of \$43/lb of U_3O_8 [47]. In 2001, the average price paid by U.S. utilities was \$10/lb U_3O_8 (\$26/kg U) [17, p. 11].

The relationship between uranium price and the contribution of uranium fuel costs to electricity costs depends on the effectiveness of fuel utilization. As was discussed in Section 9.3.4, the approximate requirement for LWRs is 200 tonnes of uranium per gigawatt-year. Then, for example, uranium at \$100/kg corresponds to a cost of \$20 million per gigawatt-year (8.76×10^9 kWh), or 0.23¢/kWh.

A useful overall equivalence for LWR uranium costs is

$$\$100/\text{lb of } U_3O_8 = \$260/\text{kg of U} \approx 0.59¢/\text{kWh.}$$

The 2001 price (\$26/kg of U) corresponds to a contribution to the cost of electricity of about 0.06¢/kWh. This is roughly 1% of the total cost of electricity from nuclear power plants. Thus, uranium would remain "affordable" even with a large increase in uranium prices. For example, uranium would cost 1¢/kWh at a price of \$440/kg of U. This is about 17 times the present uranium price, but the resulting increase in electricity costs would be small compared to the variation in U.S. electricity prices with time and location.

---

[30] The molecular weights of U_3O_8 and U are 842.09 and 238.03, respectively, and 1 ton = 0.9072 tonnes.

It should be noted that only the costs of the U_3O_8 are considered here. Fuel costs as seen by a utility are normally defined to include not only the cost of U_3O_8 but also the costs for conversion to UF_6, enrichment, fuel fabrication, and waste disposal.[31] However, these are not dependent on the grade of the original uranium ore. Rather, they make the ore cost itself relatively less significant. Typical fuel costs are now about 0.5¢/kWh (see Table 19.1). A perspective on the "affordability" of an increase in uranium prices is provided by fluctuations in natural gas prices. The cost to electric utilities of natural gas rose by $2 per million BTU between 1998 and 2001 [23, Table 9.10], corresponding to an electricity cost increase of about 2¢/kWh, with even greater price "spikes" in late 2000 and early 2001.

This comparison, and others made in Section 19.4.1, suggest that we are accustomed to accepting significant changes in energy costs. From the perspective of this past experience, uranium at $440/kg of U, or even twice that, would be "affordable," if we had no alternative. No such costs are now contemplated, but are conceivable if we turn to uranium from the sea (see Section 9.5.3).

9.5.2 Estimates of Uranium Resources

Classification of Resources

An extensive description of the world uranium industry is provided by the "Red Book" series, published by the OECD since the 1960s under the title *Uranium: Resources, Production and Demand* (see, e.g., Ref. [4]). It is common to classify resources in terms of the degree of knowledge of the location and extent of identified or expected uranium deposits. So-called conventional resources are divided in the Red Book into four groups: "reasonably assured," "estimated additional" (in two categories of differing certainty), and "speculative." This OECD terminology is mirrored in U.S. DOE publications, with "reserves" used in place of "reasonably assured" resources [17, p. 36]. Speculative resources are those estimated for regions that have not been extensively explored, but where resources are expected on the basis of geological extrapolations. Despite the possibly negative connotations of the term "speculative," it is appropriate to include these resources when seeking the most realistic estimate of the total resource base. In addition to these so-called "conventional" resources, there are unconventional resources, such as marine phosphates, as well as uranium from seawater.

The several resource categories are usually described in terms of the amounts available in different cost ranges, with the usual maximum taken to be $260/kg of U ($100/lb of U_3O_8).[32] As discussed in Section 9.5.1, there

---

[31] A charge of 0.1¢/kWh for a waste disposal reserve fund is normally counted as part of the fuel cost.

[32] The costs used in categorizing uranium resources are "forward costs," not total costs. The forward costs exclude the costs already incurred as well as, for example, income taxes and profits.

is no compelling reason to regard this cost as an upper limit for economically interesting resources. However, such a cost is so far above current levels that there is little motivation for a critical evaluation of uranium resources above \$260/kg. In fact, for most categories, the OECD tabulations only include resources up to \$130/kg [4].

Magnitude of Terrestrial Uranium Resources

The Red Books provide some important information on worldwide uranium resources. However, the data in them are based on national reports, and major gaps can exist when a country does not report on resources in a particular category. Further, with the exception of speculative resources, no results are reported for uranium at costs above \$130/kg. With the reservation that it is an incomplete summary, we present in Table 9.5 resource estimates for those countries listed in Ref. [4] as having the largest uranium resources. Separate sums are shown for this selected group of countries and for all countries listed in the Red Book.

On the basis of the data in Table 9.5, one might infer that world resources total about 17 million tonnes, including the speculative resources. However, this is misleading for a number of reasons:

Table 9.5. Uranium resource estimates for selected countries (in million tonnes of uranium).

Country	Category A[a]	Category B[b]	Sum
Australia	0.9		0.9
Brazil	0.4	0.5	0.9
Canada	1.3		1.3
China		1.8	1.8
Kazakhstan	1.7		1.7
Mongolia	1.4		1.4
Namibia	0.3		0.3
Russian Federation	0.8	0.5	1.3
South Africa	0.5	1.1	1.6
United States	2.5	0.5	3.0
Sum of countries listed above	9.8	4.3	14.1
Sum in OECD listing[c]	11.2	5.5	16.7

[a]Category A: All indicated resources at costs up to \$130/kg of U.
[b]Category B: Speculative resources at unspecified costs above \$130/kg; for the U.S. data, this corresponds to costs up to \$260/kg.
[c]The sums differ, because the full Red Book listing includes countries not included in this table.
Source: Ref. [4].

◆ Not all countries are included in Table 9.5 or in the underlying tables of Ref. [4].

◆ Many countries presented only partial reports. For example, Australia did not report resources in Category B (speculated resources above $130/kg) for the 2001 Red Book [4], although at an earlier time, it reported about 3.9 million tonnes in this category [48, p. 29].

◆ The exclusion of resources in the $130/kg to $260/kg category (except for speculative resources) is a major omission. For the United States, there are an estimated 0.8 million tonnes in this category [17, p. 39]. Even a $260/kg cutoff may be more restrictive than necessary.

◆ Given the slack demand for uranium, there has been little motivation for continuing exhaustive studies of resources.

◆ According to adjustments included in the tables of Ref. [4], the indicated totals should be reduced by about 10% to correct for losses in the recovery of uranium.

In view of both geological and economic uncertainties, it is not clear that a meaningful total can, as yet, be established for world uranium resources. It appears, however, that the total useful resource is more than 20 million tonnes, assuming practices not very different from present ones but accepting costs up to $260/kg. The indicated total might be substantially greater if all countries provided information in each of the Red Book's resource categories

The usable resources are likely to be still larger if one goes beyond these categories. Kenneth Deffeyes and Ian MacGregor have analyzed the distribution of uranium resources in the Earth's crust in terms of the probable amounts at different grades [49]. They concluded that the distribution rises rapidly as one goes from higher to lower grades, reaching a peak at the average crustal abundance of 1–3 ppm. The region presently being utilized—above a grade of 0.1% U—is on the rising part of the distribution. Thus, by accepting costs above $260/kg, and thereby accessing lower grade ores, one can probably exploit much larger resources.

These include the so-called unconventional sources. There are large phosphate deposits that contain uranium at concentrations of 10–300 ppm U (equivalent to grades of 0.001% to 0.03%). Although it is not economically attractive to use such low-grade ores for their uranium content alone, the uranium can be a useful by-product of phosphate mining [49]. Another possibility is black shale, with grades ranging from 0.003% to 0.01%, including an estimated 5 millions tons of U_3O_8 in Chattanooga shale in the United States (equivalent to 3.8 million tonnes of U), at an average grade of about 0.006% U [7, p. 411].

A cautionary note has been sounded in a comprehensive energy analysis carried out by the International Institute for Applied Systems Analysis (IIASA), in which it was pointed out that with very low-grade uranium resources, it will be necessary to mine large amounts of ore, losing one of the advantages of nuclear power over coal. The product was pejoratively termed "yellow coal" [50, p. 120]. The IIASA report concluded that uranium at 70 ppm (corresponding to a grade of 0.008% U_3O_8) requires slightly more material handling than coal. The average uranium grade of the ore in present-day conventional mining has grades above 0.1% U_3O_8, so uranium still retains a large advantage over coal in terms of volume of material extracted.

Adequacy of Terrestrial Uranium Resources

Adopting the probably conservative resource estimate of 20 million tonnes, uranium resources would suffice for the needs of LWRs for about 100,000 GWyr at a rate of 200 tonnes of U per gigawatt-year. Present world generation from nuclear power is roughly 300 GWyr/yr. Thus, a resource of this magnitude could sustain four times the present rate of generation for 80 years. Such an increase does not appear to be imminent and, therefore, there is little pressure on uranium resources at this time.

However, if one contemplates a large nuclear expansion during the next 50–100 years, uranium supplies could become a constraint for use of LWRs that consume uranium at roughly the present rate. Reducing the resource estimate to 15 million tonnes or raising it to 30 million tonnes would not drastically change the overall qualitative conclusion that, with LWRs, there is ample conventional uranium for the next several decades, but possibly insufficient uranium over the full 21st century if the nuclear industry expands greatly and large amounts of the lower-grade resources cannot be effectively extracted.

9.5.3 Uranium from Seawater

Uranium from seawater represents a very large potential additional resource, but at a cost that may be considerably higher than that of any uranium resource considered earlier. The volume of seawater in the oceans is about 1.4×10^{21} L, with an average uranium concentration of 3.2 parts per billion (see Section 3.4.3), corresponding to 4 billion tonnes of uranium. However, the energy content is very dilute, and vast amounts of water would have to be processed to extract uranium.

The cost of seawater extraction was quoted in a 1964 estimate at \$306/lb of U or about \$700/kg U [7, p. 412]. Studies of seawater extraction have continued, particularly in Japan, and there has apparently been a substantial reduction in real (constant dollar) cost. A project under the auspices of the Metal Mining Agency of Japan with funding from the Ministry of International Trade and Industry has led to pilot-plant extraction of uranium. The status was described as follows, in a 1990 report on *Nuclear Power in Japan*:

> The plant was operated continuously from 1986–1987 to produce 15.4 kg of yellow cake (U_3O_8). Basically, the process included absorption of the uranium on substances such as titanium oxide, followed by removal with an acid, and subsequent recovery of the uranium by ion

exchange techniques. An economic assessment revealed that the cost of uranium recovered from a 1000-ton/yr commercial plant would be approximately $260/lb. Ultimately, with an improved sorbent, a goal of about $100/lb would not be impossible, but it is recognized that even this would be an order of magnitude greater than the present uranium price. Impressive university research continues. [51, p. 22]

A cost of $260/lb of U_3O_8 ($680/kg U) corresponds to an increase of 1.5¢/kWh in the cost of electricity. As discussed in Section 9.5.1, this cost—although very high by present standards—is not necessarily prohibitive.

Richard Garwin and Georges Charpak cite more optimistic estimates of the possible costs of uranium from seawater [34, p. 211]. They point to tentative suggestions from French and Japanese investigators that mention costs ranging from $18/kg U to $260/kg U. This wide range of projections and the importance of having a realistic estimate indicate the need for more intensive investigations of the practicalities of extracting uranium from seawater.

9.5.4 Impact of Fuel Cycle Changes and Breeder Reactors

The question of changes in the fuel cycle and particularly the question of breeder reactors is closely connected to the issue of resources. For the most part, we have assumed a once-through fuel cycle using uranium. If, instead, a comprehensive reprocessing plan is implemented, such as the full actinide recycle option discussed in Section 9.4.3, the situation would be somewhat relieved. In this cycle, the uranium required for a given electricity output is roughly halved, which effectively doubles the resource (measured in GWyr) and halves the cost (measured in ¢/kWh). The lifetime of a nuclear economy could be further extended if use is made of the world's large thorium resources, in a ^{232}Th–^{233}U fuel cycle.

The possibility of inadequate resources would be eliminated if the further step were taken of adopting a breeder reactor fuel cycle. With breeder reactors, the energy yield per tonne of uranium increases greatly. If, say, the increase is by a factor of 50, then it becomes acceptable to have 50 times as high a uranium extraction cost. Under these circumstances, seawater and other low-grade uranium sources offer a quasi-infinite supply of nuclear fuel.

There is no immediate pressure to make a decision on breeder reactors, because present resources, with or without uranium from seawater, appear adequate to sustain a nonbreeder nuclear economy for many decades. Within this time span, extraction of uranium from the oceans may turn out to be possible at acceptable costs or other technologies—such as fusion or forms of renewable energy—may prove to be practical as large-scale providers of additional electricity,

However, there is no certainty connected with either fusion or renewables, nor is it certain that large amounts of uranium from seawater will prove economically practical. For such reasons, breeder reactors remain an important long-term contender if there is a major expansion of nuclear power use. They are, for example, among the candidates being considered as Generation IV reactors (see Chapter 16).

References

1. Organization for Economic Co-operation and Development, Nuclear Energy Agency, *Trends in the Nuclear Fuel Cycle: Economic, Environmental and Social Aspects* (Paris: OECD, 2001).

2. U.S. Department of Energy, *Nuclear Power Generation and Fuel Cycle Report 1997*, Energy Information Administration Report DOE/EIA-0436(97) (Washington, DC: U.S. DOE, 1997).

3. Uranium Information Center, *Geology of Uranium Deposits*, Nuclear Issues Briefing Report No. 34 (Melbourne, Australia: UIC, 2001) [From: http://www.uic.com.au/nip34.htm]

4. Organization for Economic Co-operation and Development, Nuclear Energy Agency, *Uranium 2001: Resources, Production and Demand*, Joint Report of OECD Nuclear Energy Agency and IAEA (Paris: OECD, 2002).

5. Manson Benedict, Thomas H. Pigford, and Hans Wolfgang Levi, *Nuclear Chemical Engineering*, 2nd edition (New York: McGraw-Hill, 1981).

6. U.S. Department of Energy, *Statistical Data of the Uranium Industry*, Report GJO-100(83) (Grand Junction, CO: U.S. DOE, 1983).

7. DeVerle P. Harris, "World Uranium Resources," *Annual Review of Energy* 4, 1979: 403–432.

8. Luther Smith, U.S. Department of Energy, private communication (August 2002).

9. Ahmad E. Nevissi and David Bodansky, "Radon Sources and Levels in the Outside Environment," in *Indoor Radon and Its Hazards*, David Bodansky, Maurice A. Robkin, and David R. Stadler, eds. (Seattle: University of Washington Press, 1987): 42–50.

10. James J. Duderstadt and Louis J. Hamilton, *Nuclear Reactor Analysis* (New York: Wiley, 1976.)

11. U.S. Department of Energy, Office of Civilian Radioactive Waste Management, *Yucca Mountain Science and Engineering Report*, Report DOE/RW-0539 (North Las Vegas, NV: U.S. DOE, 2001).

12. Ronald Allen Knief, *Nuclear Engineering: Theory and Technology of Commercial Nuclear Power*, 2nd edition (Washington, DC: Hemisphere Publishing Company, 1992).

13. Allan S. Krass, Peter Boksma, Boelie Elzen, and Wim A. Smit, *Uranium Enrichment and Nuclear Weapon Proliferation* (London: Taylor & Francis, 1983).

14. J.H. Tait, "Uranium Enrichment," in *Nuclear Power Technology, Volume 2, Fuel Cycle*, W. Marshall, ed. (Oxford: Clarendon Press, 1983): 104–158.

15. R.E. Leuze, "An Overview of the Light Water Reactor Fuel Cycle in the U.S," in *Light Water Reactor Nuclear Fuel Cycle*, R.G. Wyneer and B.L. Vondra, eds. (Boca Raton, FL: CRC Press, 1981).

16. Uranium Information Center Ltd, *Uranium Enrichment*, Nuclear Issues Briefing Paper 33 (June 2003). [From: http://www.uic.com.au/nip33.htm]

17. U.S. Department of Energy, *Uranium Industry Annual 2001*, Energy Information Administration Report DOE/EIA-0478(2001) (Washington, DC: U.S. DOE, 2002).

18. Dennis Spurgeon, "Paving the Way Today for New Enrichment Tomorrow," speech at International Uranium Fuel Seminar, 2001. [From: http://www.usec.com/v2001_02/HTML/News_speeches.asp]

19. Robert Van Namen, "American Centrifuge—The Road to Market," presentation at *Fuel Cycle 2003*, [From: http://www.usec.com/v2001_02/HTML/news.asp]

20. "New Mexico Will Host the $1.2 Billion U Enrichment Plant," *Nuclear News* 46, No. 11, October 2003: 64.

21. "USEC Ends Funding of Research on SILEX Process," USEC News Release, April 30, 2003.

22. F.J. Rahn, A.G. Adamantiades, J.E. Kenton, and C. Braun, *A Guide to Nuclear Power Technology* (New York: Wiley, 1984).

23. U.S. Department of Energy, *Monthly Energy Review, March 2003*, Energy Information Administration Report DOE/EIA-0035(2003/03) (Washington, DC: U.S. DOE, 2003).

24. U.S. Department of Energy, Office of Civilian Radioactive Waste Management, *Final Environmental Impact Statement for a Geologic Repository for the Disposal of Spent Fuel and High-Level Radioactive Waste at Yucca Mountain, Nye County, Nevada*, Report DOE/EIS-0250 (North Las Vegas, NV: U.S. DOE, 2002).

25. U.S. Department of Energy, *Integrated Data Base Report—1996: U.S. Spent Nuclear Fuel and Radioactive Waste Inventories, Projections and Characteristics*, Office of Environmental Management Report DOE/RW-0006, Rev. 13 (Oak Ridge, TN: Oak Ridge National Laboratory, 1997).

26. U.S. Department of Energy, Energy Information Administration, *Detailed United States Spent Nuclear Fuel Data, as of December 31, 1998*. [From: http://www.eia.doe.gov/cneaf/nuclear/spent_fuel/ussnfdata.html]

27. U.S. Department of Energy, *World Nuclear Capacity and Fuel Cycle Requirements 1993*, Energy Information Administration Report DOE/EIA-0436(93) (Washington, DC: U.S. DOE, 1993).

28. U.S. Department of Energy, *Spent Nuclear Fuel Discharges from U.S. Reactors 1993*, EIA Service Report SR/CNEAF/95-01 (Washington, DC: U.S. Department of Energy, 1995).

29. U.S. Department of Energy, *Characteristics of Potential Repository Wastes*, Report DOE/RW-0184-R1, Volume 1 (Oak Ridge, TN: Oak Ridge National Laboratory, 1992).

30. Jagdish K. Tuli, *Nuclear Wallet Cards* (Upton, NY: Brookhaven National Laboratory, 2000).

31. Edwin Kolbe, output of the Oak Ridge National Laboratory's ORIGEN-ARP program (private communication to the author, October 10, 2002).

32. S.F. Mughabghab, *Neutron Resonance Parameters and Thermal Cross Sections*, Part B: $Z = 61$–100, Volume 1 of *Neutron Cross Sections* (New York: Academic Press, 1984).

33. J-Y Barre and J. Bouchard, "French R&D Strategy for the Back End of the Fuel Cycle," in *Future Nuclear Systems: Emerging Fuel Cycles & Waste Disposal Options*, Proceedings of *Global '93* (La Grange Park, IL: American Nuclear Society, 1993): 27–32.

34. Richard L. Garwin and Georges Charpak, *Megawatts and Megatons* (New York: Alfred A. Knopf, 2001).

35. Luther J. Carter, *Nuclear Imperatives and Public Trust: Dealing with Radioactive Waste* (Washington, DC: Resources for the Future, 1987).

36. *Nuclear Power Issues and Choices*, Report of the Nuclear Energy Policy Study Group, Spurgeon M. Keeny, Jr., Chairman (Cambridge, MA: Ballinger, 1977).

37. National Academy of Sciences, *Management and Disposition of Excess Weapons Plutonium*, Report of the Committee on International Security and Arms Control, John P. Holdren, Chairman, (Washington, DC: National Academy Press, 1994).

38. Frans Berkhout and Harold Feiveson, "Securing Nuclear Materials in a Changing World," *Annual Review of Energy and the Environment* 18, 1993: 631–665.

39. International Energy Agency, *Nuclear Power in the OECD* (Paris: OECD/IEA, 2001).

40. Organization for Economic Co-operation and Development, Nuclear Energy Agency, *Nuclear Energy Data 1995* (Paris: OECD, 1995).

41. Pierre M. Chantoin and James Finacune, "Plutonium as an Energy Source: Quantifying the Commercial Picture," *IAEA Bulletin* 35, no. 3, 1993: 38–43.

42. National Academy of Sciences, *Management and Disposition of Excess Weapons Plutonium, Reactor-Related Options*, Report of the Panel on Reactor-Related Options for the Disposition of Excess Weapons Plutonium, John P. Holdren, Chairman (Washington, DC: National Academy Press, 1995).

43. "Surplus Plutonium: Two Reports Assess DOE's Pu-to MOX Plan," *Nuclear News* 45, no. 4, April 2002: 69–70.

44. U.S. DOE Nuclear Energy Research Advisory Committee and the Generation IV International Forum, *A Technology Roadmap for Generation IV Nuclear Energy Systems*, GIF-002-00 (December 2002). [From: http://gif.inel.gov/roadmap/pdfs/gen_iv_roadmap.pdf]

45. U.S. Department of Energy, Office of Nuclear Energy, Science, and Technology, *Report to Congress on Advanced Fuel Cycle Initiative: The Future Path for Advanced Spent Fuel Treatment and Transmutation Research* (January 2003). [From: http://www.ne.doe.gov/AFCI_CongRpt.pdf]

46. Massachusetts Institute of Technology, *The Future of Nuclear Power, An Interdisciplinary MIT Study*, John Deutch and Ernest J. Moniz, Co-chairs (Cambridge, MA: MIT, 2003).

47. John Geidl and William Szymanski, "United States Uranium Reserves and Production," in *Future Nuclear Systems: Emerging Fuel Cycles & Waste Disposal Options*, Proceedings of *Global '93* (La Grange Park, IL: American Nuclear Society, 1993): 1187–1193.

48. Organization for Economic Co-operation and Development, Nuclear Energy Agency, *Uranium 1993: Resources, Production and Demand*, Joint Report of OECD Nuclear Energy Agency and IAEA (Paris: OECD, 1994).

49. Kenneth S. Deffeyes and Ian D. MacGregor, "World Uranium Resources," *Scientific American* 242, no. 1, January 1980: 66–76.

50. International Institute for Applied Systems Analysis, *Energy in a Finite World: A Global System Analysis*, Wolf Häfele, program leader (Cambridge, MA: Ballinger, 1981).

51. Ersel A. Evans, "Nuclear Fuel Cycle," in *JTEC Panel Report on Nuclear Power in Japan*, K.F. Hansen, Chairman (Baltimore, MD: Japanese Technology Evaluation Center, Loyola College, 1990).

10

Nuclear Waste Disposal: Amounts of Waste

10.1 Categories of Nuclear Waste

10.1.1 The Nature of the Problem

The term *nuclear waste* embraces all residues from the use of radioactive materials, including uses in medicine and industry. The most highly radioactive of these are the spent fuel or reprocessed wastes from commercial nuclear reactors and reactors that produced plutonium for nuclear weapons. The other wastes have much lower levels of activity and have far less potential to cause harm, although the establishment of sites for the disposal of these wastes has encountered public opposition in many parts of the United States.

Plutonium production for weapons has ended in the United States and Europe, at least for the time being. For the United States and Russia, in particular, the emphasis is now on getting rid of excess weapons plutonium, not on producing more. This does not totally end the creation of weapons wastes because some countries are still expanding their nuclear-weapon inventories. However, the creation of weapons wastes is mostly a thing of the past. What remains is the technical problem of coping safely and effectively with the existing weapons wastes and the associated site contamination.

Wastes from commercial nuclear reactors raise more critical issues, because the amounts are greater, their production continues, and the perception of the hazards they create affects the public's willingness to accept nuclear power. This waste initially consists of the spent fuel removed from the reactor. It contains a wide array of radionuclides formed by fission of uranium and plutonium and by neutron capture in these and other elements. Some of the radionuclides are long-lived and must be isolated from the environment for an extended period.

The spent fuel is commonly referred to as "waste," although some of the radionuclides have potentially useful applications and it has been argued that spent fuel is a resource, not a waste product. Nonetheless, the attribute of spent fuel that is of greatest current significance is the waste aspect, not the resource potential, and its designation as waste conforms to the dominant official and informal usage.

10.1.2 Military and Civilian Wastes

As discussed in Chapter 2, the first nuclear reactors were those built during World War II to produce plutonium for weapons. In order to extract the plutonium, it is necessary to reprocess the spent fuel, first converting it to liquid form. The residue remaining after the plutonium and uranium (and sometimes other elements) are extracted constitutes the *reprocessed wastes*. Reprocessing increases the volume of the residue and puts it in a form that can more readily escape into the environment. This residue was originally stored as liquids in large underground, single-walled tanks.

Given the pressures of wartime development, there was no well-engineered, long-term plan for the permanent disposal or storage of these wastes. Weapons production continued and increased after World War II, with large programs at the DOE's facilities at Hanford (Washington) and Savannah River (South Carolina), but disposal plans were still not developed in a timely fashion. As a result, there were mishaps, including large leaks from some tanks at Hanford during the 1970s, and concerns arose about possible further leaks and conceivable chemical explosions.

In recent years, intense consideration has been given to methods for isolating and disposing of these wastes, but the excessively casual earlier treatment has complicated matters. To date the wastes have caused no known harm to human health, and it is not clear that there is a realistic prospect of future harm. However, their ultimate disposal is a complex and costly matter. Most of the wastes have not yet been turned into a stable, solid form, although conversion into glass rods is being carried out at the Savannah River nuclear center in South Carolina and a vitrification plant is now under construction at Hanford in Washington.

These wastes constitute the bulk of the *military wastes*. Nuclear reactors used on naval vessels constitute a second, smaller source of the military wastes, but the radioactivity levels and physical volumes involved are considerably less than those from weapons production [1, p. 44].

Civilian or *commercial wastes* are those produced by reactors built for commercial electricity generation. The amount of radioactivity produced in this manner to date is much greater than that for the military wastes, because more reactors have been involved, operating over longer total periods. The volume is less, however, because the wastes have remained as solid fuel rods. Overall, they are easier to handle than the reprocessed military wastes, being in compact solid form. The focus here will be on commercial wastes, because their successful disposal is crucial to the future of nuclear power.

The issues relating to military wastes are irrelevant to nuclear power in terms of most of the technical problems of waste handling. They do have a practical relevance, however, in that much of the public does not differentiate between the two categories of waste. Even if the distinction is recognized, it is common to conclude that errors and problems in one area imply a strong possibility of similar errors and problems in the other. Of course, there is a difference in the histories. Military wastes not only were reprocessed and put in liquid form, but they were reprocessed in an atmosphere where speed was thought to be urgent and the dangers small. Now, the situation is reversed: The civilian spent fuel has stayed in solid form, speed has played little part in the planning for its handling, and there is great attention to possible dangers.

10.1.3 High- and Low-Level Wastes

Nuclear wastes are sometimes divided into *high-level* and *low-level* waste, along with a separate category of *transuranic waste*:

- **High-level waste (HLW)** is the highly radioactive fission and neutron-capture product of the nuclear fuel cycle. It may be in the form of either spent fuel or liquid and solid products from the reprocessing of spent fuel.[1] (In some alternative definitions, spent fuel is in a category by itself. Many DOE tabulations reserve the term HLW for reprocessed wastes alone, and designate spent fuel separately.)
- **Transuranic waste (TRU)** is waste that does not qualify as high-level waste but that contains more than 0.1 μCi of long-lived ($T > 20$ years) transuranic alpha-particle emitters per gram of material. The term *transuranic* refers to elements of atomic number greater than 92. These are formed in neutron-capture chains, starting with uranium. The TRU waste is a residue of DOE military programs, with relatively small total activity. It is being transferred to a special repository that has been built in New Mexico, the Waste Isolation Pilot Plant (WIPP).
- **Low-level waste (LLW)** is the remaining radioactive material that arises from the operation of nuclear facilities, as well as residues from medical and industrial use of radionuclides.[2]

We will restrict the discussion here to consideration of high-level wastes, because the amounts of activity are the greatest for these and the potential hazards the greatest.

---

[1] This definition, with the inclusion of spent fuel, corresponds to the official definition expressed in 10 CFR 60.2 of the Code of Federal Regulations [2, Part 60].

[2] Wastes with relatively low levels of activity are categorized on the basis of their activity per unit volume, with different NRC disposal criteria for each class (see Part 61 of Ref. [2]).

Table 10.1. Inventories of commercial and military nuclear wastes in the United States.

	Commercial	Military	
	Spent Fuel	Spent Fuel	HLW[a]
Planned for Yucca Mountain (MTHM)	63,000	2,333	4,667
Projected inventory (MTHM)[b]	105,000	2,500	12,600
YM plan, as fraction of projected inventory	0.60	0.93	0.37

[a]The high-level waste (HLW) here refers only to reprocessed waste; the total includes a very small amount of reprocessed commercial fuel.
[b]The projected inventory for commercial fuel corresponds to the continued operation of existing reactors, each with a license renewal for an additional 10 years [3, p. A-14]. (Operations end in 2046.)
Source: Ref. [3, p. A-7].

10.1.4 Inventories of U.S. Nuclear Wastes

Inventories of U.S. commercial and military wastes are summarized in Table 10.1. They include the wastes that are planned for the Yucca Mountain repository as well as further amounts expected to be produced. The commercial wastes constitute 90% of the total slated for Yucca Mountain, measured in terms of the mass of the initial uranium used to generate the wastes (in MTHM). In general, they are the "hottest" wastes, so they contain more than 90% of the total activity [3, p. A-9]. When the reprocessed military waste (identified as HLW) is vitrified, the total mass of the material is about four to five times the mass of the fuel itself due to the addition of the glass materials— although the additional nonradioactive material is of little concern.[3]

The spent fuel from military programs makes a relatively minor contribution to the overall inventory and its mass is increasing only slowly. With the cessation of new nuclear weapons manufacture, there are no appreciable additions to the HLW. In fact, its total radioactivity is decreasing due to natural decay, and its volume is decreasing as liquid wastes are solidified [4, Chapter 2]. In contrast, the amounts of commercial spent fuel are increasing rapidly. Virtually all of it is presently stored at the reactor sites.[4] Roughly two-thirds of the fuel is from PWRs and one-third from BWRs.

At the present level of nuclear generating capacity, about 2100 tonnes of spent fuel are discharged each year from U.S. commercial reactors. Future rates of spent fuel production will depend on the growth or contraction of nuclear capacity. Projections made in the 1990s were often based on a scenario in which there would be no new orders for reactors, and reactor shutdowns

---

[3] The projected total mass of the solidified waste is 58,000 tonnes, compared to the 12,600 MTHM in the fuel [3, p. A-40].

[4] A very small amount is stored at a site in Morris, Illinois, where a reprocessing plant had been planned but was not completed.

would start to be numerous by 2010. For example, in a 1992 estimate, the cumulative spent fuel inventory in 2030 was projected to be 87,700 MTHM [5, 14]. A more recent estimate, reflected in the Environmental Impact Statement for Yucca Mountain published in 2002, assumed that existing plants would continue to operate and all have their licenses renewed for 10 years. The estimated inventory in this case is an eventual 105,000 MTHM in 2046, as shown in Table 10.1 [3, p. A-16].

These projections could be too low if the recent wave of applications for 20-year license renewals continues (and the renewals are implemented), or if there is a revival of nuclear reactor construction. They could be too high if nuclear power is phased out precipitously. Nonetheless, they set a scale for storage needs over the next few decades. In any projection, the total exceeds the capacity that is now allocated for spent fuel in the one repository being planned—namely the Yucca Mountain repository. The proposed repository is limited by congressional mandate to a total of 70,000 tonnes of waste (measured in MTHM), of which 63,000 tonnes is allocated to spent fuel (see Section 9.5.1). This total is expected to be reached in about 2010, the year that the repository is nominally scheduled to begin to receive wastes.[5]

The remaining capacity is allocated to military wastes, including spent fuel and reprocessed HLW. This latter waste is still in many different forms, including liquid, sludge, and salt cake, but it is expected that almost all of it will be incorporated into solid form by 2030. A variety of solid forms are possible, but present plans call for incorporating virtually all of it into glass [3, §A.2.3]. It would be contained in about 22,000 canisters. The mass of this vitrified material will about 58,000 tonnes. Of this, 44,000 tonnes would come from Hanford wastes and be in the form of borosilicate glass with a density of about 2.8 tonnes/m^3. As with the civilian wastes, the planned Yucca Mountain capacity will not suffice for the full amount of the defense wastes (see Table 10.1).

10.1.5 Measures of Waste Magnitudes

The inventories of wastes have been described earlier in terms of mass. However, several different sorts of measure can be used to describe the "amount" of nuclear wastes:

♦ *Mass.* The most common mass measure for nuclear waste is the mass of the uranium in the initial fuel, more broadly designated as metric tonnes of initial heavy metal (MTIHM or just MTHM) (see Section 9.3.1). The fuel is held in cylindrical fuel rods, usually made of zircaloy, which are grouped in assemblies. The total mass of an assembly includes the masses of the UO_2, the zircaloy cladding of the fuel rods, and the metallic structure that

---

[5] The limit of 70,000 MTHM is a statutory one at present, not a physical one, and the possibility of accommodating up to 119,000 MTHM at Yucca Mountain has been suggested in DOE documents [6, p. 2-85].

Table 10.2. Physical characteristics of representative LWR
fuel assemblies.

Parameter[a]	BWR	PWR
Fuel elements per assembly		
Fuel element array	8×8	17×17
Number of fuel rods	63	264
Number of other elements[a]	1	25
Mass per assembly (kg)		
Uranium	180	460
Uranium oxide (UO_2)	210	520
Zircaloy cladding	100	110
Hardware	9	26
TOTAL	320	660
Assembly dimensions		
Length (m)	4.5	4.1
Width and depth (m)	0.14	0.21
Nominal volume (m^3)	0.086	0.19
Fuel fraction		
Uranium as percent of total mass	56	70
UO_2 as percent of total volume	24	27

[a] These are for control and instrumentation; in the BWRs, control
rods are between assemblies rather than incorporated in them.
Source: Ref. [3, vol. II, p. A-25].

holds the fuel rods in place. Typical amounts are indicated in Table 10.2.
For reprocessed wastes the mass can be specified either as the mass of
the initial heavy metal (again in MTHM) or as the mass of the actual
reprocessed product.

♦ *Volume.* The volume of the fuel can be inferred from the UO_2 mass and
density (about 10 tonnes/m^3). The volume of the assembly is considerably
greater, if based on the outer dimensions of the assembly, due mostly to
the spacing between the fuel rods.

♦ *Radioactivity.* The mass and volume of nuclear wastes are small, especially
if compared to the residues of coal-fired generating plants. The special
property of nuclear waste is its radioactivity. Correspondingly, the wastes
are often described in terms of the activity (in curies or becquerels) taken
either for the radionuclides individually or for their sum. This sum is only
a crude measure because the radionuclides differ in the types of particles
emitted (alpha or beta particles and, in many cases, gamma rays), their
energy, the radionuclide half-lives, and the possibility of their reaching the
biosphere. Nonetheless, it provides some overall perspective.

♦ *Heat output.* The heat output is another crucial measure, because of the
need to avoid overheating the wastes themselves and the repository in

which they are placed. As discussed in more detail in Section 10.2.3, something on the scale of 6 kW of heat are produced per megacurie of activity.

10.2 Wastes from Commercial Reactors

10.2.1 Mass and Volume per GWyr

Spent Fuel

The amount of spent fuel discharged per year varies greatly among reactors depending on their size, type, and operating experience. The annual discharge for all U.S. reactors from 1990 through 1998 averaged 2100 MTHM or about 200 MTHM per reactor [7]. The electrical output during this period averaged about 70 GWyr per year [8], corresponding to an average spent fuel discharge of 30 MTHM/GWyr.

The fuel is packaged in fuel assemblies when it is inserted into the reactor, and these assemblies are kept intact when the fuel is removed from the reactor. Parameters reflecting the physical characteristics of representative BWR and PWR fuel assemblies are given in Table 10.2.[6] Overall, as reflected in the table, the BWR assemblies are smaller than the PWR assemblies. Within each category, there is considerable variation in the dimensions, and the parameters for individual assemblies may differ substantially from the representative values of Table 10.2. For example, the mass can vary by as much as 25% from the indicated 660 kg for PWRs [6, p. 3–12]. Thus, the parameters of Table 10.2 are approximate indicators, presented for purposes of orientation. The assembly mass, considering the PWR case for specificity, is about 70% uranium, 9% oxygen (in UO_2), and 21% metal. The UO_2 fuel, with a density of about 10 g/cm^3, represents only about one-quarter of the assembly volume, as taken from its outer dimensions.

Typical amounts discharged per gigawatt-year of electrical output are summarized in Table 10.3, based on the average BWR and PWR burnups anticipated for the spent fuel to be deposited at the Yucca Mountain repository. For LWRs as a whole, the mass of the UO_2 fuel ranges from about 30 to 40 tonnes/GWyr corresponding to a volume, at a density of 10 tonnes/m^3, of 3–4 m^3. The total assembly mass is in the neighborhood of 50 tonnes/GWyr and the total volume is about 13 m^3. In general, the BWR numbers are at the top of these ranges and the PWR numbers at the bottom, due to the differences in average fuel burnup (see Table 10.3).

---

[6] These parameters have been used in DOE documents as reference points for at least 20 years. They appeared in 1980 in Ref. [9, Table 3], have been used in the DOE's Integrated Data Base Report series (e.g., Ref. [1]), and were used again in the 2002 EIS with only minor modifications [3].

Table 10.3. Spent fuel discharges for light water reactors.

Parameter	BWR	PWR
Average burnup (GWd/t)	33.6	41.2
Fuel per GWyr (MTHM)[a]	34.0	27.7
Fuel per assembly (MTHM)[b]	0.18	0.46
Assemblies per GWyr	189	60
Mass per GWyr (tonne)[b]		
Uranium (MTHM)	34.0	27.7
Uranium oxide fuel	39	31
Cladding and structure	21	8
Total assembly	60	40
Volume per GWyr (m^3)[b]		
Uranium oxide fuel	4.0	3.1
Total assembly	16	11
Planned for Yucca Mountain (approx.)		
Number of assemblies (1000s)	127	94
Mass of spent fuel (10^3 MTHM)	22	41

[a]Calculated from burnup, for assumed thermal efficiency of 32%.
[b]Based in part on parameters of Table 10.2.
Source: Burnup from Ref. [3, p. A-13]; Yucca Mountain plans from Ref [6, p. 3–17].

Yucca Mountain plans call for storing 63,000 MTHM of spent LWR fuel, corresponding to 71,500 tonnes of UO_2 fuel. The average burnup for the entire inventory is expected to average 38.6 GWd/t (see Section 9.3.1). PWR fuel constitutes about 64% of the total mass of the fuel and accounts for about 69% of the electricity.

Reprocessed Waste from Commercial Spent Fuel

There are no present plans to reprocess commercial spent fuel in the United States. However, reprocessing was long assumed as part of the U.S. waste program, it is being actively carried out in France and elsewhere, and may at some future time play a role in the U.S. waste disposal program (see Section 9.4.2).

10.2.2 Radioactivity in Waste Products

Activity of Spent Fuel

When a reactor is just shut down, there are extremely high levels of activity. The actual activity at the moment of shutdown depends on the recent history of the reactor operation and is dominated by the decay of the short-lived radionuclides. These radionuclides, especially those with half-lives of the order of hours or days, are important for the heat budget of the reactor immediately after shutdown, and their role is crucial in reactor accidents in which there is difficulty in maintaining the flow of cooling water (or other coolant). However, they are not part of the waste disposal problem, because their activity falls to negligible levels while the spent fuel is still in the cooling pools at the reactor sites.

After 1 year, the total radioactivity of the spent fuel is about 1.3% of the activity at discharge, and after a total of 10 years it has decreased by more than another factor of 5 [10]. Ten years can be considered to be roughly the time at which the waste disposal problem begins, because the earlier high levels of radioactivity and heat production make it unfavorable to attempt to dispose permanently of the wastes much sooner than 10 years after removal from the reactor.

Table 10.4 shows the radionuclides most responsible for the activity in the wastes for the period from 1 year to 1,000,000 years after removal from the reactor.[7] During the first few centuries, the most important contributors to the total activity are two beta-particle emitters: strontium-90 (^{90}Sr) and cesium-137 (^{137}Cs). Each has a half-life in the neighborhood of 30 years, and each decays into a short-lived unstable product, approximately doubling the total activity.[8] More particularly, ^{90}Sr decays to yttrium-90 (^{90}Y) ($T = 64$ h) and ^{137}Cs decays 95% of the time to an excited state of barium-137 (^{137}Ba) that emits a 0.66-MeV delayed gamma ray.[9]

During the first few decades, ^{241}Pu is also an important radionuclide in terms of activity. At successive later times, the leaders include ^{241}Am (a decay product of ^{241}Pu), ^{240}Pu, ^{239}Pu, ^{99}Tc, and ^{237}Np. At 1,000,000 years, the activity is spread among many radionuclides, but these are largely concentrated in decay series that are close to being in secular equilibrium. Thus, ^{233}U, whose abundance is built up largely from the decay of neptunium-237 (^{237}Np), heads a series of nine radionuclides, for a total activity of about 300 Ci. Similarly, thorium-230 (^{230}Th) heads a series of 10 radionuclides, for

---

[7] The data of Table 10.4 are for a PWR with a specific set of operating parameters, as described in the table. The activity and thermal output of the fuel varies somewhat from one set of parameters to another, even when normalized to a fixed electrical output.

[8] For ^{137}Cs, the doubling is not exact because 5% of the decays are to the stable ground state of ^{137}Ba.

[9] The gamma-ray emitting excited state of ^{137}Ba is an isomeric state, designated as ^{137}Ba^{m}, with a decay half-life of 2.55 min. Usually, gamma rays are emitted within less than 10^{-9} s following beta or alpha decay and are not considered an additive contribution to the activity (i.e., to the number of decays). However, delayed emission from isomeric states is taken to be a separate activity. A similar situation occurs in the decay of ^{93}Zr (see Table 10.4).

Table 10.4. Activity of selected radionuclides in spent fuel versus time since discharge of fuel from reactor (for PWR, burnup of 40 GWd/t, ^{235}U enrichment of 3.75%, thermal efficiency of 32%).

Nuclide	Half-life (years)	Activity (per GWyr)[a]						
		1 yr MCi	10 yr MCi	100 yr MCi	1000 yr kCi	10^4 yr kCi	10^5 yr kCi	10^6 yr Ci
^{85}Kr	10.8	0.28	0.16					
^{90}Sr	28.8	2.5	2.0	0.23				
^{90}Y	$\ll 1$	2.5	2.0	0.23				
^{93}Zr	1.53 M					0.04	0.040	26
^{93}Nb^m	16.1					0.04	0.039	26
^{99}Tc	211 k				0.46	0.45	0.33	17
^{129}I	15.7 M							1.0
^{137}Cs	30.1	3.6	2.9	0.37				
^{137}Ba^m	$\ll 1$	3.4	2.8	0.35				
^{230}Th	75.4 k					0.01	0.040	15[b]
^{233}U	159 k						0.015	33[b]
^{237}Np	2.14 M					0.04	0.042	31
^{238}Pu	87.7	0.11	0.11	0.05				
^{239}Pu	24.1 k	0.010	0.010	0.010	10	5.7	0.58	
^{240}Pu	6.56 k	0.016	0.016	0.016	15	5.7		
^{241}Pu	14.3	4.4	2.8	0.037				
^{241}Am	432	0.012	0.06	0.14	33			
^{244}Cm	18.1	0.11	0.08	0.003				
SUM[c]		17	13.1	1.43	57	13.9	1.09	150[b]
TOTAL[c]		74	13.9	1.44	60	14.9	1.82	674

[a]To convert to activity per MTHM, use for this case: 28.5 MTHM/GWyr.
[b]The nuclides ^{230}Th and ^{233}U head radioactive decay series in approximate secular equilibrium, with totals of 10 and 9 radionuclides, respectively. Including these series would add about 400 Ci to the sum (they are included in the total).
[c]The "sum" is the sum for radionuclides listed in the column (before rounding off); the "total" is the sum for all radionuclides, including those not listed in this table.
Source: Half-lives from Ref. [11]; the symbols k and M denote 10^3 and 10^6, respectively. Activities are based on ORIGEN calculation [10].

a total activity of about 150 Ci. Together, these two series account for about two-thirds of the total activity at 1,000,000 years.[10]

Much of this data, in different units, is plotted in Figure 10.1. The rise in some of the abundances is due to the decay of precursor radionuclides, as in the decay sequence

$$^{241}\text{Pu} \rightarrow {}^{241}\text{Am} \rightarrow {}^{237}\text{Np} \rightarrow {}^{233}\text{Pa} \rightarrow {}^{233}\text{U}$$

---

[10] The ^{230}Th series includes six alpha-particle emitters and four beta-particle emitters; the ^{233}U series includes six alpha-particle emitters and three beta-particle emitters.

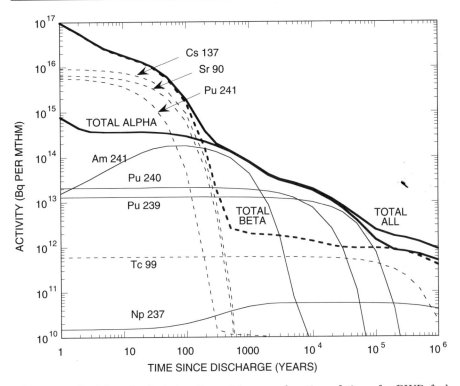

Fig. 10.1. Activity of selected radionuclides as a function of time, for PWR fuel (same parameters and data as for Table 10.4). For individual radionuclides, dotted lines correspond to beta-particle emitters and solid lines to alpha-particle emitters. The curves for ^{90}Sr and ^{137}Cs include the activity of their shorter-lived daughter products: ^{90}Y and ^{137}Ba.

The Impact of Reprocessing

The nature of the long-term waste storage problem is substantially changed if reprocessing is undertaken to remove plutonium and uranium from the waste. Plutonium removal would also reduce the contributions from americium-241 (^{241}Am) and neptunium-237 (^{237}Np), because most of their abundance in the spent fuel derives from the decay of ^{241}Pu. To use the extraction of plutonium to reduce the abundance of ^{241}Am and ^{237}Np, it is necessary to reprocess the fuel promptly, before an appreciable fraction of the ^{241}Pu ($T = 14.3$ years) has decayed. A further step would be to remove all of the actinides, including neptunium, americium, and curium, during reprocessing, as discussed in Section 9.4. This would remove all of the alpha-particle emitters from the wastes, along with their beta-decaying progeny (see Figure 10.4).

It is also possible to remove and package separately the strontium and cesium, although this is not an essential goal of reprocessing. The strontium and cesium can be encapsulated, and the integrity of their containers is of importance for only a few hundred years, up to perhaps 600 years, because their activity drops by roughly a factor of 1000 in each 300 years.

Although the separation of these radionuclides from the wastes reduces the activity of the main body of the wastes, it involves far more handling of radioactive material than does direct disposal of the spent fuel. It also creates new streams of radioactive material. The plutonium can be returned to reactors to provide additional energy and be consumed in the process. However, the increased availability of plutonium for diversion to use in weapons has caused objections to the reprocessing of spent fuel and the extraction of plutonium, as discussed in other chapters.

10.2.3 Heat Production

The handling of the nuclear wastes is significantly complicated by the heat generated in the decay of the radionuclides. On occasion, it has been suggested that the heat from the wastes could be used constructively as a heat source (e.g., for direct warming of arctic installations). However, until there is more confidence in our ability to retain the wastes safely within their containers, it is unlikely that such applications will be considered prudent. The main interest in the heat generation at present is in the demand that it creates for cooling the waste containers, with either water or air, and in the problems that the total thermal output create for a waste repository.

The heat generation per unit activity depends on the energy carried by the emitted particles. A 1-MCi source emits 3.7×10^{16} particles per second. An order-of-magnitude estimate of the heat generation per curie is obtained by assuming an average energy deposition of 1 MeV per disintegration. This corresponds to a deposition of 5.93×10^3 joules per second. Thus, for orientation purposes, the following conversion factor is useful:

$$1 \text{ megacurie} \rightarrow 5.93 \text{ kW (at 1 MeV per disintegration).}$$

Taking 1 MeV per disintegration as a rough average, 1 MCi corresponds to a heat generation on the order of 6 kW.

This is an overestimate for most beta decays and a substantial underestimate for all alpha decays. In a typical beta decay, the maximum beta-particle energy is about 1 MeV, but there is a distribution of energies, and, on average, more than half of the energy is carried off by neutrinos.[11] Thus, typical energy depositions for beta decay are well under 1 MeV. For all alpha-particle

---

[11] A precise determination of the energy deposition in beta decay requires (1) the establishment of the actual average beta energy, (2) a correction for the energy from gamma rays produced in beta decay to an excited state; and (3) a (very small) correction for the energy lost by the escape of beta particles and gamma rays from the waste into the surroundings.

activities of interest, the alpha-particle energy exceeds 4 MeV, and this energy is all captured in the fuel within 0.1 mm of the site of the decaying nuclide.

For the first several decades after the fuel is removed from the reactor, the most important activities are beta decays. At later times, alpha-particle activities become the more important, unless the alpha-particle emitters are removed in reprocessing. The actual heat output for current inventories of wastes is about 4 kW/MCi for spent fuel and 3 kW/MCi for high-level wastes from reprocessing. Both categories include samples that are broadly distributed in the times since their discharge from the reactor, within the limitation that the greatest possible time is less than about 60 years.

The heat output and activity of spent fuel are displayed in Figure 10.2 as a function of time since discharge, for fuel from 1 GWyr of reactor operation.[12] The activity and heat output fall off in roughly the same fashion, but the fractional drop is more rapid for the activity because, on average, the beta-particle emitters have shorter half-lives and less energy deposition than

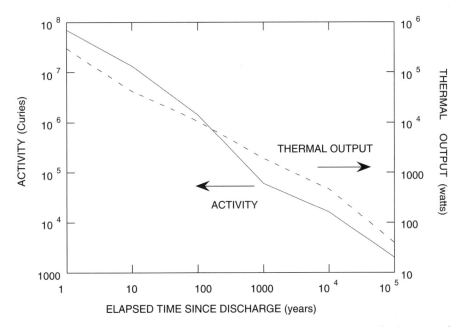

Fig. 10.2. Decay of spent fuel from 1 GWyr of PWR operation, for burnup of 40 GWd/t (28.5 MTHM): activity and thermal output as a function of time since discharge.

---

[12] The data are taken from Ref. [12, Tables 2.4.6 and 2.4.7]. The specific case shown is for a burnup of 40 GWd/t for a PWR with an initial ^{235}U enrichment of 3.72% and a fuel mass of 29 MTHM. A normalization from energy in gigawatt-days thermal (GWd) to a net electrical output of 1 GWyr is made using a thermal efficiency of 31.8%.

do the alpha-particle emitters. The heat output is 4.3 kW/MCi after 1 year and rises to 31 kW/MCi after 1000 years. At 1000 years, about 95% of the activity is due to three alpha-particle emitters: ^{239}Pu, ^{240}Pu, and ^{241}Am (see Table 10.4). The subsequent drop at still later times of the thermal output per unit of activity is due to a gradual decrease in the dominance of alpha-particle emitters.

Spent fuel from PWR operation has a heat output of 1.5 kW/MTHM after 10 years of cooling, 0.37 kW/MTHM after 100 years, and 0.07 kW/MTHM after 1000 years [3, p. A-24]. From some perspectives, these are not very great heat outputs, and after 10 years or so, spent fuel can be cooled by convective airflow. However, the combined output of a large mass of fuel creates a significant heat source, and the "thermal load" becomes a major consideration in the design of an underground repository that holds tens of thousands of tonnes of fuel. One way to reduce the peak temperature reached in a repository is to postpone the placement of wastes in it. For example, in Swedish plans for a waste repository, a long preliminary cooling time is planned, using interim storage. Other ways are to use ventilation and/or place less fuel per unit area (see Section 11.2.4).

10.3 Hazard Measures for Nuclear Wastes

10.3.1 Approaches to Examining Hazards

Exposures from Direct Contact

The spent fuel assemblies are placed, with remote handling, into cooling pools. They eventually are to be transferred from the cooling pools into protective canisters which are placed in heavy casks, again with remote handling. The canisters and casks provide substantial shielding—essentially as much as one wants if a price is paid in weight. Therefore, radiation exposure from close contact with spent fuel assemblies is not a critical safety issue, assuming proper handling during the operations that precede their placement in the casks.[13]

The high radiation levels from unshielded spent fuel provide important protection against theft by people who do not have the elaborate equipment and facilities required to handle the fuel assemblies. This aspect is discussed briefly in Section 17.4.2. However, in considering the radiation hazards from nuclear wastes, the usual focus is not on exposures of terrorists or others from direct contact. It is on the possible exposure of the general public many centuries hence, through the escape into the biosphere of radionuclides from waste repositories.

---

[13] See Section 11.1.4 for a brief discussion of exposures from fuel during transportation.

Hazards from Wastes in a Repository

The ultimate measure of the hazards created by nuclear wastes is the dose—or spectrum of doses—received by people who may ingest or inhale radionuclides that escape from the repository. Calculating this dose requires as a first step the evaluation of mechanisms by which the waste containers might be damaged, permitting the escape of radionuclides. Then, for each radionuclide, it is necessary to consider its amount in the repository as a function of time, the rate at which it would escape from damaged containers, its subsequent movement from the repository site to the biosphere, its pathways for entering the human body, and the dose resulting from its ingestion or inhalation.

This is a complex calculation, with many uncertainties. Nonetheless, such calculations are central to evaluating the safety of a repository, and these calculations have been carried out by the DOE for the proposed Yucca Mountain repository in a long series of iterations of so-called Total System Performance Assessments (TSPAs). These are discussed below in Section 12.4. In the remainder of this chapter, we will examine two ways of getting some perspective on the magnitude of the problem—one in terms of a simple hazard measure and the other in terms of an even simpler criterion, the total activity. Such approaches have the advantage of using reasonably transparent calculations, but the major drawback of looking at only a small part of the problem that is addressed in a full-scale TSPA.

10.3.2 Comparisons Based on Water Dilution Volume

Motivation for Introducing the Water Dilution Volume

Although the amount of activity (i.e., the number of decays per unit time), gives some indication of the importance of a radionuclide from the standpoint of risk to human health, activity per se is not the crucial quantity. In addition, the radiation dose for radionuclides deposited in the body depends on the type and energy of the emitted particles, on whether the nuclides are quickly excreted from the body upon ingestion (or inhalation) or retained, and, if retained, for how long and in what organs.

Just these considerations go into determining the *maximum permissible concentration* of radionuclides in water.[14] The maximum permissible concentration is established as the maximum level acceptable for drinking water. Such levels have been tabulated both by the International Commission on Radiation Protection (ICRP) and by the U.S. Nuclear Regulatory Commission.[15]

---

[14] The maximum permissible concentration is closely related to the annual limit on intake (ALI). (See Section 4.4.3.)

[15] The NRC standards, effective 1994, are specified in Appendix B to §20.1001–20.2402 of Ref. [2].

The potential hazard represented by a given radionuclide is frequently indicated in terms of the *water dilution volume* (WDV). The water dilution volume (in cubic meters) is the amount of water required to dilute the radionuclide to the maximum permissible concentration. It equals the amount of the radionuclide being considered (in becquerel or curie) divided by the maximum permissible concentration (in Bq/m^3 or Ci/m^3). There is no implication here that the radionuclide *will* be so dissolved. However, the WDV is used in an attempt to provide an overall indication of the relative hazards associated with the different radionuclides in the wastes.

Water Dilution Volume for Spent Fuel

To compare the hazards from the various radionuclides in the waste, it has been common to plot the WDV values for individual radionuclides as a function of time since discharge of the fuel. An example of such a plot is given in Figure 10.3, copied from a 1983 National Academy of Sciences (NAS) study [13].[16] The results are presented in terms of the amount of water (in m^3) needed to dilute the wastes in 1 MTHM of spent fuel.[17] Although the activity falls by a factor of more than 1000 in the first thousand years (see Figure 10.1 or Table 10.4), the WDV falls more slowly. This is in part due to the fact that the longer-lived nuclides are alpha-particle emitters. Typically, alpha particles are more energetic than beta particles, by a factor of 5–10, and have a 20 times greater relative biological effectiveness.

Over long time periods, ^{237}Np ($T = 2.14 \times 10^6$ years) becomes the most important of the waste nuclides in Figure 10.3, which used data for ^{237}Np from a 1979 report of the International Commission on Radiological Protection (ICRP 30). However, more recent reports of the ICRP and EPA quote a ^{237}Np risk that is smaller by something of the order of a factor of 100 (see Table 4.4). Adopting the newer values would substantially change the appearance of Figure 10.3 after about 50,000 years, with ^{237}Np playing a smaller role and the combined WDV reduced.

To illustrate the meaning of the water dilution volume with a specific example, consider ^{137}Cs. Ten years after discharge, the ^{137}Cs activity is 8.2×10^4 Ci/MTHM for the reactor parameters used in the NAS study [13, pp. 28 and 31]. For an annual limit on intake (ALI) of 10^{-5} Ci of ^{137}Cs and an annual water intake of 0.73 m^3, the maximum permissible ^{137}Cs concentration would

---

[16] This study is more formally a study by the National Research Council, but for the most part in this book, we will identify such studies as NAS studies—in conformity with widespread practice. The National Research Council is an agency of the National Academy of Sciences, the National Academy of Engineering, and the Institute of Medicine.

[17] This fuel was assumed to have a burnup of 33 GWd/t.

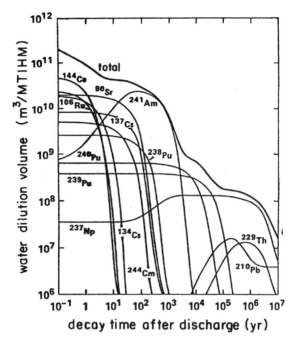

Fig. 10.3. Illustration of use of water dilution volume: WDV of radionuclides in PWR spent fuel, as a function of time, based on ICRP values for maximum permissible concentrations (in m^3/MTHM). [Adapted from Ref. [13, p. 32].]

be 1.37×10^{-5} Ci/m^3 corresponding to a WDV of 6×10^9 m^3/MTHM—in rough agreement with Figure 10.3.[18]

Comparisons of WDV of Waste and Initial Uranium

In attempts to obtain perspective on waste disposal hazards, it has been common to compare the WDV (or an equivalent measure of relative toxicity) of the spent fuel and of the uranium ore initially mined to fuel the reactor. In the 1970s, it was frequently pointed out that after several hundred years, the radiotoxic hazard measure (equivalent to the WDV) of the wastes fell below that of the original uranium ore. This was cited as demonstrating that the hazard posed by the wastes is small.

However, the WDV of the wastes was substantially increased when the focus shifted in the 1980s from reprocessed wastes, with the plutonium removed, to spent fuel. In addition, an ICRP reevaluation of maximum permissible con-

---

[18] The occupational ALI for ingestion of ^{137}Cs is 10^{-4} Ci, corresponding to a dose limit of 50 mSv/year. At the time of the NAS report, the dose limit for the general public was 5 mSv/year, corresponding to an ALI of 10^{-5}Ci. The assumed water intake is 2 L/day.

centrations for various radionuclides had the effect of decreasing the WDV for the parent uranium ore and increasing the WDV at long times for the wastes. Specific changes included an increase by a factor of 8 in the maximum permissible concentration for ^{226}Ra, a major contributor in the uranium ore, and decreases in the maximum permissible concentrations for several important actinides in the wastes, including a factor of 340 reduction for ^{237}Np (see Section 4.5.3) [13, p. 31].

As a consequence of these changes, the calculated time for the spent fuel toxicity to fall to the level of the parent ore increased greatly. Depending on which set of parameters was adopted, this time for spent fuel ranged from 10,000 years to over 1 million years.[19] In contrast, for reprocessed wastes and using the parameters set forth by the U.S. Nuclear Regulatory Commission at that time, the crossover was at about 500 years [13, p. 37].

Despite its suggestiveness, the comparison made earlier is deficient on a number of grounds:

♦ A comparison to the parent ore represents an arbitrary constraint that appears to be based on the premise that use of uranium for nuclear power should not increase the total hazard from uranium. However, this is not a useful standard for comparison, because the doses from ingestion of natural uranium and its radioactive progeny are very small and do not represent a significant hazard.[20]

♦ The use of the WDV ignores the fact that the chance of ingestion is not the same for all radionuclides. If the waste containers are penetrated by water, different radionuclides will have very different rates of entering the water and of eventual transport to the biosphere. For example, ^{99}Tc has a small WDV in view of its slow rate of decay (long half-life) and its small energy release per disintegration. However, the transport of technetium through the ground surrounding a potential repository site is unusually rapid. In comparing the relative hazards from ^{99}Tc and any other radionuclide, it is necessary to consider not only the WDV for each but also the rates at which they move from the waste package to the point of human consumption.

♦ The biological parameters used to determine the WDV for the various radionuclides are uncertain. The NAS reported results based on both ICRP and U.S. Nuclear Regulatory Commission parameters. The two sets differed substantially at the time, and such parameter sets are likely to continue to change as the behavior of radionuclides in the body is further studied. The large changes cited above for ^{237}Np illustrate the difficulty, albeit these probably are of an exceptionally large magnitude. It should

---

[19] The smaller time corresponded to the values of the maximum permissible concentration then being used by the NRC; the longer time corresponded to values used by the ICRP.

[20] The more significant pathway for exposure from these radionuclides is through inhalation of radon.

be noted, however, that this problem is not unique to the WDV approach, but exists for any attempt to estimate radiation hazards.

10.3.3 Comparisons of Activity in Spent Fuel and in Earth's Crust

As indicated earlier, the comparison of hazards should take into account the full behavior of each radionuclide before and after it enters the human body. In short, a proper investigation of the nuclear waste hazards cannot avoid the complexities of a full TSPA calculation. One nonetheless may be able to get a simple perspective by comparing the magnitude of the activities of the radionuclides in the wastes to those found in the ground. A comparison of this sort is made in Figure 10.4.

In Figure 10.4, the activity of the wastes emplaced in a geologic repository is compared to the activity in the Earth's crust down to a depth of 1 km. A layer of 1 km is chosen for comparison, because a repository is expected

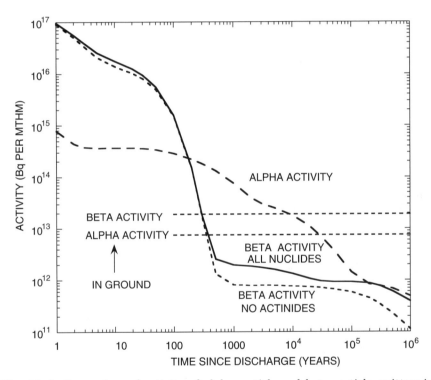

Fig. 10.4. Comparison of activity of alpha-particle and beta-particle emitters in spent fuel and in ground. The beta-particle activity is shown with and without actinides removed. The alpha-particle activity is only shown for the all-nuclides case, because it derives almost entirely from the actinides. [Spent fuel: 1 MTHM from PWR (same parameters and data as for Table 10.4). Ground: 0.005 km^3 of typical soil (see text).]

to be at least several hundred meters below the surface of the Earth. The abundances of natural radionuclides in typical soil are given in Table 3.4. The activity is due primarily to ^{40}K, ^{87}Rb, ^{232}Th, and ^{238}U, together with long chains of decay products for the latter two. In Figure 10.4, we use as a basis of comparison the activity for an area of 0.005 km^2 (5000 m^2), down to a depth of 1 km. This area was chosen as the "share" corresponding to 1 tonne of spent fuel, on the assumption that a potential total of 120,000 tonnes (almost twice the current plan) might eventually be deposited at Yucca Mountain, where an area of 600 km^2 has been assigned to the Yucca Mountain Project as the "withdrawal area" (see Section 12.2.2).

The activities in the spent fuel and in the ground are shown separately in Figure 10.4 for alpha-particle and beta-particle emitters. For the most part, alpha emitters give larger doses per unit of ingested activity, because in radioactive emissions the alpha particles are more energetic than the beta particles and they give a higher equivalent dose per unit energy deposited (see Section 3.2.2).[21]

It is seen in Figure 10.4 that the beta-particle activity of the spent fuel falls below the reference level within 300 years, whereas the alpha-particle activity does not fall below its reference level until about 30,000 years. (This difference between the time scales for alpha and beta emitters and the implications for reprocessing have already been discussed in Section 10.2.2 and seen in Figure 10.1.) Removal of the actinides from the wastes essentially eliminates the alpha-particle activity, and after 500 years it somewhat reduces the beta-particle activity.

One should be cautious, however, in interpreting the results. If a different volume of ground were chosen as the reference, the results of the comparison would be shifted. There is no definable "appropriate" volume to use as the reference, and some people find any comparison of this sort irrelevant.

However, the comparison demonstrates that the wastes are not being placed in an otherwise radiation-free environment and that the large inventory of radionuclides does not *in itself* establish that the wastes represent an unusual hazard. The high concentration of activity in a limited location makes the situation of the wastes very different from that for natural radioactivity, but whether this means more (or less) of a hazard depends on the effectiveness of the repository in immobilizing radionuclides. This again points up the need for detailed analyses, specific to the repository in question. It then is appropriate to compare the calculated radiation doses from the repository to the doses of several millisieverts per year resulting from radionuclides in the ground (see Table 3.5).

---

[21] This generalization is an oversimplification because the biological effects of a given radionuclide also depend greatly on its retention in the body and on the organs it reaches. The data in Table 4.3 show that, on the whole, the generalization is valid, but they also indicate a major exception, namely ^{90}Sr, which has a relatively high rate of absorption from the stomach into the blood system with eventual concentration in bones.

References

1. U.S. Department of Energy, *Integrated Data Base for 1992: U.S. Spent Fuel and Radioactive Waste Inventories, Projections and Characteristics*, Report DOE/RW-0006, Rev. 8 (Oak Ridge, TN: Oak Ridge National Laboratory, 1992).
2. *Energy, U.S. Code of Federal Regulations*, title 10 (1993).
3. U.S. Department of Energy, Office of Civilian Radioactive Waste Management, *Final Environmental Impact Statement for a Geologic Repository for the Disposal of Spent Nuclear Fuel and High-Level Radioactive Waste at Yucca Mountain, Nye County, Nevada*, Report DOE/EIS-0250 (North Las Vegas, NV: U.S. DOE, 2002).
4. U.S. Department of Energy, *Integrated Data Base—1996: U.S. Spent Fuel and Radioactive Waste Inventories, Projections and Characteristics*, Report DOE/RW-0006, Rev. 13 (Oak Ridge, TN: Oak Ridge National Laboratory, 1997).
5. U.S. Department of Energy, *World Nuclear Capacity and Fuel Cycle Requirements 1992*, Energy Information Administration Report DOE/EIA-0436(92) (Washington, DC: U.S. DOE, 1992).
6. U.S. Department of Energy, Office of Civilian Radioactive Waste Management, *Yucca Mountain Science and Engineering Report, Technical Information Supporting Site Recommendation Consideration*, Report DOE/RW-0539 (North Las Vegas, NV: U.S. DOE, 2001).
7. U.S. Department of Energy, Energy Information Administration, *Detailed United States Spent Fuel Data*, Table 2 (2002). [From: http://www.eia.doe.gov/cneaf/nuclear/spent_fuel/table2.html]
8. U.S. Department of Energy, *Monthly Energy Review, March 2002*, Energy Information Administration Report DOE/EIA-0035(2002/03) (Washington, DC: U.S. DOE, 2002).
9. A.G. Croff and C.W. Alexander, *Decay Characteristics of Once Through LWR and LMFBR Spent Fuels, High Level Wastes, and Fuel Assembly Structural Material*, Report ORNL/TM-7431 (Oak Ridge, TN: Oak Ridge National Laboratory, 1980).
10. Edwin Kolbe, output of the Oak Ridge National Laboratory's ORIGEN-ARP program (private communication to the author, October 10, 2002).
11. Jagdish K. Tuli, *Nuclear Wallet Cards* (Upton, NY: Brookhaven National Laboratory, 2000).
12. U.S. Department of Energy, *Characteristics of Potential Repository Wastes*, Report DOE/RW-0184-R1, Volume 1 (Oak Ridge, TN: Oak Ridge National Laboratory, 1992).
13. National Research Council, *A Study of the Isolation System for Geologic Disposal of Radioactive Wastes*, Report of the Waste Isolation Systems Panel (Washington, DC: National Academy Press, 1983).

11

Storage and Disposal of Nuclear Wastes

11.1 Stages in Waste Handling

11.1.1 Overview of Possible Stages

The main stages in nuclear waste handling, as indicated in Chapter 9, are as follows:

- Storage of spent fuel in cooling pools at the reactors.
- Dry storage of spent fuel at reactor sites.
- Reprocessing of spent fuel.
- Interim storage of reprocessed waste or spent fuel at centralized facilities.
- Permanent disposal of spent fuel, reprocessed waste, or residues of transmutation, by placement in repositories or by other means.
- Transportation of spent fuel or reprocessed waste as it moves through the stages above.

It is unlikely that a coherent waste management plan would include all of these stages. Facilities for on-site dry storage or off-site interim storage are needed only if there are delays in implementing the later stages of disposal. In fact, they were not part of the early thinking about waste disposal, when it was expected that reprocessing would be the norm, followed by relatively prompt geologic disposal.

The U.S. situation is now quite different from that envisaged earlier. Reprocessing has been abandoned for U.S. commercial wastes, there are no U.S. repositories for long-term storage, and there are no firm plans for interim storage. In practice, the spent fuel has remained at the reactor sites. Innumerable plans for waste management have been developed over the past 40 years, but, as yet, no centralized facilities have been built in the United States or in most

253

other countries, although in the United States the Department of Energy is seeking to establish a permanent repository at Yucca Mountain in Nevada.

A few countries have been more active than the United States in some aspects of waste handling. For example, France routinely reprocesses spent fuel, and Sweden has started the operation of an interim storage facility as part of its long-term plan for disposal of spent fuel. However, no country has put into operation a repository for long-term storage of high-level commercial wastes. Although site investigations have been carried out in many countries, the earliest target date is that of the United States, namely to open the Yucca Mountain repository in 2010.

For many years, the United States "nuclear establishment" felt no urgency about the waste disposal problem because it appeared to be easily solvable and not technically interesting. It is now obvious that the problem is not easy to solve, but there is deep disagreement as to whether the main difficulties are technical or political.

11.1.2 Storage of Spent Fuel at Reactor Sites

Limitations of Cooling Pools

Originally, it was expected that the spent fuel from nuclear reactors would be held in cooling pools at the reactor sites for a brief time and that reprocessing would be carried out after about 150 days. Instead, in the United States, the fuel has remained at the reactor sites, and some pools have held fuel for over 20 years. The capacity of these pools is limited, and although no reactor has had its operation stopped by a shortage of cooling pool space, individual reactors have faced a severe squeeze.

The capacity for storage of spent fuel assemblies at the reactor site can be increased to a modest extent by modifying the geometric arrangement of the assemblies in the cooling pool. A much larger expansion can be achieved by using dry storage. By 2002, a substantial number of nuclear plants had begun use of dry storage. In the absence of alternatives, this fraction will continue to grow as plants run out of excess capacity in their cooling pools.

Dry Storage of Spent Fuel at Reactor Sites

In U.S. dry storage systems, the spent fuel rods are transferred to special casks when the total activity and the heat output are reduced enough for air cooling to suffice.[1] This solves the problem of limited cooling pool capacity and is an option that can be implemented pending decisions on the establishment of centralized facilities. It also defers the contentious issue of transportation

---

[1] In some other countries, and for the fuel from the U.S. reactor at Fort St. Vrain, the spent fuel is placed in a concrete-enclosed vault, without massive individual casks, but with a more elaborate overall structure [1, p. 9].

of nuclear wastes (see Section 11.1.4). The dry storage casks are cooled by natural convective airflow, without pumps. They must be licensed by the Nuclear Regulatory Commission, which by 2003 had approved 15 different cask designs [2].

The first such casks were installed at the Surry power station in Virginia in 1986 [3]. Other utilities gradually followed suit. For example, the Palisades reactor in Michigan (a 768-MWe PWR) was near the end of its remaining capacity by the end of 1991, and at the time it was concluded that operation of the reactor could not continue beyond 1993 without alternative storage [4, p. 15]. Dry storage was proposed and approved, using casks that were designed to hold 24 PWR assemblies. The first casks were loaded in May 1993 [5].[2] The site was authorized to have up to 25 casks, enough to hold the fuel that will be discharged through the year 2007.

A diagram of the cask used at the Palisades facility is shown in Figure 11.1. The fuel assemblies are transferred underwater in the reactor cooling pool to a steel canister (called a basket). The canister is then brought to a decontamination area, where it is pumped dry, filled with helium, and sealed with redundant welded lids. The sealed canister is placed in the storage cask, which consists of an outer cylinder with a 29-in.-thick concrete wall and a steel inner liner that is at least 1.5 in. thick. The canister is cooled by natural air convection in a gap between the canister and the cask liner.

In its first years, on-site storage often attracted significant legal and political challenges, although the first facility, at the Surry reactor, was installed in 1986 without any significant opposition. Legal efforts were made to stop on-site dry storage at the Palisades reactor in 1993, but these failed in the Michigan courts. However, in Minnesota, the state courts ruled that proposed dry cask storage at the Prairie Island nuclear plant required legislative approval, which was eventually granted in May 1994 under a plan that allowed a gradual installation of storage casks, but under conditions that appear to point toward a long-term phasing out of nuclear power in Minnesota [7; 1, p. 50].

In the years since, a large and growing number of nuclear plants have installed dry cask storage systems [8, p. 210]. Overall, it appears that on-site dry storage is becoming sufficiently routine that it is usually does not face major local opposition, as had been the case for the Prairie Island reactor. In contrast to the centralized storage discussed in Section 11.1.3, it has the advantage of avoiding controversies over transportation of nuclear wastes and charges that one locality is being asked to accept a disproportionally large share of an onerous national burden. With heightened concern about terrorist

---

[2] The reactor core at the Palisades plant has 204 assemblies [4, p. 15], and one-third of the core is discharged every 15 months [private communication from Mark Savage of Consumers Power]. Thus, each cask holds roughly the output for one-half year of operation. Details of the cask construction come primarily from Ref. [6].

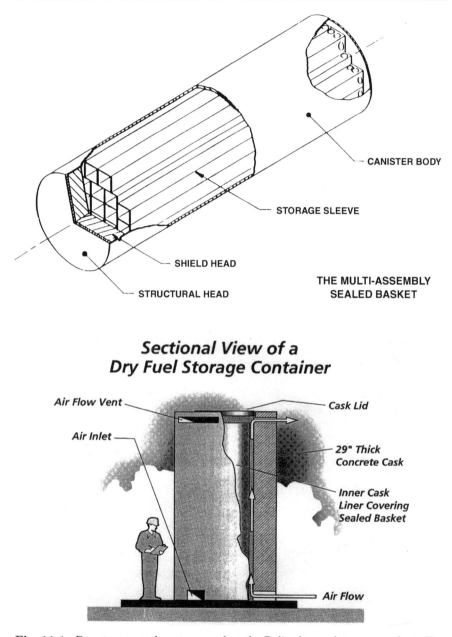

Fig. 11.1. Dry storage cask system used at the Palisades nuclear power plant. Top: canister (basket) for holding fuel assemblies; bottom: storage cask for containing sealed canister. (Adapted from diagrams provided by Consumers Power.)

threats, it may gain some political support from the perception that the spent fuel is more secure in dry casks than in cooling pools.[3]

An extensive survey of the interim storage issue, in a joint report by groups from Harvard University and Tokyo University, endorses dry cask storage in the following terms:

> Dry storage technologies, especially dry casks, have been increasingly widely used in recent years. The combination of simplicity, modularity, and low operational costs and risks offered by dry cask storage systems make them highly attractive for many storage applications [1, p. x].

This report does not put forth dry cask storage as the only acceptable option nor does it take a crisp position on the choice between on-site and centralized storage. It seems to accept as reasonable a continuation of the growing U.S. employment of on-site storage and the plan in Japan to establish a centralized facility [1, p. xxi]. This eclectic attitude toward technical solutions probably derives from the view that the main difficulties are institutional not technical, and it, therefore, is reasonable to take the path of least institutional resistance. As expressed in the report:

> Interim storage of spent nuclear fuel is technically straightforward. The key problems that have made it difficult to provide adequate interim storage arise from the difficulties of gaining political acceptance for such arrangements by the potentially affected publics, and from the complex web of legal and institutional constraints related to management of spent fuel and nuclear wastes [1, p. ix].

11.1.3 Interim Storage of Waste or Spent Fuel at Centralized Facilities

General Considerations

In parallel with the move toward on-site dry storage in the United States, there have been continued efforts to establish one or more centralized facilities. Much of the impulse for this has come from utilities that do not wish to remain responsible for the spent fuel over long time periods. They point to the provisions of the Nuclear Waste Policy Act which stipulates that the DOE would begin to remove spent fuel from reactor sites by January 1998. With neither a permanent nor an interim storage facility, the DOE cannot comply with this requirement, and DOE compensation to the utilities has been the subject of lawsuits by several utilities and of negotiations between

---

[3] For example, a prominently featured article in the magazine *Newsweek* in June 2002 singled out the cooling pools as an area of great vulnerability—at reactors that the author otherwise pictured as safe—and pointed to dry casks as the "way out" [9].

the DOE and reactor operators [10]. Retaining custody for spent fuel would be a particular burden for the reactor operator after the reactor is shut down.[4]

Pending a truly permanent solution, an intermediate step is to store the wastes at centralized interim facilities. These have been variously referred to as *monitored retrievable storage* (MRS) or *away-from-reactor* (AFR) sites. There are many MRS designs. In typical schemes, the spent fuel assemblies would be placed in steel canisters (before shipping) and the canisters stored at the MRS site either aboveground in the open in heavy concrete shielding casks (which have been called *silos*) or in enclosed vaults with less shielding for the individual canisters. Cooling would be provided by natural air convection or, in the vault case, perhaps by forced air (see, e.g., Ref. [1, p. 9]).

Aside from relieving the burden on utilities, it has sometimes been argued that even if on-site storage is safe, storage at a specially designed and located central repository (or repositories) would be safer still. In either case, there does not appear to be a significant safety issue, except for the possible vulnerability of the facility to terrorist attack.[5] The Nuclear Waste Technical Review Board discussed the terrorist matter in 1996, without giving much prominence to it or indicating a preferred approach:

> Intuitively, it would seem easier and more economical to install an effective protective system at one centralized facility than installing multiple systems at reactor sites. But it also could be argued that a single facility with a large stockpile of spent fuel might be a more tempting and visible target. Until more analyses have been performed, it is premature to assert that either an at-reactor or centralized storage would be more exposed to theft or sabotage. [11, p. 20]

We will return to the general issue of terrorism directed at nuclear facilities in Section 17.5.

As MRS was originally conceived, several such facilities would be situated in the United States. However, no MRS facilities have been approved as yet, due to local opposition in candidate states—for example, the successful blocking by Tennessee in the 1980s of a proposed storage facility at Oak Ridge [8, p. 231]. In some other countries, it is considered to be an integral and desirable stage in the waste disposal process, because interim storage allows the fuel to cool before permanent disposal. Sweden, in particular, has a centralized interim facility that uses underground water-filled pools rather than dry cask storage, and the plan has been to let the fuel cool for 40 years before placing it in a geologic repository [12, p. 49]. This facility has a spent fuel capacity of 5000 MTHM, and was already two-thirds full by the late 1990s [8, p. 209].

---

[4] This argument would lose some of its force if, when the reactor is taken out of service, it is replaced by a new reactor at the same general site.

[5] See discussion of general safety issues in the context of the proposed PFS facility in the following subsection.

Efforts to Find a MRS Site in the United States

Members of Congress made a number of proposals in the late 1990s to establish an interim storage site at Yucca Mountain, culminating in legislation passed in March 2000 that called for an interim facility which "would begin receiving wastes at Yucca Mountain as soon as possible after NRC granted a construction permit for a permanent underground repository" [10, p. CRS-5]. However, this legislation was vetoed by President Clinton on the grounds that this step should not be taken before a decision was made on a permanent repository at Yucca Mountain.

This effort followed a number of unsuccessful attempts to site an MRS facility elsewhere. In the absence of a federal plan for an MRS facility, some Indian tribes and utilities saw an opportunity to site an MRS facility in an area under Indian control. Such initiatives were encouraged by a DOE program of small study grants and by the efforts of a so-called Nuclear Waste Negotiator, who was given the task of finding potential hosts for a MRS sites among states or Native American tribes [13].[6] No states showed any interest, but a number of tribes did pursue the possibility.

It is not difficult to understand the difference between the viewpoints of states and tribes. For example, the Mescaleros Apache tribe in New Mexico—the group that showed the greatest initial interest in an MRS facility—reportedly estimated that "the project could generate as much as $2.3 billion in revenue over its 40-year lifetime" [14]. Such an amount could be a great attraction to at least some of the Mescaleros, who numbered 3400, but is not a decisive inducement if the financial benefits are spread over an entire state. The Mescaleros applied for and received in October 1991 a $100,000 DOE grant to study the desirability of "hosting" a MRS facility on its reservation [15]. However, in the face of disagreements within the tribe and opposition from New Mexico state authorities, the project was ultimately abandoned.

On the other hand, the Skull Valley Band of Goshute Indians in Utah retained an interest in pursuing the MRS option. With its cooperation, a utility consortium, known as Private Fuel Storage (PFS), applied to the Nuclear Regulatory Commission in 1997 for a license to construct an installation on the Goshutes land which could hold 40,000 tonnes of spent fuel [10]. To be implemented, the proposal requires NRC authorization and approval from three other U.S. federal agencies: the Bureau of Indian Affairs, the Bureau of Land Management, and the Surface Transportation Board. These groups issued a Final Environment Impact Statement (FEIS) for the facility in December 2001 [16].

In the PFS proposal, the spent fuel is placed in up to 4000 storage casks standing vertically in the open on concrete pads. Each cask would hold a sealed stainless-steel canister designed for 68 BWR assemblies or 24 PWR assemblies. In either case, a cask fully loaded with fuel assemblies would hold slightly over

---

[6] The position of Nuclear Waste Negotiator was established in 1987 amendments to the Nuclear Waste Policy Act of 1982.

10 MTHM. The assemblies are to be loaded into the canisters at the reactor site. An individual canister is placed in a thick-walled transportation cask for shipment to the MRS facility, where it would be transferred to a storage cask. The planned storage cask is a concrete structure surrounding a stainless-steel lining, with a total wall thickness of about 2.5 ft [17, p. 3].

The radiation risks created by this facility are small. No leakage of radionuclides from the casks is to be expected, because the fuel is solid and enclosed within the sealed canister and thick cask. In consequence, the NRC judges that "no discernible radioleakage is credible" [16, pp. 4–46]. Gamma rays that penetrate the casks are the only expected source of exposure. The maximum dose for an individual working at the boundary of the facility for 2000 h/yr is estimated by the NRC to be 0.06 mSv/yr [16, pp. 4–47]. The dose at the nearest residence in the area, about 3.2 km from the site at present, is estimated to be 0.0004 mSv/yr. Either of these levels is small compared to the natural background level of over 2 mSv/yr.

The Governor of Utah is on record as opposing the PFS facility. Thus, even if it receives federal approval, it faces local political opposition and court challenges. The project received an at least temporary setback in March 2003 when the NRC's Atomic Safety and Licensing Board (ASLB) concluded that the probability that a U.S. military plane or "jettisoned ordnance" would crash into the site was 4.29×10^{-6} per year, which exceeded the stipulated maximum probability of 1×10^{-6} per year for a credible accident [18]. This ruling was appealed to the Commission (i.e., the NRC commissioners) by both PFC and the NRC staff. The Commission put these appeals in abeyance and, instead, directed the ASLB to analyze the consequences of such crashes (which had not been part of the ASLB's earlier review), prior to a decision by the NRC on approval of a license for the facility.

11.1.4 Nuclear Waste Transportation

Waste Transportation Plans

The prospect of large amounts of spent fuel being transported from the reactor sites to a centralized location—whether an interim facility or a permanent repository—has led to a debate over the dangers that might result. The shipments, from many parts of the country, would have to pass through numerous governmental jurisdictions. Unless a public consensus develops that the dangers posed by these shipments are small, political and legal challenges could complicate the implementation of any transportation program.

The safety of the shipments depends on the characteristics of the wastes and their containers. As discussed earlier, the wastes are in solid form, mostly pellets of spent fuel contained in assemblies of fuel rods.[7] The assemblies

---

[7] In addition to spent fuel, there will be shipments of vitrified high-level wastes from the weapons program. The amounts are smaller, in terms both of mass in MTHM and radionuclide content per MTHM.

are to be placed into transportation casks at the reactor sites. The casks are massive structures that are designed to provide adequate shielding to keep the external radiation levels low and to be rugged enough to withstand potential transportation accidents. Accidents are expected to be rare, but if they occur, the cask, the fuel rod cladding, and the solid form of the fuel rod are the defense against the release of of radionuclides.

A major study of nuclear waste transportation hazards was carried out in 2000 for the Nuclear Regulatory Commission by the Sandia National Laboratory [19]. Results of the Sandia work are reflected in the DOE's Final Environmental Impact Statement (EIS) for Yucca Mountain issued in 2002 [20]. We outline in the following subsections some of the specific plans that are described in these documents. It is to be expected that as the planning for Yucca Mountain progresses, details of these plans and of the supporting analyses will be changed, but it seems unlikely that the changes will alter the fundamental picture.

Transportation System and Cask Design

The shipment of the transportation casks is to be made by a combination of truck, train, and barge. The 2002 EIS discussed the transportation plans in terms of two possible scenarios, characterized as *mostly rail* and *mostly truck*. In the mostly rail case, some use is made of trucks for reactor sites where rail transport is not readily available. In this scenario, it will also be necessary to build a branch line to the Yucca Mountain site.

The truck casks are smaller and a truck-only system would require more shipments than a mostly rail system (see Table 11.2). The DOE estimates that the transfer of 63,000 tonnes of commercial spent fuel and 7000 tonnes of defenses wastes (the amount authorized for Yucca Mountain) will require about 53,000 truck shipments (in the mostly truck case) [20, p. J-11]. These shipments would be spread over 24 years (optimistically, from 2010 to 2033) at an average rate of 2200 shipments per year. This means an average of six shipments per day would converge on the repository site from different parts of the country.

The physical arrangement for shipping is sketched in Figures 11.2 and 11.3, which present "artists concepts" of the configuration for truck and rail shipments. For truck shipments, a single cask would be placed on a trailer that is elongated to reduce the radiation level in the driver's cab. In train transport, a single cask would be placed on a flatcar.

The casks are designed to provide radiation shielding to protect people in the vicinity of the casks, including both the truck drivers and the general public. They must also protect the fuel assemblies against damage in case of an accident. A number of alternative designs have been under consideration, and until final decisions are made, their expected features are described by "generic" designs. Table 11.1 gives dimensions for generic truck and rail casks. The fuel is surrounded by three concentric metallic protective layers: a

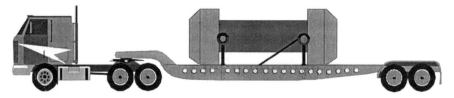

Fig. 11.2. Artist's conception of transportation cask and carrier for truck transport; total length = 18 m (56 ft). (From Ref. [20, p. J-13])

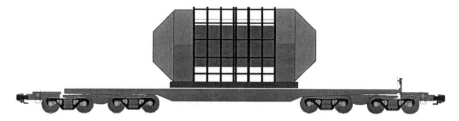

Fig. 11.3. Artist's conception of transportation casks and carrier for train transport; total length = 21 m (66 ft). (From Ref. [20, p. J-13]).

Table 11.1. Dimensions of generic transportation casks (in inches).

	Truck	Rail
Total cask dimensions		
Length	205	200
Outer diameter	37	89.5
Inner diameter	13.5	65
Thickness of concentric shells		
Outer protective layer (steel)	0.25	0.25
Neutron shield[a]	4.5	4.5
Outer steel wall	1	2
Gamma-ray shield (lead)[b]	5.5	4.5
Inner steel wall	0.5	1
Total (all shells)	11.75	12.25
Number of assemblies per cask		
PWR assemblies	1	24
BWR assemblies	2	52

[a]The indicated neutron shield in Ref. [19] is water, but a solid material may be substituted (e.g., borated polyethylene).
[b]In another version of the generic truck cask, the gamma shield is made of depleted uranium (3.5 in.), the inner diameter is 18 in. and the cask can hold three PWR assemblies or seven BWR assemblies.
Source: Ref. [19, p. 6.3].

stainless-steel liner, a lead or depleted uranium layer for gamma-ray shielding, and a stainless-steel outer shell. In addition, a neutron shield surrounds the shell, and it, in turn, is protected by a relatively thin metallic outer layer. Uranium is chosen as a possible gamma-ray shield because of its high density (19 g/cm^3) and high atomic number. The casks have "impact limiters" at the front for protection in case of collisions.

Radiation Exposure in Routine Transportation

The radiation exposure outside the containers is due to gamma rays and neutrons that penetrate the shielding material that surrounds the spent fuel assemblies. The most important radiation component is the 662-keV gamma-ray from cesium-137, although gamma rays from the shorter-lived cesium-134 ($T = 2.06$ yr) and cobalt-60 ($T = 5.27$ yr) also are significant for fuel that has been only recently removed from the reactor. The flux of these gamma rays is greatly reduced by absorption in the walls of the cask.

The casks must meet requirements set forth in NRC regulations which stipulate maximum dose rates of 0.1 mSv/h (10 mrem/h) at a distance of 2 m from the edge of the transport vehicle and 0.02 mSv/h in spaces that might normally be occupied by the truck or train crew [21, §71.47]. The specified maximum dose rate is lower for workers than for the general public, because their exposure continues for many hours, whereas it is brief for the latter. An individual who works 1000 h/yr at the maximum dose rate would receive an annual dose of 20 mSv, which is a sizable fraction of the overall U.S. occupational limit of 50 mSv/yr (see Section 4.2.2.).

These are limiting, not average, doses. Aside from drivers and other workers who must be near the casks for extended periods, the annual doses are very small. Doses at various distances are calculated starting with the assumption that the dose at 2 m is at the regulatory limit of 0.1 mSv/h (10 mrem/h). (Usually, the actual dose would be less and, therefore, the calculated doses at other distances represent upper bounds rather than typical values.) On this basis, a person at 30 m ($\approx$ 100 ft) from the vehicle is calculated to receive a dose of under 0.002 mSv/h (0.2 mrem/h), which is negligible under any normal circumstances of exposure involving moving vehicles [20, p. J-38].

In the case of truck transport, a car directly adjacent to the vehicle might be at a distance of less than 2 m and the dose could be higher than 0.1 mSv/h. Although the dose would be unimportant in typical "passing" situations, it might be a concern for a person caught in a prolonged traffic jam. The DOE estimates a total dose of 0.16 mSv for a maximally exposed individual caught in a traffic jam [20, p. 6-40].[8] With this in mind, it may desirable to devise a plan that has routes, times of travel, and escorts to avoid such a situation, although this is a not a major dose on a one-time basis. More importantly, it

---

[8] The assumed time of close proximity is not indicated.

is important to keep workers who may be repeatedly exposed from spending long periods of time directly adjacent to these casks.

Criteria for Cask Performance in Accidents

Probably the greatest cause of public concern is the possibility of an accident that releases radioactive material into the general environment. To guard against this, NRC regulations require that transportation casks be able to survive the following sequential "hypothetical accident tests" [22, p. 10]:

◆ *Impact*: a 30-ft drop onto an "unyielding surface" in a orientation to cause maximum damage. (A 30-ft drop gives only a 30-mph speed, but actual collisions are rarely with unyielding surfaces, so the NRC considers that the test simulates a higher-speed collision.)
◆ *Puncture*: A 40-in. drop onto a 6-in. diameter steel shaft.
◆ *Fire*: A 30-min exposure to an all-engulfing fuel fire at 800°C.
◆ *Submersion*: Being held held underwater for 8 h.

A measure of "survival" in these tests is for the dose rate near the cask surface (1 m away) not to rise to more than 1 rem/h (10 mSv/h) and that there be negligible gas leakage.

The DOE has tried to demonstrate that these requirements are more than met by performing a series of tests in which transportation casks survived violent collisions and fire under conditions more severe than stipulated by the NRC. In the words of a DOE document, the tests included the following:

(1) A flatbed truck loaded with a full-scale cask was smashed into a 700-ton concrete wall at 80 miles an hour.
(2) A cask was broad-sided by a rocket-propelled 120-ton rail locomotive traveling 80 miles an hour, and
(3) A transportation container was dropped 2000 feet onto soil as hard as concrete, and was traveling 235 miles an hour on impact [23, p. 5].

Overall, the NRC requirements and DOE tests are intended to assure that the casks are so rugged that there will be little or no release of radionuclides in the vast majority of possible train or truck accidents.

Probability and Consequences of Transportation Accidents

The chance of a nuclear waste transportation accident is estimated by the DOE from the average accident rates in normal U.S. interstate commerce where an event is counted as a truck "accident" if there is a death, injury, or need to tow away one of the vehicles [20, p. J-70].[9] Thus, typical small "fender-

---

[9] This may be a conservative assumption given the attention these shipments are likely to receive.

Table 11.2. Estimated accident probabilities and consequences in transportation of nuclear wastes.

	Mostly Truck	Mostly Rail
Shipments during 24 years		
Number of shipments made by primary mode	52,786	9,646
Number of shipments made by secondary mode[a]	300	1,079
Total number of accidents (24 years)	66	8
Maximum reasonably foreseeable accident		
Frequency per year, urban	2.3×10^{-7}	2.75×10^{-7}
Frequency per year, rural	1.3×10^{-6}	1.7×10^{-6}
Collective dose in urban accident (person-Sv)	11	99
Collective dose in rural accident (person-Sv)	0.02	0.16
Dose to maximally exposed individual (mSv)	30	290

[a]The secondary modes are rail in the mostly truck case and truck in the mostly rail case.
Source: Ref. [20], number of shipments, p. J-11; number of accidents, p. J-69; maximum accident, pp. J-61 and J-62.

benders" are not included. The average rates, for truck and rail transport, are given in Table 11.2. On this basis, a total of 66 accidents are expected assuming the mostly truck mode and 8 accidents for the mostly rail mode, during the 24 years when the shipments are made [20, p. J-69]. The Sandia analysis indicates that there is less than a 0.01% chance (per accident) that these accidents will involve enough damage to cause the release of any radionuclides [19, pp. 7-74 and 7-76].

To examine the possible consequences of a "worst-case" accident, the EIS excludes accidents that are so unusual that they are estimated to have less than 1 chance in 10 million of occurring in 1 year. From the remaining more plausible accidents, the report selects the case with the greatest adverse consequences. This is defined as the "maximum reasonably foreseeable accident" (MRFA). It is described as "a long-duration (many hours), high-temperature fire that would engulf a cask" [20, p. J-60].[10] The consequences of such an accident would obviously be greater in an urban environment than in a rural one due to the higher population density in the former. The chance of an MRFA occurring in an urban area is estimated in the EIS to be under 3×10^{-7} per year in both the truck and rail scenarios [20, p. J-62].

For the mostly rail MRFA, some radionuclides would be released, but the fractional amounts would be appreciable only for the highly volatile components. Over 80% of the krypton-85 (^{85}Kr) would be released, but because

---

[10] According to reports quoted in Ref. [20], the Baltimore Tunnel Fire in July 2001, which burned at 1500°F "for up to 5 days," had conditions similar to the MRFA.

krypton is an inert gas, its release has relatively little health impact. The radionuclide that is most troubling in nuclear reactor accidents, iodine-131—which is responsible for the childhood thyroid cancers at Chernobyl and motivates the distribution of potassium iodide pills near reactors—is not present in the wastes, because its half-life is only 8.02 days and virtually none is left after the wastes have cooled for several years or more. The moderately volatile radionuclide ^{137}Cs is present in large amounts, but the NRC analysis estimates that only about 0.002% of the inventory would be released in the postulated fire [19, p. 7-76].

The case with the highest expected radiation exposure is an urban train accident. The chance of an urban train MRFA is estimated to be about 3 in 10 million (3×10^{-7}) per year. The expected collective population dose is 100 person-Sv. According to standard estimates of the effects of low-level radiation exposures—which, as discussed in Chapter 4, are speculative—this corresponds to five eventual cancer deaths. Thus, the calculated average total of deaths from radiation exposure in these accidents is about 10^{-6} per year.

Unless these accident analyses are grossly in error, the risks from nuclear waste transportation are negligible compared to the risks that society routinely faces from a multitude of sources, including ordinary traffic accidents. The perhaps surprisingly low risk follows from the relatively few shipments (on the overall scale of commercial activity), the solid form of the material, and the ruggedness of the surrounding casks.

11.2 Deep Geologic Disposal

11.2.1 Multiple Barriers in Geologic Disposal

The handling of nuclear wastes is in the first instance the responsibility of the country in which the wastes are produced. Although there have been suggestions for the establishment of international repositories (see Section 11.3.1), most countries are proceeding on the basis of using a site within its own borders. In all countries that are engaged in active planning, the favored solution has been to place the wastes in deep geologic repositories. Protection against the escape of radionuclides into the biosphere is then provided by a number of barriers, with the overall set of barriers commonly divided into the *engineered system* and the *natural system.*

The *waste package*, auxiliary components such as a shield or backfill, and the configuration of the repository together constitute the engineered system. The waste package consists of the solid waste (the spent fuel assemblies or the resolidified products of reprocessing) and the surrounding protective containers. These are usually concentric cylinders made of materials that are chosen because of their ability to resist corrosion and prevent water from reaching the waste material. In some designs, additional protection against water intrusion is provided by a protective shield above the canister, and entry or escape of water may be hindered by backfill surrounding the waste package. The engineered system cannot be designed independently of the natural environment, because factors such as the water flow rate, the water chemistry, and the heat conductivity of the medium strongly influence the choice of waste package design and the repository configuration.

The natural system is the surrounding rock through which water would move to the repository, and from the repository to the biosphere. It includes the rock out of which the repository is excavated. A good repository site is one for which the location and type of rock (a) prevent or limit the flow of water into the repository, (b) provide geochemical conditions favorable for a low rate of corrosion of the waste package and low solubility of radionuclides in the event of entry of water, (c) slow the outward migration of water to the biosphere, (d) retard the motion of major radionuclides so that they move more slowly than the water, and (e) are at low risk of future disruption by earthquake, volcano, erosion, or other natural phenomena. Together, these attributes provide a series of natural barriers.

Repositories may be in rock in a *saturated zone*, lying below the groundwater table, or in an *unsaturated zone*, lying above the water table. Except in arid climates, the water table usually lies too close to the ground surface to permit having a geologic repository in the unsaturated zone, and in almost all countries the planned repositories are in the saturated zone. The proposed Yucca Mountain site, in the Nevada desert, is an exception, and the repository would be in the unsaturated zone, well above the water table.

In the saturated zone, the gaps and pores in the rock are filled with water, although the site may still be suitable if the movement of water through the rock mass is at a slow rate. In the unsaturated zone, the pores hold less water but seepage of rain water introduces some moisture into the rock, and the environment of a repository in the unsaturated zone is unlikely to be completely dry.

Different countries have placed different emphases on the engineered and natural barriers. For many years, although this has now changed, primary emphasis in U.S. planning had been on the properties of the site, whereas in Sweden, which has a very ambitious plan for the waste package, reliance was placed on the engineered safeguards. In more recent designs, the waste package for the Yucca Mountain repository has been made more rugged and has become a longer-lasting barrier (see Section 12.3.2).

11.2.2 Alternative Host Rocks for a Geologic Repository

A large number of different types of rocks have been considered for waste repositories. There is no single overall "best" choice, as evidenced by the different choices made by different countries (see Section 11.4). Among the physical factors that go into the consideration of a particular rock formation are the extent to which water entry would be inhibited, its retardation of the flow of any escaped radionuclides, and its behavior when heated by the repository wastes. Rocks that have been considered as candidates for repositories include the following (see, e.g., Ref. [24]):

◆ *Bedded salt.* Bedded rock salt was the initial candidate of choice. The existence of a salt bed was taken as evidence that there had been no water intrusion for many thousands of years. Further, salt has high thermal conductivity, which would limit the temperature rise of the wastes. Salt melts at relatively low temperatures, and the waste would eventually be surrounded by a tight resolidified mass of salt. On the negative side, salt brine is highly corrosive and may attack the canister. There had been a suggestion in the early 1970s to use a saltbed in Lyons, Kansas, but this plan was canceled following strong local opposition and the realization that prior oil drilling had made water intrusion possible.[11]

◆ *Salt domes.* Under some circumstances, the pressure on a thick bed of salt will cause some of the salt (which, in general, has a lower density than the surrounding rock) to break through the overlying material and rise upward to form a salt dome. One advantage of salt domes over bedded salt is a generally lower water content (see, e.g., Ref. [26, p. 201]). The Gorleben site, a waste disposal site under consideration in Germany, is a salt dome.

◆ *Granite.* Granite and similar rocks (granitoids) are very abundant. They are stable and generally homogenous, with low permeability to water movement. However, they are susceptible to fractures, which could provide paths for relatively rapid water flow. Granite is the choice in Sweden and Canada.

◆ *Basalt.* Basalt is an alternative rock formation, although a National Research Council review has suggested that "a major reason for considering basalt for repositories is its abundance in federal land near Hanford, Washington and the Idaho National Engineering Laboratory (INEL) and not its overall favorable characteristics" [24, p. 155].

◆ *Tuff.* Tuff is the residue of material blown out of exploding volcanoes. At high temperatures, some of the material fuses to form "welded tuff," a material of low permeability. Tuff, both welded and unwelded, is the rock type at Yucca Mountain. Tuff can be highly fractured, and a study of the fracture structure is an important component of the Yucca Mountain waste repository site characterization.

The suitability of a particular site depends not only on the type of rock but also on location-specific aspects, including the history of past human disturbance of the region (as at Lyons, Kansas), the thickness of the available rock layers, and the absence of valuable mineral resources. In practice, an

---

[11] The Waste Isolation Pilot Plant (WIPP), in a bedded salt site in New Mexico, was completed in 1991, to be used for military transuranic wastes only. Local opposition imposed long delays in the actual transfer of waste to the repository, but the facility received a "final design certification" from the EPA in 1998 and began receiving waste shipments in March 1999 [25].

important further consideration is the political acceptability of the site. As discussed in Chapter 12, an essentially political decision was made in the United States to concentrate efforts on the Yucca Mountain tuff site.

11.2.3 Motion of Water and Radionuclides Through Surrounding Medium

Travel Times for Water and Waste

The main mode of transfer of radionuclides to the biosphere from a repository that is deep underground is through movement in the groundwater. The possible paths depend on the configuration of the surrounding land and of underlying aquifers. Transfer to the biosphere would occur if an aquifer were tapped by deep wells, presumably by diggers who were unaware of the radioactive contamination. Typical travel distances from the repository to the biosphere were taken in a National Academy of Sciences (NAS) study of geological disposal to be in the neighborhood of tens of kilometers (see, e.g., [24, p. 253]).

The rate of water travel depends on the medium. If the medium is uniform, motion is through small spaces between grains. Fractures in the medium offer more rapid paths for water flow. The overall travel speed through fractures depends on the size of the fractures and the degree to which they form a connected network. For careful planning, it is necessary to map the fracture structure of a potential repository site to avoid zones with a high density of fractures and to understand the extent to which the fractures that cannot be avoided could speed the escape of radionuclides.

Water travel time estimates are presented in Table 11.3 for hypothetical "reference" repositories in a number of unfractured geologic media. The values listed are mean values from Ref. [24, p. 253]; maximum and minimum estimates are also given in Ref. [24], which in many cases differ widely from these means. Thus, the values are more useful for general orientation than as realistic estimates for actual sites. For a given site, the detailed hydrological properties must be determined by specific "site characterization" studies.

Retardation Factors

The ions of many substances travel through the ground more slowly than does water due to processes designated, in their totality, as *sorption*.[12] These processes include the adsorption of ions on the surface of the medium and ion exchange between ions in the water and ions in the host rock. These processes retard the ions without permanently sequestering them. The retardation is parameterized in terms of a retardation factor R, which is defined as the ratio of the velocity of groundwater (v_{water}) to the average velocity of the ions (v_{ion}),

---

[12] Here, we follow in part the discussion in Ref. [27, p. 62ff].

Table 11.3. Retardation factors for radionuclides in several geologic media and travel times to biosphere for reference repositories. [Note: There are large uncertainties in these numbers (see text).]

| Element | Atomic Number | National Research Council (NAS) | | | | | EPRI[a] |
		Granite	Basalt	Clay[b]	Salt	Tuff	Tuff
Water travel time (yr)							
Mean path length (km)		10	14[c]		100	6[d]	
Linear velocity (m/yr)		0.01[e]	0.9		2	5.7	10
Travel time (1000 yr)		1000	15		40	1.2	
Retardation factors, R^f							
Strontium (Sr)	38	200	200	200	10	200	250
Technetium (Tc)	43	5	5	5	5	5	1
Iodine (I)	53	1	1	1	1	1	1
Cesium (Cs)	55	1000	1000	1000	10	500	250
Lead (Pb)	82	50	50	50	20	50	
Radium (Ra)	88	500	500	500	50	500	
Uranium (U)	92	50	50	200	20	40	20
Neptunium (Np)	93	100	100	100	50	100	10
Plutonium (Pu)	94	200	500	1000	200	200	500
Americium (Am)	95	3000	500	800	1000	1000	500

[a]The EPRI reports present sorption coefficients K_d; the retardation factors are rounded-off values, calculated from K_d using Eq. (11.3).
[b]Also refers to soil and shale.
[c]Corresponds to smaller of two estimates of distance.
[d]The path and travel time are for travel in saturated tuff to a nearby well.
[e]Velocity not given in Ref. [24]; inferred from distance and time.
[f]These are values termed "suitably conservative" in Ref. [24].
Sources: NAS, Ref. [24, pp. 147 and 253]; EPRI, Ref. [28, Section 7.3].

and is equal to the ratio of the total number of ions per unit bulk volume N to the number per unit volume n being carried by the water (i.e., the number momentarily *not* trapped in the rock):

$$R = \frac{v_{\text{water}}}{v_{\text{ion}}} = \frac{N}{n}. \tag{11.1}$$

This expression can be rewritten as

$$R = 1 + \frac{N - n}{n} = 1 + K_d \frac{\rho(1 - \epsilon)}{\epsilon}, \tag{11.2}$$

where ϵ is the porosity of the medium, ρ is the density of solid rock, and K_d is the *sorption coefficient* or *distribution coefficient*, defined as the ratio of the number of ions per unit rock mass attached to the rock, $(N - n)/\rho(1 - \epsilon)$, to the number per unit water volume in the water, n/ϵ.[13] In Eq. (11.2), K_d depends on the chemical properties of the ion–water–rock system and ρ/ϵ on the physical properties of the rock.

Eq. (11.2) sometimes is approximated by the expression [24, p. 195]

$$R = 1 + 10\,K_d, \tag{11.3}$$

where K_d is in units of cubic centimeters per gram. This is a conservative approximation, because commonly $\rho > 2$ g/cm^3 and $\epsilon < 0.2$. Thus, Eq. (11.3) often underestimates the retardation factor.[14]

Determination of K_d, and hence R, depends on experimental measurements. In the words of Konrad Krauskopf,[15] the experimental results "show a discouraging lack of agreement from one laboratory to another" [27, p. 64]. Table 11.3 gives estimates taken from the 1983 NAS study that were considered by Krauskopf to be "suitably conservative for predicting the performance of conceptual repositories" [24, p. 147]. They fall between the extremes in a range of values presented in later work of Krauskopf [27, p. 65]. In the absence of reproducible, site-specific data, these data are mainly useful for providing a general overall orientation for considering different host rocks and radionuclides.

For more reliable information, site-specific studies are necessary. For the U.S. Yucca Mountain site, the situation is complicated by the presence of different types of tuff with different sorption coefficients K_d. Even for the same type of tuff, there are large variations between the maximum and minimum reported values (e.g., Ref. [29, p. 4-323]). The last column of Table 11.3 gives summary results for saturated tuff from an Electric Power Research Institute (EPRI) report published in 2002, based on DOE data. Although there are wide uncertainties, the newer results are in qualitative agreement with the 1983 values of Ref. [24]—for example, a large retardation for plutonium and no retardation for iodine. However, there are some substantial differences, notably lower retardation factors in the newer data for technetium and neptunium.

However, even approximate data illustrate the relevance of the retardation factors in understanding the role of individual radionuclides:

- *Plutonium.* It is often believed that ^{239}Pu ($T = 24{,}100$ yr) is a major long-term threat. However, its large retardation factor (in all media) may provide a long enough time delay so that it will decay to a low level before reaching the biosphere.

---

[13] The porosity of the medium is the fraction of volume that is "empty" (or filled with water) (i.e., the pore volume per unit bulk volume).

[14] It may be noted that for $\rho = 2$ g/cm^3 and $\epsilon = 0.2$, the coefficient of K_d in Eq. (11.2) is 8, not 10. However, in more typical cases, the coefficient will exceed 10.

[15] At the time, Professor of Geology at Stanford University and Chairman of the Board on Radioactive Waste Management of the National Research Council.

◆ *Americium.* The high retardation factor for americium makes ^{241}Am ($T =$ 432 yr) of relatively little concern, although it looms large for the first several thousand years in terms of activity.

◆ *Technetium.* The low retardation for technetium makes ^{99}Tc ($T = 211{,}000$ yr) a relatively important player, although its importance is limited by its relatively low total activity and its small release of energy per disintegration.

These inferences from Table 11.3 are not well-established points and in view of the uncertainties in the quoted retardation factors and in the overall models in which they are used, they are useful mainly as qualitative guides.

The actual situation may be considerably more complicated than suggested in this simple picture of travel through a rock matrix. For example, it is possible that at high concentrations of dissolved material, the adsorption capacity of the rock may become saturated, reducing the retardation factor. In view of the large effect of retardation as a modifier of the motion of radionuclides, it is important to determine the water travel time and the retardation factors in the environment of the specific waste repository being considered, with attention to issues of water chemistry and solute concentration. It is also necessary to consider possible accelerated pathways due to transport of nuclides on colloids or through fractures in the rock.

11.2.4 Thermal Loading of the Repository

One of the key decisions in the planning of a geologic repository is the choice of the desired temperature profile for the repository region, as a function of time and location. The temperature profile is controlled by the density with which the heat-dissipating wastes are placed in the repository, the time delay before their placement, and the ventilation—if any. The overall plan constitutes the "thermal loading strategy." For a fixed waste inventory, this strategy is important in determining the area of the repository.

The heat output for typical spent fuel 25 years after discharge from the reactor is 990 W/MTHM for PWR fuel and 820 W/MTHM for BWR fuel. After another 100 years, the heat output is about one-third of these values [20, p. A-24]. The thermal load can be expressed in terms of an (initial) *areal power density*, in kilowatts per unit area, or an *areal mass loading*, in tonnes per unit area. For example, at a heat output of 1 kW/tonne, an areal mass loading of 2 tonnes/acre corresponds to an areal power density of 2 kW/acre.

Delaying the burial of the fuel decreases the initial power density, due to the decay of short-lived radionuclides, and limits the peak temperatures reached by the fuel and its surroundings. The temperature history of the repository also depends on the spacing between canisters (i.e., on the areal mass loading), and on the extent to which natural or forced ventilation is used.

Depending on the thermal load, one can achieve either a *below-boiling* regime or an *above-boiling* regime. In the former case, the temperature in the rock surrounding the repository always remains below the boiling point of water. In the latter case, the temperature rises above the boiling point, perhaps for hundreds or thousands of years. An advantage of the higher-temperature option is that the heat will vaporize water and drive it away from the wastes, thereby protecting the waste containers. It also means that less area is required for the repository. An advantage of the lower-temperature regime is that it keeps the rock in a better known and more predictable condition and one with less thermal stress.[16] It also avoids possible combinations of high temperature and high humidity in which corrosion of the waste containers may not be adequately understood and may be accelerated.

A decision as to the optimal thermal loading must consider the temperature dependence of the properties of the wastes and container materials, including their corrosion rates, solubility, and sorption coefficients. It also requires an evaluation of the heat and fluid flow through the repository under different thermal conditions. The problem is complex, because the repository temperature influences the movement of water and steam and may also impact the physical structure of the rock. The effects are coupled, with water and steam providing mechanisms for heat transfer.

There have been substantial differences in the thermal loading strategies adopted by different countries. The original "baseline" United States plan for Yucca Mountain was to design the repository to be in the above-boiling mode for roughly 300–1000 years [30, p. 28]. However, studies are being made of other thermal loading approaches for the Yucca Mountain repository, ranging from those in which there is no boiling to ones in which repository temperatures remain above the boiling point for more than 10,000 years. Issues of thermal loading at Yucca Mountain are discussed further in Section 12.3.5.

In Sweden, the plan has been to operate the repository in the below-boiling mode, keeping the maximum temperature no higher than 80°C, even for the hottest canister [31, pp. 3–5]. The temperature is to be held down by having a low areal mass density for the fuel and by allowing the fuel to cool for at least 40 years before placing it in the repository.

11.2.5 The Waste Package

Relation of the Waste Package to Its Environment

The waste package—and for reprocessed wastes, the waste form—is selected to limit corrosion of the canister and leaching of the waste. The rate of corrosion or leaching of any particular material depends on the chemical composition of the water attacking it, and, therefore, on the type of rock through which the water has traveled. Therefore, the choice of waste packaging materials must be tailored to the chosen site.

---

[16] It also makes possible the use of a heat-sensitive backfill, such as bentonite.

Waste Form

If the wastes are disposed of as spent fuel, there is no option as to the waste form. The fuel is taken as it comes, usually in the form of pellets of UO_2 contained in long, thin-walled fuel rods. These are held in the reactor in assemblies, each with up to several hundred fuel rods (see Section 8.2.3). The assemblies remain the basic physical unit in the handling of the spent fuel.

For reprocessed waste, a good deal of attention has been given to finding an optimal waste form. The basic criterion is resistance against leaching under conditions of varying temperatures and of possible radiation damage from alpha particles emitted by waste products. The preferred waste form has long been borosilicate glass. As discussed in Section 9.4.2, France is currently converting reprocessed liquid wastes into borosilicate glass and has recently expanded its reprocessing facilities. Similarly, the United Kingdom built a large vitrification plant in the 1990s for the same purpose and the United States is converting its reprocessed defense wastes into borosilicate glass. This glass form is preferred because there is experience with its properties and it is not difficult to produce. It has long been the choice in the planning of most countries that have reprocessed wastes [32].

At the same time, there has been a search for better alternatives, and many have been suggested. Among these are varieties of ceramics, including one that mimics natural rocks, called Synroc. However, ceramics may be more difficult to make on a large scale than borosilicate glass, and there has not been very vigorous exploration of alternatives to borosilicate glass.

Components of the Waste Package

The waste package consists of the spent fuel or reprocessed waste plus the surrounding protective container, which is typically in the form of one or more concentric cylinders. The overall requirement is that the container provide protection against physical damage (e.g., from falling rock in the drift) and resist corrosion. Commonly, there is an inner cylinder, or canister, with metal panels that are arranged in a honeycomb structure to hold the individual fuel assemblies.[17] The panels also provide mechanical support, facilitate heat transfer to the container wall, and absorb neutrons to prevent the development of criticality. The same functions could be performed by filling the canister with a powder or sand [33, p. 63]. An outer cylinder, sometimes called the overpack, was originally envisaged as the heavier and physically stronger barrier. Now, however, in the most recent U.S. designs, the main function of the outer cylinder is to provide corrosion resistance while the inner cylinder provides the main structural strength (see Section 12.3.3).

There are many candidate materials for the waste container. Metals are currently favored. Among these, particular consideration has been given to

---

[17] A diagram of this structure is given in Figure 12.3.

copper, steel, and a variety of alloys based on titanium, nickel, or iron. The most crucial property is the resistance of the material to corrosion under the expected repository conditions. Corrosion may be either generalized, with uniform corrosion over the entire surface, or localized at pits or cracks. In many cases, this localized corrosion is the more troublesome phenomenon, and there have been intensive, but, as yet, not fully conclusive, efforts to establish rates for both general and localized corrosion for the various combinations of metal and the surrounding environment.

Placement of the Waste Packages

A typical underground repository design is a large cavern, honeycombed with tunnels, called "drifts." In the early thinking about Yucca Mountain, for example, each tunnel was to have a series of vertical boreholes for emplacing the waste packages; a cylinder containing spent fuel assemblies or solid reprocessed waste would be placed in each borehole and the top portion of the borehole refilled and sealed. More recently, the Yucca Mountain planning has been based on larger canisters, placed horizontally on the floor of the tunnel, sitting on rail tracks. This simplifies their handling, provides flexibility in moving them, and makes retrievability easier if desired in the future. These two configurations are depicted schematically in Figure 11.4.

In some designs, there is no extensive further protection against contact with water, other than the natural dryness of the environment plus the effects of repository heating. Alternatively, the cavity around the package can be filled with a *backfill* to impede water movement. A common choice is bentonite, a

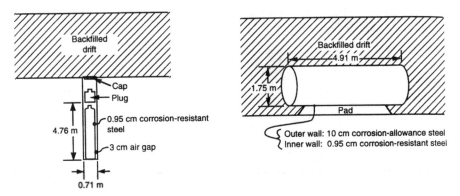

Fig. 11.4. Illustrative sketch of alternative containers and emplacement geometries for deep geologic disposal of spent fuel or reprocessed wastes. Left: thin-walled container in vertical borehole; right: thick-walled container in horizontal drift. The backfill in this illustration is coarse rock. (Note: Recent designs for the Yucca Mountain have not included backfill.) [From Ref. [34, Figure ES-3].]

material made largely of clays. Bentonite swells when water enters it, impeding the flow of water toward or away from the waste containers. Further, it adsorbs many radionuclides, reducing their migration to a rate even slower than that of the water itself. However, bentonite may not be effective if subjected to high temperatures, and thus it may be more suitable for the relatively cool environment of the planned Swedish repository than for Yucca Mountain if plans for a hotter environment are adopted. Materials known as *buffers* can also be added to the backfill to condition the chemical composition of any water moving through the backfill.

Planned Lifetime of the Waste Package

If the wastes are placed in a sufficiently impenetrable container, then the waste problem has been solved. The most ambitious plan in this regard was the one originally adopted by Sweden. The design called for putting the spent fuel in thick-walled canisters, using a copper outer wall to take advantage of the low corrosion rate of copper in the water of the host rock, which is granite. One design specified concentric cylinders, the inner with a 5-cm-thick steel wall and the outer with a 5-cm-thick copper wall [35, p. 16]. The entire canister is 88 cm in diameter and 5 m long.

The canisters in the Swedish plan are to be placed in vertical holes in tunnels in the repository and be surrounded on all sides, including top and bottom, by bentonite in the form of blocks or powder. The calculated lifetime of the waste package, under ordinary conditions, is over 1,000,000 years [31, p. 5-50]. There are scenarios that give earlier releases of radionuclides (e.g., the disruption of individual canisters due to rock motion produced by a growth of glaciers). Nonetheless, the engineering system is expected to provide very substantial protection for periods of time far in excess of 1000 years.

In contrast, early United States planning was predicated on a package life-time of only 1000 years, with the natural geologic barrier providing protection thereafter. The plan was for thin-walled containers, in part because the containers were not expected to be subjected to high pressures, because thin walls suffice for corrosion resistance in a dry environment and because the cost is less with less material [36, p. 4]. However, in all recent planning for Yucca Mountain, relatively thick containers have been adopted which are expected to keep their integrity for 10,000 years or more. Thick container walls offer important advantages. They can provide resistance against corrosion for long times, as well as shield against the radiation from the enclosed wastes. The reduced radiation level means less difficulty in keeping exposures of the workers low. It also reduces dissociation by radiolysis of any surrounding water into oxygen and hydrogen, both of which can contribute to corrosion. Together, a rugged container and a favorable environment are intended to provide "defense in depth" in geologic disposal.

11.3 Alternatives to Deep Geologic Disposal

11.3.1 Variants of Geologic Disposal

Alternative Geologic Sites

In addition to excavated caverns, which everywhere remain the adopted approach for planning purposes, several other sorts of geologic site have been considered [37, p. 1.16], although none is an active contender at the moment:

- *Deep-borehole disposal.* The wastes would be placed in holes at depths of several kilometers in crystalline rock, such as granite. This option has not been given a great deal of attention. As evaluated, for example, in a 2001 NAS report, drilling the numerous boreholes probably entails higher costs than those for excavated caverns and the option faces difficulties in sealing the boreholes and maintaining retrievability [38, p. 123]. However, consideration of the deep borehole approach was strongly endorsed in the 2003 MIT study *The Future of Nuclear Power* on the grounds that travel of radionuclides from deep underground to the surface would be very slow and that "vast areas of crystalline basement rock are known to be extremely stable, having experienced no tectonic, volcanic or seismic activity for billions of years" [39, p. 56]. Suitable rock deposits are so common that the study even mentioned the possibility of locating boreholes at or near reactor sites. This recommendation by a prestigious group may foreshadow greater future consideration of deep-borehole disposal.
- *Rock melt.* High-level wastes in liquid form would be put into an underground cavity where, at high concentrations and confined volumes, they would melt the surrounding rock. Subsequent cooling and solidification, perhaps after about 1000 years, would trap the wastes in a well-sealed environment. Of course, the wastes would not be as well trapped at early times, while the material is liquid. Further, it would be difficult if not impossible to retrieve the wastes, even at times shortly after disposal.

International Approaches

The stricture that each country should dispose of its own wastes is driven primarily by the expected unwillingness of potential "hosts" to accept foreign wastes. Nonetheless, there have been suggestions for international solutions:

- *Remote islands.* Given local opposition to waste disposal sites near populated areas, it has been suggested that underground repositories be placed on remote islands. Such a plan would face difficulties in identifying a geologically suitable site, overcoming environmental objections, and solving the problems that remoteness would create for construction, waste transportation, and provision of security against intruders. This approach does not appear to have gathered much favor.

◆ *The ARIUS Project.* The Association for Regional and International Underground Storage (ARIUS) was formed in 2002 "to facilitate progress toward multinational solutions" [40]. The founding members were organizations from Belgium, Bulgaria, Hungary, Japan, and Switzerland. The goal of ARIUS is to identify and develop a site, in Europe or elsewhere, for accepting wastes from a number of countries. There have been some indications that Russia might be willing to host a site, but such hopes are subject to national vetoes. ARIUS is a successor to the Pangea project, which had sought to establish an international repository in a remote location in Australia, but which was abandoned in the face of Australian objections.

11.3.2 Subseabed Disposal

Deep-seabed or *subseabed* disposal (SSD) provides a technically interesting alternative to geologic disposal (see, e.g., Refs. [41] and [42]). At present, it is not a viable option, because SSD is banned by an international agreement commonly known as the London Dumping Convention. Nonetheless, it warrants consideration, because the Convention could be modified if a consensus were to develop that SSD is practical and and poses no significant environmental threats.

Main Features of Subseabed Disposal

The deep seabed, at places where the ocean is several thousand meters deep, has been formed from the deposition of sediments over millions of years. The seabed is in the form of a water-saturated clay layer, on the order of 50 m thick. Its physical properties as a site for high-level waste burial were described in favorable terms in a 1994 National Academy of Sciences study on the disposition of plutonium from dismantled nuclear weapons:

> The deep ocean floor in vast mid-ocean areas is remarkably geologically stable; smooth, homogenous mud has been slowly building up there for millions of years. The concept envisioned for HLW [high-level wastes] was to embed it in containers perhaps 30 meters deep in this abyssal mud, several kilometers beneath the ocean surface...the mud itself would be the primary barrier to release of the material into the ocean, because the time required for diffusion of radionuclides through this mud would be very long. [43, p. 200]

About 30% of the ocean floor ($\sim 10^8$ km^2) is composed of sediments of this sort, and one goal of SSD studies is to identify specific regions that have been geologically stable for long time periods, in the range of 10^7 years or more [44]. An individual repository might be 10^4 km^2 in area.

The canisters containing nuclear wastes could be emplaced in this sediment either by free fall through the ocean, by forcible injection, or in predrilled holes. The canisters might be individually emplaced, with one canister per penetration of the ocean floor (e.g., Ref. [45, p. 2]) or stacked, with several canisters apiece in holes drilled deep into the seabed. The holes would be sealed either mechanically or by natural processes in the sediment. Once the canisters are in place and buried, they are in a uniform environment over wide regions, so that the durability of the canisters and the rate of migration of radionuclides through the clay will not vary greatly from place to place.

Estimates of the lifetime of a canister in the environment of the enclosing mud range from "several centuries" [41, p. 28] to "a few thousand years" [43, p. 200]. The lifetime obviously would depend on the material used and its thickness.[18] After a canister is breached, protection is provided by the slowness of movement of many radionuclides through the clay and subsequent dilution in the oceans. According to one summary:

> Experiments conducted as part of an international research program concluded that plutonium (and other transuranic elements) buried in the clays would not migrate more than a few meters from a breached canister after even 100,000 years. [46]

This description does not encompass all radionuclides, but it holds for some of the most important ones. More complete information on the movement of radionuclides in the sediment will be important to an informed decision on SSD.

Detailed decisions on the canister design and capacity and on the spacing of canisters in the seabed await further studies of details of the subseabed environment and the long-term behavior of canister materials. An order-of-magnitude picture of the scale of the problem can be obtained if one assumes that each canister contains about 1 MTHM. Thus, a large repository, holding in the neighborhood of 100,000 MTHM in the form of spent fuel or reprocessed wastes, would require 100,000 canisters. This capacity would be comparable to that of Yucca Mountain, although somewhat greater.

The required area is not well established, but appears to be small compared to the area of potentially favorable ocean floor sites. If the canisters were placed 200 m apart, an area of 4000 km^2 would accommodate 100,000 canisters.[19] A much smaller area was suggested in an early report by the Nuclear Energy Agency of the OECD (NEA) which stated that for 100,000 canisters, "a minimum site area should be 100 km^2" [48, p. 15].

The attraction of subseabed disposal lies partly in the remoteness of the repository from the human environment. The main pathway of radionuclides to people would be through the consumption of seafood contaminated with radionuclides. Estimated doses from these radionuclides are small. The NEA report estimated the peak individual dose to be about 10^{-7} Sv/yr (0.01

---

[18] Presumably, multilayer canisters are possible if needed.

[19] A review by Edward Miles indicated a possible separation of between 100 m and 300 m [47, p. 9].

mrem/yr) for a repository holding vitrified HLW waste from 1000 GWyr of output [48, p. 222].[20] The peak dose would be reached after about 20,000 years. In a later summary, Robert Klett of Sandia National Laboratories found, as a "best estimate," a dose of 5×10^{-10} Sv/yr, reached after about 100,000 years, for a repository with 100,000 MTHM [45, p. 11].[21] The large difference between this estimate and the earlier one was not addressed, although Klett's "least favorable" estimate was only slightly below the NEA estimate. However, for all of these estimates, the calculated peak dose is far below the levels of natural background and of any radiation protections standards.

Any renewed consideration of subseabed disposal would require additional research, including study of the durability of potential containers and the rate of motion of radionuclides through the clay. This information would help determine the container design and make possible better calculations of the radiation doses received by organisms living near the ocean floor and by people who eat food that might be contaminated by the radionuclides.

Prospects for Subseabed Disposal

Whatever the merits of subseabed disposal from a technical standpoint, its implementation faces institutional obstacles that have sufficed to terminate most research on SSD. A major international project—participated in by the United States and most of the other leading nuclear countries (not the former Soviet Union)—was inaugurated in 1976 to study the feasibility of SSD. This work was led by the Seabed Working Group under the auspices of the Nuclear Energy Agency of the OECD. The project was ended in the mid-1980s due to growing opposition in the United States and international pressures against "ocean dumping" (see, e.g., Ref. [47]).

In the opinion of subseabed advocates, the program was abandoned by the DOE and opposed by the U.S. nuclear industry because it would detract from the focus on the Yucca Mountain project for geologic disposal. As described by Edward Miles, one of the American leaders in the SSD programs:

> Essentially, we were killed because the science was impeccable and our results were too good. DOE's public story was that we were terminated as the result of budgetary cutbacks, but DOE lied. We were killed

---

[20] At the same rate of escape from the canister, spent fuel would give a somewhat greater dose, but the increase would be small because the chief contributor is technicium-99 (^{99}Tc) in seaweed, and technicium is not removed in reprocessing. Plutonium makes a much smaller contribution to the calculated dose due to its retardation in the clay.

[21] It is not clear in this reference whether the assumed wastes are in the form of spent fuel or vitrified HLW. In either case, assuming the specified mass refers to the initial uranium content, the repository would represent the output of roughly 3000 GWyr.

because our results fed the fires of the NIMBY syndrome and thereby threatened the future of Yucca Mountain. [47, p. 17]

In short, if subseabed disposal was seen as a viable alternative to Yucca Mountain, the opponents of Yucca Mountain would have had additional ammunition.

Even were this cause of opposition to disappear—when a decision for or against a Yucca Mountain repository is irrevocably made—subseabed disposal would face a major further obstacle because of the international protections of the oceans embodied in the London Dumping Convention. When subseabed disposal was first being considered, its proponents believed that the Convention would not apply, because the disposal was in a region "*beneath* the seabed," not in the ocean itself [44]. The issue became a matter for prolonged negotiations until the United States decided in 1995 to support applying the dumping ban to subseabed disposal [47]. This led to the adoption in the 1996 Protocol to the Convention of a provision which specifically defined "dumping" to include "any storage of wastes or other matter in the seabed and the subsoil thereof" [49].

This decision against SSD reflects a worldwide visceral opposition to ocean dumping of any material deemed potentially harmful. The fact that the London Convention specifically adopts the pejorative term "dumping" in its title may reflect and contribute to a negative feeling about the whole concept, and no country appears to be ready to risk the worldwide criticism that would arise should it decide to pursue subseabed disposal.

Nonetheless, the Convention may continue to be a subject of controversy. Issues of "ocean dumping" might also arise in the context of the sequestration of carbon dioxide either in the deep ocean or in subseabed aquifers. In principle, subseabed disposal of either nuclear wastes or carbon dioxide could contribute to the alleviation of climate change by limiting the buildup of carbon dioxide in the atmosphere. Viewed in that light, and if there is strong evidence that there are no adverse environmental effects, subseabed disposal might eventually gain political acceptance. It is also possible that countries with less available land area than the United States (or Russia and Canada) may one day press for subseabed disposal.

11.3.3 Partitioning and Transmutation of Radionuclides

Much of the calculated long-term waste hazard comes from a limited set of radionuclides, with half-lives ranging from hundreds of years to a million or so years. Exposure of selected radionuclides to high fluxes of neutrons, produced in a nuclear reactor or by an accelerator, could transmute them into nuclides with much shorter or much longer half-lives (including stable nuclei). Either outcome could reduce the long-term radioactivity by so large an amount that although the ultimate wastes might still be placed in a geological repository, the nature of the problem is sufficiently changed for this approach to be considered to be an alternative to conventional geologic disposal. Interest in these possibilities led the U.S. DOE to sponsor a National Academy study whose results are published in *Nuclear Wastes: Technologies for Separations and Transmutation* (NWTST) [50]. We draw heavily on that report in the present discussion.

A first step in the transmutation of the wastes is the separation of the fuel into different waste streams. Separation technologies were first used to extract plutonium for nuclear weapons, after converting the spent reactor fuel into liquid form. A similar approach is used to retrieve plutonium for recycling for use in other reactors, as is being done in the French reprocessing program. However, the extraction would not be limited to plutonium and uranium if a major goal is radionuclide destruction. After extraction, the radionuclides in question would be exposed to high neutron fluxes.

This could eliminate some of the long-lived fission products. For example, technicium-99 (^{99}Tc) ($T = 211,000$ yr) can be eliminated by bombarding it with neutrons. The capture product ^{100}Tc ($T = 16$ s) quickly decays to rubidium-100 (^{100}Rb), which is stable. There is no urgency in removing the rubidium, because further neutron captures continue to ^{101}Rb and ^{102}Rb, which are also stable. The net result is the conversion of ^{99}Tc into stable rubidium isotopes [50, p. 50]. Iodine-129 ($T = 15.7 \times 10^6$ yr) can be converted into stable xenon isotopes by similar processes.

Separation and transmutation (S&T) could, more importantly, also reduce the actinide abundances. Here, the situation is complicated, because there is competition between fission, which eliminates actinides, and capture, which transmutes them into heavier actinides. For example, ^{239}Pu is destroyed in neutron bombardment through fission or transformed to ^{240}Pu ($T = 6564$ yr) through neutron capture. The competition between fission and capture continues to ^{241}Pu ($T = 14.29$ yr) and beyond. At each stage, the fission-to-capture ratio depends on the neutron energy spectrum and is generally higher for fast neutrons than for thermal neutrons. Neptunium-237 (^{237}Np) ($T = 2.14 \times 10^6$ yr) is another key component of the wastes after long periods of time (see Section 10.3.2).[22] Successive neutron captures in ^{237}Np leads to the production of ^{239}Np, which decays with a 2.4-day half-life to ^{239}Pu.

If, for example, one inserts into a reactor a number of fuel rods containing plutonium (and not uranium), the combination of capture and fission processes will decrease the plutonium content and total actinide content of the sample. This means less long-lived radioactivity in the waste and particularly less alpha-particle activity, as discussed in Chapter 10 (see Figure 10.4).

The NWTST report describes a large number of technologies for separation and transmutation of the various radionuclides. The methods for producing the required fluxes of neutrons fall into two broad categories:

---

[22] ^{237}Np is produced in reactors by neutron capture, starting from ^{235}U, and in the wastes from the beta decay of ^{241}Pu followed by the alpha-particle decay of americium 241 ($T = 432$ yr).

- *Critical nuclear reactors.* These are reactors in which the chain reaction is self-sustaining, as is normal in reactors. Many sorts of reactor can be used for transmutation, including—but not limited to—light water reactors (LWRs) and liquid metal reactors (LMRs). They differ in that the neutrons in LWRs are quickly slowed to very low ("thermal") energies, whereas in the LMRs, the neutron spectrum is dominated by higher-energy ("fast") neutrons. Fast reactors have an advantage over thermal reactors for transmutation of actinides because the fission-to-capture ratio is generally higher for fast neutrons than for thermal neutrons. Thus, the proponents of several of the Generation IV reactors stress the actinide destruction aspect, in each case based on a fast reactor in a fuel cycle in which the spent fuel is reprocessed and some or all of the actinides are recycled (see Chapter 16).

- *Accelerator-driven reactors.* In this system, particle accelerators are used to produce high-energy protons which create spallation neutrons on striking a suitable target. These neutrons, in turn, bombard fissile material (such as fissile isotopes of uranium or plutonium) and produce additional neutrons, in a subcritical chain reaction which would die out were the accelerator source shut off.

For either category of reactor, the usual plan would be to generate electricity at the same time as the various radionuclides are being destroyed (and others produced). Several separate streams of radionuclides would be extracted from the fuel and the output of each stream would be returned to the reactor for transmutation or be converted to solid form for disposal by incorporating it into a glass or ceramic matrix. The goal is to have waste material which has much less long-lived activity than is in the spent fuel from the once-through fuel cycle.

The focus of the NWTST committee was on existing wastes. It concluded that it "found no evidence that applications of S&T have sufficient benefit for the U.S. HLW program to delay the development of the first permanent repository for commercial spent fuel" [50, p. 2]. It endorsed the key feature of the present policy: the placement of spent nuclear fuel in a geologic repository. It suggested, however, that "fuel retrievability should be continued to a reasonable time (on the order of 100 years)" and further suggested:

> A sustained, but modest, and carefully focused program of research and development over the next decade could prepare the technical basis for advanced separation technologies for the radionuclides in spent LWR fuel and for decisions on the possible applications of S&T as part of the more efficient future use of fissionable resources. [50, p. 10]

The main motivation for pursuing S&T activities is here indicated as resource conservation, rather than protection of health or the environment.

However, the S&T approach is likely to be more relevant to future wastes than to the relatively limited amount of existing wastes. A long-sustained major increase in nuclear power use may encounter problems of limited waste disposal capacity and uranium resources. A fuel cycle that breeds and consumes plutonium can provide very large amounts of power without a commensurate increase in the long-term activity of the waste product. This activity can be further reduced if some of the waste streams are appropriately separated and irradiated with neutrons in a fast reactor.

Two concerns remain, however. The handling of streams of highly radioactive liquid wastes may result in the occasional escape of radionuclides and exposure of workers or the general population. In addition, more threateningly, the separated plutonium could be diverted for weapons use by either government authorities or terrorists. To address some of these dangers, reactor and fuel cycle concepts have been developed in which the entire operation would be carried out in a set of facilities located at a single site. The Integral Fast Reactor program was designed to do this (see Section 16.5.1). Although this specific project has been abandoned, the general approach remains of interest.

As yet, there are no well-defined plans to proceed in this direction, beyond the ordinary reprocessing pursued in France and elsewhere. However, many different sorts of reactors, some coupled with accelerators, have been suggested as suitable to use for transmutation. As indicated earlier, these considerations are prominent in the design of possible Generation IV reactors (see Chapter 16).

The transmutation issue also arises in the different context of disposing of plutonium from dismantled nuclear weapons. One option for dealing with this plutonium is to combine plutonium and uranium oxides to form a mixed-oxide fuel (MOX), which can be substituted for some or all of the UO_2 fuel used in reactors (see Section 9.4.2). In this case, most of the ^{239}Pu is destroyed in ordinary reactor operation and the remainder becomes admixed with ^{240}Pu and fission products in such a way as to greatly lessen its utility as a weapons fuel. We return briefly to this use of transmutation in Section 18.3.3.

Extraterrestrial Disposal with Rockets

In a scheme that most observers consider frivolous, nuclear species that are particularly pernicious, in terms of half-life or biological impact, would be extracted from the waste and propelled into space, perhaps into the Sun or out of the solar system. There are no serious plans to adopt such a strategy. In view of the possibility of rocket failure, it would be difficult to establish that disposal via rocket creates less danger than does deep underground storage. Nonetheless, if convincing methods could be advanced for ensuring that failed rockets would drop their payloads without damage and then be recovered, this would provide a means for truly permanent disposal.

Polar Disposal

It was at one time suggested that a waste canister placed on a polar ice shelf would burrow down into the ice by the action of its own heat and disappear harmlessly. This suggestion was not taken very seriously even before issues of protection of the Antarctic came to the fore, and it is not a plausible option today.

11.3.4 Summary of Status of Alternatives to Geologic Disposal

Nuclear planning in the United States—and in all other countries as well— is based on the assumption that ultimate disposal of high-level wastes is to be in deep geologic repositories. It is difficult to make a definitive technical assessment of the relative advantages and disadvantages of this approach and of subseabed disposal. Either method holds the promise of being able to keep human radiation doses to a very low level. Retrieval of the wastes is more difficult for subseabed disposal than for geologic disposal. Whether this lesser accessibility is seen as an advantage or disadvantage depends on whether the focus is on possible future beneficial uses of the radionuclides or on the dangers of theft or accidental intrusion.

Advocates of one or another approach often couch their case in terms of adequate technical acceptability and superior political acceptability. Geologic disposal is politically difficult because of opposition from people living in the vicinity of any contemplated site. Obtaining international agreement on ocean disposal is possibly even more difficult, because of widespread opposition to anything that might be termed "ocean dumping." Thus, deep geologic disposal remains the leading prospect. If the obstacles to both approaches appear too formidable, and if continued interim storage is deemed unsatisfactory, there may be a turn toward transmutation of existing wastes, even without an expansion in nuclear power use. Such a turn, if it occurs, could be regarded as an act of prudence or of desperation.

11.4 Worldwide Status of Nuclear Waste Disposal Plans

The planning process for geologic disposal has everywhere moved slowly and no country, other than perhaps the United States, is likely to have an operating repository before the year 2020. Plans for geologic disposal are summarized in Table 11.4 for most of the OECD users of nuclear power. Outside the OECD, Russia and India are planning to reprocess spent fuel [51]. Except for the United States and Finland, all of the plans appear to be tentative and it is probably unwise to consider any of the plans as firm at least until such time as actual repository construction begins.

Aside from the commonality of geologic disposal, there are marked differences in approach. At least four different types of host medium are being

Table 11.4. National plans for disposal of spent fuel or high-level wastes from commercial nuclear reactors, for selected OECD countries (as of 2001).

Country	Earliest Date for Repository	Candidate Host Rock	Substantial Reprocessing?	Tentative Site
Canada	2025	Granite	No	
Finland	2020	Granite	No	Okkiluoto
France	2020	Clay, granite	Yes	
Germany	2020[a]	Salt	Yes[b]	Gorleben
Japan	2030–2045	Granite, sedimentary rock	Yes[b]	
South Korea	No target		No	
Sweden	2020	Granite	No	
Switzerland	2050	Granite, clay	Yes	
United Kingdom	2040		Yes	
United States	2010	Tuff	No	Yucca Mountain

[a]A date of 2008 is cited in Ref. [52, p. 186], but meeting such a schedule would require an abrupt weakening of the opposition to the site.
[b]Reprocessing of wastes from Germany and Japan has been carried out in France and the United Kingdom.
Sources: Ref. [52, p 182–192]; Ref. [53, Chapter 3], and Ref. [54, p. 50].

actively considered, in part reflecting differences in geologic conditions. In an aspect where there is freedom of choice, there is a division between disposing of the spent fuel directly or of reprocessing it and disposing of reprocessed vitrified wastes.

The United States, which is furthest along in preparations to open a repository, plans to put spent fuel in a repository in tuff by 2010. Germany has long been considering a site in a salt dome at Gorleben, and, for a number of years, a target date of 2008 was quoted for this site. However, strong opposition to the site has put the schedule in serious doubt, and the German waste disposal plans will not take any definite shape until the political controversies have been resolved.

France is pursuing a program in which the high-level wastes are the vitrified residues of reprocessing, currently in the form of borosilicate glass. They are to be put in a deep geologic site after several decades of cooling. The expected date for this placement is about 2020. Sites in clay and in granite are being investigated. Representatives of the French Commission for Atomic Energy have spoken with great confidence of the efficacy of their program:

> ...vitrification offers a stable and safe solution, long-since mastered, to package high-activity wastes. The behavior of this embedding.... lead to predict a minimum lifetime of 10,000 yr without significant change. The mean package integrity lifetime will most likely reach a million years....

The know-how necessary to design, construct, and implement deep sites is classical and available. [55, p 31–32]

Nonetheless, despite this apparent confidence, the French authorities plan prior tests in underground laboratories before proceeding with the final repository.

The repository target dates indicated in Table 11.4 are relatively far in the future. They range from 2010 for the United States to 2050 for Switzerland. With so long a time interval before the actual completion of a repository, it is possible that many of these plans will change. In the United States, for example, there have been changes in the waste package design in the last few years and, as the Yucca Mountain project moves ahead, it is likely that detailed design revisions will continue to be made while the DOE seeks NRC approval to begin construction. For countries with longer lead times, it is even more probable that design changes will occur. However, these are likely to all be changes in detail. In the larger picture, long-term geologic disposal continues to appear to be the universal choice.

References

1. Matthew Bunn, John P. Holdren, Allison Macfarlane, Susan E. Pickett, Atsuyuki Suzuki, Tatsujiro Suzuki, and Jennifer Weeks, *Interim Storage of Spent Nuclear Fuel*, A Joint Report from the Harvard University Project on Managing the Atom and the University of Tokyo Project on Sociotechnics of Nuclear Energy (Cambridge, MA: Harvard University, 2001)
2. U.S. Nuclear Regulatory Commission, "Dry Spent Fuel Storage Designs: NRC Approved for General Use" (2003). [From: http://www.nrc.gov/waste/spent-fuel-storage/designs.html]
3. Betsy Tomkins, "Onsite Dry Spent-Fuel Storage: Becoming More of a Reality," *Nuclear News* 36, no. 15, December 1993: 35–41.
4. U.S. Department of Energy, *Spent Nuclear Fuel Discharges from U.S. Reactors 1991*, Service Report SR/CNEAF/93-01 (Washington, DC: U.S. DOE, 1993).
5. "Late News in Brief," *Nuclear News* 36, no. 8, June 1993: 22.
6. "Dry Cask Storage of Spent Nuclear Fuel" and related information documents, Consumers Power, Palisades Nuclear Plant, South Haven, Michigan (1994).
7. "Legislature: Prairie Island ISFSI Can Proceed," *Nuclear News* 37, no. 8, June 1994: 26.
8. Allison Macfarlane, "Interim Storage of Spent Fuel in the United States," *Annual Review of Energy and the Environment* 26, 2001: 201–235.
9. Jonathan Alter, "At the Core of Nuclear Fear," *Newsweek*, June 24, 2002: 40.
10. Mark Holt, *Civilian Nuclear Waste Disposal; Updated March 8, 2002*, CRS Issue Brief for Congress IB92059 (Washington, DC: Congressional Research Service, Library of Congress, 2002).
11. Nuclear Waste Technical Review Board, *Disposal and Storage of Spent Nuclear Fuel—Finding the Right Balance, A Report to Congress and the Secretary of Energy* (Arlington, VA: NWTRB, 1996).

12. Nuclear Waste Technical Review Board, *Sixth Report to the U.S. Congress and the U.S. Secretary of Energy* (Arlington, VA: NWTRB, 1992).
13. M.V. Rajeev Gowda and Doug Easterling, "Nuclear Waste and Native America: The MRS Siting Exercise," *Risk: Health, Safety, & Environment* 9, no. 3, 1998: 229–258.
14. "Utilities Sign to Support Mescaleros Storage Project," *Nuclear News* 38, no. 9, August 1995: 84.
15. "Late News in Brief," *Nuclear News* 34, no. 14, November 1991: 25.
16. U.S. Nuclear Regulatory Commission, *Final Environmental Impact Statement for the Construction and Operation of an Independent Spent Fuel Storage Facility on the Reservation of the Skull Valley Band of Goshute Indians and the Related Transportation Facility in Tooele County, Utah*, NUREG-1714 (Washington, DC: NRC, 2001).
17. Private Fuel Storage, *Response to Questions About the Operation of the Private Fuel Storage Facility: A Report to the Citizens of Utah* (Salt Lake City: PFS, 2001).
18. U.S. Nuclear Regulatory Commission, "In the Matter of Private Fuel Storage, L.L.C," Commission Order CLI-03-05 (May 28, 2003).
19. J.L. Sprung, et al., *Reexamination of Spent Fuel Shipment Risk Estimates*, Report NUREG/CR-6672, SAND2000-0234 (Albuquerque, NM: Sandia National Laboratories, 2000).
20. U.S. Department of Energy, Office of Civilian Radioactive Waste Management, *Final Environmental Impact Statement for a Geologic Repository for the Disposal of Spent Fuel and High-Level Radioactive Waste at Yucca Mountain, Nye County, Nevada*, Report DOE/EIS-0250 (North Las Vegas, NV: U.S. DOE, 2002).
21. *Energy, Code of Federal Regulations*, title 10 (rev. January 1, 2002).
22. U.S. Nuclear Regulatory Commission, *An Updated View of Spent Fuel Transportation Risk, Discussion Draft* (2000). [From: http://ttd.sandia.gov/nrc/modal.htm]
23. U.S. Department of Energy, Office of Public Affairs, *Spent Nuclear Fuel Transportation* (2002). [From: http://www.ocrwm.doc.gov/wat/pdf/snf_trans.pdf]
24. National Research Council, *A Study of the Isolation System for Geologic Disposal of Radioactive Wastes*, Report of the Waste Isolation Systems Panel (Washington, DC: National Academy Press, 1983).
25. Mary Kruger, *A Message from the Director, The WIPP Bulletin*, EPA 402-N-99-001 (Washington, D.C: EPA, 1999).
26. A. G. Milnes, *Geology and Radwaste* (London: Academic Press, 1985).
27. Konrad B. Krauskopf, *Radioactive Waste Disposal and Geology* (London: Chapman & Hall, 1988).
28. Electric Power Research Institute, *Evaluation of the Proposed High-Level Radioactive Waste Repository at Yucca Mountain Using Total System Performance Assessment, Phase 6* (Palo Alto, CA: EPRI, 2002).
29. U.S. Department of Energy, Office of Civilian Radioactive Waste Management, *Yucca Mountain Science and Engineering Report, Technical Information Supporting Site Recommendation Consideration*, Report DOE/RW-0539 (North Las Vegas, NV: U.S. DOE, 2001).
30. Nuclear Waste Technical Review Board, *Fifth Report to the U.S. Congress and the U.S. Secretary of Energy* (Arlington, VA: NWTRB, 1992).

31. Swedish Nuclear Power Inspectorate (SKI), *SKI Project–90*, SKI Technical Report 91:23 (Stockholm: SKI, 1991).

32. J. L. Zhu and C. Y. Chan, "Radioactive Waste Management: World Overview," *IAEA Bulletin* 31, no. 4, 1989: 5–13.

33. Organization for Economic Co-operation and Development, *The Status of Near-Field Modelling*, Proceedings of a Technical Workshop (Paris: OECD, 1993).

34. M. L. Wilson et al., *Total-System Performance Assessment for Yucca Mountain—SNL Second Iteration (TSPA-93)*, Report SAND93-2675 (Albuquerque, NM: Sandia National Laboratories, 1994).

35. Swedish Nuclear Fuel and Waste Management Company (SKB), *Activities 1993* (Stockholm: SKB, 1993).

36. R. D. McCright, *An Annotated History of Container Candidate Material Selection*, Report UCID-21472 (Livermore, CA: Lawrence Livermore National Laboratory, 1988).

37. U.S. Department of Energy, *Management of Commercially Generated Radioactive Waste, Final Environmental Impact Statement*, Report DOE/EIS-0046F (Washington, DC: U.S. DOE, 1980).

38. National Research Council, *Disposition of High-Level Waste and Spent Nuclear Fuel, the Continuing Societal and Technical Challenges*, Report of the Committee on Disposition of High-Level Radioactive Waste Through Geological Isolation (Washington, DC: National Academy Press, 2001).

39. Massachusetts Institute of Technology, *The Future of Nuclear Power, An Interdisciplinary MIT Study*, John Deutch and Ernest J. Moniz, Co-chairs (Cambridge, MA: MIT, 2003).

40. Charles McCombie and Neil Chapman, "Regional and International Repositories, Not If, But How and When," paper presented at World Nuclear Association Annual Symposium, 2002. [From: http://www.arius-world.org]

41. Edward L. Miles, Kai N. Lee, and Elaine M. Carlin, "Nuclear Waste Disposal Under the Seabed," *Policy Papers in International Affairs*, No. 22 (Berkeley: Institute of International Studies, University of California, 1985).

42. R. Kilho Park, Dana R. Kester, Iver W. Duedall, and Bostwick H. Ketchum, eds., *Wastes in the Ocean, Vol. 3: Radioactive Wastes and the Ocean* (New York: Wiley, 1983).

43. National Academy of Sciences, *Management and Disposition of Excess Weapons Plutonium*, Report of the Committee on International Security and Arms Control, John P. Holdren, Chair (Washington, DC: National Academy Press, 1994).

44. Charles D. Hollister, D. Richard Anderson, and G. Ross Heath, "Subseabed Disposal of Nuclear Wastes," *Science* 213, 1981: 1321–1326.

45. Robert D. Klett, *Performance Assessment Overview for Subseabed Disposal of High Level Radioactive Waste*, Sandia Report SAND93-2723 (Albuquerque, NM: Sandia National Laboratories, 1997).

46. Charles D. Hollister and Steven Nadis, "Burial of Radioactive Waste Under the Seabed," *Scientific American* 278, no. 1, January 1998: 60–65.

47. Edward L. Miles, "Personal Reflections on an Unfinished Journey Through Global Environmental Problems of Long Timescale," *Policy Sciences*, 31, 1998: 1–33.

48. Organization for Economic Co-Operation and Development, Nuclear Energy Agency, *Seabed Disposal of High-Level Radioactive Waste, a Status Report on the NEA Coordinated Research Programme* (Paris: OECD, 1984).

49. International Maritime Organization, *Convention on the Prevention of Maritime Pollution by Dumping of Wastes and Other Matter (London Convention 1972), compilation of the full texts of the London Convention of 1972 and of the 1996 Protocol thereto*, Document LD.2/Circ. 380 (London: IMO, 1997).

50. National Research Council, *Nuclear Wastes, Technologies for Separations and Transmutation*, Report of the Committee on Separations Technology and Transmutation Systems, Norman C. Rasmussen, Chair (Washington, DC: National Academy Press, 1995).

51. Uranium Information Centre Ltd, "Waste Management in the Nuclear Fuel Cycle," *Nuclear Issues Briefing Paper 9* (Melbourne: UIC, 2003). [From: http://www.uic.com.au/nip09.htm]

52. International Energy Agency, *Nuclear Power in the OECD* (Paris: OECD/IEA 2001).

53. U.S. Environmental Protection Agency, *Environmental Radiation Protection Standards for Yucca Mountain, Nevada; Draft Background Information Document for Proposed 40 CFR 197*, Report EPA 402-R–99-008 (Washington, DC: EPA, 1999).

54. Organization for Economic Co-operation and Development, Nuclear Energy Agency, *Trends in the Nuclear Fuel Cycle* (Paris: OECD, 2001).

55. J-Y Barre and J. Bouchard, "French R & D Strategy for the Back End of the Fuel Cycle," in *Future Nuclear Systems: Emerging Fuel Cycles & Waste Disposal Options*, Proceedings of *Global '93*, (La Grange Park, IL: American Nuclear Society, 1993): 27–32.

12

U.S. Waste Disposal Plans and the Yucca Mountain Repository

12.1 Formulation of U.S. Waste Disposal Policies

12.1.1 Brief History of Planning Efforts

Military and Civilian Wastes

Planning for the disposal of nuclear wastes in the United States has undergone many changes—in both administrative leadership and overall strategy—in the almost six decades since the wastes began to be generated. The power to make decisions has moved from the tight control of a single federal body into an arena in which many governmental and even nongovernmental groups have significant, and sometimes conflicting, roles. On the technical side, the basic assumption that all of the wastes would be reprocessed has been abandoned, and the originally favored geologic formation for disposal, bedded salt, has been bypassed. It has been difficult to find disposal sites because public trust in the federal government and its scientists is insufficient to overcome local fears. Some of this history is sketched in this section, along with a summary of the current institutional status of nuclear waste disposal.

Responsibility for the handling of nuclear wastes in the United States initially resided with the Atomic Energy Commission (AEC), which was established in 1946 to take charge of all aspects of the U.S. nuclear energy program, including nuclear weapons and civilian nuclear power. Although nuclear waste management came under the AEC's purview, it was not addressed as a matter of high priority. The problems did not appear to be very great, and there were more pressing and technically interesting matters that demanded attention. On the military side, there were the difficulties of simultaneously building U.S. weapons capabilities, coping with the dangers created by the nuclear arms race between the United States and the USSR, and restraining the spread of nuclear weapons to other countries. On the nonmilitary side, there were the challenges of fostering peaceful nuclear research throughout the world and establishing a nuclear electric power industry.

It was recognized that it would be necessary to handle wastes emerging from both the civilian and military programs, but there was implicit confidence that this could be done successfully when the need became more pressing. However, the low priority originally given to the matter has led to problems with the defense wastes, as mentioned in Section 10.1.2, and to long delays in developing a program for the civilian wastes.

Overview of Civilian Waste Disposal Plans

In the first decades of civilian nuclear power, there was a consensus that the spent fuel from the civilian reactor program would be reprocessed as soon as reprocessing plants were built (see Section 9.4.2). The plutonium and uranium would be extracted for use in other reactors, and additional isotopes possibly extracted for specialized applications in medicine and industry. The remaining wastes were to be solidified, probably in the form of borosilicate glass, and eventually transferred to a federal repository for permanent disposal.

It was generally expected that the repository would be in bedded salt. The use of a bedded salt repository had been suggested by a National Academy of Sciences (NAS) panel as early as 1957, and this recommendation was reaffirmed in a 1970 NAS study which spelled out the advantages of salt, although it also suggested some further studies and indicated the importance of taking care in the selection of specific salt sites [1].

The first repository site selected by the AEC was in salt beds located near Lyons, Kansas. It was designated in 1970 in response to pressures from Idaho's governor and senators to find a location to receive wastes that had been recently transferred to the National Reactor Testing Site near Idaho Falls.[1] The NAS report gave a qualified endorsement of the Lyons site, but emphasized the need for further studies (e.g., the identification of possible holes from earlier oil or gas well drilling). As studies proceeded, in accordance with the

---

[1] A summary account of this action and subsequent developments has been given by Luther Carter [2]. This reference is an excellent source of information on historical aspects of waste management up to the mid-1980s. Later developments, with a special emphasis on technical issues related to Yucca Mountain, have been given in an authoritative review by Paul Craig [3].

NAS recommendation and spurred by local objections to the repository, it was found that prior commercial activity had created cavities and boreholes in the salt that compromised the site's safety. Fear grew that it was vulnerable to both physical collapse and the intrusion of water. A combination of possible technical difficulties and growing opposition within Kansas doomed the site, and in 1971 the AEC abandoned plans for its development.

Since the early 1970s, the U.S. nuclear waste program has been struggling to find an acceptable solution. There have been innumerable studies and plans, but every specific proposal has encountered intense opposition, especially from people in the vicinity of the proposed facility and their political representatives. The Atomic Energy Commission was itself abolished in 1974. Its roles of developing nuclear energy and of regulating it were separated and given to two newly formed organizations. As their names suggest, the Energy Research and Development Administration (ERDA) assumed the development role and the Nuclear Regulatory Commission (NRC) assumed the regulatory one. In turn, ERDA was replaced by a new cabinet-level entity in 1977, the U.S. Department of Energy. The Environmental Protection Agency (EPA) in 1976 was given the responsibility for developing radiation protection standards for nuclear wastes [4, pp. 1–3].

The organizational changes did not lead to expeditious progress in the formulation and implementation of waste disposal plans. The separation of the NRC from ERDA, and later the DOE, was made in part to forestall imprudent actions. That goal may have been achieved, but the goal of moving forward with a coherent policy was not met. This led Congress in 1987 to short-circuit the decision-making process and designate Yucca Mountain as the sole site to be investigated as a potential waste repository site. The subsequent intensive investigations reached a crucial point in January 2002 when the DOE declared the site to be suitable and recommended that the President formally approve it. He did so, with the subsequent support of Congress, setting the stage for further technical examination of the site and a new round of political and legal debates (see Section 12.2.1).

12.1.2 Organizations Involved in Waste Management Policy

Federal Agencies

Three federal agencies have particularly important roles in nuclear waste management, as defined by Congress:

◆ *Department of Energy (DOE)*. The Department of Energy, through its Office of Civilian Radioactive Waste Management (OCRWM), has primary responsibility for planning, building and operating waste management facilities and for research that provides a technical basis for evaluating waste management options and problems. In this work, it makes extensive use of private contractors and outside advisory groups.

- *Environmental Protection Agency (EPA).* The EPA has overall responsibility for establishing standards for radionuclide releases and radiation exposures arising from the operation of waste management facilities.[2]
- *Nuclear Regulatory Commission (NRC).* The NRC has the responsibility for licensing waste management facilities and for establishing standards for the construction and operation of these facilities. Its standards must be consistent with EPA standards.[3]

Beyond this trio, important roles are played by the U.S. Geological Survey and the Department of Transportation.

Federal Advisory Bodies

In addition to the main federal agencies with administrative responsibilities, there are a large number of groups and organizations with official advisory status. Some of these report to the individual agencies, some to Congress, and some to the public at large. They include bodies with broad responsibilities, of which nuclear waste studies are a relatively small component, as well as a number of boards and committees established specifically to advise on nuclear waste issues. These include the following:

- *National Academy of Sciences (NAS).* The NAS was established by congressional charter in 1863 as a self-governing body to provide scientific advice to the federal government. The National Academy of Engineering and the Institute of Medicine were established more recently under the NAS charter. The National Research Council is a joint arm of these three organizations and it serves as the organizing entity for academy studies, which are published by the National Academies Press (formerly the National Academy Press). The NAS has carried out many studies on waste disposal, and under the terms of the Energy Policy Act of 1992, it had a special role in making recommendations for the regulations for the Yucca Mountain nuclear waste repository (see Section 13.2.3) [7, Section 801].
- *Nuclear Waste Technical Review Board (NWTRB).* The NWTRB was established under the terms of the 1987 Amendment to the Nuclear Waste Policy Act (NWPA) as an independent entity within the Executive branch. It is appointed by the president from a group of persons nominated by the National Academy of Sciences as "eminent in a field of science or engineering, including environmental sciences" [8]. Its mandate is to "evaluate the technical and scientific validity of activities undertaken by the Secre-

---

[2] These regulations are set forth in Parts 191 and 197 of Title 40 of the U.S. Code of Federal Regulations [5].

[3] The NRC regulations are contained in Title 10 of the U.S. Code of Federal Regulations [6], including parts dealing with geologic disposal at Yucca Mountain (Part 63), nuclear waste transportation (Part 71), and temporary storage at the reactor or in a monitored retrievable storage facility (Part 72).

tary [of Energy]," with a focus on the activities necessary to establish a waste repository. Results of its work are contained in a series of Reports to Congress and the Secretary of Energy and are also made available to the public.

◆ *General Accounting Office (GAO).* The GAO is an arm of Congress that reports to it on fiscal and managerial aspects of federal programs. This mandate gives the GAO the scope to study the overall progress and effectiveness of these programs.

◆ *Advisory Committee on Nuclear Wastes (ACNW).* The ACNW was established by the Nuclear Regulatory Commission in 1988 to provide independent advice to the NRC on matters pertaining to the NRC's responsibilities for licensing and regulating future nuclear waste facilities.

States and Indian Tribes

It is widely feared that in attempting to address the national nuclear waste disposal problem, the federal government is willing to ignore local interests. To address this concern, Congress included in 1982 legislation extensive requirements for consultation with affected states and Indian tribes. Although, in the end, Congress retained a prerogative for federal overriding of state objections, the states were given both time and funds to formulate possible objections.

In addition to the powers explicitly granted by Congress, the states and Indian tribes have access to the courts, and on frequent occasions they have taken legal action to forestall federal actions. Thus, for example, the States of Minnesota, Texas, and Vermont joined in a suit objecting to EPA waste repository standards. Since the designation of the Yucca Mountain site, Nevada has used the courts and other means in an effort to stop the project. At one stage, local denial of water permits forced the Yucca Mountain administration to truck water from California to carry out some preliminary investigations of the site. Following congressional approval of the site in 2002, in the face of formal objections from the Governor of Nevada, the State continued its opposition through suits challenging actions taken by the EPA, NRC, and DOE in connection with the Yucca Mountain project.[4]

Private Individuals and Organizations

Private individuals and groups also have taken an active interest in waste disposal policy and have input through public hearings, normal political processes, and legal actions. Public opposition has, for example, prevented the investigation of potential granite repository sites in Maine and Michigan.

To date, most intervention by individuals and private organizations, aside from industry organizations, has been directed toward preventing the devel-

---

[4] The suits were originally scheduled to be heard in September 2003 [9, p. 64], but the hearings were subsequently delayed.

opment of specific waste disposal facilities. To counter this trend, the Office of Civilian Radioactive Waste Management (OCRWM) has undertaken extensive programs to make information available to the public and to schools, in the hope that the information provided would allay concerns over waste disposal studies and plans.

12.1.3 Congressional Role in the Site-Selection Process

Nuclear Waste Policy Act of 1982

In response to the failure of the federal waste program to reach any clear and final decisions, Congress enacted the Nuclear Waste Policy Act (NWPA) of 1982, which was intended to put the program on a new footing [10]. Key provisions of this act were as follows:

- *Geologic sites.* Geologic repository sites were confirmed as the leading choice for nuclear waste disposal.
- *Designation of candidate sites.* The Secretary of Energy was directed to nominate five sites for initial study. Following study and consultation, the secretary was to recommend to the president by January 1, 1985, three of the five sites for intensive study as possible sites of the first waste repository. This study process is termed site characterization.
- *Recommendation of selected site.* On the basis of information developed in the site characterizations and on recommendation of the DOE, the president was called upon to recommend to Congress by 1987 one site as qualified for development of a repository.
- *Nuclear Regulatory Commission.* The NRC was called upon to approve or disapprove an authorization for construction of a repository at the selected site by January 1, 1989.
- *Nuclear waste fund.* A Nuclear Waste Fund was established by assessing a fee of 1.0 mill per kilowatt-hour (0.1 ¢/kWh) on electricity generated at nuclear reactors. This fund is to be used to cover the expenses of waste disposal, including the construction of a repository.
- *States and Indian tribes.* Throughout the process, there are extensive mandated consultations with the affected states and Indian tribes, and opportunities for them to indicate disapproval. However, in the end, the authority to make decisions was retained by the president and Congress.

The responsibility of the EPA and NRC for regulating disposal facilities for civilian radioactive wastes was confirmed in this act [10, Section 121]. The EPA's mandate is to promulgate standards for protection of the offsite environment from releases of radioactive materials from repository sites. The NRC's mandate is to establish technical criteria for approving the construction, operation, and closure of repositories. These criteria "shall not be inconsistent" with the EPA standards.

The above-outlined schedule was for a first repository. The goal of this schedule was to have an operating repository by 1998. A parallel schedule, with a small lag, was also specified for a second repository.

The first step, the DOE's selection of the three choices for site characterization, was completed in May 1986, after lengthy hearings and well behind schedule. The sites chosen were a basalt site at Hanford, a bedded salt site in Texas, and a tuff site at Yucca Mountain in Nevada. The Hanford and Yucca Mountain sites were on federal land, the former part of the Hanford Reservation in Washington and the latter overlapping the Nevada Test Site, which was used for nuclear weapons tests. These recommendations engendered a great deal of controversy, including lawsuits initiated by the selected states.

With the difficulties encountered at each stage, it became apparent that the schedule outlined in the NWPA of 1982 would not be met. Not only was there no prospect that decisions would be reached on a schedule that would permit the president to recommend a single site in 1987, but it was also unclear if a decision could be made that would hold up against future lawsuits challenging the decisions and the process by which they were reached.

Nuclear Waste Policy Amendments Act of 1987 and Yucca Mountain

The matter was resolved, in a legislative sense at least, when Congress adopted the Nuclear Waste Policy Amendments Act of 1987, which designated Yucca Mountain as the *sole* site for characterization for a possible high-level waste repository [11]. Thus, instead of having three site characterization studies proceeding in parallel in the face of local opposition, there would be a single study, in a desert location where land-use conflicts seemed minimal and where it appeared that "deep, visceral opposition...[was]...lacking" [2, p. 423].

This congressional decision did not resolve all disputes. Local opposition grew, and the subsequent history has been one of continued conflict between federal authorities and representatives of Nevada. It did, however, establish the federal agenda for waste disposal planning for the following 15 years, and perhaps far beyond.

12.2 The Planned Yucca Mountain Repository

12.2.1 Schedule for the Yucca Mountain Project

The path to be followed by the Yucca Mountain project was set forth in November 1989, when the Secretary of Energy issued a "Reassessment of the Civilian Waste Management Program" [12]. The revised plan included a prolonged period of "iterative scientific investigations of the site to examine its suitability" [13, p. 2]. If found suitable, the schedule called for the opening of the repository by 2010, following a series of intermediate administrative steps set out in the NWPA of 1987: a recommendation by the Secretary of Energy, an up-down decision by the President, an opportunity for the state of Nevada to object, and a final decision by Congress. If Congress approves, then the DOE would prepare an application to the NRC for authorization to begin construction of the repository. Finally, after construction is completed, the DOE would need another NRC license to authorize it to begin to accept the wastes. The repository would be licensed to accept 70,000 MTHM, 90% of which would be spent fuel from commercial reactors (see Section 10.1.4).

Following extensive initial studies, the DOE issued a draft Environmental Impact Statement in 1999, as per schedule, and followed this with a series of major summary documents, including a *Preliminary Site Suitability Evaluation* in July 2001 that foreshadowed a favorable recommendation [14] and a *Science and Engineering Report* in February 2002 that gave the detailed rationale for a positive recommendation [15].[5] During the first part of 2002, in sequence, the Secretary of Energy and the President made positive recommendations, the Governor of Nevada submitted a notice of disapproval, and Congress voted to override Nevada's objections. The next step is for the DOE to prepare a detailed license application to the NRC requesting authorization to begin construction of the repository. The target date set for submitting the application was December 2004. Review of the application by the NRC staff and hearings before the NRC's Atomic Safety and Licensing Board are expected to take 3 years, moving the target date for construction authorization to 2007. Another NRC license will be required to authorize the acceptance of wastes. If this proceeds on schedule, the DOE hopes the repository could be ready to begin to receive wastes by 2010. During this time, it is to be expected that Nevada and other opponents will pursue further legal and political means to stop the project.

If repository construction is approved and completed and a further NRC license to receive wastes is also issued, the DOE can begin emplacement of the wastes. If the repository opens at the nominally scheduled date of 2010, transfer of the 70,000 tonnes of wastes is expected to continue until about 2033 [16, p. 1-9]. It is not necessary that the full repository be completed at the time that the first wastes arrive. In fact, it can be argued that a gradual buildup of the capacity could allow the project to gain from experience as it expands.

The earliest that the repository could be closed, under expected regulations, would be about 2060, but it might be kept open for up to several hundred years. Keeping the repository open for an extended period would facilitate cooling of the repository, allow long-term monitoring of repository

---

[5] A preliminary version of the *Science and Engineering Report* was published in May 2001 for public comment [16]. These reports are just the tip of an enormous iceberg that includes a large array of comprehensive reports and extensive subsidiary analyses of specific topics. Many of these documents have been made available on the website of the Yucca Mountain Project (www.ymp.gov).

performance, and offer the opportunity to retrieve waste packages either to correct defects or extract valuable radionuclides.

12.2.2 Physical Features of the Site

Location and Size of Repository

Yucca Mountain is located in a very dry, desert area of southern Nevada, roughly 160 km (100 miles) northwest of Las Vegas and 80 km northeast of Death Valley in California. The Yucca Mountain site is on federal land, adjacent to the Nevada Test Site, which was used in the past for nuclear weapons testing, and the Nellis Air Force Bombing Range. Part of the site would extend into land assigned to these facilities. Figure 12.1 displays both the location of the site and the orientation of the repository at the site. The total area assigned to the Yucca Mountain project, termed the land withdrawal area, is about 600 km^2, over two-thirds of which comes from land of the Nevada Test Site and Nellis Air Force Range [17, pp. 3-11].

Although political considerations played an important part in the final selection of the Yucca Mountain site, the location has long been viewed as having some special advantages, as suggested in the following DOE summary:

> In 1976, the director of the USGS [U.S. Geological Survey] identified
> a number of positive attributes in and around the Nevada Test Site

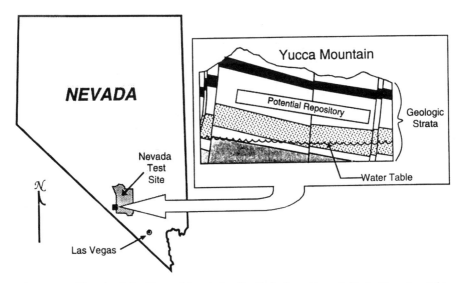

Fig. 12.1. Sketch of the Yucca Mountain site. Left: Location of site in Nevada; right: orientation of the repository at the site. The different shadings represent different layers of tuff, with different hydrological properties. (From Figure ES-1 of Ref. [18].)

that would make positive contributions to geologic disposal, including multiple natural barriers, remoteness, and an arid climate (McKelvey 1976). In 1981, a USGS scientist documented that the water tables in the desert Southwest are among the deepest in the world, and the geologic setting includes multiple natural barriers that could isolate wastes for "tens of thousands to perhaps hundreds of thousands of years."[16, pp. 1-14]

This recommendation led to the DOE's consideration of disposal in the unsaturated zone above the water table, although the repositories being considered in other countries are all below the water table.

The proposed placement of the repository is in a rock layer of welded tuff, known as the Topopah Spring Tuff unit. This formation is up to 375 m thick [16, pp. 1-28]. The water table in this region lies about 500–800 m below the surface [16, p. xxxiv]. The repository would be placed about 200 to 500 m below the surface and roughly 300 m above the water table. The main pathway by which radionuclides might reach the outside environment is by downward flow from the repository to the saturated zone and then by lateral flow in the saturated zone to an area where the water might be extracted through wells.

The rock at Yucca Mountain consists of layers of different types of tuff. A major distinction is between *welded* and *unwelded* tuff. Welded tuff is made of material that was at high temperatures and then cooled relatively slowly, leading to the formation of a dense, relatively nonporous mass [16, pp. 4-28]. It is extensively broken up by fractures. Unwelded tuff results from a different thermal history, is more porous, and has fewer fractures. Water flow in fractures is more rapid than in the bulk rock, or *matrix*.

The repository is to be honeycombed by an array of horizontal tunnels, called *drifts*, in which the wastes are placed. The eventual size of the subsurface area occupied by the repository and the layout of the tunnels will depend on decisions as to thermal loading of the repository (see Section 12.3.5). In the high-temperature mode, the area would be about 1150 acres (4.7 km^2) with about 58 drifts. In the low-temperature mode, either a larger area or greater ventilation would be used.

Emplacement Drifts

The planned drifts in which the wastes will sit are about 5.5 m in diameter and are expected to have a center-to-center spacing of 81 m [16, p. xxxvii]. Their walls gain support from steel hoops, or *sets*, spaced about 1.5 m apart. A welded-wire screen is placed against the rock behind the steel sets, and the sets and screen are bolted to the rock [16, p. 2-96]. The total length of the drifts is planned to be about 56 km [16, p. 2-83]. This is sufficient to hold the complement of roughly 10,000 waste packages and canisters (including those for defense wastes), each about 5 m in length and spaced with rather small

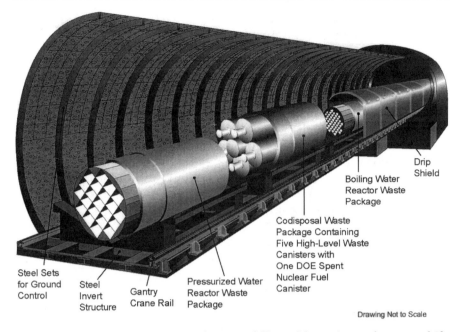

Drip
Shield

Boiling Water
Reactor Waste
Package

Codisposal Waste
Package Containing
Five High-Level Waste
Canisters with
One DOE Spent
Nuclear Fuel
Canister

Pressurized Water
Reactor Waste
Package

Steel Sets
for Ground
Control

Steel
Invert
Structure

Gantry
Crane Rail

Drawing Not to Scale

Fig. 12.2. Schematic illustration of proposed Yucca Mountain emplacement drift, showing cutaway views of the waste packages and the support structure. (From Ref. [15, Figure 3.3.])

gaps between the ends of the packages.[6] A schematic picture of the drifts and waste package is shown in Figure 12.2.

Permanent rails will be put into the drifts, and the waste packages will be moved on special flatbed railcars pushed by locomotives. Final placement of the waste package would be accomplished by a special lifting mechanism, or *gantry*, that also can ride along the rails and transfer the waste package to a permanent resting place, or *emplacement pallet*, in the drift. Both the locomotive and gantry are remotely controlled because the radiation levels and temperatures in the drifts preclude human entry. Retrieval of the waste package, should this later be deemed necessary or desirable, would be accomplished by reversing the emplacement process [16, p. 2-109 to 2-130].

12.2.3 The Waste Inventory

The capacity of the Yucca Mountain repository is limited by the Nuclear Waste Policy Act of 1982 [10, Section 114] to 70,000 tonnes of fuel or, more specifically, to 70,000 tonnes of heavy metal (MTHM). Of this total, 90%

---

[6] The spacing and other aspects of the layout could be modified if it is decided to operate the repository in the low-temperature mode.

(63,000 MTHM) will be spent fuel from commercial reactors and the remainder will be vitrified high level wastes (HLW) and spent fuel from defense programs (see Section 10.1.5). As indicated above, this will be sufficient to accommodate the commercial spent fuel produced up until about 2010. The 63,000 tonnes of spent fuel correspond to an electrical output of about 2100 GWyr or 19 trillion kWh.[7]

Beyond that, additional storage space will be needed both for spent fuel and for the HLW that is not included in the 7000-tonne allotment for military wastes. This could require the development of a second repository at a different location or an increase in the allowed capacity of the Yucca Mountain repository. The latter possibility is envisaged in Yucca Mountain planning documents, with the potential inclusion of the 105,000 MTHM indicated in Table 10.1 and perhaps as much as 119,000 MTHM [16, p. 2-85].

12.2.4 The Nuclear Waste Fund

The fee of 0.1 ¢/kWh charged against nuclear-generated electricity now produces a revenue of roughly $750 million/yr for civilian waste management—essentially the Yucca Mountain project. From FY 1983 through FY 1999, the fees totalled $8.3 billion [19, p. 3]. The expenditures have lagged far behind. This relatively low rate of spending on Yucca Mountain has been criticized as having impeded progress on important parts of the program including, e.g., the study of the waste packages [20, p. 6].

The difference between income and expenditures has created a large balance in the Nuclear Waste Fund, with the balance substantially increased by credits for interest income. In consequence, the balance in the fund amounted to over $9 billion in FY 2000 (in 2000 $), even after past expenditures for Yucca Mountain [19, p. 19]. According to a DOE projection made in 2001, expenditures will continue to substantially lag income until FY 2007, when they will rise sharply [19, p. 19]. This roughly corresponds to the time at which the DOE hopes to receive a construction permit for the repository.

The costs of the Yucca Mountain project as a whole, including transportation costs, are to be assigned in part to the civilian program (73%) and in part to defense nuclear waste programs (27%) [19, p. 10]. Based on an estimated future cost of $26 billion for the civilian program from 2000 to 2042 and anticipated income from fees and interest, the Nuclear Waste Fund is projected to have a balance at the end of FY 2042 of anywhere between $9 billion and $46 billion (all in 2000 $), depending on future economic conditions including interest rates. Further expenditures will still be required after 2042 to monitor and eventually close the repository

The key question posed in the DOE study from which these figures come is the adequacy of the 0.1 ¢/kWh fee. The overall conclusion, as of 2001, was that the fee "is sufficient at this time," based on the present repository design

---

[7] Based on an average burnup of 38.6 GWd/t and a thermal efficiency of 32%.

and the assumptions the authors made as to interest rates and inflation [19, p. 18]. This cost estimate ignored the possibility of license extensions for existing reactors or construction of new reactors, and it assumed the continued production of spent fuel by existing reactors during their originally licensed lifetimes. This led to an estimated eventual total of 83,800 MTHM of civilian fuel for placement at Yucca Mountain, not the presently authorized 63,000 MTHM.

This economic model was reasonable for a static situation. Possible additional nuclear generation would increase both the revenue and the costs. It does not seem likely that such changes would alter two main qualitative conclusions that may be drawn from the studies to date: (1) a fee of 0.1 ¢/kWh is probably adequate, and (2) doubling the fee to 0.2 ¢/kWh would almost certainly be adequate to cover unexpected costs and even more ambitious repository or waste package designs. Thus, waste disposal appears to be a relatively minor component of nuclear electricity costs.

12.3 Protective Barriers in Repository Planning

12.3.1 The Protection Requirement

Dose Limit

The basic requirement in repository design is to keep excessive amounts of radionuclides from reaching the biosphere, where humans could come in contact with them. This requirement is defined more specifically in regulations established by the EPA that set for 10,000 years a dose limit of 0.15 mSv/yr (15 mrem/yr) for the "reasonably maximally exposed individual" (RMEI) [5, §197.20]. (This standard will be discussed further in Section 13.2.5). The exposed person is assumed to live about 18 km from the repository in a farming community that obtains its water from wells drilled into the aquifer that carries the Yucca Mountain wastes. The RMEI obtains drinking water (2 L/day) from the contaminated source and has a diet similar to that of people in the region today, including consumption of food from local farms that use contaminated water for irrigation.

Assumptions as to Water Flow

The EPA imposes the requirement that the dose calculations be based on the assumed dilution of radionuclides in a plume of water that has a volume of 3000 acre-ft/yr.[8] This amount is judged by the NRC to correspond to the anticipated water demand of a community of "up to 100 individuals, living on 15 to 25 farms" [21, p. 55734]. It is a little less than the 4000 acre-ft that

---

[8] 1 acre-foot = 3.26×10^5 gal = 1234 m^3.

the EPA estimates could be "removed annually without seriously depleting the aquifer" [22, p. 32111].

The calculated dose for the RMEI is inversely proportional to the volume of water in which the radionuclides are carried. If a smaller flow is assumed, the dose would be higher than it is for 3000 acre-ft/yr, but presumably fewer people could be supported by the water. The number of people that could be supported by 3000 acre-ft/yr of water depends on their lifestyle, but as discussed in Sections 13.2.5 and 13.3.2, speculations about future changes in society cannot be part of official Yucca Mountain planning. With this limitation, the envisaged future Yucca Mountain neighbors will be mostly farmers, and their living patterns and need for irrigation water will limit the directly affected population to about 100 people.[9]

12.3.2 Defense-in-Depth

The requirement that the environment be protected for at least 10,000 years—and in some views for much longer—imposes demands on the engineered and natural protective barriers for times that greatly exceed those of most of the pertinent scientific experience. In particular, we have no direct experience of the corrosion resistance of materials for thousands of years under conditions of high temperature and diverse chemical environments. It also is impossible to know future climates with certainty (witness intermittent ice ages) or to predict with absolute confidence water flows in the regions above and below the repository. Informed inferences can be made about the various aspects of the system, but complete certainty cannot be established.

In this situation, it is important to have a system in which there are many layers of protection, so that if one—or even several—of the protective barriers proves to be less effective than expected, the overall system would remain adequately safe. This is defense-in-depth. In a highly stylized example, if there are five *independent* barriers, each with a 1% chance of failing, there is only 1 chance in 10 billion that all will fail. Of course, in the actual world of waste disposal, neither "failure" nor "nonfailure" are all-or-nothing matters, and the numerical illustration is far too simplified. Nonetheless, it illustrates the basic attribute of multiple barriers. For nuclear waste repositories, these are commonly divided into engineered and natural barriers (see Section 11.2.1), although there is substantial interplay between the two.

Overall, the defense against excessive doses from radionuclides depends on delay, decay, and dilution. The escape of the radionuclides into the biosphere is delayed by the various barriers. During this time, the level of activity

---

[9] If the no-change stipulation is ignored and no water is allocated to irrigation, the 3000 acre-ft of water/yr could support several thousand people. However, a good part of currently calculated Yucca Mountain doses comes from ingestion of food grown with contaminated irrigation water, and if this food were to be replaced by "imported" food, the doses would be lower than those in the calculations described in Section 12.4.

continually decreases due to natural radioactive decay. When remaining radionuclides can reach the biosphere, they are diluted by the water in which they are carried, further reducing the potential doses.

12.3.3 Engineered Barriers

The Waste Package

As discussed in Section 11.2.5, in the first years of planning for the Yucca Mountain repository, the goal of waste canister design was to contain the radionuclides for 1000 years. This was a relatively modest goal but sufficed if enough reliance could be placed on the natural barriers for long-term protection. However, after the early 1990s, the DOE planning shifted to the goal of a robust waste package, designed to retain its integrity for much longer than 1000 years. This shift was consistent with a defense-in-depth strategy and followed continued prodding from the NWTRB—starting with their first report in 1990 [8, p. 40]. Current waste package designs all call for enclosing the fuel assemblies in very durable, thick-walled cylindrical structures. Details of these designs have evolved with time and may be expected to continue to change as final plans are made for the repository. Figure 12.3 shows the main features of the waste package for PWR assemblies.

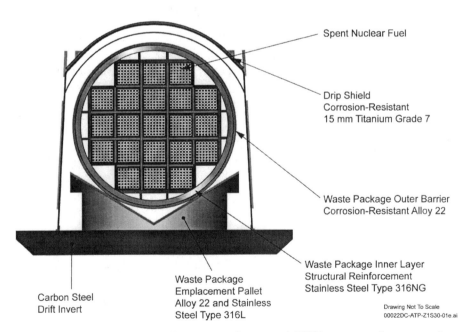

Fig. 12.3. Cross-sectional illustration of proposed PWR waste package, together with drip shield (above) and support structure (below). (From Ref. [16, Figure 3-1].)

Table 12.1. Contents and dimensions of typical waste packages for PWR and BWR assemblies.

Property	BWR	PWR
Contents of waste packages		
Typical number of assemblies per package	44	21
Average spent fuel mass per assembly (MTHM)	0.177	0.43
Average spent fuel mass per package (MTHM)	7.8	9.0
Mass of empty waste package (tonnes)	28	26
Mass of loaded waste package (tonnes)	42.5	42.3
Dimensions of waste packages		
Outer length (m)	5.2	5.2
Outer diameter (m)	1.7	1.6
Number of waste packages		
Total spent fuel, approximate (MTHM)	22,400	40,600
Number of waste packages, approximate	2,900	4,600

Note: Some small discrepancies between these data and those of Table 10.2 result from differences in the sources used in the cited references.
Source: Ref. [15, pp. 3–15].

Slightly different waste packages are being considered for the various types of wastes (i.e., spent fuel from PWRs and BWRs and defense high-level wastes). Some details of the PWR and BWR waste packages, as of 2002 when the DOE made its site recommendation, are given in Table 12.1. On average, about 8 or 9 MTHM are contained per waste package and about 7500 packages are needed for the planned 63,000 tonnes of civilian spent fuel. As seen from Table 12.1, they are large structures, each with a mass of about 42 tonnes—mostly the metal of the container and support structures.

The main components of the planned waste package include the following [16, Section 3]:[10]

♦ An outer cylinder of a corrosion resistant alloy, known as Alloy 22, made primarily of nickel, molybdenum, and chromium. The planned thickness is 2.0–3.0 cm. This layer provides the fundamental protection against corrosion. Alloy 22 was selected for its low rate of corrosion over a wide range of temperatures and humidities [16, pp. 3–21].

♦ A stainless-steel inner cylinder that provides structural strength. The indicated wall thickness is 5–6 cm.

♦ A support structure for the spent fuel assemblies made with a grid of steel plates arranged to provide (in the PWR case) 21 "slots" for the assemblies. These plates may contain boron to absorb neutrons and prevent criticality.

♦ Control rods of boron carbide, as an option to prevent criticality in lieu of borating the plates of the support structure.

---

[10] Quoted values for specific dimensions vary slightly even within a single section of Ref. [16], presumably reflecting continued changes in the designs being considered.

◆ Helium gas to fill the package before it is sealed. The helium helps to conduct heat from the fuel rods to the outer cylinders and—being an inert gas—does not cause oxidation or other degradation of the waste package materials.

◆ Lids of Alloy 22 and steel, at each end of the cask, that are welded to the inner and outer cylinders. These provide a tight seal for the enclosed helium gas.

Drip Shield

The waste packages are to be protected against the dripping of water from the walls of the drift by a "drip shield" which covers the top and sides of the waste package. The protective surfaces of the drip shield are to be made from 1.5-cm-thick sheets of a titanium alloy chosen for corrosion resistance (Grade 7) [16, p. 2-153]. In addition, the drip shield provides protection against rocks that may fall from the walls of the drift. The drip shield will not be installed initially, but is to be put in place before the repository is closed.

Invert

The cask sits on a support structure termed the "invert." Should the walls of the cask be breached, any water carrying radionuclides would drip downward to the invert. Some plans call for filling the gaps around the base of the invert with "ballast" material (e.g., crushed tuff) that would adsorb the radionuclides and impede their flow into the rock formations below the drifts [16, p. 2-151].

Multiple Barriers in the Engineered System

Overall, the engineered barriers include the drip shield, the Alloy 22 and steel walls of the waste package, the zircaloy cladding of the fuel rods, the solid form of the fuel pellets, and the ballast material in the invert region. None of these is counted on to suffice as the sole protection—although in some models the Alloy 22 cylinder will hold its integrity for well over 10,000 years—but together they are predicted to delay the release of radionuclides for very long time periods.

Evolutionary Character of the Design

It should be emphasized that the design of the waste package and remainder of the engineering barriers, as described earlier, represents a "snap-shot" of DOE thinking as of late 2001 and early 2002, at the time that the formal site recommendation was being made. This thinking reflected the results of evolving investigations and analyses that had been carried out for more than a decade, but the design details are not to be considered final. One may expect further changes both before and after the repository is constructed. However, this design is an important milestone. It establishes a specific option to examine in further analyses and critiques—by the DOE, NRC, NWTRB, and others.

12.3.4 Natural Barriers

Study of Geological Properties of the Site

A first step in the study of a potential repository, such as the Yucca Mountain repository, is *site characterization*, the purpose of which is to evaluate the suitability of the site in terms of its geological properties.[11] It requires study of the properties of the rock and the characteristics of the climate and rock geology that will determine the amounts and chemical composition of water moving through the region.

The movement of water into the repository and radionuclides away from the repository depends on the physical and chemical properties of the surrounding rock (tuff), including the number and size of fractures that provide paths for relatively rapid flow. These and related matters are analyzed by observation, experimental tests, and theoretical modeling. The investigations have included extensive on-site studies at Yucca Mountain, including the extraction of rock cores from the region of the repository and the excavation of exploratory tunnels in the rock.

Another aspect of site characterization is the consideration of the likelihood and potential consequences of violent disruptive events—in particular, volcanoes and earthquakes.

Seepage of Water into Repository Drifts

One of the initial attractions of the Yucca Mountain site was its dry environment [23]. The present average annual precipitation in the Yucca Mountain region is 19 cm/yr [24, p. I-14]. Almost all of this water returns to the atmosphere by evaporation or transpiration by plants, before it penetrates deep into the ground. The deep penetration is termed the *net infiltration* and the subsequent downward motion of the water is termed *percolation*. The fraction of the precipitation that results in net infiltration varies with humidity and precipitation rate, but, overall, the net infiltration now averages about 0.5 cm/yr.

Water passes through three tuff layers in moving from the surface to the repository: the welded Tiva Canyon unit, the unwelded Paintbrush unit, and the welded Topopah Spring unit in which the repository is located. The average percolation flux at the level of the repository is roughly equal to the net

---

[11] Site characterization is mandated and broadly defined in the Nuclear Waste Policy Act of 1982 [10].

infiltration rate, with some local differences due to lateral motion of the water [16, p. 4-68]. There is less time variation in the percolation flux than there is in the precipitation due to averaging effects, particularly in the unwelded tuff of the Paintbrush unit.[12] However, local departures from the spatial average can be caused by "focused flow" in fractures, carrying more water to some drifts than to others.

Actual seepage of the percolation flux into the drifts is reduced by capillary action in the rock, which holds some of the water in the rock matrix and diverts it around the drift cavity. Thus, seepage into the drift is less than the percolation flux [26, p. 187]. This sort of capillary barrier has been suggested as a possible explanation of the durability of the paintings in caves in France and Spain that have survived for tens of thousands of years in a climate much wetter than that of Yucca Mountain [16, p. 4-59]. Even when the capillary barrier is not fully effective, some of the moisture that reaches the wall of the drift may be removed by evaporation. Overall, therefore, the flux of water falling onto the drip shield can be expected to be less than the percolation flux.

Capillary action is most effective in reducing the fraction of the percolation flux that gets into the drift when the percolation flux is low and the pores in the rock contain little water. In fact, there is probably a percolation flux threshold—not well established—below which no seepage occurs [25, p. 4-5]. Conversely, capillary action is less effective when the percolation flux is elevated. Therefore, seepage into the drifts may be concentrated at locations where the flux is particularly high due to focused flow.

The long-standing belief that water could move only very slowly through the unsaturated zone was challenged in 1996 by the reported discovery of excess chlorine-36 (^{36}Cl) in some regions well below ground level. ^{36}Cl has a half-life of 3.0×10^5 years and the only plausible origin of a ^{36}Cl excess is fallout from nuclear weapon tests.[13] This finding suggested that water has moved through fracture pathways more rapidly than had been previously assumed. However, there is serious question as to the validity of the ^{36}Cl observation, originally found by Los Alamos National Laboratory (LANL) investigators. Later studies, carried out by the U.S. Geological Survey (USGS) with the support of analytic facilities at the Lawrence Livermore National Laboratory (LLNL), did not find the ^{36}Cl excess [27]. In addition, the USGS authors point out that studies of other isotope and isotope ratios show no sign of a fast path

---

[12] Departures from the time average may be created by "episodic flow" that could occur if water is trapped by capillary action and then suddenly released when the head of water rises sufficiently. However, there is some doubt as to the likelihood of this effect in the Yucca Mountain context [25, pp. 4-5].

[13] Very small amounts of ^{36}Cl are produced by the interaction of cosmic rays with argon in the atmosphere, but the ratio of this natural ^{36}Cl to the more abundant isotopes of chlorine ($A = 35$ and $A = 37$) is well known. The excess, now present throughout the atmosphere but not in pre-1945 samples of chlorine, was produced in bomb tests by neutron capture in the abundant isotope ^{35}Cl.

for water percolation into the unsaturated zone. They suggest, as a possible explanation for the difference between their results and the LANL results, that the LANL equipment may have been contaminated with ^{36}Cl prior to the measurements.

By the time of the DOE's 2002 recommendation to proceed with the Yucca Mountain Project, there already were serious doubts as to the existence of the "bomb-pulse" ^{36}Cl. Nonetheless, in the interests of conservatism, the model used at the time treated the bomb-pulse ^{36}Cl as a real effect and incorporated its implications into the calculation model [15, p. 4-76]. Only a limited part of the repository was assumed to have these fast flow paths and, therefore, their inclusion and their possible future omission do not greatly change calculated repository conditions.

Future Climates

Consideration of precipitation and seepage cannot be limited to the present climate conditions, because repository performance has to be analyzed for 10,000 years and to some extent up to 1 million years. The DOE analysis of Yucca Mountain adopts climate models from the U.S. Geological Survey in which it is assumed the "modern climate" lasts for another 600 years, followed by a "monsoon" climate that continues until about the year 4000, and then by a "glacial transition" climate [24]. The mean annual precipitation for these climates is 19 cm, 30 cm, and 32 cm, respectively; the mean net infiltration is 0.5 cm, 1.2 cm, and 1.8 cm, respectively [16, p. 4-78]. Of course, assumptions about future climates and infiltration rates are speculative, but the DOE studies assume that future percolation rates will be higher than the present ones.

Overall, the present dry environment provides an important initial protection, especially because the fractional infiltration and seepage rates are both low when the precipitation is low. The barriers are expected to be less effective in the future, assuming that the climate becomes wetter.

Movement of Radionuclides to Saturated Region

Despite the low seepage rates and the protections provided by the engineered barriers, some waste packages will eventually be breached by percolating water, and radionuclides in the fuel will be dissolved out and carried into the underlying rock. This also is an unsaturated region and the subsequent movement of the radionuclides will depend on the supply of water. The predominant flow path is downward and the radionuclides must be carried several hundred meters to the saturated zone below (i.e., to the water table). Most of the host rock in this region is unwelded tuff.

The travel time in the unwelded tuff is much slower for many radionuclides than for water, due to sorption in the matrix.[14] There is great variation among radionuclides with respect to sorption. In particular, isotopes of technetium, iodine, and chlorine are not at all retarded, isotopes of uranium and neptunium are retarded substantially, and isotopes of plutonium and americium are greatly retarded (see Section 11.2.3).

Transport of Radionuclides in Saturated Region

Radionuclides that reach the water table below the repository have to migrate about 20 km to reach the point where they are likely to be taken up into the biosphere [16, p. 4-310]. The first part of this path is in fractured tuff and the latter part is in alluvium.[15] The speed at which the radionuclides are transported by the flowing water is reduced by sorption of the radionuclides in the tuff and in the alluvium, as well as by migration back and forth between the fractures and the rock matrix. It can be accelerated if the radionuclides adhere to small particles (i.e, *colloids*), that move through the water. There is some evidence, for example, that the movement of plutonium at the Nevada Test Site has been accelerated by its attachment to colloids [28].

12.3.5 The Thermal Loading of the Repository

The importance of thermal loading strategy for a geologic repository was discussed in Section 11.2.4. Yucca Mountain planning during most of the 1990s was based on the assumption that the repository would operate in an above-boiling mode. However, more recently, increased consideration has been given to the possibility of operating at lower temperatures.

Background on thermal issues is provided in *White Paper: Thermal Operating Modes*, a study document prepared for the DOE and issued in February 2002. Parameters for two very similar high-temperature operating modes (HTOM) and eight low-temperature operating modes (LTOM) were presented, providing an overview of the options [29, p. 4]. In the HTOM modes, the average maximum temperature at the surface of the waste packages reaches about 160°C, while for the eight LTOMs, with one exception, the average maximum temperatures is below 85°C.

Low temperature can be achieved by some combination of longer ventilation (either forced, natural, or both), a higher airflow rate, and a lower thermal loading density. The latter can be achieved by spacing the waste packages fur-

---

[14] There is also some sorption on the walls of fractures and resulting retardation of radionuclide movement, but the *Science and Engineering Report* indicates that this "is not included in the TSPA transport evaluations because of limited data and the conservative nature of this approximation" [16, p. 4-286].

[15] Alluvium is defined in Ref. [16] as "sedimentary material (clay, mud, sand, silt, gravel) deposited by a stream or running water."

ther apart within the drifts, having a greater separation between the drifts, or by putting less fuel into each waste package. Depending on the options chosen, the repository footprints may increase from the 1150 acres (4.7 km^2) in the HTOM to as much as 2500 acres (10.1 km^2) in the LTOM, but most of the indicated LTOM scenarios have areas in the rough neighborhood of 1600 acres (6.5 km^2).

The advantages of the LTOMs include the reduction in the uncertainties in the impact of heat on the rock structure and the avoidance of combinations of temperature and humidity that might promote corrosion of the waste package. The chief advantage of the HTOMs is that they would "limit the potential for contact between water and the waste packages for time periods of hundreds to thousands of years" [15, p. xlvi]. Other advantages include a smaller repository area, a modestly lower cost, and a shorter period over which forced (and, in some cases, natural) ventilation is needed. The calculated doses are about the same for the two modes, but the uncertainties in predicting the performance of the repository system are probably less for the LTOM.

No obviously clear advantage has been found by the DOE for one mode over the other, and it had made no final decision as of mid-2003 as to which to adopt. The NWTRB, on the other hand, has expressed serious doubts about the HTOM (see Section 12.5.1). There may be no urgency in reaching a decision because the density of thermal loading and the length of time that forced ventilation will be maintained can be changed even after the repository is opened. Even in the HTOM, it is planned to have forced ventilation for 50 years after emplacement of the wastes begins, and this time could be extended. The DOE intent has been to have "flexible" design, so that either mode could be adopted if a clear preference emerges.

As enunciated in a letter from Margaret Chu, the Director of the OCRWM, to the NWTRB on September 6, 2002, the DOE decided to base its license application (and TSPA-LA) on the HTOM, but the LTOM "will be carried forward with the objective of minimizing impacts on the overall schedule if this option is selected" [30, p. 75]. In the version of the HTOM that was envisaged, the above-boiling conditions would be maintained near the drifts, keeping water away from them, but the region midway between the drifts would be below boiling, providing a path for water to drain downward past the drifts.

12.4 Total System Performance Assessments

12.4.1 The TSPA Approach

The Purpose of the TSPAs

The ultimate goal of the repository investigations is to determine the radiation doses that potentially may be received by people in the vicinity of the repository. Studies of the properties of engineered and natural features of the repository provide the data that are needed to determine the range of possible doses. Toward this end, extensive data have been collected—both at the site and in the laboratory—and voluminous analyses have been been made of the "features, events, and processes" (collectively termed FEPs) that govern repository performance.

The results of the studies of individual aspects of the repository are used as inputs to a *total system performance assessment* (TSPA), that provides estimates of the future performance of the repository. A very specific requirement is to compare the performance of the repository to the EPA standard: a limit of 0.15 mSv/yr over the next 10,000 years (see Section 12.3.1 and Section 13.2.5). The TSPA is used to determine a probability distribution for the dose to the RMEI as a function of time. Compliance with the EPA's standard requires that the mean of this probability distribution lies below the regulatory limit.

The development of the TSPAs for Yucca Mountain is not a static or one-shot effort. One positive feature of the TSPA approach is the opportunity it provides for iterative improvements in the design of the repository system. There has been a long series of TSPAs, and, in principle, each TSPA uncovers gaps in knowledge and suggests directions for further study. It also may uncover weaknesses in the design and suggest improvements. Thus, the broad purpose of the succession of TSPAs has been not only to evaluate the safety of a particular repository plan but also to provide guidance for improved designs.

The Succession of the TSPAs

Before the Yucca Mountain site was designated, a variety of preliminary analyses were carried out that were more modest than a full-scale TSPA. For example, the EPA in 1985 published a study covering a number of possible sites as a prelude to setting standards for future high-level waste repositories [4]. After the designation of the Yucca Mountain site, more detailed performance assessments were undertaken, starting with an initial study, TSPA 1991, carried out by the Sandia National Laboratories [31]. Numerous further TSPAs followed, supported by reports on specific aspects of the project. These were carried out in collaboration with the DOE's national laboratories and private contractors.

A milestone in the continuing series of analyses was the 1998 Viability Assessment for Yucca Mountain [32]. This study concluded that "Yucca Mountain remains a promising site for a geologic repository," while indicating that possible design improvements were being evaluated [32, p. 36 of Overview]. Another key stage in the process was reached with the TSPAs that served as the basis for the DOE's decision in early 2002 to recommend Yucca Mountain as a repository site. These TSPAs provided the background for some of the main documents supporting the recommendation, including the *Yucca Mountain Science and Engineering Report* (SER), issued originally in May

2001 [16] and in revised form in February 2002 [15]. The revised SER makes particular reference to three TSPA versions, developed in succession in 2000 and 2001: (1) the TSPA-SR, which was used in the preliminary SER, (2) the Supplemental TSPA Model (henceforth referred to here as TSPA-Sup), and (3) the revised Supplemental TSPA model (henceforth referred to here as TSPA-Rev).

The presidential and congressional approval of moving ahead with Yucca Mountain set in motion the next phase of repository studies. These are expected to culminate in a new TSPA that is to be developed before the end of 2004 as a key part of DOE's application to the NRC for authorization to begin repository construction. Officially, this is designated as the Total System Performance Assessment for the License Application (TSPA-LA). It is unlikely that it will be the end of the series of analyses, because, inevitably, new information and understanding will be developed during the period that the NRC studies the application.

In parallel, the Electric Power Research Institute (EPRI) has carried out its own TSPAs, drawing on DOE's work for some (but not all) of the input data but using a different analysis.[16] The EPRI studies provide a detailed and largely independent overview and have been reported in a series of extensive reports, including ones in February and December of 2002 [25, 33].

The Probabilistic Nature of the DOE TSPAs

In implementing a TSPA, certain properties of the repository can be taken as given conditions. In particular, for a repository that will hold 63,000 tonnes of commercial fuel and 7000 tonnes of reprocessed defense wastes, there are only small uncertainties in the inventory of radionuclides because the history of the fuel assemblies, including their enrichment and burnup, is well known.

Other parameters can vary widely but are under the control of the system designers. These include thermal loading of the repository and the physical construction of the canisters. Still other parameters are physical quantities that depend on the geological and chemical properties of the site and the waste, over which one has little or no control and about which knowledge is often incomplete. These include, for example, the future rainfall in the region, the structure of the rock, the rate of motion of water, and the solubility and sorption coefficients for the various nuclides.

The DOE TSPAs are carried out in a probabilistic manner, using the so-called *Monte Carlo* method. The calculation is based on an overall picture, or model, of the significant processes that determine the performance of the repository. Ingredients of the model can be described in terms of parameters, for example those that describe the corrosion rates for the waste container

---

[16] The Electric Power Research Institute was founded in 1973 as an independent consortium to provide research information of interest to the electric utility industry. It is funded primarily by the industry.

or the movement of individual radionuclides through the relevant rock formations. Many of these parameters are not well known; each such parameter is assigned a distribution function that specifies the relative probabilities of different numerical values for the parameter. A single "realization" of the model is a calculation in which one particular set of parameter values is used. Different realizations employ different parameter sets, where the likelihood of choosing a particular parameter value is determined by the probability distribution for that parameter.

One result of an individual realization might be, for example, the dose from ^{237}Np as a function of time. Many such realizations are carried out, with the parameters each time randomly chosen on the basis of the distribution function weightings. A probability distribution for the ^{237}Np dose is obtained by combining the results of a large number of these individual realizations. The "average" dose is commonly represented by the mean of the probability distribution (i.e., the "mean dose"). The median of the distribution function provides an alternative reference point. The mean usually exceeds the median by a substantial amount in these calculations, because realizations that give high doses impact the mean more than do realizations that give low doses.[17]

The focus of the TSPA is on finding the mean dose due to all radionuclides combined. The calculations also determine the mean dose for each radionuclide individually, as well as the 5th, 50th (median), and 95th percentiles in the probability distributions. However, they are not used to establish a "worst-case" dose, because picking the least favorable (plausible) value for each parameter would result in a high dose with an exceedingly low probability of occurrence. It would, however, be of interest to calculate—in analogy with what was done for waste transportation—"a reasonably foreseeable maximum dose," but this has not as yet been done for Yucca Mountain (see Section 11.1.4).

Classes of Scenarios Considered

Three main classes of scenarios are addressed in the DOE documents:

* *The nominal scenario.* This is the basic reference scenario, "representing the most plausible evolution of the repository" in the absence of disruptive events [15, p. 4-8]. A variety of parameter choices are explored within the context of this scenario.
* *Disruptive scenarios.* Three particular types of disruptive events have received attention: volcanic activity, seismic activity, and human intrusion. Their potential implications will be considered in Section 12.4.3.
* *Excluded scenarios.* This category consists of scenarios that have been "screened" out of consideration as too unlikely to require full-scale TSPAs. Scenarios in this category include (1) a future rise in the water table, such

---

[17] For example, the mean of 1, 3, 10, 30, and 100 is 29, which considerably exceed the median, which is 10.

that the repository region would become submerged, and (2) a future criticality event in which a chain reaction develops in the spent fuel. Each was hypothesized by individual scientists as a possible threat to the repository and it became necessary for the DOE to investigate its plausibility. On the basis of its own investigations and outside inputs, the DOE has concluded that neither represents a significant danger, and it does not plan to pursue detailed further examination of them [15, p. 4-415ff].

12.4.2 The DOE Nominal Scenario

Components of the Model

The TSPA is based on the results of subsidiary models used to analyze the various features of the repository and examine the effectiveness of the various barriers. A listing of the model components therefore closely parallels the list of natural and engineering barriers. As described in the DOE's Science and Engineering Report, the main model components are as follows [15, p. 4-439]:

♦ *The flow of water in the unsaturated zone.* This embraces questions of climate, infiltration of water into the ground, movement of water to the repository region, and seepage into the drifts.

♦ *The environment in the vicinity of the repository.* The relevant environmental considerations include the temperature in the region surrounding the repository, temperature effects on the flow of water, and the chemical composition of the water seeping into the drifts.

♦ *The degradation of the waste packages, drip shields, fuel rod cladding, and fuel pellets.* These model components consider the effectiveness of the main engineered barriers.

♦ *Transport of radionuclides out of the drifts and in the unsaturated and saturated zones.* These components consider the movement of the various radionuclides through the invert ballast, downward to the saturated zone, and their subsequent flow and transport to the biosphere.

♦ *Biosphere behavior.* This component considers the transfer of radionuclides into the water and food supplies and the resulting radiation doses.

Most of these components require the development of models for systems that are only partially understood, and there are many uncertainties in the results obtained from these models. The exploration of the uncertainties is a significant aspect of the TSPA itself.

Pathways for Radiation Exposure

The calculated radiation doses all come from the contaminated water extracted from the aquifer. Part of the dose comes directly from drinking this water. Another part comes indirectly when this water is used for irrigation and the radionuclides are taken up by plants (and animals feeding on the plants) and enter the food chain. In the DOE calculations, these two inges-tion pathways account for most of the dose [34, p. 3-569]. Other pathways include inhalation of dust (from the contaminated soil) and direct gamma-ray radiation from the soil.

At one time, considerable attention was given to a gaseous release path in which ^{14}C in the form of carbon dioxide would move rapidly upward through the repository into the atmosphere.[18] The resulting individual radiation doses are very small, but the collective dose for a world population of 10 billion people over a period of 10,000 years created a potential violation of an earlier EPA standard. The issue is now generally recognized to be unimportant from a safety standpoint, but remains interesting as an instance of a conflict between common sense and regulatory self-entrapment (see Section 13.2.2).

Calculated Doses

The main results of the DOE's TSPAs are encapsulated in families of curves of the dose rate versus time for the entire set of realizations of the nominal scenario. The dose rates are calculated in these studies for up to 100,000 or 1,000,000 years, although the time period over which EPA standards apply is 10,000 years. Results of the TSPA calculations that were the basis for the site recommendation are plotted in Figure 12.4, which shows the calculated mean dose as a function of time over 1 million years. The TSPA-Rev calculation applies to the RMEI, while for the other three curves it applies to the average member of the critical group.

The dose depends on the concentration of radionuclides in the water, and this, in turn, is determined by the volume of water that carries the radionu-clides. The curves of Figure 12.4 are based on an assumed volume of 2000 acre-ft/yr. Subsequently, the EPA and NRC have mandated that an annual volume of 3000 acre-ft (4 million m^3) be used as the basis for calculations (see Sec-tion 12.3.1). Thus, the doses in Figure 12.4 should be reduced 33%.

For the first 10,000 years, the mean dose remains under 2×10^{-4} mrem/yr (2×10^{-6} mSv/yr) in all of the calculations. The difference between the TSPA-SR calculation, for which the mean dose in this time period is so small as to be off-scale, and the other three calculations is that a "more conservative approach" was taken for the latter three cases and it was assumed that several waste packages failed at an early time due to improper heat treatment [15, p. 4-463]. The largest dose found in any of the realizations in the TSPA-Rev calculation was less than 2×10^{-3} mrem/yr—about 0.01% of the EPA's limit of 15 mrem/yr [15, p. 4-461].

A major contributor to keeping the calculated doses low is the predicted performance of the drip shields, which are expected to remain intact for several tens of thousands of years [15, p. 4-464]. Degradation of the waste packages tends to lag drip shield degradation, although a few early failures may precede

---

[18] The ^{14}C is produced primarily by neutron bombardment of ^{14}N, which is an impurity in the fuel and cladding.

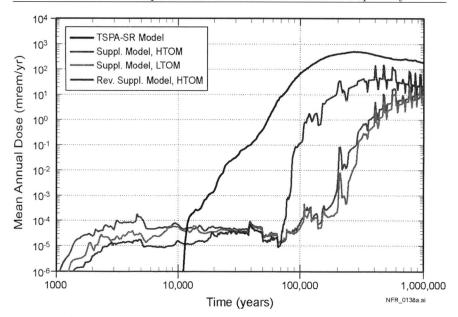

Fig. 12.4. DOE calculations of the mean annual dose over a 1 million-year period: the top curve (at 100,000 years) is for TSPA-SR, the bottom two curves are for TSPA-Sup (for both LTOM and HTOM cases, and the middle curve is for TSPA-Rev (HTOM case only). Note: These curves are based on an annual water "demand" of 2000 acre-ft; taking this demand to be 3000 acre-ft, as specified by the NRC and EPA, would reduce the plotted doses by 33% (see text). (Reproduced from Figure 4-180 of Ref. [15].)

the drip shield degradation due to manufacturing defects or corrosion in a humid atmosphere [16, p. 4-195 ff.].

After 10,000 years, the drip shields and waste packages eventually degrade, releasing radionuclides to the region below the repository. The slowness of the movement of many of the radionuclides from the repository to the biosphere creates additional delays in the occurrence of an appreciable dose. The doses rise more slowly in the TSPA-Sup and TSPA-Rev calculations than in the earlier TSPA-SR calculation, because, for the later calculations, the drip shields were assumed to remain intact longer and "more realistic" values were assumed for the solubility of some of the radionuclides, including plutonium and neptunium [15, p. 4-464].

At 100,000 years, the calculated mean dose is roughly 100 mrem/yr in TSPA-SR and 0.1 mrem/yr in TSPA-Rev. It continues to rise for several hundred thousand years more. Maxima are reached of about 500 mrem/yr after 250,000 years in TSPA-SR and of about 100 mrem/yr after 400,000 years in TSPA-Rev. The fluctuations in several of the curves of Figure 12.4, especially after 100,000 years, are due to assumed changes in precipitation.

Table 12.2. Fractional contribution of leading radionuclides to calculated mean dose (in percent), in the TSPA-SR nominal scenario.

Radionuclide	Half-life (years)	Sorption Properties[a]	Time after Closure (years) 25,000	50,000	75,000	100,000
^{99}Tc	2.11×10^5	L	65	66	12	3
^{129}I	1.57×10^7	L	26	16	3	1
^{237}Np	2.14×10^6	M	4	14	67	69
^{239}Pu	2.41×10^4	S	3	3	16	25
Other			2	1	2	2

[a]Sorption properties are based on Table 4-35 of Ref. [16]: L = low, M = moderate, S = strong.
Source: Figure 4-182, of Ref. [15].

The timing of these climate changes is not known and, therefore, the specific locations of the peaks in the fluctuation pattern have no significance, but the maxima and minima of these fluctuations suggest a rough envelope for the range of possible doses.

Table 12.2 gives the fractional contribution of various radionuclides to the mean dose at far future times after repository closure. It is seen that four radionuclides account for almost all of the calculated dose. In the first time period, the weakly sorbing fission products are dominant, but when ^{239}Pu and ^{237}Np have time to reach the biosphere, they dominate. The calculated doses from ^{237}Np, however, are probably overestimates (of more than a factor of 10)—if other aspects of the model are correct—because the DOE calculations have used old values for neptunium toxicity, which are much higher than newer ones adopted by the EPA and ICRP (see Section 4.5.3).

These results have all been for a high-temperature scenario. The curves of Figure 12.4 indicate that there is little difference in the doses for the HTOM and LTOM modes, over the full period from 1000 to 1,000,000 years.

Sensitivity Studies

In carrying out the above-described analyses, key parameters are characterized by probability distributions for their numerical values, as described in Section 12.4.1. Each realization of the calculation uses a value chosen randomly for each parameter (with a weight determined by its probability distribution). The sensitivity of the final results to the values chosen for a particular parameter can be explored by assigning a fixed extreme value to it and repeating the rest of the calculation as before.

Results of such calculations for the TSPA-SR analysis are presented in Ref. [15, p 4-496ff] for four cases in which individual key parameters are set at their 5th and 95th percentile values (one at a time) and the rest of the analysis is carried out unchanged. The selected parameters are (1) the stress responsible for stress corrosion cracking at the welds that seal the lids to the canister, (2) the general corrosion rate of Alloy 22, (3) the rate of water infiltration at the repository site, and (4) the seepage rate as determined by the extent of focused flow of water onto drifts. In none of these cases did the dose rate at 10,000 years reach 1 mrem/yr. The dose rate at 100,000 years was somewhat changed, but only for the case of an increase in the general corrosion rate did it rise as high as 200 mrem/yr (compared to 60 mrem/yr in the TSPA-SR base case) [15, p. 4-497]. Thus, the sensitivity of the final result to the selected parameters is not great enough for the qualitative results to be changed, even when the magnitudes of the parameters are significantly changed.

Multiple Barrier Analyses

The basis of a defense-in-depth strategy is to have multiple barriers, such that if some do not perform as expected, other barriers will suffice to prevent serious adverse consequences. In order to examine the robustness of the system in the face of the failure of some of the barriers, a wide variety of cases of partial barrier failure were considered for the TSPA-SR calculation [15, p. 4-523ff]. For the natural barriers, these included greater infiltration of water, more seepage into the drifts, and less effective barriers against radionuclide movement in the unsaturated and saturated zones.[19] For the engineered barriers, they included degradation of the drip shield, waste package, and cladding. The highest doses were found in the case of waste package degradation, simulated by setting parameters describing corrosion rates and quality of the welds at their 95th percentile (adverse) values. For this barrier failure, the 10,000-year dose rose to about 0.03 mrem/yr and the 100,000-year dose to about 800 mrem/yr [15, p. 4-530].

Overall, these results suggest that failures of individual barriers do not compromise the ability of the repository system to meet the EPA standards at 10,000 years, although, in some cases, they would significantly raise the 100,000-year dose. It may be noted that at 100,000 years, the dose for un-degraded conditions is much higher in the TSPA-SR calculation than in the later TSPA-Rev calculation, and the same is presumably true for degraded conditions (see Figure 12.4).

12.4.3 Disruptive Scenarios

The effectiveness of the various above-discussed barriers makes it difficult to find plausible ways—in the broad context of the nominal scenario—for there to be large releases of radionuclides to the biosphere during the first 10,000

---

[19] These calculations were reported in Ref. [15] for the TSPA-SR case, but it is plausible to assume that the same qualitative conclusions would follow for TSPA-Rev as well.

years. However, disruptive processes create possibilities for breaching all of the normally independent barriers in one event. The likelihood of such events is not great. As put into context in a report by the U.S. Geological Survey:

> The probabilities of occurrence for things like a repository-piercing volcanic eruption, a canister-breaking rockfall induced by earthquake ground motion, or an accidentally accumulated critical pile of exposed radionuclides have been calculated to be very low...

> Most Earth scientists will know that Yucca Mountain resides in the Basin and Range Province of the Western United States, generally considered to be a region of active tectonics. As such, there has been considerable concern in the Earth-science community about the seismic and volcanic hazards to which Yucca Mountain might be exposed. In fact, by the standards of active tectonics, Yucca Mountain and environs has been a surprisingly inactive place, at least for the past 500,000 years or so. Existing geologic, geomorphic, tectonic, paleoseismic, and volcanic signatures all point to very low rates of crustal deformation and landform modification in the vicinity of Yucca Mountain during this period. [35, p. 19]

The likelihood and effects of these disruptive events are considered in the next subsections. Extremely improbable events are not considered. As specified by the EPA in its standards for Yucca Mountain, the DOE "shall not include consideration of very unlikely features, events, or processes, i.e., those that are estimated to have less than one chance in 10,000 of occurring within 10,000 years of disposal" [5, § 197.36]. For events that occur randomly in time, this means the exclusion of events with estimated rates below 10^{-8}/yr.

Volcanic or Igneous Activity

Igneous activity, also referred to as volcanism, involves the transport of molten rock (magma) through the Earth's crust. The passage of magma through the Yucca Mountain repository could lead to the release of radionuclides in two ways [15, p. 4-476]:

◆ *Eruptive volcanism.* In an eruptive event, magma can pass through the repository region and rise to the Earth's surface. The hot magma would damage the waste packages, entrain radionuclides in the material that is emitted into the atmosphere, and fall back to the ground as ash. People in the vicinity are therefore immediately exposed to radiation.

◆ *Igneous intrusion.* Some or all of the magma passing through the repository may remain well below the surface of the ground. However, any waste packages that are traversed by it will be damaged, and radionuclides will become more available for transport to the biosphere by water. The required transport downward and then laterally through the saturated zone introduces substantial delays—as in the nominal case—and the potential exposures are delayed.

In these igneous events, the magma flows in long, narrow streams that are expected to destroy waste packages over a cross-sectional area that averages roughly 50 m [16, p. 4-452]. Several such flows will pass through the repository per event. The expected influx will be not massive enough to impact more than a small fraction of the waste packages. A TSPA that includes igneous events considers the number of waste packages impacted, the amounts of radionuclides swept up by the magma, and the movement of the radionuclides through the ground and (for eruptive events) in the atmosphere. On this basis, a *conditional dose* for eruptive events is calculated (i.e., the dose expected *if* a volcano occurs). For an event that can occur randomly in any year, the conditional dose becomes smaller as the radioactivity decreases with time. For the earliest time considered, stated as 100 years after closure of the repository, the mean conditional dose was calculated in the volcanic version of TSPA-SR to be approximately 13,000 mrem/yr [36, p. 3-48].[20]

However, the likelihood of such an event is small and the EPA standard applies not to the conditional dose, but to a probability weighted dose. The probability of a disruptive igneous event was estimated by an expert panel advising the DOE to be 1.6×10^{-8}/yr [15, p. 4-476], with volcanoes resulting 77% of the time. Thus, the estimated frequency of these events was just above the 10^{-8}/yr level below which they could be ignored. The contribution to the probability-weighted dose from a single year is the product of the conditional dose for that year and the event probability (e.g., 2×10^{-4} mrem/yr at 100 years after closure). The total probability-weighted dose in any year includes the contributions from all past years, because radionuclides deposited by a past volcanic event will remain in the environment for many decades. Thus, although the annual conditional dose drops with time, the probability-weighted dose may rise when contributions from past years are added.[21]

The probability-weighted dose for igneous events has been calculated in TSPA-SR for the first 50,000 years and in TSPA-Rev for the first 100,000 years. The two results differ in detail, but in neither case does the mean probability-weighted dose reach 0.2 mrem/yr [15, p. 4-484]. Eruptive events are the main source of dose during the first part of this time period and intrusive events dominate later. It is to be noted that during the first 10,000

---

[20] A calculation for a period "100 years after closure" has a clear meaning only if one knows when closure occurs. It may be guessed that this calculation presumed closure takes place roughly 75 years after waste disposal begins.

[21] Neither the conditional dose nor the probability-weighted dose includes contributions from the immediate inhalation of ash from the volcanic cloud. This could add significantly to the first-year dose in case of an eruption, but does not make any contribution to the dose in future years and, therefore, adds very little to the probability-weighted dose [36, p. 3-48].

years the maximum probability-weighted dose from igneous activity, which in TSPA-Rev is close to 0.1 mrem/yr for the first thousand years, far exceeds the dose in the nominal scenario without igneous activity (see Figure 12.4).

As the DOE was working in 2001 toward its conclusion to recommend the Yucca Mountain site, a disagreement between the DOE and NRC came to general public attention. The NRC estimated the igneous event probability to be between 10^{-8} and 10^{-7}/yr, suggesting that the probability-weighted dose may have been underestimated by the DOE. Although even in the NRC estimate the calculated dose fell below the EPA limit, this was a disturbing disagreement.

One step toward improved analyses was the formation by the DOE of the Yucca Mountain Igneous Consequences Peer Review Panel, which issued its final report in February 2003 [37]. This panel of experts was asked to "review the technical basis used to analyze the consequences of igneous events" [37, p. 1]. Its report concluded that the overall conceptual model adopted in the DOE's TSPA was "adequate and reasonable," but it recommended a more careful quantitative analysis of magma flow into drifts. It discussed the so-called "dog-leg" scenario, in which magma might flow through a drift and then vent to the surface through a fracture, damaging a large number of waste packages in the drift. However, it judged the probability of this event to be small and concluded that the additional risks implied by this scenario were "more than offset by the level of conservatism built into the existing estimates" [37, p. 77]. One of the TSPA assumptions that the panel thought to be too conservative was the assumption that the contents of damaged canisters would be widely dispersed into the environment.

The panel suggested further studies, including investigations of changes in waste package design and repository layout that would take into account the possibility of igneous intrusion. Overall, the report neither suggested that volcanic activity represented a major threat to the Yucca Mountain repository nor did it dismiss the matter as fully settled. Presumably, the issue will receive further attention during the preparation and evaluation of the DOE's license application to the NRC for authorization to construct the repository.

Seismic Activity

In principle, earthquakes could damage the repository by shaking the ground or rupturing faults. However, shaking is less at repository depths than at the surface. Several types of events were ruled out on the basis of their very low probability of occurrence: major changes in the local hydrology, dislodging and falling of large rocks in the drifts, and a large displacement of faults within the repository [15, p. 4-447]. The one seismic effect that was included was shaking sufficient to damage the fuel cladding, which was assumed to occur at a rate of 1.1×10^{-6}/yr. Rather than consider this part of a disruptive scenario, this effect was included in the TSPAs for the nominal scenario by adding it to the probability of fuel cladding failure. All in all, seismic effects appear to be unimportant for the Yucca Mountain repository.

Human Intrusion

There is little basis on which to estimate the likelihood and scale of human intrusion into the repository. The DOE human intrusion scenario, as mandated by the EPA in its standards for Yucca Mountain, assumes that the intrusion occurs accidentally, during drilling for water [5, §197.26]. A drill bit is assumed to penetrate a single waste package and the hole is continued down to the water table. Finally, the borehole is left unsealed, providing a path for radionuclides to reach the saturated zone.

Intrusions were studied that take place 100 years and 30,000 years after the repository is closed. The calculated maximum mean dose for the 100-year case is 0.005 mrem/yr in the TSPA-Rev calculation, and occurs after 875 years [15, p. 490]. However, it is unlikely that such an event would occur. Even ignoring the continued knowledge of the repository's location, inadvertent drilling into a waste package is unlikely at so early a time because of the difficulty of drilling into the titanium drip shield and the waste package with a drill bit designed for rock. The drillers would recognize that they had struck a "foreign" object. Thus, accidental human intrusion is unlikely during the first 30,000 years, before the drip shields and waste canisters have become so degraded that the drill operator will not recognize their presence. At later times, when inadvertent intrusion is deemed less unlikely, the activity of the wastes and potential doses are reduced.

12.4.4 EPRI's TSPA Calculations

Method of Calculation

The Electric Power Research Institute (EPRI) has been studying the Yucca Mountain repository in parallel with the DOE and has published its own TSPA results. These studies—as is also the case for the DOE studies—are a continuing sequence of analyses that reflect changes in Yucca Mountain design planning and in the understanding of key processes. Relatively recent studies include Phase 3 in 1996, Phase 4 in 1998, Phase 5 in 2000, and Phases 6 and 7 in 2002.

The broad outlines of the EPRI calculations are similar to those of the DOE, in that the same processes must be taken into account. However, rather than select parameter values with a probabilistic Monte Carlo method, as done by the DOE, the EPRI calculations use a logic tree, characterized by *nodes*, where choices are made for the values to assign to important parameters whose actual values are not well established, and by *branches* coming off each node that represent the choices made. EPRI's Phase 7 TSPA has a sequence of five nodes corresponding to five aspects of repository performance: infiltration rate (three branches), amount of focused flow (two branches), seepage fraction (two branches), radionuclide solubility and waste form degradation (three branches), and retardation in radionuclide flow (three branches) [33, p. 2.2].

At a node, each branch is assigned a value of the pertinent parameter and the probability that the branch will be taken. For example, at the infiltration node, probabilities of 5%, 90%, and 5% were assigned to "low," "moderate," and "high" infiltration rates, respectively, with net infiltration rates for the three cases of 1.1, 11.3, and 19.2 mm/yr during an assumed 1000-year "greenhouse" period. Together, these branches and nodes define 108 unequally weighted sets of assumptions—for each of which a curve of dose versus time is calculated. From these, weighted average dose is determined as a function of time.

EPRI Results and Comparisons to DOE Results

In the evolution of EPRI studies, the Phase 7 TSPA, published in December 2002, reported doses that were considerably higher than those of the previous EPRI studies, in particular Phase 6. The main changes going from Phase 6 to Phase 7 were a correction in the radionuclide release from regions impacted by focused flow, the addition of a branch with seepage into a higher fraction of the repository, and a more extensive consideration of diffusive release of radionuclides from the waste containers [33, 38].[22]

Earlier differences between the EPRI and DOE results were substantially reduced by these changes as well by changes in the DOE results (from TSPA-SR to TSPA-Rev). For the crucial initial 10,000-year period, both calculations give miniscule doses: under about 10^{-5} mrem/yr (10^{-7} mSv/yr) in the DOE's TSPA-Rev calculation (normalized to a flow of 3000 acre-ft/yr) and under about 2×10^{-4} mrem/yr (2×10^{-6} mSv/yr) in the EPRI Phase 7 calculation [33, p. 3-10].[23] The higher dose in the EPRI calculation resulted from the assumption that one container and 2.5% of the fuel rod cladding was defective when initially placed in the repository. Given the arbitrary nature of such an assumption and the small magnitude of both results, it is neither interesting nor meaningful to try to determine the reason for differences between the DOE and EPRI results at 10,000 years. (It is important to know if there are plausible ways in which both estimates could be grossly in error.)

---

[22] Some descriptions of the transport of radionuclides from failed containers to the ground below assume that the radionuclides are carried only by flowing water (*advective release*). However, if there is a connected film of water, radionuclides can be carried from a region of high concentration to a region of lower concentration by diffusion, even for nonflowing water (*diffusive release*).

[23] The doses reported here are calculated for a plume size of 3000 acre-ft/yr, as mandated by the EPA and NRC.

At 1 million years, the DOE TSPA-Rev calculation (again normalized to 3000 acre-ft/yr) gives a dose that is about 20 times that for the EPRI calculation (about 20 mrem/yr versus 1 mrem/yr). John Kessler of EPRI attributes this remaining difference between the DOE and EPRI partly to EPRI's assumptions of longer travel times for the radionuclides in the saturated zone and better performance of the cladding and container [38]. In addition, although not specifically commented on by Kessler, the DOE calculation assumes a response–dose coefficient for ^{237}Np that is about 13 times as great as the more up-to-date value used by EPRI (see Table 4.4).

The EPRI analysis particularly investigated the multiple-barrier aspect of the repository performance. In one part of this analysis, EPRI considered the hypothetical case in which no credit is given for the engineered barriers (i.e., no drip shield, no waste container, and no effective fuel rod cladding). Even in this case, the dose was within the 15 mrem/yr limit for the first 10,000 years [33, Figure 4-10].[24] During the first million years, the dose approached 50 mrem/yr for only a brief period at 100,000 years. The results are much less favorable if reliance is placed on the engineered barriers alone, presumably because with no natural barriers the large flow of water on the drip shields and waste containers would cause many of the fuel rods to be breached at an early time, with no subsequent delay in the transfer of radionuclides to the biosphere.

12.5 Resolving Questions About the Repository Performance

12.5.1 Evaluations of Yucca Mountain Analyses

The Need for Continued Studies

The decisions taken by the President and Congress to proceed with Yucca Mountain did not end the study phase of the project. Although the assessments carried out by the DOE and EPRI indicate that the doses for the Yucca Mountain repository will be far below any level of concern for the first 10,000 years and will not be very high at later times, some technical experts are not satisfied that these assessments are well enough grounded to provide a firm basis for confidence in the repository's performance (e.g., Ref. [39]). In general, they do not argue that the repository will be unsafe. Rather, they suggest that we still cannot be adequately confident that it will be safe. Part of this is a matter of perception: How sure is sure enough? Beyond that, however, are specific technical concerns related to the difficulty of predicting perfor-

---

[24] The reported dose at 10,000 years was 5 mrem/yr for the drinking water path only, with a plume size of 750 acre-ft/yr. The drinking water path represents about 10% of the total dose in the EPRI calculations. Thus, normalized to 3000 acre-ft/yr, the reported dose of 5 mrem/yr should be multiplied by 2.5.

mance of a complex system far into the future. (Some of the specific issues are discussed in Section 12.5.3.)

There is an interval of roughly 2.5 years between the congressional approval in 2002 and the scheduled submission of a license application to the NRC at the end of 2004. The DOE is using this time to make intensive further studies, including continuing investigations of the repository geology, the behavior of the engineered barriers, and chemical properties of the radionuclides. This will culminate in a revised TSPA (the TSPA-LA) and possibly changes in aspects of the repository design. As this work proceeds, there will be further inputs from the NRC, NWTRB, EPRI, special panels established by the DOE and, perhaps, by Congress, and innumerable other groups and individuals.

Part of the approach to improving both the repository design and the analyses of its performance has been to have reviews of the DOE's work by outside, independent bodies. The EPRI studies represent one form of independent analysis of the overall Yucca Mountain issue. Comments from groups that specifically focus on the DOE's approach to the problem are presented in the next subsections.

Comments from the Nuclear Waste Technical Review Board

The NWTRB has been making periodic reports on the progress of the Yucca Mountain repository since 1990. The function of the NWTRB is to assess the technical status of the project and make suggestions for improvement. A report issued in early 2002 expressed a number of reservations about the confidence that could be placed in then-existing assessments:

> When the DOE's technical and scientific work is taken as a whole, the Board's view is that the technical basis for the DOE's repository performance estimates is weak to moderate at this time...

> Gaps in data and basic understanding cause important uncertainties in the concepts and assumptions on which the DOE's performance estimates are now based. Because of these uncertainties, the Board has limited confidence in current performance estimates generated by the DOE's performance assessment model.

> The limited confidence is not an assessment of the Board's level of confidence in the Yucca Mountain site. At this point, no individual technical or scientific factor has been identified that would automatically eliminate Yucca Mountain from consideration as the site of a permanent repository.... [40, p. 1]

In short, the NWTRB's view in early 2002 was that although the repository *may* be quite satisfactory, the DOE analysis had not fully established the case.

The Board reiterated its oft-expressed view that there could be greater confidence if the repository operated in the low-temperature mode and recommended that a complete comparison be made between the LTOM and

HTOM (see Section 11.2.4 and Section 12.3.4.). It expressed particular concern about "uncertainties related to the performance of waste package materials under high-temperature conditions."[25] The NWTRB's continuing concerns about the HTOM were explained in considerable detail in a report issued in November 2003 [41]. An accompanying letter expressed the Board's belief that with the high-temperature design "widespread corrosion of the waste packages is likely to be initiated" [42]. Margaret Chu (Director of the OCRWM in the DOE) responded that the DOE would carry out an extensive review of the NWTRB report, but that its "analyses do not suggest such results." She looked forward to continued dialogue between the DOE and the NWTRB [42].

Comments From Peer Review Panels

At the request of the DOE, the Yucca Mountain TSPA program—more specifically, the TSPA-SR—was reviewed by a 10-member international review team (IRT) in a study that was organized by the Nuclear Energy Agency of the OECD and by the International Atomic Energy Agency and carried out in the second half of 2001 [43]. The report made a large number of suggestions for improving the detailed information that goes into the TSPA as well as for making its results more transparent. However, its overall conclusion was positive:

> While presenting room for improvement, the TSPA-SR methodology is soundly based and has been implemented in a competent manner. Moreover, the modelling incorporates many conservatisms, including the extent to which water is able to contact the waste packages, the performance of engineered barriers and retardation provided by the geosphere.

> Overall, the IRT considers that the implemented performance assessment approach provides an adequate basis for supporting a statement on likely compliance within the regulatory period of 10000 years and, accordingly, for the site recommendation decision. [43, p. 63]

For the future, the IRT recommended that more attention be paid to understanding the repository system, as distinct from the present emphasis on demonstrating that it satisfies regulatory requirements over a 10,000-year period. This entails two aspects [43, p. 59]:

♦ An analysis should be made using "realistic" parameters, based on best estimates, "regardless of whether this can be demonstrated with reasonable assurance." In short, they suggest using the best numbers, not those that can most readily withstand criticism.

---

[25] For example, in response of January 24, 2002 to questions from Senators Reid and Ensign [30, Appendix F].

♦ The TSPA approach should be modified to provide disaggregated results, so that the reader would see not only final radiation dose numbers but also intermediate results that indicate the performance of each barrier.

The report also called for additional laboratory data and study of the analytic models used, in order to reduce uncertainties in the results.

A review of the more specific topic of waste package materials was carried out by an expert Waste Package Materials Performance Peer Review Panel, which was established at the request of the DOE in 2001. The Panel, as expressed in its Executive Summary, indicated that "the current waste package design is likely to meet the performance criteria for the repository, if some technical issues are favorably resolved" [44, p. 1]. Although for most conditions the Panel concluded that Alloy 22 has "excellent corrosion resistance," it identified several types of possible corrosion under unfavorable environmental conditions. It also raised a warning flag about titanium as a drip shield. Overall, it indicated a need for further analysis and testing on a variety of specific potential corrosion problems. The Panel also addressed the choice between different temperature modes for the repository, but expressed no clear preference.

The significance of these peer review panels is not in any up-down verdict on the repository. No such clear verdicts were expected or were presented. However, the reports are valuable in providing the DOE, and advisory bodies such as the NWTRB, with advice and information that can point the way to improved designs and improved understanding of the repository performance.

12.5.2 Continuing Technical Issues

Technical issues for further study include, but are not limited to the following:

♦ *Thermal loading strategy.* Evaluation of the relative merits of the high- and low-temperature operating modes remains an important part of the DOE agenda. Calculated doses are about the same for the two modes. The HTOM may have some advantages in terms of overall economy and the LTOM in terms of less uncertainty about waste package corrosion.

♦ *Durability of the waste package.* There is no direct experience on which to base predictions about the corrosion resistance of the waste package components over times of 10,000 years or longer. The problem is particularly complicated by uncertainties in the behavior of these components over a wide range in temperatures and in the chemical composition of the water that may reach the waste packages.[26] Extrapolations must be made

---

[26] It may be noted that the uncertainties extend in both directions; there have been some suggestions that the engineered barriers are overdesigned as well as suggestions of possible deficiencies.

from tests performed over much shorter periods and from the behavior of natural analogs. Waste package studies are likely to continue for many years because even if the present designs are satisfactory, better ones are always in principle possible (in terms of safety or cost) and there is no compelling reason to use the same versions over the full period of waste emplacement.

♦ *Motion of water into the repository.* Understanding repository performance depends in part on understanding the motion of water in the overlying unsaturated zone. One uncertainty in this understanding would be reduced if the reality of the ^{36}Cl bomb pulse could be definitely established—one way or the other (see Section 12.3.4).

♦ *Movement of radionuclides in saturated zone.* The movement of escaping radionuclides from the repository to the biosphere depends on the solubility of the elements (or compounds) involved, the sorptive properties of the rock, and the possible role of fractures and colloids in speeding the radionuclide motion. Improving the necessary chemical and geological understanding has been an ongoing process, but critics maintain that the knowledge is still inadequate (see, e.g., Ref. [39]).

♦ *Volcanic effects.* Volcanic activity represents a possible mechanism for large release of radionuclides. The NRC and DOE both believe that the likelihood of damaging events is small but have in the past disagreed on the absolute likelihoods. Efforts to establish the probabilities and consequences will continue.

12.5.3 Further Institutional Measures

Interactions with the NRC

The DOE requires NRC approval in order to obtain a license authorizing it to begin construction of the Yucca Mountain repository and, later, another license authorizing it to open the repository for the receipt of wastes. In this process, Congress has called for extensive interactions between the two agencies, rather than an arm's-length relationship. Thus, there was a long period of consultations between the NRC and DOE prior to the DOE's decision in 2002 to seek construction authorization, and these interactions will continue as the license application is prepared by the DOE and later analyzed by the NRC.

This preliminary examination of the DOE's evolving plans by the NRC staff and advisory groups has led to questions that have been formalized by the definition of a large array of Key Technical Issues (KTIs). The KTIs are to be resolved as the process moves forward, with Agreements as to the scope of further work required of the DOE.[27] This extensive NRC involvement in the studies of the repository provides a continuing check on the DOE's analyses.

Science and Technology Program

The DOE established a Science and Technology (S&T) Program in 2002 to consider some of the scientific issues, with a mandate to take a longer-term perspective than has been the case in many of the past Yucca Mountain studies. As described by Robert Budnitz, a leader in the S&T program, the goal is not to obtain results that will influence the DOE's license application to the NRC in 2004 or even the waste packages first used in 2010.[28] However, it is likely to influence the waste packages in later years. Making an analogy to the evolution of airplanes from the 707 in the 1960s to the newer models today, he suggests that improved waste packages will be utilized as time progresses, although the DOE's initial license application will be couched in terms of a specific design.

Budnitz also envisages other future repository improvements. In another analogy, with nuclear reactors this time, he says:

> Take a reactor, for example like Diablo Canyon in California where I've been the past 35 years, it's been there for 15 or 20 years, and the design is 30 years old. There are amendments every week, every month, that say we want to do this a little better, and they're approved because they're efficient or they're better or they're safer, or whatever. I'm sure that's going to go on here, too, but we're not contemplating any specifics because who knows what's going to come.

This can be viewed as an informal expression of an aspect of the strategy now being put forth as "adaptive staging." In particular, it represents a recognition that it is neither necessary nor desirable to decide everything about the repository in advance. Adaptive staging is discussed further in Section 13.5.

12.5.4 Overview of Yucca Mountain Prospects

As the process of studying Yucca Mountain continues during the period that the license application is being prepared by the DOE and analyzed by the NRC, it is probable that details of the scientific understanding will be im-

---

[27] The interactions between the DOE and NRC have been discussed in detail by EPRI in Ref. [33], where it is argued that some KTIs are not significant as far as risks are concerned and that their resolution should not be a prerequisite for licensing [33, Chapter 5].

[28] These remarks of Budnitz are contained in testimony at the NWTRB meeting in May 2003 [45].

proved and the design may be modified. However, the thrust of the DOE's eventual conclusions are predictable, because it is highly unlikely that there will be any changes large enough to reverse the DOE's favorable evaluation of Yucca Mountain.

During this time, viewpoints contesting the DOE position may attract more public attention than those supporting it. Serious criticisms require careful consideration, because they may uncover difficulties and lead to improvements in the repository design or—in the extreme case—demonstrate that it is imprudent to proceed with the Yucca Mountain repository, at least as now envisaged. However, it is relevant to note that the repository performance, as presently assessed by the DOE and EPRI, far exceeds the EPA requirements and that the repository safety is based on multiple barriers. Weaknesses in one feature, should they emerge, need not compromise the overall system. Thus, moderate degradation of performance could still leave a "safe" repository, as measured against EPA's standards. Whether these standards are appropriate or too strict (or too lax) is another matter—as discussed in Chapter 13—but they provide the only widely accepted reference point for judging the repository's overall acceptability.

A favorable DOE conclusion on this matter does not guarantee NRC approval, nor is it possible to exclude further intervention by Congress or the courts that would slow or terminate the project. These considerations create pressures on the DOE to establish a case for Yucca Mountain safety that will stand up under intense and often sceptical scrutiny. However, no matter how carefully and expertly these analyses are performed, uncertainties will remain in predicting the behavior of the complex repository system over many thousands of years. Perspectives relevant to establishing policies in the face of such uncertainties are considered in the following chapter.

References

1. National Research Council, *Disposal of Solid Radioactive Wastes in Bedded Salt Deposits*, Report of the Committee on Radioactive Waste Management (Washington, DC: National Academy of Sciences, 1970).
2. Luther J. Carter, *Nuclear Imperatives and Public Trust: Dealing with Radioactive Waste* (Washington, DC: Resources for the Future, 1987).
3. Paul P. Craig, "High-Level Nuclear Waste: The Status of Yucca Mountain," *Annual Review of Energy and the Environment* 24, 1999: 461-486.
4. U.S. Environmental Protection Agency, *High-Level and Transuranic Radioactive Wastes, Background Information Document for Final Rule*, Report EPA 520/1-85-023 (Washington, DC: U.S. EPA, 1985).
5. *Protection of the Environment, U.S. Code of Federal Regulations*, Title 40 (2001).
6. *Energy, U.S. Code of Federal Regulations*, Title 10, Parts 51 to 199 (1993).
7. 102d Congress, *Energy Policy Act of 1992*, Public Law 102-486, Title VIII–High Level Radioactive Waste (October 24, 1992).

8. U.S. Nuclear Waste Technical Review Board, *First Report to the U.S. Congress and the U.S. Secretary of Energy* (Washington, DC: NWTRB, 1990).

9. "Late News," *Nuclear News* 45, no. 13, December 2002: 64.

10. 97th Congress, *Nuclear Waste Policy Act of 1982*, Public Law 97-425 [H.R. 3809] (January 7, 1983).

11. U.S. Department of Energy, Office of Civilian Waste Management, "Congress Amends Nuclear Waste Policy Act of 1982," *OCRWM Bulletin*, Report DOE-RW-0153 (December 1987/January 1988): 1–3.

12. U.S. Department of Energy, Office of Civilian Waste Management, "DOE Submits Report to Congress on Reassessment of the Civilian Radioactive Waste Management Program," *OCRWM Bulletin*, Report DOE-RW-0227 (November/December 1989): 1–3.

13. U.S. Department of Energy, Office of Civilian Waste Management, "Secretary of Energy Approves Project Decision Schedule," *OCRWM Bulletin*, Report DOE/RW-0317P (July/August 1991): 4–9.

14. U.S. Department of Energy, Office of Civilian Radioactive Waste Management, *Yucca Mountain Preliminary Site Suitability Evaluation*, Report DOE/RW-0540 (North Las Vegas, NV: U.S. DOE, 2001).

15. U.S. Department of Energy, Office of Civilian Radioactive Waste Management, *Yucca Mountain Science and Engineering Report Rev. 1, Technical Information Supporting Site Recommendation Consideration*, Report DOE/RW-0539-1 (North Las Vegas, NV: U.S. DOE, 2002).

16. U.S. Department of Energy, Office of Civilian Radioactive Waste Management, *Yucca Mountain Science and Engineering Report, Technical Information Supporting Site Recommendation Consideration*, Report DOE/RW-0539 (North Las Vegas, NV: U.S. DOE, 2001).

17. U.S. Department of Energy, Office of Civilian Radioactive Waste Management, *Final Environmental Impact Statement for a Geologic Repository for the Disposal of Spent Nuclear Fuel and High-Level Radioactive Waste at Yucca Mountain, Nye County, Nevada*, Report DOE/EIS-0250 (North Las Vegas, NV: U.S. DOE, 2002).

18. M. L Wilson, et al., *Total-System Performance Assessment for Yucca Mountain—SNL Second Iteration (TSPA-93)*, Report SAND93-2675 (Albuquerque, NM: Sandia National Laboratories, 1994).

19. U.S. Department of Energy, Office of Civilian Radioactive Waste Management, *Nuclear Waste Fund Fee Adequacy: An Assessment*, Report DOE/RW-0534 (Washington, DC: U.S. DOE, 2001).

20. U.S. Nuclear Waste Technical Review Board, *NWTRB Special Report to Congress and the Secretary of Energy* (Washington, DC: NWTRB, 1993).

21. U.S. Nuclear Regulatory Commission, "10 CFR Parts 2, 19, 20, 21, etc., Disposal of High-Level Radioactive Wastes in a Proposed Geologic Repository at Yucca Mountain, Nevada; Final Rule," *Federal Register* 66, no. 213, 2001: 55732–55816.

22. U.S. Environmental Protection Agency, "40 CFR Part 197, Public Health and Environmental Protection Standards for Yucca Mountain, NV; Final Rule," *Federal Register* 66, no. 114, 2001: 32074–32135.

23. U.S. Department of Energy, *Environmental Assessment Overview, Yucca Mountain Site, Nevada Research and Development Area, Nevada*, Report DOE/RW-0079 (Washington, DC: U.S. DOE, 1986).

24. U.S. Geological Survey, *Simulation of Net Infiltration for Modern and Potential Future Climates*, Report ANL-NBS-HS-000032 REV 00 (Denver, CO: USGS, 2000).

25. Electric Power Research Institute, *Evaluation of the Proposed High-Level Radioactive Waste Repository at Yucca Mountain Using Total System Performance Assessment, Phase 6*, Report 1003031, J. Kessler, Project Manager (Palo Alto, CA: EPRI, 2002).

26. G. S. Bodvarsson, et al., *Unsaturated Zone Flow and Transport Process Model Report*, Report TDR-NBS-HS-000002 REV00 (Las Vegas, NV: TRW Environmental Safety Systems, Inc., 2000).

27. J. B. Paces, Z. E. Peterman, L. A. Neymark, G. J. Nimz, M. Gascoyne, and B. D. Marshall, "Summary of Chlorine-36 Validation Studies at Yucca Mountain, Nevada," Proceedings of the 10th International High-Level Radioactive Waste Management Conference, Las Vegas, NV, 2003: 348–356.

28. A. B. Kersting, et al., "Migration of Plutonium in ground water at the Nevada Test Site," *Nature* 397, January 1999: 56–59.

29. Bechtel SAIC Company, LLC. *White Paper, Thermal Operating Modes* (Las Vegas, NV: Bechtel, 2002)

30. U.S. Nuclear Waste Technical Review Board, *Report to the Congress and the Secretary of Energy, January 1, 2002 to December 31, 2002* (Washington, DC: NWTRB, 2003).

31. R. W. Barnard, et al., *TSPA 1991: An Initial Total-System Performance Assessment for Yucca Mountain*, Sandia Report SAND91-2795 (Albuquerque, NM: Sandia National Laboratories, 1992); reprinted 1992.

32. U.S. Department of Energy, Office of Civilian Radioactive Waste Management, *Viability Assessment of a Repository at Yucca Mountain*, Report DOE/RW-0508 (North Las Vegas, NV: U.S. DOE, 1998).

33. Electric Power Research Institute, *Integrated Yucca Mountain Safety Case and Supporting Analysis, Phase 7*, Report 100334, J. Kessler, Project Manager (Palo Alto, CA: EPRI, 2002).

34. J.F. Schmitt, et al., *Biosphere Process Model Report*, TDR-MGR-MD-000002 REV 00 (Las Vegas, NV: TRW Environmental Safety Systems, Inc., 2000).

35. Thomas C. Hanks, Isaac J. Winograd, R. Ernest Anderson, Thomas E. Reilly, and Edwin P. Weeks, *Yucca Mountain as a Radioactive-Waste Repository, A Report to the Director, U.S. Geological Survey*, Circular 1184 (Denver, CO: USGS Information Services, 1999).

36. Bechtel SAIC Company, LLC. *FY01 Supplemental Science and Performance Analyses, Volume 2: Performance Analyses*, Report TDR-MGR-PA-000001 REV 00 (Las Vegas, NV: Bechtel, 2001).

37. Emmanuel Detournay, Larry G. Mastin, J.R. Anthony Pearson, Allan M. Rubin, and Frank J. Spera *Final Report of the Igneous Consequences Peer Review Panel* (Las Vegas, NV: Bechtel SAIC, 2003).

38. John Kessler, "Electric Power Research Institute Performance Assessment and Barrier Analyses," transcript of the Fall 2002 Meeting of the Nuclear Waste Technical Review Board (Las Vegas, NV: September 10, 2002): 253–259.

39. Rodney C. Ewing and Alison MacFarlane, "Yucca Mountain," *Science* 296, no. 5568, April 26, 2002: 659–660.

40. U.S. Nuclear Waste Technical Review Board, *Report to the Congress and the Secretary of Energy, January 1, 2001 to January 31, 2002* (Washington, DC: NWTRB, 2002).

41. U.S. Nuclear Waste Technical Review Board, *An Evaluation of Key Elements in the U.S. Department of Energy's Proposed System for Isolating and Containing Radioactive Waste* (November, 2003). [From: http://www.nwtrb.gov]

42. Letter from Michael Corradini (Chairman of NWTRB) to Margaret Chu (Director of OCWRM), November 25, 2003; letter from Chu to Corradini, December 17, 2003. [From: http://www.nwtrb.gov]

43. OECD Nuclear Energy Agency and the International Atomic Energy Agency, *An International Peer Review of the Yucca Mountain Project TSPA-SR"* (Paris: OECD, 2002).

44. J.A. Beavers, T.M. Devine, Jr., G.S. Frankel, R.H. Jones, R.G. Kelly, R.M. Latanision, and J.H. Payer, *Peer Review of the Waste Package Material Performance: Final Report* (Las Vegas, NV: February 28, 2002).

45. Robert Budnitz, "Science and Technology Program Update," transcript of the Spring meeting of the Nuclear Waste Technical Review Board (Washington, DC: May 14, 2003): 382–409. [From: http://www.nwtrb.gov]

13

Policy Issues in Nuclear Waste Disposal

13.1 The Importance of the Nuclear Waste Disposal Issue

13.1.1 The Centrality of the Issue

The Congressional Research Service, in an issue brief on nuclear waste disposal, compactly described a common assessment when it noted that "nuclear waste has sometimes been called the Achilles' heel of the nuclear power industry" [1, CRS-1]. It is a critical point of vulnerability for nuclear power, because without an acceptable solution for waste disposal, it appears irresponsible to continue the further use of nuclear power.

However there is a large range of opinions as to the intrinsic seriousness of the issue. At one pole, it is believed that the waste problem has not been solved, very possibly will not be solved in the predictable future, and represents a hazard that will persist for many thousands of years. At the other pole, it is believed that from a technical standpoint, the problem is essentially solved, with multiple ways of assuring safe disposal, and that the only really difficult problem is that of public perception.

For the first group, the main cause for concern about nuclear wastes is their long-term effects, i.e., their impact on future generations. There is also concern about the possible hazards in the transportation of the wastes, but this is generally viewed as a less weighty matter. The importance attached to the long-term issue was eloquently expressed in 1973 by Allen V. Kneese, of Resources for the Future:

Here we are speaking of hazards that may affect humanity many generations hence and equity questions that can neither be neglected as inconsequential or evaluated on any known theoretical or empirical basis. This means that technical people, be they physicists or economists, cannot legitimately make the decision to generate such hazards. Our society confronts a problem of great moral profundity; in my opinion it is one of the most consequential that has ever faced mankind. In a democratic society the only legitimate means for making such a choice is through the mechanisms of representative government. [2]

Key points in this statement are (1) waste disposal hazards are long-lasting and substantial, (2) the magnitude of the hazards cannot be evaluated, (3) we face an important moral responsibility, and (4) the decision on the course to follow should be made by representatives of the public, not by the "experts."

Those who hold the second, or optimistic, opinion do not contest the question of moral responsibility. However, they are frustrated by the emphasis placed on the issue, because they believe that the concern is highly exaggerated. This frustration is reflected in a statement by Sir John Hill, then Chairman of the UK Atomic Energy Authority:

I've never come across any industry where the public perception of the problems is so totally different from the problems as seen by those of us in the industry.[1]

Reflecting these different viewpoints, it is common to hear from responsible members of the public that their chief concern about nuclear power is waste disposal, whereas, in contrast, most physicists who are unenthusiastic about nuclear power tend to put waste disposal far down on their list of worries, well below those of weapons proliferation.[2]

In preceding sections, some of the technical aspects of nuclear waste disposal have been considered. In this chapter, emphasis will be given to policy aspects.

13.1.2 General Considerations in Nuclear Waste Disposal

It is universally agreed that nuclear wastes must be disposed of in a "safe" manner. However, defining the criteria for safety raises diverse questions, including the following:

- *The nature and level of the standard.* Should the basic protection standard be a limitation on dose or on calculated cancer risk? At what level should the protection be set? Should it apply to individuals or collectively to local or global populations?

---

[1] Statement attributed to Hill in Ref. [3, p. 65].

[2] This assessment derives from personal conversations, not from a systematic survey.

- *The time period of concern.* How far into the future do our responsibilities extend? Are we to be concerned about people 100 years hence, 1000 years hence, 1,000,000 years hence? Is a future death as much of a concern as a death tomorrow?
- *Our picture of our descendants.* Is it reasonable to assume that technological capabilities will continue to advance and, if so, should this enter into our consideration of the appropriate level of protection?
- *The context for considering nuclear waste hazards.* In thinking about intergenerational responsibilities, should nuclear waste disposal hazards be considered in isolation or in relation to other hazards?
- *The decision-making process.*[3] What procedures should be followed in reaching decisions on a waste disposal program? To what extent should society defer to nuclear professionals? To what extent should views of the public be sought and respected? How should the balance of power in making decisions be apportioned between the federal government, as reflected by federal agencies and Congress, and local interests, as reflected by state governments and citizen groups?
- *Intergenerational and intragenerational equity.* How should we compare the benefits and dangers to future generations *(intergenerational equity)* with those to people in our own and other countries today *(intragenerational equity)*? What risks and sacrifices should be accepted by our generation in order to protect future generations?

These questions have no crisp answers and there is no accepted methodology for converging upon common conclusions. In formulating U.S. standards, the EPA gives some consideration to general issues of this sort, but in the end, it must make arbitrary decisions influenced by a desire to achieve maximum acceptability in the face of widely divergent viewpoints and pressures. In practice, once the EPA has established standards it is natural to compare expected repository performance to those standards, and let the examination of the broader questions slip into the background. Nonetheless, these questions remain pertinent.

13.2 EPA Standards for Nuclear Waste Disposal

13.2.1 The Original Formulation of 40CFR191

The Nuclear Waste Policy Act of 1982 gave the EPA the responsibility to "promulgate generally applicable standards for protection of the general environment from offsite releases from radioactive material in repositories"

---

[3] This general issue has been extensively explored in a study carried out by an international panel under the leadership of the U.S. National Research Council: *Disposition of High-Level Waste and Spent Nuclear Fuel: The Continuing Societal and Technical Challenges* (see Section 13.4.1) [4].

[5, §121]. Proposed standards were set forth by the EPA in 1985 in Title 40, Part 191 of the Code of Federal Regulations [6].[4] Originally, these standards were to become effective as of November 18, 1985. Key provisions of 40CFR191 included the following:

◆ Over a period of 1000 years, no member of the general public is to receive a dose in excess of 25 mrem/yr due to releases from the repository. (This limit was specified in Subpart B of 40CFR191.)
◆ Over a period of 10,000 years, releases of radionuclides from the repositories to the outside environment must conform to specified limits. These limits were set so that the expected rate of fatal cancers in the surrounding population would be under 1000 during the 10,000-year period (i.e., an average of less than one cancer fatality per decade). [8]

When these provisions were established, it was believed that they could readily be satisfied in a well-designed repository. At the time, the EPA was focusing on repositories in formations saturated with water and concluded that there would be no excessive releases for water transport of any of the radionuclides. The possible escape of ^{14}C ($T = 5730$ years) in the form of gaseous carbon dioxide was overlooked [9]. Those analyses were later superceded by studies that considered the possible motion of gases through the unsaturated rock above the Yucca Mountain repository, and an appreciable possibility was found for ^{14}C to escape in amounts exceeding the 40CFR191 limits.

13.2.2 The ^{14}C Problem

The recognition that excessive amounts of ^{14}C could escape from the Yucca Mountain repository appeared for a time to put the entire project in jeopardy. A 1993 TSPA concluded that, "When considering the 10,000-year time period, virtually all (greater than 99.99%) of the release to the accessible environment is the result of ^{14}C" [10, p. 4-16]. Ignoring ^{14}C, these and other analyses indicated that the repository would easily meet the standards. Considering ^{14}C, it very possibly would not.

Nuclear reactors produce ^{14}C by neutron reactions with ^{17}O in the cooling water and with ^{14}N impurities in the fuel and cladding. For a 1000-MWe reactor, this leads to the release to the atmosphere of about 10 Ci/yr and the accumulation of ^{14}C in spent fuel assemblies [11]. The total amount of ^{14}C expected in the Yucca Mountain spent fuel is about 91,000 Ci [12, p. 9-4]

This is much less than the amount of natural ^{14}C present in the environment. About 4×10^4 Ci of ^{14}C are produced annually in the atmosphere by the interaction between cosmic-ray neutrons and ^{14}N, leading to an accumulated global ^{14}C inventory of 3×10^8 Ci. There is a continual interchange of carbon among atmosphere, oceans, and biosphere. Most of the carbon is

---

[4] Some of the history of the formulation of these regulations and the methodology behind them are discussed in a EPA publication on "Background Information" [7].

in the oceans, including most of the ^{14}C. The atmospheric inventory, in the absence of nuclear weapons tests, is about 5×10^6 Ci [11].[5] Living plants continually draw ^{14}C from the atmosphere during photosynthesis. This exchange ceases when they die, make ^{14}C dating an effective tool for studying the age of objects that contain carbon.

Radiation exposure from natural ^{14}C arises from the ingestion of food, which inevitably contains ^{14}C as part of the carbon present in all organic matter. The resulting dose to individuals is about 0.015 mSv/yr (1.5 mrem/yr) [12, p. 9-7], corresponding to a global collective dose of roughly 1×10^5 person-Sv/yr.[6] According to the linearity hypothesis for radiation effects, this dose implies about 5000 cancer fatalities per year or roughly 50 million fatalities from *natural* ^{14}C during the 10,000-year period being considered for waste repositories.

If the entire inventory of 91,000 Ci of ^{14}C escaped from the Yucca Mountain repository before decaying, it would represent an addition of about 0.03% to the full global ^{14}C inventory. The ^{14}C disperses around the Earth's atmosphere and almost all of it is gradually taken up by the oceans and plants. EPA calculations indicate that the added average individual dose over 10,000 years would be 3×10^{-6} mSv/yr (0.0003 mrem/yr) [12, p. 9-8].[7] However, although the individual dose is miniscule, the collective dose for a world population of 10 billion would be about 30 person-Sv/yr or about 3×10^5 Sv over 10,000 years, corresponding to a cancer toll of roughly 15,000 if one again adopts a strict application of the linearity hypothesis. Thus, if the carbon dioxide escapes from a large fraction of the waste packages, the calculated fatalities exceed the stipulated limit of 1000.

This created a substantial stumbling block in the consideration of Yucca Mountain, although many critics thought it unreasonable to be concerned about an increased dose which is about one-millionth of the natural dose (3 mSv/yr), with an individual risk amounting to about 1.5×10^{-10} per year.

Here, one is caught between the implications of a seemingly significant collective dose and obviously trivial individual doses. One also is faced with having to decide whether 15,000 deaths "matter" during a time when one might expect a total of one trillion (10^{12}) deaths from all causes.[8] To escape from these difficulties, it has often been suggested that a de minimis (or "negligible incremental dose") principle should be adopted in which one would ignore very small doses or very small incremental individual risks. This level

---

[5] Weapons testing increased the atmospheric concentration substantially, but with the cessation of U.S. and Soviet aboveground testing, the concentration has been returning toward the equilibrium value through mixing with the oceans.

[6] This collective dose is found by multiplying the average annual individual dose (0.015 mSv = 1.5×10^{-5} Sv) by the world population (6×10^9).

[7] Of course, to the extent that the waste containers retained their integrity and contained the carbon dioxide, the dose would be less.

[8] This assumes an average lifetime of 100 years for 10 billion people.

might be set, for example, at 1 mrem/yr (a risk of 5×10^{-7} per year). However, no such principle has been adopted in the United States by either the EPA or the Nuclear Regulatory Commission (see Section 4.4.4).

An alternative approach is to bypass the issue by not calculating the collective dose. If the collective dose is ignored, there is no need to decide formally upon an appropriate de minimis level or sort out the overall conceptual dilemma. This was the approach that was ultimately adopted by the EPA in 2001 (see Section 13.2.4). It amounts to a tacit acceptance of the viewpoint of a de minimis principle, with no explicit need to specify where the cutoff level lies.

13.2.3 The Overturn of 10CFR191

The EPA's proposed standards under 40CFR191 also encountered legal difficulties. One problem was the apparent inconsistency of setting a 10,000-year limit for general protection and a 1000-year limit for individual protection. The National Resources Defense Council (NRDC), along with several states and several other environmental organizations, brought suit against the EPA on grounds that included the claimed inconsistency between these time criteria and the fact that the EPA had not considered the applicability of the Safe Drinking Water Act to their standards. In 1987, the First Circuit Court of Appeals ruled in favor of the NRDC and its associates, and parts of 40CFR191 were "vacated and remanded" [13].

Before either the judicial challenge or the ^{14}C embarrassment was resolved, Congress, in the Energy Policy Act of 1992, temporarily suspended the EPA's authority to set standards for Yucca Mountain [14, Sec. 801]. It commissioned a study under the auspices of the National Academy of Science (NAS) to advise on "reasonable standards for protection" for the Yucca Mountain repository and instructed the EPA to delay formulation of new standards until after receipt of the recommendations of the NAS panel.[9] One of the specific concerns was "the release limits for carbon-14," as cited in a letter from Senator Bennett Johnson to the chairman of the National Academy committee (see Ref. [16, p. 136]).

13.2.4 The NAS Recommendations

Two Explicit Questions

Two of the questions that the NAS was called upon to address were (1) "whether a health based standard based on doses to individual members of the public. . . . will provide a reasonable standard for protection" and (2)

---

[9] More officially, this was a committee of the National Research Council, an arm of the NAS. However, following common practice, including that of the EPA [15], we refer to this as an NAS study and report.

"whether it is possible to make scientifically supportable predictions of the probability.... of human intrusion over a period of 10,000 years." The results of the NAS study were published in the summer of 1995 in the report *Technical Bases for Yucca Mountain Standards* [16]. In brief, the answer to question (1) was *yes* and the answer to question (2) was *no*. However, the significance of the report extended beyond the answers to these specific questions.

The Recommended Nature of the Health Standard

Protective standards for a nuclear waste repository can be couched in alternative ways, with limits set on dose, risk, or the release of radionuclides. They could apply to the maximally exposed individual, a limited set of selected individuals, or the world population. Such limits are coupled. For example, the limits that the EPA had earlier proposed on radionuclide releases were intended to satisfy a global limit on fatalities.

The NAS panel recommended "...the use of a standard that sets a limit on the risk to individuals..." from radiation exposure [16, p. 4]. A plausible choice, as implied in the NAS discussion of precedents, would be an incremental risk in the neighborhood of 10^{-6} to 10^{-5} per year [16, p. 5]. Such a risk range can be translated into a radiation dose limit, based on an assumed dose–response relationship, but risk is here the fundamental criterion.[10]

The panel also recommended that the limit apply to members of a "critical group" rather than to a hypothetical maximally exposed individual or to a very broad population. The critical group for which the dose would be calculated should be "representative of those individuals in the population who, based on cautious, but reasonable, assumptions have the highest risk resulting from repository releases" [16, p. 53]. This group is intended to be "relatively small," without great variations in dose within the group.

Duration of Protection

In previous EPA standards, the longest period of protection had been set at 10,000 years on the grounds that the risk decreases with time as radionuclides decay and that projections beyond 10,000 years would be highly uncertain. However, studies of the Yucca Mountain site suggest that the maximum radiation doses are reached after several hundred thousand years, because at earlier times the bulk of the radionuclides will be kept out of the accessible environment by the repository's engineered and natural barriers (see Chapter 12). With such calculations in mind, the NAS report suggested that the time horizon be extended to the neighborhood of one million years. Specifically,

---

[10] For example, for a cancer fatality risk factor of 0.05 per sievert (see Section 4.3.4) a risk limit of 10^{-5} per year would correspond to a dose limit of 0.2 mSv (20 mrem) per year. If the scientific consensus on the risk factor changes, the dose limit would be recalculated, but the underlying risk limit need not be reevaluated.

it recommended that the "assessment be conducted for the time when the greatest risk occurs, within the limits imposed by long-term stability of the geologic environment" [16, p. 7], with the suggestion that the appropriate time frame for consideration is "on the order of 10^6 years" [16, p. 9]. Assessment of compliance—the extent to which the repository would actually satisfy the established standards—was deemed "feasible for most physical and geologic aspects of repository performance" over this time scale.

Negligible Incremental Dose and ^{14}C

With specific reference to the ^{14}C issue, the report affirmed the belief "that the concept of a negligible incremental dose can be extended to risk and can be applied to Yucca Mountain" [16, p. 60]. It suggested as a "starting point" for consideration a level that had previously been suggested by the National Council on Radiation Protection and Measurements—a risk of 5×10^{-7} per year, which translates to a dose of 0.01 mSv/yr (1 mrem/yr). At this level, in the words of the report, "the effects of gaseous ^{14}C releases on individuals in the global population would be considered negligible."

The Problem of Human Intrusion

The NAS report concluded that it was not possible to predict either the likelihood that systems to prevent human intrusion into the repository will be successful over extended time periods or the probability that intrusions will actually occur. However, the report recommended that the consequences of such intrusions be studied. For specificity, it was suggested that the focus be on the integrity of the repository following the drilling of a single borehole through a waste canister and into an underlying aquifer. The calculated risk to the critical group from such intrusion should be no greater than from "normal" long-term releases from the repository.

13.2.5 EPA's 2001 Standards: 40CFR197

Stipulations of the Standards

The role of the NAS panel was advisory, but the EPA retained the authority to formulate the ultimate regulations and was not bound by the panel's recommendations. After prolonged consideration, the EPA put forth in August 1999 its proposed standards for Yucca Mountain, to be embodied in 40CFR197. Following a period for comment and consideration, the final standards appeared in June 2001, in a version little changed from the 1999 proposed version [15]. The heart of the standard is the following requirement:

> The DOE must demonstrate, using performance assessment, that there is a reasonable expectation that, for 10,000 years following dis-

posal, the reasonably maximally exposed individual receives no more
than an annual committed effective dose equivalent of 150 microsiev-
erts (15 millirem) from releases from the undisturbed Yucca Mountain
disposal system. [6, §197.20]

This limit includes the dose from "all potential pathways." The reasonably
maximally exposed individual (RMEI) is assumed to live in the neighborhood
of the Yucca Mountain site "above the highest concentration of radionuclides
in the plume of contamination," have a "diet and living style representative
of those who now reside in the Town of Amargosa Valley" (a nearby town),
and drink 2 L of water per day from local wells [6, §197.21]. In addition to the
dose limit on individual exposure, the EPA established a separate standard
for groundwater that includes limits on concentrations of radionuclides and a
40 μSv (4 mrem) limit on the annual dose from drinking water [6, §197.30].

Several features of these standards are to be noted, with special reference
to comparisons to the NAS recommendations:

◆ *The dose limit.* The adopted limit of 0.15 mSv/yr is less than the
0.25 mSv/yr limit that was in the original version of 10CFR191 and that
was also favored by the Nuclear Regulatory Commission. However, either
level will be satisfied for the required 10,000 years according to the TSPA
calculations made by the DOE.

◆ *The nature of the limit: dose based.* Despite the NAS recommendation
favoring a risk-based limit, the EPA adopted a dose-based limit. The limit
of 0.15 mSv/yr corresponds (in the standard linearity-based calculation)
to a risk of 7.5×10^{-6} per year, which is within the range suggested by
the NAS committee. Risk is of more fundamental concern than dose itself,
but from a regulatory standpoint, it is simpler to judge compliance with
a dose standard than with a risk standard, especially when the dose–risk
relationship is controversial.

◆ *Time period for the regulation.* The EPA retained the 10,000-year period
for compliance, rather than extend the period to something like one million
years, as suggested by the NAS. The EPA chose 10,000 years in large part
because of the uncertainties involved in calculations covering longer time
periods and the resulting difficulties in applying a regulatory standard [15,
p. 32096ff].

◆ *Time period for calculation.* Although the regulatory limit applies for only
10,000 years, the DOE is required to calculate the peak dose (for the
RMEI) "that would occur after 10,000 years but within the period of
geologic stability...as an indicator of long-term disposal system performance" [6, §197.35]. In practice, the DOE's TSPA calculations now extend
to one million years. It is not clear what use the EPA expects should be
made of these results.

◆ *Collective dose and* ^{14}*C.* There is no requirement to determine collective dose or collective impact and, thus, no regulatory reason to consider
^{14}C. Nonetheless, the EPA's discussion of the omission of collective dose

shows great ambivalence [15, p. 32094ff]. In the 1999 proposed version of 40CFR197, although no standard on collective dose (or total fatalities) was imposed, the explanatory document recommended "that DOE calculate the collective dose without truncation" [17, p. 46992]. In the final version, this recommendation was dropped. However, the EPA explicitly declined to adopt the concept of negligible incremental dose, as suggested by the NAS, although it agreed that calculating the effects of a 0.0003 mrem/yr increment in dose is "uncertain and controversial" [15, p. 32095]. The EPA discussion did not acknowledge that many people have stronger views, namely that such a calculation goes beyond the realm of the reasonable.

♦ *The nature of future populations.* The standard explicitly states that "the DOE should not project changes in society...or increases or decreases of human knowledge or technology."

The EPA standard makes an explicit attempt to define "reasonable expectation," but despite those efforts, the words appear open to a wide range of interpretations. This may be contrasted with the requirements spelled out in the earlier standard, 40CFR191 (§191.13): The likelihood of exceeding the radionuclide release limits "shall" be less than 10% and the likelihood of exceeding these limits by a factor of 10 "shall" be less than 0.1%. This earlier formulation may have suggested a precision in estimation that is hard to achieve, but it did define an identifiable target. It also placed stress on the importance of being quite sure that, even if the limits were exceeded, they would not be greatly exceeded.

Reactions to EPA's Standards

The proposed standards elicited criticisms from a variety of sources, including key bodies involved in establishing and enforcing radiation protection standards [18]. The Board on Radioactive Waste Management (BRWM) of the National Research Council—the body responsible for the earlier NAS study—particularly criticized the adoption of a separate groundwater standard, in addition to the overall dose limit [19]. In other areas of clear disagreement, it reaffirmed the NAS preference for a risk-based standard and for the abandonment of collective-dose calculations. The most tangible result of these criticisms of the EPA's proposals was the decision by the EPA to drop the requirement for a collective-dose calculation.

Given the criticisms and disagreements, it might appear that there are important differences in the views of the major organizations involved with waste disposal standards. However, from a broader perspective, these disputes involve secondary details, and there is agreement on more fundamental matters. In particular, all of the official organizations involved accept the basic premises that are now guiding waste disposal policy in the United States and other countries:

- People many thousands of years hence should be protected from radiation doses at about the same level as people are now protected from radiation from nuclear power facilities.
- In arriving at such criteria, it is assumed that future generations will be no more capable than our own in guarding against the effects of radiation exposures.

It is not clear whether these premises are accepted in the belief that they have inherent merit or in an effort to forestall contentious disputes. However, although widely accepted, these premises warrant further examination and are considered in more detail in the following section.

13.3 Responsibilities to Future Generations

13.3.1 The General Recognition of the Problem

As we have seen, there is virtually universal agreement that the long lifetimes of radionuclides in the nuclear wastes impose a responsibility upon us to handle the wastes in a manner that will protect the interests of future generations. This view is reflected in the quotation from Kneese cited in Section 13.1.1 and in many statements by other individuals and groups. The responsibility is a moral one, and the evolving recommendations and regulations seek to give concrete substance to the moral demands. However, defining this responsibility, which is essentially the question of respecting "intergenerational equity," is very difficult.

A comprehensive review of efforts made in this area was prepared by David Okrent in 1994 [20] and in a revised and extended form in 1999 [21]. As he summarized the difficulties:

> Within the United States and across the world, there has been and is today only limited, piecemeal consideration of how society should deal with those of its activities that have the potential to pose long term risks far into the future. The absence of accepted broad philosophical guidelines to deal with such issues makes decision-making that much more difficult. [21, p. 877]

The failure to achieve accepted guidelines is not for want of effort. Numerous individuals and groups have attempted to address the issue of intergenerational responsibilities. However, the issues do not lend themselves to easy resolution, especially in the light of ambiguities as to the nature of the hazards, the level of hazard to be avoided, and the balance to be sought between the claims of the present generation and those of future generations.

A viewpoint for considering intergenerational responsibilities and rights has been suggested in compact form by Edith Brown Weiss:

> Each generation receives a natural and cultural legacy in trust from previous generations and holds it in trust for future generations. This relationship imposes upon each generation certain planetary obligations to conserve the natural and cultural resource base for future generations and also gives each generation certain planetary rights as beneficiaries of the trust to benefit from the legacy of their ancestors. [22, p. 2]

If the term "cultural legacy" is interpreted broadly, it includes not only our intellectual, moral, scientific, and artistic heritages but also the institutions and facilities that are intrinsic to a modern society. Of course, reasonable as this formulation appears, it hardly provides a blueprint for policy implementation.

In a broad summary of the issue, the OECD issued a statement of "A Collective Opinion of the Radioactive Waste Management Committee of the OECD Nuclear Energy Agency" [23]. Emphasis was placed on respecting both intergenerational and intragenerational equity. With respect to intergenerational equity, it called attention for balance between benefits and risk:

> In the case of nuclear energy production and the management of radioactive wastes, as with various other aspects of industrial activity, the balance between the benefits which are enjoyed by present and future generations through sustained technological development, and the liabilities which may be imposed on future generations over a long period, must be carefully scrutinized. [23, p. 7]

Overall, the Committee concluded that "the geological disposal strategy can be designed and implemented in a manner that is sensitive and responsive to fundamental ethical and environmental considerations" [23, p. 5].

A rather detailed analysis of these matters was published by the National Academy of Public Administration (NAPA) [24]. Its study affirmed the obligation of each generation to future generations *(trustee principle)*, in particular not to deprive future generations of "the opportunity for a quality of life comparable to its own" *(sustainability principle)*. This obligation is somewhat qualified by an affirmation of the obligation to the living and near-future generations, with priority to be given to "near-term concrete hazards" over "long-term hypothetical hazards" *(chain of obligation principle)*. Risking irreversible or catastrophic harm should be avoided unless there is "some compelling countervailing need" *(precautionary principle)*.

The NAPA formulation attempts to establish a balance between the present and future. The last two principles have particular applicability to the issue of nuclear waste disposal. The "chain of obligation" principle can be used as an argument for moving ahead with nuclear power, on the grounds that addressing the immediate hazards of pollution from fossil fuels and of energy shortages deserves priority over the more remote risks from possible repository failures. In fact, the panel explicitly states that "the Chain of Obligation Principle strongly prefers the near future over the distant future, as long as no injustice is done" [24, p. 12]. At the same time, the "precautionary principle" implies an obligation to be sure that waste disposal creates no possibilities of catastrophic damage (i.e., that future populations will not suffer "significant irreversible harm").

Overall, nuclear power has conflicting implications in the context of intergenerational responsibilities. It may be creating problems for the future through its production of nuclear wastes. On the other hand, it may contribute to the welfare of future generations by saving fossil fuels, reducing environmental threats from global climate change, and helping to sustain the industrial base needed for the continuance of modern societies.

13.3.2 Picture of Future Generations

Almost all groups that consider waste disposal standards decline to make any assumptions as to the nature of future societies and, by default, assume them to have the same characteristics as society today. This general approach is explained succinctly in a study on nuclear waste disposal carried out by the Electric Power Research Institute:

> In order to avoid endless speculation regarding the technology, physiology and socio-economic structures of future communities, the primary assumption is made that the activities and characteristics of the population exploiting the well water are similar to those of present-day communities. [25, p. 8-5]

The same position had been previously enunciated by the OECD [23, p. 8] and the EPA [15, p. 32096].

However, whatever its rationale in terms of convenience or expediency, the no-change assumption is manifestly implausible. In other contexts, scenarios for the future assume technological progress that will soon make the world almost unrecognizable, just as it would have been difficult, probably impossible, for people living before 1800 to envisage today's technological world. Progress is likely to continue such that within several hundred years, society will again have a radically advanced set of scientific and technical capabilities. In such a context, it might be only a relatively modest step to develop cures for most cancers and to routinely monitor water quality—together removing any significant danger in the unlikely event of significant escape of radionuclides into the biosphere.

These would be small accomplishments compared to those contemplated by Freeman Dyson in *Imagined Worlds*:

> In the next hundred years we will probably have human settlements on the moon and on a few nearby planets. Perhaps also on Mars. We will see genetically engineered plants and animals adapted to the colonization of various asteroids and planets. For example, we may see the Martian potato....[26, pp. 152–153]

On a time-scale of a thousand years, the genetic differences between human populations may be increased by effects of natural selection or genetic engineering.... During the next thousand years there will be many opportunities for experiments in the radical reconstruction of human beings. Some of these experiments may succeed.... We may hope that one group of our descendants, those who cling to our old human heritage, those who are loyal to our natural human shape and genetic endowment, will be allowed to remain here in possession of our planet....[26, pp. 154–158]

These may be grand and unprovable visions, but they provide a striking counterpoint to the assumed no-change scenarios.

Of course, change does not necessarily mean advances, and other scenarios assume that there could be a discontinuity in civilization so that future generations will be living at a very low technological and scientific level—a return to a more "primitive" epoch. However, Bernard Cohen and others have pointed out:

If there should be a collapse of civilization, the residual radiation would be the least of the problems. There would surely be mass starvation as the world would have to shrink back to its pre-industrial revolution level. [27]

The same point has been made by William Sutcliffe:

It is sometimes argued that we cannot depend on the progress of science and technology to protect future generations from the hazard of a nuclear waste repository. In fact this hazard depends on society losing track of the repository location. The short rejoinder is that if the repository information is lost, the society, or what's left of it, will have much more pressing problems than those associated with nuclear waste. The cause of loss of information, e.g. melting ice caps, ice age, nuclear war, comet strike, etc., will kill more people and cause more disruption and dislocation than all nuclear waste in a repository. More generally, if science and technology do not progress in a number of important areas, e.g., the development of energy, food and water supplies, tensions will be certain and wars will be likely. In any case nuclear waste would not be the major concern of such a future society. [28]

In short, the radiation protection standards established by the EPA would be irrelevant either in the world of technological advances or in a world of technological collapse. They would only be appropriate if in 10,000 years and beyond, the world is balanced just at the point where we are today—not advanced enough to be able to cure cancer but still unscathed by global disasters that would dwarf any concerns about our nuclear waste repositories.

Why do responsible authorities base their standards on so implausible a picture of the future? It is in part because there is no firm footing for moving to any other basis. There may also be a component of moral unease. If one assumes (as seems quite plausible) a cancer cure within a few hundred years, then it should suffice to contain the wastes for, say, 1000 years. Yet, it offends our sensibilities to say that it is acceptable to contaminate the world 1000 years hence on the grounds that people will then be easily able to deal with the contamination.

The demand for almost complete confinement of the wastes for at least 10,000 years may lack a firm rational basis—and therefore an unequivocal ethical basis—but it does, at the moment, have a persuasive aesthetic basis and, even more importantly, it probably has a compelling political basis. In recognition of this, the EPA must set high standards for the repository. However, in trying to assess the implications of a possible—even if unlikely—major breach after thousands of years, it would be reasonable to take into account that whatever harm might ensue under present technological conditions, it is likely to be smaller under future technological conditions.

For perspective, it is to be noted that in other areas, we tacitly assume a great deal about social continuity and the likelihood of technological advances. The entire organization of society in cities is predicated on the stability of water supplies and of the means to transport food long distances. The casualness with which the world entered the fossil fuel epoch in the 19th century, consumed fossil fuels as if there were no tomorrow in the 20th century, and is continuing a fossil-fuel-dominated energy economy in the 21st century represents a high level of either irresponsibility or confidence that new energy technologies will bail us out as fossil fuels supplies are exhausted over the next several hundred years.

13.3.3 Discounting with Time

In doing a cost–benefit analysis of a project, it is common to discount the future benefits on the grounds that (quite apart from inflation) a dollar today is worth more than a dollar in the future. Thus, if a discount rate of 5% is used, then it is worth investing a dollar today only if the benefit in 10 years is at least $1.63.[11]

This sort of calculation works reasonably in the short term and when applied to things whose value can be quantified. However, applying such a quantitative approach to the long-term issue of waste disposal leads to senseless results. Even were we to assume a discount rate of only 0.5% per year—one-tenth of that used above—an expenditure of $1 today is equivalent to an expenditure of 4.6×10^{21} in 10,000 years. If we assume a human population of 100 billion at that time (one would hope it will be much less), then the expenditure would be equivalent to $46 billion per person. This is perhaps 1000 times the "value" of a human life today. Thus, by this logic, it would not be worth spending $1 today to save the entire human race in 10,000 years.

---

[11] Note: $1.05^{10} = 1.63$.

This is a manifestly absurd conclusion, whether or not we can pinpoint the flaws in the logic or premises.

Once it is seen that a discount rate as small as 0.5% per year leads to absurdity, it is tempting to conclude that the proper discount is zero. It is a short step then to conclude that we should consider a human life 10,000 years hence to be as valuable to us as a life today and that we should make as much effort to save it as we would for any (anonymous) individual today. In this spirit, the International Atomic Energy Agency enunciated the following principle:

> The degree of isolation of high-level radioactive waste shall be such that there are no predictable future risks to human health or effects on the environment that would not be acceptable today...the level of protection to be afforded to future individuals should not be less than that provided today. [29]

This principle appears to be fully accepted by the U.S. EPA and all other agencies involved in setting U.S. waste management policy. It has the attraction of being conceptually simple and seemingly highly responsible. In contrast, a position that embraces some sort of discount of future lives faces at least two major difficulties: (1) it is hard to avoid the appearance of moral insensitivity and (2) any decision to employ a discount requires a debatable and inherently arbitrary decision as to the magnitude or character of an "acceptable" discount.

Nonetheless, there have been advocates of discounting. Thus, in a 1977 report of the National Research Council, the following rationale was offered for some degree of discounting:

> Discounting does not reflect lessened concern for future generations, but rather the opportunity to forego investments far in advance of the life to be saved in order to have more resources available closer to the time at which the life is actually saved. Future generations may not want us to have squandered resources to reduce their mortality now. If they will have more efficient opportunities to save lives, then society should do them a favor by discounting them even more. [30, pp. 238–239]

This is essentially an efficiency, or cost-effectiveness, argument for discounting.

A more general argument can be made on the basis of the remoteness of a world 10,000 years hence. Several aspects of this remoteness are pertinent:

- The world will be very different in 10,000 years, and our efforts at protecting or improving its environment are likely to be misdirected.
- Technological progress will probably provide easy solutions to difficulties that seem to be significant today, including the dangers of radiation exposures.
- Our greatest intergenerational obligation is to set the stage for sustaining progress over the next several centuries. When the understood needs of

immediately following generations are in conflict with the unknown needs of far-distant generations, the century-scale needs should be given priority.

♦ It is inconsistent with virtually all our practices in other areas, and perhaps with human nature, to be deeply concerned about small risks for people separated from us by 10,000 years. As individuals, our concern extends perhaps as far as our great-grandchildren—but hardly much beyond.

This last point, in its suggestion of attenuated concern, is the most problematic because it suggests irresponsibility or callousness, although attenuated concern for distant people is inherent in the actual approach to human affairs taken by most of us. We have more in common with people 10,000 miles away than we have with people 10,000 years away, and our responsibilities to them might be thought to be greater. Yet, despite some efforts to help, there is a de facto acceptance by the developed nations of thousands of deaths from environmental and man-made disasters in the developing countries of Asia and Africa. Smaller-scale events, involving tens or hundreds of deaths, are often just ignored.

All this said, one must be conscious of being on a slippery slope. For 10,000 years, should one discount by a factor of two, ten, one hundred, one million? There is no firm guide and one is naturally drawn back to the simplicity of a factor of one—namely, no discounting. However, this may be more a form of mental laziness than evidence of a heightened ethical sensitivity. Fortuitously, we may be able to evade the issue, because it appears that there is a technological fix, namely a repository judged to be good enough to make discounting irrelevant.

13.4 Special Issues in Considering Waste Disposal

13.4.1 The Decision-Making Process

Attitudes Toward Science and Experts

Immediately after World War II, experts—and scientists, in particular—were held in high regard in the United States. U.S. military successes in the war and the technological accomplishments that contributed to those successes helped to form a prevailing view that scientists and engineers knew what they were doing and could be trusted in terms both of their competence and their intentions. There was a broad, if not unanimous, trust in the "establishment" of which scientists had become a key part.

By the beginning of the 21st century, the situation had dramatically changed. A number of signposts or milestones mark the growing distrust of what might be called the scientific–industrial–military complex. The publication of *Silent Spring* by Rachel Carson in 1962 put the dangers of chemical pollution before the public in dramatic form. The 1964 movie *Dr. Strangelove* showed technological and military prowess gone mad. The Three Mile Island nuclear accident in 1979, the Bhopal chemical accident in 1984, and the Chernobyl accident in 1986 dramatized the fallibility of technology. The prolonged legal battles over asbestos and tobacco provided examples of corporations that appeared indifferent to public health and experts who gave incorrect assurances that there was nothing to worry about. A more general distrust lies in the gulf that developed between science and other parts of our culture, described as early as 1959 by C.P. Snow in *The Two Cultures and the Scientific Revolution*, and which has arguably increased since.

These events or phenomena all contribute to a fear of the fruits of science and technology and an unwillingness to take the "expert" at his or her word, especially if the expert is from industry or the government. Of course, the picture is not monolithic. Many of the spectacular advances in medical technology and the computer revolution have been accepted with enthusiasm, and science is often heralded as the key to future economic progress. Thus, it is mixed and inconsistent picture.

The Role of the Public

The lessening confidence in experts has been accompanied by an increased call for public participation in decisions on technical matters, at least in the case of nuclear waste disposal. Some scientists do not welcome this trend. They probably retain a residual feeling that technical issues are too specialized for the layperson to understand, and decisions on technical matters should therefore be left to technical people. Overall, however, there is general adherence to the principle that policy decisions on matters that may impact society as a whole should be reached through broad consultation with members of the public. In the context of waste disposal planning, it is sometimes affirmed that the public should participate as full partners in the decision-making process.

This approach is advocated in part as a pragmatic recognition of political realities and in part as an expression of a broader ethical principle. The pragmatic view has been expressed succinctly:

> Involving the public does not guarantee success, but not involving the public just about guarantees failure.[12]

Similarly, an international committee convened by the Board on Radioactive Waste Management (BRWM) of the U.S. National Research Council also strongly endorsed public participation:

> Successful decision-making is open, transparent, and broadly participatory. National waste disposition programs in democratic countries cannot hope to succeed today without a decision-making process that facilitates choices among competing social goals and ethical consid-

---

[12] Attributed to Jerry Scoville in Ref. [3, p. 319].

erations. Sufficient time must be devoted to developing this process, including the involvement of broader circles of citizens in examining the choices in an informed way. [4, p. 4]

In formal adherence to this general principle, the DOE now holds numerous public hearings and has initiated outreach programs in an effort to develop a positive view among Nevadans toward the Yucca Mountain project. In addition, there have been long time intervals between the issuance of draft versions of key documents and the final versions, and during this period there is extensive opportunity for public input.[13] These efforts have not succeeded in altering the state government's opposition, and it is difficult to judge how much they may have influenced attitudes of the general public or within the DOE.

The Roles of the Public and the Experts

Whatever the proper relationship between the role of the "experts" and of the "public," a few generalizations can be made:

- ◆ No major technological project, and particularly no nuclear project, is likely to be implemented without public support.
- ◆ The leadership of experts is essential to planning and carrying out projects in a safe and effective manner.
- ◆ The input of objective "outside" experts is important for providing independent evaluations of major projects.
- ◆ Whether or not specific public concerns have a valid scientific basis, they should be addressed directly, objectively, and informatively.
- ◆ Even the best experts are fallible. Attention to public criticisms—which are usually buttressed by inputs from dissident experts—can help avoid errors and oversights.

The above-suggested guidelines are directed toward the utilization of expert opinion in the decision-making process. It may be well to recognize, however, that the public's evaluation of conflicting opinions is influenced by the desire to believe one side or another. With respect to the specific matter of nuclear waste disposal, projects such as Yucca Mountain are more likely to be deemed "unsafe" when nuclear power is seen to be unnecessary. They are more likely to be accepted as "safe" when nuclear power is seen as essential.

Local Interest, National Interest, and Global Interest

Strands in public attitudes during the past few years have been reflected in two phrases: the bumper sticker: "Think globally, act locally" and the cliché "Not

---

[13] It is not always clear what is meant by the term "the public." The public input includes contributions from industry and environmental groups and there is no obvious distinction between "the public" and "special interests."

In My Back Yard" (NIMBY). No doubt there have been instances when local citizens allowed their global consciences to outweigh their backyard interests, but when a choice must be made, the NIMBY impulse usually prevails. People generally put forth their local claims with little embarrassment and with little suggestion from others that they should hold to a higher, global standard.

It is not clear if there is any guiding philosophy or even institutional structure to decide the proper balance between national and local interests, except on an ad hoc, case-by-case basis. If the president and Congress decide that Yucca Mountain is needed for the national, or even global, welfare, does the country have the right to run roughshod over the wishes of the State of Nevada? Is it a violation of intragenerational equity for the federal government to go ahead with Yucca Mountain, disregarding local preferences? There is even a problem in deciding the meaning of "local." If the majority in a county in the immediate vicinity of Yucca Mountain favor the repository— as they might for economic reasons—does this trump the objections of people in Las Vegas who might fear for the tourist industry?

The analogous issues were rarely raised during the many years that nuclear weapons tests were conducted in Nevada, both aboveground and belowground, but the perceived needs of national security then outweighed any local concerns. Further, the general political context in which decisions are made and implemented is now greatly changed.

Congress and Federal Agencies

In considering the proper role of experts and of the public, the key part played by Congress should not be forgotten. In practice, the actual determining factor in U.S. nuclear waste policy has been Congress, which, at times, has shown an interest, even an eagerness, to be involved in the details of nuclear waste policy formulation. It has on occasion acted cavalierly, as it did in 1987 when it threw out the elaborate procedures of the NWPA of 1982 and designated Yucca Mountain as the one location for site characterization. It moved more circumspectly in its later call upon the National Academy of Sciences to frame recommendations for waste disposal standards.

More recently, in 2002, it took a decisive step when it approved going ahead with a license application for the Yucca Mountain repository in support of the President's position and in opposition to the position of the Nevada governor.[14] Of course, in reaching its decisions, Congress has hearings in which varied testimony is received and it is always sensitive to pressures from its constituents. However, the process on occasion has become brisker and less deliberative than envisaged in the somewhat theoretical discussions of the

---

[14] This became partially, but not completely, a partisan issue. On the key Senate vote (7/9/02) supporting the Yucca Mountain repository by a 60 to 39 margin, Republicans favored it 45 to 3, whereas Democrats (and one Independent) opposed it 36 to 15.

appropriate interplay between the public and experts. In the end, Congress is the key player and it listens to whomever it wishes.

The actual setting of standards is carried out by federal agencies that have limited mandates. The EPA and NRC are charged with protecting the public from specific dangers. It would be outside their mandates to consider the dangers from global warming when framing standards for nuclear radiation. The standards are set with the goal of controlling the risks of a given technology without weighing alternative risks. It might be well to have an agency with an overarching responsibility to consider energy and environmental issues in a broad context. This, in fact, was the role of the old Atomic Energy Commission (AEC). However, here one runs into the problem that led to the termination of the AEC in 1974, namely the danger of having one-and-the-same entity be the developer of a technology and the judge of its safety.

Thus, a "new AEC" will not be the solution. Nonetheless it would be desirable to develop an institutional mechanism for the consideration of energy and environmental matters on a broad and sustained basis, as a guide to the president and Congress. In the end, however, the balancing of the various factors falls to them by default.

In some instances, the ultimate arbiter is the courts. If there is a defect or ambiguity in the formulation of the relevant regulations or statutes, the courts can intervene. There has been a long history of legal battles relating to nuclear power in general and nuclear wastes in particular. One example was the judicial overthrow of parts of 40CFR191. More recently, the state of Nevada has mounted vigorous legal challenges to the federal efforts to develop the Yucca Mountain repository. Current lawsuits are not likely to be settled before late 2004, and these or new lawsuits may continue well beyond 2004.

13.4.2 Technological Optimism and Its Possible Traps

One problem in the search for a solution to the waste disposal issue may have been excessive technological optimism. There has been a willingness to impose and accept very rigorous standards of protection, lasting for 10,000 years, because the federal regulatory agencies and many nuclear advocates have deemed them to be achievable. Given the defense-in-depth of the waste package and the geological environment, there were optimistic expectations that fulfillment of very demanding requirements could be assured. However, the emergence of the "carbon-14 problem" illustrates the dangers of pointlessly stringent standards, as were set in the original version of 40CFR191 (see Section 13.2).

The new Yucca Mountain standards appear achievable, and nuclear proponents accept them with optimistic confidence. However, it may be difficult to establish a clear scientific consensus that this assurance is absolute. Some scientists believe that the behavior of metals and of geological sites cannot be conclusively demonstrated over so long a time period. Thus, the establishment of standards for 10,000 years could lead to the eventual failure of the project, should a court conclude that the requisite level of proof has not been met. However, no such technical ambiguity is expected by most nuclear advocates, and there has been no visible pressure to revisit the issue and ask if a relaxed standard would still be a responsible one.

In setting standards, a federal agency must find a compromise that will be broadly acceptable in the political arena. The temptation in such a situation is to move to stringent standards, in order to demonstrate responsibility and to reassure an important part of the public constituency. The chief brake that comes into play is that provided by practicality. If it is seemingly impossible or unreasonably expensive to meet a given standard, there is then an incentive to step back and ask if the standard is appropriate. However, if achieving a high goal seems technically feasible, there is a pressure to move the standard to the level of that goal. That is where technological optimism may create difficulties.

13.4.3 A Surrogate Issue?

Discussions of nuclear waste disposal take place in a climate of widespread concern about nuclear energy and opposition to it. The concern stems in part from the fear of specific nuclear mishaps—in particular, radiation exposure from reactor accidents, waste transportation accidents, and escape of radionuclides contained in the nuclear wastes. Despite the difficulties involved, these concerns could, in principle, be resolved by convincing evidence that the nuclear facilities are built and operated very safely.

However, other objections to nuclear power involve more general and elusive considerations. These include (a) the links between nuclear power and nuclear weapons, (b) the feeling that nuclear power epitomizes the excesses of big industry, big government, and centralized control, and (c) the belief that large-scale industrial activity usually carries environmental liabilities. These concerns are of a fundamental nature, involve judgments that are not readily amenable to resolution by discussion, and provide their own bases for opposing nuclear power. On the other hand, although these objections involve profound considerations, they are not ones that appear to resonate strongly with the general public.

Under these circumstances, opponents of nuclear power can find in waste disposal an effective issue through which the growth, and even continuance, of nuclear power can be discouraged. It has been suggested that for some opponents of nuclear power, the waste disposal issue is more a tool than a driving substantive concern. The then-chairman of the Sierra Club, Michael McCloskey, is quoted by Luther Carter in the 1987 book *Nuclear Imperatives and Public Trust* as expressing the view:

> I suspect many environmentalists want to drive a final stake in the heart of the nuclear power industry before they will feel comfortable in cooperating fully in a common effort at solving the waste problem....

Their concern would arise from the possibility that a workable solution for nuclear waste disposal would make continued operation of existing plants more feasible, and even provide some encouragement for new plants. [31, p. 431]

In short, waste disposal is a convenient surrogate issue to use in place of other issues that are harder to argue effectively but may be more fundamental.

If this assessment is correct, it suggests that some opponents of nuclear power will oppose any waste disposal plan, independent of its intrinsic merits. This obviously complicates the matter of implementing a project such as Yucca Mountain—or any alternative long-term solution—especially given the existing administrative and legal mechanisms for creating delays.

13.5 Possible Approaches to Nuclear Waste Disposal

13.5.1 A Step-by-Step Approach

Suggested Flexible Approach (1990)

In 1990, when the U.S. nuclear waste disposal program appeared to be floundering, the BRWM of the National Research Council issued a "position statement" entitled *Rethinking High-Level Radioactive Waste Disposal* [32]. This document concentrated on institutional issues, not technical ones, and made an array of still pertinent suggestions. It started with the overall assessment:

There is no scientific or technical reason to think that a satisfactory geological repository cannot be built. Nevertheless, the U.S. program, as conceived and implemented over the past decade, is unlikely to succeed. [32, p. vii]

The difficulty, in this view, was the demand imposed by "historical and institutional conditions" in the United States that the entire waste disposal facility be planned in advance, with rigid specifications, before one can move ahead. In the view of this report, achieving the required geological certainty for a period of 10,000 years, as required under current guidelines, is essentially impossible, and the demand for this certainty is undesirable. A more constructive approach would be a flexible one, carried out in an admittedly exploratory spirit. It would then be possible to adjust the disposal plan as new information is obtained, and it would be accepted that there never can be absolute certainty.

Overall, the central message of the 1990 position statement is that it is prudent to proceed with waste disposal on the basis of current knowledge, without demanding complete assurance for the future. In essence, this is because the stakes, in terms of risks, are not as great as commonly assumed. The BRWM position included the following points:

◆ Quantitative prediction of the geological characteristics of the repository "far into the future stretches the limits of our understanding." Some aspects will remain uncertain, no matter what efforts we make to obtain information, although "uncertainty does not necessarily mean that the risks are significant" [32, pp. 2–3].

◆ The perception that waste disposal carries a possibility of "catastrophe"— analogous to dangers from potential reactor accidents—is incorrect because the energies involved are "far lower" and therefore the risks are "much lower" [32, p. 16].

◆ Rather than seek "perfect knowledge" in advance, we should follow the model of mining which is "fundamentally an exploratory activity" where plans are modified when unanticipated conditions are encountered [32, p. 2].

◆ A plan that calls for spelling out everything in advance and "getting all of the needed measurement and analysis on the first pass, with acceptably high quality, is not likely to succeed" [32, p. 33].

◆ Keep the approach flexible so that changes can be made as new information is gained during the development of the repository. Such changes should "not be seen as an admission of error" [32, p. 33].

Suggested Stepwise Approach (2001)

The 2001 report on *Disposition of High-Level Nuclear Waste and Spent Nuclear Fuel*, prepared by the BRWM-convened international panel mentioned earlier (see Section 13.4.1), also favored a stepwise approach [4]. This report put considerable emphasis on measures needed to win public support for any program and saw a gradual, flexible approach as helpful in this direction:

> For both scientific and societal reasons, national programs should proceed in a phased or stepwise manner, supported by dialogue and analysis. . . . Decision makers, particularly those in national programs, should recognize the public's reluctance to accept irreversible actions and emphasize monitoring and retrievability. Demonstrated reversibility of actions in general, and retrievability of wastes in particular, are highly desirable because of public reluctance to accept irreversible actions. [4, p. 5]

> In practice, there are many experts who feel confident enough about the state of current knowledge to advocate moving ahead now with repository implementation in correctly chosen locations. There are others who believe that a lot more research is needed before we even decide on the feasibility of permanent safe disposal. There are also those somewhere in the middle who support continuing the process leading to disposal, provided this remains stepwise and reversible. **The consensus of the committee reflects the views of this middle group.** The consensus that a cautious stepwise approach is

the way to proceed was reached because those members who believe that the science and technology are already sufficiently mature recognize that societal processes take time. [4, p. 29]

A phased or stepwise approach to implementation of repositories can offer a proper compromise between minimizing future burdens and maximizing future choice. Properly designed and sited repositories can have a long period of monitored controls and enhanced retrievability before being converted into their final closed state. [4, pp. 112–113]

In fact, the Yucca Mountain project is now planning to delay closing the repository for up to several centuries (see Section 12.2.1).

Adaptive Staging

The ideas outlined in the previous sections were extended in a new study, again undertaken under the auspices of the National Research Council's BRWM, that was published in 2003 in the report *One Step at a Time: The Staged Development of Geologic Repositories for High-Level Radioactive Waste* [33]. The study committee recommended a version of a stepwise approach that it termed "adaptive staging," which it contrasted to the alternative, termed "linear planning." The strategy in the latter approach is to establish in advance a clear goal and a path for reaching it, and then try to follow that path. In adaptive staging, the strategy is to learn as one goes along, in a process where the "ultimate path to success and end points themselves are determined by knowledge and experience gathered along the way" [33, p. 1].

Overall, the report's version of adaptive staging has several key components: (1) the establishment of many Decision Points for the evolving process, (2) a review at each Decision Point to gauge the current situation and plan the next steps, and (3) measures at each Decision Point to make the findings "publicly transparent" and to engage in "dialogue with stakeholders" [33, p. 2].[15] A rather specific recommendation for Yucca Mountain is that waste emplacement start there with a small-scale pilot trial.

Considerable emphasis is placed in this adaptive staging strategy on building external confidence. This can be seen in the *One Step at a Time* report, as well as in a summary paper presented by Charles McCombie and Barbara Pastina, who chaired and staffed, respectively, the committee that prepared the report [34]. This paper features six highlighted attributes of adaptive staging. Four of these are related to building trust and having good communication with "all interested parties" and stakeholders: auditability, transparency, integrity, and responsiveness. Two attributes appear directed mainly toward

---

[15] The report explicitly suggests that there are different meanings attached to the term "stakeholder," but it appears to be used here to refer to nongovernmental groups or individuals that are actively interested in repository decisions [33, p. 55].

improving the repository design: systematic learning and flexibility. Even in the area of systematic learning, input from stakeholders is regarded as important. This emphasis distinguishes it from the 1990 statement which also recommended expanded communication with the public, but gave such communication a considerably less prominent role.

Staged Development in the Yucca Mountain Project

It would appear that the Yucca Mountain project has evolved into a mode that includes aspects of a phased approach, although, for the most part, the goal has been to plan as much as possible in advance. The preparation of successive TSPAs has been an iterative process, with changes made along the way. Modifications of assessments and plans are likely to continue during the licensing process and beyond, and the repository will be kept open for many decades, perhaps even centuries, permitting subsequent changes. Such changes will be viewed by many as a recognition of past error, but in the spirit of staged planning, they could be accepted as an appropriate implementation of a flexible approach.

As discussed in a paper by three Nuclear Regulatory Commission staff members, the present licensing system already provides for a type of phased approach. They point to "three decision points or phases (i.e., construction authorization, license to receive and possess radioactive waste, and license amendment for permanent closure)" [35]. These are times when the repository is extensively reassessed. However, these three decision points are fewer in number and different in spirit than those envisaged in *One Step at a Time*.

In a different aspect of the licensing process, the decision by the DOE and Congress to set in motion an application to the NRC for construction authorization at Yucca Mountain, before all technical issues were resolved, also represents a limited kind of phased approach. Further investigations and assessment were known to be needed in some areas (e.g., on questions of corrosion resistance and optimal thermal loading) and these matters will continue to be addressed as the planning moves ahead, although without the formality of well-defined individual Decision Points. The DOE may hesitate to incorporate a formal policy of sequenced Decision Points in its procedures, because it may fear that each Decision Point could become more an occasion for new delays than an opportunity to improve the design.

13.5.2 Framework for Considering Intergenerational Responsibilities

Here, we suggest some possible guidelines for considering intergenerational responsibilities, drawn in part from the above-summarized arguments:

1. No generation is obliged to protect future generations from hazards that are smaller than those that it accepts for itself.
2. Each generation must avoid setting the stage for potential future catastrophes.
3. Each generation should strive to pass on to immediately succeeding generations an improved world, including the potential to sustain such improvements for the indefinite future.

If criterion 1 is accepted, then the 0.15 mSv/yr standard set by the EPA can be seen as too stringent. It corresponds roughly to present standards for exposure from nuclear power facilities (although slightly more demanding), but it is much below the levels we accept for exposure to natural radiation or the levels at which the EPA recommends action for indoor radon mitigation. As a nation, we are indifferent to variations of as much as 1 mSv/yr in the levels of natural radiation (excluding the much larger differences in levels of exposure from indoor radon), and as a world community we are indifferent to the "excess" natural doses of many millisieverts per year received by people living in high-radiation regions such as in Kerala, India.

Criterion 2 suggests that we should pay greater attention to assessing the possibility of major harm from a repository such as Yucca Mountain, rather than focus on meeting standards at a level where the harm is modest. Criterion 3, in the narrow context of nuclear power alone, repeats the call quoted earlier for balancing the benefits to be derived from the continued use of nuclear power against the hazards created.

13.5.3 Putting the Risks into Perspective

The Lack of a Defined Danger

The ongoing controversies about nuclear waste disposal have had something of the aspect of shadowboxing, because there has been no clear picture of the danger to be avoided. This can be contrasted, for example, with the concern over nuclear reactor accidents. As analyzed for many years before Chernobyl and illustrated by Chernobyl, there are plausible, specific scenarios for very serious consequences in the case of a reactor accident. The issue then involves estimating the likelihood of such accidents, and the whole matter can be considered with reference to identifiable and substantial dangers.

With nuclear wastes, on the other hand, the discussion often is disconnected from meaningful concerns. Much of the attention is directed toward compliance with regulatory requirements. However, it is often not possible to find a convincing justification for those requirements. The ^{14}C case is a particularly egregious example of what can happen when the standard bears no relationship to a significant danger (see Section 13.2.2). In discussing the rationale for the 1985 40CFR191 regulations, one knowledgeable observer[16] stated:

> It is my belief that the EPA document never has been and is not today a health-based document. I believe it was and is a technology-forcing document, like somebody at EPA might do to encourage the best available technology. [36]

Setting demands for the "best available technology," without regard to the actual dangers avoided, may not only be misleading but also could be destructive. Instead of forcing a technology to be better, excessively ambitious demands could have the perhaps unintended effect of forcing it to be abandoned.

The new standards of 40CFR197, established by the EPA in 2001, also appear to be decoupled from society's actual concerns. As an aesthetic matter, there may be an attractive simplicity in requiring that for 10,000 years, no individual will receive a dose greater than 0.15 mSv/yr.[17] However, otherwise, it is a perplexing standard. As already discussed, society is essentially indifferent to much greater exposures today. The average dose of a person in the United States from natural sources is roughly 20 times this level and we shrug off substantial variations above and below this average. The limit on occupational exposures is still higher, at present 50 mSv/yr.

Given our casual acceptance of natural doses of about 3 mSv/yr for *everyone* today, it is hard to believe that we care if a few of the *most exposed* of our descendants receive additional doses of 0.15 mSv/yr. In the same vein, maintaining the EPA's earlier standard of 1000 possible deaths in 10,000 years (0.1 per year) from radiation would have been hard to reconcile with our acceptance of thousands of deaths per year in industrial accidents, an estimated 10,000 or more possible deaths per year from indoor radon, and close to 50,000 deaths per year from automotive accidents.[18] Overall, it is not obvious what danger such standards seek to avoid.

Standards and Meaningful Concerns

The question of standards may be looked at in two contexts. In the formal context, involving EPA regulations and the development and evaluation of TSPAs by the DOE and NRC, it is unlikely that any change will soon be made in the standard set in 2001 in 40CFR197—namely a dose limit of 0.15 mSv/yr for 10,000 years. It also appears probable that the DOE and EPRI calculations

---

[16] Terry Lash, formerly director of the Illinois Department of Nuclear Safety and appointed in 1994 as the Director of the DOE's Office of Nuclear Energy.

[17] Here, we use "aesthetic" to indicate a subjective attractiveness to some individuals, for reasons that do not need to be justified by objective arguments.

[18] The deaths from radiation are termed "possible" because of uncertainties as to the effects of radiation at the low doses rates being considered (see Section 4.3.4).

will continue to indicate that this standard will be satisfied for the Yucca Mountain repository.

However, in a context that is more relevant to our real concerns, the deeper issue is not risks at the levels of these standards but, rather, the possibility that the wastes may cause some form of disaster, whether it be seen as a medical, environmental, or ecological disaster. We want strong evidence that severe harm is very unlikely. This suggests that an effort should made in the DOE analyses to formulate scenarios for serious nuclear waste mishaps and to estimate the likelihood of their occurrence.[19] The crucial question then becomes the probability of *a significantly large* number of people receiving *significantly large* doses. Inevitably, subjective judgments will enter into the definition of "significantly large," but couching the issue in these terms would frame the problem in a fashion related to the fundamental concerns.

In this spirit, the emphasis in waste disposal studies could sensibly move from establishing the probable avoidance of minor effects to establishing a nearly certain avoidance of serious ones. Thus, in further examination of the Yucca Mountain project and other future projects, special attention should be given to the probability and potential consequences of plausible worst-case nuclear waste scenarios. The goal would be to make sure that no large risks have been overlooked.

Comparison of Intergenerational Burdens

The consideration of nuclear waste issues is unusual in the attention given to questions of intergenerational responsibility. This would be understandable were the risks uniquely large. However, the analyses to date suggest that they are not. For example, for a situation somewhat similar to that of nuclear waste disposal, Okrent identifies cancer risks from chemically polluted sites that are orders of magnitude greater than those allowed for nuclear wastes, assuming that institutional memory is lost and future peoples settle at those sites [21, p. 890].

In another comparison, it is to be noted that the magnitudes of the wastes differ widely for different means of electricity generation. A coal-fired plant creates about 8.5 million tonnes of CO_2 per gigawatt-year. The same electrical output from a nuclear power plant produces roughly 30 tonnes of spent fuel together with about 20 tonnes of other materials in the fuel assemblies. The CO_2 is far less dangerous than spent fuel per tonne, but in aggregate, it may create problems of global climate change that dwarf in scope the hazards from the nuclear wastes (see Section 1.2.3). Here, the difference in mass is pertinent. Isolating the relatively small mass of spent fuel from the environment for ten

---

[19] This has been done for the case of volcanic activity, with estimates of both probability and impact (see Section 12.4.4).

thousand years appears to be possible. Isolating the much greater amounts of CO_2 from the environment is problematic.[20]

The exhaustion of fossil fuels is another problem that current human activities may be creating for future generations. If nations are unable to find adequate replacements in timely fashion, there could be serious economic, political, and, possibly, military turmoil as they scramble to secure what resources are available. Here, one is contemplating millions of possible future deaths in an energy-starved world—from poverty, armed conflict, or both—in contrast to several deaths per year, or less, from a flawed repository.[21]

In view of the potential contribution of nuclear energy to reducing these problems, it is appropriate to consider both the positive and negative sides of the production of nuclear wastes. A project such as Yucca Mountain should be judged in terms of both the risks that it may create for future generations and those that it may help to ameliorate.

Summary

If the results of the current analyses carried out by the DOE and EPRI stand up upon further review and if no plausible "disaster scenarios" are found, the risks created by the wastes appear to be small. Nonetheless, it is conceivable that in the end, either technical or political considerations will prevent the implementation of the Yucca Mountain project, forcing an intensified examination of alternatives—other geologic sites, subseabed disposal, transmutation, or interim storage for an extended time period.

The last of these alternatives may be the most likely should Yucca Mountain not be opened to receive wastes. The development of interim storage facilities would meet the immediate demands of nuclear utilities that have accumulating inventories of spent fuel. In addition, for better or worse, it would postpone an ultimate resolution of the "waste disposal problem." Such a delay might satisfy those people who see no technological or political objections to interim storage as well as those who distrust any presently conceived permanent solution. Viewed broadly, however, it would decrease the prospects for nuclear power development in the United States. Conversely, sustained progress on Yucca Mountain would improve the prospects.

---

[20] See Section 20.2.2 for a brief discussion of CO_2 sequestration.

[21] If, for example, 1000 people received an average dose of 1.5 mSv/yr (10 times the EPA standard), the resulting population dose would be 1.5 person-Sv/yr. If one applies the linearity hypothesis in estimating the effects of low-level radiation, this corresponds to under one death per decade.

References

1. Mark Holt, *Civilian Nuclear Waste Disposal*, CRS Issue Brief for Congress (Washington, DC: Congressional Research Service, March 2, 2002).
2. Allen V. Kneese, "The Faustian Bargain," *Resources*, no. 44, September 1973: 1–5.
3. Riley E. Dunlap, Michael E. Kraft, and Eugene A. Rosa, eds., *Public Reactions to Nuclear Waste: Citizens' Views of Repository Siting* (Durham, NC: Duke University Press, 1993).
4. U.S. National Research Council, *Disposition of High-Level Waste and Spent Nuclear Fuel: The Continuing Societal and Technical Challenges* (Washington, DC: National Academy Press, 2001).
5. 97th Congress, *Nuclear Waste Policy Act of 1982*, Public Law 97-425 [H.R. 3809] (January 7, 1983).
6. *Protection of the Environment, U.S. Code of Federal Regulations*, title 40.
7. U.S. Environmental Protection Agency, *High-Level and Transuranic Radioactive Wastes: Background Information Document for Final Rule*, Report EPA-520/1-85-023 (Washington, DC: U.S. EPA, 1985).
8. "Environmental Standards for the Management and Disposal of Spent Nuclear Fuel, High-level and Transuranic Radioactive Wastes," *Federal Register*, 50, no. 182, 1985: 38066–38089.
9. U. Sun Park and Chris G. Pflum, "Requirements for Controlling a Repository's Release of Carbon-14 Dioxide; the High Costs and Negligible Benefits," in *High Level Radioactive Waste Management*, Vol. 2, Proceedings of the International Topical Meeting, (La Grange Park, IL: American Nuclear Society, 1990): 1158–1164.
10. Robert W. Andrews, Timothy F. Dale, and Jerry A. McNeish, *Total System Performance Assessment—1993: An Evaluation of the Potential Yucca Mountain Repository* (Las Vegas, NV: INTERA, Inc., 1994).
11. Benjamin Ross, "The Technical Basis for Regulation of Gas-Phase Releases of Carbon-14," in *EPRI Workshop 1—Technical Basis for EPA HLW Disposal Criteria*, Proceedings of conference held in September 1991, EPRI report TR-100347 (Palo Alto, CA: EPRI, 1993): 159–172.
12. U.S. Environmental Protection Agency, *Environmental Radiation Protection Standards for Yucca Mountain, Nevada. Draft Background Information for Proposed 40 CFR 197*, Report EPA 402-R-99-008 (Washington, DC: U.S. EPA, 1999).
13. *National Resources Defense Council, Inc. v. U.S. Environmental Protection Agency*, 824 F.2d 1258 (U.S. Court of Appeals, 1st Cir. 1987).
14. 102d Congress, *Energy Policy Act of 1992*, Public Law 102-486, title VIII—High Level Radioactive Waste (October 24, 1992).
15. U.S. Environmental Protection Agency, "40 CFR Part 197, Public Health and Environmental Protection Standards for Yucca Mountain, NV; Final Rule," *Federal Register* 66, no. 114, 2001: 32074–32135.
16. National Research Council, *Technical Bases for Yucca Mountain Standards*, Report of Committee on Technical Bases for Yucca Mountain Standards (Washington, DC: National Academy Press, 1995).

17. U.S. Environmental Protection Agency, "40 CFR Part 197, Environmental Protection Standards for Yucca Mountain, Nevada; Proposed Rule," *Federal Register* 64, no. 166, 1999: 47976–47016.

18. Richard A. Kerr, "Science and Policy Clash at Yucca Mountain," Science 288, 2000: 602.

19. Letter to Carol M. Browner, EPA from Michael Kavanaugh, Chair and John Ahearne, Vice Chair, Board on Radioactive Waste Management, National Research Council (November 26, 1999).

20. David Okrent, *On Intergenerational Equity and Policies to Guide the Regulation of Disposal of Wastes Posing Very Long Term Risks*, Report UCLA-ENG-22-94 (Los Angeles, CA: UCLA School of Engineering and Applied Science, 1994).

21. David Okrent, "On Intergenerational Equity and Its Clash with Intragenerational Equity and on the Need for Policies to Guide the Regulation of Disposal of Wastes and Other Activities Posing Very Long-Term Risks," *Risk Analysis* 19, no. 5, 1999: 877–901.

22. Edith Brown Weiss, *In Fairness to Future Generations: International Law, Common Patrimony and Intergenerational Equity* (Tokyo: The United Nations University, 1988).

23. Organization for Economic Cooperation and Development, *The Environmental and Ethical Basis of Geological Disposal of Long-Lived Radioactive Wastes, A Collective Opinion of the Radioactive Waste Management Committee of the OECD Nuclear Energy Agency* (NEA/OECD, 1995). [From: http://www.nea.fr/html/rwm/reports/1995/geodisp.html]

24. National Academy of Public Administration, *Deciding the Future: Balancing Risks, Costs and Benefits Fairly Across Generations*, A Report by a Panel of NAPA, Harold B. Finger, Chairman (Washington, DC: NAPA. 1997).

25. Electric Power Research Institute, *Evaluation of the Proposed High-Level Radioactive Waste Repository at Yucca Mountain Using Total System Performance Assessment, Phase 6* (Palo Alto, CA: EPRI, 2002).

26. Freeman Dyson, *Imagined Worlds* (Cambridge, MA: Harvard University Press, 1997).

27. B. L. Cohen, "Discounting in Assessment of Future Radiation Effects," *Health Physics* 45, no. 3, 1983: 4687, as quoted in Ref. [21, p. 893].

28. William G. Sutcliffe, "Faulty Assumptions for Repository Requirements," in Proceedings of *Global '99, Nuclear Technology—Bridging the Millennia* (La Grange Park, IL: American Nuclear Society, 1999) [CD-ROM only].

29. International Atomic Energy Agency, *Safety Principles and Technical Criteria for Underground Disposal of High Level Radioactive Waste*, Safety Series 99 (Vienna: IAEA, 1989), as quoted in Ref. [16, p. 56].

30. National Research Council, *Decision Making in the Environmental Protection Agency, Volume II*, Report from the Committee on Environmental Decision Making (Washington, DC: National Academies Press, 1977).

31. Luther J. Carter, *Nuclear Imperatives and Public Trust: Dealing with Radioactive Waste* (Washington, DC: Resources for the Future, 1987).

32. National Research Council, *Rethinking High-Level Radioactive Waste Disposal*, A Position Statement of the Board on Radioactive Waste Management (Washington, DC: National Academy Press, 1990).

33. National Research Council, *One Step at a Time: The Staged Development of Geologic Repositories for High-Level Radioactive Waste*, Report of the Committee on Principles and Operational Strategies for Staged Repository Systems, Charles McCombie, ch. (Washington, DC: National Academies Press, 2003).

34. C. McCombie and B. Pastina, "Staging the Development of Geologic Repositories for High-Level Waste," Proceedings of the 10th International High-Level Radioactive Waste Management Conference (La Grange Park, IL: American Nuclear Society, 2003): 971–976.

35. John Greeves, Tim McCartin, and Bill Reamer, "Phased Licensing Approach in NRC Regulations for Yucca Mountain," Proceedings of the 10th International High-Level Radioactive Waste Management Conference (La Grange Park, IL: American Nuclear Society, 2003): 989–992.

36. Terry Lash, *EPRI Workshop 1—Technical Basis for EPA HLW Disposal Criteria*, Proceedings of conference held in September 1991, EPRI Report TR-100347 (Palo Alto, CA: EPRI, 1993): 394.

14

Nuclear Reactor Safety

14.1 General Considerations in Reactor Safety

14.1.1 Assessments of Commercial Reactor Safety

The historical record of nuclear reactor performance can be interpreted as showing that they are very safe or that they are very dangerous. The former conclusion follows if one limits consideration to plants outside the former Soviet Union (FSU). The latter conclusion follows if one focuses on the Chernobyl accident and takes it as a broadly applicable indicator.

For commercial reactors in the non-Soviet world, which account for the largest part of the reactor experience, the safety record is excellent. As of the end of 2003, these reactors had a cumulative operating experience of about 10,100 reactor-years, of which about 2870 reactor-years were logged by U.S. reactors.[1] There has been no accident in any of these reactors, including the 1979 Three Mile Island (TMI) accident, that has caused the known death of any nuclear plant worker from radiation exposure or that has exposed any member of the general public to a substantial radiation dose.

If one goes beyond Western commercial reactors, there are three exceptions to this excellent record. Two involved reactors built for military purposes and are sometimes overlooked—the 1957 Windscale accident in a British plutonium-producing reactor that led to some significant exposures and the 1961 SL-1 accident in the United States in which three army technicians died. (These accidents are described briefly in Section 15.1.) The third was much

---

[1] The number of reactor-years is extrapolated from December 31, 2002 data [1, Table 7]. It includes the contribution from commercial reactors that are no longer in operation.

greater in impact and has received far more attention: the Chernobyl accident in the Soviet Union in 1986 (see Section 15.3).

The Chernobyl reactor was graphite moderated, with a number of unusual design features, and the circumstances of that accident could not be repeated in the standard LWRs and HWRs used outside the FSU. Nonetheless, no reactor has a truly zero chance of an accident and Chernobyl demonstrated that a major reactor accident could potentially impact hundreds of thousands of people. For this reason, high importance is attached to issues of reactor safety by proponents and opponents of nuclear power alike.

Assessments of reactor safety involve estimates of both the probability and severity of accidents. In the remainder of this chapter, we will explore some of the general issues involved in achieving and evaluating nuclear safety. In the following chapter, we will look at the failures, cases where accidents did, in fact, occur. We will be interested in both their causes and consequences.

14.1.2 The Nature of Reactor Risks

Categories of Reactor Accidents

There is a large spectrum of possible consequences from a nuclear reactor accident. The most serious accident is one in which there is a large external release of radionuclides, as was the case at Chernobyl. Less harmful, but still serious, are accidents in which there is damage to the reactor core, but with no appreciable release of radionuclides to the outside environment, as at Three Mile Island. The spectrum can be extended downward to include everything from near misses to harmless breakdowns that have no actual or likely adverse consequences. These lesser mishaps are of interest primarily because of the cost of the remedial measures and lost time, and for the light they shed on the probability of more serious accidents (see Section 14.4.3 on precursor analyses).

Potential major nuclear reactor accidents fall into two main categories, each illustrated by one of the two major past accidents in power reactors, the Chernobyl and Three Mile Island accidents:[2]

- *Criticality accidents.* These are accidents in which the chain reaction builds up in an uncontrolled manner, within at least part of the fuel. In an LWR of normal design, such accidents are highly improbable, due to negative feedbacks and shutdown mechanisms. They are less unlikely in some other types of reactor, given sufficient design flaws. The 1986 Chernobyl accident was a criticality accident, although much of the energy release was from a steam explosion following the disruption of the core.
- *Loss-of-coolant accidents.* When the chain reaction is stopped, which can be accomplished quickly in the case of an accident by inserting control

---

[2] The Windscale accident does not fit into either of these categories (see Chapter 15).

rods, there will be a continued heat output due to radioactivity in the reactor core. Unless adequate cooling is maintained, the fuel temperature will rise sufficiently for the fuel cladding and the fuel to melt, followed by the possible escape of radioactive materials from the reactor pressure vessel and perhaps from the outer reactor containment. The TMI accident was a loss-of-coolant accident. There was substantial core melting, but no large escape of radioactive material from the containment.

With appropriate precautions, such as the assurance of intrinsic negative feedback and the capability for rapid insertion of control rods, a criticality accident is virtually impossible in a well-designed reactor. Therefore, almost all of the attention to reactor accidents in the United States and elsewhere is directed to the more demanding task of avoiding a loss-of-coolant accident.

In the light of possible misapprehensions, it is worth noting that a bomblike nuclear explosion cannot occur in a nuclear reactor. In a bomb, a critical mass of almost pure fissile material (^{235}U or ^{239}Pu) is brought together violently and compressed by the force of a chemical explosion, and the chain reaction develops fully within one-millionth of a second—quickly enough for much of the fuel to fission before the mass is disassembled (see Section 17.2.3). In a reactor, most of the mass is not fissile. Even in the fuel, the fissile mass is small compared to the ^{238}U mass.[3] A reactor also contains a great deal of other nonfissile material in the form of coolant, moderator (if there is a moderator other than the coolant), fuel cladding, and metal support structures.

The presence of the nonfissile material has two consequences that are pertinent to the issue of explosions: (1) The multiplication factor k in a reactor is close to unity, whereas in a bomb it approaches 2, and (2) the average time between fission generations (the mean neutron lifetime l) is greater in a reactor than in a bomb, because the most frequent neutron reactions in a reactor are elastic or inelastic scattering, not fission. As a result, the chain reaction builds up much more slowly in a reactor than in a bomb [see Eq. (7.15)].

Overall, the first "line of defense" against an explosion in a reactor is the negative feedback that prevents criticality accidents. This should suffice. However, if there are mistakes in the design or operation of the reactor and the chain reaction reaches too high a power level, there is time for the ultimate "negative feedback" to come into play—the partial disassembly of the reactor core, which stops the chain reaction after only a relatively small amount of energy has been produced (i.e., only a small fraction of the nuclei have fissioned). This is what happened in the Chernobyl accident, where most of the energy of the explosion came from chemical reactions, including steam interacting with hot metal (see Section 15.3.2). Such an accident can be very serious, but the consequences are not on the scale of the consequences of a nuclear explosion.

---

[3] The fission cross section for neutrons colliding with ^{238}U is small for neutron energies below 2 MeV and is negligible below 1 MeV (see Section 6.2.3).

Aftermath of a Reactor Accident

Nuclear accidents pose particular problems because of the persistent effects of radioactivity. The heat output immediately after the reactor is shut down is about 7% of the thermal output of an operating reactor (see Section 14.2.2). Although the activity and energy release fall rapidly with time, a serious accident can occur if this heat is not removed by the cooling system. At Three Mile Island, this continued production of heat led to the fear that the accident might progress further, with the release or ejection of radioactive material from the reactor containment. At Chernobyl, there was a very large release of radioactive material, and the dispersed debris has created problems that will last for many years.

This may be contrasted with the situation in many other sorts of accident (with important exceptions, such the Bhopal accident in India in 1984). Once a dam breaks or a natural gas facility explodes, the damage is done and society feels moderately secure in coping with the aftermath. There may be more immediate fatalities than in a nuclear accident,[4] but when the accident is over, it is usually deemed to be over, and there is little investigation of possible lingering consequences. With nuclear accidents, serious consequences may persist for a long period of time—in particular, cancers caused by both the initial exposure and the continuing exposures due to radionuclides deposited on the ground.

These factors, plus less well-defined but widely held fears, put nuclear accidents in a special category of societal concern and make it particularly urgent that they be avoided. There can be debates as to the effect of an accident on the health of the public. There is no doubt, however, that each nuclear accident has been something of a disaster for the nuclear industry.

14.1.3 Means of Achieving Reactor Safety

General Requirements

Underlying the approach to safety, for any sort of equipment, are high standards in design, construction, and the reliability of components. In nuclear reactors, concern about possible accidents has led to particularly intense efforts to achieve high standards. Individual components of the reactor and associated equipment must be of a codified high quality. As described in an OECD report:

> In the early years of water reactor development in the USA, a tremendous effort was put into development of very detailed codes and

---

[4] For example, explosions in liquid-natural-gas tanks and the associated fires killed 130 people in Cleveland, Ohio, in 1944 and 40 workers on Staten Island in New York, in 1973 [2, p. 162]; in each case, the casualties exceeded the prompt fatalities at Chernobyl (see Section 15.3).

standards for nuclear plants, and these were widely adopted by other countries where nuclear plants were initially built under US licenses. [3, p. 62]

The efforts of the United States have since been supplemented by parallel efforts by other countries and the International Atomic Energy Agency (IAEA). In parallel, a nuclear reactor safety philosophy has developed which includes a number of special features, as summarized in the succeeding subsections.[5]

Passive or Inherent Safety

A distinction is made between *active* and *passive* safety systems. An active safety system is one that depends on the proper operation of reactor equipment, such as pumps or valves. For example, active safety systems include the pumps and valves that control the water supply for emergency core cooling and the motors used to insert control rods in emergency shutdowns. Passive safety features are aspects of the system that are arranged to come into play automatically, without the action either of the operators or of mechanical devices that might fail. The gravity-driven fall of a control rod is a passive feature, although its purely passive character would be compromised if the release of the rods is initiated by an active system.

The terms "passive safety" and "inherent safety" are often used interchangeably, although some authors may intend a difference in meaning or nuance. These terms suggest that the safety of the reactor will depend on immutable physical phenomena rather than on the proper performance of individual components or correct actions by reactor operators. For example, if the thermal expansion of the reactor core provides a negative feedback, the expansion provides an inherent safety feature. In the extreme version of the concept, in a passively safe reactor all operators could become incapacitated and all external electricity and water could be shut off, and still the reactor would turn itself off in the case of an accident and gradually cool with no damage.

This terminology is widely used but has also been criticized. The objections have had several strands:

- Inherent or passive safety is a matter of degree, rather than a totally new departure. A negative temperature coefficient or a negative void coefficient is a passive safety feature and, therefore, most existing reactors already have passive safety features.
- The terms are misleading in that they seem to suggest that an accident would be *totally* impossible, whereas, in fact, one can find circumstances in which any given reactor might fail if arbitrarily improbable scenarios are permitted.
- The terms could appear to have a prejudicial aspect because they could seem to suggest that existing reactors are *not* safe.

---

[5] The discussion loosely follows the organization used in Ref. [4, pp. 9ff].

The criticisms have had some force, and to defuse them alternative words have sometimes been suggested [5]. However, whatever words are used or caveats included, the concept is clear: It is safer to rely on basic physical phenomena (e.g., gravity or thermal expansion) rather than on the consistently good performance of equipment and operators.

Redundancy

The likelihood of any sort of accident can be reduced by redundancy, which can be achieved in a number of ways:

- *Identical units of the same type.* Often, more than one pump or motor is provided to perform a given safety task, although it is only necessary that one of these operates properly. It is particularly important in such cases to avoid common-mode failures, in which one failure could simultaneously disable all of the units. To achieve this, among other demands, there must be adequate physical separation between the units and between the control systems for them.[6]
- *Diverse types of systems.* An example of diversity in reactor safety design is the provision of different types of emergency core-cooling systems, which act independently.

Defense-in-Depth

A special kind of redundancy is sometimes singled out as being "at the heart of nuclear safety." This is the reliance on *multiple barriers* or *defense-in-depth*, which is described as "a hierarchically ordered set of different independent levels of protection" [6, p. 109]. The principle of defense-in-depth is seen in considering the barriers that prevent or minimize exposures due to the release of radioactivity from a reactor:

- The UO_2 fuel pellets retain most radionuclides, although some gaseous fission products (the noble gases and, at elevated temperatures, iodine and cesium) may escape.
- The zircaloy cladding of the fuel pins traps most or all of the gases that escape from the fuel pellets.
- The pressure vessel and closed primary cooling loop retain nuclides that escape from the fuel pins due either to defects in individual pins or, in the case of an accident, overheating of the cladding.

---

[6] After the Browns Ferry fire in 1975, it was recognized that multiple wiring systems, intended for redundancy, were carried in the same cable trays and, therefore, were all disabled at the same time. A simple solution is to use different paths for redundant cabling.

- ♦ The heavy outer reactor containment, with its associated safety systems, is designed to retain radionuclides that escape through the cooling system or, in the case of a very severe accident, from the pressure vessel.

- ♦ If these systems all fail and there is a significant release of activity to the outside environment, the population can be partially protected through evacuation. However, if radiation escapes the containment, then the system has been defeated even if evacuation reduces the damage.

Steps taken to avoid the overheating of the fuel—in particular, the standard and emergency cooling systems—as well as systems to suppress overpressurization of the containment can also be considered to be part of the defense-in-depth.

These barriers against radiation exposures have been put to a severe test in only two instances. In the TMI accident, the reactor containment was highly successful. In the Chernobyl accident there was no containment, as the term is understood in Western design practice, and there was a massive release of radioactive material to the outside surroundings. Subsequent emergency evacuations reduced the exposure of people in the evacuation zone, but there was substantial exposure of the public nonetheless (see Section 15.3).

Defense-in-depth and the various forms of safety redundancy represent a sophisticated version of the view that although it is likely that *something* will go wrong, it is highly unlikely that *everything* will go wrong. Illustrating the value of redundancy (reiterating a point made in Section 12.3.2), if the causes of the failures are uncorrelated, three independent barriers that each have a 1% chance of failure provide a system in which there is only one chance in one million of overall failure.[7]

14.1.4 Measures of Harm and Risk in Reactor Accidents

The most fundamental harm in reactor accidents is that caused by radiation exposures. The extent of the harm can be alternatively measured in terms of individual radiation exposures, the collective population exposure, the number of prompt fatalities caused by intense exposures, or the number of latent cancers caused by lower radiation doses. Of these, prompt fatalities represent the most dramatic and least ambiguous effect. However, the greatest predicted health consequence is latent cancer fatalities (i.e., the eventual cancer deaths expected to occur due to radiation exposures). The doses might be received

---

[7] Another way of looking at reactor safety, also sometimes termed "defense-in-depth," is to divide it into phases of accident avoidance, accident correction or protection, and accident mitigation (e.g., Ref. [7, p. 339]). Avoidance is achieved by proper design, maintenance, and operation. Accident correction is achieved by reliable safety systems that, for example, shut the reactor down promptly and alert the operators. Accident mitigation is achieved by, for example, restoration of lost cooling, an effective containment system, and, as a last ditch measure, evacuation of the immediately surrounding population.

mostly in the first few days or in the first year following the accident, but the cancer fatalities would appear over many decades, generally starting after a latent period of 10 years.

Other harm includes physical damage to the reactor plant and contamination of the surrounding environment that may force the evacuation of large regions. Plant damage was clearly the most important direct consequence of the TMI accident and ground contamination was a major, perhaps in the end *the* major, consequence of Chernobyl.

Reactor accident risks are often analyzed in terms of the probability of two defining aspects of reactor accidents. One is the probability of reactor *core damage*—in particular, the melting of part of the core. The other is the probability of a *large radiation release*, stemming from the failure of the barriers provided by the reactor pressure vessel and the reactor containment.

The distinction between these consequences is illustrated by the TMI accident where, as discussed in Chapter 15, there was surprisingly little release of radioactive material to the environment outside the reactor containment although the damage to the reactor fuel assemblies was great. This focused attention on the accident *source term*—the inventory of radionuclides released to the outside environment, as distinct from the inventory of radionuclides in the fuel.

For ^{131}I and other iodine isotopes, the source term at Chernobyl was essentially the total initial core inventory. At TMI, it was close to zero: 18 Ci out of 64 million Ci, as reported in an American Nuclear Society study [8, pp. 1–12]. The low release of iodine at TMI was the result of the fact that there was much more cesium than iodine in the core inventory and the iodine predominantly formed cesium iodide (CsI), rather than the volatile gas I_2 (see, e.g., Ref. [8, pp. 8–9]).[8] The CsI was then trapped by dissolution in water or deposition on surfaces.

A crucial question is whether the very good performance of the containment system is generic to all LWRs or was peculiar to TMI. There have been a number of studies of this matter—for example, studies carried out under the auspices of the American Nuclear Society (ANS) [8] and the American Physical Society (APS) [9]. These studies concluded that, in most cases, the source term will be substantially less than the core inventory. If the source term is sufficiently low, then there is no "large release" of radionuclides.

Although there is no single indicator of reactor safety, in practice the most significant measure may be the core damage probability. Any instance of core damage at least raises the possibility of a significant radiation release and would inevitably deeply concern the public. Further, even with no release of activity outside the reactor containment, the cleanup expense after the core is damaged would be punitively expensive for the utility. Thus, much of the

---

[8] The differences in abundances results from the continual decay of ^{131}I ($T = 8.02$ days) during the months of reactor operation, whereas ^{137}Cs ($T = 30$ yr) kept increasing in amount.

efforts in assessing and reducing reactor risks focuses on the possibility of core damage.

14.2 Accidents and their Avoidance

14.2.1 Criticality Accidents and Feedback Mechanisms

General

In normal operation of a thermal reactor, prompt criticality is avoided. The reactivity of the system is kept low enough to make delayed neutrons crucial for criticality. Thus, even if the reactivity rises, the rates of increase of the neutron flux and of the power output are relatively slow. The magnitude of any power excursion is limited in an appropriately designed reactor by inherent negative feedbacks that come into play automatically. This gives time for the insertion of control rods, which have high neutron-absorption cross sections and will terminate the chain reaction. We consider below two major feedback mechanisms that enhance reactor safety.[9] Unless otherwise indicated, it will be assumed that the reactor considered is a standard LWR.

Fuel Temperature Feedback: Doppler Broadening

Although we have been tacitly treating the nuclei of the fuel as motionless targets undergoing bombardment by neutrons, this is not a precise description. The uranium nuclei are in thermal motion, with an average speed that increases as the temperature increases. The result is to increase the effective cross section for neutron absorption in ^{238}U if the temperature of the fuel rises, through the Doppler broadening of the absorption resonances (see Section 5.2.3). The number of neutrons available for fission is reduced, and the reactivity and the reactor power output decrease.[10] This negative feedback comes into play quickly, reversing the rise in power output as soon as the fuel temperature rises.

However, the fuel temperature feedback is not automatically negative in all types of reactors. If a fuel has relatively little ^{238}U and is primarily made of fissile material, then the main effect of Doppler broadening is to increase the rate of fission at nonthermal energies, giving a positive feedback. Thus, to keep the fuel temperature feedback negative, the fraction of fissile fuel in liquid-metal fast breeder reactors is kept below 30% [7, p. 146].

---

[9] This is not intended as a full listing of feedback mechanisms. Additional ones exist, both positive and negative (see, e.g., Ref. [7, pp. 145ff]), and must be taken into account in reactor design.

[10] In terms of the four-factor formula [Eq. (7.5)], the resonance escape probability, p, is reduced.

Void Coefficients

In an LWR, water is essential for moderating the reaction. If the water is removed (e.g., if there is a pipe break and insufficient replacement water is provided), the moderation will be inadequate and the reactivity will drop, because with less thermalization, there will be more loss of neutrons through absorption in ^{238}U. More voids also mean a greater escape of neutrons from the reactor. Loss of water in the reactor vessel is the limiting case of a "void."

The term *void coefficient* is usually applied to the replacement of liquid coolant by bubbles. The void coefficient is defined as the ratio of the change in the reactivity to the change in the void fraction. A negative void coefficient means that the reactivity decreases as the volume of steam bubbles increases (i.e., the void fraction increases). The loss of water leads to two effects which contribute to a negative void coefficient: (1) less effective moderation (i.e., relatively less elastic scattering of neutrons by hydrogen) and therefore increased resonance absorption of neutrons in ^{238}U and (2) more leakage of neutrons from the reactor.[11] A negative void coefficient corresponds to a negative feedback in accident situations, because the void fraction rises as the power level rises.

However, water also acts as an absorber of slow neutrons, and too much water leads to too much absorption (a low thermal utilization factor f). Were this the dominant effect, then the void coefficient would be positive. Thus, there is a competition between the moderating and absorbing roles of water, with opposite feedback signs. When the void coefficient is negative, the reactor is *undermoderated*; when it is positive, the reactor is *overmoderated*.

For BWRs, in which steam and water are both present, an increase in the steam content corresponds to less water. The moderating role is more important than the absorbing role, and an increase in steam content decreases the reactivity. Thus, the void coefficient is always negative for BWRs. In PWRs, there is usually no direct void coefficient, but thermal expansion of water has the same general effect of reducing moderation and providing a negative feedback.

The situation is more complicated for water-cooled graphite-moderated reactors, and the sign of the feedback can go either way depending on the relative amounts of water and graphite. The role of the water as moderator is less important, and the main effect of the water (aside from the intended function of cooling) can be to absorb neutrons. Loss of this water, by conversion to steam or otherwise, can increase the reactivity (i.e., the void coefficient is positive). This was the situation at Chernobyl. However, this is not intrinsic to all water-cooled graphite reactors. In particular, the N reactor formerly operating at Hanford had a negative void coefficient.

In sodium-cooled fast breeder reactors, the sodium plays only a small role as a moderator, but this moderation acts to lower the reactivity, because

---

[11] Referring to the five-factor formula [Eq. (7.4)], these feedbacks correspond to a lower resonance escape probability p and a lower nonleakage probability P_L.

the fission cross section increases with energy for neutron energies in the neighborhood of 1 MeV. Thus, the thermal expansion of the sodium or the development of bubbles reduces the moderation, increases the average energy in the neutron spectrum, and increases the reactivity. At the same time, with less sodium in the path of a potentially escaping neutron, more neutrons can escape from the reactor, reducing the reactivity. Overall, these competing effects may leave a sodium-cooled reactor with a positive void coefficient and it is important that there be counterbalancing negative feedbacks.

14.2.2 Heat Removal and Loss-of-Coolant Accidents

Decay Heat from Radioactivity

The central problem in loss-of-coolant accidents arises from the need to remove the heat produced by radioactivity during the period after reactor shutdown. The magnitude of the initial rate of heat generation can be understood in terms of the total energy release in fission, as discussed in Section 6.4.2. On average, for each fission event, about 7.8 MeV is released in beta decay and 6.8 MeV in accompanying gamma decay, for a total of 14.6 MeV out of about 200 MeV (i.e., approximately 7% of the total energy release). Strictly speaking, this result is applicable only when equilibrium has been reached between the production of radionuclides and their radioactive decay. However, the initial activity is dominated by short-lived radionuclides with half-lives of several days or less. Thus, if a reactor has been operating at full power for, say, a month, the total activity reaches a value close to its equilibrium level.

The activity just after shutdown is the same as the activity just before shutdown (treating shutdown as essentially instantaneous), and the initial thermal output from radioactive decay is 7% of the thermal output of the reactor during normal operation, or about 20% of the electric output. Thus, at shutdown of a 1000-MWe reactor, the heat output is initially about 200 MW. It drops to about 16 MW after 1 day and about 9 MW after 5 days [10, p. S23]. Without cooling, these heat production rates are sufficient to melt the fuel.

Core-Cooling Systems

During normal operation, reactor cooling is maintained by the flow of a large volume of water through the pressure vessel. This flow can be disrupted by a break in a pipe, failure of valves or pumps, or, in PWRs, a failure of heat removal in the steam generators. Such accidental disruptions of the normal cooling system are generically termed loss-of-coolant accidents (LOCAs). To guard against the overheating of the fuel in a LOCA, light water reactors have elaborate emergency core-cooling systems intended to maintain water flow to the reactor core.

A distinction is sometimes made between large and small LOCAs. The prototypical large LOCA is a break in the pipes carrying the primary cooling water to the reactor. In a large break, the pressure in the reactor vessel will be lost and a large amount of water will escape. The emergency core-cooling system (ECCS) then comes into play. Initially, replacement water is delivered from "accumulators" driven by nitrogen gas under pressure. Later, low-pressure pumps can provide additional water from external supplies. A large LOCA would be a dramatic event, and much of the early concern about reactor safety focused on preventing such an accident and, if prevention failed, assuring an effective and independent ECCS.

A small LOCA may occur from a leak in the primary cooling loop or, as was the case for the initiating event in the TMI accident, from a problem in the secondary cooling loop. Loss of secondary flow means that heat cannot be removed in the heat exchanger from the primary loop. In such an event, the pressure in the reactor vessel may not be relieved, and it may be difficult to establish the flow of replacement water in the complex hydraulic environment created by the mixture of steam and water at high pressures. To cope with such circumstances, the ECCS has a high-pressure injection system to provide replacement water to the reactor vessel.

The effectiveness of the ECCS for both large and small LOCAs has been the subject of many studies, starting before and intensifying after the TMI accident. In addition to calculations and theoretical analyses, there have been extensive tests, particularly the loss-of-fluid test (LOFT) program at the Idaho National Engineering Laboratory. This program was carried out from 1978 to 1985 and involved simulated accidents on a specially built 50-MWt test reactor. This was an NRC facility, but tests were also carried out there for the Nuclear Energy Agency of the OECD. Analyses of the results of these tests of system performance under simulated accident conditions have led to improvements in equipment and procedures (see, e.g., Ref. [3, pp. 39–42]).

Release of Radionuclides from Hot Fuel

If either the normal or emergency core-cooling system operates properly, there will be no damage to the reactor core in case of a reactor malfunction and no concern about release of radionuclides. However, if the cooling system fails to keep the cladding temperatures low enough to avoid melting, radionuclides will escape into the pressure vessel and into the primary cooling system.

The radionuclides include both fission products and actinides. They can be grouped according to differences in their volatility. The most volatile are the noble gases. These can diffuse out of the fuel into the fuel pins even at normal fuel temperatures. As the fuel temperature rises, damage to the fuel and the cladding causes release of additional elements, the most volatile of which are iodine and cesium. Some other radionuclides, in contrast, are quite refractory and are not released in substantial amounts even under extreme circumstances.

Thus, in one hypothetical accident, presented as an example in an NRC study, the median fission product release from the fuel rods was close to 100% for the noble gases, 6% for iodine, 1% for cesium, and 0.05% for strontium [11, p. A-34]. Although these particular values cannot be taken as precise measures of what would happen in a specific actual accident, they illustrate the main trends.

Although we have emphasized transport through the cooling system as the main avenue for radionuclide release, as was the case at TMI, there are other possibilities. In one extreme case, molten reactor fuel might settle in the bottom of the reactor vessel, melt through the vessel wall, and penetrate into the concrete base below. This scenario is sometimes referred to as the "China syndrome." The main consequence is not as extreme as the name might suggest. It comes from the generation of gases (such as CO_2 and others) in the interaction between the molten fuel and the concrete. This could produce an aerosol that carries nonvolatile radionuclides out of the fuel and into the atmosphere of the containment.

If radionuclides escape from the cooling system or from the reactor vessel, the next barrier is the containment structure. The integrity of the containment can be compromised by overpressure, most likely from the buildup of steam. To avoid this, there are containment cooling systems, either passive or active, intended to condense the steam. For example, PWRs commonly have spray systems for condensation, and BWRs have pools of water for pressure suppression. Some units also have refrigeration units. It is also possible, in the case of an excessive buildup of pressure, to release gas from the containment through valves, with filters to remove radionuclides.

14.3 Estimating Accident Risks

14.3.1 Deterministic Safety Assessment

One approach to establishing and evaluating reactor safety is to establish strict criteria for reactor design and construction and to analyze the behavior of the resulting system for a variety of postulated failures. The more demanding of these failure scenarios are termed *design basis accidents*. The reactor performance is studied through experiments and computational models to investigate whether the safety systems are adequate to cope with a design basis accident. For example, one can postulate a break in a cooling system pipe and then examine whether the emergency core-cooling systems will provide alternative cooling.

This straightforward approach is called *deterministic safety assessment* and it is useful in establishing and verifying design criteria for the reactor. A limitation of the approach is that it does not address the question of likelihoods. In particular, it does not consider the probability that the design basis accident will occur or the probability that the safety system will work as

intended. Obviously, a particular sequence of events is more serious if the initiating problems are relatively probable and the safety systems have a relatively
high probability of failing.

14.3.2 Probabilistic Risk Assessment

PRA Implementation: Reactor Safety Study, WASH-1400

Estimating risk probabilities is not an easy matter in the case of nuclear
reactors. For automotive safety, by contrast, it is relatively easy to answer
questions about the chances of a fatal accident. One merely has to look at
the annual fatality rate, subdivided, if one wishes, by type of car, road conditions, driver, and so forth. There are ample data on auto fatalities, and these
lend themselves to extensive analysis. Thus, there is reasonable quantitative
knowledge of the safety of automobiles and roads.

   With no fatal accidents and with no major accidents of any sort in light water reactors other than the Browns Ferry (1975) and Three Mile Island (1979)
accidents, overall reactor safety cannot be determined from direct accident experience.[12] (Of course, it would be unacceptable to have enough accidents to
provide meaningful statistics.) Instead, it is necessary to rely on calculations
or assessments. An early effort in this direction was an 1957 Atomic Energy
Commission study (known as WASH-740) on the possible consequences of an
accident, but for many years, there was no careful estimate of the *probability* of an accident. A major expansion of nuclear power was expected in the
1970s in the United States and throughout the world, but although there were
many intuitions as to the level of risk, there was no defensible quantitative
analysis.

   To address this issue, the Atomic Energy Commission sponsored an extensive study under the direction of Norman Rasmussen of the Massachusetts
Institute of Technology. This study was issued in draft form in 1974 and in
final form in 1975 under the institutional sponsorship of the Nuclear Regulatory Commission—which by this time had assumed the regulatory functions
of the disbanded AEC. The study is variously referred to as the Rasmussen
report, the Reactor Safety Study (RSS), and WASH-1400 [12]. It was the first
major study to combine in one analysis the probability and consequences of
accidents, in order to assess the *risk* associated with reactor accidents. It was
a limited study in that only one PWR and one BWR were analyzed in detail,
although the results were often taken to be representative of the situation for
other PWRs and BWRs.

   The RSS was controversial from the moment the first draft appeared, and
the controversies were never fully resolved. However, it is generally agreed
that the study made a very important contribution in pioneering the application of methods of *probabilistic risk assessment* (PRA) to the analysis of

---

[12] See Chapter 15 for a discussion of these accidents.

nuclear reactor safety. In later terminology, especially in international usage, this approach has also been called *probabilistic safety assessment* (PSA). The terms are often used interchangeably.[13]

In principle, this approach permits an objective estimate of the *absolute* risk of accidents, although, at present, it is widely believed that the absolute PRA numbers have large uncertainties. However, even if the data and analyses fail to establish the absolute risks precisely, they can be useful in suggesting the *relative* risks of different configurations and in pinpointing weaknesses. There are some who argue that the chief value at present of probabilistic risk assessments is in identifying places where safety improvements are needed. In this view, PRAs are more useful for improving reactor safety than for estimating it.

Although improvements in reactor equipment and advances in analysis techniques have made the detailed numerical results of the RSS obsolete, they remain of historical significance, and the report itself remains a historic milestone. Subsequent to the TMI accident, numerous steps have been taken to improve reactor safety as well as to refine the analyses, with separate analyses carried out for individual reactors. The general methodology employed in the RSS has been retained.

Event Trees and Fault Trees

The PRA tools used in the RSS were event-tree analyses and fault-tree analyses. In an event-tree analysis, one imagines the occurrence of some initiating event and traces the possible consequences. We illustrate in Figure 14.1 the event tree for studying the consequences of a major pipe break, following which the emergency core-cooling system (ECCS) must operate successfully for damage to be avoided [12, Main Report, p. 55]. The worst case in this example would be the electric power failing to operate, the ECCS not functioning, the fission product removal systems within the containment not operating, and the containment integrity being breached.

The probability that everything goes wrong in this sequence is shown in the bottom leg of the "basic tree" in Figure 14.1. It is the product of five individual failure probabilities. In the "reduced tree," shown in the bottom part of Figure 14.1, cognizance is taken of the possibility that the probabilities are not independent. In particular, the bottom leg of the reduced tree, which bypasses three steps, is based on the assumption that without electrical power, the other systems will also fail and the accident will proceed to the breaching of the containment.[14]

---

[13] For example, the NRC describes the analysis in its study NUREG-1150 as a PRA [11], whereas the (American) chairman of INSAG terms this study a PSA [13, p. 50].

[14] Figure 14.1 is a simplified version of the event trees that are actually used and is shown for illustrative purposes.

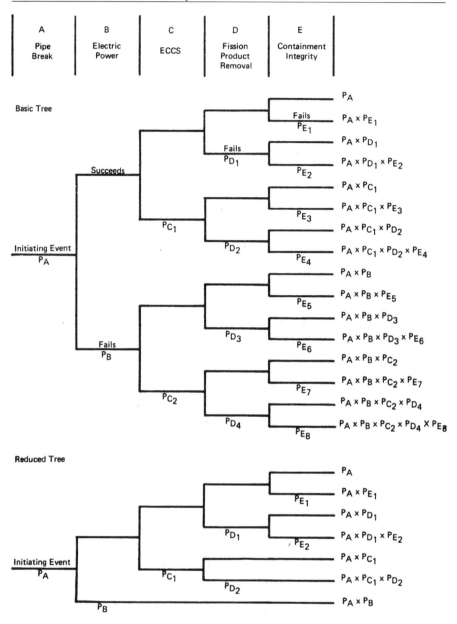

Fig. 14.1. Simplified event trees for a large loss-of-coolant accident. [For this diagram, it is assumed that the failure probabilities (P_f) are small and therefore factors of the form $1 - P_f$ are not explicitly indicated.] (From Ref. [12, p. 55].)

This is an example of a *common-mode* failure (i.e., a case in which individual failures are causally related). Such scenarios could, at least in principle, greatly increase the chance of a serious accident. It is therefore necessary, but not necessarily easy, to identify sequences in which the failure of one system enhances the likelihood of the failure of others.

What is the probability that the electric power will fail, as assumed for the event tree of Figure 14.1? That question is answered in principle by a fault-tree analysis, diagrammed in Figure 14.2. For the electric power to fail, there must be a loss of *both* the off-site AC power (the standard source) and the on-site AC power (one or more emergency generators). The loss of AC power *or* the loss of DC power (required in this case to control the AC system) would mean that the safety systems would not operate.

In many cases, the individual ingredients for the event-tree and fault-tree analyses come from an extensive database (e.g., the rate of failure of a given type of valve or motor that may be widely used outside of the nuclear power

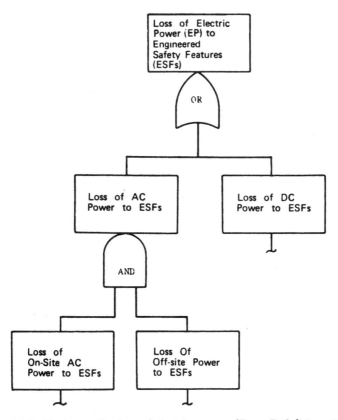

Fig. 14.2. Fault tree for loss of electric power. (From Ref. [12, p. 56].)

industry). In other cases (e.g., the probability of human error), the input numbers are likely to be only rough surmises.

Combining the outcomes of the event-tree analyses and the fault-tree analyses gives the probability for an accident scenario. Some scenarios will represent accidents with large releases of radioactivity to the environment; others will represent small releases. The overall results of the study can be embodied in graphs or tables in which the probability of an accident of a given or greater severity is displayed as a function of the severity of the accident.

The Role of Probabilistic Risk Assessment

In the original Reactor Safety Study, considerable emphasis was put on the absolute magnitude of the reactor accident risks. Uncertainties in the analysis were explicitly indicated, but there were criticisms that these had been underestimated. In NUREG-1150, a later PRA study, the issue of uncertainties was featured more prominently (see Section 14.4.2).

Despite the difficulty of making precise estimates of reactor risk with PRA techniques and the uncertainties that surround their results, they appear to offer the best available approach to risk estimation. As analysis methods are improved and input data on failure rates becomes more extensive, there can be increasing confidence in the applicability of the results. However, ambivalence remains, as reflected in comments made in a 1993 report prepared by the Nuclear Energy Agency of the OECD:

> *Probabilistic safety assessment* (PSA) is a powerful technique for providing a numerical assessment of safety. It is being increasingly used as a guide for comparing levels of safety. As such it complements the deterministic approach to safety assessment, but it is not considered as an absolute measure of safety for regulatory purposes....
>
> But the importance of PSA is not so much in the final answer that it gives for the chance of accidents. Its main value lies in the insights that are obtained in the process of the analysis. It will highlight those elements in a chain of events which contribute significantly to the probability of serious accidents—the weak links—and which if strengthened will therefore give a significant improvement in overall safety. [3, p. 63]

There appears to be little dissent from the view that PRA (or PSA) studies give useful information on relative risks and on the identification of "weak links." However, the uncertainties in the PRA estimates of absolute risk magnitude of the risks may be large, and the policies on the use of PRAs for regulatory purposes by agencies such as the NRC appears to be still evolving (see Section 14.5.1).

14.3.3 Results of the Reactor Safety Study

Summary of Results

The results of the RSS included estimates of the probability distributions for a variety of forms of harm: early fatalities, early illness, latent cancer fatalities, thyroid nodules, genetic effects, property damage, and magnitude of the area in which relocation and decontamination would be required. These results were presented in the form of graphs of the probability of occurrence as a function of the magnitude of the harm. Thus, for instance, the calculated probability of an event that would cause more than 1 latent cancer death per year was about 3×10^{-5}/reactor-year (RY); the probability dropped to 2×10^{-6}/RY for more than 100 latent cancer deaths per year [12, p. 97]. Large uncertainties were indicated for both the probabilities of the events and the resulting number of cancer fatalities.

The probability of a core melt was estimated to be 5×10^{-5}/RY, with an upper bound of 3×10^{-4}/RY, or about 1 per 3000 reactor-years of operation [12, p. 135]. The most probable cause of a core melt was found to be not a break in the large pipes providing the main cooling water but rather an accumulation of smaller failures. This was surprising in view of prior prevailing beliefs.

To provide perspective, the RSS also compared the risks from reactor accidents to those from other sorts of accidents or natural mishaps. For these other accidents, there are few data on latent effects. Perhaps the trauma of a nonfatal airplane accident increases one's chance of dying 30 years later, but this is not customarily included as a fatal consequence of airplane accidents. Thus, a direct comparison between nuclear power and other hazards is made simpler, although incomplete, if consideration is restricted to early fatalities. For a nuclear reactor accident, these would be primarily caused by very high early radiation exposures. In Figure 14.3, the annual risks from 100 reactors, as estimated in the RSS, are compared with the annual risks from other causes, such as airplane accidents and dam failures. For example, Figure 14.3 indicates an average of 1 airplane accident causing 100 or more fatalities every 3 years, whereas a nuclear reactor accident with this early toll was predicted to occur only once every 80,000 years.[15]

Responses to the Reactor Safety Study

The RSS was received very differently by different groups. Nuclear power advocates greeted it enthusiastically as a vindication of their belief in nuclear safety. It was possible to draw all sorts of dramatic comparisons from it, and these were gleefully put forth; for example, that there was less chance of

---

[15] It should be noted that with accidents of this magnitude, the consequences other than early fatalities are likely to be much more severe for the reactor accident than for the airplane accident.

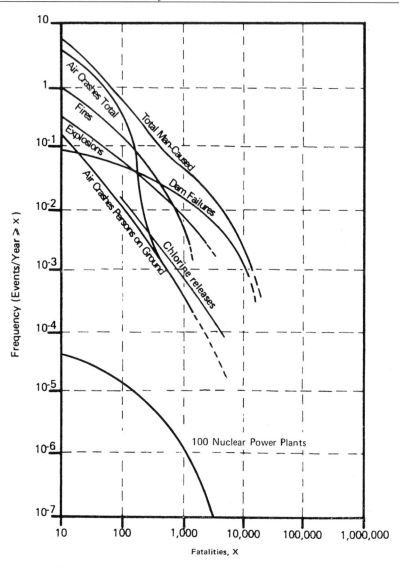

Fig. 14.3. RSS comparison of annual probabilities of accidents causing x or more (early) fatalities: 100 nuclear reactors compared to other "man-caused" events. (From Ref. [12, p. 119].)

being killed by an accident in a nearby nuclear power plant than by an errant automobile, even if you were neither in a car nor crossing a street yourself. Nuclear opponents greeted the RSS with strong criticism and even scorn. It was not surprising, in their view, that a study sponsored and carried out by the "nuclear establishment" would conclude that nuclear power was safe.

An influential critique of the RSS was done by a special review group, commissioned by the Nuclear Regulatory Commission and chaired by Harold Lewis of the University of California at Santa Barbara [14]. The main conclusions of the Lewis report were (1) the methodology used in the RSS was basically sound, (2) significant mistakes had been made, for example, in some of the statistical methods, (3) it was difficult to balance the instances of conservatism and nonconservatism, (4) the uncertainties were much greater than those quoted in the RSS, (5) the executive summary was misleading, and (6) the review group could not conclude whether the probabilities of a reactor core melt were higher or lower than those quoted in the RSS.

The Lewis report was regarded by some as a "repudiation" of the RSS, and the NRC backed away from using it as a guide for regulatory decisions. However, Lewis himself took a consistently "pro-nuclear" position in congressional testimony, stating that he felt "the plants are actually safer than stated in the Rasmussen report."[16] Lewis made this last statement, which reiterated earlier statements by him in the same vein, in May 1979, shortly *after* the TMI accident. However, TMI made such studies at least temporarily irrelevant. Quite apart from the merits and demerits of studies by academic scientists and engineers such as Rasmussen and Lewis, a significant fraction of the public concluded after TMI, and all the more after Chernobyl, that nuclear reactors were not safe enough for nearby siting. That conclusion has had a profound influence on the subsequent pace of nuclear power development.

Hindsight on RSS Predictions

It is tempting to look back with the benefit of hindsight on the RSS estimate of a core damage probability of 5×10^{-5}/RY with an upper bound of 3×10^{-4}/RY. As of the end of 2003, there had been about 2870 reactor-years of LWR operation in the United States.[17] If one uses the RSS, the predicted number of core melts through 2003 would be 0.14 with an upper bound of 0.9. The actual number of core melts was one (TMI), so the actual experience does not differ greatly from the predicted upper bound.

This comparison is not quite appropriate, however. The RSS was specific to reactors as they existed in the 1970s. Therefore, its results should be compared to reactor performance in that period before the post-TMI improvements were made. As of the end of 1979, there had been under 600 reactor-years of LWR operation in the United States, and the above "anticipated" accident rate should be reduced by about a factor of 5. Thus, one could infer that the average estimates of core melt probabilities given in the RSS were underestimates.

If one goes beyond average estimates, another interesting viewpoint emerges. As indicated earlier, the RSS studied certain reactors in detail and

---

[16] References and further quotations are given in Ref. [15].

[17] The RSS was for United States LWRs, and, therefore, it is appropriate to restrict comparison to their record.

these were taken as representative of all LWRs. In particular, the PWR analysis was based on the Surry 1 power plant, manufactured by Westinghouse. According to the RSS, using this as prototypical for all PWRs would "tend to overestimate, rather than underestimate the risk," because this was a relatively old plant and newer ones would, on average, be safer. However, as discussed in a subsequent study of the implications of the TMI accident, conducted under the auspices of the American Physical Society:

> The first reaction of many observers to the accident was that the Reactor Safety Study methodology was completely wrong because it had not predicted that type of accident would soon occur. The particular sequence...was calculated for the Surry facility...to have a frequency of once in 10^5 years. Yet...if the RSS procedures had been applied to a Babcock and Wilcox reactor like TMI-2, the methodology would have predicted a frequency of occurrence of one in 300 years. Babcock and Wilcox reactors had an operating history of about 30 reactor years. The differences stemmed from: (a) the pressure relief valve settings that caused the valve to be released before reactor scram and (b) the fact that the steam generators had a small heat capacity and dried out in ten minutes, compared with a time of about an hour calculated for the Westinghouse reactors such as Surry...if the methodology had been applied to the reactor at Three Mile Island, the plant-specific scenario differences might have been noted, modifications might have been made, and the accident perhaps avoided. [9, p. S11]

In short, the greatest mistake with the RSS analyses was the failure to apply the analyses to all reactors, individually.

14.4 Post-TMI Safety Developments

14.4.1 Institutional Responses

The TMI accident showed that the prior efforts of government agencies and reactor manufacturers to achieve safety had been insufficient. Although the consequences were limited, in that there was no large release of radioactive material to the outside environment, it was clear that further measures were needed to improve reactor safety. This was obviously felt by the public and was recognized by the nuclear industry and the U.S. Nuclear Regulatory Commission (NRC). The intensity of their subsequent efforts was increased by the fear that any accident would bring discredit to all of nuclear power. This concern is encapsulated in the phrase: "An accident anywhere is an accident everywhere."

More specifically, the accident revealed defects in physical components of the cooling system and in the systems that provided the operators with information about the status of the reactor. It also showed the need for improvements in operator training and in communication among utilities. Remedial steps were taken in all of these areas, including, in some cases, expensive and time-consuming retrofitting of existing reactors and modifications of reactors under construction. The first impact of these measures in the United States was a pause in the licensing of new reactors and even in the operation of some reactors. In consequence, the output of nuclear electricity dropped in the 3 years following TMI, before beginning a substantial rise in the mid-1980s. The last new nuclear reactor in the United States went into operation in 1996, but despite some subsequent shutdowns of older reactors, total nuclear output continued to rise during the 1990s and reached a new high in 2002.

At first, the rise was attributable mainly to additional reactors coming on line. However, since about 1990, it has been due almost entirely to an increase in the capacity factors of the reactors (see Section 2.4.2). The capacity factor provides a good overall measure of reactor performance and its rise reflects the positive impact of post-TMI improvements in equipment and maintenance procedures. As discussed in Section 14.5.3, the rise in capacity factors was accompanied by other, more direct, indicators of increased safety.

On an institutional basis, in the United States the nuclear industry established the Institute of Nuclear Power Operations (INPO) to exchange information and coordinate and monitor efforts to make reactor operation more reliable and safer. At the same time, the NRC intensified its watchdog role in setting standards and monitoring performance.

Internationally, there are long-standing reactor safety programs operated by the International Atomic Energy Agency [e.g., the IAEA's International Nuclear Safety Advisory Group (INSAG)] and by the Nuclear Energy Agency of the OECD. In addition, the World Association of Nuclear Operators (WANO) was established after the Chernobyl accident (1986) as an international counterpart of INPO [3, p. 27]. It now plays an active role in reviewing the performance of individual plants throughout the world and in facilitating the exchange of information on reactor safety.

14.4.2 1990 NRC Analysis: NUREG-1150

Analysis Procedure in NUREG-1150

A further step in the development of reactor safety analysis methods in the United States was marked by the publication in 1990 of the NRC report *Severe Accident Risks: An Assessment for Five U.S. Nuclear Power Plants*, also known as NUREG-1150 [11]. Five LWRs were analyzed in detail for this study. These are in some sense typical of LWRs in the United States, but the reported results are specific to the individual reactors.

In NUREG-1150, an explicit distinction was made between *internal* and *external* events. Internal events are those due to the malfunctioning of components of the reactor, including its control systems. External events are those initiated by things that happen outside the reactor (e.g., earthquakes).[18] External events were analyzed for only two of the five reactors.

The NUREG-1150 analysis was divided into several separate stages:

- *Accident frequency.* The goal here is to estimate the probability that the reactor core is damaged. The starting point is to identify possible initiating events and assess their frequency. For events due to internal system failures, accident probabilities were determined through a combination of event-tree and fault-tree analyses. Human error and "dependent failures" (also known as common-mode failures) were included. For events due to external hazards—including earthquakes, fire, and aircraft impacts—somewhat analogous procedures were used, but the database for the initiating events is not as good and there is a greater chance of a simultaneous failure of several components.

- *Accident progression.* Given damage to the reactor core, it is important to know what further damage occurs. Thus, probabilities were estimated for the breaching of the reactor vessel and for either breaching of the concrete containment or leaks through it.

- *Transport of radioactive material.* Given damage to the fuel and to the reactor vessel or cooling system, there will be a transfer of radionuclides to the reactor building. Release to the environment then depends on whether or not the containment fails. The noble gases are the most likely to be released, with virtually all escaping given sufficient fuel damage. For other radionuclides, the release rates depend on the volatility of the element, ranging from high for iodine and cesium and low for ruthenium and strontium. The aggregate of total releases to the environment is the source term.

- *Off-site consequences.* The radiation doses received by people outside the reactor depend on the magnitude of the source term, the movement of the plume of radioactive material in the air, and the details of the pathways by which radiation exposures occur. Doses and health consequences were calculated for a variety of assumptions as to the evacuation of the surrounding population.

- *Integrated risk analysis.* An overall integrated risk is found from the array of probabilities for each of the various stages.

As discussed in connection with the RSS, the ultimate result of this analysis is a probability distribution for the risk of occurrence versus the magnitude of the consequence, for each adverse consequence of interest. Thus, the result might be the probability distribution for exceeding various levels of population dose or of latent cancer fatalities. Although such a probability distribution cannot be fully represented by a single number, both medians and mean values are given in NUREG-1150 to provide an easily encapsulated overall

---

[18] The distinction is not clean and, customarily, loss of power from off-site sources is included as an internal event, whereas floods and fires within the plant are termed external events [11, p. 2-4].

Table 14.1. Estimated mean probabilities per reactor-year of core damage and other effects of reactor accidents, for reactors studied in NUREG-1150.

			Reactor Studied		
	Surry 1	Zion 1	Sequoyah 1	Peach Bottom 2	Grand Gulf 1
Reactor type	PWR	PWR	PWR	BWR	BWR
State located	VA	IL	TN	PA	MS
Capacity (MWe)	781	1040	1148	1100	1142
Commercial operation (year started)	1972	1973	1981	1974	1985
Internal events					
Core damage[a]	4×10^{-5}	6×10^{-5}	6×10^{-5}	4.5×10^{-6}	4×10^{-6}
Early containment failure[b]	4×10^{-6}	6×10^{-6}	7×10^{-6}	2×10^{-6}	1×10^{-6}
Individual early fatality[c]	2×10^{-8}	3×10^{-9}	1×10^{-8}	5×10^{-11}	3×10^{-11}
Individual latent cancer[c]	2×10^{-9}	3×10^{-9}	1×10^{-8}	4×10^{-10}	3×10^{-10}
External events, core damage[d]					
Seismic events, LLNL	1.2×10^{-4}			8×10^{-5}	
Seismic events, EPRI	2.5×10^{-5}			3×10^{-6}	
Fires	1.1×10^{-5}			2×10^{-5}	

[a]Number for Zion reactor reflects plant modifications after study was initiated [11, p. 7-4].

[b]A containment failure here includes both breaks in the containment structure and bypass of it. It may lead to large early release of radionuclides.

[c]The individuals considered are those within 1 mile and 10 miles of the reactor boundary for early and latent fatalities, respectively.

[d]Separate results are given for studies from the Lawrence Livermore National Laboratory (LLNL) and the Electric Power Research Institute (EPRI).

Sources: Capacity data and commercial operation dates are from Ref. [16]. Core damage data are from Ref. [11, pp. 3-4, 4-4, 5-4, 6-5, and 7-4]. Large early release data are from Ref. [11, p. 9-6]. Individual risk data are from Ref. [11, p. 12-3].

perspective.[19] Some results of the NUREG-1150 analysis are summarized in Table 14.1, for the five reactors studied.

The NUREG-1150 study is more pertinent to the present situation than the original Reactor Safety Study, because it used more advanced analysis techniques and considered reactors as they were after a period of considerable upgrading. This is a continuing process, however, and the analyses were spe-

---

[19] The mean is, in general, higher than the median, because the probability distribution for a given consequence generally has a tail extending to high magnitudes.

cific to the situation at the time they were made (in the later 1980s). Since then, conditions may be better due to further modifications in equipment and operating procedures or worse due to aging. Continual reactor-by-reactor monitoring is necessary. As discussed in Sections 14.4.3 and 14.4.4, other indicators suggest continuing gains.

Core Damage Probabilities

The mean calculated probability of core damage from internal events varied from 4×10^{-6}/RY in the best case to 6×10^{-5}/RY in the worst case, with a rough (arithmetic) average of about 3×10^{-5}/RY (see Table 14.1). The probability of core damage was considerably lower for the two BWRs than for the PWRs, although the study cautioned that it would be "inappropriate" to conclude that this was true in all cases [11, p. 8-11]. Nonetheless, some advantages of BWRs were pointed out, particularly more redundancy in the emergency core-cooling systems.

The main causes of core damage differed among the reactors [11, p. 8-3].[20] For two (Zion and Sequoyah), events with loss-of-coolant were the most important factor. For two (Surry and Grand Gulf), loss of power (station blackout) was the main factor. For one (Peach Bottom), roughly equal responsibility was placed on loss of power and failure of control rod insertion during transient disturbances.

Core damage due to external causes was considered for only two of the reactors, Surry 1 and Peach Bottom. In both cases, seismic events and fires were the only significant external sources of risk. The data of Table 14.1 might suggest that the external risks are greater than the risks due to internal failure. However, such a conclusion may be premature. For one, the risk for seismic events is highly uncertain, with two analyses considered in NUREG-1150 differing substantially. Further, the seismic risk distributions are very broad and are skewed so that the median risks are considerably lower than the mean risks [11, p. 8-6].[21]

Early Containment Failure

Radionuclides can escape into the environment due either to a breaching of the containment structure or due to valve failures, through cooling system pipes that bypass the containment and vent outside it. An emphasis is put on early failures because if the release of radionuclides is delayed, "mitigative features within the plant can substantially limit the release that occurs" (i.e.,

---

[20] A compact summary is given in Ref. [17, p. 3–7].

[21] For the LLNL analyses, which give the higher core damage probabilities, the median risk is of the order of one-tenth the mean risk: 1.5×10^{-5}/RY for Surry 1 and 4×10^{-6}/RY for Peach Bottom 2.

radionuclides may be retained inside the containment despite the containment failure [11, p. 9-5]). In each case, the early containment failure rate is estimated at under $10^{-5}/\text{RY}$.

Consequences of Accidents for Human Health

The key potential health effects of a reactor accident are early fatalities, due to very high radiation doses, and latent cancers, due to the long-term effects of smaller doses. The NRC has put forth "quantitative risk objectives" for these accident consequences (see Section 14.5.1), and the NUREG-1150 results in Table 14.1 are couched in terms that permit a direct comparison with the NRC objectives. Average early fatality risks are calculated for individuals within 1 mile of the reactor, and latent fatality risks for those within 10 miles of the reactor. Of course, the doses decrease substantially with distance, and the risks are higher in these regions than in broader surrounding areas.

The mean early fatality risk for individuals is calculated to range from under $10^{-10}/\text{RY}$ to $2 \times 10^{-8}/\text{RY}$. Even for the greater of these numbers, the risk is small—1 chance in 50 million per year. This is only 4% of the NRC objective of $5 \times 10^{-7}/\text{RY}$. For latent cancers, the highest of the calculated risks $(1 \times 10^{-8}/\text{RY})$ is only 0.5% of the NRC objective $(2 \times 10^{-6}/\text{RY})$. Therefore, unless the calculated results are greatly in error, the NRC's safety goals for individuals are satisfied by a large margin.

Seismic Risk

A striking aspect of Table 14.1 is the relative importance of seismic risks in the tabulated core damage frequencies.[22] Determination of the seismic core damage probabilities involves estimating both the probability of earthquakes of various magnitudes at the reactor site, the so-called seismic hazard, and the ability of the reactor to withstand the resulting ground accelerations.[23] The NUREG-1150 calculation used seismic hazard assessments from both Livermore and EPRI. There is no conclusive method for predicting earthquake probabilities, and both the Livermore and EPRI studies relied upon an array of expert evaluations.[24]

---

[22] The large differences between the EPRI and LLNL results prompted the NRC to commission a study on methods to be used in carrying out a probabilistic study of seismic hazards [18]. No new comprehensive analysis of seismic hazards incorporating these recommendations has been published as yet.

[23] The methods used in these analyses are described in detail in Refs. [19] and [11, Section C11].

[24] The EPRI and Livermore studies were both part of a major program undertaken in the late 1980s to assess seismic hazards, in the region of the United States to the east of the Rocky Mountains, where the large majority of the reactors are situated. These hazards are ultimately couched in terms of site-specific probability distributions for ground acceleration.

Differences in these evaluations in the two studies led to substantially different mean results for the core damage probability. The associated probability distributions are very broad. For example, for the Livermore seismic hazards, the 5th and 95th percentiles in the core damage frequency distribution differ by more than a factor of 1000 for the Surry plant, and the median probability is only about one-eighth of the mean probability [11, p. 8-6]. The large width of the distribution makes any "average" result a poorly established quantity. Subsequent to the publication of NUREG-1150, a new Livermore study of seismic hazards was carried out, which, in an overall sense, tended to move the Livermore results in the direction of the EPRI results (i.e., to lower the estimated risk [20]).

Despite the emphasis here on probabilistic risk assessments for evaluating reactor earthquake risks, this is not the chief approach adopted by the NRC. Instead, *seismic margin* methodology is being used for some present and all future reactors. This method is somewhat less demanding in terms of the analysis required and may be the most reasonable approach given the large uncertainties in estimating earthquake probabilities. The starting point is the specification of the so-called *safe shutdown earthquake* (SSE).[25] This is an earthquake whose magnitude is based on the "maximum earthquake potential" in the vicinity of the site. The reactor must be designed to shut down safely should this largest expected earthquake occur.

If a reactor can withstand an earthquake more severe than the SSE, then the reactor has a "seismic margin." The extent of the seismic margin is based on a reference earthquake, more severe than the SSE, for which there is a "high confidence of a low probability of failure" (HCLPF).[26] For example, if the SSE corresponds to a peak ground acceleration of 0.3 g at the reactor site, the seismic margin condition might be established by demonstrating fulfillment of the HCLPF condition for an acceleration of 0.45 g. The NRC requirement for new reactors will be that they demonstrate an adequate seismic margin.

Although the safety study NUREG-1150 included estimates of seismic core damage probabilities for two of the five reactors, no estimates were given of large release probabilities for seismic effects. The rationale was that if an earthquake is severe enough to damage a reactor, there will be damage to other structures such as buildings and dams, with consequences that are expected to be more severe than those from possible reactor releases [11, p. 1-4]. In the absence of meaningful estimates of the effects of this other damage on the surrounding population, the NRC lacked a reference point for considering the

---

[25] The definition and use of the safe shutdown earthquake is discussed in Appendix A to Part 100 of Ref [21]. The SSE had previously called this the *design basis earthquake*.

[26] The HCLPF criterion can alternatively be established by deterministic or probabilistic determination of failure modes. In the latter case, it is assumed to correspond to a greater than 95% confidence that the failure probability is less than 5% [19, p. 5-4].

significance of the off-site effects of nuclear reactor containment failures and did not calculate their probability for NUREG-1150.

Overall, although considerable effort has gone into making nuclear reactors "safe" against earthquakes, there has been difficulty in quantifying the level of safety. In a quite different approach, some designs for new reactors incorporate seismic isolation between the reactor and the surrounding ground, with the goal of decoupling the reactor from possible ground motion.

14.4.3 Predictions of Core Damage and Precursor Analyses

The Record Since TMI

Estimates of core damage frequency based on PRAs cannot be checked against actual experience because there has not been an LWR accident that has caused core damage since the TMI accident in 1979. In this period, there have been roughly 2300 reactor-years of operation in the United States, a record that can be taken to suggest that the core damage frequency during this period was probably less than $5 \times 10^{-4}/RY$. However, such an upper limit considerably exceeds both what is acceptable and what is predicted and, therefore, is not very useful.

The database, of course, substantially increases if one looks at all LWRs throughout the world. Again, there has been no core damage. However, it is not fully appropriate to compare worldwide experience to estimates made in NRC studies of individual U.S. reactors because, although the guiding principles are the same, differences in regulatory, construction, and operating practices may make reactors in other countries more safe or less safe than U.S. reactors.

Analysis of Precursor Events

There are continuing malfunctions of reactor equipment short of core damage, spanning a wide range of severity. These malfunctions can be viewed as *precursor* events (i.e., potentially the initial first stage in a chain of failures that, if they all occurred, could lead to core damage). Analyses of precursors provides a powerful tool for inferring the expected core damage frequency and—perhaps even more valuable—gauging progress in reactor safety.

The precursor events are identified from "licensee event reports" that reactor operators are required to submit to the NRC for each significant malfunction in reactor performance. It is possible through the PRA methodology to then calculate a "conditional core damage probability" (CCDP) (i.e., the probability that core damage would result given this first mishap). These calculations have been made in the NRC's Accident Sequence Precursor Program. An index of reactor performance is given by the sum of the CCDPs for all failures in a year divided by the number of operating reactors. This index has been variously called the "inferred mean core damage probability" [22], the "annual core damage index" [23], or—in the terminology now used by the NRC—the "accident sequence precursor (ASP) index" [24]. It provides an overall indication of progress in reducing the likelihood of core damage, although the ASP index is not a complete predictor of core damage frequency [24].[27] For example, it does not include external events. Further, year-to-year results for the index are not strictly comparable, because the detailed analysis methods and range of events included have changed with time [25].

Despite these caveats, this index provides a valuable indication of progress in reactor safety. The reported magnitudes of the ASP index for the period from 1969 through 2000 are shown in Figure 14.4 [23, 24]. The ASP index is dominated by a few events of relatively great severity (high conditional core damage probability) rather than by a large number of relatively minor events. For example, the high value of the index in 1979 was due to the TMI

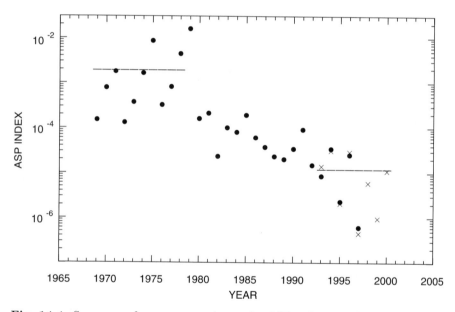

Fig. 14.4. Summary of precursor analyses: the ASP index as a function of time. The horizontal lines are averages over extended periods. [Solid circles: 1969 to 1997 (calendar years), data from Ref. [23]; crosses: 1993 to 2000 (fiscal years), data from Ref. [24].]

---

[27] The numerical values of the "mean core damage frequency" reported by Thomas Murley in 1990 [22], the "core damage index" reported by Murley in 1999 [23], and the ASP index of the recent NRC documents [24] are in good agreement, confirming that the differences are more in terminology than in basic meaning (see Figure 14.4).

accident[28] and the high excursion in 2000 was due to a single event at a two-reactor station.[29] Such events lead to large year-to-year fluctuations, although not enough to mask the overall downward trend, exhibited in averages of the ASP index taken over a number of years. This average dropped from about 2×10^{-3}/RY for the 1969–1978 period (the 10 years preceding the TMI accident) to about 1.2×10^{-5}/RY for the 1993–2000 period—an improvement of better than a factor of 100.

The retrospectively calculated pre-TMI ASP index is considerably higher than even the upper bound on core damage probability of 3×10^{-4}/RY estimated in the 1975 Reactor Safety Study. Therefore, taking the ASP index to represent an approximate core damage probability, it would appear that the reactors were less safe than then thought. Subsequent changes in equipment and operating procedures have greatly improved matters (e.g., see Ref. [26]). However, the improvement in the ASP index seems to have ended in the early 1990s, although this is difficult to interpret due to the large fluctuations during the 1990s (see Figure 14.4), in part caused by the dominance of only a few events in each year. Preliminary NRC study of the experience for 2001 and 2002, prior to the completion of ASP analyses for those years, does not suggest any marked changes, although the analysis of the Davis–Besse corrosion problem was continuing as of early 2003 (see the next section) [27].

14.4.4 Other Indications of Performance

Following the Three Mile Island nuclear accident, the U.S. nuclear industry established the Institute for Nuclear Power Operations (INPO) designed to coordinate the industry's efforts to remedy existing problems and achieve safer and more economical reactor operation. As part of this activity, INPO monitors and reports upon various performance indicators. Results comparing 2001 to early years include the following [28]:

♦ *Unplanned scrams.* Automatic shutdowns of a reactor, initiated by a failure in one of the reactor components, are called scrams. The scrams are a safety measure to avoid the development of serious accidents and the occurrence of a scram is part of proper operation of the safety systems. However, just as it is desirable to avoid the need for household circuit breakers to trip, it is desirable to minimize the need for scrams, and the number of scrams per year suggests how well a plant is operating. Measured in terms of the

---

[28] For 1979, when there were 69 reactors, the contribution to the ASP index from TMI alone (CCDP =1) was $1/69 = 0.0145$/RY; the total calculated ASP index was 0.0157/RY.

[29] The CCDP for this event was 4.5×10^{-4}. Summing for the two reactors gives, for 103 reactors, a contribution of 0.9×10^{-5} to the average for all reactors of 1.1×10^{-5}/RY for 2000 [24].

number of scrams per 7000 h of reactor operation,[30] the rates were 7.3 in 1980, 1.2 in 1990, and under 0.1 in 2001. This more reliable operation was one of the reasons for the improved capacity factors.

◆ *Radiation exposure of workers.* In the absence of a serious reactor accident, the reactor workers are the only individuals at potential risk from exposure to nuclear radiation. Some radiation exposure is inevitable among workers in a nuclear plant, but it is desirable to keep this exposure "as low as reasonably achievable." The collective exposure per reactor (i.e., the sum of individual exposures of all the workers at that unit) dropped by more a factor of 6 from 1980 to 2001.[31]

◆ *Industrial accident rate.* The rate of industrial accidents provides an indication of the overall safety of working conditions, although it has little to do with nuclear safety per se. The rate of accidents at nuclear reactors that led to lost or restricted work time or to fatalities dropped by a factor of 9 from 1980 to 2001. The rate per 200,000 worker-hours in 2001 was 0.24 for nuclear reactors compared to an average of 4.0 for U.S. manufacturing industries as a whole.

These measures of reactor performance indicate that there have been substantial safety improvements over the past 20 or 25 years. However, this does not mean that the nuclear industry or the NRC can afford to be complacent. This was forcefully brought home by the discovery in March 2002 at the Davis–Besse plant of deep corrosion in the reactor vessel head (i.e., the top portion of the reactor pressure vessel).[32] The corrosion was discovered during an inspection conducted by plant personnel while the reactor was shut down for refueling. It was caused by boric acid that leaked through small cracks in the nozzles that allow control rods to move up and down in the reactor.[33]

Both boric acid leaks and the corrosion caused by boric acid were familiar matters in the nuclear industry, but the NRC thought that it was on a scale that did not constitute a safety problem, as long as it was monitored. In fact, the inspections at Davis–Besse and other reactors were specifically required by the NRC in recognition of past boric acid leaks. However, the magnitude of the corrosion was unexpected. The cavity was about 4 in. by 5 in. and extended to a depth of approximately 6 in. This means it penetrated almost all the way through the top cover of the pressure vessel. The extent of the corrosion suggests that it might have been discovered during an earlier shutdown, given sufficient vigilance. Following this discovery the NRC required opera-

---

[30] This is equivalent to one reactor-year for a 1000-MWe reactor operating at an 80% capacity factor.

[31] Collective exposures per reactor in 2001 averaged 0.68 person-Sv for PWRs and 1.49 person-Sv for BWRs.

[32] This account of the corrosion problem is based primarily on Ref. [29].

[33] Boric acid is introduced into the reactor cooling water to adjust the reactor's reactivity through neutron absorption in boron-10 (^{10}B).

tors of other PWRs to report anew on boric acid leaks and corrosion at their plants. No comparable problems were found at other reactors. To decrease the possibility of similar future occurrences, the NRC is imposing strengthened requirements for the inspection of reactor vessel heads and is also making a broader examination of ways to assure that other reactor corrosion problems will be avoided [30].

It is of interest to determine how "near" a miss this was. There was no release of radionuclides and no damage of any sort except at the location of the leak. The NRC had not yet completed its analysis of the accident when its March 2003 report on the Accident Precursor Program was released [27]. This analysis will eventually provide an estimate of the likelihood that the accident might have led to core damage. There might have been no core damage even if the corrosion had penetrated all the way through the wall. Some reactor cooling water would have been ejected into the containment building as the pressure inside the reactor vessel was suddenly relieved, but it may have been possible to maintain cooling and avoid core damage. Pending the NRC report, it seems reasonable to conclude that in the case of the Davis–Besse incident, the overall safety system, including the inspection regime, in the end worked, and even had the inspection not come in time to avoid a reactor vessel breach, there probably would have been no major release of radionuclides to the outside environment. Nonetheless, the failure to detect and correct the corrosion promptly showed serious weaknesses in the monitoring procedures of the reactor operator and the NRC. This single event does not negate the very good and improving record of nuclear reactor performance, but should serve as a reminder against complacency.

14.5 Reactor Safety Standards

14.5.1 U.S. Nuclear Regulatory Commission Position

General Approach

Although the above discussion has emphasized probabilistic measures for specifying reactor safety, the NRC—the agency responsible for U.S. reactor safety—has been reluctant to adopt probabilistic criteria in setting reactor licensing requirements. Nuclear reactor safety is regulated using a "deterministic" approach, which, in recent years, has been modified to "risk inform" the application of the deterministic criteria. As described by the NRC in 1995:

> The NRC established its regulatory requirements to ensure that a licensed facility is designed, constructed, and operated without undue risk to the health and safety of the public. These requirements are largely based on deterministic engineering criteria. Simply stated this deterministic approach establishes requirements for engineering margin and for quality assurance in design, manufacture, and con-

struction. In addition, it assumes that adverse conditions can exist (e.g., equipment failures and human errors) and establishes a specific set of design-basis accidents. It then requires that the licensed facility design include safety systems capable of preventing and/or mitigating the consequences of those design-basis events to protect the public health and safety. [31, §IIIA]

Probabilistic risk assessment methods are used to supplement the deterministic approach by identifying both vulnerable and overprotected parts of the system:

A natural result of the increased use of PRA methods would be the focusing of regulations on those items most important to safety. Where appropriate, PRA can be used to eliminate unnecessary conservatism and to support additional regulatory requirements. Deterministic-based regulations have been successful in protecting the public health and safety and PRA techniques are most valuable when they serve to focus the traditional, deterministic-based, regulations and support the defense-in-depth philosophy. [31, §IIIA]

This is a limited use of PRA methods. The NRC has resisted suggestions to take the further step of setting probabilistic limits as part of the regulatory criteria (e.g., setting limits on the calculated core damage frequency).

Health Effects Criteria

The NRC's primary safety goals were set forth in 1983 in the report *Safety Goals for Nuclear Power Plant Operation* and were published in 1986 in a slightly revised form in the *Federal Register* [32]. This policy statement has remained the primary guide for NRC regulatory activities. The NRC staff in 2001 recommended modifications to the statement, but these were disapproved by the NRC commissioners who indicated that a change was not then timely given broader efforts underway to further "risk inform" NRC's safety regulations and the press of other demands upon the NRC [33, 34]. The 1986 statement put forth two "qualitative safety goals":

Individual members of the public should be provided a level of protection from the consequences of nuclear power plant operation such that individuals bear no significant additional risk to life and health.

Societal risks to life and health from nuclear power plant operation should be comparable to or less than the risks of generating electricity by viable competing technologies and should not be a significant addition to other societal risks. [32, § II]

The NRC also presented "quantitative risk objectives" to give a specific numerical meaning to the term "significant additional risk," but it prefaced the quantitative criteria with the cautionary comment:

> The Commission wants to make clear... that no death attributable to nuclear power will ever be "acceptable" in the sense that the Commission would regard it as a routine or permissible event. We are discussing acceptable risks, not acceptable deaths. [32, § IIIB].

Essentially, the NRC indication of objectives interpreted "significant addition" to mean an increase of 0.1% over the risk from non-nuclear sources [32, § IIIC]:

- *Prompt fatalities.* The risk to an average individual living within one mile of a plant should not exceed 0.1% of the combined average risk from all other accidents. Taking the latter risk to be 5×10^{-4} per year, the individual risk limit for a prompt fatality accident translates to 5×10^{-7} per year, as indicated in NUREG-1150 [11, p. 12-3].
- *Cancer fatalities.* The risk to the population living within 10 miles of the plant should not exceed 0.1% of the risk from all cancers. Taking the average annual cancer rate to be 19 per 10,000, the individual risk limit for cancers attributable to releases from the nuclear plant translates to 2×10^{-6} per year (again as in NUREG-1150 [11, p. 12-3]).

These were specified to be "quantitative objectives." Although they do not carry the legal force of regulations, they were deemed by the NRC to provide a "useful tool by which the adequacy of regulations... can be judged" [32, §V].

These objectives have remained unchanged in subsequent years and have been commonly referred to in safety evaluations. They were cited as "safety goal[s]" in the NRC document NUREG-1150 (e.g., Ref. [11, p. 12.3]) and were, for example, reiterated in 1998 in the context of an analysis of a proposed new reactor, the AP600 (see Section 16.3.2), with the statement: "The Commission approves the use of the qualitative safety goals, including use of the quantitative health effects objectives, in the regulatory decision making process" [35, p. 19-4].

Accident Frequency Criteria

As discussed earlier, the NRC has no official standard that specifies reactor safety requirements in terms of the probabilities for core damage or a large early release, as estimated through PRAs. However, many NRC documents have discussed the matter. Thus, a 1983 document on safety criteria stated that "the Commission has selected the following design objective" [36, p. 14]:

> The likelihood of a nuclear reactor accident that results in a large-scale core melt should normally be less than one in 10,000 per year of reactor operation. [36, p. 14]

The document went on to indicate the importance of mitigating the effects of such an accident through "containment, siting in less populated areas, and emergency planning as integral parts of the defense-in-depth concept."

It should be noted that these are "design objectives," but they were not necessarily to be incorporated in "the regulatory framework" [36, p. 15]. The terminology and formal status of this objective has been the subject of continual discussion in subsequent years. In a 2001 recommendation to the commissioners, the NRC staff suggested adopting "useful subsidiary benchmark[s]" of 10^{-4}/RY for the core damage frequency and 10^{-5}/RY for the large early release frequency (LERF) [33, §IIID]. However, as already mentioned, the commissioners declined to accept that recommendation at the time.

However, the NRC is thinking in terms of a more ambitious goal for new reactors. As early as 1986, the Commission had recommended consideration of a performance guideline for a LERF of less than 10^{-6}/RY [32, § V]. More recently, in the design certification documents for the ABWR and AP600 reactors, the NRC compared the expected performance to the "Commission's goal" of a LERF of under 10^{-6}/RY [35, p. 19-180; 37, p. 19-39]. A further indication of NRC thinking on desirable safety levels is provided in Regulatory Guide 1.174, a 1998 document discussing its approach to handling requests by a reactor operator for changes in the reactor [38]. In brief, it calls the calculated increased risk "very small" if it is less than 10^{-6}/RY for the core damage frequency and is less than 10^{-7}/RY for the large release probability.

Of course, limiting an increase to 10^{-6}/RY is less stringent than limiting the magnitude to 10^{-6}/RY. Nonetheless, this stipulation suggests that if the NRC eventually adopts a formal limit for the core damage frequency in new reactors, the number selected is likely to be more demanding than the value of 10^{-4}/RY that has appeared in past discussions.

Cost Considerations

In considering reactor safety, there is a certain temptation to ignore cost issues and say that society should "spare no cost" in its efforts to reduce the probability of a reactor accident. However, costs are not ignored for other activities (e.g., in preventive medicine, highway safety, and airplane construction, to cite a few examples) and they are not ignored for nuclear power plants. In all of these cases, one eventually reaches the point of diminishing returns.

The issue is explicitly addressed by the NRC in the spirit of cost–benefit analysis. For example, in its review of the application for design approval of the Westinghouse AP600 reactor, the NRC considered a number of possible changes beyond those already implemented in the design and compared the cost of each change to an assumed benefit of $5000 per person-rem of averted exposure [35, p. 19-255].[34] The costs for the 14 design alternatives that were considered ranged from $19,762 to $14,679,500 per person-rem [35, p. 19-270] and, therefore, none of these design changes was required.

---

[34] This figure is based on an an assigned benefit of $2000 per person-rem for health effects and $3000 per person-rem "to account for offsite property damage" [35, p. 19-251].

The number of significant figures in these numbers is not to be taken literally. The costs are somewhat uncertain and the averted risks (in person-rem) are probably considerably more uncertain. In addition, the assumed benefit of $5000 per person-rem can be disputed as being too high or too low. Nonetheless, these calculations address an essential consideration in seeking to balance the needs for safety and economy. The challenge for reactor developers is to achieve designs that will be acceptably safe and economically affordable. Cost–benefit comparisons can be a guide in considering modifications to the designs.

14.5.2 Standards Adopted by Other Bodies

Other organizations have been more explicit than the U.S. NRC in setting forth criteria for core damage frequency. The International Atomic Energy Agency has established a group specifically charged with considering matters of nuclear safety: the International Nuclear Safety Advisory Group (INSAG). In a 1999 report on safety principles, INSAG indicated a severe core damage frequency target of under 10^{-4}/RY for existing reactors [39, p. 11]. It further suggested that the application of appropriate safety principles and objectives to future plants "could lead to the achievement of an improved goal of not more that 10^{-5} severe core damage events per plant operating year."

The U.S. utility industry, through a "Requirements Document" issued by the Electric Power Research Institute in 1990, also adopted a core damage frequency limit of 10^{-5}/RY for future light water reactors [40, p. 94].

The INSAG document, in addition, suggested as an objective for future reactors the "practical elimination of accident sequences that could lead to large early radioactive releases." No quantitative meaning was assigned to the term "practical elimination" and no serious analyst will ever claim, or talk in terms of, "zero risk." However, one can speculate that if one tries to interpret the words in terms of a quantitative criterion, "elimination" might mean at least a factor of 100 beyond the core damage frequency, which would correspond to a LERF of less than 10^{-7}/RY. This is perhaps as close to zero as can be meaningfully considered.

14.5.3 Standards for Future Reactors: How Safe Is Safe Enough?

There is no universal answer to the question of "how safe is safe enough?" The acceptability of a given risk depends on circumstances, including the risks involved in alternative options. Many auxiliary factors enter, including—but not limited to—whether the risk is created by one's own actions, the actions of external institutions, or the actions of nature. Experience suggests that, at any level of numerically calculated danger, risks associated with nuclear energy are far less acceptable to the public and to most policy makers than are many other existing risks we encounter (e.g., those from automobile travel and the chemical emissions from coal-burning power plants).[35]

With these attitudes in mind, it is reasonable to conclude that the risk levels of the "useful subsidiary benchmarks" for present reactors suggested to the NRC (but not adopted by it)—of 10^{-4}/RY for core damage and 10^{-5}/RY for a large release of activity—would not meet a socially acceptable criterion of "safe enough" for a future world with, say, 4000 reactors. Taken literally, these numbers would imply a TMI-type accident every 2 or 3 years and a Chernobyl-type accident every 25 years. No matter what number of casualties is assumed for such an accident, it would not be acceptable to have a Chernobyl every few decades. The fact that the world accommodates to more severe tragedies from natural events and small-scale wars is probably irrelevant in terms of public response to nuclear power accidents.

On the other hand, a core damage frequency of 10^{-6}/RY to 10^{-5}/RY and a large early-release frequency of 10^{-7}/RY to 10^{-6}/RY might be satisfactory. Were such criteria met at the start, there would be only a small chance of either type of accident during the first decades of a large nuclear energy buildup. As discussed in Chapter 16, it may be possible to meet and exceed such standards with the new reactors that are now becoming available. It is pointless to attempt to estimate safety levels beyond a few decades, because continuing changes in nuclear reactor design—presumably with further safety improvements—would accompany a major revival of nuclear plant construction.

References

1. International Atomic Energy Agency, *Nuclear Power Reactors in the World*, Reference Data Series No. 2, April 2003 edition (Vienna: IAEA, 2003).
2. U.S. Environmental Protection Agency, *Accidents and Unscheduled Events Associated with Non-nuclear Energy Resources and Technology*, Report EPA-600/7-77-016 (Washington, DC: EPA, 1977).
3. Organization for Economic Cooperation and Development, Nuclear Energy Agency, *Achieving Nuclear Safety: Improvements in Reactor Safety Design and Operation* (Paris: OECD, 1993).
4. Uranium Institute, *The Safety of Nuclear Power Plants: An Assessment by an International Group of Senior Nuclear Safety Experts* (London: The Uranium Institute, 1988).
5. C. W. Forsberg and A. M. Weinberg, "Advanced Reactors, Passive Safety, and Acceptance of Nuclear Energy," *Annual Review of Energy* 15, 1990: 133–152.
6. International Atomic Energy Agency, *The Safety of Nuclear Power: Strategy for the Future* (Vienna: IAEA, 1992).

---

[35] Illustrating the disparity in public reactions, the death of five people due to a natural gas explosion and fire in Philadelphia on May 11, 1979, less than 2 months after TMI and in the same state, was virtually unnoticed [41, p. 930].

7. Ronald Allen Knief, *Nuclear Engineering: Theory and Technology of Commercial Nuclear Power*, 2nd edition (Washington, DC: Hemisphere Publishing Company, 1992).

8. American Nuclear Society, *Report of the Special Committee on Source Terms* (La Grange Park, IL: ANS, 1984.)

9. "Report to the APS of the Study Group on Radionuclide Release from Severe Accidents at Nuclear Power Plants," Richard Wilson, Chairman, *Reviews of Modern Physics* 57, no. 3, part II, 1985.

10. "Report to the APS by the Study Group on Light-water Reactor Safety," H. W. Lewis, Chairman, *Reviews of Modern Physics* 47, Supplement 1, 1975.

11. U.S. Nuclear Regulatory Commission, *Severe Accident Risks: An Assessment for Five U.S. Nuclear Power Plants*, Final Summary Report, Report NUREG-1150, vols. 1 and 2 (Washington, DC: NRC, 1990).

12. U.S. Nuclear Regulatory Commission, *Reactor Safety Study: An Assessment of Accident Risks in U.S. Commercial Nuclear Power Plants*, Report WASH-1400 (NUREG 75/014) (Washington, DC: NRC, 1975).

13. H. Kouts, "The Safety of Nuclear Power," in *The Safety of Nuclear Power: Strategy for the Future* (Vienna: IAEA, 1992), pp. 47–54.

14. U.S. Nuclear Regulatory Commission, *Risk Assessment Review Group Report to the U.S. Nuclear Regulatory Commission*, H.W. Lewis, Chairman, NUREG/CR-0400 (Washington, DC: NRC, 1978).

15. David Bodansky, "Risk Assessment and Nuclear Power," *Journal of Contemporary Studies* 5, no. 1, 1982: 5–27.

16. "World List of Nuclear Power Plants," *Nuclear News* 37, no. 3, March 1994: 43–62.

17. American Nuclear Society, *Report of the Special Committee on NUREG-1150, The NRC's Study of Severe Accident Risks* (La Grange Park, IL: ANS, 1990).

18. Senior Seismic Hazard Analysis Committee, R. J. Budnitz, Chairman, *Recommendations for Probabilistic Seismic Hazard Analysis: Guidance on Uncertainty and Use of Experts*, Report NUREG/CR-6372, UCRL-ID-122160 (Livermore, CA: Lawrence Livermore National Laboratory, 1997).

19. Electric Power Research Institute, *Use of Probabilistic Seismic Hazard Results: General Decision Making, the Charleston Earthquake Issue, and Severe Accident Evaluations*, EPRI Report TR-103126, prepared by Risk Engineering, Inc. (Palo Alto, CA: EPRI, 1993).

20. U.S. Nuclear Regulatory Commission, *Revised Livermore Seismic Hazard Estimates for 69 Nuclear Power Plant Sites East of the Rocky Mountains*, Draft Report NUREG-1488 (Washington, DC: NRC, 1993).

21. *Energy, U.S. Code of Federal Regulations*, Title 10 (1993).

22. T.E. Murley, "Developments in Nuclear Safety," *Nuclear Safety* 31 no. 1, 1990: 1–9.

23. T.E. Murley, "Safety Culture Indicators," MIT Safety Course (July, 1999), unpublished.

24. William D. Travers, *Status of Accident Sequence Precursor and SPAR Model Development Programs*, SECY-02-0041 (Washington, DC: U.S. Nuclear Regulatory Commission, 2002).

25. R. J. Belles, et al., *Precursors to Potential Severe Core Damage Accidents: 1997*, Report NUREG/CR-4674, ORNL/NOAC-232, Vol. 26 (Oak Ridge, TN: ORNL, 1998).

26. "Changes in Probability of Core Damage Accidents Inferred on the Basis of Actual Events," NRC staff report (forwarded to the Chairman of the NRC by James M. Taylor, April 24, 1992).

27. William D. Travers, *Status of the Accident Sequence Precursor (ASP) and the Development of Standardized Plant Analysis Risk (SPAR) Models*, SECY-03-0049 (Washington, DC: U.S. Nuclear Regulatory Commission, 2003).

28. "Performance Indicators: Another Successful Year in Performance, Safety," *Nuclear News* 45, no. 6, May 2002: pp. 28–30.

29. U.S. Nuclear Regulatory Commission, *NRC Update: Davis–Besse Reactor Head Damage* (November 2002).

30. "The Nuclear News Interview. The NRC's Brian Sheron: On Reactor Vessel Degradation," *Nuclear News* 46, no. 7, June 2003: 29–33.

31. U.S. Nuclear Regulatory Commission, "Use of Probabilistic Risk Assessment Methods in Nuclear Regulatory Activities; Final Policy Statement, *Federal Register* 60, no. 158, August 1995: 42622–42629.

32. U.S. Nuclear Regulatory Commission, "10CFR Part 50, Safety Goals for the Operation of Nuclear Power Plants; Policy Statement; Correction and Republication, *Federal Register* 51, no. 162, August 1986: 30028–30033.

33. William D. Travers, *Modified Reactor Safety Goal Policy Statement*, SECY-01-0009 (Washington, DC: Nuclear Regulatory Commission, 2001).

34. U.S. Nuclear Regulatory Commission, *Committee Voting Record, Modified Reactor Safety Policy Goal Statement* (Washington, DC: NRC, April 16, 2001).

35. U.S. Nuclear Regulatory Commission, *Final Safety Evaluation Report Related to Certification of the AP600 Standard Design,"* NUREG-1512 (Washington, DC: NRC, 1998).

36. U.S. Nuclear Regulatory Commission, *Safety Goals for Nuclear Power Plant Operation*, NUREG-0880 REV 1 (Washington, DC: NRC, 1983).

37. U.S. Nuclear Regulatory Commission, *Final Safety Evaluation Report Related to the Certification of the Advanced Boiling Water Reactor*, Report NUREG-1503 (Washington, DC: NRC, 1994).

38. U.S. Nuclear Regulatory Commission, *An Approach for Using Probabilistic Risk Assessment in Risk-Informed Decisions on Plant-Specific Changes to the Licensing Basis*, Regulatory Guide 1.174 (Washington, DC: NRC, 1998).

39. International Atomic Energy Agency, *Basic Safety Principles for Nuclear Power Plants, 75-INSAG-3, Rev. 1*, International Nuclear Safety Group Report INSAG-12 (Vienna: IAEA, 1999).

40. National Research Council, *Nuclear Power, Technical and Institutional Options for the Future*, Report of the Committee on Future Nuclear Power Development, John F. Ahearne, Chairman (Washington, DC: National Academy Press, 1992).

41. *The World Almanac and Book of Facts 1980* (New York: Newspaper Enterprise Association, 1979).

15

Nuclear Reactor Accidents

15.1 Historical Overview of Reactor Accidents

Despite the largely successful precautions taken to avoid nuclear reactor accidents, the record is not perfect. We list here the more important known reactor accidents, excluding accidents in submarine reactors and possible accidents in the former USSR and Soviet Bloc countries, other than the Chernobyl accident.[1]

The decision as to which accidents qualify as "major" accidents is somewhat arbitrary. In particular, ordinary industrial non-nuclear accidents are omitted. For example, in 1972 two workers at the Surry Power Station were fatally scalded by steam escaping from a faulty valve. This did not involve the nuclear components of the power station and therefore is not pertinent to the broader issue of nuclear reactor safety. The major past accidents are as follows:

♦ **Chalk River, Canada (1952).** There was a partial meltdown in a 30-MWt experimental reactor. The reactor was cooled by light water and moderated by heavy water. The accident was initiated by operator errors and a failure of the control rod system. This led to an elevated power

---

[1] There are reports of reactor accidents in these countries prior to the much larger Chernobyl accident (see Refs. [1] and [2]), although accounts of their course and magnitude are in dispute. In addition, there was a major release of radioactive material in an accident in 1957 at the Kyshtym nuclear complex in the Urals. The accident was a non-nuclear explosion in tanks of reprocessed radioactive wastes, not a reactor accident. It led to the reported evacuation of 10,730 people and caused a collective effective dose of about 2500 person-Sv [3, p. 116].

output and some boiling and loss of cooling water. In a typical LWR, the accident would have been mitigated by a negative void coefficient (see Section 14.2.1), but in this case the feedback was positive. The reactor was eventually shut down by draining the heavy water moderator. There were no known injuries or deaths, but the reactor core was damaged and there was escape of an unspecified amount of radioactivity [4, p. 101].

♦ **National Reactor Testing Laboratory, Idaho (1955).** The 1.4-MWt experimental breeder reactor, EBR-I, suffered a 40% to 50% core meltdown during a test in which the power level of the reactor was intentionally raised but, due to operator error, was not reduced promptly. There was little contamination of the building, no injuries occurred, and the release of radioactive material was "trivial" [4, p. 103].

♦ **Windscale, England (1957).** Overheating and fire occurred in a graphite moderated reactor used for plutonium production. The accident began in the course of heating the fuel above normal operating temperatures to release energy stored in the graphite crystal lattice.[2] This energy is a consequence of radiation damage to the graphite, a problem that arises in graphite-moderated reactors if they operate below the temperature necessary for annealing radiation damage. In this case, the heating and the energy release, although intentional, were too rapid. The reactor was shut down with control rods, but the heating had been sufficient to cause a fire in the uranium fuel and, eventually, in the graphite. The fire smoldered for about 5 days, until extinguished by flooding with water [6]. The most serious consequence was the release of about 20,000 Ci of ^{131}I ($T = 8.02$ days), which was carried by winds over much of central and southern England. The estimated consequences for England and continental Europe are 260 thyroid cancers and 13 thyroid cancer fatalities over a period of 40 years, plus 7 additional fatalities or hereditary effects [7, p. 24].

♦ **National Reactor Testing Laboratory, Idaho (1961).** Three army technicians were killed when one of them apparently rapidly removed (manually) a control rod from a 3-MWt test reactor, known as SL-1, on which they were working. Reactors of this type were intended for heating and electricity production at remote sites, and they were so primitive that the control rods could be moved by an operator standing on top of the reactor. There was a rapid increase in reactor output, followed by a steam explosion, leading to lethal levels of radiation within the reactor building. Most, but not all, of the activity was contained within the building [4, p. 109].

♦ **Fermi Reactor, Detroit (1966).** There was a partial meltdown in a 200-MWt (61-MWe) commercial breeder reactor, which was a one-of-a-kind prototype. The cause was a blockage in the flow path of the sodium coolant. There were no injuries or significant release of radioactivity, and

---

[2] The storage and release of energy in a graphite moderator is the so-called *Wigner effect* (see, e.g., Ref. [5]).

the reactor was briefly put back into operation before its final shutdown in 1973 [4, p. 32].[3]

♦ **Lucens, Switzerland (1969).** There was partial fuel melting in a 30-MWt experimental reactor due to loss of CO_2 cooling. There was severe damage to the reactor but no radiation release beyond permitted levels [4, p. 121].

♦ **Browns Ferry 1, Alabama (1975).** A fire in the electrical wiring did extensive damage to the control systems and threatened the reactor, but the reactor was turned off and cooling maintained with no radiation release and no injuries other than one individual suffering a minor burn from the fire. Despite the absence of damage to the reactor itself, this accident was of importance because it was the first major accident in a commercial LWR and demonstrated a serious vulnerability in the control systems of that period due to inadequate redundancy.

♦ **Three Mile Island, Pennsylvania (1979).** This accident is discussed in more detail in Section 15.2.

♦ **Chernobyl, USSR (1986).** This accident is discussed in more detail in Section 15.3.

Some aspects of these accidents are summarized in Table 15.1.

The only reactor accidents that caused a clearly identifiable loss of life were at Idaho Falls, where three workers died from the effects of the explosion and radiation, and at Chernobyl, where 31 operating and firefighting personnel died within about 2 months, primarily from high radiation doses. In addition, there possibly will be a small number of eventual, or "delayed," cancer fa-

Table 15.1. Major nuclear reactor accidents.

				Environmental Consequences		
Year	Reactor	Purpose	Capacity (MW)	Radioactivity Release	Prompt Deaths	Delayed Cancers[a]
1952	Chalk River	Experimental	30 (t)	Some	0	0
1957	Windscale	Pu production		Large	0	~ 13–20
1961	Idaho Falls	Test (army)	3 (t)	Small	3	0
1966	Fermi I	Demo breeder	61 (e)	Very little	0	0
1969	Lucens	Experimental	30 (t)	Very little	0	0
1975	Browns Ferry 1	Power	1065 (e)	None	0	0
1979	TMI-2	Power	906 (e)	Small	0	~ 0–2
1986	Chernobyl	Power	1000 (e)	Very large	31[b]	~ 30,000[b]

[a]Indicated cancers are possible cancer fatalities, calculated on the basis of the linear hypothesis (see Section 4.3).
[b]See Section 15.3.4 for further discussion of Chernobyl fatalities.

---

[3] For two very different assessments of the significance of this accident and the level of hazard it created, see Refs. [8] and [9].

talities from the Windscale radiation release and a large number of delayed fatalities from Chernobyl.[4] It may be noted that none of these three reactors were commercial LWRs, and except for Chernobyl, the accidents took place more than 35 years ago. Their history therefore has only limited pertinence to the present safety of commercial LWRs or of other non-Soviet commercial reactors.

Beyond reactors, the most serious nuclear accident since Chernobyl occurred at Tokaimura, Japan on September 30, 1999, at a facility for preparing reactor fuel.[5] Although this was not a reactor accident, we mention it here because a significant accident at any nuclear facility reflects unfavorably on the nuclear industry in general. The accident occurred in the course of preparing fuel for an experimental fast reactor which used uranium enriched to 18.8% in ^{235}U. In one stage of the process, in violation of authorized procedures, workers poured buckets of enriched uranium solution into a tank, apparently unaware that given the size and shape of the tank (45 cm in diameter and 61 cm high) they could create a critical mass. When they filled the tank with about 40 L of the solution, criticality was reached and there was an intense burst of gamma rays and neutrons, setting off radiation alarms. The three workers involved left the building, but all were heavily exposed and two eventually died. The only "significant" health consequences cited in an IAEA report on the accident were to these workers, although other workers were exposed to some extent, including some involved in measures taken to terminate the chain reaction [10, p. 30].[6]

15.2 The Three Mile Island Accident

15.2.1 The Early History of the TMI Accident

The Three Mile Island (TMI) accident occurred in one of two similar reactors at the Three Mile Island site in Pennsylvania.[7] The accident was in the second

---

[4] The number of fatalities is in question, given the uncertainties surrounding the effects of radiation at low doses and dose rates, but in the discussion of these accidents, we quote numbers based on the adoption of the linearity hypothesis (see Section 4.3).

[5] This summary is based largely on an IAEA report prepared shortly after the accident [10].

[6] There was no explosion, but criticality continued with a low power output for about 20 h, stabilized by thermal expansion of the fluid and the formation of bubbles. The chain reaction was terminated by draining water from a cooling jacket surrounding the tank, which reduced reflection of neutrons back into the tank, and as a precaution by injecting a boric acid solution into the tank.

[7] Extensive studies were carried out after the accident. One, referred to later as the "Kemeny report," was by a commission appointed by President Carter and

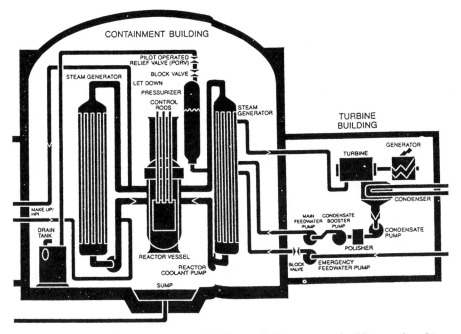

Fig. 15.1. Schematic of the TMI-2 facility, including reactor building and turbine building. Piping goes, from right to left in the diagram, through the containment building wall to the auxiliary building (not shown); piping also goes, from left to right, through the turbine building wall to the condensate storage tank and cooling tower (not shown). (From Ref. [11, pp. 86–87].)

unit, known as TMI-2. It was a 906-MWe pressurized water reactor built by Babcock and Wilcox, the smallest (in terms of number of units completed) of three U.S. manufacturers of PWRs. It had first received a license to operate at low power in February 1978 and was in routine operation at full power by the end of 1978. A schematic of the TMI-2 facility is shown in Figure 15.1 [11, pp. 86–87].

The accident started with a failure of the cooling system of TMI-2 in the early morning of March 28, 1979. The initial problem was an interruption in the flow of water to the secondary side of the steam generator. This water is the so-called feedwater. In the secondary loop, feedwater enters the steam

_____

chaired by John Kemeny, the president of Dartmouth College [11]. The second, the Rogovin Report, was by a special inquiry group instituted by the Nuclear Regulatory Commission and chaired by Mitchell Rogovin, a partner in an independent Washington law firm. The description here is drawn largely from the Kemeny Report [11] and Part 2 of Volume II of the Rogovin Report [12], as well as a further review article [13].

generator, and steam emerges to drive the turbine. The steam is condensed in a second heat exchanger (the condenser), and water is returned to the steam generator after passing through a "polisher," in which dissolved impurities are removed. The flow of water between the condenser and steam generator is maintained by the condensate pump and the main feedwater pump (see Figure 15.1).

The chain of events that led to the accident appears to have been initiated by work done to clean the polishers. In a sequence that has not been conclusively established, this operation may have caused one or more of the valves in the condensate polisher system to close, automatically shutting off (tripping) one of the condensate pumps. The tripping of the condensate pump, whatever the cause, in turn, tripped the main feedwater pumps.[8] This failure caused the emergency feedwater pumps to start automatically, in order to maintain the flow of water to the steam generator. Maintenance of feedwater flow is essential to cool the water from the reactor that flows through the primary side of the steam generator.

Up to this point, everything was "normal," in the sense that reactors are designed to handle occasional equipment failures; protection then comes from backup systems. However, the block valves in the emergency feedwater lines (there were two) were closed; according to proper operating procedures, they were supposed to be open. Indicator lights in the control room showed the closed status, but the operators at first did not notice this. Thus, no water was being fed to the secondary side of the steam generator because the pumps for the main supply were off and valves in the emergency line were closed. With no flow of water, the pressure in the steam generator rose and in response, the pilot-operated relief valve (PORV) on the "hot" side of the steam generator heat exchanger opened. The pressure excursion also caused the reactor to trip, with automatic insertion of the control rods. With the reactor turned off and the PORV open, the pressure dropped. The PORV should then have automatically closed.

At this point, there were additional equipment and design failures. The PORV did not close properly, but the control panel indicator light displayed the status of the control power to the valve (namely that it was supposedly closed), not the actual status of the valve (namely that it was open). Thus, the operators had to cope with unusual conditions in the cooling system without knowing the actual status of the valves in it. In particular, the PORV remained open for almost 2.5 h, causing a very large loss of needed cooling water.

Within 2 min after the start of the accident, the steam generators boiled dry because they had no feedwater source and there was a substantial heat output from the reactor core due to radioactive decay. Overall, the conditions of the cooling system were both unusual and confused, with the operators

---

[8] Figure 15.1 does not show the redundancy in the system. There were two main feedwater pumps and three emergency feedwater pumps.

not having correct information or sufficient training to recognize the nature of the evolving anomalies and cope with them. They did recognize that there were serious problems, and by 4:45 AM supervisory personnel began to arrive at TMI, only three-quarters of an hour after the start of the accident. By 6:22 AM, the PORV was closed, but the problems were not over. At 7:00 AM a "site emergency" was declared because there had been some release of radioactivity.

15.2.2 Evolution of the TMI Accident

Over the next few days, the accident continued to unfold, with continued difficulty in establishing proper cooling conditions. There were some small releases of radioactivity outside the plant, as well as one misinterpreted report of radiation levels that led to the incorrect belief that there had been a large release. There was a great sense of emergency both at the site and in the surrounding area, as no one was willing to give unequivocal assurances that matters were under control. This led to a recommended evacuation of pregnant women and preschool children from the immediate vicinity and a large self-initiated evacuation by individuals.

Concern reached a peak on Saturday, March 31, over the possibility of a hydrogen explosion inside the pressure vessel. As described in the subsequent Kemeny Report:

> The great concern about a potential hydrogen explosion inside the TMI-2 reactor came with the weekend. That it was a groundless fear, an unfortunate error, never penetrated the public consciousness afterward, partly because the NRC made no effort to inform the public it had erred. [11, p. 126]

Hydrogen is produced by the reaction of steam with the zircaloy cladding at high temperatures. Oxygen is formed by the breakup of water under radiation, so-called radiolysis. Together, hydrogen and oxygen can form an explosive mixture. There was fear that such an explosion could occur within the pressure vessel. Within a day or so, some NRC experts came to the conclusion that a hydrogen explosion was impossible, but this conclusion was not immediately accepted by all of the authorities. In the meantime, the hydrogen bubble had become a matter of great public concern, a concern not unambiguously dismissed by the NRC. However, by 6:00 PM on April 1, the hydrogen was removed from the bubble by "letdown, leakage, and venting" [12, p. 535]. It never had been the threat that had been believed.

The reason that the problem was not a real one was an insufficient accumulation of oxygen. In a PWR, it is normal to have some hydrogen dissolved in the water and to have continued recombination of oxygen and hydrogen. This recombination prevented the amount of oxygen from rising sufficiently to create a danger of explosion. This point is brought out in the Rogovin Report:

> Little or no oxygen was present in the bubble and a very low probability of explosion existed. The incorrect perception of an explosion hazard stemmed from contradiction among supposed experts. This perception was known or should have been known to be false by the afternoon of April 1. [12, p. 535]

It may be noted that a Babcock and Wilcox scientist had given assurances from the first that there was no problem from oxygen production [12, p. 534], but apparently this assurance did not receive much attention.

President Carter visited the site on Sunday, April 1; the hydrogen bubble itself dissipated (although not the perception of a near miss with hydrogen) and the worst of the crisis was over. However, it took more than another week for the advisory evacuation of pregnant women and preschool children to be withdrawn by the governor of Pennsylvania.

15.2.3 Effects of the TMI Accident

Core Damage and Radionuclide Releases

In retrospect, several major aspects of the Three Mile Island accident were not fully appreciated at the time and might seem to be in conflict:

♦ There was very little release of radioactivity and very little exposure of the general population. According to the Kemeny Commission, "the maximum estimated radiation dose received by any one individual in the off-site general population (excluding the plant workers) during the accident was 70 millirems.... three TMI workers received radiation doses of about 3 to 4 rems; these levels exceeded the NRC maximum permissible quarterly dose of 3 rems" [11, p. 34]. In essential agreement, the Rogovin Report found that "the maximum off-site individual dose was less than 100 mrem" [12, p. 400].

♦ The total collective dose to the 2 million people living within 50 miles of TMI was approximately 2000 person-rem (20 person-Sv).[9] From this, the Kemeny Commission estimated a 50% chance of no fatal cancers from the accident, a 35% chance of one fatal cancer, and a 15% chance of more than one [11, p. 12]. These results correspond to an average expectation of 0.7 cancer fatalities. If the 1993 NCRP risk estimate of 0.05 per sievert is adopted, then one fatal cancer is calculated for the collective dose of 20

---

[9] See Ref. [11, p. 34] and Ref. [12, p. 399].

person-Sv (see Section 4.3.4). Among these 2 million people, it is expected that 325,000 will die of cancers unrelated to TMI, so the TMI impact, if any, will be far below any detectable level.

◆ The core damage was very great. As cleanup and dismantling of the TMI-2 reactor proceeded, it was found that the core damage was greater than originally thought, and some observers have expressed surprise that the reactor vessel itself withstood the molten fuel at its bottom.

Thus, those who thought that the accident was causing, or was about to cause, large releases of radioactive material were proven to be wrong. Those who thought that matters were being exaggerated and that there was relatively little actual damage to the reactor were also wrong. The biggest surprise, however, was having these two outcomes together—great core damage and almost negligible external radionuclide releases. Only about 15 Ci of ^{131}I were released to the environment outside the containment [12, p. 358], despite an initial core inventory more than 1 million times greater. It had been commonly assumed that with core damage of this magnitude, a large fraction of the iodine would escape. Thus, the containment system, including the system to spray water into the containment to remove radionuclides, performed unexpectedly well. In the aftermath of TMI, understanding this performance—now part of what is known as the source term question—became a major issue in reactor safety studies (see Section 14.1.4).

Studies of Health Effects of TMI

The release of radioactivity from the Three Mile Island plant and the resulting radiation exposures were too small to have produced any observable effects, if one accepts official accounts of the magnitude of the releases and standard dose–response relationships. One or even 10 cancer deaths would be lost among a total of over 300,000 "natural" cancer deaths. Nonetheless, there have been persistent claims of health problems from TMI. In response to some of the early concerns, the Pennsylvania Health Secretary stated in a news release: "After careful study of all available information, we continue to find no evidence to date that radiation from the nuclear power plant resulted in an increased number of fetal, neonatal, and infant deaths. That simply isn't the case" [14]. This was based on an examination of death rates near TMI and in Pennsylvania as a whole, before and after the accident.

The Pennsylvania Department of Health carried out a later study of spontaneous abortions, in the face of a rather widespread belief among residents of the TMI area that there had been an increase in stillbirths and miscarriages [15]. The study identified 479 women living within 5 miles of the plant who were pregnant at the time of the accident. For this group, there were 436 live births, 28 spontaneous abortions or stillbirths, and 15 other abortions. The rate of spontaneous abortions and stillbirths was compared to the rate expected from earlier studies of nonexposed populations and it was found that there was no excess. In the author's words, the TMI incidence rates "compared favorably with the four baseline studies."

In a broader study of cancer rates near Three Mile Island, investigators found a statistically significant increase in cancer incidence, compared to rates at greater distances, during 1982 and 1983 [16].[10] However, this excess did not persist, and in 1985, the cancer incidence rate was slightly lower for the near-TMI group than for the more distant group. Further, the increase was seen only in cancer *incidence*, not in cancer fatalities. This lack of increase was commented on particularly for lung cancer, which progresses rapidly from incidence to fatality, as pointing to possible "screening bias." The authors concluded:

> We observed a modest postaccident increase in cancer near TMI that is unlikely to be explained by radiation emissions. The increase resulted from a small wave of excess cancers in 1982, three years after the 1979 accident. Such a pattern might reflect the impact of accident stress on cancer progression. Our study lacked a direct, individual measure of stress, however. The most plausible alternative explanation is that improved surveillance of cancer near the TMI plant led to the observed increase.

These results are consistent with the belief that there is virtually no possibility that there have been or will be observable health effects from radioactivity released in the TMI accident, given the low exposure levels. However, the post-TMI history illustrates the extent of skepticism about official reassurances in situations of possible radiation hazard. This skepticism is fed by anomalies in the data (such as the increase in observed cancer incidence in 1982). Anomalies often cannot be explained in any conclusive fashion, and the ruling out of radiation exposure as the cause may hinge on somewhat indirect arguments, such as comparisons to standard models of the time intervals between radiation exposures, cancer incidence, and cancer fatalities. The families and friends of the "victims" of the anomalies may have little incentive to accept these arguments.

These difficulties may be of only marginal interest in the case of TMI, where the weight of evidence and scientific opinion is strong, but they could assume much greater importance in evaluating the Chernobyl accident, where the exposures were very much greater and the conditions for systematic epidemiological studies are poorer. It is probable that there will be large health consequences from Chernobyl, and it is possible that some will be observable, but it may prove difficult to assess the validity of individual reports and to resolve the disagreements that will arise.

---

[10] The authors were from the School of Public Health at Columbia University, with the exception of one from the Audubon Society.

15.3 The Chernobyl Accident

15.3.1 The Chernobyl Reactors

Among reactor accidents, the Chernobyl accident in 1986 stands alone in terms of magnitude.[11] The design features of "Chernobyl-type" reactors are unique, and it is the standard assumption of nuclear analysts that a similar accident could not occur in any of the other types of reactors operating today. However, Chernobyl demonstrated the seriousness of a near "worst-case" accident. It intensified preexisting public concern about reactors of all sorts and strengthened the position of those who oppose nuclear power on safety grounds.

The Chernobyl reactor was one of the Soviet RBMK-1000 reactor series, designed to operate at a (gross) capacity of 1000 MWe. These are graphite moderated and water cooled, of a type originally used in the USSR (and with important differences in the United States) for the production of plutonium. Such reactors were also used for the generation of electricity in the USSR, dating back to a 5-MWe water-cooled, graphite-moderated reactor at Obninsk, put into operation in 1954 [18, p. 9].

At the beginning of 1986, there were four RBMK-1000 reactors at the Chernobyl site in the Ukraine, located about 130 km north of the major city of Kiev. The most recently installed reactors, completed in 1983, were Units 3 and 4, housed in a single building. At the time of the accident, all four reactors were operating and two more were under construction. The accident itself occurred in Unit 4 on April 26, 1986 at 1:24 AM. Unit 3 was turned off by the operators after almost 5 h and the nearby Units 1 and 2 were turned off after about 24 h. Units 1 and 2 were returned to operation in late 1986 and Unit 3 in December 1987. Construction was suspended and then canceled on the two reactors being built at Chernobyl [19].

Unit 2 was permanently closed, following a fire on October 1, 1991. The fire was in a non-nuclear part of the plant, and there was no release of radioactivity, but there was damage to the engine room [20]. Units 1 and 3 continued in operation beyond 1991, with the Ukrainian authorities balancing the need for their electrical output against concerns about their safety, but both were eventually shut down permanently—in 1996 and 2000, respectively. Outside Ukraine, there are 11 RBMK-1000 reactors operating in Russia (4 at Kursk, 4 near St. Petersburg, and 3 at Smolensk) and two larger RBMK reactors (1380 MWe) at Ignalina in what is now Lithuania [21].[12] In an effort to reduce the chance of another accident, steps have been taken since 1986 to improve operator training, and significant modifications have been made in the RBMK reactors themselves.

---

[11] The account of the accident progression given here is based largely on parts of Refs. [17], [18], and [24]–[27].

[12] Lithuania has agreed to shut down these plants, one in 2005 and the other in 2009, as a condition for joining the European Union [22].

Although the USSR in the past exported reactors to its neighbors, these were all PWRs, not RBMKs. Outside the USSR, the only reactor bearing some similarity to the Chernobyl reactor was the not so very similar N reactor at Hanford, which stopped operations in 1988. Future Russian plans are based primarily on LWRs, although the Chernobyl accident did not stop the completion of all RBMK reactors: One of the RBMK-1000 units now operating at Smolensk was put on-line in 1990 and a fifth unit at Kursk, for which construction began in 1985, remains under construction [23].

15.3.2 History of the Chernobyl Accident

Deficiencies in Attention to Safety

There was no crisp single cause of the Chernobyl accident. Even 6 years later, a review by a major international technical body, the International Nuclear Safety Advisory Group (INSAG), reported: "It is not known for certain what started the power excursion that destroyed the Chernobyl reactor" [24, p. 23]. The quotation may suggest more ignorance than is the case. A great deal is known about the conditions before and during the accident, even if the development of a definitive, detailed scenario has been made difficult by the complexity of the reactor conditions, the speed of unfolding of the accident, and the damage itself.

Overall, the accident was the result of a combination of design deficiencies, operator errors, and an unusual set of prior circumstances, all of which put the reactor and the operators to a test that they failed. In the INSAG view, "the accident can be said to have flowed from deficient safety culture, not only at the Chernobyl plant, but throughout the Soviet design, operating and regulatory organizations." The INSAG report listed a whole gamut of weaknesses in institutions and attitudes [24, p. 24]. In this view, with greater vigilance in both design and operations, there would have been no accident.

Design Weaknesses

Two aspects of the design have been particular targets of criticism: a positive void coefficient of reactivity and an improperly configured control rod system. Both contributed positive feedbacks, which turned an initial excursion in reactor performance into the Chernobyl disaster.

The void coefficient in a water-cooled reactor would be negative if the water acted only as a coolant and moderator. However, the water also acts as a poison, due to capture of neutrons in hydrogen. In an LWR, when water is lost or steam bubbles develop, the dominant effect is a decrease in reactivity, corresponding to the negative void feedback discussed in Section 14.2.1. However, in a Chernobyl-type reactor, most of the moderation is provided by the graphite, and the water acts mainly as a poison. When some of the cooling water is replaced by steam, there is a different balance than in an LWR between the competing feedbacks: increased resonance absorption in ^{238}U (negative void coefficient), increased neutron leakage from the reactor (negative void coefficient), and decreased absorption of thermal neutrons in H_2O (positive void coefficient). The net outcome depends on the relative amounts and arrangement of the uranium, carbon, and water in the reactor. For the Chernobyl reactor, the overall void coefficient was positive, whereas a graphite-moderated, water-cooled reactor in the United States, the Hanford-N reactor, had a negative void coefficient.

The second major defect, involving the control rods, is also related to the role of water. At the onset of the accident, most of the control rods had been fully withdrawn from the reactor to compensate for an excursion in xenon poisoning to above the normal level (see next subsection). Remarkably, the first effect of the insertion of the control rods from the full-out position was to *increase* the reactivity. This was due to a peculiarity of the RBMK-1000 control rod system, which has since been corrected. The control rods move vertically through the reactor core and are withdrawn by being lifted upward. To prevent the control rod from being replaced by water, which acts as a poison and lessens the effect of withdrawing the rod, a long graphite "displacer" was attached to the bottom of the rod. When the rod was fully withdrawn, most of the channel in the core was occupied by this graphite, and the control material was above the core. However, the graphite displacer did not completely fill the channel; instead, there was still a 125-cm column of water below the graphite displacer [24, p. 5]. The first effect of inserting the control rod, before the absorbing part of the control rod reached the core, was for the graphite to drive out the water column, increasing the reactivity by reducing the poison. The full motion of the control rods was slow, and by the time the control rod proper entered the core region, it was too late.

Since Chernobyl, there have been a number of changes to correct these defects in the RBMK reactors. These include increasing the enrichment of ^{235}U in the fuel, changing the control rod geometry so that the displacer will not displace water when inserted, and speeding up the control rod insertion [25].

Reactor Operations Prior to Accident

The accident evolved from a test that disturbed normal operating conditions. Ironically, the test was undertaken to demonstrate a safety feature of the reactor. Power for the pumps and other plant facilities normally comes from the plant's own turbogenerator units or from the off-site power grid. Should there be an off-site power outage and a shutdown of the reactor, standby diesel generators at the plant come on-line to supply power. There could be an interval of several seconds between the loss of normal power and the start-up of the diesel generators. The test was to demonstrate that the inertial coasting of the turbogenerator would provide sufficient power to operate pumps during this interval.

The first step taken to perform this test was a reduction of reactor power to one-half of its normal 3200 MWt, beginning at about 1:00 AM on April 25. One of the turbogenerators was switched off at 1:06 AM, and the power reached 1600 MWt at 3:47 AM [24, p. 53]. The remainder of the test, involving a further reduction of power, was to start at about 2:00 PM. As part of the test, the emergency core-cooling system was disconnected at 2:00 PM.

However, the reactor's power was required for the electricity grid fed by Chernobyl and instructions were given to postpone the further power reduction. The test did not resume until 11:10 PM on April 25. It was then intended to reduce the power to about 700–1000 MWt, but there was an overshoot in the shutdown and the power level dropped to 30 MWt. By about 1:00 AM on April 26, it had been brought back to 200 MWt, but the period of operation at low power caused a buildup of xenon poisoning (see Section 7.5.3), which was compensated for by removal of a large number of the control rods, more than proper under operating guidelines.[13]

At this point, there were at least two unusual circumstances: The power level for the test was lower than planned and the margin of safety, in terms of the ability to shut down the reactor with the control rods, was less than the normal operating limit. In addition, a number of safety systems had been turned off to facilitate the planned test. In hindsight, it is clear that the test should have been terminated at this point, but it was continued.

Initiation and Progress of the Accident

As the test proceeded at low power, water flow conditions were not normal, there was some decrease in steam, and the reduced reactivity caused automatic control rods to withdraw to restore the reactivity. This was a manifestation of the fact that the Chernobyl reactor operated with a positive void coefficient.[14]

This action was in itself harmless, but it raised the control rods to unusually high positions out of the core. At 1:23:04 AM, despite warning indications of the dangerous control rod configuration, the operators initiated the turbine test by shutting a valve and reducing steam flow to the turbine. The resulting changes in steam pressure and in water flow from the cooling water pumps led to a decreased water flow through the core and some boiling in the core. The displacement of water by steam caused the reactivity to rise.

In response, at 1:23:40 AM, an emergency shutdown (scram) was attempted. However, the control rods had been withdrawn too far to take im-

---

[13] After power is reduced, the decay of ^{135}I ($T = 6.57$ h) to ^{135}Xe ($T = 9.14$ h) continues, but the destruction of ^{135}Xe by neutron capture is decreased. Therefore, the amount of ^{135}Xe increases for several hours.

[14] According to Richard Wilson [26], Russian experts explained to him that this design was adopted for cost savings in establishing the graphite configuration. For power levels of 20% of normal or higher, the negative fuel temperature coefficient (the Doppler coefficient) was supposed to provide an adequate margin of safety, and at lower power levels, there were to be stringent operating regulations.

mediate effect and, due to the graphite displacers at their ends, their first effect was to increase rather than decrease the reactivity. Within 3 s there was a sharp increase in the neutron flux and power output, as the reactor went superprompt critical. Within not more than 20 s, there were two large explosions—one apparently a steam explosion that exposed the reactor fuel to the air and the other an explosion due to exothermic reactions, including the interaction of liberated hydrogen and carbon monoxide with the air.

These explosions breached the reactor building (there was no true containment) and sent burning fragments into the air, which started fires on the roof of the reactor. Firefighters from the nearby towns of Chernobyl and Pripyat arrived shortly after the accident and put out the building fires by 5:00 AM There had been a threat that the fires would spread to the other units at Chernobyl (Units 1, 2, and 3), but this was prevented by firemen working under extreme conditions of heat and radiation exposure. Remarkably, Unit 3 was not turned off until about 6:00 AM.

Although the exterior fires had been extinguished, the problem of heat generation in the reactor continued, due to (chemical) burning of the fuel and graphite in the reactor and to radioactive decay heat. These interior fires were not extinguished until May 6, following a series of attempts to quench them by dropping massive amounts of boron carbide (intended to prevent recriticality), limestone, lead, sand, and clay. In total, Unit 4 was entombed under about 5000 tons of material, and the acute phase of the accident was then over.

15.3.3 Release of Radioactivity from Chernobyl

The initial explosions and subsequent reactor fires caused large amounts of radioactive materials to be released from the reactor. The release was not all immediate, with about 24% the first day, 28% over the next 5 days, and 48% over the following 4 days [27, p. 3.9]. The release was mainly of the volatile nuclides, including the noble gases, iodine, and cesium. Much less of the non-volatile nuclides, such as strontium, escaped. Estimated release fractions and total releases are given in Table 15.2 for some of the important radionuclides.

The cloud of radioactivity from the accident during the first 2 days spread generally to the north and west and, thus, did not severely impact Kiev. In later days, when releases were lower, the winds shifted to form plumes in other directions, including to the south [28, p. 459]. The accident was not immediately made public, and the first awareness outside the USSR came from radiation measurements in Sweden and Finland. The cloud reached Sweden at about 2 PM on April 27 and was first detected about 18 h later by monitors at the Forsmark nuclear power station [18, p. 13]. This was approximately 2 days after the start of the accident itself.

Eventually, the radioactive cloud spread over most of the northern hemisphere, depositing radionuclides widely. The amounts deposited decreased with increasing distance, although the correlation with distance was not pre-

Table 15.2. Radionuclide releases from the Chernobyl accident, for selected radionuclides.

Isotope	$T_{1/2}$	Core (MCi)	Release[a] (MCi)	Fraction Released
^{85}Kr	10.8 years	0.89	0.89	1.0
^{133}Xe	5.24 days	176	176	1.0
^{131}I	8.02 days	86	48	0.6
^{134}Cs	2.07 years	4.1	1.5	0.36
^{137}Cs	30.1 years	7.0	2.3	0.33
^{90}Sr	28.8 years	5.9	0.27	0.05

[a]The actual release was less for short-lived isotopes because of decay before release; these numbers are "corrected" back to the activity at the time of the accident.
Source: Ref. [28, pp. 518–519].

cise due to wind patterns and rainfall. In general, it could be said that the fallout was substantial near Chernobyl, moderate in some other parts of Europe, and negligible in North America.

15.3.4 Observations of Health Effects of Chernobyl Accident

Overall Summary up to 2000

Subsequent to the Chernobyl accident, there have been many studies of its health impacts, and studies are likely to continue for many decades. A succinct summary of current knowledge was presented in the 2000 Report of the United Nations Scientific Committee on the Effects of Atomic Radiation (UNSCEAR) in an Overview section on "The Radiological Consequences of the Chernobyl Accident":

> The accident at the Chernobyl nuclear power plant was the most serious accident involving radiation exposure. It caused the deaths, within a few days or weeks, of 30 workers and radiation injuries to over a hundred others. It also brought about the immediate evacuation, in 1986, of about 116,000 people from the areas surrounding the reactor and the permanent relocation, after 1986, of about 220,000 people from Belarus, the Russian Federation and Ukraine. It caused serious social and psychological disruption in the lives of those affected and vast economic losses over the entire region. Large areas of the three countries were contaminated, and deposition of released radionuclides was measurable in all countries of the northern hemisphere.

> There have been about 1,800 cases of thyroid cancer in children who were exposed at the time of the accident, and if the current trend

continues, there may be more cases during the next decades. Apart from this increase, there is no evidence of a major public health impact attributable to radiation exposure 14 years after the accident. There is no scientific evidence of increases in overall cancer incidence or mortality or in non-malignant disorders that could be related to radiation exposure. The risk of leukemia, one of the main concerns owing to its short latency time, does not appear to be elevated, not even among the the recovery operation workers. Although those most highly exposed individuals are at an increased risk of radiation-associated effects, the great majority of the population are not likely to experience serious health consequences as a result of radiation from the Chernobyl accident. [29, p. 4]

Thus, two groups have been unambiguously harmed by radiation from Chernobyl: emergency workers at the site of the accident and children in a wide surrounding region who have developed thyroid cancers. No other radiation damage had been observed by the time of the UNSCEAR report in 2000. Further details on the affected groups are presented in the next two subsections.

Effects on Plant Workers and Firemen

Severe medical effects of the accident were suffered by workers at the plant and the firemen who responded to the accident. By 8:00 AM of the morning of the accident, these totaled about 600, including 69 firemen [28, p. 522]. In all, 31 died within several months of the accident: 28 from acute radiation syndrome (ARS), 2 from nonradiation injuries, and 1 apparently from coronary thrombosis [30, p. 6].[15] Most of those who died from radiation sickness also received severe skin burns from beta-particle radiation. These early deaths were exclusively among plant personnel and firemen. The latter appear to have performed in an exceedingly dedicated and self-sacrificing manner.

A total of 237 people were suspected of having ARS and this diagnosis was confirmed for 134, including the 28 who died. The deaths among the ARS group were strongly correlated with the magnitude of the radiation exposures. The fractional death rates at different exposure levels were: 0 out of 41 up to 2.1 Sv, 1 out of 50 from 2.2 to 4.1 Sv, 7 out of 22 from 4.2 to 6.4 Sv, and 20 out of 21 above 6.4 Sv [28, p. 523]. In the decade following the accident, from 1987 to 1996, an additional 14 of the original 237 patients died, but these deaths do not appear to be primarily attributable to radiation exposure [31, p. 187].

---

[15] ARS was defined by "at least minimal bone-marrow suppression as indicated by depletion of blood lymphocytes" [28, p. 488]. In many accounts of Chernobyl, the number of prompt fatalities is given as 30, apparently omitting the coronary thrombosis victim.

Childhood Thyroid Cancer

The accident led to high thyroid exposures, primarily due to ^{131}I in the cloud of radionuclides from the reactor. The relatively short half-life of ^{131}I (8.02 days) means that the dose was received within several weeks of the accident. The main pathways for this dose were through inhalation and consumption of milk or locally grown produce.[16] High thyroid cancer rates began to be seen among children in the early 1990s. Sixty-two cases were observed in 1990 and the number of cases grew steadily until about 1994, when the rate leveled off at roughly 250 per year [28, p. 545]. At first, it was suspected that the high rate of observed thyroid cancers was due to the careful search for them. However, the rate of cancer incidence is seen to be much higher for young children (based on age at the time of the accident) than for older ones, and there is no increased rate for children born after the accident [28, p. 501]. Further, the cancer rate correlates with inferred dose. Thus, the excess is well established.

Through 1998, a total of 1791 thyroid cancer cases were reported in children up to the age of 17 [28, p. 545]. A 2002 UN report states that "some two thousand cases of thyroid cancer have been diagnosed" and anticipates that "the figure is likely to rise to 8–10,000 in coming years" [32, p. 7]. Thyroid cancer is rarely fatal, but requires continued medical treatment.

15.3.5 Radiation Exposures at Chernobyl and Vicinity

Effects on Cleanup Personnel

A prolonged cleanup was carried out in the aftermath of the accident by military servicemen and civilians who were brought in to work for short periods. These were the so-called liquidators. About 600,000 people have been so designated for the years 1986 to 1989, including about 200,000 for 1986 and 1987 when the radiation levels were highest [28, p. 469]. Individual doses often reached several hundred millisieverts [28, p. 525], which is well above the U.S. occupational limit of 50 mSv in 1 year.

There have been reports of increased fatalities and sickness among the liquidators, but in the absence of comparisons to similar populations of non-exposed individuals, it is not clear that the results are meaningful [28, p. 516]. The 2000 UNSCEAR document reported no findings of increased rates of leukemia or other forms of cancer among the liquidators [28]. However, this remains an important group for continued study over the next several decades, because most radiation-associated cancers appear more than 10 years after the exposure. As summarized in the UNSCEAR report:

---

[16] Consumption of milk from cows that grazed in contaminated soil provides a quick path for transfer of ^{131}I from the ground to the human body. Use of potassium iodide tablets reduces uptake of other iodine to the thyroid, and many of the children in the Chernobyl vicinity received these tablets.

> Apart from the radiation-associated thyroid cancers among those exposed in childhood, the only group that received doses high enough to possibly incur statistically detectable increased risks is the recovery operation workers. Studies of these populations have the potential to contribute to the scientific knowledge of the late effects of ionizing radiation. Many of these individuals receive annual medical examinations, providing a sound basis for future studies of the cohort. It is, however, notable that no increased risk of leukemia, an entity known to appear within 2–3 years after exposure, has been identified more than 10 years after the accident. [28, p. 517]

In short, by the year 2000, no statistically significant radiation-caused harm was seen among the liquidators, but that does not establish that there has been no harm or that no statistically significant effects will ever be identified.

Exposure of Population in the "Affected Region"

The largest exposures of people near Chernobyl initially came from ^{131}I and other short-lived radionuclides, in part through inhalation. After several weeks, the iodine had decayed sufficiently to be a lesser contributor, and over the longer term most of the dose came from ^{134}Cs and ^{137}Cs and, later, just ^{137}Cs.

After passage of the radioactive cloud, the most important pathways for exposures were from ingested radionuclides and from gamma rays and beta particles emitted by radionuclides deposited on the ground. Ingestion is the more important in the first year, but as deposition from the atmosphere ends and the radionuclides are washed off vegetation by rain, the ground exposure becomes the more important. Overall, in terms of the long-term dose commitment (including the dose from ingested radionuclides that remain in the body), the external surface dose and the ingestion dose are roughly equal, with the former somewhat predominating.

On the day after the accident, almost 50,000 people were evacuated from Pripyat, 3 km from the reactor site [28, p. 527]. Later, the evacuation was extended to cover an "exclusion zone" around Chernobyl, which included the region within 30 km of Chernobyl plus a few outlying areas [28, pp. 472–473]. In all, about 116,000 people were evacuated in 1986 and another 220,000 people were relocated after 1986 [28, p. 453].

A series of zones have been defined around Chernobyl, in part in terms of the deposition of ^{137}Cs:

◆ *Exclusion zone.* This is the 30-km zone (cited above) from which 116,000 people were evacuated in 1986.[17]

---

[17] There has been some ambiguity about this total. Earlier estimates put it at 135,000, as reflected in Table 15.3 [28, p. 473].

◆ *Strict control zone (SCZ)*. Region where the ^{137}Cs density on the ground exceeds 15 Ci/km^2 is the region of strict control [28, p. 475]. The initial population of this region was about 270,000 [28, p. 475]. At this level, resettlement is obligatory in Ukraine, but not necessarily mandatory in Belarus and Russia [32, p. 36]. In all three countries, people have a right to resettle from regions where the density exceeds 5 Ci/km^2.

◆ *Other contaminated areas*. These are regions where the ^{137}Cs ground density is between 1 and 15 Ci/km^2. Most of the people in these regions were in areas at the low end of the range (i.e., 1–5 Ci/km^2).

The future health impacts of Chernobyl on these groups can be estimated from a study of the radiation doses. A summary of average and collective doses for the groups considered above is given in a review by Elizabeth Cardis and colleagues, prepared as a background paper for the *One Decade After Chernobyl* conference that was held in 1996 with the joint sponsorship of the IAEA and the World Health Organization (WHO) [30]. It also included data for the subgroup of liquidators who worked in 1986–1987, when the radiation levels were highest. In this paper, cancer mortality was estimated separately for solid cancers and leukemia [33]. The results are presented in Table 15.3.

The doses for the populations other than the liquidators are doses received from 1986 to 1995.[18] This represents about 60% of the total estimated long-term dose (1986–2056) [28, Table 33], suggesting that the doses in Table 15.3 underestimate the total impact of the accident. On the other hand, as the authors pointed out, no dose and dose rate effectiveness factor (DDREF) was applied in estimating the cancer fatalities (see Section 4.3.4). Thus, the calculated number of fatalities is not an underestimate if a DDREF of 2 is the "correct" factor to apply. Overall, however, it is to be remembered that there are large uncertainties in both the doses and the dose–response relation. In particular, the number of fatal cancers may be greatly overestimated, because the calculation is based on the linearity hypothesis. Thus, the 9,000 excess fatalities of Table 15.3 are the predictions of a specific model, not an assured outcome, and even in the context of the model, the results are approximate.

It is seen from Table 15.3 that the calculated number of excess fatalities is, in most cases, small compared to the normal natural rates. Even when in principle the excess is statistically significant, uncertainties in the "normal" rate may make it difficult to confirm the result by observations. The best opportunity for doing this may be for leukemia incidence among the liquidators, where a 25% excess is predicted.

The average doses for most of the people in these groups are well below the 10-year background level of 20–30 mSv (roughly 2.4 mSv/year), but the

---

[18] In such calculations, the calculated dose includes the contributions from all radionuclides, with the ^{137}Cs density taken as an indicator of the concentrations of other radionuclides. After the first year, most of the dose is due to ^{134}Cs and ^{137}Cs; after the first decade, most of the ^{134}Cs has decayed, leaving ^{137}Cs as the dominant contributor.

Table 15.3. Cumulative radiation doses (1986–1995) and calculated deaths over a 95-year period among populations in the vicinity of Chernobyl and among liquidators working in the 1986–1987 period (see text).

Group	Population size	Average dose (mSv)	Solid Cancer Deaths		Leukemia Deaths	
			Normal Rate	Calc. Excess[a]	Normal Rate	Calc. Excess[a]
Liquidators (1986–1987)	200,000	100	41,500	2,000	800	200
Evacuees (30-km zone)	135,000	10	21,500	150	500	10
Residents, SCZ	270,000	50	43,500	1,500	1,000	100
Residents, other areas	6,800,000	7	800,000[b]	4,600	24,000	370

[a]This calculated number of excess fatalities is based on the linearity hypothesis, without a DDREF.
[b]This number, given in Ref. [33] for the normal rate, appears inconsistent with the 16% normal cancer incidence rate indicated for all groups other than the liquidators.
Source: Ref. [33, p. 255].

liquidators and the SCZ residents received higher doses. By 2003, with most of the long-term dose already received, the annual dose from Chernobyl radionuclides is below natural background radiation levels, even in the SCZ.

Controversies over Health Effects from Chernobyl

It is too soon for the health effects of the Chernobyl accident to have been fully manifested, for either the workers who participated in the Chernobyl cleanup efforts or the surrounding population. Most cancers, other than leukemia, have a latent period of 10 years or longer. Further, it is impossible to attribute a given observed cancer to a particular cause. The ambiguities of statistical evidence, the dramatic nature of anecdotal evidence, and the strong incentives to reach one conclusion or another almost guarantee that there will be very different assessments of the consequences of Chernobyl.

Even at Three Mile Island, there has been some controversy over the consequences (see Section 15.2.3), and the situation is likely to be far more difficult in the Chernobyl case, especially in view of the political and economic problems in the area. These may make it difficult to obtain satisfactorily comprehensive and reliable records. It is important that vigorous efforts be made to carry out detailed health surveys and data analyses, in order to improve our understanding of the effects of prolonged exposures to radiation at low and intermediate levels.

The disagreements about Chernobyl could be reduced if there are careful epidemiological studies by a group whose legitimacy is widely accepted. There is a precedent for this in the Radiation Effects Research Foundation,[19] which has carried out continuing studies of the aftermath of Hiroshima and Nagasaki. In this vein, the 2002 United Nations report proposed the establishment of an International Chernobyl Foundation to "channel resources into health and ecological research relating to the effects of the Chernobyl accident" [32, p. 17].

15.3.6 Worldwide Radiation Exposures from Chernobyl

One of the important set of results reported at the *One Decade After Chernobyl* conference was an UNSCEAR assessment of total global population doses. These results, as presented in a paper by R.G. Bennett, are summarized in Table 15.4.

The total collective dose commitment in the northern hemisphere is estimated to be 600,000 person-Sv, where the dose commitment is calculated until 2056 (70 years after the accident). For populations that were not near Chernobyl, the accident added relatively little to the natural background. In

---

[19] This was formerly known as the Atomic Bomb Casualty Commission. It is a joint research activity of the United States and Japan.

Table 15.4. Distribution of radiation doses from Chernobyl accident (total lifetime collective dose commitment = 600,000 person-Sv).

Category	Percent
Geographical distribution	
Former Soviet Union (FSU)	36
Other Europe	53
Other, northern hemisphere	11
Time distribution	
First year	~ 33
Later years	~ 67
Contributing radionuclides	
^{137}Cs	70
^{134}Cs	20
^{131}I	6
Other[a]	4
Mode of exposure	
External radiation	60
Internal radiation	40

[a]Ascribed to "short-lived radionuclides deposited immediately after the accident."
Source: Ref. [34, p. 125].

the first year after Chernobyl, when the impact of the accident was greatest, the collective dose in Europe (outside the FSU) was about 100,000 person-Sv, corresponding to an average individual dose of 0.2 mSv for a population of about 500 million.[20] Thus, on average, Chernobyl in the first year added about 8% to the average natural radiation dose of 2.4 mSv/yr. The total per capita doses in Europe (outside the FSU) summed over the 70 years following the accident are expected to average about 0.7 mSv from Chernobyl and 170 mSv from natural radiation sources [35, p. 369]. For North America, the average dose is estimated to be 0.001 mSv in the first year and 0.004 mSv summed over all years.

If one uses a risk factor of 0.05 deaths per person-Sv, the worldwide collective dose of 600,000 person-Sv would imply a total toll from Chernobyl of 30,000 deaths spread over more than 70 years (mostly in Europe, including the FSU).[21] For the most part, these 30,000 deaths would be impossible to identify or verify. For example, about 16,000 of these deaths would be in Europe (outside the FSU). In the same time period, the European population is

---

[20] The country outside the FSU with the highest first-year dose was Bulgaria, with an average first-year dose of slightly under 0.8 mSv [34, p. 122].

[21] It should again be noted that considerable controversy surrounds the linearity hypothesis, especially when applied to the low individual doses considered here.

expected to suffer over 100,000,000 "natural" cancers.[22] The calculated 0.02% increase—if it occurs—would be undetectable.

Different perspectives on Chernobyl may be stated as follows: (a) The accident may lead to about 30,000 cancer deaths; (b) the number of cancer deaths attributable to Chernobyl is a very small fraction of those occurring "naturally"; or (c) it is inappropriate to calculate expected deaths from the collective dose, when most of the collective dose is made of individual doses that are well under 10% of the natural background. Depending on which of these formulations is taken to be the more appropriate, Chernobyl may be considered to have been either a major global disaster or no more than a serious accident, with tragic local consequences.

15.3.7 General Effects of the Chernobyl Accident

Human Consequences for People in the Affected Region

One of the major consequences of the Chernobyl accident was the fear engendered in nearby populations. An IAEA study of the consequences of the Chernobyl accident was requested in 1989 by the USSR government, apparently prompted by concerns among people living in the vicinity of Chernobyl but not close enough to have been among those originally evacuated from the exclusion zone. The regions considered were places where the ground surface concentrations of ^{137}Cs exceeded 5 Ci/km^2. It embraced an area of about 25,000 km^2, with a population of about 825,000 [37, p. 3].

This study, known as the International Chernobyl Project (ICP), was undertaken by an international committee under the sponsorship of the IAEA, with the assistance of the WHO, UNSCEAR, and other international organizations. An overview of these results was published in Spring 1991 [37]. In the populations studied, the International Chernobyl Project found no indications of adverse medical consequences from the radiation. In its conclusions it stated

> . . . .[there were] no health disorders that could be attributed directly to radiation exposure. The accident had substantial negative psychological consequences in terms of anxiety and stress. . .[37, p. 32]

This referred to health effects as of 1991.

A similar stress on psychological factors appears in a report on the *Human Consequences of the Chernobyl Nuclear Accident* that was published in 2002. This report was commissioned by several United Nations agencies and the WHO, with the goal of studying the "current conditions in which people affected by the Chernobyl accident are living" and to make recommendations for addressing their needs [32, p. ii]. It lays particular stress on the disruption

---

[22] The "over 100,000,000" figure is a crude extrapolation from Ref. [36], where an estimate of 88,000,000 deaths in Europe is given for the next 50 years.

of the lives of the people who were evacuated and the fears experienced by them and people still living in contaminated areas. Economic conditions have been poor in the countries of the former Soviet Union, and the Chernobyl accident has made matters worse due to the costs of remedial measures and the inhibitions on growing crops in contaminated areas.

While disclaiming an intent to "minimize the seriousness of the situation for health and well-being or the role played by the exposure to ionizing radiation," the report suggests giving priority to improving "basic primary health care, diet, and living conditions" [32, p. 8]. Among the adverse consequences of the Chernobyl accident, the report describes the social demoralization which has accompanied the dislocation of populations and their living with a poorly understood hazard. It suggests that "determined efforts need to be made at national and local level to promote a balanced understanding of the health effects of radiation among the public, many of whom at present suffer distress as a result of ill-founded fears" [32, p. 10].

One of the difficulties in assessing the health impacts of the radiation exposures is the generous compensation afforded to people who can establish injury and who, therefore, have an incentive to find health damage. Children in areas with a ^{137}Cs concentration of over 5 Ci/km^2 are entitled to 2 months of health holidays. With accompanying adults, in the year 2000 almost 300,000 people took holidays in Belarus and about 372,000 in Ukraine. Overall, as described in the UN report:

> [s]carce resources are allocated not primarily on the basis of medical need but rather on an individual's ability to register as a victim. The system has promoted an exaggerated awareness of ill-health and a sense of dependency, which has prevented those concerned from taking part in normal economic and social life. The pattern of behavior was described by the Kiev Conference on the Health Effects of the Chernobyl Accident...as the "Chernobyl accident victim syndrome."[32, p. 32]

Extended Effects of the Chernobyl Accident

The above-cited report suggests that Chernobyl produced a demoralization among the neighboring population, beyond that which could be directly attributed to the health effects of radiation. Chernobyl may also have had profound effects on the Soviet Union as a whole. The technological failure at Chernobyl and the attempted coverup of the accident is cited as one of the reasons for the collapse of confidence in the Soviet system. This possibility is reflected, for example, in an op-ed piece entitled "Will SARS be China's Chernobyl?" Before drawing the parallel with SARS, the author writes:[23]

---

[23] This article, by James Goldgeier, Director of the Institute for European, Russian, and Eurasian Studies at George Washington University appeared in a number of newspapers in April 2003, e.g., the *Los Angeles Times* on April 23, 2003, [38].

...a major event in the Soviet Union's downward spiral was the accident at the Chernobyl nuclear power station in April 1986. The Soviet Union initially tried to keep a tight lid on what had occurred, but radioactive fallout was not confined to Soviet territory, and European governments were quick to announce that initial reports from Moscow underplayed the danger....

Chernobyl alone would not have brought down the Soviet Union. However, the government's clumsy reaction and the growing demands for the truth on this and other issues helped stir the caldron of resentment that culminated in Soviet collapse.

The effects of Chernobyl on the stability of the Soviet Union may have been magnified by the fact that, by chance, the radiation exposures and negative impacts of the accident were greatest in two Soviet Republics that already had separatist tendencies. As described by Martin Malia, "Chernobyl, in particular, accelerated the development of Ukrainian and Belorussian separatist sentiments; and the local apparatus easily found it in their interests to espouse this sentiment against Moscow" [39, p. 440].

The accident also had a great effect on nuclear power. Like the Soviet Union, nuclear power was facing problems of its own, prior to Chernobyl. The shock of the Three Mile Island accident, 7 years earlier, had not fully dissipated and the post-TMI improvements in nuclear power plants had not yet paid off (e.g., in the higher capacity factors that began in the United States in the 1990s). Chernobyl came at a time when nuclear power was already facing economic and political difficulties, and it reinforced the existing public fears that contributed to these difficulties. There is little justification for extending the analogy and assuming that nuclear power will go the way of the Soviet Union. However, even if not a fatal setback, the Chernobyl accident was a major blow to its progress.

References

1. Zhores Medvedev, *The Legacy of Chernobyl* (New York: W.W. Norton, 1990).
2. Grigori Medvedev, *The Truth About Chernobyl* (New York: Basic Books, 1991).
3. United Nations Scientific Committee on the Effects of Atomic Radiation, *Sources and Effects of Ionizing Radiation*, UNSCEAR 1993 Report (New York: United Nations, 1993).
4. H.W. Bertini, et al., *Descriptions of Selected Accidents that Have Occurred at Nuclear Reactor Facilities*, Report ORNL/NSIC-176 (Oak Ridge, TN: Oak Ridge National Laboratory, 1980).
5. F.J. Rahn, A.G. Adamantiades, J.E. Kenton, and C. Braun, *A Guide to Nuclear Power Technology* (New York: Wiley, 1984).
6. U.K. Atomic Energy Office, *Accident at Windscale No. 1 Pile on 10th October, 1957* (London: Her Majesty's Stationery Office, 1957).

7. M.J. Crick and G.S. Linsley, *An Assessment of the Radiological Impact of the Windscale Reactor Fire, October 1957*, Report NRPB-R135 (Chilton, UK: National Radiation Protection Board, 1982).

8. John G. Fuller, *We Almost Lost Detroit* (New York: Ballantine Books, 1975).

9. Earl M. Page, "The Fuel Melting Incident," in *Fermi-1, New Age for Nuclear Power*, E. Pauline Alexanderson, ed. (La Grange Park, IL: American Nuclear Society, 1979): 225–254.

10. International Atomic Energy Agency, *Report on the Preliminary Fact Finding Mission Following the Accident at the Nuclear Fuel Processing Facility in Tokaimura, Japan* (Vienna: IAEA, 1999).

11. *Report of the President's Commission on the Accident at Three Mile Island*, John G. Kemeny, Chairman (New York: Pergamon Press, 1979).

12. U.S. Nuclear Regulatory Commission, *Three Mile Island, A Report to the Commissioners and to the Public*, Report of Special Inquiry Group, Mitchell Rogovin, Director, Vol. II, Part 2 (Washington, DC: NRC, 1980).

13. David Okrent and Dale W. Moeller, "Implications for Reactor Safety of the Accident at Three Mile Island, Unit 2," *Annual Review of Energy* 6, 1981: 43–88.

14. Pennsylvania Department of Health, "Health Department Discounts TMI as Connected to Infant Deaths," News Release (May 19, 1980).

15. Marilyn K. Goldhaber, Sharon L. Staub, and George K. Tokuhata, "Spontaneous Abortions after the Three Mile Island Nuclear Accident: A Life Table Analysis," *American Journal of Public Health* 73, no. 7, 1983: 752–759.

16. M. C. Hatch, S. Wallerstein, J. Beyea, J. W. Nieves, and M. Susser, "Cancer Rates after the Three Mile Island Nuclear Accident and Proximity of Residence to the Plant," *American Journal of Public Health* 81, no. 6, 1991: 719–724.

17. *The Accident at the Chernobyl Nuclear Power Plant and Its Consequences*, Information compiled for the IAEA Experts' Meeting, 25–29 August 1986, Vienna, Draft (USSR State Committee on the Utilization of Atomic Energy, August 1986).

18. C. Hohenemser, M. Deicher, A. Ernst, H. Hofsäss, G. Lindner, and E. Recknagel, "Chernobyl: An Early Report," *Environment* 28, no. 5, June 1986: 6–42.

19. U.S. Council for Energy Awareness, *Nuclear Power Plants Outside the United States* (Washington, DC: USCEA, 1991).

20. Yuri Kanin, "Chernobyl Fire Chronology," *Nature* 353, 1991: 690.

21. "World List of Nuclear Power Plants," *Nuclear News* 46, no. 3, March 2003: 41–67.

22. Gamini Seneviratne, "IAEA: Early Version of Review Update Released," *Nuclear News* 46, no. 6, May 2003: 53–55.

23. International Atomic Energy Agency, *Nuclear Power Reactors in the World*, Reference Data Series No. 2, April 2003 edition (Vienna: IAEA, 2002).

24. International Atomic Energy Agency, *INSAG-7. The Chernobyl Accident: Updating of INSAG-1*, A Report by the International Nuclear Safety Advisory Group (Vienna: IAEA, 1992).

25. A.A. Afanasieva, E.V. Burlakov, A.V. Krayushkin, and A.V. Kubarev, "The Characteristics of the RBMK Core," *Nuclear Technology* 103, July 1993: 1–9.

26. Richard Wilson, "Comments on the Accident at Chernobyl and Its Implications Following a Visit to the USSR on February 13–24, 1987," unpublished (1987).

27. M. Goldman, et al., *Health and Environmental Consequences of the Chernobyl Nuclear Power Accident*, Report DOE/ER-0332 (Washington, DC: U.S. DOE, 1987).

28. United Nations Scientific Committee on the Effects of Atomic Radiation, *Sources and Effects of Ionizing Radiation, Volume II: Effects*, UNSCEAR 2000 Report (New York: United Nations, 2000).

29. United Nations Scientific Committee on the Effects of Atomic Radiation, *Sources and Effects of Ionizing Radiation, Volume I: Sources*, UNSCEAR 2000 Report (New York: United Nations, 2000).

30. International Atomic Energy Agency, *One Decade after Chernobyl, Summing up the Consequences of the Accident*, Proceedings of an International Conference (Vienna: IAEA, 1996).

31. G. Wagemaker, A.K. Guskova, V.G. Bebeshko, N.M. Griffiths, and N.A. Krishenko, "Clinically Observed Effects in Individuals Exposed to Radiation as the Result of the Chernobyl Accident," in *One Decade After Chernobyl, Summing up the Consequences of the Accident*, Proceedings of an International Conference (Vienna: IAEA, 1996): 173–196.

32. United Nations Development Program, "The Human Consequences of the Chernobyl Nuclear Accident: A Strategy for Recovery," Chernobyl Report-Final-240102, A Report Commissioned by UNDP and UNICEF with the support of UN-OCHA and WHO (January 25, 2002). [From: http://www.undp.org/dpa/publications/chernobyl.pdf]

33. E. Cardis, L. Anspaugh, V.K Ivanov, I.A. Likhtarev, K. Mabuchi, A.E. Okeanov, and A.E Prisyazhniuk, "Estimated Long Term Health Effects of the Chernobyl Accident," in *One Decade After Chernobyl, Summing up the Consequences of the Accident*, Proceedings of an International Conference (Vienna: IAEA, 1996): 241–271.

34. B. G. Bennett, "Assessment by UNSCEAR of Worldwide Doses from the Chernobyl Accident," in *One Decade After Chernobyl, Summing up the Consequences of the Accident*, Proceedings of an International Conference (Vienna: IAEA, 1996): 117–126.

35. United Nations Scientific Committee on the Effects of Atomic Radiation, *Sources, Effects and Risks of Ionizing Radiation*, UNSCEAR 1988 Report (New York: United Nations, 1988).

36. Lynn R. Anspaugh, Robert J. Catlin, and Marvin Goldman, "The Global Impact of the Chernobyl Reactor Accident," *Science* 242, 1988: 1513–1519.

37. International Atomic Energy Agency, *The International Chernobyl Project, An Overview*, Report by an International Advisory Committee (Vienna: IAEA, 1991).

38. James M. Goldgeier, "Will SARS Be the Chinese Chernobyl?" *Los Angeles Times* (April 23, 2003); as presented on the website of the Council on Foreign Relations. [From: http://www.cfr.org/publication.php?id=5881]

39. Martin Malia, *The Soviet Tragedy: A History of Socialism in Russia, 1917–1991* (New York: The Free Press, 1994).

16

Future Nuclear Reactors

16.1 General Considerations for Future Reactors

16.1.1 The End of the First Era of Nuclear Power

The world is now approaching the end of the first era of nuclear power.[1] Most of the reactors built in the past several decades are still operating, and many will continue to run for a number of additional decades, including the U.S. reactors that have had their licenses renewed for another 20 years (see Section 2.4.4). However, gradually the existing reactors will be shut down, and the number of commitments for further reactors is relatively small—none in the United States as of the end of 2003 and few elsewhere.

It is not clear what will happen next. As discussed in Chapter 2, there are movements in some countries to phase out nuclear power and others continue to bar it. On the other hand, substantial efforts are underway to develop next-generation reactors that would usher in a Second Nuclear Era. Despite—or perhaps spurred by—the recent hiatus in reactor orders, reactor designers have stepped back and tried to develop plans for safer and simpler reactors, without necessarily being constrained by existing configurations.

---

[1] The terminology "nuclear eras" has been used by a number of authors, notably by Alvin Weinberg (e.g., in Refs. [1] and [2]).

There have been two general approaches to the design of new reactors.[2] *Evolutionary* reactors build upon the experience gained to date, making incremental improvements without changing the fundamental reactor design. *Innovative* reactors employ designs that are substantially different from those of existing plants. In the former case, one relies on the benefits of accumulated experience. In the latter case, reliance is placed on a revised reactor safety strategy. Some of these designs have been completed, and one new evolutionary reactor, the advanced boiling water reactor (ABWR), is already in operation in Japan (see Section 8.1.4).[3]

16.1.2 Important Attributes of Future Reactors

Safety

For an acceptable nuclear future, the entire fuel cycle—with reactors a particularly crucial part—must be safe. Calculated accident probabilities, as estimated by probability safety assessments, provide a rough guide to the anticipated reactor safety level. Reactor safety standards have been discussed in Section 14.5. For a future in which there are, say, 1000 reactors, plausible guidelines might be as follows:

Core damage frequency: less than 10^{-5}–10^{-6} per reactor-year (RY)

Large early release frequency: less than 10^{-6}–10^{-7} per RY

The stronger of these limits would imply a 1% chance of a TMI-scale accident per decade and an 0.1% chance of a Chernobyl-scale accident per decade, for a world with 1000 reactors. Given the uncertainties in all such estimates, it is reasonable to aspire to the stronger limits. In fact, reactor manufacturers believe they can do that well and better. Presumably, if a buildup to still greater numbers of reactors takes place, experience can lead to continued improvements in reactor safety, and the total worldwide risk need not rise in proportion to the number of reactors.

Economy

To make nuclear power economically competitive, it is necessary to achieve a low reactor construction cost. The new reactor designs purportedly accomplish

---

[2] The terminology is not completely codified. In IAEA usage, the designation "advanced" is used to describe reactors for which "improvements over its predecessors and/or existing designs is expected" and includes both evolutionary and innovative designs (see, e.g., Ref. [3]). It sometimes, however, has been limited to denote what we call innovative designs here.

[3] Several "advanced" PWRs that have been put into operation recently—the 1250-MWe Sizewell B reactor in the United Kingdom (1995) and the four 1450-MWe N4 reactors in France (2000)—incorporate design improvements over earlier versions and perhaps could also be termed "evolutionary."

this—while improving, not sacrificing, safety—through better planned and simpler designs and through either economies of mass production or economies of scale.

Proliferation Resistance

One of the most serious concerns involving nuclear power is its possible implications for nuclear weapons proliferation. The reactors and the accompanying fuel cycle must be designed and operated in a manner that discourages diversion of civilian nuclear fuel or fuel cycle facilities to weapons purposes, whether by national governments, subnational groups, or terrorists.

Waste Reduction and Fuel Economy

Other things being equal, it is desirable to minimize the amount of long-lived radionuclides in the nuclear wastes. In a once-through fuel cycle, some gain is obtained by having a high burnup of the fuel. A more substantial reduction in amounts of waste, and particularly of long-lived radionuclides, can be accomplished through reprocessing of the fuel and recycling uranium, plutonium, and, possibly, other actinides—and, in the limit, adopting a breeding fuel cycle.

16.1.3 Reactor Size

Current Status

A "typical" reactor today has a capacity in the neighborhood of 1000 MWe, although there are many smaller reactors—particularly older reactors and reactors in developing countries. Illustrating differences in current practice, the mean capacity of operating reactors in 2003 was about 180 MWe for India, 940 MWe for the United States, and 1070 MWe for France (see Table 2.1).

The IAEA reports a worldwide total of 32 reactors under construction as of November 2003 (see Section 2.5.3). Their mean capacity was 827 MWe. They range in size from four 202-MWe reactors in India to two 1350-MWe reactors in Japan. Many of the recently suggested designs are for reactors with capacities near 100-MWe, often with plans to site a number of reactors close together to create a large overall plant. Many other designs call for reactors with capacities well above 1000 MWe. It is not clear whether small or large reactors will be dominant in future reactor construction, and probably there will be a mix. Differing trends in reactor size are discussed next.

The Recent Trend to Larger Reactors

In the countries that have had the strongest programs of reactor construction in recent years, the new reactors are quite large [4]:

◆ In 2000, France put into operation four PWRs—the so-called N4 reactors—
each with a capacity of about 1450 MWe.
◆ Four ABWRs, each with a capacity of over 1300 MWe, are under construction—two in Japan and two in Taiwan.
◆ South Korea has two 1000-MWe PWRs under construction that are near-
ing completion.

Much of the tangible industry planning appears to continue to favor large
reactors. Japan is developing a 1700-MWe advanced BWR (ABWR-II) that
goes beyond its present 1315-MWe ABWRs [5]. Another group of Japanese
companies and Westinghouse Electric are developing the design for a 1538-
MWe evolutionary PWR that is undergoing evaluation by the Japanese licensing bodies, with the goal of achieving commercial operation of two reactors
by 2010: Tsugara 3 and 4 [6].

There has also been a joint French–German program to design a new
reactor to meet anticipated electricity needs. This is the so-called European
Pressurized water Reactor (EPR), which, at one time was being designed to
have a 1750-MWe capacity "to take advantage of economy of scale" [3].[4] In
the same vein, a representative of the French utility, Electricité de France,
stated in 1998: "The future of present or advanced light water reactors is well
defined at least until 2050, with N4 and EPR reactors in France" [7].

Two U.S. reactor manufacturers have apparently decided that economies
of scale are sufficiently important that they have put forth larger but similar
reactors as substitutes for previously planned 600-MWe reactors. Westinghouse is offering the 1090-MWe AP1000 as an alternative to the AP600 and
General Electric is replacing the SBWR by the 1390-MWe ESBWR in its
licensing and marketing efforts (see Section 16.2.2). Looking further ahead,
many of the reactors contemplated in the international Generation IV programs have capacities greater than 1000 MWe, although in some cases, there
are options for either large or small reactors within the same general design
plan (see Section 16.6.2).

Possible Future Trend to Smaller Reactors

Nevertheless, for several decades, some nuclear planners have argued for moving to smaller reactors—somewhere in the range of 100 MWe to 600 MWe.
In this concept, especially at the lower end of the capacity range, economies
of scale could be replaced by economies of mass production, with much of
the reactor being built in factories rather than on-site. Further, construction
times would be less for smaller reactors and a utility's financial commitment
reduced—both important economic considerations. The total capacity at a

---

[4] The present German government is committed to shutting down reactors, not
building new ones, so the political climate for the EPR may be complicated.
More recent EPR designs (including the new reactor for which construction has
been started in Finland) are at the 1600-MWe level.

given nuclear facility could be increased as need dictated, by clustering many small reactors at a single site with some services shared among them. Finally, most of the small-reactor designs feature passive or inherent safety, presumably decreasing the chance of an accident. If an accident does occur, the potential release of radioactivity and the financial loss are less than they would be for a large reactor.

In a 1999 paper discussing U.S. planning, William Magwood, the Director of the Office of Nuclear Energy, Science and Technology of the DOE, paid his respects to the recent large advanced light water reactors as "having a clear market for the next decade or so, mostly in Asia" [8]. However, in describing the subsequent generation of reactors, he stressed the potential role of small reactors, both for developing and developed countries. In his words,

> Transforming the process of building nuclear power plants such that they are built in a manner more like that of aircraft than aircraft carriers could provide significant economic advantages.

In this view, mass production trumps economies of scale. A number of reactor designs have adopted this approach (e.g., the proposed high-temperature gas-cooled reactors).

16.1.4 U.S. Licensing Procedures

To assure reactor safety, the nuclear plants now operating in the United States went through lengthy licensing procedures, with the design and construction of each plant subject to challenges at many stages. This caused substantial delays. It became a major gamble for a utility to order and begin construction of a reactor, and the protraction of the construction period made the reactor very costly.

In an effort to reduce the uncertainties, the NRC instituted in the late 1980s a system under which the basic design questions would be resolved and approval would be given before construction began. The manufacturer of a proposed new reactor must work with the NRC in an extensive interactive consideration of the design. If approved, the NRC issues a *standard design certification*. The first three U.S. reactor designs to have received standard design certification under this new system are the ABWR (1997), the System 80+ reactor (1997), and the AP600 reactor (1999) (see Section 16.2.2).

After a reactor has received design certification, a purchaser can apply for a *combined license* to begin construction and to operate the reactor. The application for a combined license for a reactor must address safety issues specific to the proposed site, including (a) the neighboring population distribution, (b) tornado probabilities, (c) possibilities of floods, (d) the stability of cooling water supplies, and (e) the vulnerability of the site to earthquakes [see, e.g., Ref. [9, Chapter 2]. The combined license includes requirements on "inspections, tests and analyses...that the licensee shall perform" as well as acceptance criteria required to assure that the reactor has been built in conformity with the license [10]. These conditions must be satisfied before the reactor can start to operate, but, in principle, the main design issues would have already been settled before construction begins.

Provision also exists for requesting an early site permit from the NRC even before an application is made for the construction of a reactor at that site. In this way, it would be possible, for example, to have early decisions on the seismic suitability of a number of sites before a choice is made of the site or sites to pursue for actual construction.

16.2 Survey of Future Reactors

16.2.1 Classification of Reactors by Generation

After a period of seeming indifference to the development of nuclear power, the U.S. DOE launched in 1998 a "Nuclear Energy Research Initiative" (NERI), prodded in part by a recommendation from the President's Committee of Advisors on Science and Technology [11, p. 5–13]. The DOE in 1998 also established a Nuclear Energy Research Advisory Committee (NERAC). The DOE's nuclear efforts have evolved into a multipronged program that includes three interrelated components: (1) an effort for "near-term" (i.e., by 2010) deployment of new reactors, (2) an international program to develop so-called Generation IV reactors for later deployment, and (3) grants for research on new fuel cycle and reactor concepts (under the NERI program).

A classification of reactors by "generation" has developed, especially in U.S. DOE usage, with the following structure [8, 12]:

◆ *Generation I.* These are the early small reactors, typically completed in the 1960s with capacities of under 200 MWe. Most of these reactors have been shut down.

◆ *Generation II.* These are larger reactors, with capacities ranging from several hundred to over 1000 MWe. They represent essentially all the reactors now in operation other than the ABWR.

◆ *Generation III.* These are reactors whose designs have been recently completed and that are in principle ready to enter the commercial market, given the demand. The only model that has already done so is the Advanced Boiling Water Reactor (ABWR), two of which are in operation in Japan and with several more under construction. (See Section 16.3.1 for a further discussion of the ABWR.) *Generation III+* reactors are a next stage, described as advanced systems "that can be deployed by the end of the decade (i.e., by 2010)" [12, p. 36].

◆ *Generation IV.* These are reactors that differ substantially from earlier reactors with designs intended to provide improved safety and economy. Many such reactors are now under at least conceptual consideration, but none is under construction, nor do orders appear imminent for any of them.

In present thinking, they will not be deployable until 2020 or later, so they are far-off plants—offering the possibility of highly innovative designs.

As might be expected, the actual world does not neatly match this classification system. Thus, Generation III+ has been introduced as an intermediate level, and there are a number of reactors that might be thought of as Generation III in terms of their level of design, but that are unlikely to actually be deployed until after 2010, if ever. The classification is useful for purposes of orientation, but there is a continuum in level of innovation, status of design readiness, and date of potential availability. In terms of the distinction made earlier between evolutionary and innovative designs, the former corresponds roughly to Generation III and the latter to Generations III+ and IV.

16.2.2 U.S. DOE Near-Term Deployment Roadmap

Lists of prospective reactor candidates have been developed in the recent initiatives by the U.S. DOE and in parallel international efforts. The reactors have been divided into those available for "near-term" deployment (i.e., by 2010 or 2015), and the Generation IV reactors for deployment after 2020. The U.S. thinking about the near-term was put on a formal footing in February 2002 with the announcement of the Nuclear Power 2010 program:

> The Nuclear Power 2010 program, unveiled by the Secretary on February 14, 2002, is a joint government/industry cost-shared effort to identify sites for new nuclear power plants, develop advanced nuclear plant technologies, and demonstrate new regulatory processes leading to a private sector decision by 2005 to order new nuclear power plants for deployment in the United States in the 2010 timeframe. [13]

Prior to the announcement of this goal, the DOE established a special panel, the Near-Term Deployment Group (NTDG), which in October 2001 issued a document reviewing reactor possibilities: *A Roadmap to Deploy New Nuclear Power Plants in the United States by 2010* [14]. The leading options, as put forth by the NTDG, are listed in Table 16.1, subdivided in terms of the NTDG's assessments of the prospects for their deployment by 2010.

It is not entirely clear—nor very important—how the reactors of Table 16.1 fit into the classification in terms of Generations. The ABWR and some versions of the CANDU are Generation III; the IRIS is possibly Generation IV, and the remainder are probably most appropriately termed Generation III+.

Reactors Listed in the Near-Term Roadmap

The reactors listed in Table 16.1 fall into several groups:

♦ *Evolutionary LWRs that have been design certified by the U.S. NRC.* The ABWR and the System 80+ PWR reactor became in 1997 the first reac-

Table 16.1. Reactors considered by the U.S. DOE's Near-Term Deployment Group (2001), arranged by the NTDG's assessment of their deployability.

Reactor Designation[a] (Sorted by Deployability)	Reactor Type	Size (MWe)	Main Countries[b]
Deployable by 2010			
Advanced Boiling Water Reactor (ABWR)	BWR	1350	U.S., Japan
Probably deployable by 2010			
Advanced Passive-600 (AP600)	PWR	610	U.S.
Advanced Passive-1000 (AP1000)	PWR	1090	U.S.
Pebble Bed Modular Reactor (PBMR)	HTGR	110[c]	S. Africa
Possibly deployable by 2010			
E Safe Boiling Water Reactor (ESBWR)	BWR	1389	U.S.
Siedewasser Reactor-1000 (SWR-1000)	BWR	1013	Germany[e]
Gas-Turbine Modular Helium Reactor (GT-MHR)	HTGR	288[c]	U.S., Russia
Not deployable by 2010			
International Reactor Innovative and Secure (IRIS)	PWR	100–300	U.S.
Mentioned but not evaluated by NTDG[d]			
European Pressurized Water Reactor (EPR)	PWR	1545–1750	France, Germany
System 80+	PWR	1350	U.S., S. Korea
CANDU	HWR	500–1000	Canada

[a]Deployability as estimated by the NTDG (2001) [15].
[b]The country designation is based on the history of design or development activity. The level of international participation often makes the designation ambiguous.
[c]Capacity per module; modular deployment assumed.
[d]The Near-Term Deployment Group evaluated those candidate reactors for which it received the information it had requested. Other possible candidates were mentioned but not evaluated.
[e]The SWR is being developed by Framatome ANP, a company formed in a merger between the French and German reactor manufacturers Framatome and Siemens.
Source: Refs. [15, pp. 17–21, 30–32].

tors to receive standard design certification from the U.S. NRC. They are discussed further in Section 16.3.1.

♦ *Other evolutionary designs.* The SWR-1000 is a 1000-MWe BWR, under development since 1992 by Framatome Advanced Nuclear Power (F-ANP). Its developers term it an "evolutionary" reactor with an "innovative approach to safety" [17]. In case of an accident, reactor cooling is maintained by passive systems that do not require electric power or control signals to operate, blurring the distinction between so-called "evolutionary" and "innovative" or "passive" reactors. F-ANP plans to seek design certification for the SWR-1000 from the U.S. NRC.

The EPR is a large evolutionary PWR that has been long under development in a joint French–German effort to develop a new European standard.

♦ *Advanced designs emphasizing passive safety.* The third reactor to receive NRC design certification (in 1999) was the AP600, a 600-MWe PWR designed by Westinghouse that makes extensive use of passive features in its emergency cooling systems. It and the larger AP1000 are discussed further in Section 16.3.2.

The ESBWR is a successor to General Electric's earlier 670 MWe SBWR design, where the "S" stands for safe or simple and the "E" stands for European or economical. Unlike the AP600, the SBWR was not pursued to the point of NRC design certification. In cooperation with European nuclear designers, the much larger ESBWR is being developed to take advantage of economies of scale [14, Appendix B].

The IRIS reactor is a relatively new entry into this category, with some of the novelty that is characteristic of Generation IV reactors. It is discussed further in Section 16.3.3.

♦ *High-temperature gas-cooled reactors.* The PBMR and the GT-MHR reactors are both helium-cooled and graphite-moderated reactors, but with different physical arrangements. They are discussed further in Section 16.4.

♦ *Heavy water reactors.* Previous CANDU reactors have been pressurized heavy water reactors (PHWR) of Canadian design. Future CANDU-type reactors will be available in different sizes and designs. The CANDU manufacturer, Atomic Energy of Canada Ltd (AECL), is now developing the ACR-700, which is a 700-MWe reactor that retains heavy water moderation but, unlike the traditional CANDU reactors, will use light water for cooling [18]. According to the AECL, this will give significant advances in safety and cost. The AECL is seeking licensing approval for the ACR-700 in Canada, the United Kingdom, and the United States.

The fluidity of reactor planning—in a time of few orders—can be seen by comparing the NTDG evaluations, as reflected in Table 16.1, and an evaluation of prospective new reactors made in a 1992 National Academy of Sciences study [16].[5] The NAS report in 1992 gave five reactors an overall "high" rating, in an assessment that was "mostly driven by market suitability" [16, p. 153]. Of these, one has already been built in Japan (ABWR), two have been supplanted by larger versions prior to any purchase orders (the AP600 and SBWR, supplanted by the AP1000 and ESBWR), one is no longer being considered for marketing in the United States although it is being marketed elsewhere (System 80+), and one has not been pursued by its designer (the Westinghouse Advanced PWR).

---

[5] More officially, this is a National Research Council committee, but as earlier, we follow common practice and refer to it as an NAS committee.

16.2.3 Illustrative Compilations of Reactor Designs

The Scope of Possibilities

Interest in designs for future reactors goes far beyond the limited list discussed in the previous section. When the Generation IV study began, suggestions were solicited worldwide, and in response, "nearly 100 concepts and ideas were received" [19, p. 14]. The Generation IV International Forum eventually selected six of these as the most promising systems—usually with alternative reactor options within a system—and these are considered further in Section 16.6.

Neither the list of Table 16.1 nor the list of the favored Generation IV reactors comes close to embracing all of the possibilities that have been suggested. We present in the following subsections two examples of broader lists, but as long as preliminary designs are included, no list can be complete. Most of the contemplated reactors probably will not be built, because they will be judged as not technically or economically competitive with alternatives. However, the lists illustrate a degree of ferment among nuclear designers, even at a time when there are few orders for new reactors. Most of the reactors are intended for electricity alone, but some of these and other recent designs have also been intended for special applications, such as desalination of seawater, district heating, and provision of electricity or heat to remote localities on barge-mounted reactors [3].

The broad exploration of a variety of reactor designs is in some ways reminiscent of the 1950s. As long as the activity is on paper, it is practical to think in wide-ranging ways. When actual orders are placed and construction begins, the process is much more expensive and there is a natural contraction to a smaller group of choices. The compilations of reactor designs presented here—until matched by actual construction—are of interest primarily as an illustration of the range of alternatives.

International Near-Term Deployment Group

A listing of near-term reactor possibilities that is somewhat more extensive than that of Table 16.1 was presented by the International Near-Term Deployment group (INTD) [19, pp. 25–26].[6] It identified 16 designs "whose performance is equal or better than a light water reactor performance baseline representative of Generation III." The reactors were to be deployable by 2015, a date chosen with the view that deployment by 2010 would be difficult. This group consisted of (a) the eight reactors evaluated by the NTDG (see Table 16.1) (replacing the ABWR by the ABWR II), (b) the EPR and CANDU, also listed in Table 16.1, (c) three additional large advanced LWRs, and (d) three small modular reactors.

---

[6] This was established by the Generation IV International Forum, apparently as an offshoot of its longer-range planning [19, p. 11].

IEA/NEA Compilation of Innovative Reactors (2002)

The International Energy Agency and Nuclear Energy Agency[7] published a report in 2002 with a compilation of an "Illustrative List of Innovative Reactor Designs" [20, Table 2-1]. Thirty-six reactors were listed. They can be classified in several ways:

◆ *By type.* The 36 reactors included 9 light water reactors, 9 liquid metal fast reactors, 4 gas-cooled reactors, 5 reactors for district heating only (no electrical output), 3 molten salt reactors, 2 heavy water reactors, and 4 reactors of other types.

◆ *By size.* The capacities range from 27 to 1500 MWe for reactors intended primarily for electricity generation, and from 10 to 600 MWt for reactors intended for heat production only.

◆ *By lead country.* Thirteen different countries are indicated as the lead country for one or another of the reactors, led by Japan (9), Russia (7), and the United States (5). Ten other countries, a European consortium, and the CERN laboratory, account for the remaining 15 designs.

No attempt is made in this list to classify these reactors by generation, as in U.S. DOE usage, but they mostly correspond to Generation III+ or Generation IV. It is a highly international list, made even more international by the fact that, in many cases, the work on an individual type of reactor involves more than one country.

With its emphasis on the "innovative," this list does not include the light water reactors of the sort that are being built now. Despite the interest in wide-ranging alternatives, it is quite possible that these LWRs will turn out to account for the majority of the reactors built in the near future.

16.3 Individual Light Water Reactors

16.3.1 Evolutionary Reactors Licensed by the U.S. NRC

Advanced Boiling Water Reactor

As indicated in Table 16.1, the ABWR is the new reactor that could most readily be put into operation in the United States by 2010. It is an evolutionary reactor that follows in a sequence of BWRs designed by the General Electric Company. The company's first commercial BWR was a 200-MWe reactor that went into operation in 1960. Succeeding product lines were introduced in a rather tight period from 1969 to 1978 [21]. The last of the BWRs to be completed in the United States (Limerick 2) did not go into operation until 1990, but construction on it had started in 1970 [22, Table 10]. Thus, a long

---

[7] The agencies are both associated with the Organization for Economic Co-operation and Development (OECD).

period elapsed before the ABWR was introduced, allowing ample time to design a reactor that incorporated lessons learned in the construction and operation of the earlier reactors.

The ABWR was designed by General Electric in collaboration with Japanese manufacturers and utilities [21]. Two 1315-MWe ABWRs are now operating in Japan, with one starting in 1996 and the other in 1997. Two additional ones are under construction in Japan, with operation scheduled to begin in 2005 and 2006. In Taiwan, two new ABWRs have been ordered and construction is underway, although there was some interruption due to political opposition.

The main goals in the design of the ABWR were improved safety and simplicity. Simplicity in design is intended to improve the reactor's reliability, reduce costs, and reduce construction time. The latter goal was clearly achieved in the Japanese ABWRs, which took only 4 years to move from the start of construction to the beginning of commercial operation [21]. Steps taken to enhance safety include changes in the location of pipes and pumps, a highly redundant system for providing emergency cooling water in case of a loss of the regular reactor cooling water, a redundant system for filtering gas that is released from the containment to the atmosphere in the case of an accident, and a redundant set of diesel generators to provide power if the off-site electrical power is lost.

The effectiveness of these and other measures can be estimated through a probabilistic risk assessment (PRA). A PRA provides an approximate indication of the safety level achieved by the reactor design. However, it also serves to reveal weaknesses in the design. Thus, a PRA for early versions of the ABWR suggested the core damage frequency could be substantially reduced by the introduction of additional redundancies in the safety systems. Other design improvements were implemented in response to other identified accident sequences. The PRA also suggested when the point of diminishing returns was reached [21, p. 10-2].

On the basis of the PRA for the final design, General Electric estimates the core damage frequency for internal events to be 1.6×10^{-7}/RY and the chance of a large release caused by a failure of the reactor containment to be down by a further factor of 500 [21, pp. 10-3 and 10-4]. The Nuclear Regulatory Commission reviewed the ABWR design as part of the licensing process. In its Final Safety Evaluation Report, issued in 1994, the NRC concluded:

> It is the staff's view that the mean core damage frequency for the ABWR from internal, external, and shutdown events is probably on the order of 1E-6 or less assuming the plant is constructed, maintained, and operated in accordance with the SSAR [standard safety analysis report, submitted by GE]. [9, p. 19-6]

The NRC also considered the containment failure probability for internal events and found "an extremely low likelihood of containment failure in absolute terms (i.e., on the order of 10^{-8} to 10^{-9} per year)" [9, p. 19-33]—a factor of 100 or more better than its goal of 10^{-6}/RY for new reactors (see Section 14.5.1). The dangers can also be considered in terms of individual risk. The NRC concluded that its goal of under 2×10^{-6}/RY for cancer fatalities was bettered by "several orders of magnitude," even for the upper bound in the distribution of risk estimates [9, p. 19-39].

Overall, although the NRC's numerical estimates were somewhat less optimistic than those made by General Electric, they still suggest that the ABWR design meets U.S. safety goals by a wide margin. In other arguments presented for the ABWR, General Electric projects a construction time of only 48 months, as achieved in Japan, and construction costs that are well below those of recently completed LWRs.

System 80+ Reactor

Three System 80 reactors, which are forerunners of the System 80+ reactor, were put into operation in the United States in the 1980s—the Palo Verde plants in Arizona. They were designed and built by the Combustion Engineering Company. The System 80+ reactor is slightly larger than the System 80 (1350 MWe versus about 1245 MWe) and incorporates changes for improved safety. It was designed by the combined Combustion Engineering and Asea Brown Boveri companies, but in subsequent reorganizations of the nuclear industry, it has now come under the aegis of the Westinghouse division of British Nuclear Fuels PLC (BNFL). Thus, the System 80+ reactor is now a "Westinghouse reactor."

The System 80+ reactor received a standard design certification from the U.S. NRC in 1997, making it available for construction in the United States. However, no orders for any new reactors were placed in the immediately following years and Westinghouse has now decided not to market the System 80+ in the United States, where it would be in competition with Westinghouse's AP1000 (discussed below).

In the mid- to late-1990s, South Korea put into operation four System 80 reactors that have embodied some of the System 80+ evolutionary improvements. Following upon this work, the South Korean nuclear industry has developed a 1400-MWe reactor, termed the APR 1400, that is based on the System 80+ design, although slightly larger [23]. Two such plants (Shin Kori 3 and 4) are scheduled to start operation in 2011.

Accident risks for the System 80+ reactors are estimated to be significantly less than those for the System 80 reactors. This improvement is in part attributed—by the NRC in its Evaluation Report for the license application as well as by the manufacturer—to an array of hardware improvements, including greater redundancy. For example, additional pumps were added for the systems that, in case of an accident, provide emergency cooling water to the steam generators and the reactor pressure vessel [24, Chapter 19, pp. 12, 18, and 111]. With these and other design changes, the estimated core damage frequency, as calculated in the probability risk analysis included in the application to the NRC, is 2×10^{-6}/RY compared to 8×10^{-5}/RY for the System 80 reactor [9, p. 19-17].

16.3.2 Innovative Light Water Reactors

General Considerations

The "innovative" LWRs that are under consideration are simpler, and in the first versions they were smaller, than the current generation of LWRs. In the United States, two designs led in the early development, initially with support from the DOE and the Electric Power Research Institute (EPRI): the Westinghouse AP600 and the General Electric SBWR. Both reactors were about 600 MWe in size. Special features of this category of reactors, as summarized by Forsberg and Weinberg, include the following [25, p. 140]:

♦ The emergency cooling systems are simpler and more passive, relying on large pools of water fed by gravity, rather than on flow sustained by pumps.
♦ Emergency electric power requirements are reduced so that they can be satisfied by batteries rather than emergency diesel generators.
♦ Reactor power densities are reduced.
♦ The designs have been simplified to reduce costs and sources of possible operating or maintenance error.

Although their original designs were for the 600-MWe region, both Westinghouse and General Electric have decided that economies of scale favor larger reactors. Thus, General Electric is no longer pursuing the SBWR, and, instead is proceeding with the design of a similar but larger reactor, the 1389-MWe ESBWR, as listed in Table 16.1. Westinghouse completed the NRC application process and received standard design certification for the AP600 in December 1999, but the company's main thrust appears now to be for the AP1000, for which it hopes to receive design certification and be ready to accept orders by 2005 [26].

We discuss below the AP600 and AP1000 as examples of passive LWRs because they are furthest along in U.S. licensing process. However, in the absence of actual orders, it is premature to say whether the AP1000 or the ESBWR, if either, will prove to be more successful commercially.

Design Features of the AP600

The AP600 was designed to be simple in configuration and relatively inexpensive to build.[8] Part of the savings comes from modular construction, in which components are built and to some extent assembled off-site, substantially reducing the construction time at the reactor site itself. When compared

---

[8] This discussion is based primarily on information from Westinghouse, including published documents and private communications with Ronald P. Vijuk.

to conventional PWRs, the design simplifications greatly lower the needs for equipment such as valves, control cable, and piping [27]. Among the significant improvements are a digitized control and instrumentation system and sealed pumps. All of this is intended to provide reliable and economical operation.

The AP600's reactor pressure vessel, steam generator, and much of the emergency cooling system are located within the containment vessel, a large steel tank with a 40-m diameter (see Figure 16.1). The containment is rated for a pressure of 45 psig (approximately 3 atm above the outside pressure)

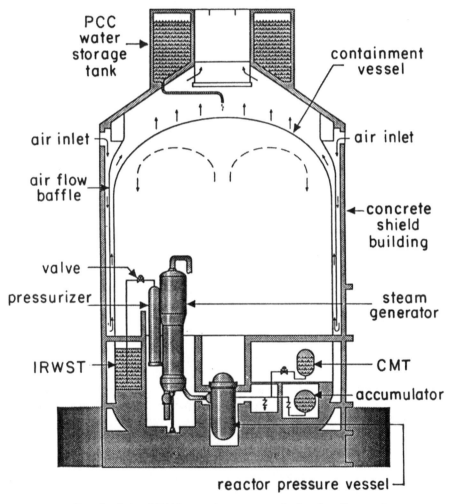

Fig. 16.1. Sketch of the AP600 containment vessel, showing the reactor pressure vessel, steam generators, and systems for emergency cooling of the containment vessel. The AP1000 has essentially the same configuration, with some increases in dimensions. (Courtesy of the Westinghouse Electric Corporation.)

and is expected to be able to survive considerably higher pressures. This containment is, in turn, surrounded, with a not very large gap, by a concrete shield building.

In the case of an accident, the two urgent goals are to maintain cooling of the reactor core and to avoid a breach in the containment that could allow the escape of radioactive material. The former is accomplished in the AP600 through passive emergency core-cooling systems. The latter is accomplished by cooling the containment vessel from the outside, to prevent a large buildup of internal pressure. The large volume of the containment coupled with the relatively low power rating of the reactor help in this regard.

We first consider a loss-of-coolant accident in the primary loop, which includes the reactor core and the input side of the steam generator. There are three types of tanks (five tanks in all) within the containment that can supply replacement water:

1. *Core makeup tanks (CMT).* There are two such tanks, each containing borated water. When the reactor pressure or water level falls below prescribed safety levels, the reactor scrams and air-operated valves open between these tanks and the pressure vessel. These are "fail-open" valves that open if the air supply is lost. The tanks are situated at a greater height than the reactor vessel, providing gravity-fed flow. It does not matter if the reactor is still under pressure, because the piping is such that the same pressure appears at the top of the core makeup tank as in the reactor vessel.

2. *Accumulator tanks.* There are two such tanks, each also containing borated water, driven by nitrogen gas at high pressure. They supplement the core makeup tanks and come into play when the reactor pressure falls below the accumulator pressure.

3. *In-containment refueling water storage tank (IRWST).* The above-described tanks provide cooling water immediately. A much larger tank, the IRWST (normally used for other purposes), provides long-term cooling in the case of an accident. Its capacity is sufficient to provide cooling for 1 h (in the AP1000) without boiling, after which it sets up a closed cooling cycle, as discussed below. A redundant set of valves is opened when the water level drops sufficiently in the core makeup tank, releasing the pressure in the reactor vessel and allowing gravity-fed flow from the IRWST into the reactor vessel.

It is to be noted that no pumps are required in this system, the flow is passively driven, and the valves are opened without operator intervention.

As the cooling process continues, it becomes self-sustaining. The reactor core sits near the bottom of the reactor vessel, well below the inlet and outlet pipes. In the event of a break that interrupts the normal cooling flow, water is boiled off and steam escapes to the interior of the containment through the break and through a depressurization system that relieves the reactor pressure in case of cooling system failures. A cycle is set up in which the steam condenses on the inside wall of the steel containment and flows down, either to the IRWST or to a pool at the bottom. Either way, it is returned to the reactor.

To avoid the buildup of high internal pressure in the containment, the outside of the containment must be cooled. This is at first accomplished with gravity-fed water from the passive containment-cooling (PCC) water storage tank, located above the containment. This wets and cools the containment. This tank water supply can suffice for 3 days. For the longer term, the containment can be cooled by an airflow established by convection in the gap between the containment vessel and the concrete shield building. However, to give a larger margin of safety, it is anticipated that water will be externally supplied to the outside of the containment.

Failures in the primary loop were considered above. If, instead, there is a failure in the secondary loop, so that the steam generator no longer serves as a heat exchanger, the water from the core is diverted through fail-open valves into an alternate heat exchanger located in the IRWST (mentioned earlier in another role) which serves as the heat sink.[9] Flow in the loop from the reactor to the heat exchanger and back is maintained by convection. As earlier, if water from this reservoir boils off, the steam condenses on the containment interior and the water drains back to the IRWST.

Safety Evaluation for the AP600

In its application to the NRC for design approval and certification of the AP600, Westinghouse projected a core damage frequency of 1.7×10^{-7}/RY for internal events. The projected frequency was four times higher if "external" events, particularly fires were included [28, p. 19-252]. The calculated probability of a large release of activity for internal events is 1.8×10^{-8}/RY [14, p. D-2].

Earthquake risks are addressed by the seismic margin approach (see Section 14.4.1). The safe shutdown earthquake corresponds to a peak ground acceleration of 0.3g. The seismic margin is established by determining for key reactor components the maximum acceleration at which there is a "high confidence of low probability of failure." Westinghouse analyses concluded that it was satisfied at 0.5g or higher [28, p. 19-74]. The failure considered includes both core damage *and* a large release, because an earthquake severe enough to cause core damage is likely to damage the containment as well.

Comparison Between the AP600 and AP1000

The motivating factor in going from 600 to 1000 MWe is cost. Westinghouse expects that construction costs would increase only modestly, compared to

---

[9] The IRWST has a capacity to provide cooling for 1 h before boiling, after which the escaping steam condenses on the interior of the containment and is returned to the IRWST [27].

the increase in output. In consequence, the expected electricity cost would drop from an estimated 4.1–4.6 ¢/kWh to about 3.0–3.5 ¢/kWh—a reduction deemed necessary for economic competitiveness.The increase in capacity over that of the AP600 is accomplished with no essential change in design but with increases in the number and length of the fuel rods, the power density in the reactor core, the size of some components, and the volume of water in the tanks that supply replacement water in case of an accident [27]. The diameter of the containment is unchanged at 40 m, but its height is slightly greater.

The improvement in simplicity and costs achieved in the AP1000, when compared to a conventional 1000-MWe PWR is indicated by large reductions in equipment requirements—36% for number of pumps, 50% for valves, 83% for piping, and 87% for electrical cables [14, p. D-9]. For example, the length of cables required is reduced from 2.77 million meters to 0.37 million meters. With the improved design, the building size is reduced by 56%.

In Westinghouse's view "the inherent safety and simplicity of the AP600 is not diminished in any way" in going to the AP1000 [27]. According to a report prepared under the auspices of the American Physical Society:

> There is a reduction in fluid flow and temperature margins compared to the AP-600 to accommodate the higher power output. The design with these reduced margins still meets present NRC regulations and the margins are still larger than those in the operating U.S. PWRs [29, Section IV(d)]

Being better than currently operating reactors would be only a modest achievement for future reactors, but Westinghouse expects to do much better. The estimated core damage frequency for the AP1000 (for internal events) is 4×10^{-7}/RY. This is about three times higher than the estimate quoted above for the AP600, but still well below any proposed standard for core damage frequency (see Section 14.5).

The IRIS Reactor

The International Reactor Innovative and Secure (IRIS) represents a relatively recent entry in the array of reactors that were reported on by the NTDG (see Table 16.1).[10] Reflecting the novelty of some of its features, it was described as "not deployable by 2010," but the international consortium led by Westinghouse anticipated that it would be deployable in the 2012–2015 time period.[11] First steps in seeking NRC licensing approval began in October 2002 [30].

---

[10] This discussion is based on the cited articles and private communications with Mario Carelli and Lawrence Conway.

[11] The IRIS consortium includes, in addition to Westinghouse, scientific or industrial representatives from organizations in Brazil, Croatia, Italy, Japan, Mexico, Spain, Russia, the United Kingdom, and United States. The U.S. participants include groups from the Massachusetts Institute of Technology, Oak Ridge National Laboratory, and the Bechtel Power Corporation.

The IRIS reactor is a small- to-medium-size PWR. In the Near-Term Deployment Roadmap, its size was indicated as ranging from 100 to 300 MWe [15, p. 17], but more recent publications indicate a planned capacity of 335 MWe in view of the "significant economic penalties" of smaller size [31]. The reactors are intended to be deployed in plants consisting of multiple 335-MWe modules. Possible configurations include one with three independent modules (1005 MWe total) and another with twin units, where each unit has two modules that share many systems exclusive of the reactors, their containments, and some safety systems (1340 MWe total). In such arrangements, the generation cost is estimated by its designers to be in the neighborhood of 3 ¢/kWh, which would make the plant economically competitive.

IRIS differs from traditional PWRs in that the steam generators, pumps, and pressurizer are located inside the reactor vessel, as shown in Figure 16.2, rather than outside. There are eight steam generators, placed in a ring close to the outer wall of the reactor vessel and sitting above the reactor core. The primary cooling water circulates up through the reactor core, through channels rising above the core, into pumps mounted above the steam generators, down through the steam generators, and eventually back again to the bottom of the core. The internal circulation of the primary cooling water eliminates the need for the large pipes that in conventional PWRs connect the reactor vessel and the external steam generators, pumps, and pressurizer (see Figure 8.1).

To contain the steam generators and pumps internally, the reactor vessel has to be larger than for other PWRs, including ones with higher capacity. It is 21 m high and 6.2 m in inner diameter. This height is almost twice that of the AP1000 reactor vessel, and the diameter is also considerably greater. However, with major components inside the reactor vessel, the IRIS containment can be relatively small. It is a spherical structure that is 25 m in diameter, topped by a cap that allows opening the reactor vessel for refueling and maintenance. Its footprint is 40% of AP1000's containment footprint.

The IRIS designers describe their approach to safety as "safety by design." For example, by placing the steam generator and related components inside the reactor vessel, there are no large pipes to carry the primary cooling water into and out of the reactor vessel. This eliminates one important class of accidents that are of concern in conventional LWRs—the large loss-of-coolant accident (LOCA), involving a large pipe break.

Smaller penetrations of the reactor vessel wall remain necessary, for example for a system to purify the primary water that directly cools the reactor core. If there is a break in one of these pipes—to consider one possible accident scenario—several features of the design serve to reduce the chance of serious damage:

◆ The pressures inside the reactor vessel and inside the containment are equalized when steam and water escape through the break into the relatively small volume of the containment vessel. The absence of an outward pressure gradient limits further outflow of water or steam from the reac-

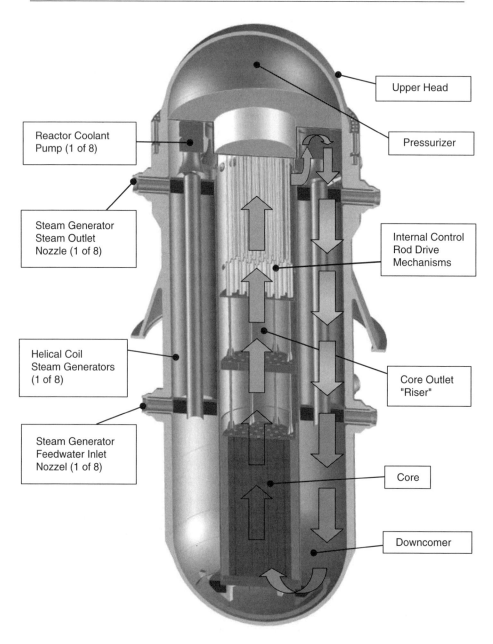

Fig. 16.2. IRIS reactor vessel: General layout of components, including reactor core and internal steam generators. Arrows indicate the flow path of the primary cooling water. (From Ref. [30], with permission of the authors.)

tor vessel. (At times, the reactor pressure even becomes lower than the containment pressure due to condensation of steam inside the reactor vessel [32].)

◆ To achieve rapid equalization of the reactor vessel and containment pressures and avoid excessive pressures in the containment, the reactor vessel pressure is rapidly reduced by steam condensation. The condensation occurs on the outer surfaces of the steam generators, which are cooled by the action of an emergency heat removal system. Relatively cool water from emergency tanks is carried by gravity flow and natural circulation into the secondary side of each steam generator. The resulting steam passes into a heat exchanger, where it is condensed, and the output water circulates back to the steam generator to continue the cycle. The cool side of the heat exchanger is a large water tank, located outside the containment. The water in this tank gradually boils off—making the atmosphere the ultimate heat sink—but the initial water supply is sufficient to last 7 days without replenishment. There are four independent heat removal subsystems for the eight steam generators. The design is such that any one of these subsystems would suffice to remove the heat produced by radioactive decay in the core.

◆ The reactor core remains covered without the need for additional water input, because steam produced by the hot reactor core condenses inside the reactor vessel and the water remains within it. The lowest pipes or other penetrations into the reactor vessel are at least 2 m above the top of the core, so the core cannot be uncovered by water flow through a pipe break.

Thus, in this accident, the core temperature is controlled simply and passively, without the need for emergency injection of water to cool the core directly.

The designers of the reactor are optimistic that these and other "safety-by-design" features will make accidents very unlikely and greatly reduce the chance of core damage should an accident occur. A probabilistic risk assessment is being carried out, in conjunction with preliminary steps in the NRC license application process.

16.4 High-Temperature, Gas-Cooled Reactors

16.4.1 HTGR Options

Planned high-temperature, gas-cooled reactors (HTGR) are carbon moderated and helium cooled. The use of carbon and helium allows the reactor to operate at higher temperatures than can be reached in LWRs, giving higher efficiencies in electricity generation—about 45% in HTGRs, compared to 35% or less for LWRs [14, p. G-1]. The high temperatures can also be an advantage in applications making direct use of the heat. For instance, one of the Generation IV reactors is an HTGR intended to operate at temperatures that are high enough for the reactor to drive efficiently a thermochemical cycle for hydrogen production.[12] Reactor safety is put forth as a further strong point by HTGR proponents.

The two leading HTGR candidates for near-term deployment are the Gas-Turbine Modular Helium Reactor (GT-MHR) and the Pebble Bed Modular Reactor (PBMR). As suggested by their names, they are both based on modular systems. Several modules, not necessarily built simultaneously, could be included in a given facility.

The GT-MHR is the latest in a series of HTGRs proposed by the General Atomics company. The PBMR is a descendent of reactors developed in the 1980s in Germany. It is being studied, in slightly different versions, by the South African company Eskom and in a collaboration between groups at the Massachusetts Institute of Technology (MIT) and the Idaho National Engineering and Environmental Laboratory (INEEL). In the Eskom version, the heated helium directly drives the gas turbine. In the MIT-INEEL version, an intermediate helium-to-helium heat exchanger is used and the turbine is driven by helium in the secondary loop.

The most obvious difference between the GT-MHR and the two pebble bed reactors (PBRs) is in the physical placement of the fuel. For the HTGRs, the basic fuel units are microspheres with a small kernel of fissile material in the center, surrounded by several concentric layers of protective materials. In the GT-MHR, the microspheres are compacted with carbonaceous material into fuel rods that are placed in channels in graphite blocks [33]. The blocks are grouped in an annular array to form the reactor core. In the PBRs, the microspheres are embedded in carbon balls that are fed into the top of the pressure vessel and removed from the bottom. Each ball is roughly 6 cm in diameter—about the size of a tennis ball.

16.4.2 Historical Background of Graphite-Moderated Reactors

The history of carbon-moderated reactors is old and mixed. The world's first nuclear reactor was the graphite moderated reactor developed by Enrico Fermi in Chicago during World War II. Reactors for plutonium production have been primarily graphite moderated, because the conversion ratio is high, and it is relatively easy to change fuel elements frequently and avoid a large buildup of ^{240}Pu. Graphite-moderated reactors can be either water cooled, as in the Hanford plutonium production reactors and the Chernobyl-type RBMK reactors, or gas cooled, as in the British CO_2-cooled reactors and in the helium-cooled Fort St. Vrain reactor in the United States. The few graphite moderated power reactors in the United States have been shut down, but some are operating in the United Kingdom, Russia, and Lithuania (see Section 8.1.4).

The safety claims for the HTGRs might seem to fly in the face of the fact that the *only* reactor accidents that have resulted in major releases of

---

[12] The Very-High-Temperature Reactor (VHTR) is designed to reach 850°C to 1000°C (see Section 16.6.2 and Section 20.3.2).

activity have been in graphite-moderated reactors, namely the Windscale and Chernobyl accidents. However, it is argued that what happened at these plants has no relevance to the planned HTGRs.

- ◆ Windscale. The HTGRs will run at higher temperature than did Windscale and there will be no buildup of stored energy in the crystal lattice (the so-called Wigner energy) because the graphite will be continually annealed. The temperature for annealing is about 350°C [35, p. 441], well below the normal graphite temperature in an HTGR.
- ◆ Chernobyl. In addition to other major design differences, the use of a helium coolant in the HTGR (rather than water, as at Chernobyl) means that loss of the coolant cannot give a positive feedback. This follows from the fact that helium has a negligible absorption cross section for neutrons and therefore, unlike the water at Chernobyl, cannot be a poison.

It might also be noted that the only significant electricity-producing HTGR in the United States, a 330-MWe prototype unit at Fort St. Vrain in Colorado, had an unusually trouble-plagued life after going into operation in 1979.[13] It was shut down in 1989 by the operating utility because it was not economical to continue to run it. The difficulties were primarily with the cooling system, and it is believed that these difficulties can be avoided in a next-generation helium-cooled reactor.

In light of the above-described history, one might imagine that the nuclear industry would shy away from further attempts to develop HTGRs. However, the arguments that the Windscale and Chernobyl experiences are not relevant to future HTGRs appear to be convincing, and there are strong believers in the HTGR as a very safe reactor for the future.

Part of this confidence is based on experience with a series of prototype pebble bed HTGRs built in West Germany. The first of these was the AVR reactor, which was put into operation in 1967 to test the HTGR concepts. It was a small reactor, only 40 MWt and 15 MWe. A 300-MWe pebble bed HTGR, the THTR-300, was put into operation in 1987. These reactors have provided experience on the behavior of HTGR fuel. However, further development work on the German HTGR systems was halted in early 1991, due to lack of commercial interest [34]. Interest was later revived by the South African company, Eskom, which has been actively considering building and marketing PBMR reactors. In addition, exploratory initiatives have recently been undertaken in Asia with the construction of small HTGRs in Japan and China: a 30-MWt prototype high-temperature test reactor (HTTR) that started up in Japan in 1998 [36] and the 10-MWt HTR-10 pebble bed reactor that started up in China in December 2000 [37].

---

[13] In addition, a much smaller HTGR—the 40-MWe Peach Bottom 1 reactor—operated from 1967 to 1974.

16.4.3 General Features of Present HTGR Designs

The Fuel Elements

The basic fuel element is a microsphere, less than 1 mm in diameter.[14] The microspheres have a small central kernel containing the fissile or fertile material, typically in oxide form. A variety of mixtures of uranium, thorium, and plutonium oxides could be used, but present planning in each case calls for the use of enriched UO_2 fuel. The fuel kernel is encapsulated within an array of concentric protective shells, made up of successive layers of porous carbon, pyrolytic carbon, silicon carbide, and pyrolytic carbon [14, p. G-3]. Microspheres of this sort have been termed TRISO particles [40, p. 1]. They were first used in a British reactor in 1967, and further experience has been obtained from tests and use in the Peach Bottom and Fort St. Vrain reactors in the United States and the AVR plant in Germany [40, p. 4].

The multicoated TRISO microspheres provide a rugged protection for the fuel and prevent the escape of radionuclides. The fuel can withstand very high temperatures. The coating reportedly does not degrade until a temperature approaching 2000°C, which is well above the maximum accident temperatures (see below).

Reliability in manufacture is another issue. According to Andrew Kadak, the leader of the PBR program at MIT, it has been shown that TRISO particles can be manufactured with a defect rate of less than 1 in 10,000, which is adequate given the small amount of uranium in each microsphere kernel [38]. This good record would have to be maintained in mass production if the HTGR technology is adopted.

Feedback Mechanisms

The reactor is protected from a runaway chain reaction by the negative feedback provided by the Doppler effect as the fuel temperature rises (see Section 14.2.1). This effect is stronger in graphite-moderated reactors than in water-moderated reactors because with a graphite moderator, more elastic collisions are required to thermalize the neutrons and there is more resonant capture in ^{238}U.[15] Thus, Doppler broadening of the ^{238}U resonances is more important. There is no possibility of a positive feedback caused by poisoning properties of the coolant, as occurred in the water-moderated Chernobyl re-

---

[14] Standard TRISO particles have diameters ranging from 0.65 mm to 0.85 mm as described for the HTGR design [33]. For the PBRs, the microspheres are indicated to be 0.9 mm in diameter [38, p. 14], [39].

[15] In terms of the four-factor formula, the resonant escape probability is lower and the thermal utilization factor higher in graphite reactors that in LWRs (see Section 7.1.2).

actor, because neutrons are not absorbed in the HTGR's helium coolant.[16]
Therefore, there is no danger of a runaway reaction as at Chernobyl.

Heat Removal

Normally, the core is cooled by the flow of helium. If the coolant flow stops
(i.e, there is a loss-of-coolant accident) the reactor will be shut down by control rods, but heat will continue to be produced by radioactivity in the core.
Because graphite is a relatively poor moderator and has a low cross section for
neutron absorption, the ratio of moderator mass to fuel mass is much larger
in an HTGR than in a LWR. The core, therefore, has a high heat capacity,
which limits the rate at which the temperature rises during the initial period
when the reactor's heat output is greatest.

In the absence of helium flow, heat is transferred by conduction and radiation from the reactor core to the reactor vessel wall and from the reactor vessel
to its surroundings. Radiation can play a part in the HTGR, which tolerates
very high temperatures, because the rate of heat transfer by radiation rises
steeply with temperature—increasing as the fourth power of temperature. The
temperature rises until the heat removal rate equals the heat production rate.
This sets a limit on the maximum temperature. According to the design analyses, the highest temperature the core can reach in a severe accident for either
the GT-MHR or PBMR is 1600°C, which is significantly below the region for
fuel damage ($\sim$ 2000°C) [14, p. H-3; 39].

Overall, the HTGRs are relatively simple systems in which safety is provided by passive mechanisms. Protection against a criticality accident is provided by negative feedback. Core damage in a loss-of-coolant accident is
avoided because passive cooling keeps the fuel temperature below the point
where the fuel microspheres are degraded.

Interaction of Fuel with Air

A concern with HTGRs is the possible entry of air into the reactor and the
resulting oxidation of the hot graphite. This possibility is discounted in the
description of the PBMR provided in the Near-Term Deployment Roadmap:

> Any concern of fire in the graphite core is avoided by showing that
> there is no method of introducing sufficient oxygen into a high-
> temperature core (> 1000 C) to achieve sustained oxidation. This

---

[16] If heating of the graphite or loss of the helium has any appreciable effect on reactivity (and such an effect is not commonly mentioned), it is to provide negative
feedback. The graphite expands slightly with heating, which makes it a less effective moderator, and loss of the helium coolant can only serve to decrease the
moderation. Both effects would reduce the reactivity, although probably only by
a small amount.

is achieved primarily by the structural design of the reactor structure and building. [14, p. G-1]

A more cautious view has been expressed by Kadak:

> [I]t is very difficult to "burn" the graphite in the traditional sense, but it can be corroded and consumed...
>
> The key issue for the pebble bed reactor is the amount of air available in the core from the reactor cavity and whether a chimney can form allowing for a flow of air to the graphite internal structure and fuel balls. Tests and analyses have shown that at these temperatures graphite is corroded and consumed but the natural circulation required for "burning" is not likely due to the resistance of the pebble bed to natural circulation flow. [38]

Continuing efforts are being made to determine if the air input can be sufficient to cause serious damage. Such studies are necessary to resolve the safety question and address criticisms such as those of Edwin Lyman. Considering the effects of a pipe break, he argues:

> While the PBMR designers claim that the geometry of the primary circuit will inhibit air inflow and hence limit oxidation, this has not yet been conclusively shown.
>
> The consequences of an extensive graphite fire could be severe, undermining the argument that a conventional containment is not needed. [41]

Here, Lyman touches on a major point, because proponents of HTGRs have sometimes viewed them as being so safe that a strong containment is not needed. However, omitting the containment is justified only if it can be unambiguously shown that the possibility of a significant radiation release from the reactor is negligible. Presumably, this will be a crucial point for the NRC to consider should it receive a license application for an HTGR. The containment issue is further complicated by concerns about terrorist attacks, which may provide an argument for a containment even for an otherwise accident-proof reactor.

16.4.4 HTGR Configurations

The GT-MHR

The overall configuration of the GT-MHR is shown in Figure 16.3. The nominal module size is expected to be 600 MWt, corresponding to an electrical capacity of 286 MWe [14, p. H-3]. This sort of system, in which hot helium drives the turbine directly, with no steam generator, was suggested in the 1980s by Lawrence Lidsky and his collaborators at MIT [42]. Lidsky et al. pointed out that this was not a new idea conceptually, but they argued that

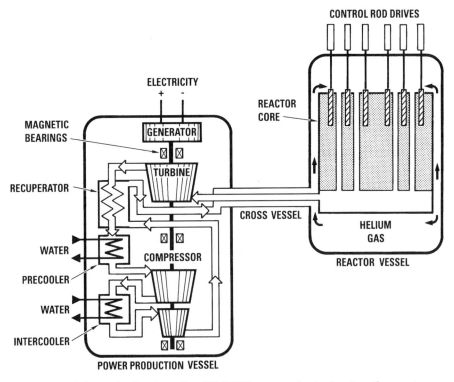

Fig. 16.3. Schematic drawing of a GT-MHR power plant, showing the reactor on the right and the turbine on the left. The turbine is driven by helium from the reactor. (Courtesy of General Atomics.)

technological developments and the push to modular design had by then made it practical.

In the GT-MHR, hot gas from the reactor drives the turbine. Gas exiting the turbine at reduced temperature and pressure passes through three heat exchangers and a two-stage compressor before returning to the reactor. The first heat exchanger (the recuperator) transfers heat from the helium leaving the turbine to the helium entering the reactor. The other two heat exchangers (the precooler and intercooler) further reduce the temperature of the gas in the compressor.[17] Partly because the recuperator is very efficient in heating the gas returned to the reactor, the overall generating efficiency is expected to be about 48%. With the high enrichment, the burnup is in excess of 100 GWd/t [14, p. H-5]. Both performance levels are well above those of present reactors.

---

[17] The nominal temperatures at various points in this cycle are as follows: turbine inlet and outlet, 848°C and 511°C; recuperator hot side inlet and outlet, 511°C and 125°C; recuperator cold side inlet and outlet, 105°C and 491°C [14, p. H-3].

The GT-MHR is now being developed in a program jointly sponsored by the U.S. DOE and the Russian Ministry for Atomic Energy (Minatom). The first construction target is for a prototype GT-MHR to be built in Russia and finished in 2009. U.S. plants would follow.

The Pebble Bed Reactor Configuration

An individual PBMR module is indicated to have a capacity in the range of 110–140 MWe, although some Eskom documents suggest a possible output of 170 MWe [43, 44].[18] Several modules would be grouped together to form a plant of the desired size. The PBMR system for using the hot helium to drive the gas turbine is similar to that depicted in Figure 16.3 for the GT-MHR, but the arrangement of the fuel in the reactor is very different.

The "pebble" of the pebble bed reactors is a graphite sphere with thousands of TRISO microspheres embedded in its matrix. Each pebble has a 5-cm-diameter core, containing about 15,000 microspheres, surrounded by a fuel-free outer graphite shell that is 0.5 cm thick, giving a total diameter of 6 cm.[19] The uranium mass is 9 g per pebble, or about 0.6 mg per microsphere. The uranium is enriched to 8% in ^{235}U.

The PBMR pressure vessel is a steel cylinder, 6 m in diameter and 20 m high, lined with graphite bricks to reflect neutrons back to the core. The core itself is 3.7 m in diameter and 9.0 m high, with a graphite central column that also serves as a neutron reflector. Present designs call for about 450,000 pebbles in the reactor. They are fed into the reactor at the top, move downward in the annular region surrounding the graphite central column, and are extracted at the bottom of the reactor. The insertion and extraction of the pebbles can be carried out while the reactor is operating, so the reactor does not have to be shut down for refueling. Each pebble makes about 10 such passages over a 3-year period. It is to be replaced when it reaches a burnup of 80 GWd/t.

The plant is designed for a lifetime of 40 years. The expected construction time is 24 months. The expected generation cost, after a "learning curve" period, is roughly 3 ¢/kWh for a plant consisting of five modules [43, p. 54]. If the necessary approvals are received from the prospective investors and the South African government, Eskom plans to build a 110-MWe demonstration plant near Cape Town, with construction starting in 2005 and commissioning scheduled for 2008.

---

[18] The Eskom and MIT-INEEL designs are similar, except for an intermediate heat exchanger in the latter. For specificity, the descriptions here refers to the Eskom PBMR [14, 39, 43, 44]. I am indebted to Tom Ferreira for Ref. [43] and additional information.

[19] The volume of a pebble's core is about 160,000 times the volume of a microsphere, so the microspheres imbedded in the graphite matrix occupy less than 10% of the core's volume.

16.5 Liquid-Metal Reactors

16.5.1 Recent United States Programs

The Integral Fast Reactor System

Liquid-metal reactors (LMRs) have had a long history, primarily directed toward the development of breeder reactors. However, in recent years, with little increase in the demand for nuclear power and a diminution in political support for it, worldwide interest in developing a breeder reactor capability has decreased. In consequence, some major facilities such as the Superphenix reactor in France have been shut down, and in the United States the Integral Fast Reactor (IFR) program at Argonne National Laboratory—which had become the focal point of U.S. work on breeder reactor systems—was dropped by the DOE in the mid-1990s.

In the IFR fuel cycle, the spent fuel from the reactor is reprocessed and the separated actinides, including plutonium and uranium, are fabricated with fresh uranium into new fuel elements that are returned to the reactor. The overall IFR system had four components: the reactor, a reprocessing plant for spent fuel, a facility to fabricate new fuel elements, and facilities for waste handling. Although the IFR project has been discontinued, some of its central ideas remain pertinent and figure prominently in the Generation IV program. These IFR components included the following:

- *Separation of chemical elements.* Spent fuel from the reactor is divided into different output streams through a series of chemical and electrorefining steps, some at high temperatures, in an overall process called *pyroprocessing* (as mentioned in Section 9.4.3). The actinides can be separated out, leaving a fission-product waste stream from which most of the long-lived components in the spent fuel have been removed.

- *Actinide burning.* The fuel cycle is closed, with the actinides returned to the reactor in new fuel elements. Most of the actinides are destroyed by fission in the fast-neutron flux of the LMR, thereby removing them from the long-term waste inventory.

- *Breeding fuel cycle.* Breeder reactors offer the prospect of virtually unlimited uranium resources. The IFR program incorporated a sodium-cooled LMR called the Advanced Liquid Metal Reactor (ALMR).

- *Passive safety.* The ALMR had passive features designed to make it it unusually safe. These features were explored in extensive safety tests at the Experimental Breeder Reactor (EBR–II). Some of the conclusions are discussed briefly in Section 16.5.2 because of their relevance to future liquid-metal reactors.

- *Colocation of facilities.* The reactor and fuel reprocessing and fabricating facilities were located in close physical proximity in the IFR system. Hence, the name "integral." The colocated facilities make plutonium diversion difficult.

Interest in the capabilities of this type of system has not disappeared and
two liquid-metal reactors are part of the current Generation IV roadmap:
the Sodium Fast Reactor and the Lead Fast Reactor. With possible future
developments in mind, we will briefly summarize here some of LMR features.
Another important aspect of the IFR—the closed, colocated fuel cycle—is
reflected in several other of the Generation IV designs.

16.5.2 Safety Features of LMRs

Sodium Coolant

The boiling point of liquid sodium at atmospheric pressure is 883°C [45,
p. 344]. Thus, it is possible to operate at atmospheric pressure and still
have a high coolant temperature, as needed for high thermal efficiency.[20]
In the ALMR design, the temperature of the sodium flowing past the core
was 500°C [46], providing a large margin of safety before the boiling point
is reached. The high-temperature that can be sustained before boiling occurs means that in the case of an accident, the reactor vessel could be cooled
primarily through radiation.

Liquid sodium requires care in handling and in the choice of metals used in
the system. Sodium (Na) reacts strongly with water and with oxygen in the air;
sodium and, more so, sodium hydroxide (NaOH) are corrosive in interacting
with some metals. However, LMR proponents argue that experience has shown
corrosion by liquid sodium to be less of a problem than is corrosion with water
in LWRs [47, p. 147]. Radioactivity is not a long-term problem with a sodium
coolant. The main activity produced in the sodium is ^{24}Na from neutron
capture in the stable isotope ^{23}Na. The half-life of ^{24}Na is 15 h, so the problem
is one of protection during and immediately after operation, rather than of
ultimate disposal.

Metallic Fuel

The fuel for the ALMR was to be metallic—an alloy of uranium, plutonium,
and zirconium [46, 48]—although oxide fuels have been used in some LMRs
(e.g., the French Phenix reactor and the U.S. Fast Flux Test Reactor). Metallic
fuels offer the possibility of very high burnup, because the problem of fuel
swelling has been successfully addressed in the alloys now being tested. Fuel
swelling is due to fission-product gases that cause the fuel to expand and
deform, with possible rupture of the cladding and escape of volatile fission
products into the coolant. In the metallic LMR fuel, after swelling reaches
about 30%, gas bubbles form and escape from the fuel [49]. The fuel pin is

---

[20] The pressure is near atmospheric at the top of the sodium pool in the reactor
vessel. There is a pressure gradient in the pool, with the pumps that drive the
sodium circulation creating a pressure head at the bottom of the reactor vessel
to drive the sodium upward through the reactor core.

designed to accommodate this swelling and has space to contain the escaped gas. With swelling no longer a major problem, it is possible to keep the fuel in the reactor for long periods and achieve a high burnup of the fissile material. For some fuels tested in EBR-II, the burnup has exceeded 18.5% of the fissile atoms [49], or about 180 GWd/t.[21]

Feedback Performance

The ALMR had an array of feedback mechanisms, including negative feedbacks from thermal expansion and a negative Doppler temperature coefficient, that would terminate the chain reaction if the sodium temperature became too high. The temperature at which this happens is well below the boiling point of sodium.

In the event of equipment failure and a reactor shutdown, heat can be removed from the core either by the normal circulation of sodium or, if the pumps fail, by convective circulation. The reactor vessel itself can be cooled, if necessary, by radiation to the surrounding containment vessel, which, in turn, is cooled by a natural convective airflow. The performance of the liquid sodium system under simulated accident conditions was evaluated in a series of demanding tests at EBR-II. In these tests, some or all of the sodium pumps were shut down, disrupting normal cooling. Normally the control rods would then be inserted, but this was intentionally disabled. The sodium temperature then rose but the rise was terminated by negative feedbacks that reduced the reactivity to zero or close to zero. In consequence, the sodium temperature oscillated as the reactivity changed with changing temperature, with no damage to the reactor.[22]

Safety Assessment

A probabilistic safety (or risk) assessment for the ALMR, carried out by the manufacturer, indicated risks well below NRC safety goals. For example, the chance that an individual in the vicinity of the plant would suffer a delayed fatal cancer as a result of an accident was calculated to be 1×10^{-10}/RY, compared to an NRC goal of 2×10^{-6}/RY (see Section 14.5.1) [46]. The analysis included both internal and external events.[23]

---

[21] The complete fission of 1 tonne of ^{235}U gives a thermal energy output of 950 gigawatt-days (see Table 9.3). Ignoring the small differences between ^{235}U and ^{239}Pu, this means, for example, that a 10% consumption (by number of atoms) corresponds to a burnup of 95 GWd/t.

[22] These and other attributes of the IFR system were discussed at greater length in the first edition [50, pp. 245–251].

[23] Protection against earthquakes is provided by a seismic isolation system, in which mechanical bearings isolate the reactor containment from Earth's motion. Use of seismic isolation is partly motivated by the fact that the ALMR operates at low

16.6 The Generation IV Program

16.6.1 Overview of the Program

Goals of the Program

The U.S. DOE inaugurated in 1999 a new program to explore possibilities for future reactors that are substantially more attractive than present reactors in terms of sustainability, economy, safety, and proliferation resistance. The contemplated reactors are the so-called Generation IV reactors. The goal has been to have the reactors ready for commercial deployment in two or three decades, with target dates set in the 2020–2030 time frame [19]. The dates are sufficiently far in the future to give time to explore a wide variety of options.[24]

Goals for the Generation IV reactors have been set forth in terms of four criteria [19, p. 12]:

- *Sustainability.* The fuel cycle should protect the environment, use resources of fissile material economically, and handle wastes in a manner that reduces the long-term burden.
- *Economy.* The reactor systems should provide electricity at competitive costs and without unusual economic risks.
- *Safety.* The reactors should operate reliably with a very low likelihood of severe core damage.
- *Proliferation resistance.* The overall nuclear system should be unattractive as a target for diversion or theft of fissile materials and should be well protected against terrorist attacks.

Of course, these same goals could have been enunciated at the start of the first nuclear era in the 1960s. Proliferation resistance, broadened to give emphasis to terrorism, now looms larger—partly because it is no longer overshadowed by concern about a nuclear exchange between the United States and the USSR and partly because of an increase in the number of countries, and possibly subnational groups, that have nuclear weapons or hope to get them.

The Generation IV effort has become an international one under the aegis of the Generation IV International Forum (GIF), which had 10 participating countries by 2002.[25]

---

pressure and, therefore, does not have as heavy a reactor vessel and pipe system as does a LWR operating at high pressures [51].

[24] Under conditions where the sense of urgency and willingness to accept risks were greater, during World War II, the first Hanford plutonium production reactors were put into operation within 2 years of the achievement of the first chain reaction.

[25] These were Argentina, Brazil, Canada, France, Japan, Korea, South Africa, Switzerland, the United Kingdom, and the United States [19, p. 11].

The Six Selected Systems

Six systems have been selected by the GIF as the ones to focus on as the most promising. These are listed alphabetically:

1. **GFR:** Gas-Cooled Fast Reactor system
2. **LFR:** Lead-Cooled Fast Reactor system
3. **MFR:** Molten Salt Reactor system
4. **SFR:** Sodium-Cooled Fast Reactor system
5. **SCWR:** Supercritical-Water-Cooled Reactor system
6. **VHTR:** Very-High-Temperature Reactor system

Parameters for these systems are given in Table 16.2. As can be seen from the alternatives indicated for some of the systems, the reactors are still more at the level of promising concepts, not fully developed designs. They probably are best looked at as illustrating the range of possibilities, rather than as an all-embracing list.

Broad Themes in the Generation IV Program

Overall, the systems have a number of specific themes, although these are not reflected in every one of the reactors types.

Table 16.2. Generation IV nuclear reactor systems, selected by GIF as being the most promising. (NA = Not Applicable)

Reactor Characteristic	VHTR	SCWR	GFR	SFR	LFR	MSR
Neutron spectrum	Thermal	Either	Fast	Fast	Fast	Thermal
Coolant	Helium	Water	Helium	Sodium	Pb or PbBi	
Moderator	Graphite	Water[a]	None	None	None	Graphite
Recycle?	No	If fast	Yes	Yes	Yes	Online
Monolithic size (MWt)	NA	3860	NA		3600	
Monolithic size (MWe)	NA	1700	NA	≈ 1500	1200	1000
Mid-size or Modular (MWt)	600	NA	600		120–400	NA
Mid-size or Modular (MWe)		NA	288	150–500	50–150	NA
Burnup (GWd/t)	150–200	45	250	150–200	100–150	
Temperature out (°C)	1000	550	850	550	550–800	700–850
Electricity production?	Possible	Yes	Yes	Yes	Yes	Yes
Hydrogen production?	Primary	No	Yes	No	At 800°C	At 850°C
Year deployable	2020	2025	2025	2015	2025	2025

[a]Moderator only for case of thermal spectrum.
Source: Ref. [19].

For each of the reactors, other than the VHTR and the SCWR operated as a thermal reactor, recycling of fuel is a basic part of the system design, and even for the VHTR it is being considered. Both aqueous reprocessing and pyroprocessing are options (see Section 9.4.2). The recycling may include all the actinides, not just uranium and plutonium. The wastes would then be just the fission products, with traces of actinides, greatly reducing the long-term activity. With the removal of the uranium, the mass of the waste is greatly reduced. The activity of the uranium itself is so low that if not returned to a reactor it could relatively easily be handled as a special category of "waste." This closed cycle has important implications for resource utilization, waste handling, and weapons proliferation. A high priority is being given to implementing it in a manner that will create little risk of plutonium diversion.

Another theme of the Generation IV reactors is their potential use for hydrogen production. Although hydrogen can be produced by any electricity generator through electrolysis, production is more efficient in thermochemical cycles at very high temperature, where the net process is production of hydrogen and oxygen from water (see Section 20.3.2). The VHTR has hydrogen production as its central goal and three of the other systems have it as a possibility.

As a whole, the Generation IV reactors seek to operate at higher temperatures, with longer lifetimes, and at higher fuel burnup rates than is usual for present-day reactors. Thus, another crosscutting theme in the Generation IV program is the development and testing of materials that will perform well under the these demanding conditions.

16.6.2 Systems Emphasized in the United States

Relative Priorities

The priorities given to these six systems are likely to vary from country to country. In the United States, as reported in a talk by Robert Versluis of the DOE Office of Advanced Nuclear Research, the primary focus as of late 2002 was expected to be on the VHTR, with a secondary focus on the SCWR and GFR and with only "nominal support of the other concepts" [52].

Very-High-Temperature Reactor (VHTR)

The VHTR is a graphite moderated, helium-cooled reactor and, thus, it is a descendant of the HTGRs discussed in Section 16.4. As indicated by its name, the purpose of the VHTR is to achieve high temperatures. These are of importance in a variety of applications, with especially high stakes involved in its potential for the production of hydrogen (see Section 20.3.2). These reactors could also be used for electricity generation, and a particular reactor could be built for one or both of these purposes. A sketch of the VHTR system

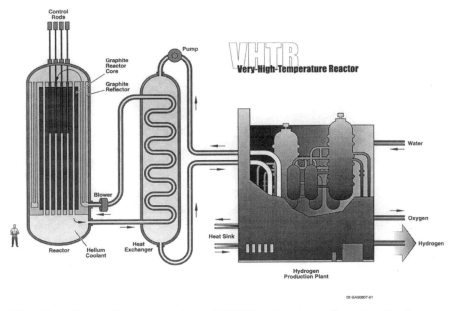

Fig. 16.4. Schematic representation of VHTR system in configuration for the production of hydrogen. (From Ref. [19, p. 54].)

is shown in Figure 16.4, in a configuration for hydrogen production only, as featured in the Generation IV Roadmap. In this configuration, hot helium gas moves from the reactor vessel into a heat exchanger, where it transfers heat to the fluid in the secondary loop, which might be helium or molten salt.

The high temperatures in the reactor place special demands on material performance. For example, the VHTR will use TRISO fuel particles, as planned for the HTGRs, but it is may be desirable to develop alternative coatings for these particles, perhaps zirconium carbide, to withstand the elevated temperatures and high burnups that are being planned. Developing these coatings and testing their performance is expected to take until 2015 [19, p. 55–58]. The planned burnup is in the region of 150–200 GWd/t. To attain this, fuel will be used with a relatively high ^{235}U enrichment—perhaps up to 20%—in a once-through fuel cycle. Burnable absorbers will be used to reduce the reactivity at the beginning of the fuel's lifetime.

The hot gas would exit the core at a temperature of at least 1000°C, and the fuel will be at still higher temperatures. The pressure vessel temperature would then be over 450°C, which is substantially higher than the pressure vessel temperatures in LWRs (300°C) or other HTGRs (400°C in the HTTR operating in Japan). Further development and testing of the materials to be used for the pressure vessel and other components (such as the heat exchangers) will be required to establish their durability under high temperatures and irradiation. As with the parallel work on the TRISO particles, completion of the development and testing is projected to take until 2015.

Supercritical-Water-Cooled Reactor (SCWR)

Conventional light water reactors use steam to drive the generating turbines. In PWRs, the steam is produced in separate steam generators. In BWRs, the steam is produced in the pressure vessel, making a separate steam generator unnecessary but requiring a pumping system to circulate the water through the core. The SCWR operates at a high temperature where water is in a supercritical state and thus is a single-phase fluid. There is no need for either the PWR's steam generator or the BWR's circulation system and, therefore, the design is significantly simpler than that of other LWRs (see Figure 16.5).

The pressure in the system is about 250 atm. The coolant enters the reactor pressure vessel at about 280°C, is heated as it passes through the core, and exits at 510°C. At this high temperature and pressure, the deliverable energy per unit mass of the fluid is high. Therefore, the turbines can be driven with

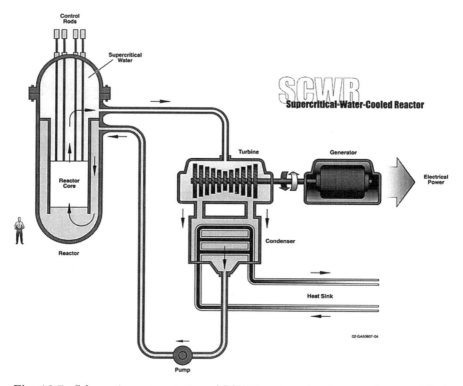

Fig. 16.5. Schematic representation of SCWR system showing use of supercritical water to drive turbines directly. The arrows show the direction of fluid flow. (From Ref. [19, p. 48].)

a smaller mass of fluid than in ordinary steam generators, and the pumps and the pipe diameters can be smaller. With the more compact reactor system, the containment building size can also be reduced.

These advantages, which together lead to a reduced cost of the plant, provide an incentive for developing the SCWR. Before such a reactor can be built, however, it will be necessary to find and test suitable materials. Supercritical water is highly corrosive and, in addition, the materials will be subjected to high levels of radiation. Some of these problems have been solved for supercritical water-cooled fossil fuel plants, which are widely used now, and these provide a base of experience. However, for reactor purposes, according to the Generation IV Roadmap summary, "no candidate alloy has been confirmed for use as either the cladding or structural material" [19, p. 49].

A key decision for the SCWR is the choice between a thermal or fast neutron spectrum. It is possible to have a water-cooled fast reactor in the case of the SCWR, because the mass of water in the reactor is relatively small and the moderation less effective than in an ordinary LWR. If a thermal spectrum is desired, additional moderation is needed.

Gas-Cooled Fast Reactor (GFR)

The GFR is a gas-cooled reactor that is designed to achieve a high burnup of fuel (up to 250 GWd/t) and destruction of actinides. Both helium and super-critical carbon dioxide (CO_2) are being considered for use a coolant [53]. With helium, the operating coolant temperature would be higher, which would be an advantage for hydrogen production, while with supercritical CO_2, a higher system efficiency can be reached due to less energy required to pump the coolant. Previous gas cooled reactors have a high ratio of carbon to uranium, which provides both moderation and the capability of withstanding high temperatures. For a fast reactor, where little or no moderation is wanted, the permissible amount of carbon is reduced, although there may be some carbon in the fuel pellets. New materials will have to be developed and tested for the fuel elements and other reactor components. This process is scheduled to continue through 2019. It is hoped that a prototype system could be built and put into operation by 2025 [19, p. 29].

16.7 Radical Nuclear Alternatives to Present Reactors

16.7.1 Fusion

The traditional nuclear alternative to fission is fusion. There has been continued study of fusion, with significant but slow progress. Fusion research and development has now become a very expensive matter, motivating international collaborations. The main hope for major further progress centers upon the building of the International Thermonuclear Experimental Reactor

(ITER), which is being planned in a collaboration involving the United States, Europe, Japan, and Russia. If ITER is built and tests with it are successful, a subsequent step would be to build a demonstration power plant.

It is not possible to say with any confidence when, if ever, this program will come to fruition. Most observers doubt that electric power from a prototype fusion reactor could come before the year 2030, and some would consider this to be an optimistic target date. Nonetheless, in principle, fusion offers the prospect of virtually unlimited electrical power, probably with fewer concerns about environmental contamination and weapons proliferation than is the case with fission power.

16.7.2 Accelerator-Driven Fission

In the early 1990s, a quite different approach was suggested for obtaining energy from fission. In this approach, a high-energy beam from a proton accelerator produces a large number of neutrons in collisions with heavy nuclei, in a process known as *spallation*. A geometric arrangement is used so that the spallation neutrons irradiate uranium or thorium, initiating fission. The configuration of fissile material is subcritical, and the number of neutrons decreases in successive fission generations. Nonetheless, this subcritical chain reaction serves to amplify the total number of neutrons above those produced by spallation alone. Thus, many neutrons are produced per initial proton, and the total energy released from fission in a properly designed reactor can considerably exceed the input energy required to run the accelerator.

There were two prominent suggestions for implementing this approach:

♦ *Los Alamos proposal.* A Los Alamos group suggested a system in which 1.6-GeV protons irradiate a heavy target (e.g., lead), spallation neutrons from the target are moderated in heavy water, and the resulting thermal neutrons produce fission in the fissile fuel (^{233}U or ^{235}U) [54]. The heavy spallation target would be in the form of molten metal, the fissile fuel in the form of a molten salt, and the fertile fuel (^{232}Th) mixed with heavy water. All of these components would flow in recirculating streams. Given a high-current proton beam and a large spallation yield, the flux of neutrons in the reactor is much higher than in normal reactors. Many of the radioactive products of fission or neutron capture can then be transmuted into products of shorter half-life, reducing the inventories of wastes that require disposal. The system assumes chemical separation of selected radionuclides from the circulating fuel and their return to the reactor for bombardment and transmutation.

♦ *CERN proposal.* A design developed at CERN is similar to conventional reactors in that the fissile and fertile materials are in solid form and the fuel and moderator are both contained in a reactor vessel [55]. However, the fuel is in a configuration that makes the assembly subcritical. Again, the system is fed by neutrons produced in proton-induced spallation reactions, either in the fuel and moderator or in a separate target. Actinides are consumed in this cycle, but there is no intent to transmute fission products.

Both systems have little danger of a criticality accident because their fuel is in a subcritical arrangement. They are envisaged for use in a ^{232}Th–^{233}U breeding cycle, which could greatly extend energy supplies. In this cycle, there is less production of long-lived actinides than in ^{238}U reactors, thereby reducing proliferation and waste disposal problems. However, for the case where the reduction is greatest, the Los Alamos approach, it is necessary to compare these gains against the possible hazards created by the chemical handling of the radioactive waste streams from the circulating fuel.

These accelerator proposals have not received the critical examination that conventional reactors have received. They clearly represent an ingenious application of accelerator technology to the extraction of energy from nuclear fission. However, it has not been established that they provide better systems than are provided by more traditional reactor approaches. As yet, no major programs have been implemented to advance these reactors beyond the present conceptual designs.

References

1. Alvin M. Weinberg and I. Spiewak, "Inherently Safe Reactors and a Second Nuclear Era," *Science* 224, 1984: 1398–1402.
2. A.M. Weinberg, I. Spiewak, J.N. Barkenbus, R.S. Livingston, and D.L. Phung, *The Second Nuclear Era: A New Start for Nuclear Power* (New York: Praeger, 1985).
3. J. Kupitz and J. Cleveland, "Overview of Global Development of Advanced Nuclear Power Plants, and the Role of the IAEA," in *Proceedings of the International Conference on Future Nuclear Systems: Global '99* (La Grange Park, IL: American Nuclear Society, 1999): document 134 [CD-ROM].
4. "World List of Nuclear Power Plants," *Nuclear News* 46, no. 3, March 2003: 41–67.
5. Toshiaki Enomoto, "Perspectives on the Future of Nuclear Power," *Nuclear News* 43, no. 12, November 2000: 64.
6. Hidaeki Suzuki, "Design features of Tsuraga-3 and -4: The APWR Plant in Japan," *Nuclear News* 45, no. 1, January 2002: 35–39.
7. H. Mouney, "A Utility View Point," in *Proceedings of the Workshop on Advanced Reactors and Innovative Fuels, Villigen, Switzerland, October 21–28, 1998* (Paris: OECD, 1999): 33–39.
8. William D. Magwood IV, "Looking Toward Generation Four: Considerations for a New Nuclear R&D Agenda," paper presented at the 1999 Summer Meeting of the American Nuclear Society, 1999.
9. U.S. Nuclear Regulatory Commission, *Final Safety Evaluation Report Related to the Certification of the Advanced Boiling Water Reactor Design*, Report NUREG-1503 (Washington, DC: NRC, 1994).
10. *U.S. Code of Federal Regulations, Energy*, Title 10, Part 52.97(b)(1) (1993).

11. President's Committee of Advisors on Science and Technology, *Report to the President on Federal Energy Research and Development for the Challenges of the Twenty-First Century* (Washington, DC: Executive Office of the President, 1997).

12. William D. Magwood IV, "Roadmap to the Next Generation of Nuclear Power Systems: A Vision for a Powerful Future," *Nuclear News* 43, no. 12, November 2000: 35–38.

13. U.S. Department of Energy, Office of Nuclear Energy Science and Technology, *Nuclear Power 2010—Overview*, 2003. [From: http://www.nuclear.gov/planning/NucPwr2010.html]

14. Near Term Deployment Group, *A Roadmap to Deploy New Nuclear Power Plants in the United States by 2010, Volume II, Main Report* (Washington, DC: U.S. DOE, 2001).

15. Near Term Deployment Group, *A Roadmap to Deploy New Nuclear Power Plants in the United States by 2010, Volume I, Summary Report* (Washington, DC: U.S. DOE, 2001).

16. National Research Council, *Nuclear Power, Technical and Institutional Options for the Future*, Report of the Committee on Future Nuclear Power Development, John F. Ahearne, Chairman (Washington, DC: National Academy Press, 1992).

17. Robert T. Twilley, Jr., "Framatome ANP's SWR 1000 Reactor Design," *Nuclear News* 45, no. 10, September 2002: 36–40.

18. David F. Torgeson, "The ACR-700—Raising the Bar for Reactor Safety, Performance, Economics, and Constructability," *Nuclear News* 45, no. 11, October 2002: 24–32.

19. U.S. DOE Nuclear Energy Research Advisory Committee and the Generation IV International Forum, *A Technology Roadmap for Generation IV Nuclear Energy Systems*, Report GIF-002-00 (2002). [From: http://gif.inel.gov/roadmap]

20. International Energy Agency, OECD Nuclear Energy Agency, and International Atomic Energy Agency, *Innovative Nuclear Reactor Development, Opportunities for International Cooperation*, a Three Agency Study (Paris: OECD/IEA, 2002).

21. GE Nuclear Energy, *The ABWR Plant General Description* (San Jose, CA: GE Nuclear Energy, 1999).

22. International Atomic Energy Agency, *Nuclear Power Reactors in the World*, Reference Data Series No. 2, April 2003 edition (Vienna: IAEA, 2003).

23. Westinghouse Electric Company, "Korea's Nuclear Strategy," *Westinghouse World View* August 2002: 7–9.

24. U.S. Nuclear Regulatory Commission, *Final Safety Evaluation Report Related to the Certification of the System 80+ Design*, Report NUREG-1462 (Washington, DC: NRC, 1994).

25. C.W. Forsberg and A.M. Weinberg, "Advanced Reactors, Passive Safety, and Acceptance of Nuclear Energy," *Annual Review of Energy* 15, 1990: 133–152.

26. R.P. Vijuk, private communication (June 11, 2002).

27. E.H. Kennedy and R.P. Vijuk, "AP1000 Overview," paper presented at CEPSI 2002, The 14th Conference of the Electric Power Supply Industry, Furuoka, Japan, 2002.

28. U.S. Nuclear Regulatory Commission, *Final Safety Evaluation Report Related to Certification of the AP600 Standard Design*, Report NUREG-1512 (Washington, DC: NRC, 1998).

29. John Ahearne, Ralph Bennett, Robert Budnitz, Daniel Kammen, John Taylor, Neal Todreas, and Bert Wolfe, *Nuclear Energy: Present Technology, Safety, and Future Research Directions: A Status Report*, Panel on Public Affairs, American Physical Society (November 2, 2001). [From: http://www.aps.org/public_affairs/popa/reports/nuclear.pdf]

30. Mario D. Carelli, "IRIS: A Global Approach to Nuclear Power Renaissance," *Nuclear News* 46, no. 10, September 2003: 32–42.

31. K. Miller and D. Paramonov, "The Economics of Iris," *Proceedings of the 10th International Conference on Nuclear Engineering, ICONE10, Vol. 2, Safety, Reliability, and Plant Evaluations; Next Generation Systems* (New York: The American Society of Mechanical Engineers, 2002): 561–565.

32. L.E. Conway, C. Lombardi, M. Ricotti, and L. Oriani, "Simplified Safety and Containment Systems for the Iris Reactor," *Proceedings of the 9th International Conference on Nuclear Engineering, ICONE9*, Nice, France (2001).

33. Malcolm P. LaBar, "The Gas Turbine-Module Helium Reactor: A Promising Option for Near Term Deployment," Report GA-A23952 (San Diego, CA: General Atomics, 2002). [From: http://www.ga.com/gtmhr/images/ANS.pdf]

34. "International Briefs," *Nuclear News* 34, no. 3, March 1991: 56.

35. F.J. Rahn, A.G. Adamantiades, J.E. Kenton, and C. Braun, *A Guide to Nuclear Power Technology* (New York: Wiley, 1984).

36. World Nuclear Association, "Nuclear Power in Japan," *Information and Issue Briefs* (July 2003). [From: http://www.world-nuclear.org/info/inf79.htm]

37. "China: First High-Temperature Reactor Goes Critical," *Nuclear News* 44, no. 2, February 2001: 34–35.

38. Andrew C. Kadak, "A Renaissance for Nuclear Energy?" *Physics and Society* 31, no. 1, January 2002: 13–17.

39. Dave Nicholls, "The Pebble Bed Modular Reactor," *Nuclear News* 44, no. 10, September 2001: 35–40.

40. *Coated Particle Fuel Technology For Modular HTGR Reactor Systems*, rev. 1, General Atomics report (August 1993).

41. Edwin S. Lyman, "The Pebble-Bed Modular Reactor (PBMR): Safety and Non-Proliferation Issues?" *Physics and Society* 30, no. 4, 2001: 16–19.

42. L.M. Lidsky, D.D. Lanning, J.E. Staudt, and X.L. Yan, "A Direct-Cycle Gas Turbine Power Plant for Near-Term Applications: MGR-GT," paper presented at the 10th International HTGR Conference, San Diego, September, 1988.

43. Pebble Bed Modular Reactor (Proprietary) Limited, *Report on the Commercialization of the Pebble Bed Modular Reactor* (March 27, 2003).

44. PBMR website: http://www.pbmr.com (2003).

45. Yeram S. Touloukian, "Thermophysics" in *A Physicist's Desk Reference, The Second Edition of Physics Vade Mecum*, Herbert L. Anderson, ed. (New York: American Institute of Physics, 1989): 336–347.

46. P.M. Magee, E.E. Duberley, A.J. Lipps, and T. Wu, "Safety Performance of the Advanced Liquid Metal Reactor," paper presented at the ARS '94 Topical Meeting—Advanced Reactors Safety, Pittsburgh, April 1994.

47. Charles E. Till and Yoon I. Chang, "The Integral Fast Reactor," *Advances in Nuclear Science and Technology*, 20, 1988: 127–154.

48. B.R. Seidel, L.C. Walters, and Y.I. Chang, "Advances in Metallic Nuclear Fuel," *Journal of Metals* 39, no. 4, April 1987: 10–13.

49. Y.I. Chang and C.E. Till, "Advanced Breeder Cycle Uses Metallic Fuel," *Modern Power Systems* 11, no. 4, April 1991: 59.

50. David Bodansky, *Nuclear Energy: Principles, Practices, and Prospects* (Woodbury, NY: American Institute of Physics Press, 1996).

51. Emil L. Gluekler (G. E. Nuclear Energy), private communication (May 9, 1994).

52. Rob M. Versluis, "Generation IV Nuclear Energy Systems," paper presented at the 2002 Winter Meeting of the American Nuclear Society, Washington, DC, November 2002.

53. Kevan D. Weaver and Hussein S. Khahil, "Gas Cooled Fast Reactor (GFR)," paper presented at the 2002 Winter Meeting of the American Nuclear Society, Washington, DC, November 2002.

54. C.D. Bowman, et al., "Neutron Energy Generation and Waste Transmutation Using an Accelerator-Driven Thermal Intense Neutron Source," *Nuclear Instruments and Methods in Physics Research* A320, 1992: 336–367.

55. R. Carminati, R. Klapisch, J.P. Revol, C. Roche, J.A. Rubio, and C. Rubbia, "An Energy Amplifier for Cleaner and Inexhaustible Nuclear Energy Production Driven by a Particle Beam Accelerator," preprint CERN/AT/93-47(ET) (November 1, 1993).

17

Nuclear Bombs, Nuclear Energy, and Terrorism

17.1 Concerns About Links Between Nuclear Power and Nuclear Weapons

Many observers believe that the most profound problem with using nuclear energy for electricity generation is the connection between nuclear power and nuclear weapons. In this view, the threat of nuclear weapons proliferation increases if the world relies on nuclear power, because nuclear power capabilities could be translated into nuclear weapons capabilities. The relative merits of renewable energy and nuclear fission energy (omitting fusion as still speculative) as eventual substitutes for fossil fuels are highly controversial, with unresolved arguments over relative economic costs, environmental impacts, practicality, and safety. However, the weapons connection is unique to nuclear fission energy and constitutes, for some people, a reason to limit or abandon it.

Giving up nuclear power would obviously avert the danger that nuclear power facilities might be diverted to weapons purposes. However, it would not avert all dangers of weapons development. It is quite possible to have nuclear weapons without nuclear power as well as nuclear power without nuclear weapons. In fact, most countries that have nuclear weapons had those weapons well before they had civilian nuclear power. Other countries that have made substantial use of nuclear power, such as Sweden and Canada, are rarely perceived to be potential nuclear weapons threats.

Nonetheless, a program in one area can aid a program in the other. The relation between nuclear power and nuclear weapons will be addressed in this and the next chapter. Special attention will be given to the rudiments of bomb technology and terrorist threats in this chapter and to past and potential national nuclear weapons programs in Chapter 18.

17.2 Nuclear Explosions

17.2.1 Basic Characteristics of Fission Bombs

Fissionable Materials for Nuclear Weapons

In a nuclear explosion, it is necessary to have a rapidly developing chain reaction. Therefore, the time interval between successive fission generations must be short, with the chain reaction propagated by unmoderated neutrons from fission. These neutrons typically have energies E_n in the neighborhood of 1 MeV. It is also necessary to have a high multiplication factor. Thus, the weapons material must have a large neutron fission cross section for $E_n \approx$ 1 MeV, and the number of neutrons emitted per fission, ν, must be relatively large. The nuclides that meet these criteria are the same as those that can serve as the fissile fuel in nuclear reactors:

- **^{235}U.** The bomb dropped at Hiroshima was a ^{235}U bomb. For reactors, uranium is typically enriched to 2–5% in ^{235}U. Uranium enriched to more than 20% is termed "highly enriched." For nuclear weapons, it is the norm to use ^{235}U enrichments in the neighborhood of 90% or higher, but it is possible to make a bomb with enrichments well below 90% (see Section 17.3). The enriched material is obtained by isotopic separation, starting with natural uranium. The minimum amount of enriched uranium needed depends on the details of bomb design, but typically quoted numbers range from 10 to 25 kg.[1]
- **^{239}Pu.** The bomb dropped at Nagasaki was a ^{239}Pu bomb. A plutonium bomb requires about 5 kg of ^{239}Pu, which must not be excessively contaminated with ^{240}Pu if the bomb is to have a reliably high yield. (The problems created by ^{240}Pu are discussed in Section 17.4.) Both ^{239}Pu and ^{240}Pu are produced in any reactor that uses ^{238}U as a fertile fuel. The plutonium is extracted from the spent fuel by chemical separation; the more demanding step of isotopic enrichment is unnecessary.
- **^{233}U.** No ^{233}U bombs are known to have been made, and there have been no apparent incentives to produce them.[2] ^{233}U could, in principle, be produced in a reactor that uses ^{232}Th as a fertile material. A bomb

---

[1] The 25-kg figure, which is somewhat higher than those cited in Section 17.2.3, follows from an equivalence indicated by the U.S. Enrichment Corporation: 500 tonnes of bomb-grade uranium corresponding to 20,000 bombs [1].

[2] It is not possible to exclude the possibility that some experimental ^{233}U bombs have been built, but, even if so, they do not represent a significant component of any nuclear weapons program.

based on ^{233}U would require uranium that is highly enriched in ^{233}U. However, the chief interest in ^{233}U is not as a bomb material but as the fissile component in a possible ^{232}Th–^{233}U fuel cycle, where ^{232}Th is the fertile component. This cycle may have advantages in protection against weapons proliferation (see also Section 18.3.3).

Types of Nuclear Bomb

The minimum mass sufficient to sustain a chain reaction is known as the "critical mass." This is the mass for which the neutrons produced in fission just balance the neutrons that are absorbed or escape. For an effective nuclear weapon, the critical mass must be assembled quickly. Two general approaches are used to achieve this, as pictured in Figure 17.1:

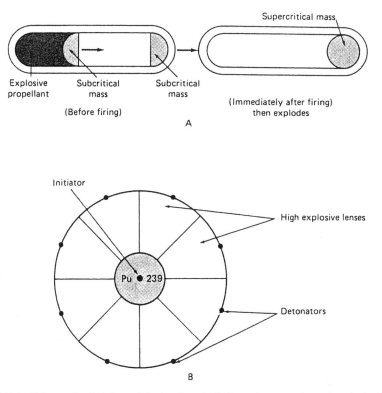

Fig. 17.1. Schematic sketches of fission bomb designs: top, gun-type bomb; bottom, implosion-type bomb. [From Paul P. Craig and John A. Jungerman, *The Nuclear Arms Race: Technology and Society*, 2nd edition (New York: McGraw-Hill, 1990), 213; with permission of McGraw-Hill.]

◆ *Gun assembly.* In the gun-type approach, two subcritical masses of fission-able material are brought together by firing one into the other, or by firing them both. This can be done in a straightforward way using a chemical propellant. The assembly speed depends on the velocity imparted to the mass or masses. As discussed in Section 17.4.1, the gun assembly method is too slow for use with ^{239}Pu bombs and is only used with ^{235}U bombs.

◆ *Implosion technique.* In the implosion approach, a given mass is changed from subcritical to critical by a very rapid compression, produced by a carefully placed set of chemical explosives surrounding the mass. The ar-rangement of the explosives and the timing of their firing require a more sophisticated design than is necessary in the gun assembly. The motivation for making this extra effort is speed of assembly. The implosion technique has been used for both ^{235}U and ^{239}Pu bombs.

Many of the nuclear weapons developed since World War II obtain a major fraction of their explosive energy from the fusion of hydrogen, in conjunction with fission. However, such "hydrogen bombs" are more complicated to build than are simple fission bombs, and although they may be more dangerous in terms of the size of the explosion produced, they are less immediate threats from the standpoint of weapons proliferation.[3]

Energy Yield of Nuclear Weapons

The energy yield of nuclear weapons is commonly expressed in kilotons (kt) or megatons (Mt) of high explosive (TNT) equivalent, where 1 kt of TNT is assumed to release 10^{12} cal (4.18×10^{12} J). The complete fission of ^{235}U in a reactor releases 8.2×10^{13} J/kg (see Table 9.3). About 86% of the energy is in the kinetic energy of the fission fragments themselves and 6% in prompt neutrons and gamma rays (see Section 6.4.2). Therefore, complete fission of 1 kg of ^{235}U would give a prompt explosive yield of about 7×10^{13} J, or 17 kt.[4] The yield for ^{239}Pu is similar.

---

[3] However, "fusion boosted" bombs might be built without a great technological effort. In a simple version of such a bomb, deuterium (^{2}H) and tritium (^{3}H) are introduced into the center of an implosion weapon, possibly as a mixture of gases or or by inserting lithium deuteride. (The lithium deuteride contains the required deuterium and serves as a source of tritium through the n + ^{6}Li → ^{3}H + ^{4}He, where the initial neutrons come from fission.) Implosion of the fissile material starts the bomb's chain reaction, creating the high temperatures that ignite the fusion reaction: ^{2}H + ^{3}He → ^{4}He + n. The importance of fusion in these bombs is not as an additional source of energy (the addition usually is relatively small) but as a source of high-energy (14-MeV) neutrons that increase substantially the fraction of the fissile material that undergoes fission and hence the bomb's explosive yield (see, e.g., Ref. [2, Section 1.5.1]).

[4] This number includes a qualitative allowance for the escape of some of the neu-trons and gamma rays.

Actual yields in nuclear weapons are less than 17 kt/kg of fissile material, because a bomb will disassemble without complete fissioning of the material. For example, the world's first nuclear bomb, used in the Trinity test in New Mexico in July 1945, is reported to have had a mass of 6.1 kg of plutonium [3, p. 127] and a yield of 18.6 kt.[5] This corresponds to a yield of 3 kt/kg of fissile material, or an efficiency of about 20%.[6] Modern bombs are more efficient, reportedly "approaching 40 percent" [6, p. 266].

17.2.2 Effects of Nuclear Bombs

Experience at Hiroshima and Nagasaki

The bomb at Hiroshima caused about 100,000 deaths, either immediately or within the first few weeks, and the bomb in Nagasaki caused a comparable, but perhaps somewhat lower, toll [7, p. 3].[7] William Schull, formerly with the joint Japanese–American Atomic Bomb Joint Casualty Commission (ABCC), reports:

> Contrary to what is commonly supposed, the bulk of the fatalities at Hiroshima and Nagasaki were due to burns caused either by the flash at the instant of the explosion or from the numerous fires that were kindled, and were not a direct consequence of the amount of atomic radiation received. Indeed, the Joint Commission [the ABCC] estimated that over half the total deaths were due to burns and another 18% due to blast injury. Nonetheless, ionizing radiation accounted for a substantial number of deaths, possibly 30%. [7, p. 12]

The delayed effects of these bombs were much less than the prompt effects. As discussed in Section 4.2.2, radiation from the bomb is believed to have been responsible for 421 excess cancer deaths during the period from 1950 through 1990, and there was no statistically significant increase in observed genetic defects among offspring of exposed individuals.

Potential Effects of Future Bombs

The destruction that would be caused by a major nuclear attack by a nation with a substantial arsenal greatly exceeds that caused by the Hiroshima and

---

[5] Richard Rhodes, in the Introduction to Ref. [4].

[6] The yields of the two bombs dropped during World War II were about 15 kt for the Hiroshima bomb and about 21 kt for the Nagasaki bomb [5, p. 15]. The Nagasaki bomb reportedly used 6.2 kg of plutonium, again corresponding to an efficiency of about 20% [6, p. 266].

[7] Schull indicates that a precise number cannot be obtained. As he explains, "the circumstances obtaining after the bombing precluded accurate counts, but the toll was obviously awesome" [7, p. 303]. Among the difficulties is uncertainty as to the number of Japanese civilians and military personnel in the cities and the number of Korean conscript laborers.

Nagasaki bombs. The impact would depend on the number and size of the weapons used, and it is perhaps impossible to gauge the effects of the bombs themselves and of the possible breakdown in society that might follow. The danger of such a nuclear exchange between the two major nuclear powers, the United States and Russia, now seems remote. There are intermittent threats that India and Pakistan might resort to the use of nuclear weapons against each other, but as of early 2004, the threat does not appear acute.

The most imminent threat at the moment for the United States, and perhaps for the world as a whole, appears to be from the possible use of single weapons by terrorist groups. Such bombs are likely to be relatively small in yield. In its Report No. 138 on *Management of Terrorist Events Involving Radioactive Material*, the U.S. National Council on Radiation Protection and Measurements (NCRP) summarized the damage that might be done by terrorist-size weapons. For example, for a 1-kt bomb, the estimated range for 50% mortality from thermal burns is 610 m (assuming no intervening structures) and the estimated range for an absorbed dose of 4 grays (Gy) (400 rads) from the initial gamma ray and neutron radiation extends to 790 m [8, p. 23].[8]

Assuming the explosion is at ground level, as distinct from the Hiroshima and Nagasaki cases where the explosions were high aboveground, there would be heavy local fallout. The downwind dose is calculated to exceed 4 Gy in the first h for a distance of 5500 m. A dose of 4 Gy is specified here, because it is the estimated level for 50% mortality in the absence of medical treatment [8, p. 20]. At smaller distances, the dose would be still greater, and medical treatment—assuming it is available—might be of no help in preventing death. Because much of the dose is due to very short-lived radionuclides, the dose rate falls very rapidly with time—by approximately a factor of 10 when the time increases by a factor of 7 [9, p. 391]. Thus, at 7 h, the dose rate is 10% of the 1-h level, and at 49 h, it is 1% of the 1-h level.

17.2.3 Critical Mass for Nuclear Weapons

Critical Mass With and Without Reflectors

A sphere of fissionable material will have a mass that is less than the critical mass if its radius is small compared to the mean free path for fission, λ. For 1-MeV neutrons in ^{235}U, $\lambda = M/\rho N_A \sigma_f = 17$ cm. (See Table 17.1 for numerical values of the fission cross section σ_f and the uranium density ρ.) Although λ sets a crude scale for the dimensions, it does not alone determine the critical radius R_c. For a reasonable estimate of R_c, it is necessary to treat the geome-

---

[8] At low doses, where one is estimating the chances of producing latent cancers, the absorbed dose in grays and the dose equivalent in sieverts are numerically equal for low-LET radiation. At high doses, often only the absorbed dose in grays is used (see Section 3.2.2).

Table 17.1. Properties of fissile materials for nuclear weapons: fission cross section, neutrons per fission, mean free path at 1 MeV, and critical mass and radius.

Material	$E_n = 1$ MeV σ_f (b)	ν	Density (g/cm^3)	λ (cm)	No Reflector M_c (kg)[a]	R_c (cm)	With Reflector M_c (kg)[b]	R_c (cm)
^{235}U	1.22	2.52	19	17	52	8.7	13–25	~6
^{239}Pu	1.73	2.95	19.6[c]	12	10	5.0	5–10	~4
^{233}U	1.9	2.53	19	11	16	5.9	5–10	~4

[a]Values of M_c identical to those listed here are given in Ref. [3] for ^{235}U, specified as "weapons-grade" (94% ^{235}U), and for ^{239}Pu, presumably in an isotopically pure form.

[b]The critical mass depends upon the reflector material and configuration. Critical masses as low as 4 kg are cited for ^{239}Pu and ^{233}U in Ref. [13, p. 908]. In Ref. [11], the critical masses for 100% pure ^{235}U and ^{239}Pu, in each case surrounded by a thick natural uranium reflector, are given as 15 kg and 4.4 kg, respectively.

[c]This density is for plutonium metal in the α-phase; for the other phase most relevant to plutonium explosives, the δ-phase, the density is 15.7 g/cm^3 [3].

Sources: σ_f and ν data for ^{235}U and ^{239}Pu are from Ref. [4, p. 20]; σ_f and ν data for ^{233}U are from Ref. [12]. M_c data are from Ref. [10, p. 24].

try in greater detail and take into account two crucial nuclear parameters: the number of neutrons produced per fission, ν, and the cross sections for elastic and inelastic scattering of neutrons. Larger ν means lower R_c, all other things being equal, and scattering, on average, increases the path length traversed by a neutron before it can escape and, therefore, decreases R_c. An approximate calculation for ^{235}U gives a critical radius R_c of about 9 cm [4, p. 74], equal to roughly one-half of the mean free path λ. The corresponding uranium mass is $M_c = 4\pi\rho R_c^3/3 \approx 60$ kg.

More accurate values of the critical mass for ^{235}U, along with values for ^{239}Pu and ^{233}U, are given in Table 17.1, assuming the fissile material to be isotopically pure, or almost so, and in metallic form. Results are given for a bare device and for one with a surrounding reflector, also known as a "tamper." The reflector returns escaping neutrons to the fissile volume by one or more scattering events; this can reduce the critical mass by a factor of 2 or 3 (see Table 17.1). Many materials can be used for the reflector (e.g., uranium and tungsten). Reflectors cited in discussions of weapons include examples with thicknesses of 4 cm of uranium [10, p. 24] and 15 cm of uranium [11, p. 412]. In addition to reflecting neutrons, the mass of the reflector (or tamper) acts to slow the dispersion of material after the explosion, thereby sustaining criticality for a longer period of time.

The critical mass for ^{239}Pu is considerably less than that for ^{235}U, a consequence of the former's higher fission cross section, σ_f, and higher number of neutrons per fission, ν. The critical mass is also small for ^{233}U, but there is no public record that such bombs have been built.

Critical Mass and Implosion

If the density of the material is increased through compression, the critical mass is reduced below the values given in Table 17.1. The mean free path for fission, λ and, therefore, approximately R_c, is inversely proportional to the density ρ. For a critical mass, the volume therefore varies as ρ^{-3} and the mass as ρ^{-2}. This is the principle underlying implosion bombs. Although the actual masses made possible by this approach remain secret, arms experts often suggest the importance of this high compression. For example, Theodore Taylor, a physicist with extensive experience in the field, cautions:

> It should not be concluded that the minimum amount of ^{235}U from which a fission explosive can be made is in the range of 10–20 kg, corresponding to the critical mass at normal density in a reasonably good neutron reflector. It is well known that "implosion" types of fission explosives achieve supercriticality by compression of the core substantially above normal density.... How little fissionable material can be used in a practical fission explosive depends on the knowledge, ingenuity, and skills of the explosive designers and fabricators. [11, p. 412]

Of course, the same considerations apply to plutonium bombs. In the same vein, a group of nongovernmental arms experts state: "Most weapons in the U.S. arsenal are believed to use only a fraction of a critical mass (at normal density)" [10, p. 25].

Nominal Numbers for the Critical Mass

The actual critical mass for nuclear weapons cannot be precisely stated. It depends on design variables, such as the compression achieved in implosion, the tamper used, and the isotopic purity of the material. Different designs give different results, and the nature of advanced designs and the resulting sizes are not made public by weapons builders. However, it is useful to have a nominal number to provide an approximate measure of the amount of material needed. One source suggests the following nominal numbers for pure material in metallic form: 10–11 kg for ^{235}U and 5 kg for ^{239}Pu [13, p. 907]. The numbers are useful for purposes of qualitative orientation, even if advanced technology may make possible the building of weapons with less material.

The possibility of making bombs with considerably smaller amounts of material has been suggested in a study published by the National Resources Defense Council (NRDC), assuming that the bomb builders have sufficient technical capability and a willingness to settle for a relatively small bomb [14]. The amounts quoted in the previous paragraph correspond, in the NRDC summary, to the amounts needed for a bomb with a 10-kt yield built using "low technology." On the other hand, according to the NRDC authors, as little as 1 kg of ^{239}Pu or 2.5 kg of uranium highly enriched in ^{235}U would suffice for a 1-kt bomb if the builders had a "high technical capacity." One important factor in reducing the required mass is the achievement of high compression during the implosion of the bomb material.

The masses suggested in the NRDC study are on the low end of published estimates. A somewhat higher mass is implied in a 1994 National Academy of Sciences Report on plutonium handling, which states that "[S]everal kilograms of separated weapons-grade plutonium...would be enough to build a nuclear weapon [15, p. 29].

Settling on a best estimate for the amount of material needed to produce a bomb of a given size at a given level of technology requires information that is currently classified. Furthermore, designations such as "high" and "low" technical capacity are not well defined. Regardless of which of the published numbers are used, it is clear that highly damaging weapons can be made with relatively small amounts of fissionable material. For the present purposes, it suffices to use the nominal numbers mentioned earlier: about 10 kg for weapons-grade uranium (although considerably larger numbers are often employed in the literature) and 5 kg for weapons-grade plutonium.

17.2.4 Buildup of a Chain Reaction

In a nuclear weapon, dispersion of the fuel begins before the developing chain reaction reaches its maximum design level. For an effective bomb, this buildup should occur in a time that is short compared to the time for appreciable disassembly of the material. This means that the time between successive fission generations must be short. The chain reaction in a bomb therefore relies on fast neutrons, rather than on the thermal neutrons that propagate the chain reaction in most of today's nuclear reactors.[9]

The time rate of change in the number of fission neutrons N present in the system is

$$\frac{dN}{dt} = \frac{k - \beta - 1}{\tau} N = \alpha(t) N, \tag{17.1}$$

where k is the effective multiplication factor, β is the delayed neutron fraction (see Section 6.3.2), τ is the mean time between successive fission generations, and $\alpha = (k - \beta - 1)/\tau$.[10] For a reactor, the effective multiplication factor k is close to unity. For a bomb, it will be larger. As the configuration of the bomb

---

[9] The discussion in this section draws to a considerable extent on *The Los Alamos Primer*, which originally appeared as notes of lectures delivered at Los Alamos by Robert Serber during World War II and was reissued in 1992 with explanatory notes and emendations by Serber [4]. It provides a simple introduction to the physics of nuclear weapons.

[10] This is essentially the same as Eq. (7.14), with some minor changes in notation. The delayed neutron fraction β is subtracted because the explosion is over before any appreciable number of delayed neutrons are emitted.

material changes during implosion, k and α vary with time, as suggested by the notation in Eq. (17.1).

The mean time between successive fissions is $\tau = \lambda/v$, where λ is the mean free path of the neutrons for fission and v is the average neutron velocity. A typical fission neutron has an energy in the neighborhood of 1 MeV and, therefore, a velocity of 1.4×10^9 cm/s. As indicated in Section 17.2.3, $\lambda = 17$ cm for a 1-MeV neutron in ^{235}U. Therefore, the time between successive fission generations in ^{235}U is $\tau \approx 10^{-8}$ s. Similar calculations for ^{239}Pu also give $\tau \approx 10^{-8}$ s.[11]

It follows from Eq. (17.1) that the fission rate in a chain reaction rises as $e^{\alpha t}$. For example, the rate doubles in a time t if $e^{\alpha t} = 2$ or $\alpha t = 0.693$.[12] In 1 kg of fissile material, there are about 2.5×10^{24} nuclei. If one assumes—as a simplification made for qualitative orientation—that there is a doubling in each generation, it would take 80 generations to go from one fission in the first generation to 1.2×10^{24} fissions in the last, and thereby consume all the fuel. It would take only a few more generations to fission 10 times this amount or to make up for a slightly lower multiplication rate. Thus, under these conditions, the chain reaction would develop completely in a time on the order of one-millionth of a second—a time short compared to the time for the material to physically disperse and become subcritical.

17.3 Uranium and Nuclear Weapons

It is considerably simpler to make a bomb using enriched uranium than to make one using plutonium, and uranium may be becoming the material of choice for countries or groups that want to build a bomb with minimal effort and chance of detection. The fission cross section, σ_f, and the average number of neutrons per fission, ν, are both somewhat smaller for ^{235}U than for ^{239}Pu, making the critical mass larger (see Table 17.1). However, premature detonation caused by neutrons from spontaneous fission (SF)—discussed in Section 17.4 for plutonium—is not a problem with uranium, because the SF rates are low for ^{235}U and ^{238}U. Therefore, a gun-assembly uranium bomb is practical. With a greater technical effort, it is possible to make a uranium implosion weapon, and less uranium is needed than for a gun-assembly bomb. For example, China's first nuclear test weapon was a ^{235}U implosion bomb (see Section 18.2.1).

---

[11] A time of 10^{-8} s was termed a "shake" in the jargon adopted early in the U.S. atomic bomb program.

[12] Neglecting β, which is small, the doubling time is $0.693\,\tau/(k-1)$. If the multiplication factor k is 2, the doubling time is $0.693\,\tau$. The relationship between the doubling time and τ is analogous to the relationship between the half-life and mean life in radioactive decay.

Enriched uranium is used for a variety of purposes—including electric power generation, ship propulsion, and bombs. The terms *low-enriched uranium* (LEU) and *highly enriched uranium* (HEU) denote ^{235}U enrichments of 0.71–20% and greater than 20%, respectively; HEU with an enrichment above 90% is *weapons-grade uranium* [16, p. 12]. Uranium bombs can be made over a wide range of enrichments, but the mass of uranium required is greater for lower enrichment. For example, with a good reflector, Taylor indicates that the critical mass for 60% enrichment is 22 kg of ^{235}U (37 kg of U), whereas only 15 kg of ^{235}U (and U) are required at 100% enrichment [11, p. 412]. Not only is very highly enriched uranium preferable for building a compact bomb, but it requires more separative work to obtain a 37-kg critical mass at 60% enrichment than to obtain the smaller critical mass (roughly 18 kg) at 90% enrichment.[13] Thus, high enrichments are used in ^{235}U bombs. The bomb at Hiroshima was apparently built with uranium enriched to about 89% [4, p. 22]. Nuclear weapons now are reported to use uranium with an enrichment of at least 90% [17, p. 417].[14]

Uranium bombs could be obtained by proliferators or terrorists by theft of loosely guarded weapons or of highly enriched uranium that could be used to make a bomb. At a higher level of technical effort, if natural or low-enriched uranium can be procured, it can serve as the feedstock for a uranium-enrichment facility. The construction and operation of an enrichment facility would probably have to be done with government leadership or at least sanction, rather than by action of a terrorist group operating in defiance of the local government.

From the standpoint of countering proliferation, disposing of surplus weapons-grade uranium is much simpler than disposing of surplus weapons-grade plutonium. As discussed in Section 18.3.3, there has been considerable indecision in selecting the optimal way for disposing of plutonium from dismantled nuclear weapons. For uranium, on the other hand, it is a simple matter. The uranium can be "denatured" by mixing it with uranium that has a low concentration of ^{235}U and the resulting LEU can be used as reactor fuel. There are ample supplies of uranium that can be used for denaturing—depleted uranium tails from the enrichment process, uranium extracted from spent fuel, or natural uranium. In consequence, considerable amounts of surplus HEU from dismantled nuclear weapons have already been handled in this way (see Section 18.3.4). There are no comparable sources of material that could be used to denature plutonium.

---

[13] At 60% enrichment, the separative work is 125 separative work units (SWU)/kg or 4600 SWU for 37 kg (assuming natural uranium feed and 0.003% enriched tails); this may be compared to 193 SWU/kg and about 3500 SWU total for 18 kg of 90% enriched uranium. (See Section 9.2.2 for a discussion of separative work and defining equations.)

[14] Discrepancies in the amounts of uranium reported as necessary for a critical mass probably are due to differences in assumed design details.

17.4 Plutonium and Nuclear Weapons

17.4.1 Explosive Properties of Plutonium

Different Grades of Plutonium

Making a nuclear bomb with plutonium requires an expertly arranged implosion and is more difficult than making a gun-type uranium bomb. The techniques to do this were developed during World War II and it can be viewed as a standard method, preferably using plutonium with a high isotopic concentration of ^{239}Pu. At one time, there was a question as to whether it would be possible to manufacture a workable bomb using plutonium that has a relatively high concentration of ^{240}Pu. It is now known that it is possible to do so, although with some difficulty. In view of the prominence of plutonium issues in the consideration of proliferation of nuclear weapons, it is of interest to examine some of the technical details involved.

The isotope ^{240}Pu has a half-life of 6564 years. Its primary decay mode is by alpha-particle emission, but it also sometimes decays by spontaneous fission. This fission produces neutrons that may cause premature initiation of a chain reaction in a bomb before the fissile material has been fully compressed. This is known as *predetonation*. With predetonation, the weapon gives a much smaller explosion than designed; in other words, it "fizzles."

During the World War II atomic bomb program, the discovery that ^{240}Pu had a high rate of spontaneous fission came as a surprise and appeared briefly to be a major threat to the development of a ^{239}Pu bomb.[15] Part of the solution to this problem was simple. If the fuel is withdrawn from the reactor before long exposure, there is insufficient time for a large concentration of ^{240}Pu to form; this results in "weapons-grade" plutonium. The plutonium from a reactor, operated in a normal cycle, has a much higher ^{240}Pu concentration and is termed "reactor-grade" plutonium; it is unsuitable for the production of efficient, reliable weapons. However, it is well accepted that sophisticated designers could make some sort of weapon even with reactor-grade plutonium (see, e.g. Ref. [18, p. 259]).

Information on the issues involved has gradually entered the unclassified open literature. A brief review by J. Carson Mark, formerly director of the Theoretical Division at Los Alamos National Laboratory, has put some of the key information in a generally available form [3]. This subsection draws heavily from his exposition. As a preliminary, some information on different grades of plutonium is summarized in Table 17.2, based on information from Mark's paper [3].

The ^{240}Pu arises from neutron capture in ^{239}Pu. Initially, starting with enriched uranium fuel, the ^{239}Pu content rises linearly with fuel burnup and the ^{240}Pu content rises quadratically. As the burnup progresses, the description of the rates of rise becomes less simple, but the ratio of ^{240}Pu to ^{239}Pu

---

[15] See, e.g., Ref. [4, p. 55], for a brief discussion of this surprise.

Table 17.2. Properties of different grades of plutonium: isotopic abundances, neutron emission from spontaneous fission (SF), and decay heat from radioactive decay.

	Grade of Plutonium		
	Super	Weapons	Reactor[a]
Fraction (by mass)			
^{238}Pu		0.0001	0.013
^{239}Pu	0.98	0.938	0.603
^{240}Pu	0.02	0.058	0.243
^{241}Pu[b]		0.0035	0.091
^{242}Pu		0.0002	0.050
SF neutrons (per g-s)	20	66	360
Decay heat (W/kg)	2.0	2.3	10.5

[a]For LWR fuel with a 33-GWd/t burnup, reprocessed after 10 years.
[b]Includes ^{241}Am.
Source: Ref. [3].

continues to increase. For the PWR fuel considered in Ref. [3], the burnup is 33 GWd/t and the isotopic abundance of ^{240}Pu is about 24% (see Table 17.2), which is typical of the abundance in present-day spent fuel (see Table 9.2).

The neutron emission rate from spontaneous fission is about 910 $(g\text{-}s)^{-1}$ for ^{240}Pu [3, p. 115], contributing about 220 neutrons per second per gram of reactor plutonium. ^{238}Pu and ^{242}Pu also have sizable spontaneous fission rates and the total neutron emission rate is 360 $(g\text{-}s)^{-1}$ for reactor plutonium [3, p. 122]. If one assumes a bomb with a total plutonium mass of 6 kg, this means an emission of 2×10^6 neutrons per second. The number of neutrons from spontaneous fission is considerably less in weapons-grade plutonium and still less in "supergrade" plutonium (see Table 17.2), and the probabilities of premature detonation are correspondingly reduced.

Although ^{240}Pu creates problems due to spontaneous fission, it does not greatly increase the required mass of ^{239}Pu for criticality. In this, there is some difference between ^{240}Pu in a ^{239}Pu bomb and ^{238}U in a ^{235}U or ^{233}U bomb. For ^{238}U, the fission cross section even at 1 MeV is low and a fast-neutron chain reaction in ^{238}U is impossible.[16] In contrast, the fission cross section for ^{240}Pu, which is very small at thermal energies, is about 1.5 b at 1 MeV and a chain reaction is possible in pure ^{240}Pu [3, p. 115]. Because ^{238}U contributes little to the fission yield but absorbs neutrons, a ^{238}U contaminant means that a greater mass of ^{235}U is required for criticality. However, with its substantial fission yield, a ^{240}Pu contaminant raises only slightly the required mass of ^{239}Pu [11, p. 412]. For example, one relatively recent estimate specifies

---

[16] A simple early discussion of this point is reprinted in Ref. [4, p. 21].

that a bomb can be built with either 5 kg of weapons-grade plutonium or 8 kg of reactor-grade plutonium, presuming the implosion to be correctly implemented [19, Appendix 3].

Predetonation of Bombs and Energy Yield

The second part of the solution to the problem of spontaneous fission was to bring together a critical mass very quickly (see, e.g., Ref. [4, p. 55]). This was accomplished by the implosion method, in which a plutonium sphere is symmetrically compressed by a shock wave caused by detonating chemical explosives placed around the sphere. The critical mass is inversely proportional to the square of the density, and compression can change the mass from subcritical to supercritical.

In a typical bomb, the implosion shock wave has a speed of about 5000 m/s and the implosion proceeds at a rate such that the bomb goes from initial supercriticality to full compression in a period of about 10^{-5} s [3, p. 118]. As the plutonium is compressed, the multiplication factor k rises and a chain reaction, if initiated, develops more rapidly. In a "ideal" bomb, ignition is triggered when the material is fully compressed, and there is then a rapid chain reaction that encompasses a reasonably large fraction of the plutonium.[17]

However, neutrons from spontaneous fission can trigger a chain reaction before the plutonium is fully compressed, at any time after the material becomes supercritical ($k > 1$). The energy generated by fission then reverses the implosion, the system expands, the multiplication factor drops, and the chain reaction eventually ceases. If the reversal occurs relatively early, when the compression and multiplication factors are low, the chain reaction will not be rapid enough to have embraced much of the plutonium before it is terminated.

According to Mark, the disassembly of the bomb begins when there have been about e^{45} fissions (3.5×10^{19} fissions) [3, p. 117]. The corresponding energy release is about 10^9 J, equivalent to 0.2 kg of TNT. However, the fission rate continues to increase during the disassembly, as long as $k > 1$. The total fission yield therefore comes primarily from fission events that occur after the expansion begins. It is determined by the magnitude of the parameter α [see Eq. (17.1)] at the turning point. Based on a simplified model for the rate of increase of k and α during compression, the minimum value of α at the turning point can be found, corresponding to smallest "fizzle." For this case, the estimated fizzle yield is about 2.7% of the full design yield. For a 20-kt bomb, the minimum fizzle yield is about 0.5 kt, which is still a sizable explosion.

---

[17] Triggering can be produced, for example, by breaking a thin barrier between a radioactive source of alpha particles and an adjacent beryllium target. Reactions between the alpha particles and the ^{9}Be nuclei will produce a burst of neutrons.

Dependence of Outcome on ^{240}Pu Content and Implosion Speed

In this analysis, the minimum fizzle yield is independent of the fraction of ^{240}Pu in the plutonium. However, the ^{240}Pu content does affect the probability of predetonation. In reactor-grade plutonium, with 360 spontaneous fission neutrons per gram-second, there will be 2×10^6 spontaneous fission neutrons per second in a 6-kg bomb.[18] This makes it likely that a chain reaction will be initiated within the first microsecond, assuring predetonation.

With these general assumptions, a more detailed calculation shows that with reactor-grade plutonium, there is a 70% chance that the yield will be less than 10% of the full design yield (or nominal yield) and a near certainty that the yield will be less than one-half the full yield (see Appendix of Ref. [3]). For weapons-grade plutonium with 6% ^{240}Pu, on the other hand, the neutron rate is only 3×10^5 neutrons per second, and there is a 50% chance of achieving more than 40% of the design yield. With 1% ^{240}Pu, there is a 90% chance that the yield will be more than 40% of the design yield and an 80% chance of it being the full design yield.[19]

The above discussion brings out the qualitative factors relating to predetonation, but it is doubtful that it can be used for quantitative estimates of yield. One important consideration is the speed of implosion. For faster implosions, the probability of obtaining the full design yield increases. In addition, the fizzle yield itself will be greater, which may account for typical statements in the literature suggesting that fizzle yields will be "in the kiloton range" [18, p. 259]. An authoritative National Academy of Sciences report has stated in regard to reactor-grade plutonium:

> [E]ven with relatively simple designs such as that used in the Nagasaki weapon—which are within the capabilities of many nations and possibly some subnational groups—nuclear explosives could be constructed that would be assured of having yields of at least 1 or 2 kilotons. Using more sophisticated designs, reactor-grade plutonium could be used for weapons having considerably higher minimum yields. [15, p. 4]

For slower implosions, predetonation is more probable and the explosive yield will be less. According to Mark (presumably taking somewhat representative velocities), at the still slower speeds of a gun-type weapon in which the critical mass is created with an assembly velocity of 300 m/s, the minimum fissile yield would be at least a factor of 30 less than that for an implosion weapon with a shock-wave velocity of 5000 m/s [3, p. 119]. In addition, the high probability of predetonation with a gun assembly makes it unlikely that

---

[18] The 6-kg size was used in Ref. [3, p. 127], as representing an early estimate for the Trinity device.

[19] The plutonium used for the Trinity test has been inferred to have had about 1% ^{240}Pu, on the basis of predictions by Robert Oppenheimer of the yields and the sort of analysis outlined here [3, p. 127].

the yield would much exceed the minimum. That is why plutonium weapons use implosion.

17.4.2 Reactor-Grade Plutonium as a Weapons Material

Difficulties in Use of Reactor-Grade Plutonium

It is clear, from the above-reviewed arguments and from a long history of less detailed but authoritative statements about the possibilities, that reactor-grade plutonium can be used to make an explosive device that would release a substantial amount of energy. This would be enough to create an explosion that would do great damage due to the blast itself, the heat and radiation produced in the chain reaction, and the radionuclides dispersed in its aftermath.

It does not seem very likely that such a project would be undertaken by a national government. A country with access to reactor-grade plutonium could probably arrange to have a domestic reactor irradiate fuel for a shorter time, yielding weapons-grade plutonium. This would require a deliberate decision by the national authorities, but one could imagine such a situation arising after a change of government or changes in relations with neighboring countries.

A large terrorist organization might be a more likely user of reactor-grade plutonium. Such an organization, if it wanted plutonium, might take whatever was accessible. Further, although uncertainties in the explosive yield could make a weapon unsuitable for national military purposes, such uncertainties would probably be less important for terrorist purposes.

However, although possible in principle, an effort to turn reactor-grade plutonium into an explosive device faces formidable obstacles:

- If the plutonium has not already been separated from the fission products in the spent fuel, a chemical separation must be carried out with very radioactive material. A small reprocessing facility would be needed to accomplish this.
- The plutonium must be carefully machined and shaped, and then assembled in a subcritical geometry together with the surrounding reflector. A mistake at this point could create a microfizzle, probably disastrous to the people making the mistake.
- The explosive must be arranged properly to obtain a rapid and symmetric implosion. If the implosion is not rapid, the yield will be small.
- In assembling the device, precautions must be taken to avoid overheating the surrounding explosive. Reactor-grade plutonium produces heat at a rate of 10.5 W/kg (see Table 17.2). A tight blanket of chemical explosives could cause the system to overheat, with a possible chemical explosion resulting [3].

The barrier provided by the high level of radioactivity of the spent fuel is often taken to be an adequate protection against the use of this material.

For example, 15 years after a fuel assembly is removed from a reactor, the activity at a distance of 1 m from its center is about 20 Gy/h [15, p. 151].[20] A lethal dose is about 4 Gy. Therefore, handling this material without elaborate equipment would be a fatal activity, although this might not be a decisive inhibition.

A typical traditional assessment was given by Frank Barnaby, past executive secretary of the Pugwash Conferences on Science and World Affairs:

> Spent reactor fuel elements are so radioactive as to be self-protecting. It would [be] extremely hazardous for people to handle them without large remote-handling equipment. However, when the fuel elements have been reprocessed and the plutonium separated from the radioactive fission products, the plutonium is in a form that can be relatively easily handled [20, p. 1]

This implies a high threshold against diversion of spent fuel. The radioactivity was thought to be a sufficient safeguard against any terrorist group, including a subnational group that lacks a sure geographical base. With the emergence of suicidal attacks as a common terrorist option, the equation is somewhat changed, although the radiation might be so quickly disabling as to thwart even suicidal groups. A national effort could surmount this threshold by developing reprocessing facilities with remote-handling capabilities. It could also, if desired, improve the effectiveness of weapons derived from a stock of reactor-grade plutonium by isotopic enrichment of the material—although here, as discussed earlier, it would be simpler to obtain spent fuel that had been irradiated for only a short time.

Reprocessing of Spent Fuel

The difficulties of making nuclear weapons from reactor plutonium are significantly reduced if the reactor fuel has already been reprocessed and the plutonium extracted. This is an argument against reprocessing. Without reprocessing, there is no separated reactor plutonium available for theft or diversion.

Reprocessing has been opposed in the context both of possible weapons development by nations and theft by terrorist groups. Any country with facilities for reprocessing, gained as part of a normal civilian fuel cycle, has the facilities to extract plutonium for weapons production. Thus, civilian reprocessing can be used to conceal clandestine reprocessing for weapons. Even if there is no weapons program initially, having reprocessing facilities creates the opportunity to move quickly into the weapons business. For these reasons, many informed observers who are not otherwise opposed to nuclear power

---

[20] At this time, the external radiation is dominated by gamma rays from ^{137}Cs. Therefore, the decrease in radiation dose with time is determined by the decay rate of ^{137}Cs, which has a half-life of 30.1 years.

have opposed the reprocessing of spent fuel and the extraction of plutonium. This contributed to the U.S. decision to forego reprocessing, as discussed in Section 9.4.2.[21]

While proliferation concerns helped to stop commercial reprocessing in the United States, they have not had a decisive effect everywhere. There are substantial reprocessing facilities for commercial spent fuel in France, the United Kingdom, Russia, Japan, and India, and in some cases, reprocessing facilities are being expanded.[22] It appears improbable that reprocessing will be abandoned in these countries.

Pending completion of Japan's expanded reprocessing capabilities, some of the spent fuel from Japanese reactors has been reprocessed in France and the United Kingdom and the plutonium was shipped back to Japan in the form of plutonium oxide or MOX fuel [21].

Japan has been strong in its assurances that the plutonium obtained from reactors will not be used for weapons and, to date, appears to have refrained from weapons development. However, it is uncertain for how long Japan will be content to be protected by a U.S. nuclear umbrella if it is faced with nuclear weapons in China and North Korea. Although a plutonium stockpile would speed the pace of a program to develop weapons, even with no prior stockpile Japan has the personnel and facilities to develop nuclear weapons quite quickly, should it choose to do so.

There is a contrary position on the issue of plutonium reprocessing and proliferation. In this view, the plutonium in spent fuel will remain a permanent danger *unless* the fuel is reprocessed and the plutonium consumed as reactor fuel (see, e.g., Ref. [22]). In fact, as the fission products decay, the spent fuel becomes less hazardous and thus more valuable to a potential terrorist or national weapons program. However, even if one were to accept this view, there appears to be no anti-proliferation reason to rush into reprocessing as a means of destroying plutonium, because the fuel will remain self-protected for many decades.

Even for separated reactor-grade plutonium, the difficulties outlined earlier make it most unlikely that the high school student of legend (who could fashion a bomb in the basement) exists, or would long exist if he or she made the attempt. However, the necessary equipment and expertise could presumably be brought together by a well-organized terrorist group. It might be argued that even for such a group, it would be irrational to proceed with a plutonium bomb when there are simpler alternatives for major destruction and terror. However, it is not prudent to rely on the rationality of terrorist groups. It is therefore important to have robust security measures that will prevent any diversion of separated plutonium, reactor grade or otherwise. (Terrorists threats are discussed further in Section 17.5.)

---

[21] Economic factors and difficulties with specific reprocessing facilities contributed to this decision, but these were encountered in an atmosphere where an influential body of opinion welcomed, rather than regretted, the difficulties.

[22] Some details have been given in Table 9.4 in Section 9.4.2.

17.4.3 Production of Plutonium in Reactors

The production of plutonium is accomplished most effectively if the reactor has a high conversion coefficient. This means the use of heavy water or graphite as the moderator. These are better moderators than light water for plutonium production, because their cross sections for neutron capture are low and their relatively slow moderation rates gives more time for neutron capture in ^{238}U (see Section 8.3.2). Most weapons programs have obtained their plutonium from graphite-moderated reactors; the U.S. program used both graphite-moderated reactors (at Hanford in Washington) and heavy water moderated reactors (at Savannah River in South Carolina).

However, plutonium production is unavoidable in any reactor that uses ^{238}U as a fertile fuel. Thus, any reactor is a potential source of plutonium for weapons, even if ostensibly being used for other purposes. For production of weapons-grade plutonium in a commercial reactor, standard fuel would have to be removed after a burnup period that is much shorter than normal. But even plutonium from fuel with the normal high burnup can be used for weapons, as discussed earlier. Alternatively, special uranium "targets" can be placed at selected locations in or near the reactor where they are irradiated by the neutron flux, producing weapons-grade plutonium.

A summary of plutonium production capabilities of different reactors (without special uranium targets) is given in Table 17.3, based on data from Ref. [16]. The indicated annual production for commercial operation corresponds to a 1-GWe reactor operating at a capacity factor of 100%. At a more attainable capacity factor of 80%, in the neighborhood of 300–600 kg of plutonium would be produced annually, depending on the type of reactor, corresponding to a production rate of about 0.3–0.6 kg of plutonium per GWd(t).[23] The lower numbers pertain to LWRs, the dominant reactors in the world today. Of course, the product of commercial operation is normally reactor-grade plutonium, which is not ideally suited to bomb production. The critical mass for an explosive device using reactor-grade plutonium is under 10 kg (see Section 17.4.1). Thus, each year of normal operation of a 1-GWe LWR produces enough plutonium for 30 or more small nuclear bombs.

For reliable weapons, it is desirable to have a low ^{240}Pu fraction (i.e., weapons-grade plutonium). In principle, this could be accomplished by isotopic enrichment of plutonium that has been chemically extracted from the spent fuel. A simpler approach is to reduce the burnup of the fuel.[24] A light

---

[23] In Table 17.3, the graphite reactor is fueled with natural uranium and the LWR with enriched uranium, giving the former a lower burnup, less destruction of plutonium, and a higher plutonium output for a given energy output. (The HWR fuel was not specified in the data used for Table 17.3, but probably also was natural uranium.)

[24] In addition to giving a higher percentage of ^{239}Pu in the plutonium, low burnup gives a higher plutonium output per unit energy generated than in a normal fuel cycle, because there is less destruction of ^{239}Pu by fission or neutron capture.

Table 17.3. Plutonium production rates for reactors operating in a commercial mode and for production of weapons-grade plutonium.

Moderator Material	Commercial Operation			Production of WG Plutonium		
	Burnup (GWd/t)	Prod. Rate [kg/GWd(t)]	Annual Prod.[a] (kg/GWe)	Burnup (GWd/t)	Prod. Rate [kg/GWd(t)]	Annual Prod.[a] (kg/MWt)
Light water[b]	30	0.29	330	1	0.5	0.18
Heavy water	7.5	0.50	630	1	0.9	0.33
Graphite	4	0.63	815	0.2	0.95	0.35
Graphite				1	0.86	0.31

[a]Calculated for capacities of 1 GWe (commercial) and 1 MWt (plutonium production), with an assumed 100% capacity factor.

[b]These numbers are approximate weighted averages of values reported separately in Ref. [16] for commercial PWRs and BWRs.

Source: Derived from data in Ref. [16, pp. 462–3 and 473], with burnup rates and thermal efficiencies used there.

water reactor at low burnup could be used for this purpose, but it is more efficient to use a graphite or heavy water reactor, as seen in Table 17.3.[25]

The precise amount and isotopic purity of the plutonium obtained from a dedicated reactor depends on the capacity factor achieved, the average burnup, and the location of the fuel in the reactor.[26] For a graphite-moderated reactor, the output approaches 1 kg of plutonium per gigawatt-day thermal [GWd(t)] or 1 g of plutonium per MWd(t). Thus, even a small graphite-moderated reactor can in a few years produce enough weapons-grade plutonium for a modest nuclear arsenal, at a rate of about 1 bomb per 5000 MWd(t) of operation. To achieve "super" and "weapons" grades of plutonium, as defined in Table 17.2, the burnups should be about 0.3 and 0.9 GWd/t, respectively [16, p. 463].

Some rough rules-of-thumb are useful in evaluating the potential of dedicated graphite- or heavy water-moderated reactors:

♦ About 1 g of plutonium is produced per megawatt-day (thermal).
♦ About 0.25 kg of plutonium is produced annually per megawatt of thermal capacity (at an 80% capacity factor). (Thus, a 20-MWt reactor has a production potential of about 1 bomb per year.)

17.5 Terrorist Threats

17.5.1 The Range of Terrorist Threats

Potential terrorist attacks, including attacks using nuclear bombs or carried out against nuclear facilities, have long been a matter of concern. However, the events of September 11, 2001 have greatly heightened the sense of concern. The emergence of Al Qaeda has demonstrated the existence of a large terrorist group that appears to have access to skilled and dedicated personnel and ample financial resources. The willingness of individual terrorists to die in their attacks broadens the scope of their potential operations. Thus, what may have seemed to many people to be a vague threat—or in a some cases, a threat with specific and limited goals—has become concrete and open-ended in scope.[27] A face had been put on what had previously been a faceless danger.

The impact has been particularly great in the United States, which had relatively little prior experience of terrorist attacks. The Oklahoma City bombing in 1995 was a traumatic event and smaller attacks such as those of the

---

[25] Plutonium with a high percentage of ^{239}Pu can also be obtained by placing uranium as a "blanket" near a reactor and relying on capture of escaping neutrons in ^{238}U.

[26] The burnup is not uniform over the volume of the reactor.

[27] The Irish Republican Army is an example of a group that inflicted major damage in Northern Ireland and even in London, but its goals were limited, and it was generally assumed that there were limits to the scale of damage that it desired to inflict.

Unabomber were taken as very serious matters. Nonetheless, these were generally considered to be the actions of isolated individuals, and not harbingers of worse things to happen in the future. The actions and the verbal threats of Al Qaeda have changed the frame of reference in the United States and elsewhere.

Before considering specific features of nuclear and radiological threats, it may be appropriate to consider them in a more general perspective. An examination of almost any individual aspect of society would probably show that it could not withstand a determined attack. Every activity and facility has some degree of vulnerability. Inventive terrorists have undoubtedly thought of innumerable potential targets. We list a few examples, but it must be assumed that the actual list is far broader:

♦ *Places where people assemble.* Movie theaters, football stadia, cruise ships, and legislative chambers, for example, could be attractive targets for attacks by bombs, missiles, automatic weapons, and even airplane crashes. The Chechen action in Moscow in October 2002, in which an armed group made captives of an opera house audience, illustrates an additional option: the taking of many hostages.

♦ *Choke points in transportation routes.* The destruction of bridges or the blocking of tunnels could severely disrupt routine life in the affected region.

♦ *Symbolic targets.* The destruction of the Statue of Liberty or the Washington Monument might cause few casualties, but would still represent a major psychological "victory" for the attackers.

♦ *National energy carriers.* Electric transmission lines and transformers, gas pipelines, and oil tankers offer diverse targets. Well-planned attacks could cripple parts of the economy and of everyday life.

♦ *Food and water supplies.* Poisons could be introduced into food and water supplies either on a large scale or in a seemingly random local manner, making the supplies unsafe and perhaps inhibiting their use to an extent greater than dictated by the actual danger.

♦ *Weapons of mass destruction.* Biological, chemical, or nuclear materials could be introduced into the environment—quietly or accompanied by violent explosions—in a manner that would kill or injure large numbers of people. The casualties that might be envisioned range from tens to tens of thousands, and conceivably much more.

Given the options envisaged in lists such as this, it is not clear how high a place nuclear terrorism occupies in planning by terrorist groups.

In the following subsections, we will consider threats involving nuclear materials and weapons. In doing so, we do not intend to suggest that nuclear threats represent the most likely terrorist targets or the most dangerous ones. We discuss them because they are important, whatever their place in the broader picture, and because the topic is a matter of widespread interest.

17.5.2 The Nature of the Nuclear Terrorist Threat

The influential book *Nuclear Power Issues and Choices* (NPIC)—published in 1977 as a report of a Ford Foundation-sponsored study—devoted a full chapter to "Nuclear Terrorism" [23]. The concerns that were reflected in NPIC about nuclear weapons proliferation and terrorism had already led President Ford in 1976 to call for a deferral of the reprocessing of spent nuclear fuel [24, p. 117]. Less than a year later, the new Carter administration decided to terminate support for reprocessing and it has been abandoned as part of the U.S. civilian fuel cycle, at least for the time being (see also Section 9.4.2). Assessments in the Overview chapter of NPIC remains pertinent today, some 25 years later:

> A particularly disturbing aspect of nuclear proliferation is that it could extend to subnational terrorist groups. Although a complicated nuclear weapon would be a more convenient target, a highly organized terrorist group might have the capability to fabricate a crude nuclear weapon from stolen plutonium or highly enriched uranium. Since neither of these materials is available in the present [U.S.] fuel cycle, this threat will only emerge if plutonium is reprocessed and recycled or if reactors requiring highly enriched uranium are introduced.

> ...Although it would not be as easy as sometimes suggested, it is conceivable that a well-organized group supported by knowledgeable individuals could construct a device that might have a yield equivalent to a few hundred tons of TNT....

> A terrorist group might sabotage a nuclear facility...The most serious target would be an operating nuclear reactor, where trained and knowledgeable saboteurs could cause a major accident....

> We believe that additional measures should be taken to reduce the possibility of terrorist acts to divert nuclear materials or sabotage facilities....

> Nuclear terrorism is international in scope. Terrorist acts in the United States could result from materials or devices seized abroad and smuggled into this country. The United States thus has a critical interest in the improvement of nuclear security and should encourage the development and implementation of effective physical security measures in all countries. [23, pp. 26–27]

NPIC was not a voice in the wilderness in its warnings. It reflected part of the conventional wisdom, although perhaps in a more intense form and certainly with a more negative view of reprocessing than shared by many nuclear advocates. In any case, it was universally assumed that nuclear facilities were to be guarded and that nuclear materials were to be kept out of unauthorized hands.

More recently, on October 24, 2001, with timing that seems something of a coincidence but which may also be seen as a sign of the general world situation, the U.S. National Council on Radiation Protection and Measurements (NCRP) issued NCRP Report No. 138 on *Management of Terrorist Events Involving Radioactive Material*. NCRP reports are long in the making, and this one was prepared prior to September 11, 2001. The audience for such reports is mostly a professional one, and this report was specifically addressed to people in and out of government who "share the responsibility to respond to this type of disaster" (i.e., ones that involve radiation dangers) [8, p. 3].

NCRP 138 focuses on coping with the consequences of a successful terrorist attack. However, the first priority is preventing such attacks and protections have long been in place—such as guards at nuclear power plants. Whether the protective measures taken before September 2001 would have sufficed against a determined attack cannot be established with any certainty.

Since September 11, 2001, new security measures have been taken in many areas, including enhanced security at nuclear reactors. It would be difficult to know whether or not these measures are at a level commensurate with the level of plausible challenges, even with access to all available classified information. Nonetheless, one can try to understand the nature of the dangers.

An instructive summary of the potential forms of nuclear terrorism is provided in a National Academy study *Making the Nation Safer: The Role of Science and Engineering Technology in Countering Terrorism*, published in 2002. It presents a matrix of possible radiological and nuclear threats. Chief among them are the following [25, Table 2]:

- *Nuclear bombs* that are smuggled into the country, either bombs that were originally state-owned or that are are devices "improvised" by the terrorists.
- *Radiological sources*, known as radiological dispersion devices or "dirty bombs," that are made of radioactive materials that are spread into the environment by a conventional chemical explosion.
- *Attacks on nuclear plants*, including the reactors themselves or the spent fuel at the reactor sites.

Each of these categories is considered in the following subsections.

17.5.3 Nuclear Bombs

The Magnitude of the Danger

The nuclear threat that is in a class of its own is the detonation of a nuclear fission bomb, to say nothing of a hydrogen bomb. The dimensions of the danger are brought out in the following paragraphs from a report published jointly by the Project on Managing the Atom (at Harvard's Kennedy School of Government) and by the Nuclear Threat Initiative:

> If stolen or built abroad, a nuclear bomb might be delivered to the United States, intact or in pieces, by ship or aircraft or truck,

or the materials could be smuggled in and the bomb constructed at the site of its intended use. Intercepting a smuggled nuclear weapon or the materials for one at the U.S. border would not be easy. The length of the border, the diversity of means of transport, and the ease of shielding the radiation from plutonium or highly enriched uranium all operate in favor of the terrorists. The huge volume of drugs successfully smuggled into this country provides an alarming reference point. [26, p. v]

The detonation of such a bomb in a U.S. (or any other) city would be a catastrophe almost beyond imagination. A 10-kt nuclear explosion (from a "small" tactical nuclear weapon from an existing arsenal or a well-executed terrorist design) would create a circle of near-total destruction perhaps 2 miles in diameter. Even a 1-kt "fizzle" from a badly executed terrorist bomb would have a diameter of destruction nearly half as big. If parked at the site of the World Trade Center, such a truck-bomb would level every building in the Wall Street financial area and destroy much of lower Manhattan. [26, p. 6.]

The damage done by the bomb would depend on its size, the location at which it exploded, and the fraction of the material in it that is fissioned. The above quotation provides one measure of scale. Another measure is given by the Hiroshima or Nagasaki bombs. In each of those cases, there were of the order of 100,000 deaths. However, whatever the details of bomb size or location, a nuclear explosion would be a devastating event.

Sources of Weapons and Weapons Materials

Terrorists might obtain, through theft or gift, a bomb built as part of a government-run nuclear weapons program. The above-mentioned *Making the Nation Safer* (MNS) study considered this possibility along with a broad array of other terrorist threats, both nuclear and non-nuclear. It assessed the level of danger of such diversion to be low for weapons made in the United States, Britain, China, France, and Israel because the weapons are well protected and, except for the United States, the inventories are relatively small [25, p. 42]. However, it judged the level of threat to be "medium" for Pakistan, India, and Russia—for Pakistan and India because the "political situation is unstable" and for Russia because of a "large number of weapons with poor inventory controls."

The report particularly pointed to the security of Pakistani weapons over the next several years as "problematical" [25, p. 40]. This assessment is understandable in view of the recent political history of Pakistan and the suspected presence in Pakistan of groups belonging to or sympathetic to Al Qaeda. There have also been widely expressed concerns in the United States about the security of the weapons and stockpiles of plutonium and highly enriched uranium that were under Soviet control before the collapse of the Soviet Union. In the words of the NAS report: "Theft or diversion of Russian nuclear weapons for terrorist use may represent a significant near-term threat to the United States, especially the theft or diversion of smaller, man-portable weapons" [25, p. 40].

An alternative for terrorists, if they cannot obtain a ready-made bomb, is to make one. A moderately large and technically capable group, operating with a secure base, could conceivably build a bomb if they had sufficient fissionable materials—again obtained by theft or gift. However obtained, a bomb could be delivered in a variety of ways, including missiles and by smuggling into the country. For the immediate future, it seems unlikely that any terrorist group would have the missile capabilities necessary to deliver a bomb to the United States. Further, if the missile is sent from a land base, its point of origin is likely to become known, giving the potential "host" country a powerful incentive to prevent the activity. On the other hand, a smuggled device—sometimes called a "suitcase bomb," although it could be delivered by a small plane or by ship—could have an anonymous sender.

The first line of defense is to keep fissionable materials out of the hands of unauthorized individuals or groups. Worldwide information on illicit traffic in nuclear materials is being compiled at the Center for International Security and Cooperation (CISAC) at Stanford University, in its Database on Nuclear Smuggling, Theft, and Orphan Radiation Sources (DSTO). Of course, the database can only include cases in which the traffic was detected. Access to the database is restricted, but some results have been reported by authors associated with CISAC [27].

They report a widespread pattern of "diversions" of nuclear materials from nuclear facilities in the decade from 1992 to 2002, involving primarily materials from the former Soviet Union. The reported troubling incidents and trends include the following:

♦ Three men who were trying to sell 3 kg of uranium enriched to 90% in ^{235}U were arrested by St. Petersburg police in 1994. This material was smuggled from the Electrostal plant near Moscow, the leading Russian producer of nuclear fuel.

♦ Approximately 2 kg of uranium enriched to 90% in ^{235}U disappeared from a research institute in the Republic of Georgia between 1992 and 1997, a period when there was conflict between Georgia and the neighboring Republic of Abkhazia.

♦ A fuel rod with 0.19 kg of uranium enriched to 19.9% in ^{235}U was found in 1998 in the possession of mafia groups in Italy, who "intended to sell it to an undisclosed buyer in the Middle East." It had been stolen from a research reactor in the Congo. This amount, by itself, is much too little for a bomb and would require further enrichment. However, one seizure could represent the tip of a mostly hidden iceberg.

♦ The detected traffic from the former Soviet Union, which in the early 1990s had been mostly westward into Europe, has more recently switched southward, including to Iraq, Iran, and Afghanistan. Reportedly, the number of such incidents "increased sharply during the years 1999 and 2000," totaling 29 for the 2 years.

It is not clear in these instances if the intended recipients were terrorist groups or countries interested in developing a national weapons capability (e.g., Iraq).

There is no evidence from these reports that any completed nuclear weapons have been stolen. However, the number of detected thefts of materials that could be used for nuclear weapons (i.e., enriched uranium) is substantial and one may surmise that many thefts have not been detected. This suggests the danger that terrorist groups, with or without the direct aid of friendly governments, may obtain sufficient material to make a bomb or may have already done so.

The largest potential sources for weapons materials are in countries that have nuclear weapons. Measures that can be taken to reduce the availability of this material include the following (see, e.g., Ref. [26]):

- Ensure that stockpiles of weapons and sensitive nuclear materials (weapons-grade plutonium and highly enriched uranium) are surely protected in all countries that have them, especially the countries with the largest stockpiles: the United States and Russia.
- Secure the stocks of plutonium removed from dismantled weapons.
- Reduce the stocks of highly enriched uranium, by diluting the HEU with natural or depleted uranium and using the mixture as reactor fuel. A joint United States–Russia program of this sort is currently in effect. As of the end of 2003, over 200 tonnes of HEU from Russia had been converted into reactor fuel for use in the United States (see Section 18.3.4).
- Improve security or remove plutonium and enriched uranium from vulnerable facilities around the world. The theft of material from a reactor in the Congo, mentioned earlier, suggests the wide scope of the problem.

Of course, each of these steps requires the cooperation of the country involved.

The Potential for Smuggling a Bomb

Bringing a bomb into another country, including the United States, should not be very difficult, because the bombs emit little radiation that escapes outside. This reduces the radiation risks for the person carrying the bomb and makes detection difficult. However, given sufficient effort, it may be possible to reduce the risk, by detecting some smuggling attempts and deterring others. If success is uncertain, the potential smugglers might choose easier terrorist avenues. There are three general ways to detect a nuclear bomb. We list them, plus one speculative alternative:[28]

---

[28] Here, we rely heavily on an analysis by Steve Fetter and American and Russian colleagues, written primarily in the context of monitoring compliance with arms control agreements, but relevant to terrorist bombs as well [28].

◆ *Direct detection.* Detect radiation emitted from the bomb, primarily gamma rays or neutrons.

◆ *Radiography.* Pass the bomb (in practice, the vehicle or container that may hold it) through a machine analogous to a medical X-ray machine in which the elements with high atomic numbers (uranium or plutonium) will stand out. This could be done with a high-energy X-ray generator (a small linear accelerator in the example discussed by Fetter et al. [28, p. 242]) or a neutron generator.

◆ *Induced fission.* Irradiate the bomb with neutrons and observe gamma rays or prompt and delayed neutrons emitted in the fission of uranium or plutonium isotopes.

◆ *Muon radiography.* A recent suggestion, from scientists at the Los Alamos National Laboratory, is to detect bomb materials in a vehicle by studying the passage of muons through the vehicle. Substances of high atomic number Z (e.g., uranium or plutonium) would be identified by changes in the trajectories of the muons [29].[29] This method must still be viewed as highly speculative, but it illustrates the range of possibilities being explored.

The first of these methods is termed *passive* and the next two *active*; the last does not fit neatly into either category.

Passive detection is, by far, the simplest approach, but it may not be adequate to the task, especially for uranium bombs. Given the long half-life of ^{235}U (704 million years), its specific activity is only 2.16 μCi/g (see Table 4.1). Therefore, a bomb with 10 or 20 kg of ^{235}U is a relatively weak source. ^{235}U decays by alpha-particle emission accompanied by low-energy gamma rays—most of which are stopped in the uranium itself and most of the remainder in the tamper surrounding it. For a hypothetical bomb with 12 kg of uranium (93% ^{235}U, 5.5% ^{238}U, and 1% ^{234}U), the most prominent escaping gamma ray is a 1.001-Mev gamma ray from a daughter product of the ^{238}U decay chain. The emission rate at the surface of the bomb is calculated to be only 30 gamma rays per second [28, p. 232]. Detecting such a source would require a large and expensive detector and a long detection time. This might be practical if the task was to inspect a single box in a leisurely fashion. It may be beyond any practical capabilities for coping with the large volumes of material that enter the United States each day through normal import channels.

However, a stolen uranium bomb may have a "contaminant" of an additional uranium isotope that could help in detection. It is unlikely that terror-

---

[29] Cosmic rays at the Earth's surface have a large component of high-energy muons (about 200 per square meters per second). Muons are charged particles similar to electrons but heavier. They pass through large amounts of material undisturbed, except for a gradual loss of speed and slight changes in direction due to the electrical interactions with the electrons and nuclei of the material through which they pass. Materials of high atomic number Z—such as uranium and plutonium— have nuclei with high charges (Ze) and therefore cause greater changes in the directions of muon motion than are caused by other materials.

ists would have much choice in picking the ideal bomb for smuggling purposes if they rely on bombs originally built without concealment in mind. If, as the case for most U.S. and Russian bombs, the enriched uranium came from reprocessed fuel, there will be a small amount of ^{232}U ($T = 68.9$ yr), which, after a long series of relatively rapid decays, leads to the emission of a 2.61-MeV gamma ray [28, p. 238].[30] Gamma rays of this energy are particularly penetrating and it would be difficult for the smugglers to provide shielding to prevent detection.

Passive detection is easier for plutonium bombs than for uranium bombs because, with plutonium, there is a reasonably large production of spontaneous fission neutrons and the bomb material and tamper do not stop neutrons as effectively as they stop low energy gamma rays. For a 4-kg plutonium bomb (93% ^{239}Pu and 6% ^{240}Pu), the neutron emission rate at the surface is calculated to be 400,000 per second.[31] Fetter et al. estimate that a source of this magnitude could be detected in 1 s with a hand-held neutron detector at a distance of 5 m and by a larger but transportable detector at a distance of 15 m [28]. This suggests that passive detection of plutonium bombs may be practical at defined ports of entry.

Active detection methods could be used to address the problem of low-emission uranium bombs. Radiography with high-energy X-rays or neutrons could identify suspicious objects, but there may be some difficulties in distinguishing plutonium or uranium from other elements with similar absorptive properties. Induced fission, using a pulsed accelerator to produce short bursts of neutrons, offers the opportunity to obtain a unique signature for fissionable materials through the detection of gamma rays or delayed neutrons from the fission products.

Overall, it would appear that there are technological approaches to the detection of smuggled weapons, but that an effective system would be expensive to deploy. The report *Making the Nation Safer* calls for efforts by federal agencies to "evaluate and improve the efficacy of special nuclear material detection systems that could be deployed at strategic choke points for homeland defense" [25, p. 57]. This recommendation is based on the judgment that the technology is not at a level where it would be feasible to deploy a nationwide detection system, but that an effective system could probably be put into operation at selected "choke points," such as major cargo container ports and selected roads, bridges, and tunnels.

Of course, terrorists may attempt to circumvent any such system by avoiding points where detectors are deployed. They might also detonate a bomb at an entry point (e.g., a port) prior to inspection. Thus, there is a premium

---

[30] This gamma ray comes from the beta decay of a daughter product of ^{232}U to an excited state of ^{208}Pb at 2.61 MeV.

[31] This rate exceeds the spontaneous fission rate that is indicated in Table 17.2. However, this number includes additional neutrons produced in fission and (n, 2n) reactions.

on efforts to solve the problem at the source, by protecting bombs, separated plutonium, and highly enriched uranium from theft or diversion.

17.5.4 Radiological Dispersion Devices ("Dirty Bombs")

Radiological dispersion devices (RDD), commonly referred to as "dirty bombs," could be readily built by anyone who has access to radioactive materials. The radioactivity can be introduced into the general environment by combining radioactive material with conventional chemical explosives. The purpose of using such a bomb is to spread radioactive contamination, rather than cause direct damage through the explosion itself.

Large radioactive sources can be found in numerous locations, where they are used for medical or industrial purposes, with source strengths extending to many thousands of curies. Radionuclides that might be used for dirty bombs include cesium-137 (^{137}Cs, $T = 30.1$ yr), cobalt-60 (^{60}Co, $T = 5.3$ yr), and strontium-90 (^{90}Sr, $T = 28.8$ yr). The consequences of such dispersal would depend on the size of the source and the fraction that is made airborne.

As an example, suppose that 1000 Ci of ^{137}Cs were dispersed uniformly over an area of 1 km^2. A person in the path of the cloud that carries the radionuclides would be briefly exposed through direct irradiation and inhalation of radionuclides. However, the most troubling consequence is the long-term contamination of the ground. At the hypothesized average areal density of 1 mCi/m^2, the dose to an exposed individual could be of the order of 1 mSv/day (100 mrem/day), although probably it would be closer to 0.2 mSv/day or 0.1 Sv/yr (10 rem/yr).[32]

At this level, the dose is too small to produce radiation sickness or any immediate symptoms. However, the annual dose of nearly 100 mSv would exceed any ordinary standard for exposures to the public and exceed the occupational limit of 50 mSv/yr. Questions would inevitably arise as to the need for evacuation of people in the region, the degree to which the contaminated areas should have the radionuclides removed, and the conditions under which people would be allowed to return to their homes or places of work. The NCRP suggests in Report No. 138 that evacuation is "almost always" justified to avert a dose of 500 mSv and perhaps justified to avert a dose of 50 mSv [8, p. 109].[33] If that guideline is accepted, immediate evacuation would not be necessary in the particular example presented here. The same discussion suggests that relocation is advisable if the cumulative dose would exceed 1000 mSv or if the dose would exceed 10 mSv per month over a long period of

---

[32] The dose would be at a rate of roughly 2 mSv/day for a person who is outdoors at a time before any of the cesium washed away or worked its way deeper into the ground. Under conditions considered for the Chernobyl accident and taking into account the time spent indoors, an UNSCEAR analysis finds an average dose of about 0.23 mSv/day for a deposition of 1 mCi/m^2 [30, p. 336].

[33] This recommendation is presented as a reiteration of a 1993 recommendation of the International Commission on Radiological Protection.

time. In the present example, the calculated monthly dose is close enough to 10 mSv/month to make relocation a consideration. Of course, this example is artificial in that the deposition of radioactive material would not be uniform. Rather the deposition would be concentrated in the direction of the prevailing wind and would decrease with distance. In any practical situation, the doses would vary greatly depending on location.

This suggests the complex nature of the issues that would have to be addressed. Some obvious questions: How much should be spent per person-Sv of dose averted? Should a person be forbidden to return to a home if the dose will be several hundred mSv/yr? According to standard estimates of radiation risks, at 100 mSv/yr a person would, each year, incur an additional 0.5% chance of a fatal cancer. However, there is little direct evidence as to the effects of prolonged exposures at moderate dose rates (see Chapter 4). Even if one has a clear view of what should be done at 100 mSv/yr, what should be done at higher or lower doses? If radiation "experts" reach a consensus on guidelines for these emergency conditions, how can the public be convinced to accept them—given individual reactions that may range from a cavalier dismissal of any risk where neither prompt effects nor a causative agent can be observed to a fear of even trivially small amounts of radioactivity?

Radiological devices can be used in other ways than dispersal by explosion. Another scenario has been suggested by Steven Koonin in which "a several-curie source of a long-lived isotope is stolen and covertly released throughout the business district of a major city" [31]. Such an attack could produce radiation levels that are readily detected as being well above normal background (for example, a rate corresponding to doses of 10 mSv/yr) and that might suffice to cause major disruption. For this scenario, he concludes that the "predominant effects will not be casualties, but rather psychosocial consequences and economic disruption."

The overall summary of *Making the Nation Safer* reaches the same general conclusions as does Koonin. It concludes that for radiological attacks, "few deaths likely, but potential for economic disruption and panic is high" [25, p. 46]. It expands on this point with the comment:

> [T]he likely aim of an RDD attack would be to spread fear and panic and cause disruption. Recovery would therefore depend on how such an attack is handled by first responders, political leaders, the media, and general members of the public.
>
> In general, public fear of radiation and radioactive materials appears to be disproportionate to the actual hazards. Although hazardous at high doses, ionizing radiation is a weak carcinogen, and its effects on biological systems are better known than those of most, if not all, toxic chemicals. Federal standards that limit human exposure to environmental ionizing radiation, which are based on the linear, nonthreshold dose-response relationship, are conservative and protective, and the government continues to fund R&D to improve scientific

understanding of radiation effects on biological materials. [25, pp. 61–62]

The terrorism issue lends additional importance to the studies cited in the last sentence. Radiation doses from nuclear power are so low under ordinary circumstances as to make the debates over the dose-response relationship somewhat irrelevant to decisions on the use of nuclear power. However, understanding the effects of radiation may prove essential for an intelligent response to terrorist attacks.[34]

17.5.5 Attacks on Nuclear Power Plants

Nuclear power plants have two vulnerable components: the reactors themselves and the spent fuel that is stored on-site, near the reactor. The example of September 11, 2001 raises the spectre of an airplane crash into a nuclear power plant. The containment structure enclosing the plant is designed to withstand the impact of a small plane but, in most cases, not that of a large airliner. However, even if an airliner could be hijacked, a nuclear power plant is relatively difficult to hit directly because of its low height and small "target" area. The potential results of such an impact are being studied, but the results are classified [25, p. 42]. Nonetheless, it is likely that they would vary widely, depending on the precise point of the impact and the unpredictable "good fortune" the attacker might have in disabling the cooling system and breaching the reactor vessel.

An alternative form of attack is by armed intruders. The reactor sites are guarded against such attacks and it can be assumed that the reactors could be shut down at the first sign of an attack, with no option for restarting. Then, the success of the attack would depend on the ability of the intruders to overcome the guards and disable the normal and emergency cooling systems in such a manner that they could not be restored in time to avert an eventual reactor meltdown. The worst-case scenario is a serious one, but evacuation of the surrounding population could mitigate the harm and the chances of success are dubious.

An attack could also be mounted against the spent fuel at the site. *Making the Nation Safer* gives little emphasis to this possibility. It points out that the heat output of the spent fuel is low relative to the heat output of fuel in a reactor that has just been shut down [25, p. 47]. Thus, even if a cooling pool is breached, the restoration of improvised emergency cooling was not thought to be difficult. No mention is made of the possibility of spreading the spent fuel beyond the reactor site through an explosion, possibly because it would be difficult to create an explosion that would cause a wide dispersal of the solid uranium oxide pellets that are contained in fuel rods.

---

[34] The case is similar for the aftermath of the Chernobyl accident, where issues of population resettlement remain important (see Chapter 15).

A more dangerous picture is painted in another paper on threats to spent fuel in cooling pools [32]. The authors conclude that if a terrorist attack breached a cooling pool and the water was lost, the fuel could heat dangerously. The fuel is densely packed in many of the pools, with limited opportunity for air circulation. In consequence, the fuel could melt and large amounts of ^{137}Cs could escape into the atmosphere. The suggested solution is to transfer the fuel to dry storage after 5 years in the cooling pools. This would permit a reduced density of packing of the "fresh" fuel and avoid melting.

Although *Making the Nation Safer* did not suggest an urgent need for such a transfer, the report did endorse the use of dry storage casks, expressing the view that they "are very robust and would probably stand up to aircraft attacks as well" [25, p. 47]. A similarly positive assessment of the protection provided against terrorist attacks was made for the casks to be used in the transportation of spent fuel to a waste repository [25, p. 48].

Concerns about the dangers associated with terrorist attacks on nuclear power plants have, in some cases, led to calls that they be shut down. In particular, in 2003 a vigorous local campaign was started to shut down the two Indian Point reactors located about 35 miles north of New York City. The chief expressed fear is that evacuation could not be accomplished quickly enough to protect the surrounding population. On the other hand, a group of present and past leaders in the nuclear industry and nuclear education have argued that there is little possibility of a large release of activity in case of terrorist attacks, whether by airplane or otherwise [33]. Unfortunately, without access to the now-classified studies carried out by the Nuclear Regulatory Commission, it is very difficult for outside observers to make independent assessments of the chance that a terrorist attack could succeed or the range of consequences for various attack scenarios.

Nuclear power plants (NPPs) are not unique as potential targets. As suggested earlier, many other vulnerable targets exist, such as football stadia, opera houses, and bridges—to cite a few prominent examples. To the extent that we believe that these facilities are used for desirable activities, we will be reluctant as a society to shut them down. In the same vein, it is likely that attitudes toward terrorist attacks on NPPs correlate with perceptions of the need for nuclear power. The difficulties in reaching a balanced assessment of the risks are suggested in *Making the Nation Safer*:

> The potential vulnerabilities of NPPs to terrorist attack seem to have captured the imagination of the public and the media, perhaps because of a perception that a successful attack could harm large populations and have severe economic and environmental consequences. There are, however, many other types of large industrial facilities that are potentially vulnerable to attack, for example, petroleum refineries, chemical plants, and oil and liquified natural gas supertankers. Their facilities do not have the robust construction and security features characteristic of NPPs, and many are located near highly populated urban areas.

The committee has not performed a detailed examination of the vulnerabilities of these other types of industrial facilities and does not know how they compare to the vulnerabilities of NPPs. It is not clear whether the vulnerabilities of NPPs constitute a higher risk to society than the vulnerabilities of other industrial facilities. [25, pp. 43–44]

The attention paid to nuclear power plants may seem to suggest that the dangers from them are particularly great. However, this attention also means that they are guarded unusually carefully. Such measures do not provide an absolute assurance against a successful terrorist attack. However, they may inhibit terrorists who will realize that the chances of failure are substantial and that softer rich targets exist elsewhere.

References

1. U.S. Enrichment Corporation, "Status Report: U.S Russian Megatons to Megawatts Program, Turning Nuclear Warheads into Electricity" (as of September 30, 2003). [From: http://www.usec.com/v2001_02/HTML/Megatons_status.asp]
2. Carey Sublette, *Nuclear Weapons Frequently Asked Questions*, Section 1.5.1, March 8, 1995. [From: http://www.virtualschool.edu/mon/Outlaws/faql].
3. J. Carson Mark, "Explosive Properties of Reactor-Grade Plutonium," *Science and Global Security*, 4, no. 1, 1993: 111–124. Frank von Hippel and Edwin Lyman, "Probabilities of Different Yields," *Science and Global Security*, 4 no. 1, 1993: 125–128.
4. Robert Serber, *The Los Alamos Primer, The First Lectures on How to Build an Atomic Bomb* (Berkeley: University of California Press, 1992).
5. J. Malik, E. Tajima, G. Binninger, D.C. Kaul, and G.D. Kerr, "Yields of the Bombs," in *U.S.–Japan Joint Reassessment of Atomic Bomb Radiation Dosimetry in Hiroshima and Nagasaki, Final Report*, W.C. Roesch, ed. (Hiroshima: Radiation Effects Research Foundation, 1987): 26–36.
6. Frank Barnaby, "Types of Nuclear Weapons," in *Plutonium and Security*, Frank Barnaby, ed. (New York: St. Martins Press, 1992): 264–279.
7. William J. Schull, *Effects of Atomic Radiation, A Half-Century of Studies From Hiroshima and Nagasaki* (New York: Wiley–Liss, 1995).
8. National Council on Radiation Protection and Measurements, *Management of Terrorist Events Involving Radioactive Material*, NCRP Report No. 138 (Washington, DC: NCRP, 2001).
9. Samuel Glasstone and Philip J. Dolan, *The Effects of Nuclear Weapons* (Washington, DC: U.S. DOE and ERDA, 1977).
10. Thomas B. Cochran, William M. Arkin, and Milton M. Hoenig, *Nuclear Weapons Data Book, Volume I, U.S. Nuclear Forces and Capabilities* (Cambridge, MA: Ballinger Publishing, 1984).
11. Theodore B. Taylor, "Nuclear Safeguards," *Annual Review of Nuclear Science* 25, 1975: 407–421.
12. Victoria McLane, Charles L. Dunford, and Philip F. Rose, *Neutron Cross Section Curves*, Vol. 2, *Neutron Cross Sections* (New York: Academic Press, 1988).

13. F.J. Rahn, A.G. Adamantiades, J.E. Kenton, and C. Braun, *A Guide to Nuclear Power Technology, A Resource for Decision Making* (New York: Wiley, 1984).

14. Thomas B. Cochran and Christopher E. Paine, "The Amount of Plutonium and Highly-Enriched Uranium Needed for Pure Fission Nuclear Weapons" (Washington, DC: National Resources Defense Council, 1994).

15. National Academy of Sciences, *Management and Disposition of Excess Weapons Plutonium*, Report of the Committee on International Security and Arms Control (Washington, DC: National Academy Press, 1994).

16. David Albright, Frans Berkhout, and William Walker, *Plutonium and Highly Enriched Uranium 1996, World Inventories, Capabilities and Policies* (New York: Oxford University Press, 1997).

17. Leonard S. Spector, *Nuclear Ambitions: The Spread of Nuclear Weapons 1989–1990* (Boulder, CO: Westview Press, 1990).

18. D. Albright and H.A. Feiveson, "Plutonium Recycling and the Problem of Nuclear Proliferation," *Annual Review of Energy*, 13, 1988: 238–265.

19. David Albright and Holly Higgins, "Setting the Record Straight About Plutonium Production in North Korea," Appendix 3 of *Solving the North Korean Nuclear Puzzle*, David Albright and Kevin O'Neill, eds. (Washington, DC: ISIS Press, 2000). [From: http://www.isis-online.org/publications/dprk/book/app3.html]

20. Frank Barnaby, Introduction, in *Plutonium and Security*, Frank Barnaby, ed. (New York: St. Martins Press, 1992): 1–9.

21. World Nuclear Association, *Nuclear Power in Japan* (April 2003). [From: http://www.world-nuclear.org/info/inf79print.htm]

22. Myron B. Kratzer, "'Demythologizing Plutonium," paper presented at *Global '93, Future Nuclear Systems: Emerging Fuel Cycles and Waste Disposal Options*, Seattle, September 1993.

23. *Nuclear Power Issues and Choices*, Report of the Nuclear Energy Policy Study Group, Spurgeon M. Keeny, Jr., Chairman (Cambridge, MA: Ballinger, 1977).

24. Luther Carter, *Nuclear Imperatives and Public Trust: Dealing with Radioactive Waste* (Washington, DC: Resources for the Future, 1987).

25. National Research Council, *Making the Nation Safer: The Role of Science and Technology in Countering Terrorism*, Report of the Committee on Science and Technology for Countering Terrorism (Washington, DC: National Academies Press, 2002).

26. Matthew Bunn, John P. Holdren, and Anthony Wier, *Securing Nuclear Weapons and Materials: Seven Steps for Immediate Action*, Report co-published by the Project on Managing the Atom and the Nuclear Threat Initiative (Cambridge, MA: JFK School of Government, Harvard, 2002).

27. C. Braun, F. Steinhausler, and L. Zaitseva, "International Terrorists Threat to Nuclear Facilities," paper presented at American Nuclear Society 2002 Winter Meeting, Washington, DC, November 2002.

28. Steve Fetter, Valery A. Frolov, Marvin Miller, Robert Mozley, Oleg F. Prilutsky, Stanislav N. Rodionov, and Roald Z. Sagdeev, "Detecting Nuclear Warheads," *Science & Global Security*, 1 1990: 225–302.

29. John Roach, "Cosmic Particles Could Detect Nuke Materials, Scientists Say," *National Geographic News*, March 19, 2003. [From: http://news.nationalgeographic.com/news/2003/03/0319_030319_cosmicrays.html]

30. United Nations Scientific Committee on the Effects of Atomic Radiation, *Sources, Effects and Risks of Ionizing Radiation*, UNSCEAR 1988 Report (New York: United Nations, 1988).
31. Steven E. Koonin, "Radiological Terrorism," *Physics and Society* 31, no. 2, April 2002: 12–13; statement delivered before Senate Foreign Relations Committee (March 6, 2002).
32. Robert Alvarez, et al., "Reducing the Hazards from Stored Spent Power-Reactor Fuel in the United States, *Science & Global Security*, 11, no. 1, 2003: 1–51.
33. Douglas M. Chapin, et al., "Nuclear Power Plants and Their Fuel as Terrorist Targets," *Science*, 297, no. 5589, September 2002: 1997–1999.

18

Proliferation of Nuclear Weapons

18.1 Nuclear Proliferation

18.1.1 International Treaties

The Original Non-Proliferation Treaty

Following World War II, during which the United States (aided by Britain) developed and used atomic bombs, other countries undertook their own nuclear-weapon programs. The first successful international agreement to control the spread of nuclear weapons came after President Dwight Eisenhower's "Atoms for Peace" proposal, made in 1953, under which the United States would aid countries in their pursuit of peaceful applications of nuclear energy while withholding the technology needed for weapons.[1] This proposal eventually led to the establishment of the International Atomic Energy Agency (IAEA) in 1957, with the dual role of promoting peaceful nuclear uses and establishing safeguards, including inspections, to prevent this aid from being used to develop nuclear weapons. The IAEA submits reports to the United Nations but is not an agency of the United Nations. It receives financial support from its 136 member states. The members include virtually all countries with current or prospective nuclear activities, with the exception of North Korea, which withdrew its IAEA membership in June 1994.[2]

---

[1] Relatively recent summaries of efforts to control the spread of nuclear weapons are given by Richard Garwin and Georges Charpak in *Megawatts and Megatons* [1] and by Robert Mozley in *The Politics and Technology of Nuclear Proliferation* [2].

[2] Current listings of IAEA membership are posted on the IAEA website at [http://www.iaea.org/worldatom/About/Profile/member.html].

A next step in the effort to forestall the spread of nuclear weapons was the adoption of a Treaty on the Non-Proliferation of Nuclear Weapons, commonly known as the Non-Proliferation Treaty (NPT). The first signatories, in 1968, were the United Kingdom, the United States, and the USSR. These became the "Depository Governments" when the NPT went into force in 1970, initially with the signatures of 97 countries and ratification by 47 [2, p. 47]. The preamble to the treaty expressed, as motivating purposes, the desire to[3]

♦ Prevent the wider dissemination of nuclear weapons.
♦ Make peaceful applications of nuclear technology widely available.
♦ Achieve cessation of the nuclear arms race and move toward nuclear disarmament.
♦ Seek to achieve discontinuance of test explosions of nuclear weapons.

The treaty thus attempted simultaneously to discourage military applications of nuclear energy and to foster peaceful applications.

The NPT started with and retains a crucial asymmetry between nuclear-weapon states and non-nuclear-weapon states. A nuclear-weapon state (NWS) is defined in the NPT [Article IX, Par. 3] to be one that had "manufactured and exploded a nuclear weapon or other nuclear explosive device prior to January 1, 1967." By this definition, the NWSs are China, France, the USSR, the United Kingdom, and the United States. These countries were the victors in World War II and, in consequence, are also the permanent members of the United Nations Security Council.

Upon becoming a party to the treaty, the key obligation assumed by a NWS is to do nothing, directly or indirectly, to aid nuclear weapons development in a non-NWS. Each non-NWS party to the treaty is committed not to receive or to make nuclear weapons. The non-NWSs also undertake to accept safeguards against the diversion of nuclear activities from peaceful to weapons purposes. The safeguards are to be set forth for each country in an agreement to be negotiated with the IAEA.[4] The NPT specified (in Article X) that a conference was to be held in 25 years (1995) to "...decide whether the Treaty shall continue in force indefinitely, or shall be extended for an additional fixed period or periods."

The asymmetry embodied in the NPT could exist only because it reflected a substantial asymmetry in national power as of 1970. The nonweapon states received explicit inducements in terms of a commitment by the weapon states to "pursue negotiations in good faith on effective measures relating to the cessation of the nuclear arms race at an early date and to nuclear disarmament..." (Article VI). All parties to the Treaty also agreed to the "fullest possible exchange of equipment, materials and scientific and tech-

[3] The text of the NPT is reproduced, for example, in the appendices of Refs. [2] and [3].

[4] Amendments to the treaty can be made by a majority of the parties to the treaty, but this must include the affirmative votes of *all* of the NWSs.

nological information for the peaceful uses of nuclear energy" (Article IV, par. 2).

As is to be expected, some countries have resented the inequalities in status contained in the Treaty. However, they would have gained little by abstaining unless they were contemplating weapons programs of their own, and by 1994, almost all countries had adhered to the NPT. The most important holdouts—at the time and continuing to this day—are India, Israel, and Pakistan, which are abstaining because they have nuclear-weapon programs that would be in violation of the NPT, were they signatories. Subsequently, in January 2003, North Korea withdrew from the NPT.

Extension of the NPT in 1995

Although the NPT has not been completely successful in detecting proliferation efforts, it was deemed sufficiently valuable that it was made permanent at the scheduled international conference of its signatories in May 1995 [4]. The conference, held in New York with the participation of 175 nations, was convened to carry out the prescribed 25-year review of the NPT. Although the extension of the NPT was approved without explicit dissent, there was sufficient dissatisfaction on the part of a substantial minority that approval was achieved by what amounted to general consent, rather than by a formal vote. In essence, there was agreement that a majority favored indefinite extension, and this provided the basis for the conference president to declare, without reported objection, that the extension was adopted.

The dissatisfaction stemmed in part from the belief of many of the non-NWSs that there had been insufficient progress toward nuclear disarmament, as was called for in the NPT. In addition, the Arab states objected to the fact that Israel had not acceded to the NPT. However, rather than explicitly oppose the indefinite extension of the NPT, the dissenting nations accepted the extension, and some of their concerns were addressed in companion documents. In particular, again by general consent rather than a formal vote, the conference adopted a set of 20 "Principles and Objectives for Nuclear Non-Proliferation and Disarmament." Among the key principles were the following [4]:

◆ A call for a comprehensive test ban treaty by 1996 and for a "universally applicable convention banning the production of fissile materials for nuclear weapons." The NWSs were also called upon to undertake "systematic and progressive efforts to reduce nuclear weapons globally, with the ultimate goal of eliminating those weapons."

◆ A call to encourage the "development of nuclear-weapon-free zones, especially in regions of tension, such as in the Middle East...."

◆ An affirmation that the peaceful use of nuclear energy is an "inalienable right of all parties to the treaty," together with a call for the "fullest possible exchange of equipment, materials, and scientific and technological information for the peaceful use of nuclear energy..."

This statement of principles reaffirmed both opposition to nuclear weapons and encouragement of peaceful uses of nuclear energy. Thus, as far as the NPT is concerned, there is no conflict between the maintenance and expansion of civilian nuclear power and the achievement of a world free of nuclear weapons.

Review of NPT in 2000

The NPT is to be reviewed every 5 years to monitor progress toward the implementation of the treaty objectives. A scheduled review was concluded in May of 2000 at the United Nations. Tension between the NWSs and the non-NWSs continued, but in the end, a consensus document was produced.[5] This extensive document reaffirmed the importance of the NPT and, among other items, pointed a special finger at those countries which had not as yet adhered to the NPT:

> The Conference urges all States not yet party to the Treaty, namely Cuba, India, Israel and Pakistan, to accede to the Treaty as non-nuclear-weapon States, promptly and without condition, particularly those States that operate unsafeguarded nuclear facilities.

> The Conference deplores the nuclear test explosions carried out by India and then by Pakistan in 1998. The Conference declares that such actions do not in any way confer a nuclear-weapon State status or any special status whatsoever. [6]

The countries named above were not parties to the Conference, and therefore this injunction was made in their absence.[6] The NWSs did concur in a call for a number of "practical steps" toward implementation of the principles of the NPT, including

> An unequivocal undertaking by the nuclear-weapon States to accomplish the total elimination of their nuclear arsenals leading to nuclear disarmament to which all States parties are committed under Article VI.

Although the wording is quite strong, no deadline or timetable is set for fulfilling this undertaking.

The International Atomic Energy Agency retains a central role in advancing the goals of the NPT—both in establishing and verifying safeguards against nuclear-weapons proliferation and in furthering the safe development of nuclear power.

---

[5] A detailed description of the negotiations is given in Ref. [5].

[6] Cuba subsequently acceded to the NPT, in 2002.

The Comprehensive Nuclear Test Ban Treaty

An additional step toward the limitation of nuclear weapons was the adoption of the Comprehensive Nuclear Test Ban Treaty (CTBT) by the United Nations General Assembly on September 10, 1996. It commits the parties to the treaty "not to carry out any nuclear weapon test explosion or any other nuclear explosion." It was signed by all the NWSs when it was opened for signature on September 24, 1996, as well as by many others. By the end of 2003, it had been signed by 170 states (out of a total of 193 states) and ratified by 108.[7]

However, for the CTBT to go into effect, it must be signed and ratified by each of the 44 so-called "Annex 2 states," a group that includes the countries that have power or research reactors. Among these, 41 have signed, with India, North Korea, and Pakistan the only holdouts. Thirty-two of the signatories have also ratified the CTBT, with the United States among those not yet ratifying it. Even though it is not yet officially in effect, the CTBT at least puts a moral stigma on nuclear testing and may serve to inhibit some countries, including the United States. Of course, it would be a more effective instrument if all Annex 2 states ratified it.

An important part of the CTBT regime is a large network of seismic observation points, and many of these installations are now active. They detected the May 11, 1998 nuclear weapons tests in India and the subsequent tests in Pakistan, but not the tests of the very small devices that the Indian government said were carried out on May 13, 1998 (see Section 18.2.2). The implications of this "failure" are in doubt. They have led to the charge that the seismic detection network did not provide adequate monitoring, as well as to the suggestion that the May 13 explosions were smaller than stated by the Indian government [7].

The Influence of the Treaties

The NPT and CTBT do not command the universal adherence that a fully effective nonproliferation regime should have. Nonetheless, they are taken seriously. The three major nonadherents to the NPT—India, Israel, and Pakistan—have refrained from signing the NPT in order to be free to pursue their nuclear weapons programs (see Section 18.2 for descriptions of the weapons programs of individual countries) and North Korea has oscillated in its membership along with oscillations in its nuclear program.[8] This, in effect,

---

[7] Current information on the CTBT is available on the website of the Preparatory Commission for the Comprehensive Nuclear-Test-Ban Organization, established under the terms of the treaty [http://www.ctbto.org].

[8] North Korea, which had signed the NPT in 1985, withdrew from the NPT in March 1993 but suspended its withdrawal just before the expiration of the 3-month interval before the withdrawal could officially go into effect [12, p. 287]. In January 2003, North Korea again withdrew, this time stating that the withdrawal was in effect immediately (in light of the earlier suspended withdrawal) [8].

is a tribute to the significance of the NPT and the safeguards possible under its terms.

The existence of the NPT has served to turn the spotlight on the nonsignatories. Nonetheless, world pressures have not had a decisive effect on their programs, perhaps because, at crucial times each has had powerful protectors and a certain amount of sympathy for its situation. India has been threatened by China and Pakistan, Pakistan has been threatened by India, and Israel has been threatened by the Arab world. Countries otherwise friendly to one or another of them have not been willing to condemn their possession of nuclear weapons. Thus, the USSR provided political and technical help to India in the past, and the United States, to one degree or another, has tolerated the Pakistani and Israeli nuclear programs. The same arguments of security put forth to justify nuclear weapons in the NWSs have been advanced on behalf of other fearful states.

The NPT has also had some failures even with signatories (see Section 18.2.3). Iraq had a well-advanced weapons program by the time of the 1991 Persian-Gulf war, although it was an NPT signatory with a full safeguards agreement with the IAEA. On the positive side, however, the NPT provided the legal basis after the war for the inspections carried out by the IAEA under the aegis of the United Nations.

The situation is somewhat similar with respect to the CTBT. The unwillingness of India, North Korea, and Pakistan to sign it reflects their desire to retain the weapons testing option and also a respect for the letter of the treaty. In a somewhat similar vein, China and France—which, as NWSs, were not limited by the NPT— carried out nuclear weapons tests in 1995, apparently out of a desire to complete their testing programs before the anticipated adoption of the CTBT.

18.1.2 Forms of Proliferation

Any increase in the number of nuclear weapons or in the means for making them can be considered to be a form of proliferation, whether the increase is within a country that already has weapons or extends to a new country. The end-point in proliferation is the acquisition of an effective bomb, but more rudimentary proliferation steps are also important. These range from obtaining technical advice to obtaining weapons-grade uranium or plutonium. The barrier against proliferation is substantially lowered if a country possesses facilities for enriching uranium or reprocessing spent fuel to extract plutonium, and the development of such facilities is taken as a danger signal, however much the country involved professes a peaceful intent.

Proliferation can proceed by many different routes including, but probably not limited to, the following:

1. A state with nuclear weapons (whether or not a recognized NWS) could increase the number or the effectiveness of its weapons.

2. A state with nuclear weapons could publicly transfer weapons to another state. This is what happened following the breakup of the Soviet Union and the division of weapons among Russia, Ukraine, Belarus, and Kazakhstan. But with the subsequent delivery of these weapons to Russia, the overall outcome served to reduce proliferation dangers.[9]

3. A state without nuclear weapons but with advanced nuclear capabilities could covertly or openly embark on a weapons program based on its own resources. Japan, for example, could do this in reaction to the nuclear activities of North Korea.

4. A state without nuclear weapons could utilize equipment obtained for ostensibly peaceful purposes to facilitate the development of nuclear weapons. India, Israel, and North Korea took advantage of this opportunity.

5. A state with nuclear weapons could have weapons or fissile material surreptitiously transferred to other states or groups, by government action, by action of dissident officials, or by theft. The former Soviet Union and Pakistan have been thought to be particularly vulnerable as sources for such transfers.

6. Nuclear weapons technology, including designs and specialized items of equipment, could be transferred from states, private companies, or knowledgeable individuals to other states or groups.[10]

7. Subnational groups or individual terrorists could buy or steal a completed weapon or build weapons using fissile material obtained by theft or diversion.

Initially, as nuclear weapons were being developed, the most serious dangers were thought to come from national programs, including the possible use of nuclear weapons by one or more of the NWSs. The danger from the NWSs was successfully managed during the period of the cold war and now seems diminished. However, the acquisition of nuclear weapons by several non-NWSs and the suspected nuclear ambitions of other states, including so-called "rogue nations," have kept the expansion of national weapons programs a major concern. More recently, the events of September 11, 2001 gave new prominence to the potential dangers from terrorists. While there is no public evidence that terrorists have obtained any large amounts of nuclear materials by theft or otherwise, there have been many reports of thefts of relatively small amounts

---

[9] See the paragraphs on the USSR in Section 18.2.1 for a brief discussion of the disposition of weapons in the former Soviet Union.

[10] The range of such possibilities was demonstrated in early 2004, with public revelations that designs and equipment for the production of centrifuges for uranium enrichment were shared by Pakistan with Iran, Libya, and North Korea (see Section 18.2.2). Other reports indicate that Libya received detailed instructions for bomb construction that had been prepared in China and forwarded through Pakistan.

(see Section 17.5.3). The discovery of any thefts, even small ones, raises the specter of larger, undiscovered, thefts.

It is unlikely that any attempt to classify and rank the specific threats can be fully satisfactory. Detailed information on nuclear technology is now held by many countries, commercial enterprises, and individual scientists, and the industrial capability to manufacture specialized components is widespread. The prospect remains that additional countries or subnational groups may seek to obtain weapons. Given the variety of avenues for obtaining fissile uranium or plutonium, the many sources of technical knowledge and equipment, and the potentially large array of aspirants, nuclear weapons proliferation may appear in unexpected places and forms.

18.1.3 Means for Obtaining Fissile Material

A nation that seeks to develop nuclear weapons has several options for obtaining fissile material:

♦ Isotopically separating natural uranium to obtain uranium highly enriched in ^{235}U.

♦ Operating a uranium-fueled reactor and extracting ^{239}Pu from the spent fuel.

♦ Producing ^{239}Pu or ^{233}U by irradiating externally placed "targets" of uranium or thorium in the neutron flux emerging from a reactor.

♦ Obtaining the material from another country—by purchase, friendly transfer, or theft.

Methods for enrichment were discussed in Section 9.2.2. These require moderately sophisticated equipment, but the procedures are not hazardous. The chemical extraction of plutonium from spent fuel, on the other hand, is hazardous without remote-handling equipment, because it is necessary to work with highly radioactive materials. Once extracted, the activity of the separated plutonium is relatively low, making its handling easier. A terrorist group, or even a country intent on building a nuclear weapon surreptitiously, would prefer to start with separated plutonium rather than with spent fuel, given the difficulty of building and concealing the required chemical separation facilities.

Proliferation has often been treated as a matter of safeguarding plutonium. Hence, there has been a substantial focus on reprocessing, which provides chemically separated plutonium as an end product. However, consideration of the technological options and of the actual history of countries undertaking weapons programs suggests that ^{235}U may be equally attractive, making the possession of enrichment technology an important stepping stone to obtaining weapons.

In fact, Peter Zimmerman in 1994 suggested that North Korea might be one of the *last* countries to try the plutonium route, with the next proliferators opting for ^{235}U if they have uranium available from domestic resources or outside purchase [9]. Consistent with this surmise, enrichment programs were subsequently undertaken by North Korea, Iran, and (apparently with less progress) Libya. Another observer has stressed that the "examples of Pakistan, South Africa, Argentina, and Iraq have demonstrated that enrichment is not the exclusive province of the technological elite" [10]. Iran and Libya could be added to this list as additional examples, and the opening of an enrichment plant by Brazil in 2003 further illustrates the point. A dramatic demonstration of the accessibility of enrichment technology is provided by the activities of Pakistan in obtaining and selling components and plans (see Section 18.2.2).

18.1.4 Nuclear Weapons Inventories

Table 18.1 gives estimates of the number of nuclear warheads in the arsenals of the NWSs, as made by the National Resources Defense Council (NRDC) [11]. In most cases, the number of weapons remaining in 2002 was well below the past peak number. For the United States and Russia, which had, by far, the largest arsenals and are now committed to reducing them, current totals are less than one-half of past peak totals.

A separate listing is given in Table 18.2 for those non-NWSs that are believed to have nuclear weapons, based on estimates from the Institute for Science and International Security (ISIS). The NRDC reference (used for Table 18.1) gives estimates for India and Pakistan that agree well with those of Table 18.2, but its estimate for Israel is about twice as great as that of Table 18.2, namely about 200 warheads. In contrast with the trends for the NWSs, where the inventories are in most cases decreasing, there is believed to be a continuing buildup of weapons material in the non-NWSs, as of 2003. Some details of the nuclear weapon programs in the NWSs are given in Section 18.2.

Table 18.1. Estimated number of operational nuclear warheads for nuclear-weapon states, 2002, and past peak number of warheads.

Country	Operational Warheads (2002)	Peak Number	
		Year	Warheads
United States	10,600	1966	31,700
Russia	8,600	1986	40,723
United Kingdom	200	1975	350
France	350	1991	540
China	400	1991	435

Source: From the National Resources Defense Council (NRDC), as summarized in Ref. [11].

Table 18.2. Estimates of weapons-grade fissile material inventories and numbers of warheads, at the end of 1999, for non-signers of NPT.

	Fissile Material		No. of Warheads	
Count	Pu (kg)	U (kg)	Range[a]	Median
India	240–400	little	45–95[b]	30[b]
Israel	380–640	?	75–125	100
Pakistan	2–13	580–800	30–50	40
North Korea	< 30–40			1–2

[a]The range for fissile material and number of warheads is from the 5th to 95th percentile of estimates.
[b]For India, the tabulated range in number of warheads corresponds to the number if the full plutonium inventory is converted to warheads. The tabulated median is the estimated number of warheads, which, for India, corresponds to conversion of about one-half of the plutonium inventory. (For the other countries, it is assumed that most of the inventory is now in warheads.)
Sources: Website of Institute for Science and International Security (http://www.isis-online.org) and Ref. [12].

18.2 History of Weapons Development

18.2.1 Official Nuclear-Weapon States

As of 2003, seven countries have acknowledged the possession of nuclear weapons: the original quintet of China, France, the USSR, the United Kingdom, and the United States, plus the later additions of India and Pakistan. The first five are the nuclear-weapon states (NWSs), as defined in the NPT. Other countries that have nuclear weapons, including India, Israel, and Pakistan, are not technically NWSs.

In the immediately succeeding sections, we briefly outline the early history of weapons development in the NWSs. The programs of these countries have progressed well beyond their early form, and each country now has extensive facilities for the reprocessing of plutonium and the enrichment of uranium. Fusion weapons (hydrogen bombs) have been added to fission weapons. However, the early history is pertinent to the question of weapons proliferation, because new entries—discussed in subsequent sections—may find it easiest to retrace some of the same early steps.

United States

Development of U.S. nuclear weapons was accomplished in the massive Manhattan Project during World War II. The program culminated in the successful Trinity test carried out near Alamogordo, New Mexico in July 1945. This was the world's first man-made nuclear explosion. It was quickly followed by the dropping of bombs at Hiroshima and Nagasaki in Japan in August 1945.

The Trinity test and the Nagasaki bomb used ^{239}Pu, produced in reactors at Hanford, in Washington. The fissile material for the Hanford reactors was natural uranium. The reactors were moderated with graphite and cooled with water, and plutonium was extracted from the spent fuel discharged from these reactors. The Hiroshima bomb used ^{235}U, produced in electromagnetic separators from uranium partially enriched by thermal diffusion and gaseous diffusion. The thermal diffusion process was soon abandoned, electromagnetic separation fell by the wayside, and after World War II, gaseous diffusion became the sole U.S. enrichment method [13, p. 15]. However, a centrifuge enrichment facility is now being planned in the United States for commercial reactor fuel (see Section 9.2.2).

The USSR

The Soviet Union, although allied with the United States during World War II, was excluded from the Manhattan Project. Nonetheless, it became aware of the U.S. program through surmise and information from spies. It was previously alerted to weapons possibilities by the discussion of fission in the open literature in 1939, and a Soviet nuclear bomb program was initiated during the war. Soviet scientists first proposed in 1943 a reactor fueled with natural uranium and moderated with graphite, but wartime pressures on the USSR delayed the program until the assembly of the 10-watt F-1 reactor, which went critical on December 25, 1946.[11] This was followed by the construction of a 100-MWt plutonium production reactor at Chelyabinsk that reached full power in June 1948. The first Soviet nuclear bomb was set off in a test at Semipalatinsk in Kazakhstan in August 1949.

This is similar to the path the United States had taken a few years earlier. The Soviets reached the point of a critical chain reaction at the end of 1946 and exploded a bomb less than 3 years later—a history shifted from the U.S. history by 4 years. There is some dispute as to the extent to which the information obtained from spies sped their progress, but it is highly likely that even without this information, the Soviet scientists would have been able to develop a bomb without a great delay—especially when spurred by the knowledge that the United States had succeeded.

As part of this effort, it was necessary for the Soviets to develop reprocessing facilities. A parallel program to produce enriched uranium was also undertaken in the early years. Andrei Sakharov, who went on to lead the Soviet hydrogen bomb effort, reports that as a student in 1945, he had actively speculated about techniques for isotope separation, and one can assume that senior scientists were more seriously pursuing the matter [15, p. 92].

---

[11] Richard Rhodes, in *Dark Sun: The Making of the Hydrogen Bomb*, gives an interesting account of the Soviet program, including speculations as to the extent to which the program may have been accelerated by espionage [14, Chapters 14–19].

Following the breakup of the Soviet Union, nuclear weapons remained in a number of the newly formed states, namely Russia, the Ukraine, Belarus, and Kazakhstan. Of these, only Russia appears firmly determined to keep nuclear weapons, and only Russia is accepted as the legitimate heir of the USSR as a NWS. The other three countries have acceded to the NPT, Ukraine completing the process in December 1994 [16]. They reportedly completed the transfer of all of their nuclear weapons to Russia in 1996 [2, p. 159].

United Kingdom

The United Kingdom was a partner of the United States in the Manhattan Project. Although it was a junior partner in the end, it had been ahead of the United States initially and British scientists played a major role in the joint efforts. After the war, British scientists had all the necessary information to proceed, although they were excluded from further U.S. weapons developments. The first British bomb, which used plutonium from specially built plutonium-production reactors, was tested in 1952 [17, p. 120]. The British program was also based on graphite-moderated natural-uranium reactors.

France

France, overrun by Germany in World War II, was not officially a participant in the Manhattan Project, but individual French scientists had joined an Anglo-Canadian project in Canada during the war. This program worked in loose collaboration with the Manhattan Project. Its main wartime project was the building of a heavy water reactor. After World War II, these scientists returned home and were able to contribute to a newly started French atomic energy program despite somewhat ambiguous commitments to secrecy about their war work [18, p. 65].

With a strong base of nuclear scientists, including those who had remained in France during the war, development of low-power research reactors began by 1946, starting with a natural-uranium heavy water reactor. France gradually developed plutonium-production capabilities, based on natural-uranium graphite-moderated reactors, but there was considerable political ambivalence about weapons development. By the late 1950s, momentum toward bomb development had developed, and the decision to proceed with nuclear weapons was confirmed when General Charles de Gaulle assumed power in 1958.[12] The first French ^{239}Pu bomb was set off in the Sahara Desert on February 13, 1960 [18, p. 139]. Since then, following policies established by de Gaulle, France has maintained an independent nuclear-weapons capability, including

---

[12] Garwin and Charpak cite, as a factor in the French decision, the threat made by Khrushshev during the 1956 Suez crisis to use nuclear weapons against France and England [1, p. 288]. They suggest that this illustrates the motivation a country may have for possessing its own nuclear deterrent.

a program of nuclear-weapons testing in the South Pacific that was not terminated until 1996.

China

Initial Chinese efforts to develop nuclear weapons were facilitated by a military collaboration agreement in October 1957 between China and the Soviet Union. Under this agreement, China obtained atomic bomb design information. This agreement lasted only until 1959, but it gave China a valuable head start.

The original plan was to develop both ^{239}Pu and ^{235}U weapons. However, a decision was made in 1960 to give priority to the ^{235}U program [19, p. 113], perhaps because with the loss of Soviet help the plutonium program was temporarily deemed more difficult. Gaseous diffusion plants for uranium enrichment were built and put into operation in the early 1960s. The first uranium bomb was an implosion bomb. After prior practice, the enriched uranium was machined into a sphere, with tight specifications, by a single person in a single night, according to a dramatic account of the event [19, p. 167]. The machining was accomplished in May 1964, and the bomb itself was exploded on October 16, 1964, at the Lop Nur test site in a desert region of northwest China [19, p. 185]. Although China soon moved on with plutonium-producing reactors, and very quickly to hydrogen weapons, it was unique among the NWSs in testing a ^{235}U bomb before testing a ^{239}Pu bomb.

Lag Between Nuclear Weapons and Commercial Nuclear Power

For each of these countries, the development of civilian nuclear power lagged behind weapons development. Table 18.3 compares the year in which each of these countries obtained nuclear weapons and the year in which the country first obtained power from a reactor designed for electricity production for civilian use.

Table 18.3. Comparison of years of achieving nuclear weapons and civilian nuclear electric power, for acknowledged nuclear-weapon countries.

| Country | Year of Achieving | | First Power Reactor |
	Weapon	Electric Power	
United States	1945	1957	Shippingport (60 MWe)
Former USSR	1949	1958	Troisk A (100 MWe)
United Kingdom	1952	1956	Calder Hall 1 (50 MWe)
France	1960	1964	Chinon A1 (70 MWe)
China	1964	$\sim 1992^a$	Qinshan 1 (300 MWe)

[a]Qinshan 1 achieved criticality in 1991 and produced some power in 1992, but it is not listed as having achieved "commercial operation" until 1994 (see, e.g., Ref. [20]). *Sources:* For civilian electric power, Ref. [20]; for weapons, see text.

18.2.2 Other Countries with Announced Weapons Programs

India

India began a rudimentary nuclear program in 1948, shortly after achieving independence from Great Britain. It built a small research reactor in 1956 using enriched uranium from Britain, and in 1960, it obtained a larger research reactor under a joint Canadian–Indian–U.S. program [21, p. 24]. The reactor was given the acronym CIRUS in recognition of its parentage. This was a 40-MWt natural-uranium reactor moderated with heavy water. Part of the initial fuel was provided by Canada and the heavy water by the United States. By 1962, India had indigenous supplies of uranium and a small plant for producing heavy water.

India then proceeded to develop a reprocessing facility to extract plutonium from CIRUS. It was able to use this plutonium to build a nuclear explosive device, which it set off on May 18, 1974. It is estimated that the device used about 15 kg of ^{239}Pu [21, p. 34].[13] It is termed a "device" rather than a bomb, because it is thought to have been too bulky to be used in a deliverable weapon. At the time, India claimed that the purpose of the explosion was peaceful, to study the effects on rocks. However, outside India, the explosion was considered a first step in an Indian weapons program, undertaken as a potential counter to China and, more recently, as part of a nuclear competition with Pakistan. India's potential to produce plutonium was increased in 1985 with its construction of the 100-MWt Dhvura research reactor, also heavy water moderated [12, p. 266].

Although India for many years did not acknowledge the possession of nuclear weapons, it was widely believed to have had them. It clearly had facilities that would permit a large weapons program. These included several heavy water reactors from which fuel can be continuously removed—facilitating the extraction of low-burnup, weapons grade plutonium—and reprocessing facilities. India's nuclear program came into the open in May 1998 when it carried out a series of underground nuclear tests and declared that "India is a nuclear weapon state" [22]. As explained by the Indian embassy in the United States:

> On May 11, a statement was issued by Government announcing that India had successfully carried out three underground nuclear tests at the Pokhran range. Two days later, after carrying out two more sub-kiloton tests, the Government announced the completion of the planned series of tests. The three underground nuclear tests carried out at 1545 hours on 11 May were with three different devices—a fission device, a low-yield sub-kiloton device and a thermonuclear device. The two tests carried out at 1221 hours on 13 May were also low-yield devices in the sub-kiloton range. [22]

---

[13] A mass of 15 kg is above the critical mass specified in Table 17.1, even for a bare ^{239}Pu sphere. It could have been subcritical before implosion if, for example, it was in the form a spherical shell with a sizable hollow core.

The embassy document went on to affirm India's right to have nuclear weapons, to deplore the need for nuclear weapons, and to criticize the NPT for the preferred position it gave to the NWSs—"the five countries who are also permanent members of the U.N. Security Council." It also stated that, with the completion of its May 1998 tests, India would "now observe a voluntary moratorium" on further tests.

There is some disagreement as to the number of nuclear devices involved in the tests and their yields. According to Indian sources, the May 11 explosions were with devices of 15 kt (fission), 45-kt (thermonuclear with a 15-kt fission device to start the bomb), and 0.2-kt, for a total of about 60 kt, and the May 13 test involved the simultaneous detonation of 0.3- and 0.5-kt devices. American observers thought the May 11 figure to be an overestimate, a belief based in part on analyses of data from a system of seismic detectors set up in advance to monitor compliance with the Comprehensive Nuclear Test Ban. They reported no detection of the May 13 devices [7].

However, uncertainty as to the actual size and nature of the explosions does nothing to cast doubt on the existence of a vigorous Indian nuclear weapons program. Together, the Cirus and Dhruva reactors may have the capacity to produce roughly 30 kg of weapons-grade plutonium per year [12, p. 266]. India also could extract plutonium from some of its CANDU electric-power-generating reactors, but there is no evidence that it has the motivation to do so. David Albright gives a median estimate of 310 kg for India's weapons-grade plutonium at the end of 1999—sufficient for 65 nuclear weapons [23]. India has also developed some capabilities to enrich uranium. The goal of that program is not clear [12, pp. 269–271].

In carrying out its nuclear program, India has been able to blur the distinction between peaceful uses of nuclear energy on the one hand and a weapons program on the other, having made use of research reactors and, conceivably, power-generating reactors to obtain fissile material. Perhaps this is the leading case where there might be a positive link between possession of civilian nuclear power and the development of nuclear weapons. The start appears to have come using a research reactor, rather than a power reactor, but an expanded weapons program could be partly hidden behind its power generation program, if India chooses to do so.

Pakistan

Following its defeat in a mini-war with India in 1971, Pakistan reportedly decided in 1972 to develop its own nuclear weapons [24, p. 90]. It took a dual approach toward obtaining fuel, including facilities for both uranium enrichment and plutonium extraction. However, its weapons program is believed to have been based primarily on enriched uranium produced with centrifuges.

As of 1990, Leonard Spector concluded that "Pakistan probably could deploy five to ten bombs for delivery by aircraft" [24, p. 89], but other estimates of its bomb capabilities at that time were lower [12, pp. 271–272]. It is evident, in any case, that Pakistan had a substantial program directed toward the building of nuclear weapons, which presumably provided an expanding arsenal as time progressed. In late May of 1998, less than 3 weeks after the Indian nuclear detonations, Pakistan was in a position to announce six nuclear detonations of its own: one of 25–36 kt, two of 12 kt, and three below 1 kt [25]. Albright has estimated that, as of the end of 1999, Pakistan's inventory of weapons-grade uranium was probably between 585 kg and 800 kg. His median estimate was 690 kg, which would be enough for about 39 weapons at an estimated "most likely" value of 18 kg each [23].[14]

Many countries contributed to the ultimate success of Pakistan's efforts to develop enrichment facilities, whether this was done knowingly or not. Many of these contributions were made clandestinely to circumvent export controls of the countries involved. The Pakistani program was led by Dr. Abdul Qadeer Khan, a Pakistani metallurgist who worked in the early 1970s in the Netherlands at a company associated with Urenco, a consortium set up by Germany, the United Kingdom, and the Netherlands to provide a European center for uranium enrichment.[15] Its major facility was located at Amelo, in the Netherlands. Dr. Khan had the opportunity to become familiar with Urenco's equipment and processes, and reportedly stole Urenco designs which he brought to Pakistan in 1976. Pakistan's further efforts were aided by key equipment sent surreptitiously from abroad. Shipments of specialized equipment from China and from companies in Germany were reported by Albright et al. in their 1997 book [12, p. 272]. This probably was just the tip of the iceberg, with aid to Pakistan coming from companies and individuals in a wide array of countries.

In recognition of Khan's work on the weapons program, the enrichment plant he developed was in 1981 renamed the A. Q. Khan Research Laboratories. At some point, Pakistan began to sell technical information and equipment to Iran, Libya, and North Korea, including a shipment of uranium hexafluoride (UF_6)—the feedstock for enrichment plants—to Libya. Other reports indicate that approaches were made to Iraq and Syria. Khan's prestige as the "father" of the Pakistani nuclear bomb program gave him considerable freedom in his actions, and it is not clear whether the sales and transfers were carried out with the knowledge of the Pakistani government. Khan's own

---

[14] Albright suggests a range in sizes from 12 to 25 kg per weapon. The estimated 18 kg is above the nominal value suggested in Section 17.2.3, perhaps reflecting the assumption of a not very advanced design.

[15] The account presented here of the development of the Pakistani enrichment program, including transfers to and from other countries, is based largely on reports in *The New York Times* and other media outlets that appeared in January and February of 2004 when these activities first became matters of public knowledge. Although these accounts appear to be reliable, it is very possible that further revelations will substantially amplify the picture.

motivations appear to have stemmed from both ideological attitudes and the desire for personal financial gain.

Although aspects of this program had been known to U.S. and European intelligence agencies, it did not come into public view until after the interception of a German freighter that had passed through the Suez canal in October 2003 en route from Dubai to Libya, and the seizure of centrifuge components from the ship's cargo. The components had been manufactured by a company in Malaysia using blueprints supplied by Dr. Khan. U.S. intelligence had been aware of this activity and had tracked the shipment from Malaysia to Dubai and then on its way to Libya. The Malaysian company has denied knowledge of the intended use of this equipment. The proliferation dangers may be even greater with unwitting accomplices than with conscious ones, although in this case the denial has been met with some skepticism.

Following this seizure, Libya agreed to abandon its nuclear program (see Section 18.2.4), and Pakistan came under considerable pressure to reveal and terminate its export program. In response, Khan gave a brief public television presentation admitting the sale of Pakistani nuclear technology to foreign countries, absolving the government of responsibility for his actions, and asking for forgiveness. Shortly thereafter, President Musharraf pardoned Khan, who he said would be allowed to keep his personal gains from the nuclear sales, but disclaimed broader governmental responsibility for the export of Pakistani nuclear technology. President Musharraf indicated that such exports were terminated.

In addition to its uranium-enrichment program, Pakistan has built and announced the operation in 1998 of a 50–MWt reactor at Khushab, which is reportedly moderated by graphite or heavy water. This gives Pakistan a significant capability for plutonium production [12, 23]. It is clear that whatever is happening to its export activities, Pakistan feels justified in having nuclear weapons for its own use in a world where other countries, including India and Israel, have them.

18.2.3 Countries Believed to Have or Be Seeking Nuclear Weapons

Israel

Since the 1960s, there have been reports that Israel was developing nuclear weapons. A key early step in this program was the construction of a reactor at Dimona; the reactor was fueled with natural uranium and moderated with heavy water. Constructed with the aid of France starting in the late 1950s, the reactor began operation in 1963. The reactor was nominally a 24-MWt research reactor, but a series of changes have apparently raised its capacity to at least 40 MWt and possibly to 70 MWt or even higher [12, p. 258]. The initial source of heavy water was Norway, under nominally strict controls to restrict use to peaceful purposes. Israel completed a plutonium reprocessing facility at Dimona in the 1960s, with a capacity estimated to be 15–40 kg/yr. It has been suggested that Israel has centrifuge or laser uranium-enrichment capabilities for producing highly enriched uranium and has produced tritium (for boosted fission) by irradiating lithium at its reactor [12, pp. 263–264].

With the Dimona reactor, Israel has the capacity to produce something in the neighborhood of 10 to 18 kg of weapons-grade plutonium per year [12, p. 259].[16] There has been no official acknowledgment from Israeli sources that this indeed has been done, but somewhat ambiguous statements have been interpreted as confirmation that Israel could take whatever components it has and quickly make bombs. This is probably more a semantic issue than a technical one, and most observers assume Israel has, or could have on very short notice, a substantial number of nuclear weapons. The actual number is a matter of speculation, but the consensus appears to be that it is at least in the neighborhood of 100 and is perhaps more (e.g., Refs. [2, p. 194] and [12, p. 263]). Overall, it appears that Israel has a well-developed nuclear-weapons program, started with the aid of France and continued with the acquiescence of the United States.

North Korea

North Korea illustrates how a country with a relatively small technical base may be able to go it alone in weapons development, given sufficient determination and some small initial help. North Korea began with a small light-water-moderated research reactor, its IRT reactor, received from the USSR in 1965 [26]. It had an initial capacity of 2 MWt; it was later upgraded to 4 MWt and then to 8 MWt [27]. North Korea also was helped by the USSR with small-scale reprocessing equipment, and a small amount of plutonium was extracted from the IRT reactor fuel, reportedly first in 1975 [27].

The start of a serious weapons program may have come later, in the 1980s, with the construction of a graphite-moderated, gas-cooled reactor, variously described as 5 MWe and 20 to 30 MWt (it will be referred to below as a "5-MWe reactor"). It went into operation in 1986 [12, p. 295]. It was of an elementary design, similar to designs developed in the 1940s in Britain, and may have been built without significant outside help. This reactor and the IRT reactor are both located at Yongbyon.

Although North Korea ratified the NPT in 1985, it managed to keep the 5-MWe reactor free of continuous, comprehensive IAEA inspections, and there could only be speculations about the purposes and history of its operation. An IAEA team concluded from a 1992 inspection that plutonium had been separated from spent reactor fuel in 1989, 1990, and 1991, in conflict with the

---

[16] More specifically, the estimated production is about 10.5 kg/yr at a 40-MWt capacity and 18 kg/yr at 70-MWt, in each case for operation at a capacity factor of 75% and a fuel irradiation period of three months. The estimated burnup is 400 MWd/t, about 1% that of a power reactor.

North Korean claim that the only plutonium separation had been of 100 g in 1990.[17] On the basis of assumptions as to operating times and levels, Albright et al. estimated that 7 to 9.5 kg were possibly discharged from the 5-MWe reactor in 1989. Discharges in 1990 and 1991 could have added appreciably to this, but the authors report that "US officials have repeatedly discounted [this] possibility" [12, p. 298]. Additional amounts may have come from the IRT reactor [27]. Although North Korea did not admit to having extracted any large amounts of plutonium, much less to the construction of nuclear weapons, estimates of this sort led to suggestions that North Korea had probably built one or two bombs by 1994, and perhaps more.

An additional concern was a refueling of the 5-MWe reactor in April 1994, with the removal of spent fuel from it. This newly discharged fuel created the potential for substantial additional extraction of plutonium, although, at first, it was in cooling pools and had not been reprocessed. Albright et al. estimated that about 25–30 kg of plutonium were available from this fuel, enough for about five bombs [12, p. 299].

Beyond this batch, further production from the 5-MWe reactor would give enough plutonium for roughly one bomb a year.[18] In addition, two larger graphite-moderated, gas-cooled reactors were under construction at the time: a 50-MWe (200 MWt) unit at Yongbyon and a 200-MWe (600–800 MWt) unit at Taechon. These reactors could have been completed by about 1996 and together would have given North Korea the potential to produce roughly 200 kg of weapons-grade plutonium per year (at a 70% capacity factor), although the 200-MWe reactor may have been intended primarily for electricity production [12, p. 300].

The prospect of a significant North Korean nuclear-weapons program led to protracted negotiations involving North Korea, the IAEA, the United States, South Korea, and Japan that led in Autumn 1994 to an agreement between the United States and North Korea—the so-called Agreed Framework (AF)—under which North Korea would halt construction of the two new graphite reactors and allow IAEA inspection of its nuclear facilities under the NPT. The AF included suspension of reprocessing activities. In return, assistance from South Korea, Japan, and the United States was to be given to North Korea to help it obtain two LWRs with a total capacity of about 2000 MWe, plus heavy fuel oil to provide energy pending completion of the

---

[17] The IAEA conclusion was based on the relative amounts of ^{241}Am and ^{241}Pu in material samples taken by swabbing the insides of North Korean production-handling facilities [12, p. 285]. The ratio of ^{241}Am to ^{241}Pu is a measure of when separation took place, because the ^{241}Am in these samples comes from the radioactive decay of ^{241}Pu ($T = 14.35$ yr).

[18] According to the rule-of-thumb cited in Section 17.4.3, a dedicated reactor can produce about 1 g of plutonium per MWd(t) of operation. Therefore, continuous operation of the 5-MWe reactor (~ 20 MWt) would yield about 7 kg of plutonium per year, although continuous operation is impossible and the actual output will be less.

first reactor [12, pp. 290–291].[19] It is striking that the U.S. government found it prudent to help provide nuclear reactors for North Korea at a time when it showed no interest in promoting nuclear power in the United States.[20]

Some aspects of the Agreed Framework were implemented, including a freeze on construction of the new reactors and the reprocessing of spent fuel, although with some disputes over inspection procedures. In the meantime, work began on the planned 1000-MWe reactors, built under the auspices of the Korean Peninsular Energy Development Organization (KEDO), a U.S.-sponsored consortium which is mostly funded by South Korea. The reactors are 1000-MWe PWRs and use a standard South Korean reactor as the basis for design and construction [26, p. 7].

Difficulties in the relationship that had been established under the Agreed Framework came to the fore in Autumn 2002. In October, North Korea acknowledged a program to enrich uranium, in what could be a step in another route to weapons and a violation of the spirit of the Agreed Framework. Starting in November 2002, North Korean authorities began making statements that were taken to affirm that North Korea had at least a potential nuclear-weapons program, but that were perhaps intentionally ambiguous as to the status and goals of the program. In specific steps, North Korea announced on December 12, 2002 that it was going to reactivate the 5-MWe reactor at Yongbyon which has been shut down since the 1994 agreements, and later in December, it asked the IAEA to remove the seals and monitoring equipment that were intended to bar the reprocessing of the spent fuel that had been discharged in 1994. When the IAEA declined, North Korea did this unilaterally and asked the IAEA inspectors to leave the country [8]. In January 2003, North Korea withdrew from the NPT. Thus, North Korea appeared to be trying to develop a large weapons capability, obtaining fuel from the enrichment of U and the reprocessing of spent fuel.

As of the beginning of 2004, the future of the North Korean nuclear program was still a matter of negotiation—now involving North and South Korea, China, Japan, Russia, and the United States. It was not clear whether North Korea is determined to have nuclear weapons under any circumstances or whether something like the Agreed Framework could be again put into force.

Iraq

Iraq's nuclear ambitions have expressed themselves in two, or possibly three, stages. Iraq obtained a large research reactor from France in 1976, the Osirak

---

[19] The agreement was to provide 500,000 tonnes of oil per year, enough to replace approximately 300 MWe of nuclear capacity.

[20] The 2 GWe of nuclear capacity could produce roughly 500 kg of reactor grade plutonium a year. However, reprocessing of conventional power reactor fuel (UO_2) requires additional equipment, beyond that needed for the metal-alloy fuel used in the North Korean 5-MWe reactor. It was hoped that agreed-upon IAEA inspections plus this additional technical difficulty would prevent surreptitious violations of the agreement not to reprocess spent fuel.

reactor (also known as the Tammuz-1 reactor) [2, pp. 194–196]. This was a 70-MWt reactor, fueled by highly enriched uranium.[21] With less ^{238}U in the fuel, the yield of ^{239}Pu in the core is relatively small, and the apparent intent was to place natural or depleted uranium "targets" in the neutron flux outside the core. This would have produced ^{239}Pu, reportedly in amounts sufficient for at least one bomb per year [2, p. 195]. The Osirak reactor was destroyed by an Israeli bombing raid in June 1981. Iraq was a signatory to the NPT, and its facilities were under IAEA supervision, but not under continuous IAEA inspection.

Following the destruction of the Osirak reactor, a standard view was that Iraq was "unlikely to join either the civil or military nuclear club during the rest of this century" [28, p. 42]. However, Iraq was not so easily discouraged. It reportedly made an unsuccessful effort to purchase plutonium from Italian arms smugglers and later began efforts to produce enriched uranium. By late 1989, Iraq's efforts to develop centrifuge enrichment facilities had been widely recognized [24, p. 192].

After Iraq's military defeat in early 1991, IAEA inspections revealed a previously unreported major electromagnetic separation program, an ambitious but incomplete centrifuge enrichment program, and the separation of small amounts of plutonium from a research reactor [29]. Iraq had been helped in this program, presumably unwittingly, with equipment and information from many countries, including West Germany, Switzerland, and the United States. It is apparent from a 1990 Iraqi Progress Report, obtained by the IAEA inspectors and reprinted in a report by Peter Zimmerman, that Iraqi scientists had very considerable scientific understanding and technical expertise [29].

Iraq's defeat in the Gulf War and subsequent U.N. inspections presumably destroyed Iraq's physical facilities for making nuclear weapons. After the withdrawal of U.N. inspectors from Iraq in 1998, the nuclear program may have resumed. No conclusive evidence of a nuclear-weapons program was found in the immediate aftermath of the 2003 war with Iraq, and as of the end of 2003 the nature of Iraq's nuclear efforts in the 1999–2003 period is not publicly known.

Iran

Iran has for many years attempted to develop a nuclear power program. Under the Shah, who was driven from power in 1979, this was with the cooperation of Western countries, starting with a 5-MWt research reactor from the United States in the 1960s [2, p. 198]. As part of a larger long-term program, construction was begun in 1974 at Bushwehr of two 1300-MWe West German LWRs [12, p. 354]. Iran also made contracts to obtain low-enriched uranium

---

[21] This capacity, from Ref. [2], may overstate the size of the Osirik reactor. Albright et al. termed it a 40-MWt reactor [12]. It was not Iraq's first choice. Iraq had originally asked France for a 500-MWt graphite-moderated reactor but France refused—presumably because the requested reactor was best suited for plutonium production [2, p. 195].

for its reactors. After the fall of the Shah in 1979, this program was aborted due to domestic opposition, foreign unhappiness with Iran, and bomb damage during the later Iraq–Iran war.

However, a group of trained nuclear scientists remained. Efforts to establish an ostensibly civilian nuclear program were subsequently renewed. Russia agreed in 1995 to complete one of the reactors at the Bushwehr plant, but given the wartime damage and the passage of time, a 915-MWe Russian PWR has replaced the original German reactor. At the end of 2002, it was scheduled to go into operation in 2004 [20]. Its completion may be somewhat delayed but does not seem to be in doubt. The second Bushwehr plant is still listed by the IAEA as being "under construction" (see Table 2.5), but no completion date is specified and this may only be a nominal status. However, there are reports of continuing discussions between Russia and Iraq about the construction of additional reactors [20, p. 42].

Any such discussions are proceeding against a backdrop of concern about Iran's suspected nuclear-weapons ambitions. From the first, many observers assumed that Iran—a country that is very rich in oil resources—was interested in nuclear reactors as a stepping stone to nuclear weapons, despite Iran's denial of such ambitions. By 2003, it was confirmed that Iran was building a uranium-enrichment plant and a heavy water production plant, making the concerns immediate enough to prompt a special IAEA investigation in Iran. Facilities acknowledged by the Iranian authorities and noted by the IAEA include the following [30]:

- A pilot centrifuge plant for the enrichment of uranium (ready to begin small-scale tests) and a commercial-scale plant (ready to accept centrifuges in 2005), both under construction in Natanz.
- A heavy water production plant, under construction at Arak.
- A 40-MWt reactor (the IR-40), to operate with natural uranium fuel and a heavy water moderator, on which construction is scheduled to start at Arak in 2004.

The enrichment plant, as discussed by David Albright and Corey Hinderstein, is designed to have a capacity of 100,000–150,000 SWU/yr, which is enough to produce the amount of low-enriched uranium used annually for a 1000–MWe reactor [31]. They warn that this technological capability suggests that Iran could develop a secret 10,000–SWU/yr enrichment facility, which could annually provide enough highly enriched uranium for about three bombs.[22] The development of this enrichment capability was aided by a variety of out-

---

[22] Under typical conditions, a 1000–MWe PWR consumes about 1.0 tonnes of ^{235}U per yr, in the form of low enriched uranium. At the rate of 142 kg-SWU per kg of ^{235}U (for a 3.75% enrichment), this corresponds to an annual requirement of 142,000 kg-SWU (see Sections 9.2.2 and 9.3.3). Production of 95% enriched uranium for bombs requires about 220 kg-SWU per kg of U, and a 10,000 SWU/ yr facility would provide about 45 kg of enriched uranium annually.

side players, including sources in Germany, Switzerland, China, and Pakistan, although the full dimensions of the aid received by Iran remains unknown [32].

Flexibility is added to Iran's potential weapons program by the heavy water facilities, which could be used to make weapons grade plutonium in a reactor that uses natural uranium.

Iran is not barred from having enrichment or heavy water facilities under the terms of its NPT agreements with the IAEA, although the IAEA report criticized the Iranian failure to keep the IAEA properly informed of its activities [30]. Future Iranian efforts to build weapons could be restrained by internal Iranian policy decisions, pressure from Russia, economic and political campaigns organized by the United States, or more stringent IAEA requirements. These factors are closely interrelated, with the United States trying to influence Russian policy, many interested countries having inputs to the IAEA, and Iran's own policies probably impacted by the carrots and sticks that may be offered by the outside world.

In a positive development—whose long-term implications remain to be seen—an agreement between Iran and the IAEA was signed on December 18, 2003, which took the form of an Additional Protocol to Iran's earlier agreement under the NPT [33]. As reported by Mohamed ElBaradei, the IAEA Director General, Iran agreed to "more robust and comprehensive inspections" and "as a confidence building measure" to suspend enrichment and reprocessing activities. He expressed the hope that these "welcome steps. . .will be sustained."

18.2.4 Countries That Have Abandoned Nuclear Weapons Programs

The Bridling of Weapons Ambitions

Although, by now, a troublingly large array of countries have nuclear weapons or are believed to be trying to get them, the number could be much larger if all countries with the potential ability to do so had made the attempt. Many, in the terminology of a book on the subject, had "bridled" any weapons ambitions that they may have had [34]. We discuss here the cases of Argentina, Brazil, South Africa, and members of the former Soviet Union.

Additional examples, mentioned by Albright and colleagues, are weapons programs that were pursued by Sweden in the 1950s, by South Korea in the 1960s and 1970s, and apparently by Taiwan in the 1980s [12, pp. 351 and 366ff]. The South Korean program was abandoned under "international pressure," but conceivably it could be revived, if attempts to dissuade North Korea from building a nuclear arsenal fail. Sweden gave up an "active program" under internal pressure, and it is difficult to think of any circumstances under which this program would be revived. Taiwan obtained a 40-MWt heavy water research reactor from Canada in 1969, known as the Taiwan Research

Reactor (TRR).[23] This reactor was used to produce enough weapons-grade plutonium for about 17 bombs. Most of this plutonium has been shipped to the United States, and under U.S. pressure, the TRR has been shut down and its heavy water has been sent to the United States. Thus, the Taiwanese weapons program—assuming there really was one—has been given up, at least for the present.

The bridling of these ambitions was a domestic decision for Sweden. However, in the case of South Korea and Taiwan, U.S. influence played a key role. This influence is probably less persuasive for countries that are not under an overall U.S. protective umbrella. For a protected country without aggressive goals, it is less expensive to rely on such an umbrella than to undertake to develop its own weapons.

Argentina

Argentina obtained a small research reactor in 1958 and a 320-MWe heavy water CANDU power reactor in 1974. It made its first public moves toward a weapons program in 1978, with the announcement of a planned plutonium reprocessing plant and the start on construction of a gaseous diffusion plant for uranium enrichment. Work on the reprocessing plant was terminated in 1990, apparently before it went into operation, but the gaseous diffusion plant began operating in 1988. The plant was designed for a 20% enrichment in ^{235}U but could be used for higher enrichments, with a theoretical capacity of 100 kg/yr of weapons-grade uranium [24, p. 228].

Although this program appeared originally to be directed toward a weapons program, in competition with Brazil, there was a change in the political climate in South America, and in 1990, both Argentina and Brazil renounced nuclear weapons [35, p. 177]. In December 1991, they signed a joint agreement with the IAEA for inspections [12, p. 369]. In 1994, Argentina officially ratified the Treaty of Tlatelolco, the regional South American agreement barring nuclear weapons, and it signed the NPT in 1995.

Brazil

Partly in response to Argentina's nuclear program, Brazil took an important step toward nuclear-weapons capabilities in 1979, with the start of a centrifuge uranium-enrichment plant. It was put into operation in 1988, with the ostensible purpose of producing uranium enriched to 5% or 20% in ^{235}U [24, p. 250]. Brazil has also had a very small plutonium reprocessing program. At the same time, a 626-MWe PWR was put into commercial operation in 1985. Whatever its original aspirations, Brazil gave up its nuclear-weapons

---

[23] The TRR is a duplicate of the CIRUS reactor in India, which provided the start for the Indian weapons program.

program, paralleling the decision by Argentina.[24] Brazil signed the Treaty of Tlatelolco in 1994 [12, p. 369] and the NPT in 1998 [36]. However, in a move that indicated a desire to remain independent in nuclear matters, it opened a new uranium-enrichment plant in December 2002, with the avowed purpose of providing enriched uranium for its two nuclear power plants [37].

South Africa

South Africa undertook a weapons program based on the enrichment of uranium using an aerodynamic process, developed with German help [2, p. 109].[25] Its major facility for weapons-grade uranium enrichment was built in the 1970s and reportedly had an annual output capacity of about 60–90 kg of 93% enriched uranium [12, p. 382]. This gave South Africa the ability to build a substantial arsenal.

In 1990, along with many other changes in political direction, South Africa decided to abandon its nuclear-weapons program. It shut down its plant for producing highly enriched uranium and suggested that it would join the NPT. Culminating its renunciation of nuclear weapons, South Africa signed the Non-Proliferation Treaty and entered into a safeguards agreement with the IAEA in 1991 [38]. The scope of the South African nuclear program was revealed by President F.W. de Klerk in an address to the country's parliament in March 1993, when he reported that six nuclear weapons had been built by 1990 and later dismantled [39].

Former Soviet Union: Belarus, Kazakhstan, Ukraine

The Soviet Union collapsed in December 1991, when Ukraine voted for independence and the Soviet Union, at the initiative of Boris Yeltsin, dissolved into the far looser Commonwealth of Independent States. The nuclear weapons of the former Soviet Union were located in four of these states: Russia, Belarus, Kazakhstan, and Ukraine. It was agreed that only Russia would remain a NWS and the others would transfer all weapons to Russia and become nonweapon states. By the end of 1996, all three states had completed these transfers and had acceded to the Non-Proliferation Treaty. As described by Mitchell Reiss, this compliance came about with pressure from Russia and the United States and some financial incentives from the United States—albeit with prolonged negotiations, especially in the case of Ukraine [34].

---

[24] Brazil was apparently interested in uranium enriched to 6–7% in ^{235}U for naval submarines as well as more highly enriched uranium for weapons [34, p. 50].

[25] In this process, a spiraling motion is imparted to a jet of uranium hexafluoride gas causing the lighter and heavier components to be separated by centrifugal action.

Libya

An attempt by Libya to develop nuclear weapons came to public attention in late 2003, along with an announced agreement by Libya to abandon the program and submit to inspections. The program was based on centrifuge technology for uranium enrichment which reportedly was transferred to Libya by individual Pakistani scientists [40]. This possibly was done without the knowledge of the Pakistani government. The program was abandoned after negotiations with the United Kingdom and United States, but neither the full scope of the program nor the incentives or pressures used to dissuade Libya were public knowledge as of early in 2004.

18.2.5 Summary of Pathways to Weapons

We summarize in Table 18.4 the ways in which fissile material has been sought or obtained by the various countries that have undertaken nuclear-weapon programs. (Iran is omitted as an ambiguous case.) A striking feature of the table is the reminder it provides that many of the countries that built or aspired to build nuclear weapons have first used enriched uranium rather than plutonium. This means that antiproliferation efforts must take into account uranium-enrichment facilities as well as plutonium-producing reactors.

Countries seeking fissile material for weapons need amounts that are small compared to the amounts handled in nuclear reactors. A single 1000-MWe LWR produces enough plutonium each year for close to 30 bombs, albeit perhaps not very efficient ones (see Section 17.3.2). Producing the enriched uranium to fuel a reactor for 1 year requires in the neighborhood of 140,000 kg-SWU of separative work. This amount of separative work could produce over 600 kg of 95% uranium, enough fuel for at least something in the neighborhood of 50 weapons. Thus, the size of the facilities needed to obtain material for a significant number of nuclear weapons is much less than that required for the generation of a significant amount of nuclear electricity.

18.3 Nuclear Power and the Weapons Threat

18.3.1 Potential Role of Nuclear Power in Weapons Proliferation

Peaceful Uses of Nuclear Energy

Starting with the Atoms for Peace program of the Eisenhower Administration and continuing through the formulation of the Non-Proliferation Treaty, it was believed by the U.S. government that it would be possible to keep the military and civilian aspects of nuclear energy separate and that the promise of aid with peaceful nuclear activities could help induce countries to forgo military uses. Articles IV and V of the NPT in fact obligate NWSs to aid non-NWS

Table 18.4. Initial methods used to obtain fissile material, for countries that have nuclear weapons or that are believed to have attempted to develop nuclear weapons.

Country	Material	Method for Obtaining Fissile Material
Nuclear-weapon states[a]		
United States	^{239}Pu	Natural U reactor, graphite moderated
	^{235}U	Diffusion and electromagnetic separation
Former USSR	^{239}Pu	Natural U reactor, graphite moderated
United Kingdom	^{239}Pu	Natural U reactor, graphite moderated
France	^{239}Pu	Natural U reactor, graphite moderated
China	^{235}U	Gaseous diffusion
Acknowledge weapons		
India	^{239}Pu	Natural U research reactor, D_2O mod.
Pakistan	^{235}U	Centrifuge
Probably have weapons		
Israel	^{239}Pu	Natural U research reactor, D_2O mod.
North Korea	^{239}Pu	Natural U reactor, graphite moderated
Relinquished weapons		
South Africa	^{235}U	Jet nozzle technique
Started weapons program		
Argentina[b]	^{235}U	Gaseous diffusion
Brazil[b]	^{235}U	Centrifuge
Iraq	^{239}Pu	Enriched U reactor; external U targets[c]
	^{235}U	Electromagnetic separation[d]
Libya[b]	^{235}U	Centrifuge

[a]See text (Section 18.2.1).
[b]These programs were terminated voluntarily.
[c]The reactor for this program was destroyed before it became operational.
[d]This enrichment program was terminated in 1991, following the Persian Gulf War.

parties to the treaty in pursuing peaceful applications of nuclear power. One result of these policies has been the spread of research reactors throughout the world.

It can be debated whether this policy was wise or quixotic. However, the good or the harm has been done.[26] In examining the links between commercial nuclear power and nuclear weapons, it is necessary to keep this history in mind.

---

[26] Here, we are focusing on the negative aspects of nuclear energy, namely the potential to produce weapons. There are positive practical aspects as well, aside from electricity production. Examples include the use of radionuclides in medical diagnosis and therapy and myriad industrial applications.

A wide diffusion of nuclear knowledge and technology has already taken place, extending to many countries that as yet have no nuclear power. Further, to the extent that peaceful nuclear energy has been involved in helping start weapons programs, research reactors, not power reactors, have, to date, been the apparent culprit.

Nuclear Power as a Facilitator of Proliferation

Although nuclear power has contributed little to past weapons proliferation, a major objection to it arises from the fear that it will serve to facilitate the future spread of nuclear weapons, either to additional countries or to terrorists. The efforts to establish "nuclear-free" zones, in which nuclear weapons and nuclear energy are both barred, dramatize the view that the linkage is so tight that both must go.[27] In this position—both in this dramatic form and in scholarly analyses—it is argued that nuclear power can be a stepping stone to nuclear weapons.

For example, if a country has nuclear power, obtained ostensibly for peaceful purposes, it can assert the need to have enrichment facilities to produce the slightly enriched uranium used in reactors. Thus, Brazil started up a small enrichment plant for this purpose in 2002, and Iran has been building one in conjunction with its new reactor program. In neither case is domestic enrichment essential to the reactor program, because enriched fuel commonly is imported from a few major suppliers. However, it is difficult to deny a country the right to strive for "energy independence." Similarly, Iran reportedly was building a reprocessing plant, which would enable it to recycle plutonium into a reactor. Such facilities do not give countries an immediate weapons capability, but they lower the technical barriers and ease the path to the production of weapons.

Most fundamentally, the possession of nuclear reactors helps the country's scientists and engineers gain familiarity with nuclear technology. Further, it can provide a "cover" for the importation or manufacture of equipment that could be used in bomb making. The overall case is made compactly by Harold Feiveson:

> [C]ountries or terrorist groups could divert fissile material directly from the civilian nuclear fuel cycle into nuclear explosives;
>
> ...countries aspiring to obtain nuclear weapons could use civilian nuclear facilities (power reactors, research reactors, reprocessing plants, uranium-enrichment plants, etc.) and trained cadres of nuclear scientists, engineers, and technicians as a cover and/or training ground for the dedicated acquisition of fissile material for nuclear weapons. [41, p. 11]

---

[27] Such zones have been already been declared in some U.S. cities—in particular, Berkeley, California.

Feiveson stresses the danger that—given a large global expansion of nuclear power—the technology will spread to many additional countries, whose peaceful intent cannot always be counted upon. Thus, in this analysis, more countries become proliferation threats, immediate or latent. Measures to reduce this danger are discussed in Section 18.3.4.

18.3.2 Weapons Dangers for Different Categories of Countries

Countries with Nuclear Weapons, Admitted or Suspected

The most obvious nuclear weapons threats come from countries that already possess weapons:

- The NWSs, which admit to nuclear weapons and have no plans to renounce them.
- India and Pakistan, which carried out surprise nuclear weapons tests in 1998 and have raised the specter of possibly using nuclear weapons if their political and military conflicts escalate.
- Israel, which almost certainly could quickly deploy a large number of nuclear weapons if it felt sufficiently threatened.
- North Korea, which may or may not have weapons or weapons materials and whose political and military goals are not clear.

Commercial nuclear power has played virtually no role in the weapons buildup of these countries, although small reactors obtained ostensibly for research purposes have, in some cases, been important. The five NWSs all obtained bombs through weapons programs which preceded nuclear power. Of the last four countries, Israel and North Korea have no nuclear power, for India the chief early link with peaceful uses of nuclear power was with research reactors, not power reactors, and Pakistan relied on a dedicated uranium enrichment program. As yet, there is no known case of plutonium having been diverted for weapons purposes from a civilian reactor used for electricity generation.

Aspiring Weapons States

Iraq has made a substantial effort to obtain nuclear weapons, although it has been thwarted in this, at least temporarily. Again, commercial nuclear power is irrelevant to its programs, because Iraq neither has nuclear power nor any announced plans to obtain it.

Iran is another matter. Russian steps to help it with civilian nuclear power are viewed with suspicion, because it is feared that Iran intended to use this program to further weapons development. An interest in civilian power can be just a pretext for obtaining reactors and, as a next step in the process, enrichment or reprocessing facilities. Many people in the United States would breathe easier if Iran did not have nuclear power, even those who would favor nuclear power in countries they view as benign. The feared Iranian weapons program might be undertaken surreptitiously, evading IAEA inspections, or openly with or without withdrawal from the NPT.

Overall, we would feel more secure if we could deny nuclear power to countries that we believe have aggressive intentions. The United States has taken a particularly active role in promoting such a policy, but it has only limited influence and is constrained by pragmatic considerations—as in the decision in 1994 to help provide reactors to North Korea while attempting at the time to deny them to Iran. There is no recognized international mechanism for implementing a policy of barring nuclear power selectively, especially in view of the difficulty of getting agreement as to the intent of suspected nations. In particular, there is no international body that is likely to have both the will and the means to prevent major powers, such as Russia, China, and the United States, from providing nuclear facilities to other nations if they wish to do so.

Countries with Nuclear Power But No Weapons

Many countries exhibit the other side of the coin: they have nuclear power but do not have nuclear weapons. Some have not shown any inclination toward weapons (e.g., Canada, Germany, and Japan).[28] All have strong, comprehensive nuclear power programs and could easily develop weapons with no external aid. Sweden, which also has a strong civilian nuclear power program and technological base, embarked on a sophisticated program of weapons development, but the effort was abandoned before attracting much outside public attention [42]. Beyond these, Argentina and Brazil had moved toward obtaining nuclear weapons, and South Africa succeeded, but all three have given up these programs.

There are different motivations for abstaining from nuclear weapons: matters of principle, the belief that a nuclear weapons race would be counterproductive in terms of security, the lack of a threatening enemy, and the feeling that they are already under a protective umbrella established by the United States. Economic pressures also play a part. The action of Argentina in renouncing nuclear weapons was explained in terms that have a broad relevance to developing countries. As stated by the Argentinian Under Secretary of Foreign Affairs:

> We found we were blacklisted by the international community for our aggressive policies and in the end found we had to cooperate with the netherworld of third-world countries.... The paradox was that Argentina was trying to reach high tech through its nuclear program, but because of a lack of openness Western countries made sure it would not get there. I think we have learned our lesson. [43]

---

[28] Japan's position may change if North Korea pursues the development of nuclear weapons.

Here, the premise of the atoms-for-peace policy seems to have been vindicated. For countries seeking weapons, the stick is technological denial. For those that renounce them, the carrot is technological reward, possibly including help in obtaining commercial nuclear power—although the promise of this reward has not as yet been effective with North Korea.

Countries with Neither Nuclear Power nor Nuclear Weapons

Countries that lack both nuclear power and nuclear weapons tend to be countries without a highly advanced technological base. As such, they might not be thought to pose a significant threat. However, a number of countries mentioned above (Iran, Iraq, and Libya, as well as North Korea which may already have built weapons) all could be placed in this category, and they have been viewed as potential threats. Many other countries could mount similar efforts. Those that have abstained often have lacked the desire more than the technical ability.

Zimmerman has suggested the concept of a "bronze medal" technology for weapons development [44]. To win the bronze medal, a country need only achieve the technological level reached by the United States about 60 years earlier. The nuclear weapon thus developed might not win plaudits from sophisticated weapons experts, but it could wreak great actual or threatened damage. South Africa exemplifies a success of this approach, building its six bombs using a gun design similar to that of the Hiroshima bomb and without any very great investment of money or personnel [44]. Other countries could do likewise.

It is not clear how much difference commercial nuclear power would make to the development of nuclear-weapon programs in such countries. However, given the possible connections, the guiding principle is probably the same as the unsatisfactory principle suggested earlier in the case of Iran: try to deny nuclear power to suspected enemies and support it for your friends. Obviously, such a policy, if adopted by the United States, cannot be defended in terms of any general principles of international equity. At best, it is a pragmatic response of only incomplete and temporary applicability, requiring that the United States have more influence and economic power than it probably will possess in most situations. Of course, this influence and power would be amplified if it could be exerted through the United Nations or a strengthened IAEA, but here there may be difficulty in obtaining an effective consensus as to which countries should be helped to obtain nuclear power and which should be impeded.

Subnational Groups

The main premise of the preceding subsections is that it is relatively easy for many countries to build nuclear weapons, including countries that are labeled as "developing." However, it would be a difficult matter for any subnational group that does not have a secure geographical base. The difficulty is greatest if it is necessary to start using plutonium from spent fuel or uranium that requires enrichment. In this case, the subnational group would face an almost insuperable task unless the host country acquiesces in the construction of reprocessing or isotopic enrichment facilities. This would, in effect, turn the effort into a national, rather than subnational, program.

The difficulties are reduced if it is possible to obtain separated plutonium or enriched uranium. Even then, it might be difficult to assemble the people and facilities to make a bomb without the knowledge of the host country. However, if these materials are available, the threshold is lowered. It is therefore essential that they be guarded against potential diversion (see Chapter 17 for a more extensive discussion of terrorist threats).

18.3.3 Reducing Proliferation Dangers from Nuclear Power

Technological Measures

There is no infallible technological fix to the problem of nuclear proliferation. There are many routes to nuclear weapons for determined countries that possess even a modest industrial and scientific base. However, as we have seen, the threshold is raised if there is no easy access to separated plutonium.

It may be noted in this connection that, although the use of MOX fuel described earlier will decrease the stock of separated ^{239}Pu, it develops facilities that could later be used to establish a "classical" reprocessing fuel cycle in which plutonium is extracted from LWR spent fuel for use in MOX fuel. This option was abandoned by the United States in the late 1970s due largely to proliferation concerns (see Section 9.4.2). The MIT study *The Future of Nuclear Power* reiterates this opposition 25 years later:

> The PUREX/MOX fuel cycle produces separated plutonium and, given the absence of compelling reasons for its pursuit, should be strongly discouraged in the growth scenario on nonproliferation grounds. Advanced fuel cycles may achieve a reasonable degree of proliferation resistance, but their development needs constant and careful evaluation so as to minimize risk. [45, pp. 68–69]

This opposition to MOX fuel has been rejected by other important countries—including France, Japan, and Russia—who have been unwilling to accept what they consider to be a long-term economic penalty, and the MIT study calls on the United States to "work with France, Britain, Russia, and Japan to constrain more widespread deployment of this fuel cycle" [45]. Overall, there is no unanimity—worldwide or in the United States—as to whether the risks in this fuel cycle outweigh the gains.[29]

---

[29] See Section 17.4.2 for additional discussion of reprocessing and proliferation.

Technological measures that are explicitly intended to reduce plutonium-related proliferation risks include the following:

- *Once-through fuel cycle.* Restrict nuclear power to a once-through fuel cycle, as is now the practice in the United States.
- *Colocation.* If a fuel cycle with reprocessing is deemed necessary, to extend uranium resources or to reduce the burden of actinides in the wastes, the opportunity for plutonium diversion can be reduced by confining the reactors and the reprocessing plant to the area of a single facility. This concept was at the heart of the original Integral Fast Reactor program at Argonne National Laboratory and is a major consideration in the Generation IV program for new reactors (see Section 16.6).
- *Thorium fuel cycle.* An often-suggested approach is the ^{232}Th-^{233}U cycle, in which ^{233}U is the fissile fuel and ^{232}Th is the fertile fuel. ^{233}U extracted from the spent fuel could be mixed with natural uranium to obtain uranium fuel of low ^{233}U enrichment, to be returned to a reactor. The ^{239}Pu, which is formed from neutron capture in ^{238}U, could be left as part of the highly radioactive wastes. Thus, at no stage is the material in a form that is readily usable for weapons. This cycle can operate as a breeding cycle and there may be interest in pursuing the option in the future, if the perceived need for breeder reactors increases. At the moment, however, there are no active programs to implement this cycle.
- *Self-contained reactors.* Smaller countries could get the benefits of nuclear power in the form of small self-contained units that would be delivered to them with a fully loaded reactor core. These units would be taken back by the supplying country at the end of their lifetimes—at least 10 years and probably longer.

Institutional Measures

No matter what the technical measures, a determined proliferator can develop facilities for weapons—with or without nuclear power. The most promising means of restraining nuclear proliferation is a combination of rigorous inspections, presumably by the IAEA, backed by economic pressures. These pressures could be exerted through the United Nations or, alternatively, by the United States and like-minded countries. Such mechanisms can be faulted for their initial failures in the detection and prevention of the Iraqi and North Korean nuclear programs—which, at times, proceeded vigorously despite the NPT and the IAEA inspections.

However, Iraq's nuclear-weapons program was eventually terminated by a combination of military actions and intense IAEA inspections, and strong political and economic pressures have been brought against North Korea as a result of its flouting the NPT. Iran is emerging as another test of the effectiveness of international measures to inhibit nuclear-weapon programs. These measures could be made more effective by giving the IAEA the expanded authority and resources to to carry out intrusive inspections of both acknowledged and suspected nuclear facilities.

The inhibiting power of inspections and economic coercion are not complete, but their effectiveness should not be underestimated. A number of countries have given up nuclear-weapon programs and one country (South Africa) has given up actual weapons. Although their reasons may have differed, it would appear that international pressures played an important role. The unwillingness of India, Israel, and Pakistan to sign the NPT is testimony to its potential effectiveness. It may be too late, or even undesirable, to try to roll back their programs, but the desire of these countries to avoid being in violation of the letter of the NPT (by signing but not complying) suggests that international safeguard measures are, even now, not completely toothless.

The institutional arrangements contemplated here involve an asymmetry in which some countries have the facilities for a complete fuel cycle while others are restricted, especially with respect to enrichment and reprocessing. It is difficult to justify this asymmetry on the basis of any satisfactory broad principles, but it may be the most practical course for the near future. It mirrors the asymmetry that exists with respect to nuclear weapons. With the adoption of the NPT in 1968, five countries became recognized weapons states, and they are the only current signatories of the NPT that have weapons. This has given the world some stability—but a stability that is far from complete, given the development of weapons by the three countries that have declined to sign the NPT (India, Israel, and Pakistan) and the probable development of weapons by North Korea, which has withdrawn from the NPT.

18.3.4 Nuclear Power and Moderation of Weapons Dangers

Diminishing Conflicts over Oil

So far, we have stressed ways in which nuclear power might exacerbate proliferation dangers. There are also ways in which it might help to reduce them. Conflicts between nations, including conflicts that might escalate into nuclear conflicts, are more likely given severe competition for resources or domestic unrest due to extreme economic difficulties. Were nuclear power able to alleviate these sources of conflict, it could serve to lessen the risks.

The most obvious conflicts of this sort are caused by the need for oil. One of the precipitating factors in Japan's entry into World War II was the tension between the United States and Japan over oil. In an attempt to restrain what it saw as a program of aggressive expansion, the United States sought to limit Japan's access to oil. The sequels were the attack on Pearl Harbor and the bombs at Hiroshima and Nagasaki. Looking to the future, competition for oil could again become desperate enough to lead to dangerous military confrontations, possibly including nuclear war if any of the contestants have nuclear weapons and feel driven to use them.

The wars with Iraq in 1991 and 2003 partially illustrate the point, although Iraq had not yet succeeded in building nuclear weapons and the United States had no need or impulse to use them. However, the wars serve as a reminder of the high stakes involved in Persian Gulf oil. Concern over the security of oil supplies arguably gave the United States a motivation to go to war in 1991, and in the pre-1991 period oil sales gave Iraq the resources to build a strong military machine and begin an ambitious nuclear-weapons program. The wars did not develop into major military conflicts, but that was because the United States and its allies had decisive military superiority. If Iran—or at some later date Iraq—succeeds in developing nuclear weapons or if there is a radical shift in the attitude of Russia, a future conflict over Persian Gulf oil could be disastrous.[30]

Nuclear power can to some extent serve as an alternative to oil—directly as a substitute for oil and natural gas in electricity generation (freeing natural gas to replace oil in many other applications) and less directly through increased electrification and the production of hydrogen for use as an alternative fuel. It can thereby lessen some of the pressures arising from the central role of Persian Gulf oil in the world's energy economy. Although even with the largest plausible expansion in its use nuclear power could not alone eliminate the world's heavy dependence on this oil, it could reduce the level of dependence and thereby dampen the intensity of the competition for it and the associated risks.

Disposition of Excess Fissile Material from Weapons Programs

With the end of the cold war in the 1990s, the United States and Russian governments concluded that they had more nuclear weapons and more weapons-grade fissile material than were needed for security. In fact, it could be argued that their security would be improved by reductions, because there would be less chance of the accidental or intentional misuse of the weapons or the diversion of stock-piled materials. This conclusion led to the START I and START II treaties under which each country agreed to sharp reductions in its number of operational weapons [46, p. 39].

The available amounts of fissile materials was large. Albright et al. estimated the inventories of weapons-grade plutonium to be about 85 tonnes for the United States in early 1996 and roughly 130 tonnes for the former Soviet Union at the end of 1993 [12, pp. 49 and 58]. The estimated weapons-grade uranium inventories were about eight times as large, although most of the uranium was in reserves, not in weapons [12, p. 80]. The reduction of these stockpiles makes large quantities of weapons-grade uranium and plutonium

---

[30] To date, nuclear weapons have provided deterrence, including perhaps deterrence of the United States by North Korea's possible bombs. However, if the stakes are high enough—in particular, the urgency of controlling oil supplies great enough—deterrence might not be effective.

available for disposition. One solution for the disposition of this material is to consume it as fuel in power-generating nuclear reactors—as suggested by the phrases "swords into ploughshares" or "megatons into megawatts."

This is most easily done for uranium. Highly enriched uranium can be "denatured" by mixing it with natural or depleted uranium. This provides substantial supplies of low-enriched uranium, suitable as a fuel for reactors but not usable for weapons. A 1993 agreement between the countries stipulated that the United States would purchase 500 tonnes of HEU from Russia over a 20-year period. This HEU would be diluted to LEU in Russia and then shipped to the United States. Deliveries under this program began in the mid-1990s and by the end of 2003 over 200 tonnes of HEU had been converted to LEU and delivered to the U.S. Enrichment Corporation (USEC) where its enrichment can be further adjusted, as desired by the purchasers of reactor fuel.[31] The full 500 tonnes corresponds, in a conservative estimate by the USEC, to 20,000 bombs. The fuel would suffice for 500 GWyr of reactor operations at 1 tonne of ^{235}U per GWyr (see Section 9.3.3), and perhaps more if the fuel is used more efficiently. The United States has begun a similar program to reduce its own stockpiles of HEU, on a somewhat smaller scale.

The problem of plutonium from dismantled weapons is less tractable. A 1994 National Academy of Sciences (NAS) study indicates that "50 or more metric tons of plutonium on each side are expected to become surplus to military needs" [46, p. 1]. At 50 tonnes each from the United States and Russia, there is enough plutonium for about 20,000 bombs, assuming a nominal 5 kg per bomb.

There are several alternatives for keeping this material out of circulation:[32]

- *Storage option.* The plutonium could be stored in well-guarded facilities. This is the most immediate solution, requiring the least handling. It is an economical solution, saving a potentially useful resource, but a dangerous one should there be any breach of the integrity of the storage. For these security reasons, the NAS study cautions against extending such storage "indefinitely" [46, p. 226].

- *Spent fuel option.* The plutonium could be consumed in nuclear reactors. With some choices of reactors and fuel types, the resulting spent fuel would be similar to LWR spent fuel. It would still have a significant plutonium component, but now the plutonium would be protected by the fuel's high level of radioactivity, and the ^{240}Pu contamination would make it reactor grade rather than weapons grade. With more elaborate fuel cycles, including multiple reprocessing, the amount of plutonium remaining could be

---

[31] Based on periodically updated information on the USEC website (http://www.usec.com) and private communication from Charles Yulish to the author (December 29, 2003).

[32] This summary is based on Refs. [46], [47], and [48], which give overviews of the problem of dealing with separated plutonium.

substantially reduced, but this would increase the cost and, through extra handling of plutonium, some proliferation risks [49, pp. 134–136].

♦ *Vitrification option.* The plutonium could be mixed with highly radioactive wastes and stored as waste (e.g., in borosilicate glass placed deep underground). This would put the plutonium in a relatively inaccessible form, making it difficult to reclaim for malign or benign purposes.

♦ *Borehole option.* The plutonium could be placed in deep boreholes, at depths of several kilometers. Recovery would be difficult, and perhaps prohibitive, if the boreholes were sealed with clay and concrete.

The NAS study also mentioned an array of other possibilities, including subseabed disposal, dilution in the oceans, launching into space, and transmutation in accelerators.[33] However, the study's preferred options were the spent fuel option, the vitrification option, and, contingent upon further study, possibly the deep borehole option [46, p. 143]. The spent fuel option is the only one of these that is closely connected to nuclear power. Nuclear power, in this scenario, would be used not only to generate electricity but also to consume fissile material. Many different types of reactor can be used for this purpose, although not at equal levels of effectiveness.[34]

A companion NAS study, published in 1995, focused on "reactor-related options" and concluded that the most promising options were the spent fuel option, using MOX fuel in LWRs or CANDU reactors, or the vitrification option [49, p. 3]. These reactors are not the only possible choices for the consumption of plutonium, but the NAS study concluded that the alternatives would involve additional delay and expense. With LWRs and HWRs, it would be possible to use existing reactors or relatively quickly available evolutionary modifications of them.

An alternative might have been to convert plutonium-producing fast breeder reactor (FBR) programs into plutonium-destroying reactor programs. This was suggested for the Superphenix reactor in France and the proposed U.S. Integral Fast Reactor program (see Section 16.5.1)—but both programs have been terminated. If breeder programs are revived, weapons plutonium would be a useful resource for jump-starting the breeder fuel cycle. However, with no breeder program in sight in the United States in the near future, this option would require storing the plutonium for an extended time period.

At present and for the near future, LWRs are far more numerous than FBRs and they offer the more immediate prospect for large-scale plutonium consumption. The plutonium would be used in the form of a mixed-oxide fuel (MOX) of PuO_2 and UO_2. As discussed in Section 9.4.2, most LWRs are limited to using only about a one-third fraction of MOX in the reactor core, with the remainder being ordinary uranium fuel.

---

[33] There is no sharp boundary between the spent fuel and transmutation options, because in either case, there would be *some* transmutation and *some* residual plutonium.

[34] Summaries of alternatives are presented in Refs. [46], [49], and [50].

If a full load of MOX fuel is used, the ^{239}Pu consumption rate would be expected to be roughly the same as the rate of consumption of ^{235}U, namely about 1 tonne per GWyr. Consistent with this, the 1995 NAS study suggests that a U.S. inventory of 50 tonnes of plutonium could be consumed in roughly 50 GWyr of operation with full MOX loads or 170 GWyr of operation with one-third loads, but the actual numbers depend on the fuel composition and reactor operating conditions [49, pp. 121–122].

Present U.S. plans call for this use of MOX fuel to begin within several years (see Section 9.4.2). If, as expected, this is done in commercial LWRs, the swords-into-ploughshares aspect might provide a psychological or public relations dividend for nuclear power. However, the total contribution would represent less than 1 year's generation of nuclear power at today's levels, and the real benefit would be in the destruction of weapons plutonium, not in the gain of additional reactor fuel.

18.3.5 Policy Options for the United States

No attempt will be made here to outline a nonproliferation policy for the United States. The issues are too complex and, in many cases, too removed from nuclear power to be appropriately treated. However, it is worth making a few observations:

◆ The coupling between nuclear power and nuclear weapons is weak. The issue is important and profound because the stakes are great. But in the end, nuclear power policies, and more particularly, U.S. nuclear power policies, are unlikely to have much impact—in either direction—on the nuclear-weapon programs of aspiring countries.

◆ However, weak is not zero. In a few cases (e.g., Iran), there may be a significant coupling and it would be useful from a weapons proliferation standpoint to discourage nuclear power in Iran and similar countries, or at least limit the type of facilities they have.

◆ The United States could play a role in securing greater powers for the IAEA in carrying out inspections and in making the United Nations more determined in dealing with countries that pose nuclear threats. It could also provide intelligence data to the IAEA and United Nations. Support of these international bodies does not preclude unilateral action on the part of the United States and its allies, especially in terms of sanctions and trade limitations.

◆ The United States (and other countries with existing enrichment capabilities) could provide low-enriched uranium reactor fuel on advantageous terms to countries with small nuclear-power programs, as a disincentive to their development of enrichment plants.

◆ The political or moral position of the United States and other weapon states would be strengthened if the rest of the world saw more movement in the direction of nuclear disarmament, as called for in Article VI of the

NPT, and if the United States and China (the two holdouts among the NWSs) ratified the CTBT.

◆ It would accomplish little for the United States to abandon nuclear power in the name of nonproliferation. Many other countries, particularly Japan and France, would decline to follow suit. Further, abandoning nuclear power without entirely giving up nuclear weapons might be seen as an empty gesture.[35]

◆ In attempting to influence nuclear programs elsewhere, economic pressures and incentives may be the most effective tool available to the United States. Here, economic strength is a prerequisite. Therefore, even from the standpoint of weapons proliferation, our energy policy should seek to increase U.S. economic strength and economic and political freedom of action. Arguments will still remain as to the extent to which nuclear power can contribute to overall economic strength and to a reduced dependence on oil-exporting countries.

In the end, however, it should be recognized that success in preventing the use of nuclear weapons and in limiting nuclear proliferation depends on finding prudent and effective policies in areas that have little to do with nuclear power. Specific matters to address include achieving appropriate reductions in the U.S. and Russian nuclear arsenals, encouraging responsible control over weapons and nuclear materials in the former Soviet Union, developing a spectrum of incentives and disincentives to dissuade potential proliferating states and their suppliers, and determining the appropriate role, if any, for nuclear deterrence.

References

1. Richard L. Garwin and Georges Charpak, *Megawatts and Megatons: A Turning Point in the Nuclear Age?* (New York: Alfred A. Knopf, 2001).
2. Robert F. Mozley, *The Politics and Technology of Nuclear Proliferation* (Seattle, WA: University of Washington Press, 1998).
3. National Academy of Sciences, *Nuclear Arms Control: Background and Issues*, Report of the Committee on International Security and Arms Control (Washington, DC: National Academy Press, 1985).
4. William Epstein, "Indefinite Extension—with Increased Accountability," *Bulletin of the Atomic Scientists* 51, no. 4, 1995: 27–30.
5. Tariq Rauf, "An Unequivocal Success? Implications of the NPT Review Conference," *Arms Control Today*, 2000. [From: http://www.armscontrol.org/act/2000_07–08/raufjulaug.asp]
6. United Nations, *2000 Review Conference of the Parties to the Treaty on the Non-Proliferation of Nuclear Weapons, Final Document*, NPT/CONF.2000/

---

[35] It is here assumed that the United States will not give up all nuclear weapons, if for no other reason than fear that another country would then seek nuclear dominance through a few illicit nuclear weapons.

28 (May 24, 2000). [From: http://www.ceip.org/files/projects/npp/resources/NPT2000FinalText.htm]

7. Eliot Marshall, "Did Test Ban Watchdog Fail to Bark?" *Science* 280, no. 5372, 1998: 2038–2040.

8. Paul Kerr, "North Korea Quits NPT, Says It Will Restart Nuclear Facilities" (Arms Control Association, January/February 2003). [From: http://www.armscontrol.org/act/2003_01–02/nkorea_janfeb03.asp?print]

9. Peter D. Zimmerman, private communication, March 1994.

10. Myron B. Kratzer, "'Demythologizing Plutonium," *Global '93, Future Nuclear Systems: Emerging Fuel Cycles and Waste Disposal Options*, Seattle, September, 1993 (unpublished paper).

11. Robert S. Norris and Hans M. Kristensen, "NRDC Nuclear Notebook: Global Nuclear Stockpiles, 1945–2002," *Bulletin of the Atomic Scientists* 58, no. 6, 2002: 103–104.

12. David Albright, Frans Berkhout, and William Walker, *Plutonium and Highly Enriched Uranium 1996, World Inventories, Capabilities and Policies* (New York: Oxford University Press, 1997).

13. A.S. Krass, P. Boskma, B. Elzen, and W.A. Smit, *Uranium Enrichment and Nuclear Weapon Proliferation*, Stockholm International Peace Research Institute (London: Taylor & Francis, 1983).

14. Richard Rhodes, *Dark Sun: The Making of the Hydrogen Bomb* (New York: Simon & Schuster, 1995).

15. Andrei Sakharov, *Memoirs* (New York: Knopf, 1990).

16. "Ukraine: Country Joins NPT in December Ceremony." *Nuclear News* 38, no. 1, January 1995: 45–46.

17. William Sweet, *The Nuclear Age: Atomic Power, Proliferation and the Arms Race*, 2nd Edition (Washington, DC: Congressional Quarterly Inc., 1988).

18. Bertrand Goldschmidt, *The Atomic Complex: A Worldwide Political History of Nuclear Energy*, translated by Bruce M. Adkins (La Grange Park, IL: American Nuclear Society, 1982).

19. John Wilson Lewis and Xue Litai, *China Builds the Bomb* (Stanford, CA: Stanford University Press, 1988).

20. "World List of Nuclear Power Plants," *Nuclear News* 46, no. 3, March 2003: 41–67.

21. Leonard S. Spector, *Nuclear Proliferation Today* (New York: Vintage Books, 1984).

22. Embassy of India, Washington, DC, *Evolution of India's Nuclear Policy*, May 27, 1998. [From: http://www.indianembassy.org/pic/nuclearpolicy.htm]

23. David Albright, *India's and Pakistan's Fissile Material and Nuclear Weapons Inventories, End of 1999*, Institute for Science and International Security, October 11, 2000. [From: http://www.isis-online.org/publications/southasia/stocks1000.html]

24. Leonard S. Spector and Jacqueline R. Smith, *Nuclear Ambitions; The Spread of Nuclear Weapons 1989–1990* (Boulder, CO: Westview Press, 1990).

25. Federation of Atomic Scientists, *Pakistan Nuclear Weapons*, May 27, 2000. [From: http://www.fas.org/nuke/guide/pakistan/nuke/]

26. Center for International Security and Cooperation (Stanford University) and Center for Global Security Research (Lawrence Livermore National Laboratory),

Verifying the Agreed Framework, Michael May, General Editor, Report UCRL-ID-142036 and CGSR-2001-001 (Palo Alto, CA: CISAC, 2001).

27. Jared S. Dreicer, "How Much Plutonium Could Have Been Produced in the DPRK IRT Reactor?" *Science & Global Security* 8, 2000: 273–286.

28. Peter Auer, Marcelo Alonzo, and Jack Barkenbus, "Prospects for Commercial Nuclear Power and Proliferation," in *The Nuclear Connection*, Alvin Weinberg, Marcelo Alonso, and Jack Barkenbus, Editors (New York: Paragon House Publishers, 1985): 19–47.

29. Peter D. Zimmerman, *Iraq's Nuclear Achievements: Components, Sources, and Stature*, CRS Report for Congress, Report 93-323 F (Washington, DC: Library of Congress, 1993).

30. International Atomic Energy Agency, "Implementation of the NPT Safeguards Agreement in the Republic of Iran," Report by the Director General, GOV/2003/40 (June 6, 2003).

31. David Albright and Corey Hinderstein, "Iran: Furor over Fuel," *Bulletin of the Atomic Scientists* 59, no. 3, 2003: 12–15.

32. David Albright and Corey Hinderstein, "The Centrifuge Connection," *Bulletin of the Atomic Scientists* 60, no. 2, 2004: 61–66.

33. International Atomic Energy Agency, "Iran Signs Additional Protocol on Nuclear Safeguards: Signing Takes Place at IAEA," IAEA staff report (December 18, 2003). [From: http://www.iaea.org/NewsCenter/News/2003/iranap20031218.html]

34. Mitchell Reiss, *Bridled Ambition: Why Countries Constrain their Nuclear Capabilities* (Washington, DC: The Woodrow Wilson Center Press, 1995).

35. Peter A. Clausen, *Nonproliferation and the National Interest, America's Response to the Spread of Nuclear Weapons* (New York: HarperCollins, 1993).

36. Arms Control Association, "Fact Sheets: The State of Nuclear Proliferation 2001" (2001). [From: http://www.armscontrol.org/factsheets/statefct.asp]

37. Reuters News Service, "Brazil opens uranium enrichment plant" (December 13, 2002). [From: http://www.planetark.org/dailynewsstory.cfm?newsid=19032&newsdate=13-Dec-2003]

38. "Late News in Brief," *Nuclear News* 34, no. 13, October 1991: 26.

39. "De Klerk Tells World South Africa Built and Dismantled Six Nuclear Weapons," *Nuclear Fuels* 18, no. 7, March 1993: 6.

40. Patrick E. Taylor and David E. Sanger, "Pakistan Called Libyans' Source of Atom Design," *The New York Times*, January 6, 2004: A1.

41. H. A. Feiveson, "Nuclear Power, Nuclear Proliferation, and Global Warming," *Physics and Society* 32, no. 1, 2003: 11–14.

42. Peter D. Zimmerman, private communication, May 1994.

43. Nathaniel C. Nash, "Sequel to an Old Fraud: Argentina's Powerful Nuclear Program," *The New York Times*, January 18, 1994: p. A6.

44. Peter D. Zimmerman, "Proliferation: Bronze Medal Technology is Enough," *Orbis* 38, no. 1, 1994: 67–82.

45. *The Future of Nuclear Power, An Interdisciplinary MIT Study*, John Deutch and Ernest J. Moniz, Co-chairs (Cambridge, MA: MIT, 2003).

46. National Academy of Sciences, *Management and Disposition of Excess Weapons Plutonium*, Report of Committee on International Security and Arms Control, John P. Holdren, Chair (Washington, DC: National Academy Press, 1994).

47. F. Berkhout, A. Diakov, H. Feiveson, H. Hunt, E. Lyman, M. Miller, and F. von Hippel, "Disposition of Separated Plutonium," *Science and Global Security* 3, 1993: 161–213.

48. U.S. Congress, Office of Technology Assessment, *Dismantling the Bomb and Managing the Nuclear Materials*, Report OTA-O-572 (Washington, DC: U.S. Government Printing Office, 1993).

49. National Academy of Sciences, *Management and Disposition of Excess Weapons Plutonium: Reactor-Related Options*, Report of the Panel on Reactor-Related Options for the Disposition of Excess Weapons Plutonium, John P. Holdren, Chair (Washington, DC: National Academy Press, 1995).

50. Carl E. Walter and Ronald P. Omberg, "Disposition of Weapon Plutonium by Fission," in *Future Nuclear Systems: Emerging Fuel Cycles & Waste Disposal Options*, Proceedings of *Global '93* (La Grange Park, IL: American Nuclear Society, 1993): 846–858.

19

Costs of Electricity

19.1 Generation Costs and External Costs

A major reason for the decreased interest in the building of new nuclear power plants in recent years has been the relatively high cost of nuclear power. In this section, we will consider the role of costs in electricity generation choices, particularly in the context of the situation in the United States.

The usually specified cost of electricity is the *generation cost*, namely the sum of the costs of constructing the plant, purchasing the fuel, and maintaining and operating the plant. All of these are tangible, definable costs that can be calculated by standard accounting practices. There are other costs that must be included in the price of electricity to the consumer, such as taxes and costs of transmitting and distributing the electricity, but these too are readily calculable.

However, there are additional sorts of cost related to energy use. These include, for example, the costs to society of environmental pollution and of hidden or explicit government subsidies. If one takes a very broad view, one could also include the costs of providing energy security through a strategic oil reserve or military action. These are sometimes called "social costs," or "hidden costs," but a common present usage is to refer to them as *external costs* or *externalities*.

In some models of a desirable approach, a system of taxes and fees would be created in which the external costs are all recovered and are reflected in the price of electricity. However, determining these costs would be fraught with ambiguity. What, for example, is the cost of carbon dioxide emissions? To answer this, it is necessary to know the climate changes that are produced by the emissions and the price to attach to the impacts of these changes.

Extensive debates surround these matters, and we have neither the scientific knowledge nor the accounting principles needed to resolve the disagreements.

One of the more ambitious attempts to address external costs in a comprehensive fashion is the ExternE project, undertaken by the European Commission with the collaboration of the U.S. DOE.[1] Although the ExternE study represents an interesting effort toward assessing external costs in a rational fashion, it does not—and probably cannot—resolve the crucial uncertainties.

For example, the largest cost component for the coal fuel cycle is listed as "global warming." For Germany, this cost is indicated as lying between 3 and 111 mECU/kWh—a range that is too wide to be very helpful. The estimated range for the full costs of the coal fuel cycle is 17 to 138 mECU/kWh.

For nuclear energy in Germany, the total external cost is indicated to be between 4.4 and 7.0 mECU/kWh/ The largest component (3.5 mECU/kWh) is attributed to the global effects of long-lived radionuclides, primarily ^{222}Rn emissions from milling tailings (see Section 9.2.1). The assumed harm comes from small radiation doses received by many people, calculated over 10,000 years—raising unanswerable questions as to the effects of low radiation doses and the discount, if any, that should be applied to injuries that occur thousands of years from now.[2] A further large part (0.1 to 2.7 mECU/kWh) of the total cost is attributed to the global warming consequences of fossil-fuel use in the construction of nuclear facilities and in the production of nuclear fuel. The magnitude of this "cost" will depend on the fossil fuel share of the energy economy.

Overall, the ExternE studies are useful in focusing attention on the importance of externalities. However, in view of the uncertainties in the numerical estimates and in the appropriateness of including some of the cost components, they do not provide decisive guidance on energy choices.

In practice, some of the external costs (only a small fraction) are addressed by assessing charges against utilities or by compelling utilities to take steps to reduce undesirable impacts. Thus, producers of nuclear power are charged a fee of 0.1¢/kWh to cover the costs of nuclear waste disposal. Similarly, under the Price–Anderson Act, utilities are required to contribute to a fund maintained to reimburse victims of a hypothesized nuclear accident. Coal-burning utilities have been required to install equipment to limit effluents. There also have been recurring suggestions that a "carbon dioxide tax" or a "BTU tax" be imposed on fossil fuel use.

---

[1] Results of the ExternE studies can be found through its homepage (http://externe.jrc.es). The most concise results of these studies are all-Europe overview tables of the external cost in mECU/kWh for each country and type of electricity source. [The European Currency Unit (ECU) was approximately equal to one U.S. dollar and therefore 1 mECU ≈ 0.1 cent. The Euro has now replaced the ECU.]

[2] The ExternE results quoted here are for a 0% discount rate. The study also considers a 3% discount rate, which reduces these costs from 3.5 mECU/kWh to 0.01 mECU/kWh.

Such fees, equipment expenditures, and taxes, when paid by the energy producer, automatically become a part of the producer's expenses and thus are internalized in the narrow definition of cost. The issue of the appropriate magnitude to assign to an external cost is thereby resolved by law or regulation, even if the decision is arbitrary. When an external cost is not defined by law or regulation, attempts to assign an appropriate cost carry the discussion into regions of great uncertainty, as discussed earlier.

Normally, when the costs of different options for electricity generation are compared, only generation costs are considered, and these costs dominate most of the active process of choosing among alternative technologies. However, external costs, even if not quantified, play a major part in the broad decision process. When a country bars the construction of nuclear power plants, as a number have done, that is tantamount to attaching an infinite external cost to the perceived dangers of nuclear power. Similarly, when the construction of new natural gas- or oil-fired base-load power plants was barred, as was done briefly following the 1974 oil embargo, this amounted to assigning an infinite external cost to excessive dependence on limited resources. On the reverse side of the coin, the incentives that have been given to renewable energy, in the form of tax credits, are tantamount to assigning a negative external cost.

19.2 Institutional Roles

19.2.1 Who Provides Electricity?

The main suppliers of electricity traditionally were investor-owned utilities, commonly referred to as private utilities. In addition, in many parts of the country, federal and local entities created publicly owned utilities for electricity generation, including large regional organizations such as the federal Tennessee Valley Authority in the southeast and small local bodies such as the city-run Seattle City Light in Washington.

In the early days of electricity generation, a large fraction of the national electricity supply was generated by industry for its own use. However, by 1980 all but 3% of U.S. electricity was being produced by electric utilities, reflecting the increase in overall consumption and the efficiencies of central power stations [1]. A reversal of this trend eventually came from the effects of the Public Utilities Regulatory Policy Act of 1978 (PURPA), which mandated the purchase of electricity by utilities from "qualifying facilities" at the utility's avoided cost of generation.[3] To qualify, a facility had to be of relatively small size and use renewable resources or, if the efficiency was high enough, cogen-

---

[3] The avoided cost is the cost that the utility nominally avoids by not having to produce the electricity itself. According to P.L. Joskow, these costs were often set by contract at prices much higher than the true avoided costs [2, p. 220].

eration. By 1991, 5% of the electricity distributed by the utilities came from such sources, called *nonutility generators* (NUGs) [3, p. 215].

The shift from utilities to nonutility power producers was greatly accelerated by the deregulation of the electric power industry beginning in the 1990s, particularly with the growth of *independent power producers* (IPPs).[4] The nonutility share of total electricity output was 11% in 1998 and 34% in 2002 [4, p. 97]. This growth has been accomplished largely through the sale of facilities by regulated utilities to unregulated entities.

Under a regulated regime, where utilities own their generating facilities, once a power plant is built, it has an assured market for its output in the utility's service area, limited only by the demand for electricity and the price authorized by public regulatory authorities. To prevent utilities from exploiting their local monopoly positions, their rates are controlled by state public utility commissions (PUCs). An appropriate electricity rate in this framework, intended to be fair to both the utility and the consumer, is one that covers the cost of producing and delivering the electricity plus a reasonable rate of return upon the capital invested. The utility commissions have discretion in establishing the rate of return and determining what costs are to be recognized in the rate base. Although in some case utilities, especially those with nuclear power plants, have been penalized by "prudency" reviews—which questioned the prudence of the decision to build the plant and disallowed some of the costs—the old arrangements provided utilities with a degree of financial security.

The deregulation of the electric power industry has changed the situation greatly, and in many states, the link between power production and power distribution has been broken. In response to these new market regulations, the IPPs have entered the picture. In a deregulated market, the utilities distributing power can purchase electricity from whomever they wish. This can be done via bilateral contracts between a distributor and an IPP or the distributor can purchase power on the open market from the lowest-cost source. The IPPs can also sell electricity directly to customers.

This creates particular uncertainties for the potential builder of a nuclear power plant. After a plant is built, the lower cost of an alternative source (e.g., natural gas) might require that the plant's output be sold at a price below a break-even figure. Conversely, high costs for alternatives could, in principle, leave the nuclear provider with a large profit. To avoid such a speculative situation, the reactor operator can enter into bilateral agreements to give the parties the security of more stable prices.

---

[4] Independent power producers are nonutility entities whose primary purpose is the generation of electricity for public use. The earlier NUGs were often producers that generated electricity for their own use or as part of a cogeneration facility that also supplied heating.

19.2.2 The Role of Government in Electricity Generation Decisions

Other levels of government beyond the public utility commissions can also have important impacts on electricity generation. This can be done through subsidies, regulations designed to encourage or discourage a given form of generation, or taxes and fees. For example, coal has been the beneficiary of federal incentives to railroads dating to the 1800s, and nuclear power has been the beneficiary of technology developed for nuclear weapons and naval ship propulsion. Even more directly, U.S. hydroelectric power is largely the legacy of past federal programs to build hydroelectric dams.

All three of these beneficiaries of past federal assistance now face environmentally based objections. Coal plants are being required to adopt emission controls, nuclear plants are subject to very strict safety regulation, and hydroelectric output is being curtailed in some areas to preserve stream flow. These circumstances have encouraged electric utilities and NUGs to consider other generation sources.

Environmental groups have long advocated the use of renewable resources beyond hydroelectric power for electricity generation, and there has been increasing interest in renewable resources on the part of the Electric Power Research Institute and some utilities. However, for the near future, it would appear that natural gas will be preferred by most utilities as a more immediately practical and economical alternative. It is becoming the compromise fuel of choice, aided by technological advances that have improved the efficiency of gas-fired plants, in particular in the development of gas-fired combustion turbines.

This represents a striking reversal in policy. In 1978, through the Power Plant and Industrial Fuel Use Act, the federal government banned the construction of large power plants using natural gas or oil. This action was motivated by fears of eventual fuel shortages. This ban was lifted in 1987, although there were no fundamental changes in the fuel resources.

Another, also extreme, example of the negative impact that government policies can have (in this case those of local and state government) is the Shoreham nuclear power plant, built for the Long Island Lighting Company (LILCO) in New York. The plant was completed in the mid-1980s, and in 1989—after a series of battles in the courts and in NRC deliberations—it was given an NRC license to operate at full power. However, local and state opposition prevented it from going into operation, and it has been dismantled. The resulting loss is borne, in proportions that are probably hard to assign, by LILCO, the New York state government, and individual electricity ratepayers.

Government can also express its preferences or concerns through financial penalties or incentives. The charge of 0.1¢/kWh on nuclear-generated electricity for nuclear waste disposal, is a modest example of what can be done; in this case, the fee is a charge against expected expenses rather than as a punitive measure, and it is too small to be a significant discouragement. A carbon dioxide tax, which at the present does not exist in the United States, could be an explicit discouragement to the use of fossil fuels.

On the other side of the coin, the federal government is now providing strong incentives for the development of renewable energy. Federally supported research programs have increased, and under the Energy Policy Act of 1992, an inflation-adjusted production tax credit is paid for electricity generated by wind power or a dedicated biomass facility [5, Section 1914]. A similar subsidy for the generation of electricity using renewable energy is given to nonprofit or governmental producers who do not pay taxes [5, Section 1212]. The rate for these benefits started at 1.5¢/kWh and rose with inflation to 1.8¢/kWh for the year 2003.[5]

19.3 The Generation Cost of Electricity

19.3.1 Calculation of Costs

Components of Generation Cost

The generation cost is the cost of electricity as it leaves the plant and is sometimes referred to as the "busbar cost."[6] The cost to the customer is higher than the generation cost because it also includes the costs of transmission, distribution, other utility costs such as billing, and any special pricing provisions established by the utility.[7] The generation cost is generally described in terms of cost per kilowatt-hour (kWh). It is the sum of two components:

1. *Production cost.* This includes operations, maintenance, and fuel costs, where operations and maintenance (O&M) costs are sometimes grouped together as "operating costs." The operating costs may include, in addition, costs of capital expenditures made after the plant has gone into operation. The largest component of nuclear fuel costs in the United States is for enrichment; the cost of the uranium itself is only about one-third of the total fuel cost [6, p. 161].
2. *Capital charges.*[8] These provide for the payment of costs of constructing the plant, including the return on the capital expended in constructing the plant, plus taxes.

---

[5] The benefits were scheduled to be phased out at the end of 2003, but if past precedent is followed, Congress will renew them.

[6] The *busbar* is the conductor which connects the output of the generator to the input of the transformer that feeds the transmission lines. Thus, the busbar cost is the cost of electricity as it leaves the plant.

[7] Examples of special provisions include bulk rates to some large commercial users and residential rates that depend on the amount of use and, in some cases, time of use.

[8] The equivalent terms "fixed charges," "capital-related charges," and "carrying charges" are sometimes used in place of "capital charges."

Calculation of Capital Charges

If a plant could be built instantaneously, its cost would be the sum of the base construction cost, also known as the engineering, procurement and construction cost, together with a contingency cost. This constitutes the "overnight cost." However, a plant cannot be built instantaneously and so-called time-related charges must be added, namely interest on the investment and the escalation of costs during the period of construction. For example, in a study by the DOE-sponsored Near Term Deployment Group (NTDG), the time-related charges were estimated to add roughly 30% to the investment cost of the plant, for an assumed construction time of 42 months [7, p. 4-10].

The investment in the plant is the sum of the overnight cost and the time-related charges. The shorter the construction time, the lower the total investment cost. One of the reasons why nuclear plants completed in the 1980s and 1990s had very high capital charges was that the construction time was prolonged, a situation made worse by the high interest rates prevailing in that period. An important goal for future plants is to keep the construction time short.

The investment cost is repaid by a stream of income received during the amortization period of the facility, which is usually 40 years for a nuclear power plant. This stream is commonly calculated as a series of constant payments known as the capital charges or *fixed charges*. The annual fixed charges C are given by the relation:

$$C = 100 \times \frac{RI}{fH}, \tag{19.1}$$

where f is the capacity factor, $H = 8760$ is the number of hours in a year, R is the *fixed charge rate*, and I is the investment in the plant. It is common to express I in dollars per kWe and C in cents per kWh; the factor 100 is the conversion between dollars and cents.

The annual fixed charges are analogous to a mortgage payment rate and provide for payment of the construction costs and a return on the investment. The fixed charge rate is determined by the *discount rate* and the number of years over which the payments are calculated. The discount rate is the sum of the inflation rate and the real interest rate. If inflation is ignored, the discount rate equals the interest rate. The discount rate defines the present value of a future payment P. For example, if the discount rate is 10%, a payment P expected in 15 years has a present value of $P \times 1.10^{-15} = 0.24\,P$. To provide a 40-year income stream that has a present value equal to the investment cost, requires a fixed charge rate of 0.025 at a zero discount rate, 0.0583 at a 5% discount rate, and 0.102 at a 10% discount rate.[9]

Capital charges (or annual fixed charges) are the largest component of nuclear power costs, and future nuclear competitiveness depends on keeping the

---

[9] The fixed charge rate is here calculated from the expression $R = d/[1-(1+d)^{-N}]$, where d is the discount rate and N is the number of years of payment.

investment cost low and the capacity factor high. These are the basic determinants of capital charges. However, the discount rate is also important. For
example, for a 40-year amortization period the calculated fixed charge rate is
75% higher for a 10% discount rate than for a 5% discount rate. The U.S. Near
Term Deployment Group study assumed a discount rate of 12% [7, p. 4-20],
whereas the Nuclear Energy Agency of the OECD bases its calculations on
5% and 10% discount rates [6].

The discount rate is a construct of the investment environment, and discount rates at a given time vary with interest rates and perceived risk.[10]
Since nuclear plants are perceived to carry a higher risk than alternatives,
these rates will usually be higher than for alternative types of power plants.
However, in comparing different energy alternatives, a uniform discount rate
is usually assumed.

19.3.2 Recent Trends in Electricity Prices

The history of electricity prices in the United States since 1960 is shown in
Figure 19.1, which gives average prices expressed in both current (or nominal)
dollars and constant (or real) 1996 dollars [9, p. 241]. Although the price

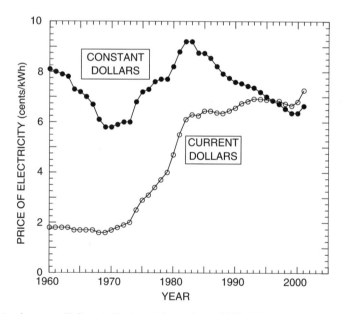

Fig. 19.1. Average U.S. retail electricity prices, 1960–2001, expressed in current
dollars and in constant 1996 dollars. (From Ref. [9, Table 8.6].

---

[10] The discount rate also depends on the inflation rates, but here we are considering
costs in constant dollars and ignoring inflation.

in current dollars has quadrupled, real electricity prices decreased over this four-decade period. Despite the overall decrease, there have been substantial swings in the real electricity price, especially the rise in the decade following 1973—although these variations were less than those for oil and natural gas. The largest swing in real prices was from 5.8¢/kWh in 1969 to 9.2¢/kWh in 1982.

Different classes of customers are charged at different rates. Thus, in 2002, the overall average price was 7.21¢/kWh, whereas the prices averaged 8.46¢/kWh for residential customers and 4.88¢/kWh for industrial consumers [10, Table 1d]. There are also large regional differences. For example, in Washington state, where "old" hydroelectric power served to keep the price down, the average residential electricity price was 6.3¢/kWh, whereas the price in New York and California was more than twice as great. In assessing the importance of differences between the costs of competing sources of electric power, it is pertinent to keep in mind the changes with time of more than 3¢/kWh and the variations across the country of 6¢/kWh.

19.3.3 Costs of Nuclear and Fossil Fuel Electricity Sources

Recent Trends in Comparative Costs

For nuclear power to be of interest to commercial companies in a competitive electricity market, its generation costs must be as low as those of coal or natural gas. In the past, the main cost competition has been between nuclear power and coal. From 1975 to 1982, the generation costs of nuclear- and coal-powered plants were about the same, but from 1982 to the late 1980s, nuclear costs rose rapidly while coal costs decreased slightly, both measured in current dollars.[11] The rise in nuclear costs was due to expensive retrofitting of operating plants and increases in the costs of bringing new plants on line, in part because the construction times were greatly extended. In the meantime, the price of coal fuel was decreasing.

Since 1989, nuclear generation costs have dropped. The production costs, which include operation, maintenance, and fuel, rose (in constant 2002 dollars) from 2.48¢/kWh in 1981 to 3.48¢/kWh in 1987 and then gradually dropped to 1.71¢/kWh in 2002 [12]. Production costs for coal continued to decrease after the late 1980s, although not by as large a margin as nuclear costs.

In the meantime, natural gas began to appear as an environmentally and economically attractive alternative to coal or nuclear power. Advanced gas-fired turbine systems, in which the turbines are driven by hot natural gas rather than by steam produced by the combustion of gas (or any other fuel), offer a number of important advantages for future power plants: high thermal efficiency (especially in combined-cycle systems), low CO_2 emissions per unit

---

[11] These data, with references, are given more fully in the first edition of this book [11, pp. 302–303].

of electricity generated (compared to coal-fired steam plants), short construction times, and low capital costs. In 2002, natural gas accounted for 18% of total electricity generation, only a modest increase from its 13% share 10 years earlier, in 1992 [14, p. 99]. However, it dominates plans for new additions to electrical generating capacity. Thus, the forecast in the U.S. DOE's *Annual Energy Outlook 2003* anticipates that natural gas will account for 80% of new U.S. generating capacity added between 2001 and 2025 (although most of it will not be for base-load generation) and about one-half of the increase in generation, with most of the rest of the increase coming from coal [15, pp. 68 and 131].

Major Factors Impacting Costs of Gas, Coal, and Nuclear Electricity

The major question mark surrounding the attractiveness of natural gas for electricity generation is that of natural gas prices. The prices have been quite volatile in the past, and we may face both volatility and a long-term increasing trend if there is increasing natural gas use for electricity generation and other applications. Figure 19.2 displays simulated fuel costs in constant 1996 dollars for coal and natural gas at the prices prevailing from 1973 to 2001, assuming thermal efficiencies that represent "traditional" performance (33%) and improved future performance for gas-fired combined cycle plants (54%) [16,

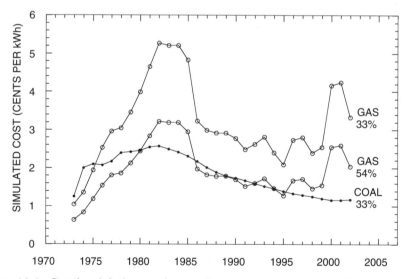

Fig. 19.2. Simulated fuel cost of natural gas and coal for electricity generation, 1973–2002, using actual U.S. costs per BTU in constant 1996 dollars and assumed thermal efficiencies of 33% (coal and gas) and 54% (gas only).

p. 67].[12] Even at 54% efficiency, natural gas at the prices of the early 1980s would mean fuel costs of over 3¢/kWh.[13]

Coal prices are comparatively stable and, in fact, have been gradually declining for 20 years. Coal is therefore is likely to remain the least uncertain option for electricity generation, unless it is penalized because of its CO_2 emissions. This penalty could be in the form of a tax on the amount of CO_2 emitted or an indirect "tax" in the form of a requirement that the emitted CO_2 be sequestered so that it does not enter the atmosphere.

Nuclear power costs are dominated by the construction cost. The key to having a low generation cost is a low investment cost and a high capacity factor. High capacity factors (near 90%) have been achieved in present nuclear plants and are presumably achievable for future plants. The investment cost is more uncertain. The overnight construction cost can be reduced by careful design and economies of scale—or, for plants made up of small modular units, economies of mass production. Time-related charges can be held down by rapid construction. The NTDG projection assumes a 40-month construction time. Based on the experience in Japan, where construction times averaged about 50 months for the 30 plants connected to the grid since 1983, this is a plausible target [13, Table 14]. However, it would mark a great change from the most recent U.S. experience (an average of 149 months for 40 plants that were connected to the grid since 1983). Finally, as discussed in Section 19.3.1, the discount rate is a major factor in determining the capital charges, and nuclear generation costs are therefore sensitive to interest rates.

Comparative Cost Estimates

Several estimates of future electricity costs are presented in Table 19.1 for new U.S. plants going into operation in 2010.[14] Costs projected by the U.S. DOE's Energy Information Administration are given for advanced coal, advanced gas, nuclear, and wind generation. For nuclear plants, two estimates are made, a reference case and an advanced cost case which assumes that the construction costs are reduced in a manner "consistent with goals endorsed by the DOE's Office of Nuclear Energy" [15, p. 71].

Table 19.1 also includes cost estimates for an Advanced Light Water Reactor (ALWR) and a combined-cycle gas-turbine plant, made in a detailed

---

[12] The costs (in current dollars) per BTU were found from Ref. [14, Table 9.10]; conversion to constant dollars was based on the implicit price deflator from Ref. [9, p. 353].

[13] For example, the 1982 peak of $5.10 per million BTU translates at 54% efficiency to a cost of 3.2¢/kWh.

[14] In accordance with common practice, the cost reported in this table, as well as in the MIT study discussed later, is the *levelized cost* (i.e., an assumed constant annual cost of electricity that over the lifetime of the plant would provide a stream of income sufficient to cover its construction and operation).

Table 19.1. Cost of generating electricity in new plants in 2010, as estimated by the Energy Information Administration (EIA) and by the Near-Term Deployment Group (NTDG) (in cents per kWh).

	Advanced Coal	Advanced Gas		Reference Nuclear	Advanced Nuclear		Wind
	EIA	EIA	NTDG	EIA	EIA	NTDG	EIA
Capital costs	3.51	1.23	1.25	5.00		3.16	4.06
O&M costs	0.45	0.13	0.20	0.74		0.50	0.82
Fuel costs	1.04	3.22	2.82	0.46		0.50	0.00
Total costs	5.00	4.59	4.27	6.2	∼ 5.2	4.16	4.88

Source: The EIA data are from Refs. [15, pp. 69 and 71], and [17]; the NTDG data are from Ref. [7, p. 4-21]. The costs are in constant dollars: 2000 dollars for the NTDG data and 2001 dollars for the EIA data.

analysis by the NTDG, as part of its Roadmap for the deployment of new nuclear plants by 2010 [7, Section II-4]. The competitive position of nuclear power depends on the construction costs, as well as the financing arrangements as reflected in the discount rate. The NTDG estimate in Table 19.1 was based on a base construction cost of $1000/kWe for the ALWR and a 12% discount rate. For a construction cost of $1200/kWe, the total nuclear cost increases by 0.6¢/kWh to 4.73¢/kWh.

An independent set of cost estimates has been made in the MIT study *The Future of Nuclear Power*, which compares nuclear, coal, and natural gas sources for a "base case" (which represents "current perceived costs") and for alternative assumptions [18, Chapter 5]. The estimated generation costs (in 2002 dollars) for plants with a 40-year lifetime and a 85% capacity factor are as follows:

◆ *Nuclear.* Base case cost = 6.7¢/kWh. This cost could decrease to 4.4¢/kWh if (a) the overnight capital cost dropped from $2000/kWe (base) to $1500/kWe, (b) the construction time dropped from 5 years (base) to 4 years, and (c) the costs of capital were the same for nuclear plants as for coal and gas plants, instead of higher due to greater uncertainties (as in the base case).

◆ *Coal.* Base case cost = 4.2¢/kWh. This cost would rise if a carbon tax were imposed (e.g., to 5.4¢/kWh for a tax of $50/tonne of carbon).

◆ *Natural gas.*[15] Base case cost = 4.1¢/kWh (at "moderate" gas prices and a 47% thermal efficiency). If the cost of gas, expressed as a 40-year levelized cost per million BTU, changed from moderate ($4.42) to low ($3.77) or high ($6.72) the electricity cost would change to 3.8¢/kWh or 5.5¢/kWh, respectively. A carbon tax would further raise the cost; a more efficient use of fuel would lower it.

---

[15] A combined-cycle combustion turbine is assumed.

The MIT report terms the nuclear cost reductions "plausible but unproven" [18, p. 41]. They roughly correspond to estimates made in the NTDG study, and there is no essential conflict between the NTDG estimates and those of the reduced-cost MIT nuclear scenario. Coal costs would be strongly influenced by the imposition of a carbon tax or high costs of carbon dioxide sequestration. Natural gas costs are determined almost entirely by uncertain future gas prices.

19.4 Costs and Electricity Choices

19.4.1 The Role of Cost Differences

The estimated generation costs for the four main options considered in Table 19.1 are mostly between about $4\text{¢}/\text{kWh}$ and $5\text{¢}/\text{kWh}$. The MIT estimates overlap this range, although the base nuclear estimate is higher. Overall, the differences among the sources are probably less than the uncertainties in the estimates for each source and are small compared to the variations in the average U.S. cost of electric power over the past several decades and to regional variations (see Section 19.3.4). Nonetheless, a utility or IPP planning a new facility is forced by competitive realities to pay attention to even small differences in generation costs. A difference of $1\text{¢}/\text{kWh}$ can be a decisive difference to a utility, at least if the difference is stably maintained.

It is an open issue, however, whether such differences are important enough to be the determinants of long-term national policies for electricity generation. They will be the determining factor if "letting the marketplace decide" is national energy policy. Such a course seeks to minimize governmental interference and avoid the inefficiencies of poorly conceived government programs. However, environmental considerations or long-term energy security needs may argue for not adopting the solution with the lowest generation costs—i.e., for taking external costs into account, in least in a qualitative fashion.

The short-term economic penalty of such a course might be appreciable, but not necessarily prohibitive. Electricity sales in 2002 were 3.5 trillion kWh and sales to the residential sector were 1.3 trillion kWh [14, Table 7.5]. An extra $1\text{¢}/\text{kWh}$ corresponds to an additional national electric bill of $35 billion and an average extra cost of a little over $100 per household. These are not trivial additional costs, but, nonetheless, are of a magnitude that we are accustomed to accept. As seen in Figure 19.1, the real cost of electricity over the past 40 years has varied from a little under $6\text{¢}/\text{kWh}$ to a little over $9\text{¢}/\text{kWh}$—a variation from the center of about $1.5\text{¢}/\text{kWh}$. Much greater swings occur for oil. The United States uses about 7 billion barrels of oil per year [14, Table 1.7]. The cost of oil rose by more than $10 per barrel (equivalent to about $0.25 per gallon) from 1999 to 2000 [14, Table 9.1], corresponding to a change in the national bill of roughly \$70 billion per year, about half of which is in import costs.[16]

19.4.2 Leveling or Tilting the Playing Field

Past Government Intervention

Allowing the marketplace to decide is closely related to the often expressed calls for a "level playing field." However, this nonintervention ideal does not correspond to past history, which, as discussed earlier, has seen frequent government intervention. Examples of intentional or incidental tilting of the playing field by government action have included the following:

- The incentives given in the mid-1800s to promote railroad construction, which facilitated the shipment of coal throughout the nation.
- The program of the Rural Electrification Administration, established in 1935, which brought electricity to farms and displaced the numerous small windmills that had been providing power [19, p. 152].
- The hydroelectric dams built by the federal government in the mid-1900s.
- The launching of nuclear power, including the research on nuclear energy during and after World War II and the pioneering development of nuclear reactors for the U.S. Navy.
- The barring of the use of oil and natural gas in new electricity generation, under the terms of the Power Plant and Industrial Fuel Use Act, from the time of its passage in 1978 to its repeal in 1987.
- The inhibitions placed upon nuclear power, often by local governments, including the shutdown of the newly completed Shoreham plant in New York in 1989 (see Section 19.2.2).
- The tax credit of 1.5¢/kWh for renewable energy established in the 1992 Energy Policy Act.
- The limitations on emissions from coal-fired power plants.
- The diplomatic and military efforts of the U.S. government to secure oil supplies.

These examples illustrate the past role of government actions in influencing the production and use of energy in the United States. They have been motivated by a mix of considerations, including the welfare of the economy, protection of the environment, and concern about energy supplies. In some cases, these actions involved costs in an obvious fashion, as with the tax credit. In other cases, cost accounting played little or no part, as in the early development of nuclear energy.

Nuclear Power's Need for Federal Support

Given present conditions, it appears unlikely that a new nuclear power plant will be built in the United States in the near future without federal action

---

[16] Net imports represented 53% of petroleum supply in 2002 [14, Table 1.7].

to reduce some of the financial risks of such a project. Even if nuclear power had, say, a 1¢/kWh cost advantage over the main competing source—most likely natural gas—that might not be enough to encourage a reactor order. The strength of public and government opposition or support will also play a part.

Nuclear power is unique in the intensity of the opposition it induces, and any new nuclear project would start under a cloud of uncertainty. The Shoreham experience is a reminder of the possibility that even a completed nuclear plant might not be permitted to go on-line if local or state officials turn out to be sufficiently opposed. Therefore, if the federal government believes it is important to reinvigorate nuclear energy, it may have to intervene to shield the first nuclear projects against such financial risks. Larry Foulke, then Vice-President Elect of the American Nuclear Society, reflected a common view among nuclear advocates when he suggested the following:

> The government could assume extraordinary costs associated with delays due to the acts of government or the acts of the public (as a consequence of government actions) through standby credit facilities. Through these facilities, the government would agree to carry interest payments resulting from construction delays caused by changing government requirements and not contractor faults. Such standby credit facilities could also offer a "make whole" provision under which the government would take ownership of the plant and repay both the lender and equity-holder in the event that "acts of the government" and "acts of intervenors" (that could result from government actions) prevent plant commissioning. [20, p. 36]

In the same article, Foulke suggested three additional measures to help a revival of nuclear plant construction: help in meeting the high costs of the first plants, a guarantee of government purchase of some of the power output if private demand is less than expected, and recognition of nuclear energy's "national benefits such as environmental quality, energy security, and the burnup of weapons-grade fissile material."

In short, it appears that Foulke is asking that the playing field be tilted in favor of nuclear energy. No doubt his supporters would respond with the counterclaim that these measures would merely level the playing field by addressing some of the unique, noneconomic obstacles that nuclear power faces. Determining the magnitude and direction of the tilt or tilts would be highly controversial. It might also be irrelevant if one rejects the "playing field" metaphor.

If the focus is not on the competitive situation of the producers, but on perceived national benefits, costs change from being a determinant of policy to being a tool. This is a form of "social engineering"—an approach that is variously viewed as discredited or as irresistibly tempting (or both). If an electricity source is preferred on grounds other than generation costs, the government can use carrots or sticks to favor its use (e.g., by relieving some of its costs or adding to the costs of its competitors). We discuss several of the possible mechanisms for doing this in the following subsection.

19.4.3 Mechanisms for Encouraging or Discouraging Electricity Choices

Moratoria or Prohibitions

The most direct way of discouraging the use of a particular electricity source is to bar it. This has been done for new nuclear power plants in a number of countries in Europe. As mentioned earlier, the United States, in the (since-repealed) Power Plant and Industrial Fuel Use Act of 1978, banned new power plants that would use oil or natural gas for base-load generation.

Tax Credits

Renewable energy has received benefits in the form of tax credits, beginning with credits established by the federal government and California in the 1970s. These lapsed, but the federal Energy Policy Act of 1992 instituted an inflation-adjusted tax credit that started at 1.5¢/kWh and has risen, with inflation, to 1.8¢/kWh in 2003 [15, p. 20]. It favors wind power and some biomass producers (see also Section 19.2.2). These credits have been extended on an annual basis since their original expiration date in 1999.

In principle, such benefits could also be provided for nuclear power, on the grounds that it too can substitute for fossil fuels. However, congressional approval of tax credits to promote nuclear power may be difficult to obtain, because nuclear power elicits stronger opposition than does renewable energy. Further, the justification of helping a fledgling industry does not apply.

Carbon Tax

A direct way of discouraging fossil fuel use, and even discriminating among fossil fuel uses, is the imposition of a "carbon tax"—a levy on fossil fuel use based on the magnitude of the CO_2 emissions. Such a tax would automatically give a competitive gain to both nuclear and renewable sources. For example, a tax of $50 per tonne of carbon emitted would have the following consequences:

◆ An increase in the cost of electricity of 1.3¢/kWh for a coal-fired plant operating at 33% efficiency and of 0.46¢/kWh for a gas-fired plant operating at 54% efficiency.
◆ An increase of $0.12 per gallon of gasoline.
◆ Tax revenues of about $80 billion per year, at the emission rates of the year 2000—ignoring a possible decrease in consumption as prices rise.

The magnitude of the tax rate (in $ per tonne C) could be set at whatever level appears appropriate. A possible approach would be to phase it in with scheduled increases, in a staircase model. The tax could be made revenue neutral by reducing other taxes, or the revenue could be used, for example, for nonfossil energy projects.

19.4.4 Reactor Longevity

A private reactor operator—whether a utility or an independent power producer—has a responsibility to provide an adequate return to the investors within a reasonable period of time. The amortization period for nuclear power plants is typically 40 years. However, prospective investors today have a shorter time horizon. A discount rate of 12%, as adopted in the NTDG Roadmap, reflects the need to compensate investors relatively quickly. For a 40-year stream of payments at this rate, 90% of the original present value is recovered within 20 years. If the plant continues to operate beyond 40 years, it becomes very profitable to the people who then own it, but this prospect may be of little interest to the original investors.

Long-lived plants may be of interest to society as a whole however, although looking ahead to distant times raises the previous encountered issue of discounting future developments. As discussed in Section 13.3, at any nonzero discount the "present value" of avoiding the impact of nuclear wastes in 10,000 years is close to zero. Yet, we conclude that it is worth spending large sums today to reduce risks in 10,000 years. Similarly, in calculating the external costs of nuclear power, the ExternE study applies a zero discount rate to the cost of radiation doses from mill tailings (see Section 19.1). The question of discounting also arises in considering the implications of long-lived nuclear power plants.

If nuclear plants operate for 100 years and their capital costs are amortized in 40 years or less, future users will be able to obtain power from plants whose main costs have already been paid. The period of 100 years may seem surprisingly long. However, the surge of applications for operating license renewals suggests that most of the present generation of U.S. reactors will be able to operate for 60 years. With forethought as to potentially vulnerable points, it is plausible that the next-generation reactors will last 100 years, albeit with replacements or upgrades of individual components.

With these long lifetimes, the future price of power can be much below the original price. In the words of Alvin Weinberg, "time annihilates capital costs" [21]. However, although this is true from the standpoint of the future consumer, the original investors face nonannihilated costs in the present. The present value of a benefit that is delayed for 40 years is small for the investor and perhaps for society as a whole. In short, if one applies a standard discount rate the bonanza of future cheap power is not an important gain.

However, the long-lived plant is a very valuable if one follows the approach taken for nuclear waste disposal—essentially of assuming a zero discount rate.

This can be looked at in terms of intergenerational justice, a consideration that was raised earlier in the context of waste disposal (see Section 13.3). In considering reactor longevity, Weinberg reiterates a point he had first made in 1985:

> [C]heap electricity from a fully amortized reactor is a gift that the generation that pays for the plant bestows on future generations. To a degree, this gift to future generations compensates for the burden of geologically sequestered wastes. [21]

In this view, reactors become part of the overall infrastructure that is required to meet the essential needs of society—in this case the need for electric power—and their construction is a long-term national investment.

References

1. Marc Gervais, *Electricity Supply and Demand into the 21st Century* (Washington, DC: U.S. Council for Energy Awareness, 1991).
2. P.L. Joskow, "The Evolution of Competition in the Electric Power Industry," *Annual Review of Energy* 13, 1988: 215–238.
3. U.S. Department of Energy, *Annual Energy Review, 1992*, Energy Information Administration Report DOE/EIA-0384(92) (Washington, DC: U.S. DOE, 1993).
4. U.S. Department of Energy, *Monthly Energy Review, March 2003*, Energy Information Administration Report DOE/EIA-0035(2003/03) (Washington, DC: U.S. DOE, 2003).
5. 102d Congress, *Energy Policy Act of 1992*, Public Law 102-486, October 24, 1992.
6. Nuclear Energy Agency and International Energy Agency, *Projected Costs of Generating Electricity, Update 1998* (Paris: OECD/IEA, 1998).
7. Near Term Deployment Group, *A Roadmap to Deploy New Nuclear Power Plants in the United States by 2010, Volume II, Main Report* (Washington, DC: U.S. DOE, 2001).
8. International Energy Agency, *Nuclear Power in the OECD* (Paris: OECD/IEA, 2001).
9. U.S. Department of Energy, *Annual Energy Review 2002*, Energy Information Administration Report DOE/EIA-0384(2002) (Washington, DC: U.S. DOE, 2003).
10. U.S. Department of Energy, "Electric Sales and Revenue 2002," Energy Information Administration report (2003). [From: http://www.eia.doe.gov/cneaf/electricity/esr/esr_tabs.html]
11. David Bodansky, *Nuclear Energy: Principles, Practices, and Prospects* (Woodbury, NY: American Institute of Physics Press, 1996.)
12. Nuclear Energy Institute, "Average U.S. Nuclear Industry Production Costs, (1981–2002). [From: http://www.nei.org/documents/Production_Costs_1981_2002.pdf]

13. International Atomic Energy Agency, *Nuclear Power Reactors in the World*, Reference Data Series No. 2, April 2003 Edition (Vienna: IAEA, 2003).

14. U.S. Department of Energy, *Monthly Energy Review, June 2003*, Energy Information Administration Report DOE/EIA-0035(2003/06) (Washington, DC: U.S. DOE, 2003).

15. U.S. Department of Energy, *Annual Energy Outlook 2003, With Projections to 2025*, Energy Information Administration Report DOE/EIA-0383(2003) (Washington, DC: U.S. DOE, 2003).

16. U.S. Department of Energy, *Annual Energy Outlook 2000, With Projections to 2020*, Energy Information Administration report DOE/EIA-0383(2000) (Washington, DC: U.S. DOE, 1999).

17. Alan Beamon, Energy Information Administration, U.S. DOE, private communication, January 31, 2003.

18. Massachusetts Institute of Technology, *The Future of Nuclear Power, An Interdisciplinary MIT Study*, John Deutch and Ernest J. Moniz, Co-chairs (Cambridge, MA: MIT, 2003).

19. D. R. Smith, "The Windfarms of the Altamont Pass Area," *Annual Review of Energy* 12, 1987: 145–183.

20. Larry R. Foulke, "The Status and Future of Nuclear Power in the United States," *Nuclear News* 46, no. 2, February 2003: 34–38.

21. Alvin M. Weinberg, "On 'Immortal' Nuclear Power Plants," paper presented at the Winter Meeting of the American Nuclear Society, Washington, DC, November, 2002.

20

The Prospects for Nuclear Energy

20.1 The Nuclear Debate

20.1.1 Nature of the Debate

Debates about nuclear energy have sometimes appeared to have the aspect of a religious war, especially in the 1970s and 1980s when an expanding nuclear enterprise came into conflict with a growing antinuclear movement. Nuclear energy was discussed as if it were intrinsically good or evil, and for many of the protagonists, it became instinctive to oppose or support it.

That debate has since become more muted. This does not mean that the basic issues are settled. Rather, it means that less attention is being paid to nuclear energy. Fossil fuel energy has been plentiful, energy problems are not at the center of the world's attention, and nuclear energy is not at the center of what consideration is given to energy. In discussions of U.S. energy policy, nuclear power is sometimes just ignored.

However, as discussed in Chapter 1, the world's heavy dependence on fossil fuels creates serious problems, and nuclear power can help in addressing them. So too, at least in principle, can a number of alternatives, but none of these can be fully counted on to serve as a large-scale source of additional energy. Each either is too limited in the scope of its potential expansion—as in the case of hydroelectric power—or lacks the proven ability to provide energy in the amounts that will be required. Some, particularly wind power, show promise (see Section 20.2.3), but it is premature to rely on wind or other alternatives to provide a sure solution. Under these circumstances, nuclear power remains a contender as a needed contributor to future energy supplies.

Ingredients in the decisions that will be made about the utilization of nuclear power include the following:

◆ The demand for energy—particularly energy in the form of electricity.

◆ The practicality and the environmental, economic, and national security implications of the alternative energy sources.

◆ The success that nuclear power has in avoiding reactor accidents, safely handling nuclear wastes, and reducing reactor construction costs.

◆ Assessments of the relationship between nuclear power and nuclear weapons and terrorism.

It is impossible to know how these considerations will play out. They raise technical and economic questions that are, in principle, resolvable. However, they also involve issues where the answers depend primarily on individual surmises and values and are perhaps impossible to resolve by "objective" analysis.

National policies toward nuclear energy evolve in part with advice from organizations, within and outside government, that study the issues in a somewhat neutral and technically oriented spirit and in part in response to pressures from groups and individuals with strong and often unyielding positions for or against nuclear power. There is also an element of chance in the political processes by which the policies are ultimately determined. Broadly speaking, "conservatives" tend to lean to the "pro-nuclear" side, whereas "liberals"—especially when allied with "greens" as in Germany—tend to lean against it. Nuclear power is rarely the central issue in elections, and political parties may gain power for reasons independent of nuclear or other energy considerations. Nonetheless, the electoral results may have—as an incidental effect—a crucial impact on energy policy.

Thus, in discussing the prospects of nuclear power, we face two major sources of uncertainty. We do not know how the alternative energy contenders will compare on technical, economic, and environmental grounds. We know even less how public and political attitudes will evolve. There are also differences among countries that sometimes have no clear explanation. It is easy to understand why Norway has no nuclear power while Sweden has employed it extensively. The answer lies in Norway's ample hydroelectric resources that have been providing over 99% of its electricity [1]. However, it is hard to find such straightforward explanations for Italy's abandoning of nuclear power while France was emphasizing it, or the difference between substantially nuclear Switzerland—which in 2003 referenda voted against giving up nuclear power—and nuclear-free Austria.

In the remainder of this chapter, we will discuss some of the factors that will influence the future development of nuclear power. However, at every turn, it will be necessary to recognize that there are large uncertainties on both the technical and political sides.

20.1.2 Internal Factors Impacting Nuclear Power

The future acceptability of nuclear energy, which we restrict to energy from nuclear fission here, will depend, in part, on internal factors—the strengths and weaknesses of nuclear power itself. Key factors are as follows:

- *Nuclear accidents.* The sine qua non for the acceptance of nuclear power is a long period of accident-free operation, worldwide. Any major nuclear accident will heighten fears of nuclear power and each decade of accident-free operation helps to alleviate them.
- *Reactor designs.* For nuclear power to be attractive, next-generation reactors must be manifestly safe and also must be economical to build. These could be either large evolutionary reactors, of the sort recently built in France, Japan, and South Korea, or smaller reactors that may be a better match to markets of modest size.
- *Waste disposal.* The completion of integrated and fully explained waste disposal plans would encourage people to believe that the problem is "solved." In particular, smooth progress with the Yucca Mountain project would suggest that waste disposal problems are surmountable. However, for a large expansion of nuclear power, it will be necessary to demonstrate the ability to handle the wastes from many more years of reactor operation.
- *Resistance to proliferation and terrorism.* For nuclear power to be acceptable, its facilities must be well protected against terrorists and the nuclear fuel cycle must be proliferation resistant.
- *Assessments of radiation hazards.* Most professionals believe that public fears of radiation—and, in particular, radiation from nuclear power—are out of proportion to the actual risks. A more realistic understanding of the dangers would, in this view, lessen some of the opposition to nuclear power.

20.1.3 External Factors Impacting Nuclear Energy

Verdicts on the "internal factors" discussed earlier will be influenced by perceptions of need. Here, factors external to nuclear power determine the apparent need. These include the following:

- *Energy and electricity demand.* Economic expansion and population growth act to increase the demand for additional energy, including nuclear energy. Effective conservation measures reduce it.
- *Limitations on oil and gas resources.* The need for alternatives is enhanced if these resources are seen to be inadequate.
- *Global climate change.* If the increasing concentration of carbon dioxide in the atmosphere looms large in the public consciousness as an environmental threat, then the pressures to find alternatives to fossil fuels will intensify. Complicating the equation is the prospect of carbon sequestra-

tion, which, at least in principle, offers the possibility of "carbon-free" coal. (We will return briefly to this prospect in Section 20.2.2.)

♦ *Renewable energy.* The technical and economic feasibility of renewable sources and assessments of their environmental impacts are critical to judging the need for nuclear power.

♦ *Fusion energy.* If the hopes for fusion energy are fulfilled, the need for alternatives will be greatly lessened.

Some of these matters have already been discussed in Chapter 1. Others are discussed further in subsequent sections of this chapter.

An additional factor, which might be called "external," is that of government initiative. In principle, government decisions are shaped by a weighing of the internal and external factors cited earlier, and by the public's views on these issues. The public's attitude is decisive when the future of nuclear power becomes a referendum issue. However, a government can sometimes act on its own, resolving complexity with a decisive action. As alluded to earlier, the importance of the Green party to the governing coalition led to a decision to phase out nuclear power in Germany. The U.S. Congress, in this instance at the urging of the President, took critical action in favor of the Yucca Mountain project in 2002. These decisions were not inevitable. For example, a different U.S. administration might have put off indefinitely a decision on Yucca Mountain. Firm government leadership, obviously in autocratic states but also in democracies, can play a role in deciding the future of nuclear power—effectively determining by fiat how all the contributing factors should be balanced.

20.2 Options for Electricity Generation

20.2.1 Need for Additional Generating Capacity

The worldwide demand for electricity will almost certainly increase substantially in the coming decades. The increase will partly be driven by an expansion in conventional uses, as world population grows and the underdeveloped countries strive to raise their presently very low per capita use of electricity. It may also be driven by the expanded use of electricity in relatively new applications—such as the production of hydrogen and the desalination of water. The demand for electricity would also grow if fossil fuels are replaced by electrical power (from non-fossil-fuel sources) in heating and transportation.[1]

Successes in conservation may diminish the need for additional electricity but are unlikely to eliminate it. The amount of electricity that will be required—or wanted—cannot be projected with any certainty, but it appears

---

[1] This has been contemplated for transportation using hydrogen as an intermediary, but it could, in part, be done directly, for example with electrified mass transportation.

probable that over the next 50 years, there will be a doubling or tripling in world electricity demand, with a still greater increase not excluded (see Section 20.3.1). However, although it is valuable to have a sense of scale, it is neither important nor possible to pin down an accurate estimate. With any plausible estimate, it is clear that a great deal of additional generating capacity will be needed to meet new demand and replace existing equipment.

The potential role of nuclear energy in providing the needed energy is considered in Section 20.3. As a prelude, in the remainder of this section, we review briefly the main alternative energy sources—particularly natural gas, coal with carbon sequestration, and renewable energy. These various sources, nuclear and non-nuclear, can be viewed either as competing options or as complementary ones.

20.2.2 Fossil Fuels with Low CO_2 Emissions

Natural Gas

The combustion of coal in electricity generation was responsible in 2001 for about one-third of the U.S. man-made CO_2 emissions (see Section 1.2.3). These emissions would be greatly reduced if natural gas were to be substituted for coal. There are two gains: (1) The CO_2 production using natural gas is about 56% of the production using coal, at the same thermal energy output (see Section 1.2.3). (2) Gas-fired combined-cycle combustion turbines can operate at a thermal efficiency of over 50%, compared to a typical efficiency of 33% for coal-fired steam plants today. Therefore, with modern natural gas plants, the CO_2 output per kilowatt-hour is less than 40% of what it is with standard coal-fired plants.

There are two major difficulties with this solution (as previously discussed in Section 1.2): (1) The carbon dioxide output is reduced but not eliminated, as it would be with nuclear or renewable sources;[2] and (2) the world supply of natural gas, at moderate prices, may prove to be too limited to sustain the contemplated expanded uses of natural gas—in particular, its use as a large-scale replacement for coal. Nonetheless, whether or not switching from coal to natural gas can provide a major long-term solution, it has already been helpful in some countries. For example, total CO_2 emissions in the United Kingdom have dropped in the past decade largely due to a switch from coal to North Sea natural gas [2].

Sequestration of Carbon Dioxide

The reduction of carbon dioxide production would be less of a priority were it possible to sequester the carbon dioxide after it is produced (i.e., to capture

---

[2] It is often pointed out that fossil fuel energy is used in constructing the non-fossil-generating facilities. However, at most, this is a small "correction."

it before it enters the atmosphere and permanently dispose of it in a secure location). The amounts involved are large. For each GWyr of coal-fired electric power, there is a release of about 8.5 million tonnes of CO_2, or, in the units that are commonly used, 2.3 million tonnes of carbon. For the year 2002, U.S. electricity generation from coal amounted to 220 GWyr, corresponding to over 0.5 gigatonnes of C (GtC) and almost 2 Gt of CO_2.

One possibility for storing this CO_2 is to capture it at the generating plant and move it in pipelines to locations where it can be pumped underground. Sam Holloway, of the British Geological Survey, suggested in a review of underground sequestration that the volume of caverns or mines is insufficient to handle the amounts of CO_2 that are produced, but that there is a larger capacity in porous sedimentary rocks or "reservoir rocks" [3, p. 149]. The CO_2 can be injected by pumping it into wells drilled in the rock. The CO_2 then permeates the rock, where it displaces some of the water that is commonly present. For deep rocks, this water is usually saline and the formations are termed saline water aquifers. At high temperatures and pressures (above $31.1°C$ and 72 atm), CO_2 becomes a supercritical fluid with a specific gravity of 0.2–0.9. This condition can be reached deep underground. Even at the low end of this range, the density is more than 100 times that of gaseous CO_2 (at standard temperature and pressure), greatly reducing the required storage volume [3, p. 151].

The total holding capacity of the potential sites for CO_2 is not well established. The global capacity has been estimated by Ted Parson and David Keith to be approximately 200–500 GtC for depleted oil and gas reservoirs, 100–300 GtC for deep coal beds, and 100–1000 GtC for deep saline aquifers [4]. World CO_2 production from fossil fuel combustion is now about 6.6 GtC/yr [2, p. 234], so the capacity might suffice for many years. However, more experience and study are needed to judge the difficulties of large-scale extraction, transportation, and storage of CO_2, as well as to gauge storage capacity.

There is already some experience with both pipeline transport of CO_2 and small-scale sequestration. For example, at the Sleipner West natural gas field in Norwegian North Sea waters, substantial amounts of CO_2 are extracted from the natural gas—which otherwise is mostly methane. This waste CO_2 is injected into sandstone below the seabed.[3] Holloway concludes that this formation "appears to be an excellent repository for CO_2" [3, p. 156]. However, it is only receiving 1 million tonnes of CO_2 per year, which is about one-eighth the output from 1 GWyr of coal plant operation.

An alternative to injecting CO_2 into the ground is pumping it into the deep ocean. The natural content of inorganic carbon in the ocean is about 40,000 GtC, primarily in the form of carbonic acid (H_2CO_3) and bicarbonate and carbonate ions (HCO_3^- and CO_3^{2-}) [5, p. 178]. The injection of several billion

---

[3] This experience illustrates the efficacy of a "carbon tax." When this sequestering project was inaugurated in 1996, Norway had a carbon tax of \$170/tonne of carbon and the project was a response to the tax [4].

tonnes of CO_2 annually would not change the total carbon content of the ocean appreciably, although the local concentrations will be increased near the points of injection. The CO_2 may be injected from pipelines extending from the shore or from tankers carrying liquified CO_2 [6, p. 3–4]. Locally, the water would become more acidic. Further research is needed to explore both injection techniques and environmental impacts.

It may also be possible to convert the CO_2 into a solid, e.g., calcium bicarbonate [$Ca(HCO_3)_2$] or magnesium carbonate ($MgCO_3$). Klaus Lackner suggests that with this approach, the CO_2 would be sequestered "safely and permanently" in a form that requires less volume than is needed in the other various burial approaches [7]. Conversion to solid form would, in principle, make possible the handling of vast amounts of CO_2.

Sequestration would be most practical where the source of the CO_2 is localized, as in a power plant. It has the attraction of offering a possible way to exploit the very large resources of coal that are available and the possibly large resources of natural gas. However, at best, assuming that it proves practical, it is a clumsy approach, requiring the handling of enormous amounts of material. The eventual cost of large-scale carbon sequestration is not well known, but estimates are in the rough neighborhood of $100/tonne of carbon, with an apparent uncertainty of more than a factor of 2 [3, 4]. A cost of $100/tonne corresponds to 2.7¢/kWh for a coal-fired power plant. This would not be a prohibitive cost if there were no alternatives, but would be a significant penalty in a competitive market.

An adequate evaluation of the potential role of sequestration cannot be made without further investigation of the economic costs and environmental impacts, including consideration of the long-term stability of the storage under normal and abnormal (e.g., an earthquake) conditions. As a step toward testing the potential of sequestration in the context of electricity generation, the U.S. Secretary of Energy announced in February 2003 a one-billion-dollar plan to build a prototype coal-fired plant that would produce hydrogen and generate 275 MWe of electricity [8]. The plan calls for the sequestering of over 90% of the CO_2 produced by the plant. This project was expected to take about 10 years to design, construct, and evaluate.

20.2.3 Renewable Sources

Overview of Renewable Sources

In principle, renewable energy sources offer an alternative that avoids the CO_2 produced by fossil fuels and the radionuclides produced by nuclear power. The terms "renewable energy" and "solar energy" are sometimes used interchangeably, but renewable energy includes the nonsolar sources of tidal and geothermal power. Tidal power ultimately is based on gravitational forces, with the tides arising from the motions of the Earth and Moon. Geothermal power is ultimately derived from the decay of long-lived radionuclides in the interior of the Earth. These are nonsolar, renewable sources. However, although both have been used to some extent, especially geothermal power, neither appears to be a leading candidate for a major expansion.

Solar energy sources are both direct and indirect. Direct sources, in which sunlight is converted to energy with no intermediary, include photovoltaic devices, systems for heating fluids with focused light to generate electricity, and growth of biomass to use as a fuel. Indirect sources include hydroelectric power, where solar energy evaporates water that eventually returns as rain, and wind power where solar energy heats the atmosphere to produce air currents.

Direct Use of Solar Energy

All direct uses of solar energy for electricity generation suffer from the "dilute" nature of the solar source. The average flux of solar energy at the surface of the Earth is about 200 W/m^2. Thus, it requires about 5 km^2 to collect 1 GW of incident solar energy. The area required for electricity generation depends on the efficiency of conversion from solar energy to electricity.

One potential source of electricity is biomass, used as a fuel in a steam turbine plant. The main source of biomass now used in electricity generation is wastes, including wastes from the forest product industry. However, the amounts of such wastes are limited. A major increase in biomass use for electricity generation would require dedicated biomass plantations and adequate supplies of water and fertilizer. As estimated by David Hall and colleagues, the "practical maximum yields" of biomass in temperate climates corresponds to an annual average efficiency of about 1% for conversion from solar energy to chemical energy in the plants [9, p. 600]. The thermal efficiency for biomass combustion is unlikely to reach the 33% efficiency achieved for coal, and some of the plantation area must be used for nonproductive purposes such as roads. Thus, an optimistic estimate—probably unrealistically optimistic—of the area required for biomass production of electricity is 2000 km^2/GWe. More typical estimates are roughly a factor of 2 higher, and some are still higher.[4] In any estimate, however, the demands on land and water are high.

Solar photovoltaic power is suitable for use in remote locations, but, at present, it is too expensive to be a candidate for supplying large amounts of power to the electric grid. If the cost is eventually reduced to an acceptable level, the land requirement would be substantial, but much less than for

---

[4] For example, an estimate made by Eric Larson for a large plantation in Brazil assumes a yield of 1000 dry tonnes of biomass/km^2, 20 GJ/tonne, and a 60% use of land. This corresponds to 1.2×10^{13} J/km^2 of thermal energy per year or an average electric output of about 0.12 MWe/km^2, which translates to 8000 km^2/GWe [10, p. 579]. In *Global Energy Perspectives*, projections of future biomass energy yields of 4–10 toe (tonnes oil equivalent) per hectare are indicated, corresponding to roughly 2000–6000 km^2/GWe at a 33% thermal conversion efficiency [11, p. 82].

biomass. It would depend on the solar flux at the site chosen, the efficiency of the photovoltaic cells and associated electronics, the orientation of the solar panels with respect to the sun (including possible tracking of the sun), and the fraction of the land occupied by the panels. If one assumes an overall module efficiency (i.e., for conversion of sunlight to usable electricity) of 10%, an average solar flux of 250 W/m², and a 50% coverage of the ground by the solar modules, then the area required would be 80 km²/GWe. This is only a crude, order-of-magnitude estimate and future systems may be more efficient. However, this may not be an appropriate way of looking at the matter. For the near future, solar photovoltaics are most likely to be used in niche markets, extending down to arrays as small as rooftop modules for individual homes, rather than in very large arrays. Thus, the area occupied may not be an immediate concern.

Neither of these direct sources, or other less conspicuous candidates, should be excluded from consideration, but among renewable sources of electrical power, they at present appear less promising than wind (see later subsection). This is reflected in the differences in the rates at which new facilities are being installed.

Hydroelectric Power

Hydroelectric power is the most important renewable energy source, and it dominates the renewable contribution to electric power generation. It has played an important role in many countries, including the United States, where in the past it accounted for a large share of all electricity generation (e.g., 32% in 1949 [12]). Some major new hydroelectric dams are still being developed, notably China's project on the Yangtze River, which is expected to provide an annual output of about 10 GWyr. However, the most desirable sites throughout the world have been exploited and there is an increasing awareness of the negative impacts of the displacement of people and the interruption of natural stream flows. In fact, in the United States, it appears that there is more interest in removing dams that interfere with the migration of fish than in building new ones. Hydropower provided 7% of U. S. electricity in 2002, and although the amounts vary somewhat from year to year with water conditions, it appears unlikely that hydroelectric capacity will be significantly increased in the United States or that it could make a major contribution to meeting the world's need for additional energy sources.

Wind Power

Wind is a rapidly growing source of electricity in some countries, particularly in Denmark, and the available resource is large. For the OECD as a whole, wind energy output rose from 0.5 GWyr in 1992 to 3.9 GWyr in 2001, an average rate of increase of 25% per year [1]. Some studies indicate that the wind resources in the United States are adequate to produce more electricity than is generated today from all sources.[5] However, wind still makes only a very small contribution in most of the world—0.3% of U.S. electricity generation in 2002 and 0.4% of generation in all OECD countries in 2001 (see Table 1.2). Its exploitation was originally impeded by growing pains in developing reliable and economically competitive wind turbines, and it still faces some environmental objections and the problem of the intermittency of a wind-dependent output.

The power delivered by a wind turbine is proportional to the area swept out by the turbine and the cube of the wind speed.[6] The power varies as the wind speed varies, sometimes dropping to zero. The capacity factor is the ratio of the actual output over the year to the output were the turbine to operate continuously at its rated power. Typical values are in the range of 30% to 40%.

A sense of scale can be obtained by considering a specific example of a large turbine that might be placed in the upper Midwest, where there are extensive areas with favorable wind conditions—in particular, a 1500-kW unit, with a rotor diameter (i.e., a turbine blade diameter) of 77 m [14]. For this case, the indicated capacity factor is about 0.38 and the annual output is about 0.57 MWyr. In the envisaged arrangement, there would be an array of these turbines with about six turbines per square kilometer.[7] To provide 1 GWyr of power each year would require about 1750 turbines, each with blades extending to about 90 m above the ground and occupying almost 300 km² in all.[8] Only a small fraction of this area is preempted by the turbine facilities themselves, and most of it is available for farming or grazing.

Wind power appears to have considerable promise, but until it grows further it may not be possible to have a good picture of its actual potential or of possible difficulties that might be created by a truly large-scale exploitation

---

[5] For example, a study at the Pacific Northwest Laboratories indicates that with "moderate exclusions" on the land that can be used, and using winds of "class 4" and higher, the annual potential is about 630 GWyr [13]. This is roughly 40% more than total U.S. generation in 2002.

[6] The so-called power in the wind P_W equals the product of the volume of air passing through the turbine area per second (Av) and the kinetic energy carried per unit volume ($\rho v^2/2$), where A is the area swept out by the turbines and ρ is the density of air. (It is assumed that the turbine is properly oriented to face the wind.) Thus, $P_W = \rho A v^3/2$. The power output of the turbine can be expressed as $P = C_p P_W$, where C_p is the *coefficient of performance*. Its theoretical limit, known as the *Betz limit*, is $C_P = 16/27 = 0.593$; actual systems reach about 0.4.

[7] The contemplated arrangement would have staggered rows of turbines with a spacing of $7d$ within a row and of $4d$ between rows, where d is the rotor diameter [15]. Thus, each turbine would require an area of $28d^2$, or 0.166 km².

[8] This estimate corresponds to about 0.11 km²/MWe of capacity, which falls in the middle of the range suggested in a recent review article of 0.04–0.32 km²/MWe [16, p. 166].

of wind resources. One obvious problem is that of intermittency. In addition, there is conflict between those who value it as a clean resource and those who consider it a visual blight, especially in otherwise scenic off-shore locations, and too noisy.

20.2.4 Fusion

Fusion energy is, in principle, a long-term alternative to fission energy and renewable energy. In the long run, success in developing fusion would have profound effects on the need for the other options. However, as suggested in the brief discussion in Section 16.7.1, it is too soon to base energy planning on any specific timetable for the successful deployment of fusion, or even to say with absolute confidence that it will become an important energy source in a predictable future.

20.3 Possible Expansion of Nuclear Power

20.3.1 Projection of Demand

Demand for Electricity

In planning for electricity growth, the time horizon is on the scale of decades, and the year 2050 has been selected in some recent publications as a target date for estimates. Such projections, although highly speculative, are useful in suggesting the scale of efforts required to meet future demand.

A number of different scenarios were analyzed in a joint study by the World Energy Council (WEC) and the Institute for Applied Systems Analysis (IIASA). The estimates for world electricity generation in 2050 range from about 2600 GWyr to 4800 GWyr, corresponding to about 155% to 285% of the generation in 2000 [11, p. 88]. The higher projection is described as corresponding to a scenario with "ambitiously high rates of economic growth and technological progress" [11, p. 7], although its average annual electricity growth from 2000 to 2050 (2.1%) is well below the rate of the previous 20 years (3.1%). The lower projection is for an "ecologically driven" scenario with "incentives to encourage energy producers and consumers to utilize energy more efficiently" [11, p. 9a]. It is described as a "normative" or "prescriptive" scenario, rather than a predictive one. In the "middle course," or the case deemed most likely in this family of scenarios, the electricity generation in 2050 is 3500 GWyr, corresponding to an average annual growth of 1.5% from 2000 to 2050.

The WEC/IIASA scenarios all give growth rates that are more conservative than those of the U.S. DOE, which projects in its "reference case" an annual average increase of 2.7% per year in world electricity consumption from 1999 to 2020 [17, p. 188]. Were this growth rate to continue until 2050, it would mean almost a quadrupling of electricity consumption from 2000 to 2050.

These several scenarios suggest that there is likely to be a doubling (1.4% per year) or tripling (2.2% per year) in global electricity demand from 2000 to 2050, although the increase could be substantially more or somewhat less.

Conservation

Conservation, especially in the form of higher efficiency in energy use, can reduce the demand for electricity. It has already contributed importantly through the introduction of more efficient lighting, refrigeration, and motors. Further exploitation of efficient technologies can make major additional contributions. However, conservation measures are already presumed in the scenarios discussed above. For example, the WEC/IIASA scenarios assume for the OECD countries improvements in energy intensity [i.e., decreases in the ratio of primary energy to Gross Domestic Product (GDP)] averaging 1.1% per year in the two higher growth scenarios and 1.9% per year in the low-growth case [11, p. 36].[9]

Demand for Nuclear Power

Given the uncertainties in the world demand for electricity and the even greater uncertainties in the future acceptance of nuclear power, any estimate of nuclear power use in 2050 is highly speculative. However, we consider here several estimates for 2050 that suggest the possible scale of nuclear capacity and generation if there is to be a "large" expansion of nuclear power:

◆ In the highest of the WEC/IIASA projections discussed earlier, annual electricity generation in 2050 was projected to be 4800 GWyr. If we arbitrarily assume that nuclear power provides one-half of this, then annual nuclear generation would be 2400 GWyr.
◆ William Sailor and colleagues projected an annual nuclear output of 3300 GWyr in a scenario designed to stabilize CO_2 emissions at twice the preindustrial level [18].
◆ Harold Feiveson, in pointing to the proliferation dangers raised by a "robust" nuclear expansion, took 3000 GWe in 2050 as a benchmark for what would be necessary to "make a dent in global warming" [19]. At a capacity factor of 90%, this would mean an annual nuclear output of 2700 GWyr.

For specificity in the following discussion, we will use the last of these estimates, which lies between the other two: a nuclear capacity of 3000 GWe and an output of 2700 GWyr. Of course, a nuclear capacity of anything from

---

[9] Data are reported only for the overall improvement in energy intensity, without a separate indication for electricity.

1000 to 5000 GWe would be a major expansion above the present world level of about 360 GWe.

If nuclear electricity supplies one-half of the total, then electricity generation from all sources would be at an annual rate of 5400 GWyr. This exceeds the higher of the WEC/IIASA projections cited earlier, but it would still not provide lavish electricity supplies. At this rate, world electricity use (for an assumed population of 9 billion) would be at a per capita rate of 0.6 kWyr/yr (i.e., consumption at a rate of 0.6 kWe per person). This is approximately twice the present world rate, two-thirds of the present average OECD rate, and 40% of the present U.S. rate [1, 2].

For the United States, which already uses electricity at a rate much above the world average, we will assume that total electricity generation rises in step with population, from 434 GWyr in 2000 to about 650 GWyr in 2050 (an average increase of 0.8% per year).[10] Assuming again a 50% contribution from nuclear power, this would mean an output of 325 GWyr or a U.S. capacity of about 360 GWe.

The speculative nature of any such number should be emphasized. The actual electricity consumption will depend on end-use efficiency improvements and, as suggested earlier, possibly expanded use in a variety of applications, including electronic equipment, building and industrial heating, mass transportation, hydrogen production, and desalination of ocean water. If we look backward and note that U.S. electricity use rose 12-fold from 1950 to 2000 (greatly outstripping the population growth), the 50% rise contemplated here for the next 50 years is a modest projection. The specification of a 50% share for nuclear power is in no sense a prediction. It does, however, identify a target to consider in judging what sort of expansion might be achievable.

Possible Additional Applications

The preceding discussions have mentioned the possible expanded use of nuclear energy in two important applications, namely the production of hydrogen and the desalinization of seawater. Each addresses limitations in the availability of a key resource. Hydrogen is a potential substitute for oil in transportation and desalinization offers a remedy for regional scarcities of water. They are discussed in the immediately following subsections.

In each application, depending upon the method used, the main energy input can be in the form of either electricity or heat and can be derived from either nuclear or non-nuclear sources. The nuclear community has shown considerable interest in these possibilities, as illustrated, for example, by a meeting on *Nuclear Production of Hydrogen* organized by the Nuclear Energy Agency of the OECD [21] and a report on *Introduction to Nuclear Desalinization,*

---

[10] A "fairly constant" population growth rate of 0.8% per year for 2001–2025 was projected in the U.S. DOE's *Annual Energy Outlook 2003* [20, p. 50], and here we arbitrarily project the same growth rate to 2050.

A Guidebook published by the International Atomic Energy Agency [22]. If either application is implemented on a large scale, the demand for electricity and nuclear energy could be substantially increased.

20.3.2 Production of Hydrogen

Methods of Hydrogen Production

Hydrogen is sometimes referred to as the energy source of the future. This is a misnomer, or at least misleading. Hydrogen is not a fundamental energy source, in that there is very little hydrogen on Earth in a pure elemental form. Most of the hydrogen is trapped in water (H_2O) or in hydrocarbons, and energy must be provided to produce it in elemental form.

Once produced, hydrogen has the advantage of having a very high heat of combustion per unit mass: 142 megajoules per kilogram (MJ/kg) for molecular hydrogen (H_2) compared to 46 MJ/kg for gasoline and 55.5 MJ/kg for methane (the main ingredient of natural gas). On the other hand, it has the disadvantage of being a gas except at extremely low temperatures, with a density that is one-eighth that of methane (at the same temperature and pressure) and therefore a lower energy content per unit volume.

Hydrogen is now used in large amounts in the chemical industry [e.g., in the production of ammonia (NH_3) and in oil refining] [23, p. 232]. Its most interesting large-scale prospective use is as an automotive fuel (see the following subsection). The main means of producing hydrogen today is the steam reforming of methane. In this process, steam (H_2O) and methane (CH_4) combine to form carbon monoxide (CO) and hydrogen (H_2), and the carbon monoxide combines with water to form carbon dioxide and more hydrogen. The net reaction is

$$CH_4 + 2H_2O \rightarrow CO_2 + 4H_2.$$

Any process based on fossil fuels has the disadvantage of producing CO_2, although this environmental liability, in principle, could be addressed by sequestering the CO_2.

The simplest alternative to producing hydrogen from methane (or coal) is electrolysis of water, a process in which the water is dissociated into hydrogen and oxygen. Although practical, this method is relatively expensive in energy. The efficiency of electrolysis systems is indicated by Joan Ogden to be about 70–85% [23]. If electricity for electrolysis is generated at, say, 40% efficiency, this corresponds to a system efficiency of about 30% in going from the original fuel to hydrogen.

Greater efficiency is provided by a wide range of thermochemical cycles that use water as the only feedstock and produce hydrogen and oxygen as the end products. They consume less energy than electrolysis for the same hydrogen output. One possible thermochemical cycle is the I–S process which uses sulfuric acid (H_2SO_4) and hydrogen iodide (HI) in a cycle in which each neither is ultimately consumed and in which the net reaction is just the breakup of water into hydrogen and oxygen (see, e.g., Ref. [21]):

$$H_2O \rightarrow H_2 + (1/2)O_2.$$

The I–S cycle is not the only possible choice, but it appears to now be a favored one.[11] One limitation is that it requires a temperature of at least 800°C and operates more efficiently at still higher temperatures.

One of the Generation IV reactors now under consideration is particularly intended for the production of hydrogen: the Very-High-Temperature-Reactor (VHTR) (see Section 16.6.2). The estimated production capacity of a 600-MWt plant is over 2 million normal m^3 per day [25, p. 54].[12] This is equivalent to roughly 26 million MJ per day or 300 MW in the form of hydrogen fuel, corresponding to a 50% efficiency for the conversion of the reactor's thermal energy.

When there is a poor overlap between the time the electricity is produced and the time when it is needed, hydrogen production can help to address the mismatch. Thus, for nuclear reactors, which are most effectively used as base-load sources that always run at peak capacity, the output could be used for hydrogen production (probably by electrolysis for most reactors) when the demand is low—typically at night. Renewable sources such as wind and solar photovoltaic power are inherently intermittent. Again, hydrogen can be produced when there is excess output. The hydrogen then serves as an energy storage medium, for use in electricity generation or for other purposes.

The DOE included in its FY2004 budget proposal, issued in January 2003, $4 million to inaugurate a new Nuclear Hydrogen Initiative, intended to study the use of nuclear energy to produce hydrogen. A focus will be R&D on thermochemical cycles at high temperatures [26, p. 87–89]. If this program moves forward, it may lead to the construction of a VHTR at the Idaho National Engineering and Environmental Laboratory. In undertaking this initiative, the DOE is not abandoning the investigation of other approaches to hydrogen production. It is a much smaller initial commitment than the coal-based initiative announced in February 2003 (see Section 20.2.2). There are also possibilities for producing hydrogen using renewable energy sources. Therefore, although nuclear power may be used in the future to produce hydrogen, there are many competitors. If nuclear power is used, it may be to provide heat rather than electricity.

Hydrogen as a Fuel

A major attraction of hydrogen as a fuel is its cleanliness: Combustion of hydrogen leaves no "waste product" other than water (H_2O). As such, it has

---

[11] An earlier review (1976) listed nine such cycles, although it did not include the I–S cycle [24, p. 287].

[12] 1 normal m^3 (Nm^3) has a mass of 0.090 kg and a combustion energy of 12.8 MJ.

been urged as an automotive fuel in vehicles powered by hydrogen fuel cells, as well as a source of electricity or heat for the home (see, e.g., Ref. [27]). The use in cars has attracted particular interest. If hydrogen could replace gasoline as the fuel to drive cars, it would both reduce the world's dependence on petroleum and eliminate cars as a source of pollution.

The enthusiasm shown by the federal government for the "FreedomCAR" initiative to develop vehicles powered by fuel cells led to the formation of a committee within the American Physical Society (APS) to "examine what is reality and what is unsupported optimism" [28]. The study concluded that for the initiative to be successful, it will be necessary to have large reductions in the costs of fuel cells and to solve the problem of hydrogen storage on vehicles. For storage, it suggested the most likely method will be compressed gas, at perhaps a pressure of 700 atm, although other methods may also prove practical.

One of the concerns that the report addressed, if only briefly, is that of safety. It cited a number of tests that suggested that "H_2 is, if anything, safer than gasoline or jet fuel" [28, p. 21]. Hydrogen is helped in this regard by its lightness, which causes it to dissipate quickly if released. Part of hydrogen's bad reputation derives from the Hindenburg accident in 1937, when the hydrogen-filled German dirigible was destroyed in a spectacular fire as it was about to land at the beginning of a visit to the United States (in New Jersey). Recent studies appear to clear hydrogen of the blame for the fire and instead implicate an "extremely flammable" paint used on the dirigible's skin.

The end-use energy efficiency of the hydrogen fuel cell is more than twice that of a gasoline-driven car. The APS study indicated that a hypothesized hydrogen fuel cell vehicle would use energy at a rate equivalent to 82 miles per gallon (mpg) of gasoline, whereas the rate with a gasoline engine (in the "probable" case) would be 38 mpg—presumably with other changes in the car's design to improve its "mileage." This is an increase in *end-use* efficiency. If the energy efficiency of hydrogen production is 30%, as it might be with electrolysis, the hydrogen advantage disappears when gauged in terms of primary energy input. However, the use of hydrogen is not motivated by the prospect of efficiency gains. It is motivated by the desire to replace oil as a primary energy input. If the hydrogen is made from natural gas rather than nuclear power—as is likely to continue to be the case for at least the near future—there would be a reduction in primary energy use, at the expense of some CO_2 emissions (albeit less than with a gasoline engine) and a possible strain on natural gas supplies.

Gasoline consumption at the rate of 82 mpg corresponds to an end-use energy of 1.60 MJ per mile.[13] Providing electricity to produce hydrogen for motor vehicles would require a substantial increase in generating capacity. Total motor gasoline demand in the United States is about 8.7 million barrels per day, corresponding to an energy of about 17 EJ per year [12, p. 152]. Considering the hypothesized "82-mpg" hydrogen car rather than today's av-

---

[13] The heat content of gasoline is 5.21 MBTU per barrel or 131 MJ/gallon [29].

erage 22-mpg car, the energy requirement is reduced to about 4.7 EJ in the form of hydrogen. An annual electrical input of about 200 GWyr would be needed to produce this hydrogen by electrolysis at an electricity-to-hydrogen energy conversion efficiency of 75%. For comparison, one can note that total U.S. petroleum product use in 2002 averaged 19.3 million barrels per day and electricity generation totaled 438 GWyr. Thus, the substitution (using electrolysis) would involve a reduction in petroleum consumption of about 45% and an increase in electricity generation of about 45%. If the electricity is provided by nuclear reactors it would be necessary to more than triple the present nuclear capacity. If a thermochemical cycle is used instead of electrolysis, the demand for nuclear energy would be somewhat reduced.

For this hypothetical scenario to materialize, it will be necessary not only to solve problems of fuel cell costs and hydrogen storage but also to develop economical and reliable cars to use the fuel cells and to develop a distribution infrastructure analogous to that provided by present-day gasoline filling stations. Some test hydrogen fuel cell vehicles have been built [30, p. 316] and a number of long-distance pipelines have been in use for several decades carrying hydrogen at high pressure [23, p. 248], but establishing such an infrastructure would be a formidable challenge.

Despite the seeming attractiveness of this hydrogen scenario as a project for the future, for the near term the most effective way to reduce gasoline consumption is to build vehicles that are more fuel efficient than those in the present automobile and truck fleet (for example, by reducing weight and using hybrid gasoline–electric engines). Replacing the present passenger car fleet with 45-mpg cars, without changing fuels, would cut gasoline consumption in half. However, for the long run, if it can be achieved, the ultimate rewards of a hydrogen-based program to displace gasoline would be great—whether the hydrogen is produced with nuclear or renewable sources.

An even more ambitious hydrogen scenario has been envisaged by Chauncey Starr, who has suggested a "Continental SuperGrid" that would make possible an energy economy based on hydrogen and nuclear energy [31]. In this picture, nuclear plants would be used to produce electricity and hydrogen. Both would be transmitted throughout the country in pipes that would carry liquid hydrogen in an inner pipe and electricity in a surrounding superconductor. The liquid hydrogen would both cool the superconductor and serve as a fuel when extracted from the pipeline system. This version of a hydrogen economy is put forth as a project for the 21st century, not for the next decade or two. Nonetheless, work today on suitable nuclear reactors and on superconducting cables would help to test the practicality of the concept and, to the extent it is practical, to help launch it.

20.3.3 Desalination of Seawater

Many parts of the world are faced with water shortages, as populations and standards of living rise and groundwater resources are depleted. Desalination of seawater offers a solution that is being increasingly employed. An IAEA document published in 2000 reported that in 1997 there were about 12,500 desalination plants in the world operating or under construction, with capacities ranging from the very small to over 400,000 m^3 per day [22, Section 2.4].[14] Total world capacity was given as 23 million m^3 per day. There is steady growth in this output, and a capacity of about 38 million m^3 per day for 2010 has been projected in another IAEA paper [32]. Even this output, however, would be less than 0.5% of total world water withdrawal, so desalination is still of local rather than global importance [33, p. 374]. The largest facilities are in Saudi Arabia, but plants exist in many other countries, including the United States.

The main techniques for desalination are distillation, which requires mostly heat energy, and reverse osmosis, which requires mostly electrical energy to drive pumps. Reverse osmosis is the least costly approach. The production of 1 m^3 of water in large-scale reverse-osmosis plants is estimated to cost about $1 and require about 6 kWh of electricity [22, pp. 64 and 154–155]. To get a sense of scale as to the implications of these costs, we can consider a hypothetical example given in a summary of IAEA nuclear desalination studies [34]. A 300-MWe plant (which would generate 6.1 million kWh per day at an 85% capacity factor) would be used to supply 1 million m^3 of water per day to a population of 3–4 million people. This corresponds to a per capita supply of about 0.3 m^3 (80 gal) per day or a little over 100 m^3 per year, at a cost of about $100 per person per year.

For comparison, it may be noted that the average annual per capita consumption of water for household purposes in the 1980s was estimated to be about 260 m^3 in the United States, 94 m^3 in Europe, and 30 m^3 in China [33, Table H1]. Household use of water is small compared to the use in industry and agriculture, and to provide the full water needs of a country, the household use numbers should be multiplied by a factor that typically would be between 5 and 20.

At these rates of water use and cost, desalination provides an expensive way to obtain water, but—as seen by its growing use—not a prohibitively expensive one. It would be "affordable" in the United States in regions of water shortages, especially if the high prices led to reductions in the use rates. Total U.S. water use is about 2000 m^3 per year per person (mostly for agriculture), or about 6×10^{11} m^3 for the entire country. If 10% of this water were to be eventually supplied by desalination—to take an arbitrary number for purposes of illustration—this would require about 40 GWyr of electrical energy. This would be a significant increment to electricity demand, but still a modest one compared to potential other sources of increased electricity use.

Worldwide, there may be something of a mismatch at present between nuclear power and desalination. The need for desalination is now greatest in countries of the Middle East, where oil and gas are unusually plentiful, and in underdeveloped regions, where nuclear generation would probably not be

---

[14] 1 cubic meter = 264 U.S. gallons.

affordable and where the amount of water locally needed is small compared to the potential output using nuclear power. However, it is not necessary to think in terms of large, dedicated nuclear reactors providing electricity for desalination alone. The waste heat from nuclear (or other) power plants could be used for distillation processes. Alternatively, for reverse osmosis, electricity could be obtained as part of the output of a general-purpose plant.

Some specifically nuclear desalination facilities have been in operation and others are being undertaken. The largest facility was in Kazakhstan, where some of the power from a 135-MWe breeder reactor (since shut down) was used to produce 80,000 m^3 of potable water per day [35]. A demonstration desalination facility is being installed at a heavy water reactor at Kalpakkam, India that will produce 6300 m^3 per day [32]. In Japan, desalination facilities supply fresh water for use at 10 reactors in amounts of 1000–3000 m^3 per day [35]. In a different variant, the possibility has been explored by Russia and China of small reactors for desalination that would be mounted on barges and which could presumably be moved where needed [22, p. 46].

To date, desalination has made only a relatively small contribution to water supplies, and the nuclear energy desalination projects are to be viewed more as demonstrations of feasibility than as significant producers. For the longer run, however, desalination is likely to be more extensively exploited as a means of providing fresh water. Adequate energy resources are the key to making this possible, and nuclear energy could provide part of this energy in some countries.

20.3.4 Possible Difficulties in Nuclear Expansion

The Pace of Reactor Construction

An expansion to 3000 GWe of nuclear capacity in 2050 may seem a very ambitious goal, especially when at present there is little reactor construction in the world. For a future world population of 9 billion people, the hypothesized per capita nuclear output would correspond to about 40% that of France in 2000. Most of the French increase in nuclear generation occurred in a 20-year period starting in about 1977. It should be possible for the world to achieve less than one-half the present French per capita nuclear output in twice the time (say, from 2010 to 2050), given the conviction that the goal is desirable and if adequate resources are dedicated to the task.

The comparison to France can also be considered in terms of the size of the economies involved, rather than population. World Gross Domestic Product (GDP) in the year 2020 is estimated to reach about 60 trillion dollars (in 1997 U.S. dollars) and the 1990 French GDP was 1.3 trillion dollars [17, Table A3]. This would make the world economy almost 50 times the size of the French economy, each taken somewhere in the middle of expansion. A target of 3000 GWe of nuclear capacity is 48 times the present French nuclear capacity of 63 GWe. Thus, the suggested world expansion is about the same as the French expansion, if normalized to the size of the respective economies. It would appear achievable in 40 years, which is twice the time of the French expansion.

In the above discussion, it was hypothesized that U.S. nuclear capacity might be about 360 GWe in 2050. This would mean an average addition of about 9 GWe of new capacity per year for 40 years—a plausible target for a revived nuclear industry.[15] It may be too conservative a target, given that it is based on a projected rate of rise in electricity consumption of less than 1% per year, or too high a target in that it assumes the nuclear fraction of electricity generation rises from 20% to 50% of the total. However, it provides a sense of scale.

Uranium Resources

An immediate concern in contemplating such an expansion is the uranium supply. World uranium resources were estimated in Section 9.5.2 to be about 20 million tonnes, enough for 100,000 GWyr of reactor operation for present reactors. This would suffice to sustain a linear buildup to 2700 GWyr in 2050 and roughly another 15 years of continued operation. However, it would make little sense to bring reactors on line in 2050 that would run out of fuel in 2065.

The hypothesized expansion could be sustained, however, if greater land resources of uranium are found at acceptable costs, if it proves practical to extract uranium from seawater, if thorium resources are exploited, or if fuel cycles are adopted that make more effective use of uranium. The most uranium-efficient fuel cycle is the breeder cycle. The prospect of breeders elicits enthusiasm in some circles, because it offers a virtually unlimited energy source. It raises concern in others, because it may allow more ready access to plutonium for nuclear weapons. A decision on breeder reactors could be deferred for a long period of time, even if a substantial nuclear expansion begins (see Section 9.5.4). However, ultimately it is an important, high stakes decision.

Nuclear Wastes

The nuclear waste problem can be considered in the context of the U.S. experience. The Yucca Mountain repository has a planned capacity of 70,000 metric tons of heavy metal (MTHM). Typical spent fuel output is now about 30 MTHM/GWyr. The contemplated expansion to an annual 325 GWyr would mean, were there no changes in reactor performance, a U.S. spent fuel output of about 10,000 MTHM per year. Thus, one Yucca Mountain scale repository would be needed every 7 years. However, future reactors may have twice the

---

[15] We assume that all of today's reactors would reach the end of their operating lifetimes by 2050.

burnup, in which case the mass of spent fuel would be halved.[16] In that case, the radioactivity and heat output per unit mass will increase and it may be desirable to have a longer period of predisposal cooling, either in on-site dry storage casks or at centralized off-site facilities.

Depending on the size of the nuclear expansion and the fuel burnup achieved, a future expansion in the United States might require one "Yucca Mountain" every 5 to 20 years—if we continue with a once-through fuel cycle. The actual Yucca Mountain project pays for itself, through the 0.1¢/kWh fee paid by the reactor operators. Therefore, this would be economically affordable. Finding geologically satisfactory sites may be possible, given the large array of plausible sites under consideration before expediency led to the selection of Yucca Mountain. However, the large amount of spent fuel in an open fuel cycle creates incentives to use fuel cycles that reduce the amount of fuel and particularly the amounts of actinides that require disposal. Burning the actinides in fast reactors, as discussed briefly in Section 9.4.3 and Section 16.6.1, would reduce the magnitude of the waste disposal problem.

The global problem is similar, but on a larger scale. It may be desirable to internationalize waste disposal, with countries that have large and geologically suitable areas accepting wastes from other countries. An upsurge in the world use of nuclear energy might also eventually motivate a reconsideration of subseabed disposal (see Section 11.3.2). However, any of these approaches would have to overcome strong and perhaps insuperable opposition, unless the expansion in nuclear energy use is accompanied by a substantial change in public attitudes toward waste disposal hazards.

Weapons Proliferation

The most serious objection to nuclear power, in the view of many technical people, is its link to the spread of nuclear weapons, either to additional countries or to terrorists, as was discussed at some length in Section 18.3. A large worldwide expansion could increase proliferation risks, because the greater the number of countries with nuclear power, the greater the number of actual or latent proliferators (see, e.g, Ref. [19]). For example, more countries could assert the need for uranium-enrichment facilities, ostensibly for low-enriched uranium for civilian reactors but potentially easing the path to high-enriched uranium for weapons. An expansion could also increase the risk of plutonium diversion or theft by encouraging the reprocessing of spent fuel and the use of breeder reactors.

These risks can be reduced by the adoption of appropriate technical and institutional measures. With a large nuclear expansion, the task of inspection and monitoring would be more demanding. However, a greater economic stake

---

[16] Here, we are considering once-through fuel cycles only. Some of the Generation IV reactors have much greater burnups, and in most cases, recycling the fuel is integral to the planning.

in nuclear power might make the world community more willing to give the IAEA the resources and authority that are needed to make it an effective monitor.

20.4 Regional Prospects for Nuclear Power Development

20.4.1 World Picture

There is no substantial expansion of nuclear power underway at this time outside Asia (see Chapter 2). Many countries in Europe could undertake a program of nuclear reactor construction, but most lack the political impulse. Exceptions include Finland and, if announced plans materialize, Russia. The Finnish program will perforce be small, perhaps restricted to a single reactor, whereas the Russian program potentially could be much larger. The significance of the Finnish reactor will be one of example; its impact on the world economy will be small.

Little use of nuclear power is being made in Africa and it does not appear that most African countries are in a position to undertake a substantial nuclear program. However, one of the more highly touted of the new reactors—the Pebble Bed Modular Reactor—is being developed in South Africa, with ambitions for a world market. This appears to be something of an anomaly, and for the near future South Africa is likely to be a nuclear exception in a largely non-nuclear continent. In the western hemisphere, Canada has been the one country other than the United States with a substantial nuclear program, including vigorous efforts to compete in the world market for nuclear reactors.

20.4.2 United States

The Decline in U.S. Leadership

The United States was the world pioneer in nuclear energy and, by virtue of the size of its economy, is still the world leader in total nuclear power generation. However, it is not the leader either in the fraction of electricity that comes from nuclear power, or in rate of growth. The U.S. share of world nuclear generation was 50% in 1975 [36] but had dropped to 30% by 2002, and U.S. DOE projections suggest that this share will continue to slip [17, p. 186]. Light water reactors of U.S. design provided the model initially followed by most countries, including France and Japan, but many countries now have strong reactor design and construction capabilities of their own.

This is not to say that nuclear power is unimportant in the United States or that the technical capabilities of the industry are gone, but the United States is losing the clear primacy it once had, and the size of the nuclear industry and of the university programs that educate future nuclear engineers have both diminished. One assessment of the impending situation was given about 10 years ago by an American participant at an international nuclear conference held in the United States in 1993, who suggested that the conference might be remembered as marking the passing of the "mantle" of nuclear power leadership from the United States to France and Japan [37]. This trend does not seem to have been reversed in the following decade, although the United States has taken a significant initiative with respect to new reactor designs in launching the Near-Term Deployment and Generation IV programs, which are now international in scope (see Section 16.2).

Projections for Future Growth

The long-standing uncertainty about the future of nuclear power in the United States is illustrated by alternative DOE projections made in 1993 for the growth of nuclear power up to the year 2030 [38, p. 9]. Three scenarios from these projections are presented in Figure 20.1: (1) a "no new orders" scenario, in which nuclear capacity decreases as existing reactors are phased out; (2) a "lower reference" case in which there is a cautious resumption of nuclear expansion; and (3) an "upper reference" case that assumes a vigorous economy and use of nuclear power *and* coal for new generation. U.S. nuclear capacity in the year 2030 was projected to be 5 GWe, 119 GWe, and 168 GWe in the three scenarios, respectively.

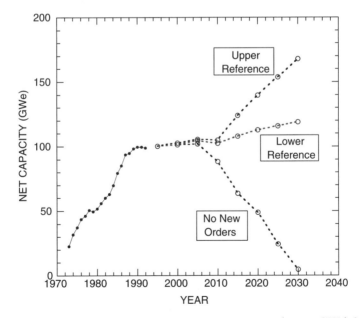

Fig. 20.1. Projected United States nuclear power capacity (in net GWe) for 1995–2030 in three 1993 DOE scenarios, together with actual capacity in prior years (1973–1992).

These projections have stood up reasonably well for the first 10 years since they were made, with their anticipation of little change before 2005. Beyond that, the spate of license extension applications (see Section 2.4.5) makes the "no new orders" estimate look unlikely, while a pessimistic or "lower reference" projection made today would not show the increase of Figure 20.1. The rapid increase shown in the "upper reference" projection appears achievable, as would an even greater increase, but the actual future for the United States remains a matter of public and industry choice.

The stated policy of the U.S. DOE is to encourage nuclear growth. More specifically, the Roadmap (discussed in Section 16.2.3) seeks an order by private industry for a new reactor by 2005, with the goal of deploying it by 2010. This is intended by the DOE to be the first stage in a revival of nuclear power in the United States, to be followed by the construction and deployment of Generation IV reactors by 2030. On the other hand, the DOE's Energy Information Administration projected in early 2003 that no new nuclear plants will be put into operation by 2025 "because natural gas and coal-fired units are projected to be more economical" [20, p. 70]. The validity of this assessment may in large measure depend on measures taken by the DOE itself.

Institutional Issues

Some observers, including most advocates of nuclear power, believe that the crucial issues in the United States are more institutional than technical or even economic. The division of authority and initiative among many levels of government has made it difficult to adopt and implement policies that would permit rapid development of nuclear power (or even its prompt curtailment). Important roles are played by the president, Congress, the courts, and a host of federal agencies, including the Nuclear Regulatory Commission, the Department of Energy, and the Environmental Protection Agency. There is also a confusing division of powers between the federal government, the states, and, in some cases, individual cities or counties. With many opportunities for de facto vetoes, smooth progress in any direction requires a strong consensus. This does not now exist with respect to nuclear power, and existing public opposition remains a problem for the nuclear industry.

One step has been taken to reduce institutional difficulties, namely the streamlining of the licensing process so that once a plant is approved, the only remaining requirement, in terms of NRC procedures, will be the meeting of the *original* specifications. This change is reflected in new regulations, but it will not be known if this process will indeed be smooth until it is tested with an actual license application. The DOE has indicated an interest in helping to cover the costs of the first attempt of this sort.

Even if the public and institutional climate are supportive at the time a reactor is ordered, potential purchasers of nuclear reactors remain fearful that changes in the policies of the federal or state governments could make it difficult to put the reactor into operation when it is completed. The precedent of the Shoreham reactor in New York is not forgotten (see Section 2.4.1). Hence, a prudent utility is likely to be hesitant about ordering a large nuclear reactor. The hesitancy might be somewhat less for smaller reactors, especially modular reactors, because the investment is less and the lead time between an order and reactor operation is expected to be shorter. Even so, problems remain if only because the economies of modular reactors are not realized until a sizable number of them have been ordered and deployed.

It is not clear whether the measures being considered by the DOE and Congress to encourage nuclear power will suffice to overcome industry hesitation. Thus, Larry Foulke, in outlining what would be needed to reduce uncertainties (see Section 19.4.2), warned:

> The DOE's expectation that industry will lead in the introduction of new nuclear technologies is not valid. This means that the DOE is the logical leader in the development and demonstration of advanced reactor schemes with the necessary financial support. [39, p. 38]

In short, nuclear power in the United States may face difficulties without federal measures to aid prospective purchasers of reactors in meeting some of the costs and risks of being pioneers in a nuclear revival.[17]

20.4.3 Asia

In Asia, three countries obtain a substantial fraction of their electricity nuclear power and continue to build new reactors: Japan, South Kor Taiwan. In each of these countries, there is some degree of tension energy planners who would like to build nuclear plants and nuclear o who have had some success in slowing expansion efforts. One succ opponents has been in Taiwan, where the two reactors listed as b construction have had an on-and-off (and now, apparently, on) political power has shifted.

The construction programs in India and China are large com is going on elsewhere in the world (see Table 2.5), but both starting from a small base and the currently committed ex compared to the overall energy needs of India and China. It they will be able to obtain the capital needed for an expansic substantial impact on their energy economies. China may b tion mark. Although its present program is still small for this could change. China's economy is growing rapidly coal is contributing to serious pollution. With a centrali could be a decision to undertake a major nuclear expa economic growth and increased access to the necessa a sign of considerable flexibility in this program, of

---

[17] The U.S. Senate took steps in this direction in June
    islative outcome was not settled during 2003.

struction in China at the end of 2001, two reactors were coming from each of four suppliers divided among France, Canada, Russia, and China [40].[18]

The two remaining Asian countries with reactors that have been under construction are Iran and North Korea. These reactors are intertwined with weapons proliferation issues, as discussed in Chapter 18. In the case of North Korea, the reactors were proposed as an antiproliferation measure, to induce North Korea to forego the use of reactors that would lend themselves to the production of weapons-grade plutonium. In the case of Iran, the power reactor is widely viewed as a proliferation threat, with Iran's nominally civilian nuclear program suspected of being a cover for developing a weapons program.

At present, Asia is just beginning to catch up with the United States and Europe in the scale of use of nuclear power.[19] However, it may soon take the lead. This was predicted by Dr. Kunihiko Uematsu, a Japanese scientist serving as Director-General of the Nuclear Energy Agency of the OECD, who described in 1993 an expansive future for nuclear power in Asia. Citing the existing programs in Japan, Korea, Taiwan, and China and the planning and studies then underway in Indonesia, Thailand, and the Philippines, he suggested:

> . . .nuclear power generation in this region will soon reach the level of the OECD's European and North American regions. This is a striking example of the general shift in the world's energy pattern from the traditionally developed countries of the OECD to other parts of the world. . .with the increasing importance of nuclear power in this part of the world, **the future development of this energy source may no longer be spearheaded by the traditionally developed countries of Europe and North America.** [41, p. 20]

Boldface, as shown, was used in the published paper, indicating the significance the author attached to the point being made.

A more whimsical statement of the anticipated Asian leadership was made in 2002 by Dr. Chang Kun Lee, a commissioner of South Korea's Atomic Energy Commission and then the current chairman of the International Nuclear Societies Council:

> As far as power reactor deployment is concerned, the advanced nations bounded out of the starting line and hopped sprightly along at the pace of a rabbit while we Asian countries plodded along at the slow crawl of the turtle. At the moment, however, the Western nuclear rabbit is taking a nap under a roadside tree (hung with limp moratorium banners) while the Asian nuclear turtle is still toddling along on the road carrying the nuclear seed.

---

of these reactors were connected to the electricity grid in 2002.
generation in 2002 was 58 GWyr in Asia, 92 GWyr in the United States,
05 GWyr in Western Europe [29].

> You could say that Asia is keeping alive a "nuclear technology shelter," keeping the flame burning…these former students of nuclear technology in Asia will be ready to pay back their previous teachers in the West with state-of-the-art technical know-how and new or next generation hardwares. [42]

It is possible to question the applicability of the "turtle–rabbit" analogy. It is also possible that Asian countries will go through the same sequence experienced by Western countries—early enthusiasm followed by strong and often paralyzing public opposition. Nonetheless, the implications of his talk are clear, that, at least for the moment, the "mantle" of nuclear leadership has passed to Asia.

20.5 Issues in Nuclear Decisions

20.5.1 Categories of Issues

Resolvable Issues

Contentious as nuclear disagreements are, some of the key issues are basically technical, and, in principle, conscientious people can eventually reach a common understanding. In particular, there are strong disagreements as to the safety of nuclear reactors and nuclear waste disposal, but it is possible to localize the points of disagreement and, with enough study and patience, it should be possible to resolve them. If one chooses to be optimistic, one can look forward to an eventual convergence of views or at least to the reaching of a consensus that, even if short of unanimity, provides objective policy guidance.

Even the question of the effects of low levels of radiation can be discussed quantitatively, despite its being sometimes considered as beyond the reach of scientific analysis. Upper limits can be put on the possible rate of cancer fatalities, and one can look forward to the day when a better understanding of the underlying biological mechanisms or more comprehensive epidemiological studies can shed light on the validity of the linearity hypothesis, the possible existence of a threshold for radiation damage, and the reality of hormesis (see Chapter 4).

More Intractable Issues

If those were the only sorts of issues involved, the nuclear policy debate would be less difficult, notwithstanding the skepticism with which many people react to the conclusions of "expert" consensus. However, there are two nontechnical issues that cannot be authoritatively decided but that raise profound questions concerning nuclear power. These are the issues of weapons proliferation and of defining what might be called—for want of a better description—a "desirable society." In the former case, any conclusions are largely a matter of political guesswork. In the latter case, they involve personal philosophical or aesthetic viewpoints—with no good way to resolve differences. A dominant position may emerge on each of these issues, but, in the end, the positions may not amount to more than people voting their instincts. (These issues are discussed further in the next two subsections.)

Perceptions of Need

Overhanging all of these considerations is the question of need. Logically or not, the perception of the dangers of nuclear power correlates with the perception of the need for it, including judgments as to the promise of the alternatives. Of course, considerations of danger and need are appropriately linked when a cost–benefit analysis is being made—even an informal one. They are not appropriately linked when an estimate is being made of the absolute risk. It is therefore important to guard against having extraneous views on the desirability of nuclear power influence assessments on technical issues—for example, estimates of radiation effects, reactor accident probabilities, and the effectiveness of the various barriers at the planned Yucca Mountain repository.

20.5.2 Proliferation Risks and Nuclear Power

Some of the detailed issues bearing on the connection between nuclear power and proliferation of nuclear weapons have been discussed in Chapter 18 and earlier in this chapter. Two contrasting assessments can be made as to the nature of this connection. In one view, any country with nuclear power has a headstart as a potential proliferator. Whether or not it has nuclear weapons at the moment, possession of nuclear power makes its path to nuclear weapons easier—in terms of both professional expertise and the procurement of materials and equipment. Further, it will have more soft targets for terrorist theft. The way to reduce this threat is to phase out nuclear power. It would also be desirable to eliminate the nuclear weapons in the countries that already have them, but, even if this cannot be accomplished immediately, a phasing out of nuclear power would reduce the number of potential proliferators. Just stopping the expansion of nuclear power to new countries would avoid adding additional potential proliferators.

An opposing view is that it is too late to adopt this strategy. The argument for phasing out nuclear power is tantamount to an argument that it would be better had nuclear fission been impossible. However, by now the genie is irrevocably out of the bottle. Thirty-one countries have nuclear power, 2 countries without nuclear power have nuclear weapons (Israel and North Korea),[20] and over 20 additional countries have research reactors [43]. Even if

---

[20] The North Korean case is ambiguous, in that it is not certain that it has actual weapons.

all countries gave up their equipment, the technical knowledge would remain, and a country could at any time try to develop nuclear weapons in secret. Further, a decision to phase out nuclear power would be on a country-by-country basis. It would presumably be led by the most "socially responsible" countries, but with no assurance that the less "socially responsible" countries would follow suit. A preferable course, in this view, would be for the "responsible" countries to provide leadership in the use of nuclear power and use their influence to establish fuel cycles that reduce proliferation risks and to strengthen international mechanisms to uncover and discourage proliferation efforts.

An additional argument advanced for nuclear power is that it can help to reduce the need for oil and the likelihood of military conflicts over oil, including potential nuclear conflicts. Thus, even if nuclear power increases the opportunities for developing weapons, it reduces the incentive to use them.

Each of these two overall views has a degree of plausibility, but their implications are contradictory. The choice between them cannot be determined by an orderly, analytic decision-making process, because no matter how the assessments are fleshed out, they appear in essence to only be educated guesses. No amount of new data would establish which assessment—or guess—is the more realistic.

20.5.3 Nuclear Power and a Desirable Society

Feelings About Material Development

Attitudes toward nuclear power are also influenced by aesthetic or philosophical positions on the nature of a desirable world. Is it better for us (i.e., humans) and the planet to have copious energy supplies or is it preferable for energy limitations to restrain unbridled material development? Individual answers to this rather vague question appear to influence the frame of reference in which people view energy issues.

We live with a mix of conflicting attitudes toward technology. On the one hand, we embrace many of the conveniences and applaud some of its fruits (e.g., in medicine and in reducing the drudgery of household chores). On the other hand, at least some people see attractions in a life that is less dependent on and encumbered by mechanical and electrical devices. Also, as we become more dependent on modern technology, we lose some of our sense of control. It is no longer possible for the average "handy" person to fix a car in case of a breakdown, which may or may not be outweighed by the decreased frequency of breakdowns.

Electricity is at the heart of modernization, and the discomfort that critics may have about its impact has been suggested by Amory Lovins:

> In an electrical world, your lifetime comes not from an understandable neighborhood technology run by people you know who are at

your own social level, but rather from an alien, remote, and perhaps humiliatingly uncontrollable technology run by a faraway, bureaucratized, technical elite who have probably never heard of you. Decisions about who shall have how much energy at what price also become centralized....[44, p. 55]

Whatever unease people may feel about electricity—and for electricity itself the unease appears to be less general than Lovins suggests—the concern is intensified for nuclear power. It is indeed a remote technology, with reactor development having become an international enterprise carried out by a small "technical elite" in the employ of large companies. The nuclear industry tries to put a human face upon itself, but it has a difficult task.

Since Lovins wrote those words, the idea of "globalization" has entered the popular culture. Nuclear power can be considered as the epitome of an enterprise controlled by very large corporations with international scope. It offers the world a somewhat uniform product that is expensive to install, difficult to understand, and dominated by highly industrialized countries. The concept of globalization and the objections to it are not well codified, but it is easy to expect that those who have a general distaste for globalization will have a special dislike for nuclear power.

Human Population and Impact

An underlying matter that—consciously or not—may figure in the nuclear debate is our feeling as to the desirability of satisfying the energy demands of a world population that was 2.5 billion in 1950, was 6 billion at the beginning of the 21st century, and appears headed to 9 billion or more in 2050. Nuclear power is pointed to as an aid in meeting these demands. However, some may take that as a curse instead of a blessing. It raises the question of the size of the population that we would welcome. A quotation from John Stuart Mill, written in 1848 when England and the world were much less densely populated, is pertinent today. As quoted by Joel Cohen in *How Many People Can the Earth Support?*, Mill contended:

> A population may be too crowded, though all be amply supplied with food and raiment. It is not good for man to be kept perforce at all times in the presence of his species.... Solitude, in the sense of being often alone, is essential to any depth of meditation or of character; and solitude in the presence of natural beauty and grandeur, is the cradle of thoughts and aspirations which are not only good for the individual, but which society could ill do without. [45, p. 397]

This attitude resonates strongly today, at least among the prosperous.

Considerations of the world's possible population are sometimes couched in terms of the "carrying capacity of the Earth." As discussed by Cohen in the book cited earlier, the carrying capacity depends not only on the material resources—such as land, water, and energy—but also on the degree of crowding that we are willing to tolerate. By providing more ample material resources, nuclear energy can help to sustain a larger population. This is one of the common arguments for nuclear energy—it will enable the world to adequately support more people. However, this can also be taken as an argument against nuclear power.

One can juxtapose the image of a densely populated world that makes extensive use of nuclear energy with one that relies exclusively on renewable energy. Carrying capacity estimates based on energy considerations were made in 1994 by David Pimentel and collaborators [46] and by Paul Ehrlich and collaborators [47]. Each group concluded that an optimal global population for a sustainable future is under 2 billion. The argument is made most explicitly in the Pimentel paper. The authors envisage a world in which fossil fuels have been exhausted and solar energy is the only sustainable energy source. They take 35 quads of primary solar energy as the maximum that could be captured each year in the United States. Assuming that the present average per capita U.S. energy consumption is halved through conservation, the 35 quads would suffice for a population of about 200 million. For the world as a whole, the total available energy in this picture would be about 200 quads. If the world per capita energy consumption were to converge to the new U.S. average, this would support a population of somewhat over 1 billion, which Pimentel et al. interpret as meaning that "1 to 2 billion people could be supported living in relative prosperity."[21]

One need not accept the details of this argument, including the maximum energy assumed for solar energy and the assumed absence of any nonrenewable energy sources. Nonetheless, it suggests alternatives of a densely populated, energy-rich world with nuclear power or a sparsely populated, energy-poor world without it. If energy limitations were accompanied by a shrinking of the world population to 2 billion without social upheavals and deep poverty, this might be a benign scenario. However, it is hard to envisage a peaceful transition of this sort on a quick enough time scale to be germane to the anticipated future energy problems.

Somewhat akin to the concern about excessive population is the concern about mankind's impact on the environment. At a time when cold fusion was being cited as a potentially unlimited source of energy, Albert Bartlett suggested that "if an abundant source of low-cost energy could be found it may be the worst thing that has ever happened to the human race" [49]. The specifically cited danger was the temperature rise accompanying untrammeled expansion of energy consumption, but a broader concern in this argument is that if mankind has unlimited options, the options may be exercised in ways that would severely damage the environment.

The overall argument, although perhaps between debaters passing in the night, is between those who most fear that without nuclear power there will

---

[21] This section, and some of the neighboring sections, are closely based on Ref. [48].

not be enough energy to support the world's population at an "adequate" level and those who most fear the encouragement of population growth and the resulting damage to the environment and the quality of individual lives. This is a second issue that defies "objective" debate.

20.5.4 The Road to Decisions

One Path or Many

A variety of solutions to the world's energy problems are on the table, and each has its enthusiasts and detractors. In the background, and complementing all of the solutions, is conservation. Reduction of wasteful or inefficient uses of energy can substantially reduce the demand for energy. However, at any plausible degree of conservation, world energy consumption will rise—and even without a rise in consumption, replacing present fossil fuel use is desirable. The options for the required energy supply and their attractions when viewed optimistically, include the following:

1. *Coal and carbon sequestration.* The world's coal resources are large and widely distributed. With carbon sequestration, they can be used cleanly, assuming that more tractable emissions such as sulfur dioxide are also eliminated.
2. *Fusion.* Fusion energy, if it can be mastered, represents an ultimate solution, as an almost unlimited resource with few negative impacts.
3. *Natural gas.* The resources of unconventional natural gas may be very great, and the pollution from natural gas is small compared to that from coal (without sequestration).
4. *Nuclear energy.* Energy from nuclear fission is already a major carbon-free electricity provider and, with new fuel cycles, could provide energy for the indefinite future, assuming that waste disposal and other contentious issues are resolved.
5. *Renewable energy.* The solar energy falling upon the Earth far exceeds any possible needs. Its capture, if achievable despite the diluteness of the source, would provide clean energy in perpetuity.

However, there are powerful reasons to question these optimistic assessments and the degree to which we can rely on the listed options. In most cases, they involve a tremendous buildup of technologies whose practicality and impacts have not been tested on a large scale. This caution applies to carbon sequestration, fusion, extraction of natural gas from unconventional sources or of uranium from seawater, and each of the expandable forms of renewable energy.[22] The most tested of these options, nuclear power, is also the one that elicits the most fears and opposition.

---

[22] Here, we are assuming that hydroelectric power and the use of biomass for electricity generation have only limited potential for expansion.

A choice among the options can be made in either a decisive or exploratory fashion. A decisive choice would be to now pick the winner, or winners, and abandon the others. An exploratory choice would be to pursue all plausible options, until their comparative merits are clearer. The case for a single path was made forcefully by Lovins in *Soft Energy Paths*. He argued that attempting to pursue several paths simultaneously would be a distraction and impede implementing the proper one—a mix of conservation and renewable energy [44]. The case for multiple paths reduces to the maxims of not putting all of one's eggs in one basket or of hedging one's bets. The argument is reflected picturesquely in the advice: "When you come to a fork in the road, take it."[23] If one thinks of a person in a car, the advice is intentionally absurd. However, for an army traversing unfamiliar terrain, it can make sense to explore multiple roads.

Even in the exploratory approach, de facto decisions are made by the pace at which efforts are invested in one option or another. A minimal investment is to "keep the option open." Sometimes, in recent years, the dispute over nuclear power has been reduced to one between those who want to phase it out and those who favor keeping the option alive—either because they foresee the day when the need will be recognized or because they are truly undecided about the eventual need. In the United States, the federal policy in recent years has been to keep the option open and, more recently, to encourage it, but to make only relatively minor investments in it.

Differences Among Countries

Although all countries are impacted by some of the same economic factors and resource pressures, it is not to be expected that they will all reach the same decisions. The differences in the nuclear policies of different countries can arise from basic aspects of their physical environment or from the political and economic character of their societies.

In terms of its environment, Japan is in a particularly difficult situation. It is poor in fossil fuel resources and it has a population density that is roughly 12 times that of the United States, limiting its options for use of renewable energy.[24] For Japan, nuclear power offers a path to partial energy independence that cannot be obtained in any other way. In contrast, the United States and Canada are much richer in fossil fuels and have large areas that could, in principle, be used for renewable energy. Thus, they are not under the same pressures as Japan.

---

[23] This advice is attributed to Yogi Berra, the American baseball player and putative source of many pithy phrases.

[24] Japan, in 2001, obtained 59% of its electricity from fossil fuels, 31% from nuclear power, 8% from hydroelectric power, and 2% from biomass and other renewable sources [1].

The effects of differences in physical circumstances are illustrated by the previously mentioned example of Norway and Sweden, which in many ways are similar in attitudes and sociology. Sweden, with somewhat limited hydroelectric resources, has reluctantly continued to use nuclear power for roughly 40% of its electricity, despite the planned shutdown. Norway, in contrast, uses its abundant hydroelectric power to provide virtually all of its electricity and has no nuclear power. Its per capita electricity consumption is 50% greater than that of Sweden [1].

Differences in political mood, in economic opportunities, and in national institutions can also play an important role. The more legal and political avenues nuclear opponents have for contesting nuclear development, the more difficult it is to proceed with nuclear power. As several commentators have pointed out, the difficulties are greater in a country with a federal government than in a country where decisions are made by a centralized national government (e.g., Ref. [41]). A federal government offers many opportunities to raise objections, and the objections are put forth in an atmosphere in which local concerns can take precedence over national priorities. The United States is quite vulnerable in this regard, with important prerogatives held by the states and with a system of checks and balances within the federal government.

It is to be expected that differences in their objective situations and institutions, as well as possibly transient differences in popular attitudes, will continue to lead countries to differing choices. Thus, even were the United States to abandon nuclear power, there is no reason to expect that Japan and France would follow suit. Other countries (e.g., China and India), may wish to accelerate their use of nuclear power, but be held back by a lack of capital. In the end—despite globally common technology, fuel markets, and environmental concerns—decisions on energy policy will be largely national decisions.

Constituencies For and Against Nuclear Power

In reaching a national decision as to the future of nuclear power, the role of a constituency is important. At present, there is a determined and effective constituency against nuclear power, including most environmental organizations.[25] There has been the image of a comparably active and determined constituency for nuclear power, namely the nuclear industry. However, with the decrease of nuclear reactor construction, the nuclear industry has shrunk, and this has not been a valid image for some years. To be sure, there is continuing activity in the operation and improvement of existing power plants and in the completion of a few others, plus some prospect of possible future reactors. This sustains interest on the part of both utilities and manufactur-

---

[25] Here, we will focus on the United States, but the general points have broader applicability.

ers. But the total scale of development is relatively small, and the utilities in countries like the United States are more interested in trouble-free operation of existing reactors than in building new ones. At present, there is no powerful and vocal constituency for the further development of nuclear power.[26]

There are, however, two potential enlarged constituencies: the technical community and the environmental community. For the most part, engineering and scientific organizations and their members support nuclear power, and if energy issues become pressing, there might be a greater sense of urgency in this support.

However, the emotional drive behind any position in the nuclear controversy is heightened when there are important environmental concerns. At present, the "environmental movement" is largely opposed to nuclear power, although with different degrees of finality in the opposition. The movement is not monolithic and there are many strands. From a somewhat extreme standpoint, the fundamental difficulty with nuclear power, or any technology that facilitates increased use of energy, is that it increases the potential impact of humans upon the natural environment—impacts that are likely, in this view, to be undesirable. Those who share this fear will always oppose nuclear power.

Other parts of the environmental movement would welcome a truly clean energy source to replace fossil fuels. Over the next years, some environmentalists might turn to nuclear power in preference to fossil fuel combustion, if they conclude these are the actual alternatives. If such a revisionist view of environmental priorities takes hold, it could provide the impetus for a nuclear revival that may not come from industry or government initiatives alone.[27]

20.5.5 Predictions and their Uncertainty

Summary of Factors Impacting Nuclear Power

As discussed earlier, the factors that will determine whether nuclear power moves ahead or regresses include the following:

- The safety record of existing reactors, the progress of the Yucca Mountain repository, and the perceived safety of next-generation reactors.
- The level of concern about global climate change, oil or natural gas shortages, and the world's dependence on Persian Gulf oil.
- The perceived prospects of renewable energy, carbon sequestration, and fusion.

---

[26] However, if the federal government is sympathetic, the influence of any constituency is amplified, as was the case for conservation during the Carter presidency in the United States and may be the case for nuclear power during the present Bush administration.

[27] Author's note: This thought appears also in the 1996 edition; since then, I have become aware of an organization Environmentalists for Nuclear Energy (EFN), based in France, that has been founded by Bruno Comby [50].

♦ Judgments as to the extent to which nuclear power contributes to or detracts from national and world security.

♦ Attitudes toward technology, globalization, and growth.

♦ The economic competitiveness of nuclear power and the nature of government intervention (e.g., tax credits or carbon taxes).

♦ The orientation of individual governments as they evaluate the issues and their vigor in facilitating the adoption of one or more of the competing technologies.

Given this array of factors—many involving highly controversial or ambiguous questions—it is not possible to know how the balance of forces will affect nuclear power's evolution over the next few decades. Presently, construction of new reactors is confined largely to Asian countries, but there are renewed government expressions of interest in the United States and Russia. Although there is no hint that sudden changes are in the offing, there is no assurance that any industrialized country, wherever it now lies in the spectrum, will maintain its present energy policies over prolonged time periods—whether the changes are in the direction of phasing out nuclear power, expanding its use, or adopting it for the first time.

There are no absolute barriers to a return to a rapid growth in nuclear power. There are nuclear suppliers in North America, Europe, and Asia who are eager to build the reactors if the demand develops. The question is not whether a major expansion of nuclear power is possible, but whether it is desirable. Predicting what will appear to be desirable 10 years hence, or even 5 years hence, is very problematic, especially if the predictions attempt to embrace all countries.

A Past Failure of Prediction

It is interesting to look back almost 30 years and examine the prescience of predictions made then. Conveniently for this purpose, a conference was held in Paris in 1975, with the complacent title *Nuclear Energy Maturity*. The underlying premise of the conference was that nuclear power had arrived, and that it remained to consider how to proceed so that nuclear power could "...represent a long-term-solution, that is for thousands of years rather than the few decades set by the uranium supply required by the 'proven' reactors" [51, p. x].

This long-term issue was addressed in a panel on the Role of Breeders. One speaker gave projections for future generation in the "Western World" (for this purpose, much the same as the OECD countries). In a variety of scenarios, western capacity was projected to be 700–1000 GWe in 1990 and 2000–4000 GWe in 2005 [51, p. 328]. In actuality, total *world* capacity was only 320 GWe in 1990 and there is no possibility that it will reach even 500 GWe by 2005.

These were not atypically optimistic projections. Similar projections were presented by other speakers [51, pp. 319 and 322]. There was at least one dissenting voice [51, p. 324], but it seems to have been a voice in the wilderness. There was a clear consensus that the world was moving into a period of very substantial nuclear expansion.

The failure of this prediction carries two cautionary reminders:

◆ Looking into the recent past does not enable one to see the future. There is a possibility that we are repeating this mistake today, in taking nuclear power's sluggishness of recent years as an indicator of future sluggishness.
◆ The proponents of a newly evolving technology can have an unduly enthusiastic picture of future prospects and may underestimate the difficulties. That was true for nuclear power in 1975. It could be true for some of the emerging technologies today.

Competing Considerations

In the end, policies on nuclear power will depend on judgments of the relative risks of using it or of trying to do without it. With it, we may face risks of radioactive contamination from reactor accidents or waste disposal. Without it, we may face increased risks from climate change and energy shortages. In both cases, there are risks of nuclear bomb manufacture and use. Conclusions as to the magnitude of these risks and how they balance are likely to vary from country to country, given different national circumstances and internal political forces. Depending on the conclusions reached, nuclear power could shrink over the next several decades and remain important in only a few countries, or it could expand substantially in much of the world.

References

1. International Energy Agency, *Energy Balances of OECD Countries 2000–2001* (Paris: OECD/IEA, 2003).
2. U.S. Department of Energy, *International Energy Annual 2001*, Energy Information Administration Report DOE/EIA-0219(2001) (Washington, DC: U.S. DOE, 2003).
3. Sam Holloway, "Storage of Fossil Fuel-Derived Carbon Dioxide Beneath the Surface of the Earth," *Annual Review of Energy and the Environment* 26, 2001: 145–166.
4. E. A. Parson and D. W. Keith, "Fossil Fuels Without CO_2 Emissions," *Science* 282, no. 5391, 1998: 1053–1054.
5. James F. Kasting and James C.G. Walker, "The Geochemical Carbon Cycle and the Uptake of Fossil Fuel CO_2," in *Global Warming: Physics and Facts*, Barbara Goss Levi, David Hafemeister, and Richard Scribner, eds., AIP Conference Proceedings 247 (New York: American Institute of Physics, 1992): 175–200.
6. U.S. Department of Energy, Office of Fossil Energy, *Carbon Sequestration: State of the Science*, draft (February 1999).
7. Klaus S. Lackner, "Carbonate Chemistry for Sequestering Fossil Carbon," *Annual Review of Energy and the Environment* 27, 2002: 193–232.

8. U.S. Department of Energy, "Abraham Announces Pollution-Free Power Plant of the Future," *Techline* (February 27, 2003). [From: http://www.netl.doe.gov/publications/press/2003/tl_futuregen1.html]

9. David O. Hall, Frank Rosillo-Calle, Robert H. Williams, and Jeremy Woods, "Biomass for Energy: Supply Prospects," in *Renewable Energy: Sources for Fuels and Electricity*, Thomas B. Johansson, Henry Kelly, Amulya K. N. Reddy, and Robert H. Williams, eds. (Washington, DC: Island Press, 1993): 593–651.

10. Eric D. Larson, "Technology for Electricity and Fuels from Biomass," *Annual Review of Energy and the Environment* 18, 1993: 567–630.

11. Nebojša Nakićenović, Arnulf Grübler, and Alan McDonald, eds., *Global Energy Perspectives* (Cambridge: Cambridge University Press, 1998).

12. U.S. Department of Energy, *Annual Energy Review 2002*, Energy Information Administration Report DOE/EIA-0384(2002) (Washington, DC: U.S. DOE, 2003).

13. D. L. Elliott, L. L. Wendell, and G. L. Gower, *An Assessment of the Available Windy Land Area and Wind Energy Potential in the Contiguous United States*, Report PNL-7789/UC-261 (Richland, WA: Pacific Northwest Laboratory, 1991).

14. Mark Z. Jacobson and Gilbert M. Masters, "Exploiting Wind Versus Coal," *Science* 293, no. 5534, 2001: 1438.

15. Mark Z. Jacobson, private communication, August 30, 2001.

16. Jon G. McGowan and Stephen R. Connors, "Windpower: A Turn of the Century Review," *Annual Review of Energy and the Environment* 25, 2000: 147–197.

17. U.S. Department of Energy, *International Energy Outlook 2002*, Energy Information Administration Report DOE/EIA-0484(2002) (Washington, DC: U.S. DOE, 2002).

18. Willian C. Sailor, David Bodansky, Chaim Braun, Steve Fetter, and Bob van der Zwaan, "A Nuclear Solution to Climate Change?" *Science* 288, no. 5469, 2000: 1177–1178.

19. H. A. Feiveson, "Nuclear Power, Nuclear Proliferation, and Global Warming," *Physics and Society* 32, no. 1, January 2003: 11–14.

20. U.S. Department of Energy, *Annual Energy Outlook 2003, With Projections to 2025*, Energy Information Administration Report DOE/EIA-0383(2003) (Washington, DC: U.S. DOE, 2003).

21. Organization for Economic Co-operation and Development, Nuclear Energy Agency, *Nuclear Production of Hydrogen*, First Information Exchange Meeting, 2000 (Paris: OECD, 2001).

22. International Atomic Energy Agency, *Introduction to Nuclear Desalination, A Guidebook*, Technical Report Series No. 400 (Vienna: IAEA, 2000).

23. Joan M. Ogden, "Prospects for Building a Hydrogen Energy Infrastructure," *Annual Review of Energy and the Environment* 24, 1999: 227–279.

24. D. P. Gregory and J. B. Pangborn, "Hydrogen Energy." *Annual Review of Energy* 1, 1973: 279–310.

25. U.S. DOE Nuclear Energy Research Advisory Committee and the Generation IV International Forum, *A Technology Roadmap for Generation IV Nuclear Energy Systems*, Report GIF-002-00 (2002). [From: http//gif.inel.gov/roadmap]

26. U.S. Department of Energy, Office of Budget, *FY 2004 Congressional Budget, Energy Supply/Nuclear Energy*, February 2003. [From: http://www.mbe.doe.gov/budget/]

27. T. Nejat Veziroğlu, "Hydrogen Technology for Energy Needs of Human Settlements," *International Journal of Hydrogen Energy* 12, no. 2, 1987: 99–129.

28. Craig Davis, Bill Edelstein, Bill Evenson, Aviva Brecher, and Dan Cox, "Hydrogen Fuel Cell Vehicle Study," Report Prepared for the Panel on Public Affairs (POPA), American Physical Society (June 12, 2003). [From: http://www.aps.org/public_affairs/popa/reports/fuelcell.pdf]

29. U.S. Department of Energy, *Monthly Energy Review, March 2003*, Energy Information Administration Report DOE/EIA-0035(2003/03) (Washington, DC: U.S. DOE, 2003).

30. Supramaniam Srinivasan, Renault Mosdale, Phillippe Stevens, and Christopher Yang. "Fuel Cells: Reaching The Era of Clean and Efficient Power Generation in the Twenty-First Century," *Annual Review of Energy and the Environment* 24, 1999: 281–328.

31. Chauncey Starr, "National Energy Planning for the Century," 2001 Winter Meeting of the American Nuclear Society, Reno, Nevada (November 2001).

32. T. Konishi and B. M. Misra, "Freshwater From the Seas: Nuclear Desalination Projects Are Moving Ahead," *IAEA Bulletin* 43, no. 2, 2001: 5–8.

33. Peter H. Gleick, ed., *Water in Crisis, A Guide to the World's Fresh Water Resources* (New York: Oxford University Press, 1993).

34. Juergen Kupitz, "Nuclear Energy for Seawater Desalination: Updating the Record," *IAEA Bulletin* 37, no. 2, 1995: 21–24.

35. World Nuclear Association, "Desalination" (April 2003) [From: http://www.world-nuclear.org/info/printable_information_papers/inf71print.htm]

36. U.S. Department of Energy, *International Energy Annual 1983*, Energy Information Administration Report DOE/EIA-0219(83) (Washington, DC: U.S. DOE, 1984).

37. Unidentified floor discussant at Global '93, Future Nuclear Systems: Emerging Fuel Cycles & Waste Disposal Options, 1993.

38. U.S. Department of Energy, *World Nuclear Capacity and Fuel Cycle Requirements 1993*, Energy Information Administration Report DOE/EIA-0436(93) (Washington, DC: U.S. DOE, 1993).

39. Larry R. Foulke, "The Status and Future of Nuclear Power in the United States," *Nuclear News* 46, no. 2, February 2003: 34–38.

40. International Atomic Energy Agency, *Nuclear Power Reactors in the World*, Reference Data Series No. 2, April 2002 edition (Vienna: IAEA, 2002).

41. Kunihiko Uematsu, "The Outlook for Nuclear Power," in *Future Nuclear Systems: Emerging Fuel Cycles & Waste Disposal Options*, Proceedings of *Global '93* (La Grange Park, IL: American Nuclear Society, 1993): 18–23.

42. Chang Kun Lee, "A Nuclear Perspective from Asia," 2002 Winter Meeting of the American Nuclear Society, Washington, DC, November 2002.

43. International Atomic Energy Agency, *Nuclear Research Reactors in the World*, Reference Data Series No. 3, December 1997 edition (Vienna: IAEA, 1998).

44. Amory B. Lovins, *Soft Energy Paths, Toward a Durable Peace* (San Francisco: Friends of the Earth International, 1977).

45. Joel E. Cohen, *How Many People Can the Earth Support?* (New York: W. W. Norton & Co., 1995).

46. David Pimentel, et al., "Natural Resources and an Optimum Human Population," *Population and Environment, A Journal of Interdisciplinary Studies*, 15, no. 5, May 1994: 347–369.

47. Gretchen C. Daily, Anne H. Ehrlich, and Paul R. Ehrlich, "Optimum Human Population Size," *Population and Environment, A Journal of Interdisciplinary Studies*, 15, no. 6, July 1994: 469–475.

48. David Bodansky, "Nuclear Power and the Large Environment," *Physics and Society* 29, no. 1, January 2000: 4–6.

49. Albert A. Bartlett, "Fusion and the Future," *Physics and Society* 18, no. 3, July 1989: 11–12.

50. Bruno Comby, *Environmentalists for Nuclear Energy*, English edition, revised by Berol and Shirley Robinson (Paris: TNR Editions, 2001).

51. Pierre Zaleski, Editor-in Chief, *Nuclear Energy Maturity, Proceedings of the European Nuclear Conference* (Oxford: Pergamon Press, 1976).

A

Elementary Aspects of Nuclear Physics

A.1 Simple Atomic Model

A.1.1 Atoms and Their Constituents

Before 1940 scientists had identified 92 elements. These were commonly arranged in the classical *periodic table*, which organized the elements into groups with similar chemical properties. The "last" element in this table was uranium, and for many years it seemed as if this table provided a full representation of matter. Then, with new facilities and insights, attempts were made to produce elements beyond uranium, the so-called *transuranic* elements. These efforts eventually proved successful, starting with the production and identification of neptunium in 1940. Subsequently, more than 20 additional elements have been produced and identified.

There is no fundamental distinction between the "original" 92 elements and the later "artificial" elements. They all were made in the original cosmic processes of nucleosynthesis. The very heaviest elements were unstable and changed relatively quickly into lighter components. Therefore, they are not found on Earth, although some of them can be recreated in the laboratory, lasting for times ranging from a fraction of a second to many thousands of years.

The smallest possible amount of an element is a single *atom*. Each atom is made up of a central *nucleus* surrounded by one or more electrons. The electrons are much smaller in mass than the nucleus, and their distances from the nucleus are much larger than the radius of the nucleus. This is somewhat analogous to the configuration of the solar system, with a heavy central sun orbited by a number of relatively light and distant planets.

The nucleus has two constituents of roughly equal mass, each much more massive than the electron: the *neutron* and the *proton*. Generically called *nucleons*, they differ in that the neutron is neutral while the proton carries a positive charge equal in magnitude to the negative charge of the electron. This equality has been established with great precision from the neutrality of bulk matter. An un-ionized atom is neutral because it contains equal numbers of electrons (outside the nucleus) and protons (inside the nucleus).

It is now recognized that neutrons and protons are not fundamental "building blocks" of matter, but are themselves composed of still more elementary entities, called quarks. We will not discuss quarks nor most of the other so-called *elementary particles*. Despite their importance to the understanding of the earliest origins and ultimate structure of matter, their existence can be ignored in considering the processes important in nuclear reactors, such as radioactivity, neutron capture, and nuclear fission.

A.1.2 Atomic Number and Mass Number

The chemical properties of an element are determined by the number of electrons surrounding the nucleus in an un-ionized atom, which in turn is equal to the number of protons in the nucleus. This number is the element's *atomic number*, Z. Each element can be identified in terms of its atomic number. Thus,

Z = atomic number = no. of protons in nucleus = no. of electrons outside.

The "natural" elements range from hydrogen ($Z = 1$) to uranium ($Z = 92$). Beyond that, the next elements are neptunium ($Z = 93$) and plutonium ($Z = 94$), both produced in nuclear reactors. A list of elements through $Z = 110$ (darmstadtium) is given in Table B.3 of Appendix B.

Nuclei with a given number of protons need not have the same number of neutrons, although for the most part the spread in the number of neutrons for a given element is rather narrow. The total number of nucleons in a nucleus is the *mass number* of the nucleus and is customarily denoted by A:

$$A = \text{mass number} = N + Z,$$

where N denotes the number of neutrons in the nucleus.

A.1.3 Isotopes and Isobars

For a given element (i.e., same Z), nuclei with different numbers of neutrons (i.e., different A) are called *isotopes* of the element. Nuclei with the same mass number A but different atomic number Z are called *isobars*. Different isotopes of an element are virtually identical in chemical properties (although small differences may arise from their different masses and therefore their different mobilities). In nature, the relative abundances of different isotopes are usually closely the same for different samples of an element, because it is mainly the chemical properties of the atom that determine the atom's history on Earth.

To specify fully a nuclear species, both Z and A must be given. For example, the most abundant isotope of carbon ($Z = 6$) has mass number 12. It is called "carbon twelve" and is conventionally written ^{12}C. In this notation, it is not necessary to specify that $Z = 6$, because the symbol C, by identifying the element as carbon, defines the atomic number to be 6. Nonetheless, sometimes the redundant notation $^{12}_{6}$C provides a useful reminder. The term *nuclide* is used to denote a particular atomic species, as characterized by its atomic number and atomic mass number; the term *radionuclide* denotes a radioactive nuclide.

A.2 Units in Atomic and Nuclear Physics

A.2.1 Electric Charge

In the International System of Units (SI), the unit of charge is the Coulomb, itself defined in terms of the unit of current, the ampere. In atomic and nuclear physics, it is usually more convenient to express charge in terms of the magnitude of the charge of the electron, e, where[1]

$$e = 1.6022 \times 10^{-19} \text{ Coulombs.}$$

The charge of the electron is $-e$, and the charge of the proton is $+e$. The atomic number of carbon is 6, and therefore the charge on the carbon nucleus is $6e$, or 9.61×10^{-19} Coulombs.

A.2.2 Mass

The SI unit of mass is the kilogram (kg). It also remains common in atomic and nuclear physics to express mass in grams (g)—i.e., in the centimeter-gram-second (cgs) system—when mass must be expressed in macroscopic terms. For most purposes in nuclear physics, however, it is more convenient to express mass in terms of the *atomic mass unit* (u), which is defined so that the mass of a hydrogen atom is close to unity.

More precisely, the atomic mass unit is defined in the so-called unified scale by the stipulation that the mass of the neutral ^{12}C atom is precisely 12 u. (Earlier, a scale based on oxygen, not carbon, had been used, and this can occasionally cause confusion, especially as the numerical values for the

---

[1] Numerical values for general physical constants and for nuclear masses and half-lives are taken from Ref. [1].

Table A.1. Masses and rest mass energies.

Quantity or Particle	Symbol	Mass or energy	
		u	MeV
Atomic mass unit		1 (exact)	931.4943
Electron	e	0.0005486	0.5110
Proton	p	1.0072765	938.2723
Neutron	n	1.0086649	939.5656
H atom ($A = 1$)	^{1}H	1.0078250	938.7833
C atom ($A = 12$)	^{12}C	12 (exact)	11177.9

two scales are close.) With this definition:

$$1 \text{ u} = 1.66054 \times 10^{-24} \text{ g} = 1.66054 \times 10^{-27} \text{ kg}.$$

Although on this scale, the masses, or atomic weights, of the ^{1}H atom (M_H), the proton (m_p), and the neutron (m_n) are all close to unity, they are not exactly unity nor equal to each other (see Table A.1 for actual values). The mass of the electron is $m_e = m_p/1836$.

A.2.3 Avogadro's Number and the Mole

It is convenient in chemistry and physics discussions to introduce the *gram-molecular weight* or *mole* as a unit for indicating the amount of a substance. For any element or compound, a mole of the substance is the amount for which the mass in grams is numerically equal to the atomic (or molecular) mass of the substance expressed in atomic mass units. For example, the mass of one mole of ^{12}C is exactly 12 g, and the mass of one mole of isotopically pure atomic hydrogen (^{1}H) is 1.0078 g.

The number of atoms (or molecules) per mole is termed *Avogadro's number*, N_A. It has the numerical value

$$N_A = 6.02214 \times 10^{23} \text{ per mole.}$$

The mass of an atom (in grams) must equal the mass of one mole of the substance (in grams) divided by N_A. Therefore,

$$1 \text{ u} = \frac{1}{N_A} \text{ g} = 1.66054 \times 10^{-24} \text{ g}$$

as stated above. It is experimentally more practical to determine an accurate value for N_A than to make an absolute determination of the mass of an individual atom. For that reason, N_A is the primary experimental number, and the mass corresponding to 1 u, expressed, for example, in grams, is a derived result.

A.2.4 Energy

Energy is expressed in joules (J) in SI units, but it is much more common in atomic and nuclear physics to express energy in electron volts (eV), kilo-electron volts (keV), or mega-electron volts (MeV), where 1 eV is the energy gained by an electron in being accelerated through a potential difference of 1 volt.[2] This energy is $q\Delta V$, with $q = e$ and $\Delta V = 1$. Thus,

$$1 \text{ eV} = 1.6022 \times 10^{-19} \text{ J}.$$

In atomic physics, the convenient unit is usually the eV. In nuclear physics, where the energy transfer per event is much higher, the more convenient unit is the MeV.

The average energy of an assembly of particles is sometimes expressed in terms of temperature. Thus, the neutrons responsible for initiating fission in typical nuclear reactors are termed "thermal neutrons." Their kinetic energy distribution is the distribution characteristic of a gas at the temperature of the reactor core. In particular, the mean translational kinetic energy of a molecule in a gas at temperature T is $\frac{3}{2}kT$, where T is expressed in degrees Kelvin (K) and k is the Boltzmann constant:

$$k = 1.381 \times 10^{-23} \text{J/K} = 0.862 \times 10^{-4} \text{ eV/K}.$$

The product kT is a characteristic temperature for a wide range of phenomena. A convenient reference point, or memory aid, is that at room temperature (taken to be $T = 20°\text{C} = 293$ K), $kT = 0.0253$ eV $= \frac{1}{40}$ eV.

In a gas at temperature T, there is a broad distribution of kinetic energies, and a small fraction of the particles have kinetic energies well in excess of the $\frac{3}{2}kT$ average. Fusion in stars and in (still experimental) fusion reactors is initiated by positively charged nuclei in the high-energy tail of the energy distribution. High energies are important in fusion because the higher the energy the better the prospect of overcoming the Coulomb repulsion between the positively charged interacting particles.[3] For neutron-induced fission, on the other hand, there is no Coulomb repulsion acting on the neutron.

A.2.5 Mass–Energy Equivalence

The equivalence of mass and energy is basic to nuclear and atomic physics considerations. The energy E associated with the mass m is $E = mc^2$, where c is the velocity of light. Expressing mass in kilograms and velocity in meters/sec, the energy equivalent of 1 atomic mass unit is

---

[2] 1 keV $= 10^3$ eV; 1 MeV $= 10^6$ eV.

[3] Coulomb repulsion is the name commonly given to the repulsion, governed by Coulomb's law, between objects that carry charges of the same sign.

$$E = (1.66054 \times 10^{-27}) \times (2.9979 \times 10^8)^2/(1.6022 \times 10^{-13}) = 931.5 \text{ MeV}.$$

Expressed in terms of energy, the mass of the electron is

$$m_e c^2 = 0.511 \text{ MeV}.$$

Often it is simply stated that the electron mass is 0.511 MeV, with no distinction made between "mass units" and "energy units." In Table A.1, the particle masses are expressed in both MeV and atomic mass units (u).

A.3 Atomic Masses and Energy Release

A.3.1 Atomic Mass and Atomic Mass Number

In describing a nuclear species—for example, ^{238}U—one could specify either the mass of the nucleus or the mass of the atom (the nucleus plus the electrons). It is virtually universal practice to specify atomic mass. The mass M of an atom (expressed in atomic mass units) is not exactly equal to the mass number A of the atom, except for ^{12}C, where the equality is a matter of definition. However, the numerical difference between M and A is small. This is because the constituents of the atom—the neutrons, protons, and electrons—have masses quite close to 1 u or quite close to zero. Although the atomic mass is slightly less than the sum of the masses of the constituent particles, it is not very different. Tabulations of atomic masses are often expressed in terms of the *mass excess* Δ, where $\Delta = M - A$.[4] For ^{238}U, for example, $\Delta = 47.304$ MeV or 0.05078 u.[5] Correspondingly, the atomic mass of ^{238}U is 238.05078 u.

A.3.2 Isotopes and Elements

It is necessary to keep in mind the distinction between the atomic mass M_E of the element, as it occurs with its natural mixture of isotopes, and the mass M_i of any particular isotope. The atomic mass of an element is given by

$$M_E = \Sigma f_i M_i, \tag{A.1}$$

where f_i is the fractional abundance of the isotope, by number of atoms. For example, carbon has two stable isotopes: ^{12}C and ^{13}C, with atomic masses M_i

---

[4] In this expression, although A is ordinarily defined as a dimensionless integer, we attach to it the same units as the units of M, rather than further encumber the notation.

[5] It might be argued that Δ has been defined in units of mass, not energy. These quantities, however, are physically equivalent, and it is common to use the symbols Δ and M to represent either mass or the equivalent energy, as context requires.

equal to 12.00000 u and 13.00336 u, respectively. For every 10,000 C atoms on Earth, 9890 are ^{12}C and 110 are ^{13}C, giving fractional abundances f_i of 0.9890 and 0.0110, respectively.[6] By Eq. A.1, it follows that $M_E = 12.0110$ u.

It is possible to specify relative abundances of the constituents of a sample of matter in terms of the relative number of atoms (or molecules) or in terms of the relative masses of the constituents. It is usual to specify isotopic abundances f_i in terms of the relative number of atoms. However, an important exception in the context of nuclear power occurs in describing the isotopic abundances of uranium isotopes. These are often specified in terms of the fraction by mass. (This distinction is discussed further in Section 9.2.2.)

Elemental abundances, on the other hand, are usually specified in terms of relative masses. For example, the abundance of uranium is commonly specified in terms of parts per million (ppm). An abundance of 2 ppm means that there are 2μg of uranium per gram of rock. Again there is an exception. For gases, relative abundances are sometimes specified in fraction by volume, which is equivalent to specifying the relative number of molecules.

A.3.3 Binding Energy, B

The mass of a nucleus is almost, but not exactly, equal to the sum of the masses of the constituent neutrons and protons. The difference is the *nuclear binding energy*, B. Thus, for a nucleus characterized by mass number A and atomic number Z,

$$M_{nuc} = Zm_p + (A - Z)m_n - \frac{B}{c^2}. \tag{A.2}$$

The binding energy B represents the total energy that would be required to dissociate a nucleus into its constituent neutrons and protons.

Neglecting the binding energy of the electrons in the atom,[7] the atomic mass M is equal to the sum of the masses of the nucleus and the surrounding electrons: $M = M_{nuc} + Zm_e$. Similarly, the mass of Z hydrogen atoms is $ZM_H = Zm_p + Zm_e$. Therefore, adding Zm_e to both sides of Eq. A.2, the atomic mass can be expressed as

$$M = ZM_H + (A - Z)m_n - \frac{B}{c^2}. \tag{A.3}$$

Rewriting Eq. A.3, the nuclear binding energy is given by

$$B = [ZM_H + (A - Z)m_n - M]c^2. \tag{A.4}$$

---

[6] We here ignore small variations that may occur in fractional isotopic abundances. These can be caused, for example, by differences in the temperature at the times when different samples of a given material were formed.

[7] Although this neglect inserts some error in Eq. A.3, if B is strictly interpreted as a *nuclear* binding energy, the error is small because electron binding energies in the atom are much smaller than nucleon binding energies in the nucleus.

Eq. A.4, or the equivalent preceding equations, can be taken to be the definition of the binding energy.

For calculational purposes, in view of the manner in which nuclear data are tabulated, it is useful to rewrite Eq. A.4 in terms of the mass excess Δ, in effect subtracting Z, $A - Z$, and $-A$ in successive terms within the bracket of Eq. A.4. It is also convenient to ignore the difference between mass units and energy units and drop the factor c^2 in Eq. A.4. Then the binding energy can be written

$$B = Z\Delta_H + (A - Z)\Delta_n - \Delta, \qquad (A.5)$$

where $\Delta_H = 7.289$ MeV, $\Delta_n = 8.071$ MeV, and Δ is expressed in MeV.

The average binding energy or binding energy per nucleon, B/A, provides a measure of the stability of a nucleus. In general, more stable configurations have higher values of B/A. As an example, we calculate the binding energy per nucleon for one of the more tightly bound of the nuclei, ^{56}Fe ($Z = 26$ and $\Delta = -60.601$ MeV). From Eq. A.5,

$$\frac{B}{A} = \frac{(26 \times 7.289) + (30 \times 8.071) - (-60.601)}{56} = 8.79 \text{ MeV/nucleon.} \quad (A.6)$$

A.3.4 Energy Release in Nuclear Processes

The total energy of the system remains unchanged in any nuclear process, in accord with the principle of conservation of energy. Energy will be released in the process if the total binding energy is greater for the final nuclei than for the initial nuclei. Equivalently, energy is released if the total mass of the final nuclei is less than the total mass of the initial nuclei. The energy release, often denoted by the symbol Q, is given by the mass difference:

$$Q = (\Sigma M_i - \Sigma M_f)c^2, \qquad (A.7)$$

where the summations are taken over all initial masses M_i and final masses M_f. Eq. A.7 is used in Section 6.4.1 to calculate the energy release in fission.

A.4 Energy States and Photons

One of the great breakthroughs of early 20th-century physics, embodied in the Bohr model of atomic structure, was the recognition that atoms can exist only in certain states or configurations, each with its own specific energy. In the simple Bohr picture of the hydrogen atom, different states correspond to electron orbits of different radii, each with a well-defined energy equal to the sum of the kinetic and potential energies of the electron in its orbit. This very simple mechanical picture has been subsequently modified by quantum mechanics, but the basic point remains that the possible configurations of an atom correspond to a limited set of states. The discrete energies associated with these states are the allowed energy levels of the atom. The same rule holds for nuclei, although we are not yet able to account theoretically for the exact energy levels of nuclei with as great precision as we can for atoms.

Thus, in broad terms, each atom or nucleus can exist in a state of lowest energy, the so-called *ground state*, or in one or another state of higher energy, the so-called *excited states*. With a few exceptions, the excited states are short-lived; that is, they quickly emit their excess energy and the system (atomic or nuclear) reverts to its ground state. The energy lost by the atom in a transition from one excited state to a lower one (or to the ground state) is commonly carried off by electromagnetic radiation.[8] For nuclei, a typical time for a transition from an excited state to a lower state is in the neighborhood of 10^{-12} sec, although very much longer and somewhat shorter lifetimes are also possible.

When an atom (or nucleus) in a state of initial energy E_i makes a transition to a final state of lower energy E_f, the energy carried off in electromagnetic radiation is

$$E_{rad} = E_i - E_f. \tag{A.8}$$

Throughout the 19th century, light and (when recognized) other forms of electromagnetic radiation were thought to be properly described by waves. One of the revolutionary new insights of early 20th-century physics was the recognition that light also has particle-like properties. In particular, the electromagnetic radiation corresponding to a single atomic (or nuclear) transition is carried in a single discrete packet, called a photon.

The energy of the photons for a given transition is simply related to the wavelength, or frequency, of the associated radiation:

$$E = h\nu = hc/\lambda, \tag{A.9}$$

where λ is the wavelength, ν is the frequency, and h is a universal constant known as Planck's constant: $h = 6.626 \times 10^{-34}$ joule-sec. Thus, in the transition of Eq. A.8, the photon energy is

$$h\nu = E_i - E_f. \tag{A.10}$$

Visible light is associated with transitions involving the outer electrons of atoms or molecules, with photon energies in the neighborhood of several eV (3 eV corresponds to $\lambda = 4130$ Å $= 0.413 \times 10^{-6}$ m). X-rays correspond to transitions involving the inner electrons of atoms, with typical energies of 1 to 100 keV, depending on the atomic number of the atom. Radiative transitions

---

[8] There are two classes of exceptions to this: (a) sometimes the energy can be transferred to an electron in a process known as internal conversion, still leaving the nuclide unchanged; and (b) in some cases an excited state can decay by emitting a particle, such as a beta particle or neutron, thereby changing the atomic number or the mass number of the nuclide.

between nuclear levels typically involve energies in the neighborhood of 100 to 10,000 keV. The photons from nuclei are called gamma rays. There is no difference between these groups of photons other than their energy. In fact, it is possible, although not common, to have gamma rays with energies lower than those of typical x-rays. The names "x-ray" and "gamma ray" date to the times of the original discovery of the then-mysterious radiations. In principle, there is no need for different terms to distinguish between photons from atomic transitions (x-rays) and photons from nuclear transitions (gamma rays), but the terminology is retained, perhaps because it provides a reminder of their physical origin.

A.5 Nuclear Systematics

Nuclei are categorized as being stable or unstable. Loosely speaking, stable nuclei are those that remain unchanged forever. Unstable nuclei decay spontaneously into lighter nuclei on a time scale characteristic of the particular nuclear species. This time scale can be expressed in terms of the half-life of the species, defined to be the time interval during which one-half of an initial sample will decay (see Section A.7.2). If the half-life for decay is greater than some (undefined) small fraction of a second, the process of decay is called radioactivity (see Section A.6). Half-lives of different species vary from much less than a second to many billions of years.

The concept of "stability" is not an absolute one. The heaviest "stable" nuclide is bismuth-209 (^{209}Bi), with $Z = 83$ and $A = 209$. There is some evidence, however, that it decays with a half-life in the neighborhood of 10^{18} years. (This is stable enough for most purposes, as the age of the universe is only on the order of 10^{10} yr.) It should also be noted that some current theories suggest that the proton itself is not stable, but if the proton does decay at all, it decays at an extraordinarily slow rate (a half-life of more than 10^{31} yr). Of course, such slow decays have no relevance to the processes of radioactivity that are of interest here.

Most of the nuclides found in nature that are lighter than ^{209}Bi are stable. However, there are exceptions, such as potassium-40 (^{40}K) and rubidium-87 (^{87}Rb), which are both long-lived residues of stellar nucleosynthesis processes, as well as carbon-14 (^{14}C), which has a relatively short half-life ($T = 5730$ yr) but is produced continuously in the atmosphere due to cosmic rays.

Above ^{209}Bi, continuing up to ^{238}U ($Z = 92$, $A = 238$), the nuclei found in nature are not stable. The reason that some are still here is either that they have very long half-lives themselves, as in the case of ^{232}Th, ^{235}U, and ^{238}U, or they are progeny of these nuclei. Above $Z = 92$, a considerable number of nuclei have been artificially produced, and the properties of some of them are well established. As one goes higher and higher, the half-lives of the nuclei tend to decrease. Some of the nuclei above uranium have half-lives of thousands or even millions of years [e.g., ^{237}Np ($T = 2.14 \times 10^6$ yr) and

^{239}Pu ($T = 2.41 \times 10^4$ yr)], but at very high atomic numbers ($Z > 104$), most of the elements have no isotopes with half-lives as long as 1 min.

Most of the mass of the universe is concentrated in hydrogen (mostly ^{1}H), in helium (mostly ^{4}He), and in nuclei with even values of Z and with $A = 2Z$, starting with carbon ($Z = 6$) and continuing through calcium ($Z = 20$), i.e., ^{12}C, ^{16}O, ... ^{40}Ca. At higher atomic numbers, the stable isotopes have $A > 2Z$, i.e., more neutrons than protons in the nucleus. At each mass number through $A = 209$ (other than masses $A = 5$ and $A = 8$, where there are no stable nuclei), there are one or more stable nuclei and a host of nuclei that are unstable against beta-particle emission (see Section A.6.1). The stable nuclei are clustered about a trajectory, depicted in Figure 3.1, which follows the line $A = 2Z$ up to $A = 40$, and continues up to $Z = 83$, $A = 209$. At higher A, all nuclei are unstable for decay by either beta-particle or alpha-particle emission, but some of the alpha-emitting nuclei have long half-lives and are included in Figure 3.1.[9] Most of the naturally radioactive isotopes lie between $Z = 82$ and $Z = 92$, with ^{40}K the most important exception.

The binding energy per nucleon (see Eq. A.6) is close to 8 MeV over most of the range of stable nuclei. It is zero for ^{1}H and is small for the very lightest atoms, but is above 7.4 MeV for all stable nuclei from ^{12}C to the top of the periodic table. It rises from values below 8 MeV for the lighter nuclei to a broad peak at about 8.8 MeV near $A = 60$ and then falls gradually to 7.57 MeV at ^{238}U. A plot of the binding energy per nucleon, B/A, as a function of mass number A is presented in Figure A.1.

The lightest and heaviest nuclei are less tightly bound, while intermediate nuclei are more tightly bound. This suggests two paths for liberating energy: form intermediate nuclei by combining very light nuclei or by splitting very heavy ones. These are the respective processes of *fusion* and *fission*. The difficulty, to the extent there is one, is in accomplishing these processes in a controlled fashion.

A comprehensive 1999 compilation of stable and unstable nuclides includes over 3,000 entries [1], with many different entries for most values of atomic number (many isotopes) and for most values of atomic mass number (many isobars). The number of listed entries can be expected to grow as investigations continue of nuclei with extreme combinations of A and Z. However, only a limited number of these nuclides are of interest from the standpoint of nuclear energy. The most important of these are: light nuclei ($A \leq 12$), which may be targets or products in fusion or may serve as moderators in fission reactors, neutron-rich medium-mass nuclei formed as fission products ($76 \leq A \leq 160$), and heavy nuclei used as fuels for fission or produced by neutron capture ($232 \leq A \leq 246$) in a fission reactor.

---

[9] The trajectory depicted in Figure 3.1 is sometimes referred to as the "valley of beta stability," because if a three-dimensional picture is envisaged, with Z and A as axes in the horizontal plane and atomic mass as a vertical axis, this trajectory would follow a valley of minimum mass at each value of A.

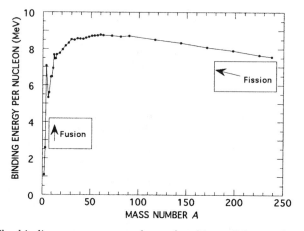

Fig. A.1. The binding energy per nucleon of stable nuclei, as a function of mass number. At very high A, the plotted nuclei are not stable against alpha-particle decay but are stable against beta decay.

A.6 Radioactive Decay Processes

A.6.1 Particles Emitted in Radioactive Decay

Many nuclei are stable in the sense that they normally remain unchanged for exceedingly long periods of time—in fact, "forever" as far as it has been possible to study them. Other nuclei are unstable. In its most common form, the instability is exhibited by the phenomenon of *radioactivity*, in which the nucleus spontaneously emits an *alpha* (α) particle or *beta* (β) particle, often accompanied by the emission of one or more *gamma* (γ) rays.

The phenomenon of radioactivity was discovered in the 1890s, first in uranium and later in other natural elements. Beginning in the 1930s a host of additional radioactive elements have been produced in nuclear accelerators and nuclear reactors. All three of the radiations can cause the blackening of a photographic plate or the discharge of an electroscope. This is accomplished by ionization of the medium, and alpha particles, beta particles, and gamma rays are referred to as *ionizing radiations*, distinguishing them from other radiations such as radio waves. Other than the shared ionizing property, the three rays are very different. We will not examine the history by which the properties of these rays was established, but the properties themselves will be briefly described.

Alpha particles are nuclei of helium-4 atoms (^{4}He). Thus, an alpha particle has a mass of about 4 atomic mass units (u) and is positively charged, with a charge of magnitude $q = Ze = 2e$. Compared to the other radiations, alpha particles can penetrate only a small distance in matter. The most energetic of the alpha particles emitted from radioactive nuclei are stopped after passing through less than 10 cm of air, i.e., about 10 mg/cm^2 of matter or about 0.1 mm of a material whose density if 1 g/cm^3 (such as water).[10] The penetration distance is strongly dependent on the initial alpha-particle energy, and for lower-energy alpha particles, the penetration will be even less.

Beta particles are electrons. As such, they have a mass far smaller than that of the alpha particle. Except for a very few cases, the beta particles emitted in natural radioactivity are negatively charged; they are more completely designated as β^- particles. The β^- particles are identical to the ordinary electrons surrounding the nuclei of atoms. In the 1930s positive electrons, called positrons or β^+ particles, were discovered. These are emitted from artificial radionuclides produced when positive particles, such as protons or alpha particles, combine with a nucleus to form an unstable "proton-rich" nucleus. β^+ emitters are very rare in natural terrestrial material and among the radioactive nuclei produced in fission. Typical beta-particle penetration distances are on the scale of 0.1 to 1 g/cm^2, increasing with increasing beta-particle energy.[11]

Gamma rays are photons or quanta of radiation. They have neither mass nor charge and are the most penetrating of the trio, with penetration distances typically on the scale of 5 to 20 g/cm^2, depending on the gamma-ray energy and the atomic number Z of the absorbing material.[12]

A.6.2 Alpha-Particle Emission

Heavy Nuclei as Alpha-Particle Emitters

Within a nucleus, the repulsive electric force (or *Coulomb* force) between the protons competes with the attractive force that holds a nucleus together—the *nuclear force*. As one goes to heavier nuclei, with higher nuclear charge and higher nuclear radius, the Coulomb force becomes relatively more important, because it has a longer range than does the nuclear force. When it wins out by a sufficient margin, there can be nuclear fission or, less spectacularly and more commonly for nuclei in the ground state, the nucleus can decay by alpha-particle emission.

---

[10] More generally, the distance in grams per square centimeter (g/cm^2) is the product of the distance in centimeters (cm) and the density of the material in grams per cubic centimeter (g/cm^3).

[11] In specifying typical penetration distances for alpha particles, beta particles, and (below) gamma rays, we are ignoring their very different penetration profiles. To take the extremes, alpha particles of a given energy have a quite well-defined penetration distance, while gamma rays of a given energy are very broadly spread in penetration distance. Further, the processes responsible for the particles being stopped are quite different. Nonetheless, the concept of "typical" distances is useful for a qualitative understanding of the behavior of these particles.

[12] For a high-Z material such as lead ($Z = 82$), the typical penetration distance is only 1 g/cm^2 at 200 keV and becomes much smaller at still lower energies, making, for example, thin sheets of lead very effective for stopping x-rays.

A requirement for alpha decay, or any decay, is that there be a final system whose total mass is less than the mass of the initial system. In that case, emission can in principle occur, because there will be net energy left over to provide the kinetic energy of the emitted particles. Other things being equal, the greater the available energy, the shorter will be the half-life for alpha-particle decay. The natural radionuclides that emit alpha particles are all heavy nuclei, with atomic numbers Z well over 80. For these nuclei, alpha particles will be emitted at a fast enough rate to be observed—i.e., with a short enough half-life—only if the available energy is not much below 4 MeV. Typically, alpha-particle energies lie between about 4 and 8 MeV.

Conservation Rules in Alpha-Particle Emission

A typical example of alpha-particle decay is the decay of uranium-238 (^{238}U) to form thorium-234 (^{234}Th):

$$^{238}\text{U} \rightarrow {}^{234}\text{Th} + {}^{4}\text{He}.$$

The initial mass number (238) equals the sum of the final mass numbers (234 + 4), and the initial atomic number (92) equals the sum of the final atomic number (90 + 2). These equalities are dictated by two important rules that apply in radioactive decay:

Charge: The sum of the charges of the final products equals the charge of the original nucleus.

Number of nucleons: The total number of nucleons in the final products equals the total number of nucleons in the original nucleus.

These rules are closely related to the general conservation laws that apply in all nuclear processes.

Energy Relations in Alpha-Particle Emission

In the decay of ^{238}U, about 77% of the alpha particles are emitted with a kinetic energy of 4.20 MeV and 23% with an energy of 4.15 MeV. For the 4.20-MeV alpha particle, which corresponds to a transition to the ground state of ^{234}Th,

$$E_\alpha + E_{Th} = [M(^{238}\text{U}) - M(^{234}\text{Th}) - M(^{4}\text{He})]c^2, \qquad (A.11)$$

where the mass M is the atomic mass of the species in question, and E_α and E_{Th} are the kinetic energies of the alpha particle and thorium nucleus, respectively. The 4.15-MeV alpha particle corresponds to a transition to an excited state of ^{234}Th, with an excitation energy of 0.05 MeV. The excited state then decays to the ground state with the emission of a 0.05-MeV gamma ray.

In each case the total decay energy, summed over two successive steps in the latter case, is 4.20 MeV, plus a small additional kinetic energy of the recoiling ^{234}Th nucleus. From conservation of momentum, when a stationary ^{238}U nucleus decays, the momenta p of the alpha particle and of the ^{234}Th nucleus are equal in magnitude and opposite in direction. Noting that $E = p^2/2M$, it follows that $E_{Th} = (4/234) \times E_\alpha = 0.07$ MeV. Thus the total decay energy is 4.27 MeV, as also could be found from the mass differences if numerical values are substituted in Eq. A.11.

A.6.3 Beta-Particle Emission

Neutrinos and Anti-Neutrinos

In contrast to alpha decay, where only a few discrete decay energies are possible for each nuclear species, the spectrum of emitted beta particles is continuous. All beta-particle energies are possible from zero to a fixed maximum, called the *endpoint energy*. The endpoint energy in beta decay corresponds to the mass difference between the parent atom and the residual product, as would be expected from conservation of energy. However, the average energy of the beta particles is less than one-half the endpoint energy. When this was first discovered, it was thought that there might be a conflict with the demands of energy conservation.

It was eventually realized, by the early 1930s, that the electron shared the available energy (i.e., the energy corresponding to the mass difference between the initial and final constituents), with an elusive partner, which was termed the *neutrino* (ν). In present usage, the "neutrino" emitted in β^- decay is more precisely termed an *anti-neutrino* ($\bar\nu$). The neutrino itself is emitted in the analogous process of β^+ decay. When the distinction between the neutrino and anti-neutrino is not important, they are both generically termed neutrinos.[13] In standard physics terminology, the β^- and the ν are called "particles," and the β^+ and the $\bar\nu$ are termed "anti-particles." According to very general considerations, in each beta decay one particle and one anti-particle are emitted.[14]

---

[13] Unless otherwise indicated, we will use "neutrino" as a generic term and indicate the specific species by ν or $\bar\nu$.

[14] There are two issues here. One is the trivial matter of establishing the (arbitrary) convention for choosing which is called the particle and which the anti-particle. We do not here consider the second and much more significant issue: Why are there both particles and anti-particles and what are the relationships between the particles and anti-particles? Ordinary matter is made up of particles and is termed *matter*, as distinct from *anti-matter*. There is no evidence for the existence of bulk anti-matter outside of science fiction, although individual anti-particles are observed in cosmic rays and can be created and observed under special laboratory conditions.

For many years it was common to say that the mass of the neutrino was zero. However, experimental studies during the past decade have provided convincing evidence that the neutrino has a small, nonzero mass. The precise value has not been determined, but it is less than 0.001% of the mass of the electron. Although the distinction between so small a mass and a zero mass is of fundamental importance in some areas of physics and astrophysics, it has no relevance to the understanding of phenomena related to nuclear energy.

The neutrino has zero charge and can typically pass through very large amounts of material without stopping. For example, a flux of neutrinos is not appreciably depleted in passing through the Earth.[15]

Beta Decay and Beta-Particle Energy

β^- decay occurs when a nuclide has "too many" neutrons, i.e., when the neighboring isobar of higher atomic number has a lower atomic mass.[16] In that case, it is energetically profitable for a nucleus to change a neutron into a proton, with the emission of a β^- and a $\bar{\nu}$. A typical process of this sort is

$$^{234}\text{Th} \rightarrow {}^{234}\text{Pa} + \beta^- + \bar{\nu}.$$

Here the initial and final number of nucleons is 234, the β^- and the $\bar{\nu}$ not contributing to the nucleon count. For thorium (Th), the nuclear charge is +90, while for protactinium (Pa), it is +91, so charge is conserved with the emission of a β^- particle. For the case of a nucleus with "too many" protons, β^+ emission will occur; in this case, the β^+ is accompanied by a ν.

The maximum beta-particle kinetic energy, or endpoint energy, can be calculated from the mass differences. The sum of the kinetic energies of the beta particle and the neutrino is equal to the endpoint energy (ignoring the small, and usually negligible, kinetic energy of the residual nucleus). The mean energy of the beta particles is typically about one-third of the endpoint energy, and the mean energy of the neutrinos is typically about two-thirds of the endpoint energy.

---

[15] Nonetheless, neutrinos are not "infinitely" penetrating, and if sufficiently large numbers of neutrinos pass through material, some will stop. For example, with a high-flux nuclear reactor and a moderately large nuclear detector, it is possible to observe some neutrino interactions with the material of the detector. This observation was first accomplished by Reines, Cowans and collaborators in 1953. The detection of neutrinos and anti-neutrinos—including neutrinos produced in nuclear processes near the center of the Sun—is now routine in some classes of experiments.

[16] A *free* neutron, i.e., a neutron that is not part of a heavier nucleus, is itself unstable due to β^- decay, with a half-life of 10 min. This decay is possible because the mass of a neutron exceeds that of a hydrogen atom. However, for many nuclides the mass of the product of β^- decay would be greater than the mass of the initial atom, making β^- decay impossible.

Electron Capture

The processes of β^- and β^+ emission are identical, aside from the signs of the charges. However, there is an important asymmetry. In the atoms of ordinary matter, the nuclei are surrounded by (negative) electrons. The nucleus undergoes the same change as in β^+ decay if it captures an electron from outside the nucleus (usually an electron from one of the inner shells of the atom). Thus, *electron capture* exists as an alternative to β^+ emission and in fact can take place with a lesser mass difference between the nuclei. There is no comparable alternative to β^- emission, in the absence of anti-matter atoms with positrons surrounding an anti-nucleus.

Some radionuclides decay by both β^+ emission and electron capture. For example, sodium-22 (^{22}Na) decays about 90% of the time by β^+ emission and about 10% of the time by electron capture. The two processes can be written

$$\beta^+ \text{ emission:} \quad ^{22}\text{Na} \rightarrow {}^{22}\text{Ne} + \beta^+ + \nu$$
$$\text{Electron capture:} \quad ^{22}\text{Na} + \text{e}^- \rightarrow {}^{22}\text{Ne} + \nu.$$

In both cases, the residual nucleus is neon 22 (^{22}Ne).

It is also possible to have cases where the nuclear masses are so close together that electron capture is possible, while β^+ emission is energetically impossible (e.g., ^{7}Be). For either process to occur, the initial energy of the system must exceed the final energy. The electron rest mass energy adds to the initial energy in electron capture, while it adds to the final energy in β^+ emission.

A.6.4 Gamma-Ray Emission

Ordinarily, gamma-ray emission is not a primary process in radioactive decay but follows alpha-particle or beta-particle emission. It occurs in those decays where the transition is to an excited state of the product nucleus rather than to the ground state. The excited state then gives off its excitation energy by gamma-ray emission, either to the ground state or to a lower-lying excited state. When the transition is to another excited state, the de-excitation sequence continues with further gamma-ray emission. As mentioned in Section A.4, typical half-lives for gamma-ray emission are on the order of 10^{-12} sec.

There are several reservations that should be made in partial modification of the description given above. First, some nuclei have long-lived excited states, known as *isomeric states*, with half-lives ranging from an appreciable fraction of a second to many years. The gamma-ray emission in these cases appears as a primary radioactive process, rather than as a sequel to alpha-particle or beta-particle emission, although ultimately it can be traced back to such initiating processes. Second, not all of the excitation energy is carried off by the gamma ray, because, as in the other decay modes discussed above, the nucleus carries off a small amount of kinetic energy as dictated by conservation of momentum. Third, as an alternative to gamma-ray emission, decay of an excited state can occur by *internal conversion*, a process in which the excitation energy is transferred to one of the inner electrons of an atom. Typically in such cases, gamma-ray emission and internal conversion are competing de-excitation processes. The internal conversion electrons will have an energy equal to that of the competing gamma ray, minus the energy required to remove the electron from the atom.

A.7 Rate of Radioactive Decay

A.7.1 Exponential Decay

The number of nuclei of a given radioactive species that will decay in any time interval is proportional to the number of nuclei present. The constant of proportionality is termed the *decay constant*. Thus,

$$-\frac{dN}{dt} = \lambda N, \qquad (A.12)$$

where λ is the decay constant, N the number of nuclei, and dN the change in the number of nuclei in the time interval dt. The negative sign corresponds to the decrease in number with time. Each decay channel, defined in terms of a given initial nucleus and decay mode, has its own decay constant λ. If a nucleus has more than one decay mode—for example, alpha-particle transitions to different states of the residual nucleus—the overall decay constant λ is the sum of the individual decay constants.

The number of nuclei remaining after time t is given by integration of Eq. A.12:

$$N(t) = N_0 e^{-\lambda t}, \qquad (A.13)$$

where N_0 is the initial number (at $t = 0$). As seen from Eq. A.13, radioactive decay is an exponential decay. The rate of decay, from combining Eqs. A.12 and A.13 or from differentiating Eq. A.13, is

$$-\frac{dN}{dt} = \lambda N(t) = \lambda N_0 e^{-\lambda t}. \qquad (A.14)$$

A.7.2 Mean Life and Half-Life

One way of specifying an "average" time before a nucleus decays is in terms of its *mean life*, τ. Following the standard definition of the mean value of a quantity, the mean life is given by

$$\tau = \frac{\int_0^\infty t\lambda N(t)\,dt}{N_0} = \lambda \int_0^\infty t e^{-\lambda t}\,dt, \qquad (A.15)$$

where $\lambda N(t)\,dt$ is the number of decays occurring in the time interval dt, and $N(t)$ is found from Eq. A.13. Carrying out the integration of Eq. A.15, it follows that

$$\tau = \frac{1}{\lambda}. \tag{A.16}$$

Another way of specifying an "average" time before a nucleus decays is to specify the *half-life*, T. The half-life is the time required for one-half of an original sample to decay. It is defined by the relation

$$e^{-\lambda T} = \frac{1}{2}. \tag{A.17}$$

Evaluating Eq. A.17 and substituting from Eq. A.16, we have the following relations among T, λ, and τ:

$$T = \frac{\ln 2}{\lambda} = \tau \ln 2 = 0.693\tau. \tag{A.18}$$

It is more common to characterize radionuclides by their half-lives than by their mean lives. Sometimes the ambiguous term "lifetime" is used; it usually means the half-life.

A.7.3 Nuclei Remaining after a Given Time Interval

Given the half-life, or equivalently, the decay constant, it is a simple matter to calculate the fraction of nuclei remaining after the time t. This can be done either on the basis of the definition of the half-life or by using Eq. A.13. It is particularly simple when the time t is an integral number of half-lives. For example, if $t = 3T$, $N/N_0 = 1/8$. More generally, the fraction remaining after time t is given by

$$\frac{N(t)}{N_0} = \left(\frac{1}{2}\right)^{t/T}. \tag{A.19}$$

We also have, from Eqs. A.13 and A.18

$$\frac{N(t)}{N_0} = e^{-\lambda t} = e^{-t \ln 2/T}. \tag{A.20}$$

By comparing the natural logarithms of the right-hand terms of the two equations, it can be seen that Eqs. A.19 and A.20 are equivalent.

Eqs. A.12 and A.13 are examples of statistical equations, describing the average behavior of an assembly of radionuclides. As long as N is large, a statistical description gives accurate results. When N becomes small, the fluctuations about the expected statistical average become more significant, and the time of the last decay, when N goes from 1 to 0, cannot be specified with any precision. It also is a time that has no significance, except perhaps as part of a conceptual game of asking when the activity is "all gone."

It is rarely of importance to consider the statistical variations in radioactive decay, but they can nonetheless be simply described. The expected number of decays in a given time interval can be designated as $\bar{n}$. If $\bar{n}$ is a reasonably large number, the probability that the actual number of decays will be within the interval from n to $n + dn$ is given by the *normal* or *gaussian* distribution:

$$P(n)\,dn = \frac{1}{\sqrt{2\pi\sigma^2}}e^{-(n-\bar{n})^2/2\sigma^2}\,dn, \tag{A.21}$$

where the parameter σ is the *standard deviation*. If $\bar{n}$ is sufficiently large, say greater than 100, then to a good approximation

$$\sigma = \sqrt{\bar{n}}. \tag{A.22}$$

In 68% of the trials, n will lie within 1 standard deviation of $\bar{n}$, and in 95% of the trials it will lie within 2 standard deviations. For example, if $\bar{n} = 1000$, then $\sigma = 31.6$, and in 95% of the cases n will fall within 6% of $\bar{n}$. In most problems of interest in radioactive decay, the relevant $\bar{n}$ is much greater than 1000, and the fluctuations about $\bar{n}$ are of no interest. Ignoring them is rarely a significant omission.[17]

A.7.4 Decay Chains

In many cases, the product of a radioactive decay is another radioactive nucleus. In this case, the decay of the first nucleus is still given as in Eq. A.13:

$$N_1(t) = N_{10}e^{-\lambda_1 t}. \tag{A.23}$$

For the next generation nucleus,

$$dN_2/dt = \lambda_1 N_1 - \lambda_2 N_2. \tag{A.24}$$

The solution of Eqs. A.23 and A.24 is

$$N_2(t) = N_{10}\frac{\lambda_1}{\lambda_2 - \lambda_1}(e^{-\lambda_1 t} - e^{-\lambda_2 t}). \tag{A.25}$$

We can consider several limiting cases for Eq. A.25, applicable after a time t sufficient for $\lambda_i t \gg 1$, where λ_i is the larger of the two decay constants:

$$\lambda_1 \gg \lambda_2\ (T_2 \gg T_1): \quad N_2(t) = N_{10}e^{-\lambda_2 t}, \tag{A.26}$$

and

$$\lambda_2 \gg \lambda_1\ (T_1 \gg T_2): \quad N_2(t) = \frac{\lambda_1}{\lambda_2}N_1(t). \tag{A.27}$$

---

[17] On the other hand, statistical fluctuations can be very important in epidemiological studies of the health effects of radioactivity.

The second of these limiting cases, given in Eq. A.27, is of importance in the consideration of the naturally occurring radioactive decay chains (see Section 2.3.2). Each of these chains is headed by a very long-lived radionuclide. After its decay, a chain of faster decays follows. By an extension of the previous analysis, the asymptotic condition is approached where

$$\lambda_1 N_1 = \lambda_2 N_2 = \lambda_3 N_3 = \ldots \tag{A.28}$$

The situation described by Eq. A.28 is known as *secular equilibrium*. At secular equilibrium, for any individual species, the same number of nuclei are added per unit time as decay, aside from the very slow variation governed by the original parent.

References

1. Jagdish K. Tuli, *Nuclear Wallet Cards* (Upton, N.Y.: Brookhaven National Laboratory, 2000).

B

General Tables

Table B.1. Selected physical constants.

Quantity	Symbol	Value (in SI units)
Speed of light in vacuum	c	2.9979×10^8 m/sec
Elementary charge	e	1.6022×10^{-19} C
Avogadro's number	N_A	6.0221×10^{23}/mol
Electron mass	m_e	9.1094×10^{-31} kg
Atomic mass unit	u	1.6605×10^{-27} kg
Boltzmann constant	k	1.3807×10^{-23} J/K
Planck constant	h	6.6261×10^{-34} J-sec

Table B.2. Selected conversion factors and approximate energy equivalents.

Length		
1 foot (ft)	0.3048 m	
1 mile (mi)	5280 ft	1.609 km
Area		
1 square mile (mi^2)	640 acres	2.590 km^2
1 hectare (ha)	10^{-2} km^2	2.471 acres
1 acre	4.047×10^{-3} km^2	4.356×10^4 ft^2
1 barn (b)	10^{-24} cm^{-2}	
Volume		
1 liter (l)	10^{-3} m^3	0.2642 gal
1 cubic foot (ft^3)	28.32 l	7.481 gal
1 gallon (gal)	0.1337 ft^3	3.785 l
1 barrel	5.615 ft^3	42 gal
Mass		
1 pound (lb)	0.4536 kg	0.0005 ton
1 tonne	1.102 ton	2205 lb
Pressure		
1 bar	10^5 N/m^2 [10^5 Pa]	0.9869 atm
1 atmosphere (atm)	1.013×10^5 N/m^2	14.7 lb/in^2
Energy		
1 electron-volt (eV)	1.6022×10^{-19} J	
1 million electron-volts (MeV)	1.6022×10^{-13} J	
1 British thermal unit (BTU)	1055 J	
1 International Table calorie	4.1868 J	
1 exajoule (EJ)	10^{18} J	0.948 quad
1 quadrillion BTU (quad)[a]	10^{15} BTU	1.055 EJ
1 megatonne oil equiv. (Mtoe)[b]	10^{16} IT calories	0.04187 EJ
1 kilowatt-hour (kWh)	3.600×10^6 J	3412 BTU
1 gigawatt-year (GWyr)	3.1536×10^{16} J	8.76×10^9 kWh
Energy equivalents		
1 atomic mass unit (u)	931.49 MeV	
1 electron mass (m_e)	0.5110 MeV	
1 kWh (e) at 33% efficiency	10,340 BTU (thermal)	
1 fission event (^{235}U)	200 MeV	
1 tonne ^{235}U	$\approx$ 1 GWyr(e)	

[a]Unit widely used in U.S. DOE publications.
[b]Unit widely used in OECD publications.

Table B.3. List of elements, $Z = 1 - 110^a$.

Z	Symbol	Name	Z	Symbol	Name	Z	Symbol	Name
1	H	Hydrogen	38	Sr	Strontium	75	Re	Rhenium
2	He	Helium	39	Y	Yttrium	76	Os	Osmium
3	Li	Lithium	40	Zr	Zirconium	77	Ir	Iridium
4	Be	Beryllium	41	Nb	Niobium	78	Pt	Platinum
5	B	Boron	42	Mo	Molybdenum	79	Au	Gold
6	C	Carbon	43	Tc	Technetium	80	Hg	Mercury
7	N	Nitrogen	44	Ru	Ruthenium	81	Tl	Thallium
8	O	Oxygen	45	Rh	Rhodium	82	Pb	Lead
9	F	Fluorine	46	Pd	Palladium	83	Bi	Bismuth
10	Ne	Neon	47	Ag	Silver	84	Po	Polonium
11	Na	Sodium	48	Cd	Cadmium	85	At	Astatine
12	Mg	Magnesium	49	In	Indium	86	Rn	Radon
13	Al	Aluminum	50	Sn	Tin	87	Fr	Francium
14	Si	Silicon	51	Sb	Antimony	88	Ra	Radium
15	P	Phosphorus	52	Te	Tellurium	89	Ac	Actinium
16	S	Sulfur	53	I	Iodine	90	Th	Thorium
17	Cl	Chlorine	54	Xe	Xenon	91	Pa	Protactinium
18	Ar	Argon	55	Cs	Cesium	92	U	Uranium
19	K	Potassium	56	Ba	Barium	93	Np	Neptunium
20	Ca	Calcium	57	La	Lanthanum	94	Pu	Plutonium
21	Sc	Scandium	58	Ce	Cerium	95	Am	Americium
22	Ti	Titanium	59	Pr	Praseodymium	96	Cm	Curium
23	V	Vanadium	60	Nd	Neodymium	97	Bk	Berkelium
24	Cr	Chromium	61	Pm	Promethium	98	Cf	Californium
25	Mn	Manganese	62	Sm	Samarium	99	Es	Einsteinium
26	Fe	Iron	63	Eu	Europium	100	Fm	Fermium
27	Co	Cobalt	64	Gd	Gadolinium	101	Md	Mendelevium
28	Ni	Nickel	65	Tb	Terbium	102	No	Nobelium
29	Cu	Copper	66	Dy	Dysprosium	103	Lr	Lawrencium
30	Zn	Zinc	67	Ho	Holmium	104	Rf	Rutherfordium
31	Ga	Gallium	68	Er	Erbium	105	Db	Dubnium
32	Ge	Germanium	69	Tm	Thulium	106	Sg	Seaborgium
33	As	Arsenic	70	Yb	Ytterbium	107	Bh	Bohrium
34	Se	Selenium	71	Lu	Lutetium	108	Hs	Hassium
35	Br	Bromine	72	Hf	Hafnium	109	Mt	Meitnerium
36	Kr	Krypton	73	Ta	Tantalum	110	Ds	Darmstadtium
37	Rb	Rubidium	74	W	Tungsten			

^{a}These are the elements that have been named by the International Union of Pure and Applied Chemistry, as of 2003. Additional elements (up to $Z = 117$) are believed to have been produced in the laboratory, and await official naming.

Table B.4. Properties of selected radionuclides.

Radio-nuclide	Half-life^a	Main decay mode	Chief source(s) and special aspects
^{3}H	12.33 yr	β^-	cosmic rays; neutron reactions^b
^{14}C	5730 yr	β^-	cosmic rays; neutron reactions with ^{14}N, ^{17}O
^{40}K	1.277×10^9 yr	β^-	natural (0.0117%)
^{60}Co	5.271 yr	β^-	neutron capture in ^{59}Co
^{85}Kr	10.77 yr	β^-	fission product
^{87}Rb	47.5×10^9 yr	β^-	natural (27.83%); fission product
^{90}Sr	28.79 yr	β^-	fission product
^{90}Y	64.0 h	β^-	β^- decay of ^{90}Sr
^{99}Tc	0.211×10^6 yr	β^-	fission product
^{129}I	15.7×10^6 yr	β^-	fission product
^{131}I	8.021 d	β^-	fission product
^{135}I	6.57 h	β^-	fission product (parent of ^{135}Xe)
^{135}Xe	9.14 h	β^-	fission product; β^- decay of ^{135}I
^{137}Cs	30.07 yr	β^-	fission product
^{137}Ba^m	2.55 m	β^-	β^- decay of ^{137}Cs^c
^{210}Pb	22.3 yr	β^-	^{238}U decay chain
^{222}Rn	3.82 d	α	^{238}U decay chain (gas)
^{226}Ra	1600 yr	α	^{238}U decay chain
^{232}Th	14.05×10^9 yr	α	natural (100%)
^{233}U	0.159×10^6 yr	α	neutron capture in ^{232}Th and β^- decay
^{234}U	0.245×10^6 yr	α	natural (0.0054%); neutron capture in ^{233}U
^{235}U	0.704×10^9 yr	α	natural (0.7204%)
^{238}U	4.468×10^9 yr	α	natural (99.2742%)
^{239}U	23.45 m	β^-	neutron capture in ^{238}U
^{237}Np	2.14×10^6 yr	α	(n,2n) in ^{238}U and β^- decay; α decay of ^{241}Am
^{239}Np	2.36 d	β^-	Neutron capture in ^{238}U and β^- decay
^{238}Pu	87.7 yr	α	neutron capture in ^{237}Np and β^- decay
^{239}Pu	24110 yr	α	neutron capture in ^{238}U and two β^- decays
^{240}Pu	6564 yr	α	neutron capture in ^{239}Pu
^{241}Pu	14.29 yr	β^-	neutron capture in ^{240}Pu
^{242}Pu	0.373×10^6 yr	α	neutron capture in ^{241}Pu
^{243}Pu	4.96 h	β^-	neutron capture in ^{242}Pu
^{241}Am	432.2 yr	α	β^- decay of ^{241}Pu

(*continued*)

Table B.4. *Continued*

Radio-nuclide	Half-life[a]	Main decay mode	Chief source(s) and special aspects
^{243}Am	7370 yr	α	β^- decay of ^{243}Pu
^{244}Cm	18.10 yr	α	neutron capture in ^{243}Am and β^- decay

[a]m = minutes; h = hours; d = days; yr = years.

[b]Tritium is produced, for example, in neutron reactions with ^{6}Li and ^{10}B and as a third fission product in a small fraction of fission events.

[c]In the β^- decay of ^{137}Cs, 94.4% of the decays are to an isomeric state of ^{137}Ba that decays to the ground state with the emission of a 662-keV gamma ray; 5.6% are to the ground state.

Sources: Half-lives and isotopic abundances from Jagdish K. Tuli, *Nuclear Wallet Cards* (Upton, N.Y.: Brookhaven National Laboratory, 2000); decay of ^{137}Cs from Richard B. Firestone *et al.*, *Table of Isotopes*, Eighth Edition (New York: John Wiley, 1996).

Acronyms and Abbreviations

ABB	ASEA/Brown Boveri
ABWR	Advanced Boiling Water Reactor
ACNW	Advisory Committee on Nuclear Wastes [U.S. NRC]
ACR	Advanced CANDU Reactor
AEC	Atomic Energy Commission [U.S.]
AECL	Atomic Energy of Canada Limited
AF	Agreed Framework [U.S. and North Korea]
AFR	Away from reactor
AGCR	Advanced Gas-Cooled Graphite-Moderated Reactor (also AGR)
AIF	Atomic Industrial Forum
ALARA	As low as reasonably achievable
ALI	Annual limit on intake
ALMR	Advanced Liquid Metal Reactor
ALWR	Advanced Light Water Reactor
ANS	American Nuclear Society
AP600	Advanced Passive PWR (or Advanced PWR)—600 MWe (approx)
AP1000	Advanced Passive PWR (or Advanced PWR)—1000 MWe (approx)
APS	American Physical Society
APWR	Advanced pressurized water reactor
ARS	Acute radiation syndrome
ASLB	Atomic Safety and Licensing board [U.S. NRC]
ASP	Accident sequence precursor
AVLIS	Atomic vapor laser isotope separation
AVR	Arbeitsgemeinschaft Versuchs Reaktor
BEIR	Biological Effects of Ionizing Radiations

BNL	Brookhaven National Laboratory
BRWM	Board on Radioactive Waste Management [National Research Council]
BTU	British thermal unit
BWR	Boiling water reactor
CANDU	Canadian deuterium uranium (reactor)
CCDF	Complementary cumulative distribution function
CCDP	Conditional core damage probability
CE	Combustion Engineering
CERN	Centre European de Recherche Nucleaire
CF	Capacity factor
CFR	Code of Federal Regulations [U.S.]
CO_2	Carbon dioxide
CRBR	Clinch River Breeder Reactor
CTBT	Comprehensive Nuclear Test Ban Treaty
DDREF	Dose and dose-rate effectiveness factor
DOE	Department of Energy [U.S.]
EBR	Experimental Breeder Reactor
ECCS	Emergency core cooling system
ECU	European Currency Unit
EIA	Energy Information Administration [U.S. DOE]
EIS	Environmental impact statement
EJ	Exajoule (10^{18} joules)
EPA	Environmental Protection Agency [U.S.]
EPR	European Pressurized Water Reactor
EPRI	Electric Power Research Institute
ERDA	Energy Research and Development Administration [U.S.]
ESBWR	European Simplified Boiling Water Reactor
eV	Electron volt
FBR	Fast breeder reactor
FFTF	Fast Flux Test Facility
FFTR	Fast Flux Test Reactor
FGR	Federal Guidance Report [U.S. EPA]
FSU	Former Soviet Union
FY	Fiscal year
GA	General Atomics
GAO	General Accounting Office [U.S.]
GCR	Gas-cooled reactor
GDP	Gross domestic product

GE	General Electric
GENII-S	Generation II Hanford dosimetry computer code (variant)
GeV	Billion (giga) electron volts
GFR	Gas-Cooled Fast Reactor
GIF	Generation IV International Forum
GT-MHR	Gas Turbine Modular Helium Reactor
GtC	Gigatonnes of carbon [10^9 tonnes]
GW	Gigawatt [10^9 watts]
GWd/t	Gigawatt-days (thermal) per tonne
GWDT	Gigawatt-days thermal
GWe	Gigawatt (electric)
GWt	Gigawatt (thermal)
GWyr	Gigawatt-year
HCLPF	High confidence of low probability of failure
HEU	Highly enriched uranium
HLW	High-level (radioactive) waste
HPS	Health Physics Society
HRE	Homogeneous Reactor Experiment
HTGR	High temperature gas-cooled reactor
HTOM	High-temperature operating mode
HTTR	High Temperature Engineering Test Reactor [Japan]
HWLWR	Heavy-water-moderated light-water-cooled reactor
HWR	Heavy water reactor
IAEA	International Atomic Energy Agency
ICP	International Chernobyl Project
ICRP	International Commission on Radiological Protection
IEA	International Energy Agency
IFR	Integral Fast Reactor
IIASA	International Institute for Applied Systems Analysis
INEEL	Idaho National Engineering and Environmental Laboratory
INEL	Idaho National Engineering Laboratory (redesignated as INEEL)
INPO	Institute of Nuclear Power Operations
INSAG	International Nuclear Safety Advisory Group [IAEA]
INTD	International Near-Term Deployment
IPCC	Intergovernmental Panel on Climate Change
IPP	Independent power producer
IRIS	International Reactor Innovative and Secure
IRT	International Review Team [for Yucca Mountain]

ITER	International Thermonuclear Experimental Reactor
JTEC	Japanese Technology Evaluation Center [Loyola College]
keV	Thousand (kilo) electron volts
kt	Kilotonne [10^3 tonnes]
kW	Kilowatt [10^3 watts]
kWe	Kilowatt (electric)
kWh	Kilowatt-hour
kWt	Kilowatt (thermal)
LANL	Los Alamos National Laboratory
LERF	Large early release frequency
LET	Linear energy transfer
LEU	Low-enriched uranium
LFR	Lead-Cooled Fast Reactor
LGR	Light-water-cooled graphite-moderated reactor
LLNL	Lawrence Livermore National Laboratory
LLW	Low-level (radioactive) waste
LMFBR	Liquid metal fast breeder reactor
LMR	Liquid metal reactor
LNT	Linear non-threshold
LOCA	Loss-of-coolant accident
LOFT	Loss-of-fluid test
LTOM	Low-temperature operating mode
LWR	Light water reactor
MBTU	Million British thermal units
MCi	Megacurie [10^6 curies]
mECU	milliECU [10^{-3} ECU]
MeV	Million (mega) electron volts
MHR	Modular helium reactor
MIT	Massachusetts Institute of Technology
MJ	Megajoule [10^6 joules]
MOX	Mixed oxide [uranium and plutonium oxides]
MPa	Megapascal [10^6 pascals]
MPBR	Modular Pebble Bed Reactor
MPC	Maximum permissible concentration; multi-purpose canister
MRFA	Maximum reasonably foreseeable accident
MRS	Monitored retrievable storage
MSR	Molten Salt Reactor
MTHM	Metric tons of heavy metal (same as MTIHM)
MTIHM	Metric tons of initial heavy metal

MW	Megawatt (10^6 watts)
MWDT	Megawatt-days thermal
MWe	Megawatt (electric)
MWt	Megawatt (thermal)
NAPA	National Academy of Public Administration
NAS	National Academy of Sciences [U.S.]
NCRP	National Council on Radiation Protection and Measurements [U.S.]
NEA	Nuclear Energy Agency [OECD]
NERAC	Nuclear Energy Research Advisory Committee [U.S. DOE]
NERI	Nuclear Energy Research Initiative
NIMBY	Not in my backyard
NPIC	Nuclear Power Issues and Choices (report)
NPP	Nuclear power plant
NPT	Treaty on the Non-Proliferation of Nuclear Weapons
NRC	Nuclear Regulatory Commission [U.S.]
NRC	National Research Council [U.S.] (acronym not used in this book)
NRDC	National Resources Defense Council
NTDG	Near-Term Deployment Group
NUG	Non-utility generator
NWPA	Nuclear Waste Policy Act [of 1982]
NWS	Nuclear-weapon state
NWTRB	Nuclear Waste Technical Review Board [U.S.]
NWTST	Nuclear Wastes: Technologies for Separations and Transmutation [report]
O&M	Operation and maintenance
OCRWM	Office of Civilian Radioactive Waste Management [U.S. DOE]
OECD	Organization for Economic Co-operation and Development
OPEC	Organization of Petroleum Exporting Countries
ORIGEN	Oak Ridge Isotope Generation and Depletion code
ORNL	Oak Ridge National Laboratory
OTA	Office of Technology Assessment [U.S. Congress]
PBMR	Pebble Bed Modular Reactor
PBR	Pebble bed reactor
PCAST	President's Committee of Advisors on Science and Technology [U.S.]
PFS	Private Fuel Storage
PHWR	Pressurized heavy-water reactor

PORV	Pilot operated relief valve
ppm	Parts per million
PRA	Probabilistic risk assessment (or analysis)
PSA	Probabilistic safety assessment (or analysis)
PUREX	Plutonium and uranium recovery by extraction
PURPA	Public Utility Regulatory Policy Act of 1978
PWR	Pressurized water reactor
R&D	Research and development
RBMK	Reaktor bolshoi moshchnosti kanalnyi [high-power channel reactor]
RDD	Radiological dispersion device
RERF	Radiation Effects Research Foundation
RMEI	Reasonably maximally exposed individual
RSS	Reactor Safety Study [WASH-1400]
RY	Reactor-year
S&T	Separations and transmutation
S&T	Science and Technology
SARS	Severe acute respiratory syndrome
SBWR	Simplified Boiling Water Reactor
SCWR	Supercritical-Water-Cooled Reactor
SCZ	Strict Control Zone [Chernobyl]
SER	Science and Engineering Report [Yucca Mountain]
SFR	Sodium-Cooled Fast Reactor
SILEX	Separation of isotopes by laser excitation
SSD	Subseabed disposal
SSE	Safe shutdown earthquake
START	Strategic Arms Reduction Treaty
SWR	Siedewasser (boiling water) Reactor
SWU	Separative work unit
TMI	Three Mile Island
TRISO	Tri-isotropic
TRU	Transuranic (radioactive waste)
TSPA	Total system performance assessment
TSPA-LA	Total System Performance Assessment for the License Application
TSPA-Rev	Total System Performance Assessment—revised supplemental
TSPA-SR	Total System Performance Assessment for the Site Recommendation
TSPA-Sup	Total System Performance Assessment—supplemental

TVA	Tennessee Valley Authority
TWh	Terawatt-hour [10^9 kWh]
U.K.	United Kingdom
U.N.	United Nations
U.S.	United States
UNDP	United Nations Development Programme
UNSCEAR	United Nations Scientific Committee on the Effects of Atomic Radiation
USCEA	U.S. Council for Energy Awareness
USEC	United States Enrichment Corporation
USGS	United States Geological Survey
USSR	Union of Soviet Socialist Republics
VHTR	Very-High-Temperature Reactor
VVER	Vodo-Vodyanoy Energeticheskiy Reaktor (water-water power reactor)
WANO	World Association of Nuclear Operators
WDV	Water dilution volume
WEC	World Energy Council
WHO	World Health Organization
WIPP	Waste Isolation Pilot Plant [New Mexico]
WWER	Alternative transliteration for VVER

Glossary

Note: In many cases, the definitions given below are specific to usage in the context of nuclear energy, rather than general definitions.

Absorbed dose. The energy deposited by ionizing radiation per unit mass of matter (also known as the *physical dose*); the common units of absorbed dose are the *gray* and the *rad*. (See Section 3.2.2.)

Absorption, neutron. An inclusive term for neutron-induced reactions, excluding elastic scattering. (See Section 5.1.2.)

Abundance. *See* **isotopic abundance** and **elemental abundance**.

Accelerator. A facility for increasing the kinetic energy of charged particles, typically to enable them to initiate nuclear reactions.

Actinides. Elements with atomic numbers from $Z = 89$ (actinium) through $Z = 103$ (lawrencium); this group includes the fertile and fissile nuclei used in nuclear reactors and their main neutron capture products.

Activity. The rate of decay of a radioactive sample.

Adaptive staging. In nuclear waste repository planning, an approach in which the project proceeds in stages, with the detailed plans for each stage adjusted in response to developments in the preceding stages.

Adsorption. The attachment of molecules in a fluid to a surface in contact with the fluid.

Alloy 22. A corrosion-resistant high-nickel alloy, containing nickel, chromium, molybdenum, and tungsten.

Alpha-particle decay (or alpha decay). Radioactive decay characterized by the emission of an alpha particle.

Alpha particle. A particle emitted in the radioactive decay of some heavy nuclei; it is identical to the nucleus of the ^{4}He atom. (See Section A.6.1.)

Antineutrino. A neutral particle, with very small mass, emitted together with an electron in beta-particle decay; sometimes referred to simply as a *neutrino*. (See Section A.6.3.)

Aquifer. A permeable formation of rock that can hold or transmit water.

Assembly. A bundle of nuclear fuel rods that serves as a basic physical substructure within the reactor core. (See Section 8.2.3.)

Asymmetric fission. Fission events in which the masses of the two fission fragments are significantly different. (See Section 6.3.1.)

Atom. The basic unit of matter for each chemical element; it consists of a positively charged nucleus and surrounding electrons; the number of electrons in the neutral atom defines the element. (See Section A.1.1.)

Atomic energy. Used as a synonym for *nuclear energy*.

Atomic mass (M). The mass of a neutral atom, usually expressed in atomic mass units. (See Section A.2.2.)

Atomic mass number (A). The sum of the number of protons and the number of neutrons in the nucleus of an atom. (See Section A.1.2.)

Atomic mass unit (u). A unit of mass, defined so that the mass of the ^{12}C atom is exactly 12 u; 1 u = 1.6605×10^{-27} kg. (See Section A.2.2.)

Atomic number (Z). An integer which identifies an element and gives its place in the sequence of elements in the periodic table; equal to the number of protons in the nucleus of the atom or the number of electrons surrounding the nucleus in a neutral atom. (See Section A.1.2.)

Avogadro's number (or Avogadro constant) (N_A). The number of atoms or molecules in one mole of a substance, e.g., the number of ^{12}C atoms in 12 g of ^{12}C; $N_A = 6.022 \times 10^{23}$ per mole. (See Section A.2.3.)

Back end (of fuel cycle). In the nuclear fuel cycle, the steps after the fuel is removed from the reactor. (See Section 9.1.2.)

Backfill. Material placed in the cavities of a nuclear waste repository, after emplacement of the waste canisters.

Barn. A unit used to specify the cross section for a nuclear reaction, equal to 10^{-28} m^2 (or 10^{-24} cm^2).

Becquerel (Bq). A unit used to specify the rate of decay of a radioactive sample, equal to 1 disintegration per second.

BEIR Report. One of a series of reports prepared by the National Research Council's Committee on the Biological Effects of Ionizing Radiations.

Bentonite. A rock formed of clays which swell on absorbing water; often suggested as a backfill material for use in nuclear waste repositories.

Beta particle. A particle emitted in the radioactive decay of some nuclei; negative beta particles are electrons and positive beta particles are positrons. (See Section A.6.1.)

Beta-particle decay (or beta decay). Radioactive decay characterized by the emission of a beta particle.

Binding energy. *See* **nuclear binding energy**.

Biomass fuel. Fuel derived from organic matter, excluding fossil fuels; e.g., wood, agricultural crops, and organic wastes.

Biosphere. The zone where living organisms are present, extending from below the Earth's surface to the lower part of the atmosphere.

Boiling water reactor (BWR). A light water reactor in which steam formed inside the reactor vessel is used directly to drive a turbine. (See Section 8.2.1.)

Boron (B). The 5th element ($Z = 5$); used as a control material because of the high cross section of ^{10}B for thermal-neutron absorption.

Borosilicate glass. A glass with high silicon and boron content, used for encapsulating reprocessed nuclear wastes.

Breeder reactor (or breeder). A reactor in which fissile fuel is produced by neutron capture reactions at a rate which equals or exceeds the rate at which fissile fuel is consumed.

Burnable poison. A material with a large thermal-neutron absorption cross section that is consumed during reactor operation; inserted in reactor to compensate for the decrease in reactivity as fissile fuel is consumed and fission product poisons are produced. (See Section 7.5.2.)

Burnup. Denotes the consumption of nuclides in a nuclear reactor; particularly used to specify the energy output from nuclear fission per unit mass of initial fuel. (See Section 9.3.1.)

Cadmium (Cd). The 48th element ($Z = 48$); used as a control material because of the high cross section of ^{113}Cd for thermal-neutron absorption.

CANDU reactor. A reactor design developed in Canada, using heavy water as the coolant and moderator; also known as the *pressurized heavy water reactor* (PHWR) (Abbreviation for Canadian-deuterium-uranium.).

Canister. *See* **waste canister**.

Capacity. The power output of a reactor under the designed operating conditions, commonly expressed in megawatts or gigawatts. *See also* **net capacity** and **gross capacity**.

Capacity factor. The ratio of the actual output of electricity from a reactor during a given period (usually a year) to the output that would have been achieved had the reactor operated at its rated capacity for the full period.

Capillary action. For water in rock, the attraction between water molecules and rock surfaces.

Capture. *See* **neutron capture**.

Carbon (C). The 6th element ($Z = 6$); used as a moderator in some types of nuclear reactors because of its relatively low atomic mass and the very low thermal-neutron absorption cross section of its stable isotopes.

Carbon dioxide (CO_2). A gas present naturally in the atmosphere; the atmospheric concentrations of carbon dioxide are being substantially increased by the combustion of fossil fuels.

Centrifuge. A rapidly rotating device for separating portions of a gas or liquid that differ in density; centrifuges provide one method for isotopic enrichment. (See Section 9.2.2.)

Chain reaction. A sequence of neutron-induced nuclear reactions in which the neutrons emitted in fission produce further fission events.

China syndrome. A term applied to a hypothetical sequence of events in which molten nuclear fuel penetrates the bottom of a reactor vessel and works its way into the Earth.

Cladding. The material surrounding the pellets of nuclear fuel in a fuel rod, isolating them physically, but not thermally, from the coolant.

Collective dose. The radiation doses to an entire population, equal to the sum of the individual doses.

Colloid. A small particle (typically, 10^{-9} m to 10^{-6} m in diameter) suspended in water; in a waste repository, radionuclides can bind to colloids and travel with them.

Commercial reactor. A nuclear reactor used to generate electricity for sale to consumers.

Committed dose. *See* **dose commitment**.

Compound nucleus. An unstable nucleus formed when an incident particle combines with a target nucleus; it quickly decays by emitting some combination of neutrons, charged particles, fission fragments, and gamma rays.

Condenser. A component of a power plant in which heat energy is removed from steam, changing water from the vapor phase to the liquid phase.

Constant dollars. A unit used to compare costs in different years by correcting for the effects of inflation.

Containment building (or containment). A heavy structure housing the reactor vessel and associated equipment; it is designed to isolate the reactor from the outside environment.

Continuum region. The region, either of nuclear excitation energies or of neutron energies, in which the neutron absorption cross section varies only slowly with energy because the compound nuclear states formed by neutron absorption overlap in energy. (See Section 5.3.)

Control rod. A rod composed of a *control substance*; used to regulate the power output of a reactor and, if necessary, to provide for rapid shutdown of the reactor.

Control substance (or control material). A material in solid or liquid form, with a high thermal-neutron absorption cross section, used to control the reactivity of a nuclear reactor. (See Section 7.5.2.)

Conversion. The production of fissile material from fertile material in a nuclear reactor.

Conversion ratio (C). The ratio of the rate of production of fissile nuclei in a reactor to the rate of consumption of fissile nuclei; for newly inserted uranium fuel, C is the ratio of the rate of production of ^{239}Pu to the rate of consumption of ^{235}U. (See Section 7.4.)

Converter reactor (or converter). A reactor in which there is substantial conversion of fertile material to fissile material; sometimes restricted to reactors with conversion ratios between 0.7 and 1.0.

Coolant. Liquid or gas used to transfer heat away from the reactor core.

Cooling pool. A water tank near the reactor, in which spent fuel is cooled and kept isolated from the environment for periods ranging from several months to several decades.

Core. *See* **reactor core**.

Cosmic rays. Atomic and sub-atomic particles impinging upon the Earth, originating from the Sun or from regions of space within and beyond our galaxy.

Coulomb force. The force between electrically charged particles; described by Coulomb's Law.

Critical. An arrangement of fissile material is critical when the effective multiplication factor is unity.

Criticality. The state of being critical; this is the necessary condition for sustaining a chain reaction at a constant power level.

Criticality accident. A nuclear reactor accident caused by some or all of the reactor fuel becoming super-critical.

Criticality factor. *See* **effective multiplication factor**.

Critical mass. The minimum mass of fissile material required for a sustained chain reaction in a given nuclear device.

Cross section, neutron (σ). A measure of the probability that an incident neutron will interact with a particular nuclide; cross sections are separately specified for different target nuclides and different reactions; the cross section has the units of area and can be (loosely) thought of as an effective target area for a specific process. (See Section 5.1.2.)

Crude oil. Liquid fossil fuel found in underground deposits; it is processed (refined) to yield petroleum products.

Crust. For the Earth, the solid outer layer composed mostly of rock.

Curie (Ci). A unit used to specify the rate of decay of a radioactive sample, equal to 3.7×10^{10} disintegrations per second (3.7×10^{10} Bq). (See Section 3.3.2.)

Current dollars. The unit for expressing actual costs at a given time, with no correction for inflation.

Daughter nucleus. Alternative term for *decay product* or *progeny*.

Decay chain (or decay series). *See* **radioactive decay series**.

Decay constant (λ). The probability per unit time for the radioactive decay of an atom of a given radioactive species. (See Section A.7.1.)

Decay product. The residual nucleus produced in radioactive decay; also referred to as *daughter nucleus*.

De-excitation. A transformation in which an excited nucleus gives up some or all of its excitation energy, typically through gamma-ray emission.

Delayed critical reactor. A nuclear reactor in which the fissions caused by delayed neutrons are essential to achieving criticality. (See Section 7.3.2.)

Delayed neutron. A neutron which is emitted from a fission fragment at an appreciable time after the fission event, typically more than 0.01 seconds later. (See Section 6.3.2.)

De minimis principle. A principle that very small consequences should be ignored, for example, the effects of very small radiation doses; it is derived from the expression *de minimis non curat lex* (the law does not concern itself with trifles).

Design basis accident. A hypothetical reactor accident that defines reference conditions which the safety systems of the reactor must be able to handle successfully.

Deterministic. (a) In radiation protection, a deterministic effect is one for which the magnitude of the effect depends upon the magnitude of the dose and is approximately the same for similar individuals; *see, in contrast,* **stochastic**. (b) In nuclear accident analysis, a deterministic safety assessment focuses on the meeting of specifications that are designed to prevent failures or enable the reactor to withstand them without serious harmful consequences; *see, in contrast,* **probabilistic safety assessment**.

Deuterium (^{2}H or D). The hydrogen isotope with $A = 2$, where A is the atomic mass number.

Deuteron. The nucleus of the deuterium atom; it consists of one proton and one neutron.

Diffusion. *See* **gaseous diffusion**.

Dirty bomb. A bomb made up of conventional explosives and radionuclides, designed to create an area contaminated by dispersed radioactivity (also known as a *radiological dispersion device*).

Doppler broadening. In reactors, the increase in the effective width of resonance peaks for neutron-induced reactions, caused by thermal motion of the target nuclei. (See Section 5.2.3.)

Dose. In radiation physics, a brief form equivalent to *radiation dose.*

Dose commitment. The cumulative radiation dose produced by inhaled or ingested radioactive material, summed over the years following intake (typically up to about 50 y).

Dose and dose rate effectiveness factor (DDREF). A factor sometimes applied in estimating radiation effects, to take into account a presumed reduction in risk per unit dose when the dose or dose rate is small. (See Section 4.3.3.)

Dose equivalent (H). A measure of the estimated biological effect of exposure to ionizing radiation, equal to the product of the quality factor for the radiation and the physical dose; the common units for the dose equivalent are the *sievert* and the *rem*. (See Section 3.2.2.)

Doubling dose. The dose of ionizing radiation that leads to a mutation rate in an exposed population that is twice the mutation rate in an unexposed comparable population.

Drift. In mining, a horizontal passageway or tunnel.

Dry storage. The storage of nuclear wastes under conditions where cooling is provided by the natural or forced circulation of air.

Effective dose equivalent (H_E). An overall measure of the estimated biological effects of radiation exposure, taking into account both the type of radiation and the region of the body exposed. (See Section 3.2.3.)

Effective multiplication factor (k). The ratio of the number of neutrons produced by fission in one generation of a chain reaction to the number produced in the preceding generation; also known as the *criticality factor*. (See Section 7.1.1.)

Efficiency. *See* **thermal conversion efficiency**.

Elastic scattering. For neutrons, a reaction in which the incident neutron and the target nucleus remain unchanged and their total kinetic energy is unchanged, while the neutron's kinetic energy is reduced and its direction is altered.

Electrons. Stable elementary particles of small mass and negative charge that surround the nucleus of an atom and balance its positive charge; sometimes the term is used to include the "positive electron" or *positron*. *See also* **beta particle**.

Electron volt (eV). The kinetic energy gained by an electron in being accelerated through a potential difference of 1 volt; 1 eV $= 1.6022 \times 10^{-19}$ J. (See Section A.2.4.)

Elemental abundance. The relative abundance of an element in a sample of matter, expressed as the ratio of mass of that element to the total mass of the sample. (See Section A.3.2.)

Energy state (or energy level). For nuclei or atoms, one of the possible configurations of the constituent particles, characterized by its energy.

Enrichment. (a) The relative abundance of an isotope, often applied to cases where the isotopic abundance is greater than the level found in nature (for uranium the enrichment is usually expressed as a ratio by mass, not by number of atoms). (b) The process of increasing the isotopic abundance of an element above the level found in nature.

Epidemiological study. For radiation, the study of the association between the radiation doses received by members of a population group and the incidence of selected illnesses or deaths within the group.

Excited state. A state of the nucleus in which the nucleus has more than its minimum possible energy; usually this energy is released by the emission of one or more gamma rays. (See Section A.4.)

Exposure. In radiation physics, a brief form equivalent to *radiation exposure*.

External costs. Costs of electrical power generation attributable to indirect factors, such as environmental pollution; also referred to as "social costs." (See Section 19.1.)

Fast reactor. A reactor in which most fission events are initiated by fast neutrons.

Fast neutrons. In reactors, neutrons with energies extending from the energies of typical fission neutrons (the neighborhood of 1 MeV) down to the 10-keV region.

Fertile nucleus. A non-fissile nucleus, which can be converted into a fissile nucleus through neutron capture, typically followed by beta decay. (See Section 6.2.2.)

Fissile nucleus. A nucleus with an appreciable fission cross section for thermal neutrons. (See Section 6.2.2.)

Fission. *See* **nuclear fission**.

Fission fragment. One of the two nuclei of intermediate mass produced in fission.

Fossil fuels. Fuels derived from organic materials which decayed many millions of years ago, namely coal, oil, and natural gas.

Fractionation. The process of changing the elemental or isotopic abundances of material in a region due to preferential transport of certain elements or isotopes out of or into the region.

Fracture. A break in the underground rock.

Fragment. *See* **fission fragment**.

Front end (of fuel cycle). In the nuclear fuel cycle, the steps before the fuel is placed in the reactor. (See Section 9.1.2.)

Fuel cycle. *See* **nuclear fuel cycle**.

Fuel rod (or fuel pin). A structure which contains a number of pellets of fuel; typically a long cylindrical rod with a thin metal wall. (See Section 8.2.3.)

Fusion power. Power derived from nuclear reactions in which energy is released in the combining of light atoms into heavier ones (e.g., of hydrogen into helium).

Gamma ray. A photon emitted in the transition of a nucleus from an excited state to a state of lower excitation energy. (See Section A.6.4.)

Gas-cooled reactor (GCR). A nuclear reactor in which the coolant is a gas, typically carbon dioxide or helium.

Gaseous diffusion. An isotopic enrichment process based on the dependence on mass of the average velocity of the molecules in a gas at a given temperature. (See Section 9.2.2.)

Generation costs. The total costs of generating electrical power, including operations, maintenance, fuel, and carrying charges that cover capital costs of construction; sometimes termed "busbar cost." (See Section 19.1.)

Geologic waste disposal. Disposal of nuclear wastes in an underground site, excavated from the surrounding rock formation.

Geometric cross section. The cross section equal to the projected area of the nucleus. (See Section 5.1.2.)

Giga (G). 10^9, e.g., 1 gigawatt $= 10^9$ W.

Gigawatt electric (GWe). Electrical power produced or consumed at the rate of 1 gigawatt.

Gigawatt thermal (GWt). Thermal power produced or consumed at the rate of 1 gigawatt.

Gigawatt-year (GWyr). The electrical energy corresponding to a power of 1 GWe maintained for a period of one year; equal to 3.15×10^{16} J.

Graphite. A crystalline form of carbon; the carbon used for reactor moderators is in this form.

Gray (Gy). The SI unit of absorbed dose for ionizing radiation; equal to 1 joule per kilogram. (See Section 3.2.2.)

Gross capacity. The capacity of a generating facility as measured at the output of the generator unit (also known as "busbar capacity"); equal to the sum of the net capacity and the power consumed within the plant.

Ground state. The state of a nucleus in which the nucleus has its minimum possible energy and therefore cannot decay by the emission of gamma rays.

Ground water. Water present in an underground zone.

Half-life (T). The average time required for the decay of one-half of the atoms in a radioactive sample.

Heavy metal. An element with isotopes of atomic mass above 230; for example, thorium, uranium, and plutonium; these are the elements that are used as nuclear fuel or are produced by neutron capture and other reactions in nuclear fuel.

Heavy water. Water (H_2O) in which the hydrogen atoms are primarily atoms of ^2H (deuterium).

Heavy water reactor. A reactor in which heavy water is used as the coolant and moderator.

High-level waste. Highly radioactive material discharged from a nuclear reactor, including spent fuel and liquid or solid products of reprocessing. (See Section 10.1.3.)

Highly-enriched uranium (HEU). Uranium with a ^{235}U enrichment of greater than 20%.

Hormesis. For radiation, the theory that small radiation doses have a beneficial effect, e.g., reducing the rate of cancer induction.

Hydroelectric power. Electric power generated by the flow of water through turbines; often this is falling water that has been stored behind a dam.

Igneous activity. The movement of molten rock (*magma*) into the Earth's crust or through it to the Earth's surface or into the atmosphere.

Implosion bomb. A nuclear bomb in which criticality is achieved by the rapid compression of uranium or plutonium to high density.

Inelastic scattering. For neutrons, a neutron-induced reaction in which the two final particles (neutron and nucleus) are the same as the initial particles, but with the final nucleus in an excited state and the total kinetic energy of the neutron and nucleus reduced.

Infiltration. The penetration of water into the ground to a depth at which water does not return to the atmosphere by evaporation or otherwise.

Initial heavy metal. The inventory of heavy metal (typically uranium) present in reactor fuel before irradiation.

Interim storage. Temporary storage of spent fuel or reprocessed wastes, in anticipation of further treatment or transfer to a long-term storage facility.

International System of Units (SI units). The internationally adopted system of physical units, based on the metric system; designated in French as the Système International d'Unités.

Ionizing radiation. Radiation in which individual particles are energetic enough to ionize atoms of the material through which they pass, either directly for charged particles (e.g., alpha particles and beta particles) or indirectly for neutral particles (e.g., X-rays, gamma rays, and neutrons) through the production of charged particles.

Irradiation. The act of exposing to radiation.

Isobar. One of a set of nuclides that have the same atomic mass number (A) and differing atomic numbers (Z).

Isomer. One of a set of nuclides that have the same atomic number (Z) and atomic mass number (A), but differing excitation energies; a state is considered to be an isomeric state only if a significant time elapses before its decay.

Isotope. One of a set of nuclides that have the same atomic number (Z) and different atomic mass numbers (A).

Isotopic abundance (or fractional isotopic abundance). The relative abundance of an isotope of an element in a mixture of isotopes, usually expressed as the ratio of the number of atoms of that isotope to the total number of atoms of the element. (*See also* **enrichment**, for exception in the case of uranium.)

Joule (J). The SI unit of energy.

Kilo. 10^3, e.g., 1 kilowatt $= 10^3$ W.

Kiloton. Used as a unit of energy to describe the explosive yield of a nuclear weapon; equal to the energy released in the detonation of 1 million kg of TNT, nominally taken to be 10^{12} calories or 4.18×10^{12} joules.

Kinetic energy. Energy attributable to the motion of a particle or system of particles.

Level. *See* **energy level**.

Level width (Γ). The energy interval between points on either side of a resonance maximum where the absorption cross section is one-half the cross section at the maximum. (See Section 5.2.2.)

Light water. Water (H_2O) in which the hydrogen isotopes have their natural (or "ordinary") isotopic abundances (99.985% ^1H).

Light water reactor (LWR). A reactor which uses ordinary water as both coolant and moderator. (See Section 8.1.4.)

Linear energy transfer (LET). The rate of energy delivery to a medium by an ionizing particle passing through it, expressed as energy per unit distance traversed. (See Section 2.4.1.)

Linear-nonthreshold hypothesis (or linearity hypothesis). The hypothesis that the frequency of damage due to ionizing radiation (e.g., the rate of cancer induction) is linearly proportional to the magnitude of the dose, remaining so even at low doses. (See Section 4.3.3.)

Liquid metal reactor. A reactor in which the coolant is a liquid metal, typically molten sodium.

Low-enriched uranium (LEU). Uranium with a ^{235}U enrichment of less than 20%.

Low-level wastes. A classification of nuclear wastes that excludes high-level and transuranic wastes. (See Section 10.1.3.)

Low-power license. An authorization granted by the Nuclear Regulatory Commission for the limited operation of a nuclear reactor.

Magma. Molten rock, as found below the Earth's crust; in igneous events, magma penetrates into or through the crust.

Matrix. In rock, the bulk mass between fractures.

Mean free path (λ). For neutrons, the mean distance a neutron travels before undergoing a nuclear reaction in the medium through which it is moving.

Mean life (τ). The mean time for nuclear decay, equal to the reciprocal of the decay constant. (See Section A.7.2.)

Mega (M). 10^6, e.g., 1 megawatt $= 10^6$ W.

Megawatt electric (MWe). Electrical power produced or consumed at the rate of 1 megawatt.

Megawatt thermal (MWt). Thermal power produced or consumed at the rate of 1 megawatt.

Metallic fuel. Nuclear fuel in metallic form, as distinct from oxide form; e.g., uranium rather than uranium oxide.

Methane. The compound CH_4, which is a gas at standard temperature and pressure; it is the main component of natural gas.

Milli (m). 10^{-3}, e.g., 1 mrem $= 10^{-3}$ rem.

Millibarn (mb). A cross section unit equal to 10^{-3} barn.

Milling. The process by which uranium oxides are extracted from uranium ore and concentrated.

Mill tailings. The residues of milling, containing the materials remaining after uranium oxides are extracted from uranium ore; the tailings include radionuclides in the ^{238}U series, such as ^{230}Th. (See Section 9.2.1.)

Mixed oxide fuel (MOX). Reactor fuel composed of uranium and plutonium oxides (UO_2 and PuO_2).

Moderating ratio. A figure of merit for comparing moderators. (See Section 7.2.2.)

Moderator. A material of low atomic mass number in which the neutron kinetic energy is reduced to very low (thermal) energies, mainly through repeated elastic scattering.

Mole. The amount of a substance for which the mass in grams is numerically equal to the atomic (or molecular) mass of an atom (or molecule) of the substance expressed in atomic mass units. *See also* **Avogadro's number**.

Monte Carlo method. A probabilistic method for surveying the range of outcomes for a complicated process. A distribution of outcomes is determined from the calculation of the individual outcomes for many sets of randomly-selected values of key parameters, with the random selection weighted by the assumed probabilities of occurrence of different parameter values.

Multiplication factor. *See* **effective multiplication factor**.

Natural gas. Gaseous fossil fuel found in underground deposits, composed mostly of methane (CH_4).

Natural radionuclide. A radionuclide produced by natural processes such as stellar nucleosynthesis or the radioactive decay of a natural radioactive precursor.

Net capacity. The capacity of a generating facility measured at the input to the transmission lines carrying power away from the plant; equal to the difference between the gross capacity and the power consumed within the plant.

Neutrino. A neutral particle, with very small mass, emitted together with a positron in β^+ decay; sometimes the term *neutrino* is used generically to include the *antineutrino*. (See Section A.6.3.)

Neutron. A neutral particle with a mass of approximately one atomic mass unit; it is important as a constituent of the nucleus and as the particle which initiates fission in nuclear reactors.

Neutron capture. A process in which a neutron combines with a target nucleus to form a product from which no nucleons are emitted and that, therefore, has the same atomic number as the target nucleus and an atomic mass number greater by unity. (See Section 5.1.1.)

Noble gas. One of the elements that are chemically inert and are in gaseous form at ordinary temperatures and pressures; the noble gases are helium, neon, argon, krypton, xenon, and radon.

Nuclear binding energy. The total energy required to dissociate a nucleus in its ground state into its constituent nucleons.

Nuclear energy. Energy derived from nuclear fission or fusion, usually converted to electricity but sometimes used directly in the form of heat; it is sometimes loosely used as a synonym for *nuclear power*.

Nuclear fission. A process in which a nucleus separates into two main fragments, usually accompanied by the emission of other particles, particularly neutrons. (See Section 5.1.1.)

Nuclear force. The force that exists between every pair of nucleons, in addition to and independent of the Coulomb force between protons.

Nuclear fuel cycle. The succession of steps involved in the generation of nuclear power, starting with the mining of the ore containing the fuel and ending with the final disposition of the spent fuel and wastes produced in a nuclear reactor. (See Section 9.1.)

Nuclear power. Electrical power derived from a nuclear reactor; it is sometimes loosely used as a synonym for *nuclear energy*.

Nuclear power plant. A facility for producing electricity using energy from fission or fusion processes.

Nuclear reaction. For neutrons, an event in which a neutron interacts with an atomic nucleus to give products which differ in composition or kinetic energy from the initial neutron and nucleus. (See Section 5.1.1.)

Nuclear reactor. A device in which heat is produced at a controlled rate, by nuclear fission or nuclear fusion.

Nuclear waste package. The nuclear wastes, their container and its contents, and the immediately surrounding materials used to isolate the container.

Nuclear wastes. Radioactive materials remaining as a residue from nuclear power production or from the use of radioisotopes.

Nuclear weapon (or nuclear bomb). A weapon which derives its explosive force from nuclear fission or nuclear fusion.

Nucleon. A generic term for neutrons and protons, reflecting similarities in some of their properties.

Nucleosynthesis. The series of nuclear reactions by which the elements are formed in nature; these reactions started in the big bang and continued in stars and in the interstellar medium.

Nucleus. The very dense central core of an atom, composed of neutrons and protons. (See Section A.1.1.)

Nuclide. An atomic species defined by its atomic number and atomic mass number.

Ocean sediment. Material on the ocean floor deposited by the settling of insoluble solid material from the ocean.

Once-through cycle. A nuclear fuel cycle in which the spent fuel is stored or disposed of without extraction of any part of the fuel for further use in a reactor.

Operating costs. The costs of operating and maintaining a power plant, sometimes including the costs of capital additions and usually excluding fuel costs. (See Section 19.3.1.)

Operating license. For nuclear reactors, a legally required prerequisite for the operation of the reactor; in the United States the license is granted by the Nuclear Regulatory Commission.

Overpack. A container placed around a nuclear waste canister to provide additional protection against corrosion and physical damage.

Oxide fuel. Nuclear fuel in oxide form, as distinct from metallic form; e.g., uranium oxide rather than uranium.

Parent nucleus. A preceding radioactive nucleus in a chain of one or more radioactive decays.

Passive safety. Reactor safety achieved without dependence on human intervention or the performance of specialized equipment; e.g., achieved by the action of unquestionable physical effects such as the force of gravity or thermal expansion.

Percolation. For a nuclear waste repository, the flow of water through the rock down to the repository.

Photon. A discrete unit or quantum of electromagnetic radiation, which carries an amount of energy proportional to the frequency of the radiation. (See Section A.4.)

Physical dose. An alternative term for **absorbed dose**.

Pitchblende. An ore rich in uranium oxides.

Plutonium (Pu). The 94th element ($Z = 94$); does not occur in nature because of its isotope's short half-lives, but a key isotope (^{239}Pu) is produced in reactors for use as a fissile fuel or for nuclear weapons.

Poison. *See* **reactor poison**.

Porosity (ϵ). The fraction of the volume of a rock formation which is open, and may be filled with air or water.

Positron. A positively charged particle which is identical to the electron in mass and the magnitude of its charge; sometimes called a "positive electron." *See also* **beta particle**.

Potassium (K). The 19th element ($Z = 19$); the natural radionuclide ^{40}K in the body contributes substantially to the total human radiation exposure. (See Section 3.5.1.)

Potential energy. The energy associated with a particular configuration of a system; configuration changes that reduce the potential energy may result in a release of kinetic energy as, for example, in fission.

Precursor. (a) In radioactive decay, the parent nucleus or an earlier nucleus in a radioactive decay series. (b) In reactor safety analyses, an early malfunction in a sequence of malfunctions that could together lead to a reactor accident.

Predetonation. In a nuclear weapon, the initiation of a fission chain reaction before the fissile material has been sufficiently compacted for fission to take place in a large fraction of the material. (See Section 17.4.1.)

Pressure vessel. *See* **reactor vessel**.

Pressurized heavy water reactor (PHWR). *See* **CANDU reactor**.

Pressurized water reactor (or pressurized light water reactor) (PWR). A light water reactor in which the water is at a high enough pressure to prevent boiling. *See also* **steam generator**.

Primary cooling system. The system in a reactor which provides cooling for the reactor core, carrying heat from the core to a heat exchanger or turbine generator.

Probabilistic safety (or risk) assessment (PSA or PRA). A method for estimating accident probabilities by considering the failure probabilities for individual components and determining the overall probability of combinations of individual failures that result in harmful consequences; used interchangeably with "probabilistic safety analysis." (See Section 14.3.2.)

Production costs. In electricity generation, the sum of operating costs and fuel costs. (See Section 19.3.1.)

Progeny. Residual nuclei following radioactive decay; often called *daughter nuclei*.

Prompt critical reactor. A nuclear reactor in which the fissions caused by delayed neutrons are not essential to achieving criticality. (See Section 7.3.2.)

Prompt neutron. A neutron emitted at the time of nuclear fission, with no significant delay.

Proton. A particle with a mass of approximately one atomic mass unit and a charge equal and opposite to that of an electron; a basic constituent of all atomic nuclei.

Quality factor (Q). A measure of the relative effectiveness of different types of ionizing radiation in producing biological damage, taken to be unity for X-rays and gamma rays. (See Section 3.2.2.)

Rad. A traditional unit for the absorbed dose of ionizing radiation; equal to 100 ergs per gram or 0.01 gray. (See Section 3.2.2.)

Radiation. In nuclear physics and engineering, often used as shorthand for *ionizing radiation*.

Radiation dose. A measure of the magnitude of exposure to radiation, expressed more specifically as the *absorbed dose* or as the *dose equivalent*.

Radiation exposure. The incidence of radiation upon an object, most often a person; the term is also used in a specific sense for the dose received from incident X-rays. (See Section 3.2.1.)

Radioactive. Possessing the property of *radioactivity*.

Radioactive decay. The spontaneous emission by a nucleus of one or more particles, as in *alpha-particle decay* or *beta-particle decay*. (See Section A.6.1.)

Radioactive decay series (or chain). A sequence of radioactive nuclei, connected by the successive emission of alpha particles or beta particles and terminating in a stable "final" nucleus. (See Section 3.4.2.)

Radioactivity. (a) The phenomenon, observed for some nuclides, of spontaneous emission of one or more particles. (See Section A.6.1.) (b) The activity of a sample, commonly in units of becquerels or curies.

Radiological dispersion device. A *dirty bomb* or other mechanism used to create an area contaminated by dispersed radionuclides.

Radionuclide (or radioisotope). A nuclear species that is radioactive.

Radium (Ra). The 88th element ($Z = 88$); a natural radionuclide in the ^{238}U radioactive decay series.

Radon (Rn). The 86th element ($Z = 86$); the isotope ^{222}Rn, a product of the decay of ^{226}Ra, is responsible for a large fraction of the radiation dose received by most individuals. (See Section 3.5.1.)

Reactivity (ρ). A measure of the extent to which a reactor is supercritical ($\rho > 0$) or subcritical ($\rho < 0$). (See Section 7.3.)

Reactor. *See* **nuclear reactor**.

Reactor core. The region of a reactor in which the nuclear chain reaction proceeds.

Reactor-grade plutonium. Plutonium with the isotopic mixture characteristic of spent fuel, typically 20% or more ^{240}Pu. (See Section 17.4.1.)

Reactor period. The time during which the neutron flux and reactor power level change by a factor of e, where e (equal to 2.718) is the base of the natural logarithms. (See Section 7.3.2.)

Reactor poison. A material, produced in fission or otherwise present in the reactor, with a high cross section for absorption of thermal neutrons. (See Section 7.5.)

Reactor vessel. The tank containing the reactor core and coolant; when, as in water-cooled reactors, a high pressure is maintained within the vessel, it is alternatively known as the *pressure vessel.*

Red Book. A biennial report on uranium resources, production, and demand, prepared jointly by the OECD Nuclear Energy Agency and the IAEA.

Rem. A unit of radiation dose equivalent, equal to 0.01 sievert; historically an abbreviation for roentgen-equivalent-man. (See Section 3.2.2.)

Renewable energy. Energy that is derived from sources that will remain undiminished into the very distant future, for example, hydroelectric energy, wind energy, direct solar energy, and energy from ocean tides.

Reprocessing. The processing of spent fuel from a nuclear reactor in order to extract and separate plutonium and other selected components.

Resonance (or resonance peak). For neutrons, a local maximum in the neutron absorption cross section, arising when the kinetic energy of the neutron corresponds to the energy of an excited state of the compound nucleus.

Resonance escape probability (p). The probability that a neutron emitted in fission in a reactor will be moderated to thermal energies, avoiding resonance capture in the fuel. (See Section 7.1.2.)

Resonance region. The region of incident neutron energies within which the reaction cross section varies rapidly with energy due to discrete resonance peaks.

Retardation factor (R). The ratio of the velocity of water through a medium to the average velocity with which a given ion travels through the medium. (See Section 11.2.3.)

Retrievable storage. The storage of nuclear wastes in a repository in a manner that permits the later removal of the wastes, for example for reasons of safety or to extract selected nuclides.

Roentgen (R). A measure of radiation exposure, usually restricted to X-ray exposures. (See Section 3.2.2.)

Runaway chain reaction. A chain reaction in which the reactivity increases at a rapid, uncontrolled rate.

Saturated zone. A zone in the ground in which the voids in the rock are filled with water; this is the zone lying below the water table.

Scattering. *See* **elastic scattering** and **inelastic scattering**.

Seabed. Top layer of solid material under the ocean.

Secular equilibrium. A condition where the number of decays per unit time is virtually the same for all members of a radionuclide decay series in a sample of radioactive material. (See Section 3.4.2.)

Seismic. Relating to vibrations in the earth, particularly due to earthquakes.

Separation energy. For nucleons, the energy required to remove the nucleon from a nucleus in its ground state. (See Section 6.2.2.)

Separative work. For uranium, a measure of the difficulty of enriching the uranium to a higher ^{235}U fraction (defined in Section 9.2.2, footnote 18).

Separative work unit (SWU). A unit, with dimensions of mass, used to specify the magnitude of separative work. (See Section 9.2.2.)

Sievert (Sv). The SI unit of radiation dose equivalent. (See Section 3.2.2.)

SI unit. A unit in the *International System of Units*, widely used as the standard system of units in physics and engineering.

Slow neutrons. Neutrons of low kinetic energy, often taken to be less than 1 eV.

Sorption. Processes that retard the transport of nuclides in ground water by binding them (usually temporarily) to the rock through which the ground water flows. (See Section 11.2.3.)

Source term. In a nuclear reactor accident, the inventory of radionuclides that escape from the containment into the environment.

Specific activity. The rate of decay of a radionuclide sample per unit mass.

Spent fuel. Nuclear fuel that has been used in a reactor and has been removed, or is ready to be removed, from the reactor.

Stable nucleus. A nucleus which, in its ground state, does not undergo radioactive decay.

State. *See* **energy state**.

Steam generator. A heat exchanger used to produce steam in a pressurized water reactor. (See Section 8.2.1.)

Stochastic. In radiation protection, a stochastic effect impacts individuals randomly; the probability of a given impact on an individual (e.g., cancer induction) depends on the magnitude of the dose received. (See Section 4.1.3.)

Subcritical. For nuclear fuel, having an effective multiplication factor that is less than unity.

Supercritical. (a) For nuclear fuel, having an effective multiplication factor greater than unity. (b) For a fluid, being at a temperature above the liquid-vapor critical point, so that there is no distinction between liquid and vapor; the fluid exists in a single phase as a gas.

Symmetric fission. Fission events in which the masses of the two fission fragments are approximately equal.

Thermal conversion efficiency (or thermal efficiency). For an electrical power plant, the ratio of the electrical energy produced to the total heat energy produced in the consumption of the fuel.

Thermal energy. (a) The energy, kT, characteristic of particles in thermal equilibrium in a system at temperature T; k is here the Boltzmann constant. (See Section A.2.4.) (b) The total heat energy produced in a power plant (as distinct from the electrical energy produced).

Thermal equilibrium. A system is in thermal equilibrium when there is no net flow of heat from one part to another; in a reactor at temperature T, neutrons reach thermal equilibrium when their average kinetic energy is the same as that of the molecules of a gas at temperature T.

Thermalization. The reduction, primarily by elastic scattering in the moderator, of the energies of fission neutrons to thermal energies (below 0.1 eV).

Thermal loading. In a nuclear waste repository, the heat burden arising from the radioactive decay of radionuclides in the wastes; expressed, for example, in kilowatts per unit area.

Thermal neutrons. Neutrons with a kinetic energy distribution characteristic of a gas at thermal equilibrium, conventionally at a temperature of 293 K; this temperature corresponds to a most probable kinetic energy of 0.025 eV. (See Section 5.4.2.)

Thermal reactor. A reactor in which most fission events are initiated by thermal neutrons.

Thermal utilization factor (f). The fraction of thermal neutrons that are captured in the fuel (rather than in the coolant or structural materials). See Section 7.1.2.)

Ton. A unit of mass in the English system, equal to 2000 lbs (907 kg).

Tonne (or metric ton). A unit of mass in the metric system, equal to 1000 kg.

Total cross section (σ_T). The sum of all reaction cross sections, for a given pair of particles at a given energy. (See Section 5.1.2.)

Transmutation. The conversion of a nuclide into another nuclide in a reactor or accelerator.

Transuranic element. An element of atomic number greater than 92.

Transuranic wastes. Wastes that contain more than 0.1 μC of long-lived alpha-particle emitting transuranic nuclides per gram of material, but which are not *high-level wastes*. (See Section 10.1.3.)

Tuff. A rock formed by the compaction of the ashes of material emitted in volcanic explosions.

Uranium (U). The 92nd element ($Z = 92$); important in nuclear reactors and nuclear weapons, particularly for its fissile isotope ^{235}U.

Unsaturated zone. A zone in the ground in which voids in the rock are at least partly filled with air, rather than completely filled with water; this is the zone lying above the water table.

Vitrification. For nuclear wastes, incorporation of liquid wastes into a molten glass, which solidifies on cooling.

Void. (a) In reactors, a region (or bubble) of vapor in a coolant that normally is a liquid. (b) In rock formations, gaps in the rock matrix that can be filled by air or water.

Void coefficient. The rate of change of the reactivity with change in the void volume; a positive void coefficient corresponds to an increase in reactivity when the void volume increases. (See Section 14.2.1.)

Waste canister. A container in which solid nuclear wastes, including spent fuel, are placed for storage or transportation.

Wastes. *See* **nuclear wastes**.

Water dilution volume (WDV). The volume of water required to dilute a sample of nuclear wastes to the maximum permissible concentration for drinking water; it serves as a measure of the relative hazard posed by individual radionuclides in the wastes. (See Section 10.3.2.)

Water table. The top of the region of the ground where voids in the rock are filled with water; i.e., the top of the *saturated zone.*

Watt (W). The SI unit of power; $1 \text{ W} = 1$ joule per second.

Weapons-grade plutonium. Plutonium with isotopic concentrations that make it effective in nuclear weapons; commonly this means a ^{240}Pu concentration of less than 6%. (See Section 17.4.1.)

Weapons-grade uranium. Uranium with an enrichment in ^{235}U that makes it effective in nuclear weapons; commonly this means an enrichment of greater than 90%.

Welded tuff. *Tuff* which has been welded together by heat and pressure.

Working level month. A unit of exposure to short-lived radon daughters, defined in terms of the concentration of radon and its daughters and the duration of the exposure; one working level month corresponds roughly to an effective dose equivalent of 10 mSv.

Xenon (Xe). The 54th element ($Z = 54$); the fission product ^{135}Xe has a very high absorption cross section for thermal neutrons.

Xenon poisoning. The reduction of reactivity in the core due to a build-up of the ^{135}Xe concentration. (See Section 7.5.3.)

X-rays. Photons emitted by excited atoms or in the rapid acceleration (or deceleration) of electrons (commonly reserved for wavelengths below the ultraviolet region).

Zircaloy. An alloy of zirconium, commonly used as a cladding material in nuclear fuel rods.

Zirconium (Zr). The 40th element ($Z = 40$); a major component of the fuel cladding alloy, zircaloy.

Index